PETROLEUM AND THE
MARINE ENVIRONMENT

Under the High Patronage of
HSH Prince Rainier III of Monaco

Springer-Science+Business Media, B.V.

First published in 1981

© Springer Science+Business Media Dordrecht 1981
Originally published by Eurocean in 1981

ISBN 978-94-017-5539-9   ISBN 978-94-017-5537-5 (eBook)
DOI 10.1007/978-94-017-5537-5

CONTENTS

SESSION 3 POLAR ACTIVITIES

SESSION 6: RAPPORTEURS' SUMMARIES - GENERAL DISCUSSION
AND CONCLUSION

SESSION I

Introduction to the Conference and Opening Remarks

OPENING ADDRESS

Dr. Hanns Kippenberger
Président d'EUROCEAN
Administrateur - Directeur Général
Banque Européenne de Crédit
Bruxelles, Belgique

Your Serene Highness, Ladies and Gentlemen,

It is for me a pleasure and a privilege to welcome you to this conference on Petroleum and the Marine Environment. We are especially honoured that His Serene Highness has accepted the invitation to officially open the conference and the exhibition. The special interest which the Grimaldi Family has given to marine matters makes Monaco, in my opinion, a particularly well chosen place for such a conference.

I would like to thank first of all those who have been involved in the preparation of this conference for all their efforts that have made this conference possible and for the work they still have to carry out during and after the conference. I would in particular like to thank Mr. D.E. Lennard, Chairman of the Conference Committee and his team and the staff of Eurocean for taking the responsibility of this challenging task.

The conference has been organized by Eurocean (Association Européenne Océanique), an association of European industrial companies whose main aim is to initiate and develop ocean-linked projects and environmental questions that are of particular concern through the co-operation of its members. It has actively been supported by the International Petroleum Industry Environmental Conservation Association (IPIECA), an association which aims to facilitate the communication between the petroleum industry and governmental bodies with regard to international protection of the environment. The Oil Industry International Exploration and Production Forum (E & P Forum) which is an international organization of oil companies which advises governments and international bodies on technical matters relating to the exploration and production of oil, has also given its support. Furthermore, the recently established International Arctic Committee (Comité arctique international) has contributed a great deal to the conference. This latter organization has as its main target to improve the understanding of the problems of the Arctic area, the exploration and the preservation of its natural resources and sites and last but not least we have been supported by the United Nations Environment Programme (UNEP) Nairobi.

-   We are living today in a world in which all events are increasingly happening on a larger scale.

-   The technical progress of today offers increasing opportunities but also unfortunately involves an increasing number of dangers and disadvantages.

- The hazards associated with offshore petroleum activities that have arisen during recent years have been widely reported by the media. Deplorable accidents such as the Torrey Canyon and the Amoco Cadiz and Ixtoc I are familiar to all of us but it is certainly true that the less sensational aspects of day-to-day oil and gas exploration, exploitation and transportation are less known to the public.

Therefore, the main aim of this conference could above all be seen to establish the "state of art" in the field of petroleum exploration, exploitation and transportation and their impact on the marine environment which I hope will help to promote the understanding between the various parties involved.

I therefore would particularly like to welcome the representatives of

- governmental and international organizations

- petroleum and gas industries

- scientists and researchers

- independent consultants

- representatives from the media

I express the hope that during this week we shall be able to come to a better understanding and eventually some conclusions with regard to:

- expected further development in this area

- managing the ecological and economic consequences linked to offshore activities, the cost/benefit analysis of environmental management

- contributing to finding a common basis for future decision-making in this important field.

May I now invite Your Serene Highness to officially open the conference and the exhibition?

H S H Prince Rainer III
of Monaco

Mr. Chairman, Ladies and Gentlemen,

The recent catastrophes resulting from the damages caused by oil spills and blowouts in some of the seas of this world, have shown clearly in a dramatic evidence the urgent necessity to really "mobilize" immediately, not only our qualm of conscience but also our determination to promote technical and financial means to solve the serious problems that assail "petroleum and the marine environment".

This conference - taking place in the Principality of Monaco on this very theme - at this high level of competence and knowledge - takes a special dimension and affords me the precious opportunity to congratulate Eurocean for its laudable initiative while, at the same time, I can extend my warm welcome to all of you and all the eminent specialists coming from many countries of either governmental or non-governmental organizations or associations. I feel confident that you will find here a very favourable atmosphere for your talks and work on these shores of the Mediterranean Sea, in all times a link between the lands it separates.

What will be dealt with during your meetings will - I am sure - contribute to define the proper means and solutions in dealing with numerous and grave problems that occur from oil pollution in the seas:

- to establish the measures taken and progress made in controlling the hazards associated with offshore petroleum activities;

- to make an objective review of present and future offshore petroleum activities;

- to promote a better understanding between the various parties concerned in the development of petroleum activities.

These are the goals of your meetings in Monaco, and they are also the preoccupations that concern all those who hope - for themselves and their children - for a happier life in a clean world.

I am, therefore, particularly pleased to note - not only that your initiative but also the motivations of this conference - show clearly the concern for this future assurance of a better life.

And I trust that, above all, the important decisions you or others may take, will result in a carefully planned action for the preservation of the sea, as we should never forget that it is the source of all life.

However, the sea can no longer wait: it cannot wait to be defended and spared from the polluting aggressions of all kinds that now constantly poison it more and more!

I sincerely hope, therefore, that the conclusion you will reach will demonstrate - in all evidence - that beyond our conscience of the dangers and the damage that constantly threaten the seas, there must be a will to take action urgently to overtake hesitating intentions with feeble means to fight with little efficiency oil spills and blowouts of any importance.

I believe that the true manner to carry out this fight against pollution is a matter of agreed international planning. Pollution shows the obvious interdependence of all nations: each having to fight for the salvation of all.

For the future will come from the sea! And as a citizen of a country bordering the sea, and as President of the International Commission for the Scientific Exploration of the Mediterranean Sea, it is not only my wish but I believe my duty, to stress most emphatically the vulnerable and precarious situation of closed seas as to the effect and consequences of any form of pollution.

Undoubtedly, the specialists are fully aware of the serious dangers to which our sea is exposed, and they are already giving their full attention to the problem, being also - I am sure - thoroughly conscious of the adequate measures to take in order to fight a particular situation endangering the marine environment of the Mediterranean Sea.

But I wish, however, to avail myself of this exceptional opportunity to ask that special attention be given to the problems and the solutions to any form of petroleum accidents in these closed seas. Obviously, special techniques and appropriate accelerated methods are necessary, but also more severe measures have to be agreed upon to prevent before fighting - my hope is that they might be found and defined here during these meetings.

The sea bordering populations fully realize - with anxiety - the dramatic consequences of any oil spills or blowouts comparable to those off the coast of Brittany or in the Gulf of Mexico occurring in the Mediterranean Sea! These populations hope and expect that initiatives such as yours will determine an international cooperation between all polluters and governments in order to implement efficiency in the combat against any such eventuality.

But the problem - I feel - will only find its true solution when, from the simple citizen through to the ship owner, the captain of the ship and the industrialist, all feel really truly concerned and personally responsible for the respect and protection of the sea - this representing for each of us the inheritance of all.

In my mind, this is vitally important: it shows the duty of all free men of good will to feel concerned with these problems surrounding petroleum and marine environment as a matter of civic responsibility in a more civilized world.

But then it has been said that the aim of civilization is to make politics superfluous and science and art indispensable. Is this to be an unreachable quest? I sincerely hope not!

I have the honour to officially declare open the Petromar 80 Conference.

OPENING REMARKS

Dr. Hanns Kippenberger
Président d'EUROCEAN
Administrateur - Directeur Général
Banque Européenne de Crédit
Bruxelles, Belgique

Your Serene Highness, Ladies and Gentlemen,

INTRODUCTION

To ask a banker to address a conference on "Petroleum and the Marine Environment" is like asking a petroleum engineer to comment on economic developments and inflation.

Both pollution and inflation are most unfortunate phenomena which have a number of effects in common and this gives me the courage to stay here.

I   Pollution as well as inflation are both man-made and are the consequence of technical insufficiencies, negligence, cost reasonings, indolence and lack of discipline.

II   Both have the tendency to stay with us, they become more and more disturbing and only considerable common efforts will make it possible to reduce at least the most unpleasant and disastrous effects that they have on our physical and economic life - no doubt an improvement to strive for worldwide.

Allow me with these first opening remarks to put the question of petroleum and the marine environment into a somewhat more general perspective.

The determining factors for the political, social and economic development of our world are the constant population growth the volume of food production, the level of industrialization, the exploitation of raw materials, energy and environmental problems. Whilst the population growth and food production, as well as worldwide industrialization, will remain long-term problems, the question of the exploitation of raw materials, especially energy such as oil, coal and gas, and environmental problems have suddenly become an immediate concern in particular in Western countries.

THE CHALLENGE

ENERGY: The impact of the change of the energy situation on our political, social and economic life

The economics of our industrial world were built on the practically unlimited availability of inexpensive energy which, during the 19th century was derived from coal but since the beginning of this century has been derived from oil and gas, a development which has in particular become more evident since the early 60s.

The world energy consumption has developed as follows:

WORLD ENERGY CONSUMPTION[1]

	1948	1965	1970	1980	1990	2000
oil (in %)	30	48	52	54	45	37
gas	7.5	17	17	18	18	16
synthetic	-	-	-	-	2	4
Coal	55	28	24	18	20	24
Nuclear	-	-	-	3	7	10
Hydro and others	7.5	7	7	7	8	9
Total in %	100 %	100 %	100 %	100 %	100 %	100 %
Total (mb/d)	29.4	60	75	100	130	165

Energy consumption has more than tripled between 1948 and 1980. The world consumption of oil itself has increased approximately 6 times. The reason for the rapidly increasing share of oil in the consumption pattern was no doubt the attractive low price of $ 1.5 per barrel (1950) and $ 3/ b (1972), a price which, as a result of the price policy of the OPEC countries, has quadrupled in Autumn 1973 to $ 12/ b and since then has increased further to around $ 30/ b today. The recent increase of $ 2/ b represents an increase of the total OPEC invoice of ar. $ 20/ b.

Energy expenses in the early 1950s are estimated to have been in the order of 1-2 % of the GNP in industrialised countries whereas today this figure is estimated to be around 15 % of the GNP (10 % in 1975). A continuation of the energy consumption pattern of the early 1950s would probably have had a retarding effect on worldwide economic growth but on the other hand it would probably have avoided the very severe adjustment process that we have to go through during the coming 2-3 decades. We would possibly have been able to avoid some of the serious political, social and economic frictions we are now facing.

The political tensions in this world, apart from the philosophical differences, are also influenced by the endeavours of the Super Powers of this world to ensure a sufficient access to natural resources and especially to oil and gas which are necessary for the further development of their industries.

Also, the future economic development of each individual country and world trade primarily depend upon the availability and the price of energy.

---

[1] Taken from "World Energy Outlook" - Exxon Background Series December 1979 page 11

Since the first oil shock in 1973, and even more so since the doubling of the oil prices in 1979, the growth rate of the World Gross National Product has come down from around 5 % (1965-1973) to around 3 % (1974-1978) and will probably only slightly increase to an annual average of around 3 - 3 1/2 % during the coming decade. Lower income developing countries will suffer most (a growth rate of 1.2 % [1] instead of 1.9 %). [2] Industrial countries might reach a figure of 2.3 % [1] instead of 4 %, [2] energy exporting countries will reach 2.8 % [1] instead of 4.6 % [2] and higher income developing countries 3 % [1] instead of 5.2 % [2].

Inflation, although a well-known phenomenon in most countries today, has particularly been put under pressure through the increase of oil prices.

Just to take the example of the European Community which consumes around 55 % of its energy in the form of oil, 85 % of its oil has to be imported from outside of which 90 % comes from OPEC countries. The effect on inflation in industrial countries, stemming from oil price increases during 1979 can be estimated to have been ar. 4.5 %. The recent increase of $ 2 /b (8 % price increase) would cause an inflationary effect of ar. 0.4 - 0.5 %.

The effects on the balance of payments of a number of countries and on their foreign indebtedness have been even more detrimental.

SUMMARY OF CURRENT ACCOUNT BALANCES IN BILLION US $ 3)

	1975	1976	1977	1978	1979	1980	1981
Industrial countries	+ 16.1	- 2.6	- 6.6	29.6	- 10.9	-47.5	- 17
OPEC	+ 35	40.0	31.7	5.0	+ 68.4	115	87
Non-oil LDC	- 45.8	- 32.1	- 28.0	- 36.2	- 54.9	-68	- 78
Comecon, PRC etc.	5.3	5.3	- 2.9	- 1.6	- 2.5	- 0.5	- 8

Whereas during 1979 the industrial countries imported oil at the counter-value of around $ 206 b which represented around 19.5 % of total exports (USA - 30.2 %, Japan - 36.9 %, Germany - 15.5 %), an increasing number of less developed countries (LDCs) had to use an even greater share of their exports for oil imports: Brazil 45 % (Brazil around 60-70 % 1980) and this significantly contributed to an increase of foreign debt in all those countries which were unable to compensate this rise in imports by additional exports. This was true for a number of industrial countries but particularly for developing countries which increased their foreign debt between 1970

---

1) 1978_1990

2) 1965-1973

3) IMF Doc. 80/5 21.4.1980, page 13

and 1979 from ₰ 55 billion to ₰ 350 billion whereas Opec countries have accumulated during the same period net assets of around ₰ 225 billion which will probably increase by another ₰ 183 billion during 1980 and 1981 to ₰ 407 billion. The importance of oil imports within EC countries has increased from 1.47 % of the GNP (1973) to around 3.9 % of the GNP (1980); a figure which compares with around 3 % for the whole of all the OECD countries - all in all still a modest figure.

For all these reasons it can be expected that the change of energy and particularly oil consumption patterns as well as production structures, which have slowly started to develop during recent years, will become more important in the future. Improvement in energy efficiency, technological advancements, growth of less energy intensive industries, energy savings as well as substitution of expensive or politically uncertain types of energy by less expensive or more reliable domestic sources such as coal, nuclear, synthetic fuels and renewable sources (nuclear fusion, solar energy, OTEC) will play an increasingly important role.

Example: for energy efficiency

    Existing average consumption of US cars: 15 mpg
    New 1980 models: 20 mpg
    by 1985 the average should be: 27 mpg

    1 mpg would signify 450,000 barrels per day
    all US cars with 30 mpg would save ar. 4.5 b/d = 50 %
    of all US oil imports during 1979.

WORLD AVERAGE ANNUAL ENERGY DEMAND GROWTH RATE (excluding Comecon, PRC) [1]

	Energy	Oil
1965-73	5.5	7.7
1973-78	1.7	0.5
1978-90	2.6	1.1
1990-2000	2.4	0.3

In spite of the estimated relatively decreasing importance of oil for the whole energy supply (54 % today, an estimated 45 % in 1990 and 37 % for 2000), the estimated world oil demand will nevertheless increase from around 53.5 million b/d today (not including Comecon countries or the PRC) to ar. 60 mb/d (1990) and 65 mb/d (2000). This increase will however, exclusively be for countries other than the USA, Europe and Japan whose share today is still ar. 75 % of total world oil demand, a figure which will come down to ar. 57 % by the year 2000. Our future industrial development will, therefore, continue to depend increasingly on oil as the major energy source, but the ability to increase the production of oil is limited by:

-    the availability of undiscovered reserves

1) Exxon background series: World Energy Outlook Dec. 1979, page 10

- the rate at which new reserves are found and developed
- the price level for oil and other energy resources which would allow the exploitation of more expensive oil resources [1]

The relationship between onshore and offshore production has considerably changed during recent years:

	1968	1978	1980
Onshore production of oil	88 %	82 %	80 %
Offshore production	12 %	18 %	20 %

In the future, offshore oil production will play an increasingly important role.

Until about 1970 oil discoveries, which ranged between 10-25 bb/a, were well in excess of production. Since then the situation has been reversed (today new discoveries are around 5 bb/a below production of ar. 21 bb/a). The proven conventional oil reserves amount presently to 570 bb with expected total reserves still to be discovered amounting to a further 1000 bb (25 y estimate).

Exploration patterns will undoubtedly move to:

- more remote locations with a difficult environment: offshore, deep water, arctic climate

- higher exploitation cost areas

- developing the production of synthetic fuels: from heavy oil, oil sands, oil shales, agricultural products which could become important if considerably higher investment costs were acceptable. (Total production estimate 2 m b/d 1990, 7-9 m b/d 2000).

As mentioned before, the energy investment costs have so far been very modest:

EC energy investment	1968-72	1976-80
in percentage of GNP	1.4	1.7
in percentage of total investment	6.2	7.9

The total investment of the oil industry for new oil and gas production has amounted to ar. $ 12 b in 1975. ($ 16 b in 1979). It now looks as though the industry's total annual investment for developing new oil and gas production will move from $ 20 b (1980) to $ 50 b (1990),

---

1)

capital cost per b/d of capacity (Shell est. Mr. de Bruyn Noroil Dec. 1979)

	today	2000
low cost	$ 2000 b/d	$ 6000 b/d
medium cost	$ 8000 b/d	$ 14000 b/d
high cost	$ 20000 b/d	$ 33000 b/d

$ 70 b (2000), $ 90 b (2010) and $ 110 b (2020).[1]

The bulk of these investments will go into offshore developments (outside OPEC countries): and these will make the oil and gas industry one of the fastest growing industries in the world during the coming decade. It is not expected that the necessary finance for these investments will cause any problems even if the internal financing of oil companies might decrease from the present rate of 90 % to around 80-85 %. Banks would be happy to extend their lending portfolio to the oil industry from its present relatively modest level.[2]

THE IMPORTANCE OF MARINE SOURCES

This brings us to the role that the oceans and the seas will play in the future. So far the first function of singular economic importance has been the means of transport. This task has become increasingly more important with the impressive expansion of the world trade and shipping transport volumes.

TOTAL WORLD MERCHANT FLEET   1st January

	1954	1960	1970	1980
World Tanker Fleet				
- Number of ships	3500	4089	5688	7218
- TDW (mio)	37	61	138	338
World Total Dry Cargo and Tanker Fleet				
- Number of ships		22.636	28.141	33.429
- TDW (mio)		173	312	653

The second traditional ocean-linked activity of mankind has been fishing which today represents around 55-60 tons of fish and shellfish every year.

The development of the oceans to yield other resources in any significant quantity is only recent and started with the oil discoveries on the continental shelf in California and Texas around 1900 and has been booming since the discovery of an increasing number of offshore oil and gas fields during the last 20 years. Since then a complete reassessment of the oceans and their value for mankind has taken place. The exploration of the sea and the sea beds is only just beginning and yet we can already be assured that the geological structures beneath the sea

---

1) D. de Bruyn: "Financing Oil Development", NOROIL December 1979, page 59

2) estimated total energy investments (world) during 1980-90 are $ 1272 b (IMF estimate), of which $ 800 b for oil, $ 84 b gas, $ 258 b nuclear and $ 130 b coal. This would represent around 2 % of the GNP and 10 % of the total of the industry's investments. It is estimated that bank lending for energy will increase to around 20 % of total lending. EC estimated total energy investments: ECU 400 b 1980-1990, EC paper "Energy Policy" 20.3. 1980 page 6.

bear an enormous amount of natural resources, the importance of which we have only begun to realise (cu 15 x Ni 1500 x, Mn 4000 x and the bulk of new oil and gas discoveries). We are only now starting to develop the necessary technology to explore and exploit these resources. An impressive development has taken place during the last 20 years which could hardly have been foreseen. Apart from hydrocarbons and minerals, the ocean will in the future also play an increasing role in the production of other types of energy (currents, tides, waves, temperature differentials, wind, salinity gradients, etc.).

Apart from such resources, the sea has for a long time been the greatest source of revitalisation in the face of man's considerable degradation and distortion of the natural ecological balance and beaches have always been a place of recreation for many people.

MARINE ENVIRONMENT

The quality of the marine environment is constantly affected by an estimated volume of 30 b tons of solids carried by the rivers, wind and glaciers from the land to the oceans. The expanding world population and the growing consumption per head of materials and energy by our industralised world is unintentionally or deliberately seriously affecting the marine environment by dispersing part of the waste into the oceans. The materials affecting the marine environment in a serious way are heavy metals - very often toxic -, halogenated hydrocarbons and oil, of which only the latter is of interest to our discussions here.

Man's introduction of crude oil and petroleum products to the ocean exceeds by far the natural seepage.[1] The total loss for 1970 was estimated to be in the order of around 2 m tons p.a., a figure which increased to 4 m tons in 1975 and today it is estimated to be in the region of 6 m tons p.a.[2]

The bulk of these losses stems from transportation, coastal refineries, municipal and industrial waste and river outlets, and only a minority amount (estimated to be around 80 000 tons) is due to spills from offshore drilling and production operations.

Blowouts have seldom occurred before the Ixtoc accident and only a limited number have resulted in serious pollution (Santa Barbara Channel 1969, 6000 tons Ekofisk 1977 - 21 300 tons, Ixtoc 1979, 509 000 tons). Accidents connected with storage and transportation are of increasing importance. During 1979 altogether around 159 accidents were reported with an estimated total amount of oil entering the oceans of around 1 million tons (1978 around 109 accidents with 800 000 tons).[3]

---

1) estimated natural seepage around 100-150 000 tons per year
2) taken from :- Exxon background series: "The Offshore Search for Oil and Gas" 3rd Edition 7/78 page 13

3) According to the "Oil Spill Intelligence Report" published by the "Centre for Short-lived Phenomena".

The rise in world trade, especially the transport of petroleum and petroleum products, has considerably increased the likelihood of such accidents.

Whereas any important oil spills are cleansed by natural processes of volatilisation, biodegradation and sedimentation, the impact of large oil spills has dramatic consequences on sea life causing both ecological and touristic disasters. The damage caused by the wrecking of the Amoco Cadiz alone is estimated at about $ 300 million apart from the immeasurable damage to life, people and the environment.

ACTION CONSEQUENCES

The serious threat to the marine environment has been recognised for a long time by the public, governmental bodies, international organisations and the industry itself. This has led to the founding of the "Intergovernmental Maritime Consultative Organisation" (IMCO) by the UN and the majority of maritime states became members of the "International Convention for the Prevention of Pollution of the Sea by Oil" (1954, amended in 1962 and 1969). These treaties were drawn up as instruments for checking and reducing the rapid increase of marine oil pollution. A number of conventions followed to establish rules for specific areas: conventions of Oslo, Paris, London, Helsinki and Barcelona. These set rules and strategies in order to protect the marine environment, sometimes in the form of permanent commissions.

Although it took some time before industry was able to develop new techniques and production processes in reply to the demand for new improved environmental protection, the industry has definitely responded successfully to these new requests:

- the introduction of tankers with segregated ballast tanks, already a must for ships sailing to US ports, is in my opinion only a question of time and cost.

- industry, together with governments, has developed efficient containment and cleanup equipment (booms and numerous kinds of skimmers) and other recovery vessels used for picking up the oil from the water. The availability of a sufficient amount of equipment is still a problem.

- a number of oil-spill co-operatives have been established to respond rapidly to accidental oil discharges in their geographical areas (USA: Clean Gulf Associates, Clean Seas Inc. and Clean Atlantic Inc.).

- coast guards have trained strike teams and cleanup equipment in key centres in the US. In Europe there is, as I understand, still some work to be done before sufficient equipment is available.

The development of new oil production and transportation techniques, the production of adequate equipment to prevent oil spills or to clean up existing pollution will lead to an innovation process for new types of products. The need for such protection equipment will no doubt increase the final cost of the oil. Although I do not have any figures available, I understand that the estimated amount for anti-pollution investment, compared to the total investment of the oil industry, has increased considerably in recent years. It would be more instructive, however, to compare the sum of the running costs, depreciation and interest charges of anti-pollution equipment in order to give a better picture. The amount needed for offshore protection would presumably be rather small if one considers the problem of oil transportation separately. The inflationary effect on GNP is probably negligible,[1] apart from the positive effect of creating new jobs for new products.

CONCLUSION

This brings me to the end of my short introduction and I would like to draw the following first conclusions:

1. For the future development of our industry, oil will remain the main raw material especially for the chemical and pharmaceutical industry and oil will continue to be the most important individual source of energy even beyond the year 2000.

2. Although offshore oil represents today only 20 % of total oil production, the bulk of new oil discoveries will probably take place in offshore areas. This makes the development of new techniques necessary to enable us to work under the most hostile circumstances (deep water, rough seas, ice, etc.). The price for oil is therefore likely to go up in order to improve the present supply and demand structure.

3. In order to safeguard the oceans and seas, which will remain one of the most important sources for food, raw materials, energy and recreation for man, we have today to be aware of the serious damage we might cause to the sea water if we do not become even more conscious of the quality of the marine environment. The cost involved is minimal compared to the total investments which will be made in marine projects and compared to the advantages of a less polluted or an unpolluted ocean.

4. It will therefore remain of utmost importance:

    -   that there is a close co-operation between governmental bodies, international organisations and industry in order to establish and improve the basic rules and regulations for the protection of the sea, to obtain better information on the necessary data and to establish clear rules of conduct.

----

1) According to a Japanese macro-economic model calculation, inflation effect ar. 0.3 p.a.

-   that industry is required to continue its efforts to develop new techniques and equipment which would allow an improvement in the protection of the seas.

-   that there is a further improvement in the attitude of individuals vis-à-vis the marine environment.

I am confident that we all realise the responsibilities we have to take for the future in our own interest and in that of following generations.

DISCOURS D'OUVERTURE

M. Carpentier
Directeur Général du service de l'environnement
et de la protection des consommateurs à la
Commission des Communautés Européennes
Bruxelles, Belgique

Monseigneur, Monsieur le Président, Mesdames, Messieurs,

C'est pour moi un honneur particulier que de participer à la
séance inaugurale de cette Conférence, dans une ville si traditionnel-
lement attachée à la grande cause de la protection des mers et de
l'environnement marin comme l'atteste la présence de S.A.S. le Prince
Rainier III.

Au nom de la Commission des Communautés européennes, je désire
réitérer aux organisateurs, l'Association Européenne Océanique (EUROCEAN)
ainsi qu'à IPIECA, E & P Forum et au Comité arctique international tout
l'intérêt que celle-ci attache à la tenue de cette Conférence ainsi qu'à
l'organisation de l'exposition internationale qui l'accompagne.

Les problèmes que posent à la société d'aujourd'hui l'extraction
et le transport du pétrole d'une part, la protection de l'environnement,
d'autre part, sont devenus, en l'espace de quelques années, des problèmes
majeurs dont la solution conditionne notre vie, sinon notre survie.

Un des mérites de cette Conférence, et non des moindres, est de
permettre à des responsables de l'industrie et de la protection de
l'environnement de confronter leurs enseignements et d'échanger leurs
expériences et leurs perspectives sur le pétrole, facteur essentiel de la
vie de notre économie moderne, ainsi que sur la mer, essentielle à notre
vie tout court.

C'est une des raisons pour lesquelles la Commission des Communautés
européennes est heureuse de s'associer à cette manifestation. J'espère
en particulier que cette conférence nous aidera à mieux maîtriser les
difficultés de toutes sortes que posent les activités pétrolières et
leurs conséquences sur l'environnement marin, ainsi qu'à mieux évaluer les
rapports coûts/avantages qui en résultent pour la santé.

Nous devons, en effet, nous efforcer de déterminer, en vue d'une
meilleure prise de décision pour les années qui viennent, ainsi que d'un
partage plus équilibré des responsabilités et des charges, le jeu réel
des forces ainsi que l'interaction des facteurs qui caractérisent les
relations entre l'industrie pétrolière (compagnies pétrolières et transporteurs), les Etats, les organisations internationales, les assureurs et
"Last but not least" les consommateurs, lesquels sont tous, à la fois, à
des degrés divers, bénéficiaires mais aussi parfois victimes des opérations
mettant en jeu le pétrole et la mer.

En ce qui concerne l'ampleur du problème qui nous occupe, je me permettrai de vous rappeler, à titre indicatif, quelques chiffres qui nous aideront à situer la question dans sa véritable perspective et son évolution. Alors qu'en 1954, le volume d'hydrocarbures transporté annuellement par voie maritime de par le monde était de 258 millions de tonnes, en 1977, ce volume avait augmenté de 700%, et s'élevait à 1 milliard 700 millions de tonnes d'hydrocarbures, dont près de 500 millions transitent par la Manche et la Mer du Nord. Pour transporter cet hydrocarbure, la flotte mondiale de pétroliers a également crû dans des proportions notables. Alors qu'en 1954, cette flotte atteignait à peine 3500 navires pour une capacité totale de 37 millions de tonneaux de jauge brut, en 1977 on comptait 7000 pétroliers en service, totalisant environ 340 millions de tjb. Alors qu'en 1954, le plus grand pétrolier en service était de 30.000 tjb, aujourd'hui de nombreux pétroliers sillonnent nos mers en transportant une charge de 500.000 tjb dans leurs flancs (1). Au demeurant d'ailleurs, pétrole et armement sont souvent réunis et les sociétés pétrolières, particulièrement concentrées et puissantes au plan commercial, financier et technologique - arment leur propre flotte : en 1976, près de 40% du tonnage de la flotte pétrolière en service était sous leur contrôle contre 34% en 1973 (2).

Quant à la pollution des océans par les hydrocarbures, le dernier rapport établi par l'Oil Spill Intelligence Report nous apprend que plus de 12 milliards de litres de pétrole ont été répandus dans les mers en 1979 ou répandus en fumée. Au cours de la même année, 159 accidents majeurs ont été répertoriés contre 109 en 1978. 250 personnes ont été tuées ou ont disparu au cours de ces accidents qui ont par ailleurs entraîné des conséquences écologiques très graves. 73% de cette pollution a été causée par cinq accidents : celui du puits Ixtoc 1, la rupture d'une cuve à Forcados au Nigeria, la collision de l'Atlantic Empress et de l'Aegean-Captain près de Trinidad, la collision du Burmah Agate et d'un cargo au large de Galveston (Texas) et enfin la collision de l'Independanta et d'un cargo dans le Bosphore.

En ce qui concerne l'extraction pétrolière, la production de pétrole "offshore" est passée de 850.000 tonnes en 1954 à quelque 485 millions de tonnes en 1974. La part de la production en mer qui était de 0,12% en 1954 semble avoir maintenant atteint un palier, plus ou moins 17%, de 1972 à 1975, de la production mondiale, mais près de 24% de la production du monde libre. Toutefois, jusqu'à aujourd'hui, seule la zone des profondeurs océaniques jusqu'à 200 mètres a été systématiquement explorée ; dans la prochaine décennie, l'exploration systématique de l'ensemble des plateaux continentaux à des profondeurs beaucoup plus importantes sera entreprise sur une grande échelle. Quant aux Etats concernés par l'exploitation ou la recherche au large de leurs côtes, ils étaient en 1974, 80.

D'autres intervenants plus qualifiés que moi vous parleront de ce que cette production et ce potentiel de pollution impliquent et des problèmes que pose, quant au fond et sur le plan général d'abord et dans le monde par secteur géographique ensuite, la gestion de l'environnement marin

---

(1) Sources : OMCI - TSPP Conference papers

(2)   "   : "Les Etats et la mer" par Lucchini et Voelckel

en abordant tour à tour les aspects techniques, économiques et politiques, les objectifs de prévention et de contrôle de la pollution tant des navires que de l'exploitation offshore, la préservation des écosystèmes, etc..

Je voudrais, pour ma part, à ce stade, vous rappeler très brièvement les initiatives de la Communauté économique européenne en matière de prévention et de contrôle de la pollution des mers par les hydrocarbures. La Communauté a poursuivi plusieurs objectifs à des niveaux divers. Elle a, en premier lieu, tout en les distinguant, attaché une importance égale aux actions relatives à la prévention de la pollution et à celles relatives à la lutte contre cette pollution.

Une distinction a en outre été faite entre les différents types de pollutions, celle-ci exigeant, en effet, pour les combattre, des solutions techniques et juridiques différentes : pollution tellurique d'une part, et pollution provenant des navires d'autre part, elle-même d'origine opérationnelle ou d'origine accidentelle.

La CEE a ensuite cherché à différencier le niveau de l'action : international, communautaire ou national. La Communauté a bien compris qu'il fallait d'abord apporter son soutien politique et juridique, voire quand elle le pouvait, financier aux organisations internationales telles que le PNUE et l'OMCI qui traitent de la pollution au niveau international. Elle a ensuite agi, au niveau régional qui est le sien, dans trois directions : Elle s'efforce, en premier lieu, de rendre applicable dans la Communauté les normes de prévention et de lutte adoptées au niveau international. Je fais référence ici à la ratification et la mise en application plus rapide, voire anticipée, par les Etats membres, des conventions internationales de sécurité maritime et de prévention contre la pollution et des résolutions de l'OMCI relatives aux directives et aux procédures de contrôle sur les navires.

Une recommandation du Conseil, adoptée en juin 1978 (1) a demandé aux Etats membres de ratifier les grandes conventions internationales de sécurité maritime et de prévention de la pollution maritime, à savoir Solas de 1974 et son Protocole de 1978, MARPOL de 1978, OIT N° 147 de 1976. Une deuxième recommandation de décembre 1978 (2) prie les Etats membres de ratifier avant le 31.12.1979 la Convention internationale de 1978 sur les normes de formation des gens de mer, de délivrance des brevets et de veille, tous ces domaines ayant, en effet, une forte incidence sur les causes des catastrophes maritimes dont la majorité est souvent imputable à l'erreur soit professionnelle soit humaine.

Exception faite de la Convention Solas, entrée en vigueur, il y a quelques jours, il faut malheureusement observer que ces "recommandations" de la Communauté n'ont pas eu le résultat escompté, en ce sens que ses Etats membres n'ont pas tous ratifié ou accédé dans les délais indiqués, aux 3 ou 4 conventions que je vous ai rappelées, jugées cependant indispensables pour la sauvegarde de la sécurité et de l'environnement marin, et alors qu'il ne fait pas de doute que la ratification par un groupe d'Etats jouant un rôle prépondérant dans le domaine des transports maritimes eut pu, de façon très sensible, hâter la mise en vigueur de ces nouvelles règles internationales ; c'est d'ailleurs ce qu'a souligné de nouveau l'Assemblée

(1)   Recommandation du Conseil du 12.6.1978 (JO N° L194 du 19.7.1978

(2)   Recommandation du Conseil du 21.12.1978 (JO N° L 33 du 8.2.1979)

parlementaire européenne dans sa résolution du 14 mars 1980 invitant la Commission à proposer au Conseil la ratification de toutes les conventions internationales portant sur les constructions navales, l'entretien des navires, la qualification, la formation et les conditions de travail des équipages.

Dans le domaine des directives et procédures de contrôle des navires, la Commission a proposé, dans un premier stade, que les Etats membres appliquent obligatoirement celles qui ont fait l'objet des résolutions de 1976 et de 1977 de l'Assemblée de l'OMCI relatives aux contrôles de l'application, par les navires, des règles de sécurité et de prévention de la pollution : Conventions SOLAS 1960, Ligne des charges 1966 et OILPOL 1954). Le Conseil des Ministres des Transports n'a pas encore statué sur cette proposition.

La Communauté a élaboré, en second lieu, des législations communautaires propres, dont la transposition dans les droits nationaux des Etats membres est de droit. Deux directives ont été adoptées en décembre 1978, l'une en matière de pilotage (1), l'autre dans le domaine de l'information préalable à donner par les navires-citernes lors de leur entrée ou de leur sortie des ports de la Communauté (2).

Enfin, en troisième lieu, la Communauté s'efforce de contribuer aux actions entreprises au plan national, régional ou local, en vue de lutter contre les pollutions, après qu'un accident s'est produit. Cet objectif est celui du programme d'action de la Communauté en matière de contrôle et de réduction de la pollution des mers par les hydrocarbures, approuvé par la Résolution du Conseil du 26 juin 1978 ; ce programme, dont je suis directement responsable, vise essentiellement l'organisation, dans la Communauté européenne, de la lutte contre la pollution, surtout massive et accidentelle, en vue de rendre le plus efficace possible le front commun et solidaire que les responsables des rivages européens doivent pouvoir opposer rapidement à une pollution de ce type, pollution que les moyens technologiques actuellement disponibles ne permettent pas encore de combattre avec toute l'efficacité voulue : il s'agit d'échanges d'informations et d'expériences d'une part sur les ressources humaines et matérielles de lutte contre la pollution, d'autre part sur les navires-citernes à haut risque, en particulier des informations sur leurs conditions et leurs équipements de sécurité, les infractions dont ils se rendent coupables et les accidents dans lesquels ils sont impliqués. Dans tous les domaines que je viens de mentionner, la Commission s'est efforcée de promouvoir des actions efficaces et pragmatiques. Un autre thème a encore fait l'objet de nos préoccupations : celui de l'indemnisation, les régimes juridiques mis au point au niveau international n'étant plus toujours en rapport avec l'importance des dommages causés, y compris le dommage aux éco-systèmes.

En exécution de ce programme d'action de la Communauté, différentes études de faisabilité ont été entreprises, qui ont été menées à terme en 1979 et sont en ce moment discutées avec les experts gouvernementaux. Deux de ces études nous conduisent à penser qu'il sera utile de créer un système communautaire d'informations concernant :

1° Les différents moyens ou produits disponibles pour récupérer et traiter les nappes de pétrole en mer et faciliter le nettoyage des côtes ;

---

(1) Directive du 21.12.1978 (JO N° L 33 du 8.2.1979)
(2) Directive du 21.12.1978 (JO N° L 33 du 8.2.1979)

2° Les plans d'urgence nationaux et régionaux mis en place par les autorités compétentes des Etats membres ;

3° Les caractéristiques des navires-citernes et en particulier leur état de conformité avec les règles internationales de sécurité et de prévention de la pollution.

Des propositions dans ce sens sont en préparation.

Nous envisageons également la création d'un Comité consultatif en matière de contrôle et de réduction de la pollution qui devrait notamment servir de forum de discussion et de réflexion ainsi que d'organe de coordination dans le domaine de la lutte contre la pollution des mers par les hydrocarbures.

La Commission compte enfin proposer au Conseil d'accorder un soutien financier communautaire à des actions spécifiques en vue de renforcer la coopération et l'efficacité des équipes d'intervention et des matériels (promotion d'expériences-pilotes, formation du personnel d'alerte et d'intervention, définition et harmonisation de spécification de matériels, création ou développement de centres d'essai du matériel, développement d'équipements nouveaux, etc.).

Le dernier point de notre programme d'action concerne la recherche et l'expérimentation de moyens chimiques ou mécaniques nouveaux pour le traitement et la récupération des nappes de pétrole, qui sont également des problèmes de grande technicité particulièrement ardus à résoudre et où le soutien de la Communauté peut s'avérer particulièrement nécessaire.

Monseigneur, Monsieur le Président, Mesdames, Messieurs, j'ai voulu par ce rappel de quelques données et d'informations sur l'action de la CEE souligner, non seulement l'intérêt de cette dernière à la protection de l'environnement marin, mais encore l'enjeu de ces journées, ici à Monaco. Ces journées se révèleront, j'en suis certain, d'un intérêt passionant pour chacun de vous, quel que soit le groupe professionnel ou privé auquel il appartient, qu'il soit producteur ou consommateur, quel que soit la mer ou l'océan - du Golfe Arabe, de l'Arctique, ou de la Méditerranée - que dans son coeur il entend protéger, pour le mieux-être de nos populations parce que ces mers sont notre héritage et notre patrimoine communs.

OPENING ADDRESS

Dr. Franco Magi
Vice-Chairman, IPIECA
Environmental Coordinator, ENI, Rome, Italy

Your Serene Highness, Ladies and Gentlemen,

In 1972 during the Conference of Stockholm, called by the United Nations to face up to world pollution problems with programmed measures, it was pointed out, among other things, that the role of industry was essential in the preparation of any programme for environmental protection, both in the planning stage and in the stage of implementation.

Without repeating this well known story, I will simply recall that 1972 saw the birth of UNEP (United Nations Environment Programme), the organization of the United Nations that is continuously working on and orienting all kinds of activities in the environmental sector.

But the magnitude of tasks that UNEP must face could be eased, for example, if it were possible to reduce the number of interlocutors. This would lead to greater speed of communication and of decision making, more circumspect strategies, a greater amount of information and so forth.

A large part of the petroleum industry immediately understood the importance of the function it was called upon to perform since it was already aware of the need to avoid, in the interest of all, those errors in the formulation of policies and regulations which could have international repercussions. Because of experience acquired in facing environmental problems connected with its operations, the industry was aware of the contribution it could make, to the prevention of and the struggle against pollution.

The outcome was that in 1974 the representatives of many international oil companies and sector associations met in London and founded the "International Petroleum Industry Environmental Conservation Association" (IPIECA).

What is the IPIECA? IPIECA is cooperation, it is the exposition of the points of view of its associates on the environmental planning of the United Nations and other international organizations which are interested in the environment. At the same time it is the focal point within the petroleum industry for all of its interlocutors. It works in cooperation with and is assisted by other petroleum industry associations such as CONCAWE, OCIMF, API, etc., in the role of a co-ordinator and of a prompter, suggesting areas of research and development on environmental problems.

From the date of its foundation in 1974 until today, this voluntary organization of the international petroleum industry has grown considerably, from its few founders to the present-day composition of 25 companies or oil groups plus 8 oil industry associations which together represent the entire world of oil, from Europe to North and South America, from the Middle East to the Far East and to Africa.

My position as Vice President of IPIECA, on the one hand, and as a representative of a Mediterranean country, on the other, affords me the pleasure of speaking about the IPIECA and also about the subject of this conference, Petroleum and the Marine Environment, and even more specifically, Petroleum and the Mediterranean Environment.

In the framework of environmental problems in the Mediterranean, there is a particular situation which justifies the intense activities on an international scale which have been going on for years to protect this sea. I want to recall only a few of the particularly unfavourable conditions characteristic of the Mediterranean which make it a subject of particular attention. First of all is its characteristic of being a basin which is almost entirely closed, so much so that it is estimated that complete seawater replacement requires at least 70-80 years. The vast zones of shallow bottoms give rise to phenomena of eutrofication. There is an extreme density of some of the sources of pollution due to dense population of the coastal areas. Along the coasts of the Mediterranean there are 18 countries with a resident population of about 100 million persons, and during the summer season this number doubles.

The heavy maritime traffic, particularly of oil tankers! Recent estimates indicate that such traffic in the Mediterranean, the surface of which represents only 0.7 % of the world's oceans, witnesses the passage of 35 % of the oil transported over the seas of the world. An estimate by IMCO for 1977 shows that a good 600 million tons of oil were transported through the Mediterranean.

These, I repeat, are only a few of the reasons which have aroused concern in and for the Mediterranean area.

Today it may be said quite calmly that it is almost natural that the blame is placed upon the type of pollution that for years has been the only source taken into consideration, oil pollution coming from tanker traffic. This has been the most noticeable phenomenon and that which undoubtedly aroused popular emotion and was also, as experience has confirmed, the kind of pollution which was the easiest to control.

It was natural, therefore, that measures taken in this sector found support at IMCO, the UNO organization created in 1948 for the purpose of instituting a system of cooperation between governments for regulating international navigation.

This led to the Covenant of London of 1954, renewed in following years up to the covenant of 1973 which may be considered as one of the most important stages in the brief history of ecological protection activities. It was with this covenant, in fact, that the opportunity or rather the necessity was recognized of considering pollution no longer in a global manner but, of taking into account individual situations.

For the first time thought was given to and proposals were made for agreements of a regional type and certain areas were established where special measures should be taken. For the first time specific areas were discussed. Among these, the Mediterranean.

But we still have covenants at an international level which, though achieving notable results, cannot deal with detailed cases by suggesting or imposing individual solutions. What in the Covenant of London of 1973 was an indication, taking into account the fact that such a covenant covers only one of the numerous aspects of pollution, should have been translated into something as general as the environment, even for a small area. We are arriving at a form of broader awareness among the countries interested and affected by a single type of phenomenon. Now we have the Covenant of Barcelona, involving all countries of the Mediterranean. What was the effect? Mainly to promote the meeting of all Mediterranean countries, above all between the industrialized and the developing nations, whose differences in visualizing environmental problems were very apparent at the Stockholm Conference in 1972.

The rapprochement of the different positions, which finds its "raison d'être" in the awareness of the common difficulties to overcome, is, I would say, fundamental in order that international agreements may be effectively put into operation.

In 1976 in reply to this necessity, under the aegis of UNEP, the covenant of Barcelona was agreed upon. This was a rare example of regional accord which, with all the difficulties encountered along its course, I would consider an extremely valid instrument for regional cooperation.

Such cooperation finds a solid base in the socioeconomic programme that was given the name of "Piano Bleu" which lends support to the covenant, with the objective of carefully examining the causes of Mediterranean environmental degradation.

But there are certain specific problems to be solved for which regional type agreements are at a too general level and therefore are not adapted for the purpose. This brings us to the covenants of a bilateral type between neighbouring countries in order to make more detailed cooperative efforts on more specific problems such as common systems for quick action and for traffic regulation in certain zones considered particularly dangerous.

The Italy-Jugoslavia agreement, another between Italy and France and yet another between Italy and Greece are valid examples.

What results have been achieved? I would say that they are really important. The man in the street who does not know the problems but suffers from results of pollution, as for instance a user of the Mediterranean beaches, cannot fail to notice that year-by-year he less and less frequently finds his feet dirtied by oily or bituminous

residues. As a technician and an enthusiast on the subject I may affirm
that this is a direct result of the covenant as well as what has been
spontaneously achieved by the oil industry which has drastically reduced
pollution of the Mediterranean by oil, pollution which in the not too
distant future should be eliminated. The process certainly is not one of
the most speedy. As a matter of fact it is not sufficient to make laws
and change technologies, which of course may be done relatively quickly,
it is necessary to change the mentality of man towards a problem which
has never before existed in the history of man. And this cannot be done
in a short time. Nor should it be forgotten that there are also other
completely different forms of pollution which need to be tackled. One must
not be deceived into believing that by combatting pollution of the seas
by oil, marine pollution will be eliminated. In my opinion, these
things call for closer cooperation among people.

In any case, and whether it is by covenants at world level, or by
bilateral agreements, whatever is done, emphasis must be placed on the
term "cooperate".

I want to say a few words about this term, which means very clearly
"to work together" but which is often distorted or transformed.
Cooperate must only mean using a common force for achievement together
where singly one cannot operate.

Cooperate must signify only and always to lend a hand together,
simultaneously, to overcome an obstacle which is impossible to do
alone.

And it is in this spirit that I wish to terminate by recalling that
it is always said that pollution knows no boundaries. I think that the
time has come to demonstrate that to combat it, no boundaries should
exist.

Synopsis of Opening Address - F. Hughes, Head of Production
Shell International Petroleum Co. Ltd., and Chairman of the
E & P Forum

The theme of this introductory address to the Conference is
the Oil Industry International Exploration and Production Forum;
why and how it has been developed and its role as an industry
focal point for the dissemination of authoritative advice to
governments and inter governmental agencies on technical aspects
of oil and gas exploration and development. The Forum's
activities concentrate on the marine environment and are
controlled by an Executive Committee comprised of senior
representatives of member companies and operator organizations.
Specific aspects of the Forum's work are undertaken by a number
of sub-committees whose members are specialists in a particular
technical sphere and who are assigned to those sub-committees
by member organizations. In this way staffing appropriate to
specific specialized areas is arranged and can be varied to
suit particular needs. Currently there are some 7 sub-committees as follows:

A. Sub-Committee A on Offshore Pipelines works very closely
   with the Oil Companies' International Study Group for
   Conservation of Clean Air and Water (Western Europe)
   (CONCAWE) oil pipeline advisory group and is collecting
   statistics of pipeline spillages in the North Sea and in
   the Middle East.

B. Sub-Committee B on IMCO Activities is the liaison body
   with IMCO and provides detailed expert advice and opinions
   to IMCO Committees and Sub-Committees with the help of
   working groups drawn from member companies.

C. Sub-Committee C on Physical Oceanography and Meteorology
   looks into weather analysis and the utilisation of weather
   information and criteria by the oil industry on a world-
   wide basis and also advises on the best means of ensuring
   adequate oceanographic and weather knowledge in new
   exploration regions.

D. Sub-Committee D on Civil Liability reviews the activities
   and liaises with other bodies such as OPOL which are
   proposing methods of dealing with liability and compensation
   in the event of oil pollution. Sub-Committee D is the appointed
   Technical Advisor to the North Sea Convention on Civil Liability
   and can advise on similar Conventions in other parts of the
   world.

E. Sub-Committee E on Offshore Structures reviews the design
   codes of practice that are being produced by several groups
   in different parts of the world with a view to making
   sure that a correct balance is achieved with regard to
   safety and liability. Sub-Committee E works with IMCO and
   IACS and with regional bodies such as API and UKOOA on
   design, performance and stability of offshore oil units.

F. Sub-Committee F on Pollution Prevention and Control in
E and P operations is collecting data on oil spills from
the vicinity of production platforms, the cost of control
and of preventative measures and the causes of spills and
the risks involved and in this way backs up the work of
Sub-Committee D.

G. Sub-Committee G on Personnel Competence Standards is collecting world-wide information on what training, testing and
certification regulations and facilities exist or are mandatory with a view to producing basic requirements which
should be considered by any governmental body who wishes to
formulate regulations regarding personnel offshore.

The Forum members include the accepted industry leaders in
this field and in many cases individual members operating companies on their own behalf or on behalf of a consortium of
companies or government enterprises have the ultimate responsibility for the integrity and safe operation of facilities, the
safety of the personnel employed and the protection of the
environment from harmful effects of these activities.

Thus the industry has established an effective focal point
which can address specific problem areas and advise Governments
and Governmental bodies without impairing objectivity through
the circumstances and competitive considerations of individual
members. Moreover it serves to promote sound and safe practices
and professional standards of competence based upon the whole
range of expertise available in the industry.

OPENING REMARKS

Dr. Finn Sollie
Member of the Directorate of the International Arctic Committee
The Fridtjof Nansen-Foundation, Norway

Your Serene Highness, Mr. Chairman, Ladies and Gentlemen,

I speak here today as a representative of the International Arctic
Committee and thus as a person whose main attention is focused on
conditions and developments in that "Mediterranean of the North", the
Arctic Ocean and its chain of shallow rim seas.

I must remind you that as a result of the undiminishing demand for
energy in the world, rapidly rising prices of oil and natural gas and
growing concern about future supply, the search for petroleum resources
has now also been concentrated in the northern regions. At present,
exploration for and development of petroleum resources is in rapid
progress in several parts of the Arctic. Already, the Soviet Union is
producing some 50 per cent of its total oil output from wells in the West
Siberian Basin, in the area at the river Ob and its tributaries. More
than half of the production from this basin takes place under arctic
conditions and all of it in a region where the water flow is towards the
north, into the Arctic Ocean. In addition to oil, vast quantities of
natural gas are produced from the West Siberian Basin, and particularly
from its northernmost parts. Still concentrated in the land areas, we
may soon expect the Soviet Union to acquire the capability to move its oil
and gas development into its vast offshore areas in the North. While
the Soviet Union has made the most massive effort to develop its subarctic and arctic petroleum resources, similar moves towards northern
development are taking place in other countries as well, in Alaska and
Canada. Exploratory drilling offshore Greenland and on the Svalbard
islands so far has been disappointing. However, this summer Norway starts
offshore drilling at 71 ° North, at the edge of the Barents Sea, where
prospects are more promising.

Every indication is that in coming decades the Arctic will make an
increasing contribution to the world supply and that a large share of
that contribution will come from northern continental shelf regions.

The Arctic Ocean is an enclosed sea of enormous proportions. Its
total area is some 13 million square kilometres and half of that is
shallow water over the largest continuous continental shelf anywhere in
the world. This continental shelf is considered to be quite promising
in terms of its potential for petroleum resources in many areas.

Operations under polar conditions - on shore and, in particular, off
shore - present us with some of the most difficult challenges man has
ever been faced with in his industrial development. Extreme climatic
conditions, problems of ice, permafrost and long periods of winter
darkness together with long distances from established industrial

communities and support facilities make operations difficult, costly and risky. At the same time, the vulnerability of the polar environment and the fragile nature of its ecology call for special precautions to reduce risk and avoid damage. The safety of operations depends not only on superior technology, but equally on precise knowledge and broad understanding of the special nature of the polar regions and the natural processes at work there.

Our knowledge about the polar regions in many respects is patchy and fragmentary. Much research remains to be undertaken to improve knowledge and broad interdisciplinary cooperation is essential to coordinate efforts and to integrate results to build a safer platform for broad understanding. Furthermore, in the Arctic in particular, it is important that this understanding be helped by international cooperation: conditions in the Beaufort Sea must be understood and in some respects may be forecast on the basis of data from the East Arctic. The Arctic Ocean is an enclosed sea also in the sense that it forms one overall system. The importance of the system is further demonstrated by the fact that the Arctic may be described as the "weather kitchen" of the northern hemisphere. Through oceanic and atmospheric circulation, the Arctic has an effect upon the climate in vast regions and major disturbances in one part of the Arctic may affect the natural environment not only in all of the arctic system but also in the areas beyond.

Your Serene Highness will remember that under your guidance a meeting was held in Monaco in February last year to discuss these problems. It was a meeting of experts from various fields in countries that have a special interest in the development in the North and in its effects. The result of that meeting was the establishment of the International Arctic Committee which has become associated with the Centre Scientifique de Monaco. This is a true international group established to promote understanding, interdisciplinary and international exchange of information and exchanges between the scientific community, the business community and governments. The Arctic Committee wishes to inspire and to sustain harmonious and balanced development in the Arctic and, in this context, to cooperate with national and international organizations, institutions and authorities. Its major concern is the arctic environment.

I know of no other ocean areas in the world where international understanding and cooperation are more essential to the safe conduct of the exploration for and development of natural resources. On this basis, the Arctic Committee takes pride - and pleasure - in cosponsoring PETROMAR 80 as one significant contribution to broader and deeper understanding of the arctic environment, of its possibilities and of its problems.

Mr. Chairman, I am very happy that this conference is taking place in Monaco.

OPENING ADDRESS

                         Dr. M. K. Tolba
                        Executive Director
            United Nations Environment Programme
                         Nairobi, Kenya
given by   Dr. Stjepan Keckes
        Director, UNEP Regional Seas Programme Activity Centre
                      Geneva, Switzerland

Your Serene Highness, Ladies and Gentlemen,

    It gives me great pleasure, as Executive Director of UNEP, to extend
a warm welcome to the distinguished participants and speakers gathered
together in Monaco for the International Conference on Petroleum and the
Marine Environment. UNEP has the honour to be one of the sponsors of
the Conference. I am certain that this conference, whose aim is to assess
the present state of knowledge in the field of offshore oil and gas
exploitation and exploration and to promote understanding between the
petroleum industry, governmental and intergovernmental organizations, and
the general public, will serve as a useful complement to UNEP's own goals
and objectives and efforts in this direction.

    The United Nations Environment Programme was created by the General
Assembly of the United Nations "as a focal point for environmental action
and co-ordination within the United Nations". We interpret such
environmental action as being based upon a comprehensive transectoral
and multidisciplinary approach to environmental problems that deals not
only with the consequences but also with the causes and prevention of
environmental degradation. With specific regard to the marine
environment, UNEP seeks to provide a continuing, policy-oriented overview
of all major activities which may influence the oceans. Our aim is to
mobilize all the varied expertise required to help governments make
rational judgements on environmental risks and benefits in respect of
the oceans. For example, we are concerned with protecting marine living
resources as well as human health, with long-term questions of ecosystem
maintenance as well as immediate economic pressures.

    We are well aware that expertise is to be found in all sectors of
society, both public and private. Governments, industry, international
organizations - all stand to gain a great deal from a sharing of
experience and knowledge, from identifying and mutually strengthening
our technical competence and research capabilities. This is why I am
convinced that this meeting is so very important to all parties
involved in offshore activities.

    UNEP for its part has been directly involved, at the international
level, in several aspects of offshore petroleum exploration and
exploitation:

    First, UNEP's Industry and Environment Office, as a follow-up to a
1977 seminar on environmental conservation in the petroleum industry,
has established an environmental consultative committee for the

petroleum industry to ensure a permanent process of communication and consultation on environmental aspects of the industry and to facilitate the dissemination of the resulting information including appropriate training. This committee consists of participants from governments, industry and international institutions and meets at least once a year.

Second, UNEP's working group of experts on environment law is at present preparing global guidelines on the legal aspects of offshore mining and drilling carried out within the limits of national jurisdiction. It is our expectation that these guidelines will be submitted to the UNEP Governing Council at its ninth session in 1981.

Third, UNEP's Regional Seas Programme is dealing with environmental protection and management in eight regional seas areas and these activities must obviously take into account the exploration and exploitation of offshore resources. Dr. S. Keckes, Director of UNEP's Regional Seas Programme will be explaining our activities to you in some detail during the conference.

Finally, UNEP is deeply involved in the development of a programme of cost-benefit analysis as applied to environmental protection measures. As you are well aware, increasing oil supplies must be quantified so that meaningful decisions are possible. Techniques such as cost-benefit analysis are being applied increasingly to environmental concerns with the result that decision makers are now in a better position to evaluate the alternatives and tradeoffs. You have before you an extended paper by Mr. Yusuf J. Ahmad, Deputy Assistant Executive Director of UNEP, on the application of cost-benefit analysis for your examination and discussion.

Ladies and Gentlemen, we in UNEP are realistic believers in the ability of man to solve his environmental problems and to manage safely his environment, while simultaneously achieving a sustainable level of development that will assure maximum benefit from the finite resources of our earth. Your discussions during the conference should add to our knowledge of how to manage the oceans resources - wisely, safely and efficiently. I wish you success in your deliberations, and I look forward to benefitting from their results.

ACCIDENTS DISASTERS AND CATASTROPHES:
INEVITABLE AND UNPREDICTABLE

by Dr Anthony R. Michaelis
    Editor, Interdisciplinary Science Reviews, London.

Accidents, Disasters and Catastrophes occurred on this planet Earth long
before homo sapiens first took evasive action to escape a falling rock.
Natural disasters, such as floods, earthquakes and volcanic eruptions have
exerted an ever increasing toll of human lives as the population increased
from thousands to millions and now to thousands of millions. The definition
adopted here is ACCIDENT 1 - 1000 people dead or in imminent danger of death,
DISASTER 1000 - 1 million dead or in imminent danger of death and CATASTROPHE
over 1 million dead or in imminent danger of death.

Man-made accidents and disasters, whether caused by carelessness, neglect,
material failure or through use of equipment in an environment for which it
was never designed, have increased greatly in severity during the last few
decades. This is at least partially due to the larger target, the increased
density of human population, but basically due to the greatly increased
uncontrolled energy which is released whenever an accident occurs today. A
train crash or steam boiler explosion in the last century may have killed
tens of people , a crashing jet aircraft would today bring death to hundreds
and an atomic reactor melt-down would threaten thousands or hundreds of thousands
tomorrow.

Also novel in the disaster area is the existence of a special literature, best
labelled "Disaster Fiction" which often with great accuracy and in detail
foretells accidents and their consequences. So for example The Prometheus
Crisis, made into the film China Syndrome, accurately foretold the accident
at Three Mile Island. Fortunately this could be arrested before it reached
the truly catastrophic dimension of the Disaster Fiction, the total elimination
of all Los Angeles through radio-active fall-out. Other examples include The
Deluge, anticipating a flooding of London if the Woolwich Barrage is not
built in time and Ice-Quake, the coming of the next ice-age, following the
masses of antarctic ice surging from their supporting rock into the sea. The
value of this literature lies in the fact that it makes the reading public
aware of dangers and thus hopefully supports activities to prevent what is
preventable.

Accidents are no Acts of God. When considering the cost-effectiveness of
managing the environment, accidents are undoubtedly the worst examples of
a breakdown and of waste, costing enormous sums of money to restore the
status quo ante. As the public becomes increasingly aware of environmental
damage caused by accidents, and as the judiciary supports their claims for
vast sums of compensation, it is quite clearly in the interest of all management, whether of private or of state-owned industry, to take all possible
steps to avoid costly accidents. The most expensive research effort to find
a safer technique, the most costly instrument to give fuller information, the
up-grading of operator duties at higher wages and salaries, all these steps
are infinitely cheaper than one single component failure or the wrong action
of an untrained operator.

Accidents are no Acts of God, but fate chooses their victims. Mr Geoffrey
Brigstocke, an Undersecretary in the British Ministry of Transport and one of
my closest personal friends, had led his team to an official conference in
Moscow and was returning to London via Paris, there fatefully taking Flight 981
of Turkish Airlines leaving for London at 12.30 on 3 March 1974. No other flight
was available to him due to strike action. About 10 minutes after leaving Orly
airport, he and 345 other passengers of this flight were dead, the lower cargo
door of the ill-famed DC-10 having blown open. This led to a collapse of the
cabin floor and the destruction of the control cables to the tail section.

Fate played another tragic part in this accident as two years previously an
identical aircraft also manufactured by McDonnell Douglas had an identical
accident, its cargo door blew open, and forced a crash-landing at Windsor,Ontario;
the faulty door-mechanism was never rectified. The damages paid to the relatives
of the Turkish airline accident came to a total of $ 62 million, a sum vastly
greater than the modification of the door closing mechanism which in any case
was later compulsorily carried out on orders of the US Federal Aviation
Administration on all DC-10 aircraft. Suggestions of criminal negligence were
certainly justified and, had they been pressed by relatives, would have greatly
increased the amount of damages.

Even more costly was the negligence which led to making the name of Seveso infamous
throughout the world. Previous accidents during the manufacture of Trichlorophenol
had led to the release of dioxin; this happened on 10 July 1976 at the Icmesa plant
at Meda, near Milan, when about 2kg escaped into the atmosphere. Four years later
Hoffman La Roche, the Swiss owners, were said to have paid out-of-court damages of
$ 114 million for irrevocable damage to land, buildings and human health to those
affected in the neighbourhood of Seveso.

DESIGN FAILURES

Innumerable examples could be quoted where design failure led to an accident.
In the case of the DC-10 accident just mentioned, it was firstly the failure
of the design team of the manufacturers to construct such a simple piece of
equipment properly, secondly the human failure to close the faulty door
mechanism properly, and thirdly it was surely criminal negligence not to have
rectified a fault which had caused a previous near-accident. Faulty design can
of course also be due to ignorance. Take two modern bridges, the one over the
River Tay which collapsed with a train crossing it in 1879, 75 killed, and the
Tacoma Narrows bridge which broke up in 1940 - none killed. In both cases
wind forces led to the accidents and it has been argued that as the effect of
wind on bridges was at the time an unknown factor, it could not be included in
the engineering calculations.

Maybe so, but then the technology of landing men on the Moon was also completely
unknown in 1961 and yet it was achieved 10 years later without any loss of life
on the way to the Moon. The death of three American astronauts during a fire
on the ground in 1967 was due to negligence combined with a design fault, again
a faulty door closing mechanism. It could be rectified in time for the Moon
flight. The Apollo Project then became a highly cost-effective operation, as
it stayed within calculated limits of cost and time.

INEVITABILITY OF MECHANICAL FAILURES

Yet there were many mechanical failures during the Apollo Project. The spacecraft
and the propulsion rockets were manufactured to the highest practicable standards
and specifications called for 99.99 % perfection, a failure rate of 1 : 10 000.
There were a total of 15 million individual parts which meant that for the
specified failure rate, 1500 parts would fail during each and all Moon flights.
The answer to this was an elaborate back-up system for all vital parts but again
this could not be 100.000 %. Most spectacular was the failure of a minor
electrical relay on Apollo 13 which overheated, fused and caused the explosion
of a liquid oxygen tank; only super-human and super-computer efforts saved the
lives of the astronauts.

Now very few, if any, normal engineering components can be manufactured to 99.99 % perfection, the costs would simply be prohibitive. Consider for example the very ordinary 38-ton tanker truck travelling near Tortosa, Spain, on 11 July 1978 with a full load of liquid prolylene at the perfectly normal speed of 60 km/h. Suddenly some human or mechanical failure occurred, never fully diagnosed, and the truck went out of control, overturned, exploded, creating a crater 20 m across and 5 m deep. The explosion occurred opposite a holiday camp; 150 of the 800 sunbathing or siesta-taking campers were killed at once, hundreds more seriously injured.

Was it fate, or simply bad luck that the accident occurred just opposite a holiday camp? Or was the driver's attention momentarily distracted by a beautiful sunbather? Or did the hot afternoon sun raise the pressure of the liquid gas beyond the wrongly calculated safety limit of the vessel? Or did metal fatigue in the truck's steering gear shear a vital bolt?

Such accidents appear to be inevitable, had often occurred before and surely will occur again and again. A truck loaded with gasoline exploded in Pennsylvania in 1959, killing 11 and only 3 weeks later two liquid butane filled tank cars exploded in Georgia, killing 22 persons. In May 1950 a street car in Chicago collided with a tank truck carrying 7500 gallons of gasoline and the death roll was 34. More serious was a mass collision of 10 buses, trucks and cars which ploughed into the burning wreckage of a 22 ton liquid butane tanker truck on a Mexican highway in July 1978; over 100 killed and 150 seriously burnt, all due probably to a tyre blow-out. Such accidents are inevitable and occur far too frequently. Collision with road tankers carrying liquified gases or other highly inflammable petroleum products took place:

PLACE	TIME	KILLED	INJURED
Syria	3rd March 1962	31	39
Istanbul	5th August 1963	21	17
Mexico	14th March 1970	27	32
Karachi	10th January 1974	24	40
Mexico City	16th July 1978	11	200

How many of these could have been prevented, IF ONLY more stringent maintenance of brakes and steering of these trucks had been carried out? IF ONLY the truck drivers had been more carefully selected for their aptitude, better trained, less tired through overwork, or even if they had only been sober when working? These kind of accidents which appear at first glance so inevitable and unpredictable, could probably all have been prevented if simple factors of human and mechanical failures had been recognised and eliminated.

HUMAN FAILURES

We are not only mortal, but all of us are vulnerable to human failures, our own carelessness and the carelessness of others. As we are all drivers of motor cars, we have all had accidents, but of course they have always been the carelessness of others, never our own faults! Excessively long working hours of truck drivers are wellknown causes of accidents, and excessive alcohol consumption is probably an even more common cause of accidents. Drunken train drivers have caused many accidents, and I daresay alcoholically inclined ships' captains have also had serious accidents; however the administration of a breath analysis test to them is hardly possible soon after the occurrence of an accident.

Here again a simple consideration of cost-effectiveness would show the ready remedy. Oil drill rigs, submarines and the American Navy are dry, absolutely dry, and the US Navy has never been accused of cowardice because its officers and crews work and fight without alcoholic stimulus. If truck drivers, ships' crews and all tanker crews were forbidden alcohol, they might well demand higher wages; to accede to their demands would be highly cost-effective if it avoided one single accident.

Human failures are more frequently the causative factors in accidents as the relevant technology becomes older and more perfect. (See Tables I and II). Steam boiler explosions on land and at sea, which caused thousands of deaths in the 19th century have become a rarity, although there is always an exception. For example the explosion in 1962 of a steam boiler in the New York Telephone Company's canteen when 23 were killed. Tyre blow-outs have similarly become a rarity, yet earlier this century they were responsible for countless deaths; here again a notable exception has already been mentioned.

The introduction of new technologies is always dangerous, but in some cases determined efforts have succeeded to reduce their rate of causing mortality. Take for example aircraft accidents with more than 10 killed per accident during the last 20 years. The number of accidents during each year has stayed remarkably constant at an average of 26 a year and the total number of passengers killed per year has fluctuated from 839 to 2354, an average per year of 1240 from 1959 to 1978. The average number of deaths per accident has changed little during that period, and stands at 47 killed per accident.

This is quite a remarkable achievement when it is borne in mind that during the years from 1959 to 1978 the passenger kilometers flown by the world's airlines have increased enormously. Apart from design faults, already discussed, it is now pilot error that causes most of the present accidents. A sad example here is the DC-10 accident of Air New Zealand on their Antarctic Flight in December 1979. It was the crew's first experience of the extremely hazardous conditions of the Antarctic and for some still unexplained reason the pilot of the aircraft crashed into Mount Terebus. No survivors.

An equally sad aircraft accident occurred, again due to pilot error, in January 1966 when an Air India 707 jetliner hit a ridge 15 m below the summit of Mount Blan All 117 persons aboard the aircraft died, including India's then most famous scientist, Dr Homi Bhaba; he too was a personal friend of mine.

THREE MILE ISLAND

But of all accidents caused by human failures, the nuclear accident at Three Mile Island is so far the biggest in technological history. There, only a few miles fro Philadelphia, Unit 2 of the Nuclear Power Plant suffered from a brief failure of a minor water circulation pump. This had happened before at this and nine other nuclear stations built by the same company, Babcock and Wilcox. However at TMI, the event escalated through a series of related failures into a partial melt-down of the uranium reactor core. If this had not been arrested in time, a true disaste situation would have arisen with 1000 to 1 million persons in danger of serious radioactive contamination and possible death. The totally inadequate design of the instrumentation system of the reactor, and the failure of training the operators were found amongst the major faults by the Kemeny Commission appointed by President Carter to investigate the accident. The cost of rectifying the accident's sequels, such as de-contamination of the reactor and the site, as well as loss of earning

through stoppage of electricity production will cost the Metropolitan Edison
Corporation, the owners of TMI, a total of ₰ 1.5 thousand million.

The main causes of the accident were defined as "lack of operator training
in the handling of accidents". At the beginning there were two operators
in the control room when the first alarms sounded soon followed "by a cascade
of alarms that numbered 100 within minutes." One of the operatores recalls
in his evidence to the Commission "I would have liked to have thrown away the
alarm panel. It wasn't giving us any useful information". This must be one
of the most damning statements in the whole report. If instruments fail to
convey the relevant information about unusual conditions than either the
operators have been inadequately trained, the instrumentation system is
antiquated and useless, or as in the case of TMI, both. Modern high technology
demands that only the most superbly trained and thoroughly experienced operators
control such technology. Nothing below astronaut-standard should be acceptable
when so much is at risk.

COUNTER ACCIDENTS AND COUNTER DISASTER MEASURES

1. Human Factors

In all technological societies it is common practice, both in industrial and
military circles, to advance salary and social position with increasing
responsibilities. The Captain of a civilian jet aircraft, being responsible
for 500 lives for eight hours a day, may earn as much as ₰ 100 000 a year.
Operators of nuclear power stations at present receive about a quarter of such
salaries, although their responsibility, in case of a serious accident, could
easily extend to 5 million lives.

To counter accidents therefore, the first requirement will be to choose operators
responsible for preventing accidents to high technology, with the same exquisite
care with which the Apollo astronauts were chosen, giving them the most searching
mental and physical tests, with regular, annual checks. The high salary and
consequent high social position would lead to an exceptionally well qualified
corps of operators, who must spend much of their time at simulation exercises,
again similar to the Apollo astronauts; boredom was certainly never one of the
complaints of the astronauts.

However, highly qualified operators are not enough for high technology. Even
more important is the instrumentation which faces them during their hours of
work, whether this is the pressure in the tank of liquified gas on the back of
their truck, the pressure in its 64 tyres or merely the instruments on the
bridge of a super-tanker. None of these instrumentation systems measure up
to the emergency conditions considered here. In conservative industries, like
shipping for example, the only concessions made during the decades were to add
further instruments to the compass and the revolution counter of the propeller
shaft. To-day the view of instrument panels on a typical ship's bridge is
similar to that of Three Mile Island, with dozens of instruments conveying minor
information and only confusing any executive officer in an emergency.

The only satisfactory solution is a hierarchy of instrumentation systems,
electronically combined and displayed on a single television monitor screen
which has projected on to it a series of relevant legends and image buttons.
These can be touched by the operator's finger and thus connect him with any
component which is to be operated. This novel philosophy and technique was
developed and adopted for the biggest atomic accelerator at CERN in Geneva.
Urgent research is needed to adapt this control and executive system to nuclear
and other power stations, for tankers and other ships at sea, to trucks carrying
explosive loads, to oil refineries and chemical plants.

2. Material Factors

It is obviously "penny-wise and pound-foolish" to substitute cheaper materials
when only the safest will do. Here the usual concept is that of a relevant
"safety factor", based on previous operating conditions and cost of materials,
will satisfy conditions. In view of the ever higher costs to be met when
accidents occur in a world which has become extremely conscious of the environment
and which no longer hesitates to evoke the rigours of the law for the highest
possible compensations, it would no doubt be a good defence in a court of law to
state that "the safety factor for a particular operation has been greatly increased
after a recent review". Of course such action will cost money, but far less than
compensation of the injured and a rebuilding programme after an accident. In
certain cases, knowledge of the behaviour of materials in a novel environment may
simply be unknown at present; but then research should be carried out to investigate.
Again this would be good defence in a court of law, should the need arise.

3. Rescue Operations

Rescue is a vital part of safety and deserves as much advance planning as the prevention of accidents. Here the basic philosophy is the same, whether a gas pipe line has burst and is burning, an explosion has taken place in an oil refinery, or a natural disaster has destroyed a whole city, as for example Darwin in Northern Australia on Christmas Day in 1974. The order of rescue priorities, stated below, was given by me on a previous occasion in 1972 and was almost precisely adhered to by Major General Stratton when he took command after Cyclone Tracey had devastated Darwin.

i. Chain of Command

The appointment of a Disaster Commander, physically and mentally of supreme fitness and stamina, is the highest priority. He must have absolute power over the disposal of men and materials, communications and sypplies. It will be his responsibility to decide what will be communicated to the media and to the population in the neighbourhood of the disaster area. Access to the site by politicians will come under his control and whether help should be requested from outside and from abroad. His appointment must be made at the earliest possible opportunity after news of the accident has been received and if foreseeable, as for examples in chemical factories and oil refineries, such a Commander must be in constant readiness. He will call exercises without previous notice and will apply the lessons learnt, similar to the age-old practise of fire drills.

ii. Reconnaissance

To assess the extent of devastation, either from survivors, or by immediate sending of experts into the area, if necessary by parachute drops, is the second priority. Photographs from the air may become essential and, although voice contact with the aircraft would give the first indications, photographic records will be vital a few hours later. Such photographs will also be invaluable legal evidence when reconstruction claims are discussed by Insurance Companies.

iii. Communications

Communications with survivors and the company's own experts on the disaster site are next in order of priority. Even if all regular telephone lines have been destroyed, radio amateurs have often in the past formed the only available link with the outside world. If this is impracticable, then a communication engineer with his own portable radio equipment must join the first experts entering the disaster area.

iv. <u>Evacuation of Injured</u>

Without good communications, it will prove impossible to arrange for the evacuation of injured and living survivors - in that order - from the site. Obviously a two-way traffic flow must be established as soon as possible, ferrying out victims and bringing in return the most urgently needed emergency supplies. One would expect that 24 hours after the event, all injured had been evacuated and accommodated in Mobile Field Hospitals if the regular facilities were inadequate or had been destroyed.

v. <u>Care for Survivors</u>

Mental as well as physical care for the survivors is the next essential action if it has not already been carried out far from the site of the accident. If evacuation of all survivors has proved impossible, a constant flow of information, probably best by radio and direct from the Commander, is an absolute essential if morale is to be maintained and if panic is to be avoided. Panic has often doubled casualty figures in past accidents, and there are many cases on record when quite simple accidents have led to panic and caused the deaths of hundreds. On the physical side, water, food, clothing, heating and shelter are of course essential, and vaccination against possible epidemics should never be forgotten.

If these priorities sound exaggerated for a minor accident, such as an exploded road tanker, then one can only hope that when a major accident has occurred, such as Three Mile Island, that the above priorities are followed. One of the most serious aspects of the nuclear accident at Three Mile Island was the utter confusion which reigned on the site for days on end. There were conflicting authorities, some ordering evacuation, others countermanding them. Confused statements to the press from many different sources, and while the accident was still escalating, a great overcrowding of the control room. This by itself contributed to the severity of the confusion and hence the accident itself.

<u>FUTURE</u>

In my opinion, Insurance Companies have a grave responsibility in improving accident precautions and preventions. After all it appears to be in their own financial interest to see that the number and severity of accidents are reduced on a world wide scale. If companies like Lloyds of London increased their premiums to those who had not taken the fullest possible steps, as outlined here, an

improvement would surely soon be noticeable. The correct scientific approach
to the whole question of accidents in high-technology areas is to learn from
past experience.

As a first step, small research staff should be assembled simply to collect the
facts and apparently the only attempt to do so has been a slim paperback volume
entitled "Catastrophe! When Man Loses Control". It was prepared by the Editors
of the Encyclopaedia Britannica and published by them in 1979; they estimated that
in the 20 years from 1959 - 1978 10 000 persons a year lost their lives in man-made
accidents, specifically excluding all natural disasters. According to them the
world wide markets of private insurance companies, covering aircraft, train and
shipping accidents, fires and explosions, came to a total of $ 250 thousand million
in the 20 years.

A further step for the future is to expand and make known world wide the work of
the Australian Counter Disaster College at Macedon, Victoria, a unique organisation
in the world. Much of what is known to them and published is as applicable to
man-made accidents as it is to natural disasters and their regular seminars guided
by an expert staff of over 50 are of the highest standard. The Organisation was
established in 1974, only months before the Darwin Cyclone and comes under the
Australian Ministry of Defence. Its main function today is the development of
contingency plans and one can only hope that other countries will follow this
remarkable enterprise and send their own experts to Macedon.

From such co-operation should arise an International Rescue Organisation for natural
Disasters and an International Environmental Safety Organisation for preventing
man-made accidents. Just as it is well-known that disease is no respecter of
international frontiers, so it should be recognised that high-technology accidents
will spread their effects over continents if chemical and radio active clouds are
not halted in space and time. It will be too late to call an international expert
committee after the first nuclear disaster has occurred.

So far mankind has been undeservedly lucky. The "accident of all accidents", the
accidental explosion of an atomic bomb has not occurred, although on a number of
occasions they have been accidentally dropped from aircraft. Much publicity was
given to one such occurrence off the coast of Spain in 1964 and soon afterwards to
a similar event near Thule in Northern Greenland. No doubt similar events have
happened since then but have been shrouded in utmost secrecy. Let us form our
International Fire Brigade before the fire - not afterwards!

New Delhi 800312   Melbourne 800303

A D C   46

<u>TABLE I</u>

<u>RAILWAY ACCIDENTS   1959-1978</u> (Above 10 killed)

	<u>TOTAL</u>
JAPAN	357
USA	128
INDONESIA	149
BRAZIL	350
TURKEY	65
YUGOSLAVIA	377
ARGENTINA	321
SIERRA LEONE	13
ITALY	204
POLAND	86
SOUTH AFRICA	306
CHILE	25
D.R.GERMANY	248
CZECHOSLOVAKIA	128
SPAIN	282
F.R.GERMANY	206
FRANCE	193
INDIA	1046
NETHERLANDS	114
COLUMBIA	40
ROMANIA	70
GUATEMALA	20
ENGLAND	135
URUGUAY	30
HUNGARY	143
EGYPT	168
PORTUGAL	132
SOUTH VIETNAM	17
MEXICO	353
SWEDEN	24
SUDAN	15
PHILIPPINES	10
BURMA	102
DENMARK	14
ZAMBIA	11
C/F	5882

TABLE I (cont.)

RAILWAY ACCIDENTS 1959–1978 (Above 10 killed)

		TOTAL
	C/F	5882
BELGIUM		42
GREECE		36
SOUTH KOREA		166
AUSTRALIA		92
NIGERIA		132
PAKISTAN		57
BULGARIA		36
ALGERIA		35
CUBA		24
MOZAMBIQUE		60
MALI		20
NORWAY		27
TAIWAN		40
CAMEROON		100
KENYA		14
USSR		19
ZAIRE		22
	TOTAL:	6804 KILLED
	ACCIDENTS:	186
	36.6 deaths/accident	

TABLE II

AVIATION ACCIDENTS (Above 10 killed)

YEAR	NUMBER OF ACCIDENTS	DEATHS	AVERAGE
1959	29	979	33.8
1960	36	1242	34.5
1961	26	1193	45.9
1962	35	1461	41.7
1963	19	969	51.0
1964	22	986	44.8
1965	27	1236	45.7
1966	28	1485	53.0
1967	27	1230	45.5
1968	30	1336	41.2
1969	34	1465	43.1
1970	21	1105	52.6
1971	18	1090	60.5
1972	29	2354	81.1
1973	25	1351	54.0
1974	27	1308	48.4
1975	24	1228	51.2
1976	27	1478	54.7
1977	23	1453	63.2
1978	21	839	39.3

| 20 Years | AVERAGE: 26.4 | 25788 | 46.97 |
| | | 1250 per year | deaths/accidents |

(Both Tables compiled from Encyclopaedia Britannica publication quoted in the text)

HUMAN AND BIOLOGICAL IMPACT OF OIL DEVELOPMENT

IN POLAR AREAS

Louis REY, Ph.D.

President of the Directorate of the Arctic Committee
Centre Scientifique de Monaco - MC-Monte-Carlo

The eager competition for the world's energy supply has boosted offshore production of fossile fuels towards more and more remote areas in which circumpolar continental shelves are likely to play an important role in the future. This not only represents a formidable technological challenge due to the drastic conditions prevailing there, but might also have a major impact on the human, socio-cultural and ecological equilibria in a highly specialized, sensitive and as yet untouched environment. Since oil exploration and exploitation is currently forecasted both in southern and northern higher latitudes, the occurrence of petroleum as an industrial commodity and as a potential pollutant, has to be studied on a global basis for the Arctic and for the Antarctic.

Indeed, both areas present some common characteristics.

- Due to the slant solar illumination and its huge seasonal variations, the primary production is of the bi-modal type with a major spring bloom and it shows a cyclic sensitivity to chemical and mechanical challenge by pollu-tants.

- Biocoenoses are of limited diversity and variable population densities but show limited flexibility towards environmental changes because of their narrow resistance and capacity adaptations to low temperatures.

- The presence of extensive pack-ice of high albedo at the sea-surface buffers thermal transfers but also acts as a mechanical barrier preventing the evaporation and dispersion of low-volatiles while supporting a rather active biological life in the brine pockets of high nutrients concentration.

- The occurrence of semi-permanent subsidence phenomena over polar areas enhance large deposits of air-borne chemicals resulting from industrial and urban activities at mid-latitudes. This includes occurrence of unburnt hydrocarbons coming from ships' engines and large plants which may over-run accidental spills or regular seeps from offshore installations.

These and many other special features of polar areas need to be studied and fully understood prior to any extensive exploitation of fossile fuels. However, since the industrial developments in the northern hemisphere are likely to come on first, a major emphasis has to be placed on the specific characteristics of the Arctic Ocean, which, in many ways, differs widely from the circum-antarctic seas.

- Boreal terrritories have been the site of human settlements for millenniums and, as such, any new development there is bound to interact with traditional socio-cultural patterns and might have important political or strategic implications.

- The Arctic Ocean is an almost entirely closed "mediterranean" of assymetric structure, with stratified water masses and an eastwards surface flow, mainly derived from the North Atlantic Drift. Thus pollutants get concentrated there in a multi-layered system and are subsequently distributed in the whole Arctic basin. The polar ocean then behaves as the common "pool" of all neighbouring countries which must share the environmental responsibility.

- Contrary to the Antarctic, the fresh water inflow is of considerable importance in the north, both on the Siberian and North-American sectors and it plays a major role in the occurrence and distribution of land-based or land-deposited sediments and pollutants.

- The levels of primary production are seasonal and, for a given sun-illumination, depend on nutrients availability, ice-coverage and freshwater situation, resulting in

wide local variations of the biological productivity,
species diversity and correlative susceptibility to pollution damage.

These are some of the numerous factors which have to
be taken into account to design an ecologically sound policy
of oil development in polar areas.

CHAIRMAN'S OPENING ADDRESS TO SESSION 4 - TEMPERATE ZONE

Mr. D.P. Heath
Coordinator Environmental Conservation
Mobil Oil Corporation, New York, USA
Chairman of IPIECA
Vice Chairman of API Committee on Environmental Affairs

The classical definition of the temperate zone comprises everything between the Tropic of Cancer and the Arctic Circle in the North, and the Tropic of Capricorn and the Antarctic Circle in the South. We should therefore more properly speak of the Temperate Zones. These vast areas contain the greater part of mankind, in all its cultural and economic diversity, and sub divide into many widely differing geographical regions, each with its particular problems and environmental sensitivities. For the purposes of our session it is relevant that substantially all of the 1.6 billion metric tonnes of crude oil and products transported world wide by sea each year pass through the temperate zones, which contain most of the higher risk areas on the tanker routes, and that off shore exploration and production is accelerating in increasingly deep and difficult waters in both hemispheres.

It is therefore not surprising that this is the longest session in the conference. In opening the session I would remind myself of our objective, which I take to be a review of the environmental management of current and projected marine oil and gas production and transportation. This is a wide field of enquiry and a broad, logical methodology would be:-

1. Assessment of the hazards to the environment.
2. Developing operational procedures to avoid accidents.
3. Planning for emergencies.
4. Dealing with emergencies.

We are fortunate to have a very distinguished panel of speakers and, whether by accident or design, their papers cover the four step approach I have suggested.

On assessment we have a review of the API sponsored research on the Fate and Effects of oil on the marine environment, and a UK approach to oil pollution research and monitoring.

On operational procedures we have a paper on the role of a data buoy in satisfying environmental data requirements in any new UK exploration area, another on the importance of oceanography, and, as to the platform itself, a review of the treatment of produced water discharged from platforms in US coastal waters.

Planning for emergencies, a subject close to my heart and recent experience, is thoroughly covered by reviews of the UK and US government contingency plans: the expanding UNEP Regional Seas Programme: oil industry planning in the Mediterranean and the North Sea, and SLIKFORCAST, an oil industry sponsored activity which enlists computer aid in tracking oil spills and is, therefore, a valuable tool in contingency planning. In addition we have a paper on the dispersion and weathering of chemically treated crude oils. The use of chemicals to disperse or contain spilled oil is a subject needing much more consideration in the planning stage for emergency preparedness.

Finally, we have a paper on the actual experience of the AMOCO CADIZ accident.

The authors of our papers come from the United Nations Environment Programme, from regulatory and technical agencies of government, from scientific organisations and from the oil industry. I am myself from the oil industry and I am speaking as the Chairman of a world wide oil industry organisation, the International Petroleum Industry Environmental Conservation Association (IPIECA). The papers themselves illustrate the diversity of disciplines and organisations involved in the safe and orderly development of the marine environment. Therefore, the theme on which I would open the session is the need for cooperation between all these entities if any programme for protecting the marine environment is to succeed.

In this conference we are looking at the management of the marine environment from the national and the international viewpoints. A glance at the organisational arrangements by which this is done shows a line of parallel development in government and industry. Before the modern upsurge in environmental concern, many governments had environmental legislation and regulations, of one sort or another, dealing with such basics as public health, industrial effluents and land use. Over the past fifteen years such legislation has increased enormously, as has the size of the departments dealing with it, resulting in moves of varying success towards national centralisation and harmonisation. On the international level it would, I think, be true to say that practically all the international agencies related to the United Nations have developed environmental programmes. This profusion led to a need for international coordination - hence the establishment of the United Nations Environment Programme at the end of 1972 following the Stockholm Conference on the Human Environment.

Within the oil industry all companies have practised some form of environmental control since the beginning of the industry. On the national level, most countries with a developed oil industry establish national institutes of petroleum which continue to develop industry responses to national legislation. Many such national institutes are members of IPIECA. During the 1960's and 70's the industry responded to the internationalisation of environmental management by establishing international industry organisations. The first such was CONCAWE - the European organisation of petroleum refiners - which was founded in 1963. In 1970 the Oil Companies International Marine Forum was formed and in 1974 the Exploration and Production Forum. When the UNEP came into being in 1972 the need was felt for an oil industry association which could respond to it, and to other international organisations, and which would not itself be associated with any geographical region or specific function. The need for such an organisation was not only felt within the industry but was also requested by the UNEP, to simplify their contacts with the industry. IPIECA was there-fore founded in March 1974 to act as a focal point through which the UNEP and other international organisations could, as it were, plug into the world wide petroleum industry. IPIECA obtained category II non-governmental organisation status with the Economic and Social Council of the United Nations and has established working contacts with the UNEP and many other of the technical agencies. Among the oil industry association members, two which I have already mentioned - OCIMF and the E & P Forum - have formal non-governmental organisation status with the Intergovernmental Maritime Consultative Organisation. Our 26 oil company members represent a wide spectrum of the world's industry and between them account for the major part of the crude oil and products produced, transported, refined and marketed in the world. Between them, therefore, our members' activities cover the whole gamut of engineering, technical, scientific and operational matters touching on the management of the environment in the context of the oil industry, and they all conveniently meet on common ground in IPIECA.

Turning to marine activities, we asked ourselves the question some months ago - what should the petroleum industry's stance be on the main areas of concern? After a thorough review we produced an IPIECA Position Paper on Marine Oil Spills. This was published in a newsletter in February 1979 and copies are available to anyone who wants one.

This paper was not designed to be a passive pronouncement, or a general statement of goodwill, but as a recommendation, subscribed to by all IPIECA members, for the continuance and implementation of existing activities and cooperation in the environmental field, and for the development of future programmes. In short, an affirmation of current action and a blueprint for the future. I will briefly review the contents of the paper to set the scene for this session.

1. Paragraph 1 states the policy which we have consistently pursued, to urge countries which have not already done so to bring into force the 1969 Amendments to the 1954 IMCO Convention, and to hasten ratification and implementation of the 1979 Protocols to the 1973 MARPOL Convention, the 1974 SOLAS Convention, the 1969 Civil Liability Convention and the 1971 Fund Convention. All these Conventions have been developed by IMCO and the Oil Companies International Marine Forum has cooperated fully in the highly technical development programme. Marine transportation of oil is probably unique insofar as international standardisation is demonstrably to the benefit of everybody.

2. Paragraphs 2 - 4 recommend the continuance by governments, IMCO, the oil industry and the shipping industry of the already advanced programmes to improve safety and pollution prevention in tanker operations and in tanker movements.

3. Paragraph 5 recommends that oil companies should provide the capability for dealing with spills in static situations, such as loading and discharge terminals (including refineries), and at offshore oil operations.

In other situations we recommend that every oil company by itself or in cooperation with others should be able and prepared to marshal resources of men and materials to deal with oil spill emergencies. We therefore recommend the establishment of contingency plans and mutual aid arrangements capable of rapid mobilisation.

This is an area of high activity within the oil industry. Currently the industry is in the process of making a thorough world wide review of the contingency plans which exist around the world and the industry's capability to react in emergencies. As I am sure everybody here knows, the oil industry does in fact react in emergencies and has a high capacity to mobilise experts and equipment. There also exist innumerable industry oil spill cooperatives, many of them involving central and local government, harbour authorities and the like. Nevertheless there is always scope for improvement which, I believe, will result from the industry review we are undertaking. Naturally, there are limits to what industry can do on its own, and also on what it is allowed to do. Any major marine emergency involving a massive oil spill will usually, and properly, give rise to government involvement and ultimate control.

4. Paragraph 6 deals with government/industry cooperation. Our basic recommendation here is that the oil industry, to the extent it is allowed to do so, should work closely with governments to improve their spill management capabilities and to cooperate in clean up efforts. We identify the following specific matters on which industry can assist:-

(a) We recommend that industry should cooperate in effective planning for major oil spill contingencies. As I have said earlier, we have in this session several papers touching on the complex and difficult task of planning for major emergencies. It calls for a high degree of commitment from both government and industry and we recommend making industry resources in expertise, equipment, materials and skilled manpower available during emergency periods and for training exercises.

As in war, nobody can be exactly certain what will happen in an emergency save that, without planning, chaos is guaranteed.

(b) We recommend the industry to give advice on the effectiveness and effects of mechanical and chemical clean-up techniques. I am pleased to report that IPIECA participated in a meeting of experts to consider the application and environmental effects of oil spill chemicals convened by the UNEP at Brest in November last year. This subject has, perhaps understandably, been clouded with uncertainty and prejudice. The UNEP report on this meeting will be issued at the end of this month and an IPIECA paper will be published at the same time. It is our hope that this initiative will be pursued and will result in a better understanding of the subject and in the proper and effective use of chemicals being incorporated in national contingency planning.

(c) We recommend assisting governments in making inventories of regionally available equipment and materials. There have been many efforts to make such inventories, with varying degrees of success. They are difficult to make and, in particular, it is extremely difficult to keep them up to date. However, the

reviews being made within the industry, to which I have referred,
will undoubtedly be of assistance to governments and it is my hope
that we shall increasingly avoid the situation, which has often
arisen, in which a government faced with an emergency has very
little knowledge as to what is available, even within its own
national borders.

(d)   We recommend providing information on the methods of assessing
      the short and long-term effects of specific spills on the marine
      biota. The literature is extensive and we shall be having papers
      from Dr. Lasday, Mr. Hardy, Mr. d'Ozouville and Dr. McAuliffe
      which deal with various aspects of it. The importance of the
      subject is unquestionable, involving food, health and the web of
      life in the seas. Unfortunately, much of the literature is highly
      technical, as a glance at the 1977 report on the "Impact of Oil on
      the Marine Environment" by the Joint Group of Experts on the
      Scientific Aspects of Marine Pollution (GESAMP) will show, and
      some hasty and not always accurate conclusions have been drawn
      from it. A significant part of the research and field work has
      been sponsored by the oil industry and carried out independently
      by universities and scientific establishments, some of which will
      be touched on today.

      It is very relevant that we should be discussing this subject in
      Monaco, the home of several renowned laboratories, and with the
      UNEP Regional Seas Programme represented. I have followed closely
      the development of the Barcelona Convention and, in particular,
      the mobilisation of the many research centres in the Mediterranean.

      For the particular benefit of non-technical management in
      government and industry IPIECA and OCIMF will shortly be publishing
      a booklet on "The Fate and Impact of Oil Spills in the Marine
      Environment", which reviews the existing state of knowledge and
      assesses the risks.

(e)   We recommend providing information on damage control and salvage
      techniques for aiding disabled vessels. These subjects are
      already under intensive study by OCIMF and IMCO.

5.   Finally, we recommend regional oil company/government cooperation
      supplemented by the competent international industry organisations.
      This is being achieved by management dedication and by the expansion of
      the technical organisations.

Having briefly reviewed some of the cooperative activities which exist, I
express the hope that this conference will enhance them.

PETROMAR 80

EUROCEAN CONFERENCE

INTERNATIONAL CONFERENCE AND EXHIBITION
"PETROLEUM AND THE MARINE ENVIRONMENT"

OVERVIEW BY: DR. JOSE RAFAEL DOMINGUEZ
             CHAIRMAN SESSION 5 - TROPICAL ZONE

LADIES AND GENTLEMEN:

It is a great honour, and a pleasure for me to chair the important session on oil spills and pollution in the tropical zone.

The seven papers to be presented cover a great variety of aspects, scientific as well as technical or administrative and it
is very fortunate that the distinguished organizers of this e-
vent have been able to get leading specialists and outstanding
scientists to appear on this rostrum.

The contributions to this tropical session cover three important
areas: contingency planning; the effects of oil spills on the
marine ecology; and computer application.

In the past several years, contingency plans have developed from
simple lists of human and material resources available to deal
with an emergency, to complex plans establishing priorities in
the protection of certain areas according to their sensitivity
to spills.

Some recent examples have shown that practically no country on
its own can deal with the staggering logistical, technical and
scientific problems of a massive oil spill. National and international cooperation will have to be sought. One recent example
of international cooperation will be highlighted in this session.

Experience has shown that to deal with a 10.000 ton spill, an
investment of about 5 million dollars is required, and a combined effort of about 2.000 people. In another example of cooperation we shall hear about the way in which several oil companies established a method to provide a weather forecast system.
It is almost indispensable that such systems be made available
world-wide so that oil fleet and E&P operations can be made more
secure.

Scientific research in the field of marine ecology and especially on the effects of oil spills on tropical marine habitats is
really only in an early stage of development. Coral reefs and
mangrove swamps are apparently highly sensitive to oil spills.

Two papers in this session will cover these areas. Finally, the use of computers in the field of marine pollution by oil has already reached an impressive level of sophistication. A mathematical model for biological control and a simulation model of oil-slick trajectories are described in two different presentations.

It has been estimated that close to 90% of world-oil production comes from the tropical or subtropical zones, and this is likely to be the case for years to come, even though intensive oil exploration and production activities have extended to temperate and frigid climates.

Contrary to this, the spectacular marine spills, most of them related to tanker transport operations and well blow-outs, appear to have occurred in the more temperate or even frigid zones, possibly because of worse climatic conditions in these areas. Most accidents seem to occur near shores and harbours. Statistics also seem to indicate that they occur near the points of destination, rather than at the loading ports. Looking at the world tanker routes one notices that most high-risk areas are in the temperate regions, the highly congested North Sea passages and the entrances to the consumer areas of the U.S. North-East Coast.

However, and this must be the exception to the rule, last June the IMTOC-1 blew out in the Gulf of Mexico and a few weeks later two supertankers, the "Atlantic Empress" and the "Aegean Captain" collided in tropical waters off the coast of Tobago; both the largest spills so far on record; the latter, unfortunately with a great loss of lives.

Both accidents happened not too far away from Venezuela, a tropical country and one of the very early oil producers and exporters. Venezuela's Lake Maracaibo could be considered the cradle of offshore oil operations. The first wells in the lake were drilled in 1924.

The seven papers to be presented in session 5 represent an interest in both the scientific aspects of the effects of oil spills on tropical ecosystems, and the technical-administrative aspects required, or that should logically be undertaken, to combat an oil spill once it occurs.

The first paper in the session is a good example of where international cooperation has been successfully achieved in the Gulf Area. Human and material resources of private as well as government oil companies are pooled on a non-profit basis.

In this particular case, 17 multinational members from 7 different countries have found a way to help one another, while maintaining trust between industry, governments and the people of the region.

It is interesting to see that these organizations - they can be called Mutual Aid Organization, as in this case, or Cooperatives - tend to have similar problems; organizational problems outweighing technical ones. A very important aspect are communications and the need for a coordinator or at least a focal point is soon felt. We have seen this in the Caribbean. In most organizations of this kind the employment of a full time Director or Secretary becomes a necessity.

Another example of intercompany and international cooperation we shall hear about, is the way a weather forecast scheme was created in the Gulf. The second paper in this session is called "The Gulf Weather Forecast Scheme - 21 years on". Weather and wave forecasting are obviously of prime importance to the offshore oil industry. Existing communication facilities have been used for weather observation and reporting, and this could finally provide an adequate forecast service. Here again, close cooperation between private and Government interests has been achieved. One can imagine the importance of the information gathered during the past 22 years. The valuable bank of wind, wave, weather and allied data is of scientific significance not only for the oil industry, but for many other fields. Obviously this kind of information is an important contribution to the Oil Spill Contingency Plans being developed in many parts of the world, and specifically to the computer applications to oil slicks, which form part of these plans.

The third paper in this session presents recommendations for the development of national oil spill contingency plans with specific reference to tropical conditions. Currently, plans in the tropics seem to be less institutionalized than in temperate zones. The oil industry has generally taken responsibility for oil spill contingency planning in their own installations, and they have participated in inter-company cooperatives. However, coordination of national and international response resources for the development of regional or international plans will require several support projects, many of which require assistance by specialized environmental consultants. Background studies must be carried out to determine the potential for adverse effects on the marine environment of oil slicks or chemical dispersants or both in order to establish preferred cleanup methods. Under certain conditions or in certain geographical areas the physical impact of an oil slick may be more damaging than the chemically dispersed oil; or the contrary may be the case for instance at cooling water inlets for power plants or desalinization plants.

Papers 4 and 5 of this session describe the effects of petroleum hydrocarbons on two specifically tropical marine ecosystems: Mangrove forests and coral reefs.

Paper number 4 assesses possible oil industry effects in tropical intertidal, sub-tidal and open water systems, with special references to experience obtained in mangrove studies in Indonesia. Acute short-term detrimental effects like high mortalities of invertebrates and defoliation of the mangroves have been observed, however in longer term, relatively quick recuperation has been reported.

The author of the 5th paper, "The Effects of Petroleum Hydrocarbons on Corals" recommends the use of low toxicity dispersants while the oil is still distant from the reef in deep water. This coincides with studies carried out in the Caribbean which conclude that physical harm by smothering may be the most damaging effect of oil spills to corals. A very important aspect of cleaning operations is highlighted; i.e. the mechanical damage caused by cleanup crews and equipment. In this case it is the physical damage to the coral reef, but the same holds for damage to sandy shores, mudflats, mangrove forests and seagrass beds.

Quite a different aspect is highlighted in paper number 6 "The Development of a Model for Biological Control of Shallow Tropical Waters subjected to Petroleum Threats". The model simulates the effect of an oil slick on the underlying ecosystem studying the light penetration, and elaborates on the damaging effects upon the fish population. Finally, it presents its views as to how ecosystems survive under pollution threat, and ends up with the notion of comprehensive management of selected zones in which microsensors extrapolate readings into visual form using the differential equations developed by the authors.

Finally, in paper number 7 "Oil Slick Movements in the Arab Gulf", high pollution risk coastlines are identified by computer simulation, using statistically determined seasonal average drift vectors.

Please allow me to say a few words about oil operations and pollution control in Lake Maracaibo. A detailed presentation was scheduled for this conference, but unfortunately was not available in time.

Maybe it is worth while to know that Environmental Protection had a relatively early start in Lake Maracaibo with the initiative of the international Oil Companies, which was taken over and further developed by the nationalized industry. During a two year period before the nationalization, Battelle Pacific Northwest carried out a research program, sponsored by the Venezuelan Oil Industry, on "The Effects of Oil Discharges and Domestic and Industrial Wastewaters on the Fisheries of Lake Maracaibo". As a result of an extensive oil spill control equipment test programme, a wide array of equipment is available in the Lake Maracaibo area. There are several self propelled skimmers, one of them a seaworthy 90 feet catamaran, possibly the largest skimmer built so far, worldwide.

Environmental affairs for the Lake Maracaibo area are being carried out by a Committee formed by representatives of the four national oil company affilliates to the holding company "Petroleos de Venezuela".

These four companies are also working together on an ambitious project of a National Contingency Plan to protect the Caribbean North Coast against oil spills resulting from tankers or offshore production operations. Tanker traffic in the southern Caribbean is second only to the movements out of the Middle East. A large percentage of the Caribbean trade nowadays consists of African and Middle Eastern oil being shipped to and from the large transhipment ports in that area.

Drilling operations are being carried out in various sites of the Venezuelan continental shelf.

Consequently many ports on the Venezuelan coast can be considered high risk areas which warrant detailed plans and a preparedness to minimize the impact of spills on the marine environment. The development of a comprehensive national oil spill contingency plan is now being undertaken.

Like in other parts of the world, the many aspects of such a plan will have to be addressed by the execution of background studies in the biological, geological, meterological, chemical and other fields.

Much has been done already, however mainly in the U.S., Canada and in Europe. The tropical zones seem to have been the least studied scientifically. It has been speculated that tropical water communities might be less tolerant of pollution than temperate ones, although on the other hand, accumulation of the more toxic fractions of hydrocarbons should be less likely in tropical waters due to the quicker evaporation. Seagrass beds and certain species of mangroves and especially coral reefs appear to be highly vulnerable although very little seems to be known about the long-term biological and chemical effects of spilled hydrocarbons. The issue is complicated by the fact that scientists do not always agree on the interpretation of the results obtained in very specialized studies. For instance, difficulties seem to persist in distinguishing petroleum derived hydrocarbons from biogenic ones which occur naturally in certain tropical aquatic environments like some mangrove forests.

Many studies have been carried out in Venezuela universities and scientific institutions, a few in cooperation with foreign centers or research institutes. Our oil industry has been instrumental in promoting several of these.

The public in general has been very much aware of the problem of
Marine Pollution since the Third Conference of the United Nations
on the Law of the Sea was held in Caracas in 1974.

In the tropical zone, biological activity is greater than in any
other area and therefore the damage caused by oil spills is pro-
portionately of larger dimensions.

Again, in the tropical zones you will find the greatest population growth which in some countries reach the staggering figure
of 3,5 to 4%/year; it is also well known the dependency of the
population on the fishing industry for their protein requirements. Any demage to this proteinic sources by oil spills will
have a marked negative incidence on the population. A substantial reduction on this natural means of subsistence will not
only affect the people in the tropical zone but could have serious repercusions on the rest of the world.

So far many oil spill conferences have dealt with even more specific aspects of the subject, yet, it is our hope and sincere
belief that the papers to be presented in this section will not
only contribute to growing knowledge on the subject, but more
importantly, will stimulate further scientific investigations
through governmental and private industry action towards the
development and implementation of the necessary means to deal
collectively with such mishaps, which by their very nature often
go beyond the scope and capabilities of any one entity.

If as a result of this well organized Conference in Mónaco, we
shall find a closer cooperation and a better understanding between Nations and between their governments and the petroleum
industry in the sense that any accident or mishap that results
in the pollution of the seas is a problem to everyone, then we
should agree that we have taken a great step towards the solution of one the problems that as of today are seriously endanger
ing the very existence of the inhabitants of this planet.

We look to the past for experience, we look to the present for
actions, we look to the future for a better world.

                              Tank you.

SESSION 2

Cost/Benefit Analysis of Environmental Management

MAREES NOIRES : QUESTIONS D'INDEMNISATION

Martine REMOND GOUILLOUD
Chargée de conférences
à l'Université de Paris-I

Avec la pollution des mers, deux mondes sont entrés en collision. Depuis toujours, le monde des marins cotoyait en l'ignorant la civilisation des terriens. Ses mœurs, et les règles juridiques qui en sont le reflet, lui étaient propres. Et l'avènement de la recherche offshore depuis un quart de siècle n'avait même pas troublé cet ordre.

Quatre accidents ont bouleversé ces données, provoqué le choc de la mer et de la terre : le Torrey-Canyon en 1967, Santa-Barbara en 1969, l'Amoco-Cadiz en 1978 et IXTOC-I en 1979.

Désormais les terriens ont leur mot à dire dans le droit des marins. Ils exigent que des règles préventives les protègent contre les marées noires. Ils réclament, surtout, qu'une fois l'accident survenu, leurs dommages soient indemnisés.

L'indemnisation est sans doute la clef du problème. Mieux que la plus sévère interdiction, les montants d'indemnisation en jeu dans les récentes marées noires sont de nature à inspirer au chef d'entreprise une salutaire réflexion sur le choix de ses hommes et l'entretien de son matériel. Or dans ce domaine régnait la plus grande incertitude. Lorsque le Torrey-Canyon s'était échoué, les règles internationales permettant la réparation des dommages résultant d'une marée noire étaient inexistantes : la nappe de pétrole se diluant sur des zones soumises à des compétences différentes, provoquant des dommages de nature inconnue jusqu'alors, défiait les systèmes juridiques les plus raffinés. De plus, l'accident pouvait se produire en haute mer, dans un espace qui n'était pas sans loi mais où la loi était sans force. Le droit maritime, enfin, contribuait à rendre le problème insoluble : à la victime qui lui réclamait une indemnité, le propriétaire de navire opposait le jeu de la fortune de mer qui lui permettait de se dégager de toute obligation en abandonnant la valeur du navire et du fret, souvent même une simple épave, aux mains de ses créanciers.

Le Torrey-Canyon a eu raison de cette carence. Les années soixante-dix ont vu s'élaborer un mécanisme d'indemnisation ingénieux, assurant une redistribution du risque sur les milieux professionnels en cause, armateurs et pétroliers. Suivant une Convention adoptée en 1969, le propriétaire du navire est tenu responsable des dégâts dans la limite de quinze millions de dollars (1) ; au-delà, et jusqu'à trente millions de

---

(1) Convention de Bruxelles du 29 novembre 1969 : la limitation exacte de responsabilité du propriétaire est de 2.000 F Poincaré par tonneau de jauge toute, avec un plafond de 210.000 F Poincaré. Ces chiffres on été, en 1976, transformés en droits de tirage spéciaux du FMI. Le propriétaire

67

dollars, un Fonds pétrolier alimenté par les industries importatrices d'hydrocarbures devait prendre le relais (2). L'industrie s'est associée à l'effort d'indemnisation des victimes, à titre intérimaire et complémentaire : deux plans volontaires, dits TOVALOP et CRISTAL, y ont été consacrés (Cf. infra.) (3).

A leur tour enfin, les opérateurs offshore, forts de l'expérience des transporteurs, ont mis sur pied un système d'indemnisation : peu après l'accident d'Ekofisk, une Convention de Londres du 17 décembre 1976 était adoptée, tandis qu'un plan OPOL, contribution volontaire de l'industrie, venait jouer dans l'offshore le rôle des plans TOVALOP et CRISTAL. En somme, il était permis de penser que, malgré quelques failles, le mécanisme international d'indemnisation étant en place, le financement assuré, le problème était réglé.

Survient l'Amoco-Cadiz et le système s'effondre. Devant l'ampleur de la catastrophe, les plafonds d'indemnisation fixés par les conventions internationales paraissent totalement inadaptés : 160 millions de francs (4) sont disponibles, mais c'est sur 800 millions de dollars que portent les demandes (5).

Quelques mois plus tard, l'éruption du puits IXTOC-I dans le Golfe de Campêche, fournit un sinistre contrepoint au désastre de l'Amoco-Cadiz dans le monde de l'offshore. Etrange répétition de l'histoire : de même qu'en 1969 l'accident de Santa-Barbara avait montré aux pétroliers qu'ils se trouvaient, comme les transporteurs d'hydrocarbures en 1967, confrontés à un vide juridique, de même IXTOC-I vient, en 1979, faire écho à l'Amoco-Cadiz, avec un sinistre d'ampleur catastrophique (6).

Ainsi, une deuxième génération de problèmes s'est-elle fait jour. De nouvelles questions se posent, de nouveaux mécanismes devront être mis en place. Parmi ces questions, certaines intéressent surtout l'auteur du

---

est exonéré dans certains cas (ex. évènement naturel imprévisible et irrésistible, fait intentionnel d'un tiers) ; en revanche, sa faute personnelle le prive de toute limitation.

(2) Convention de Bruxelles du 18 décembre 1971. Le Fonds est, en outre, appelé à se substituer au propriétaire du navire insolvable ou dispensé de la réparation par certains cas d'exonération (cf. infra I.1).

(3) TOVALOP : Tanker Owners Voluntary Agreement regarding Liability on Pollution, signé en 1969 ; CRISTAL : Contract Regarding an Interim Supplement for Tanker Liability, 1971.

(4) Francs français. Cette somme correspond au jeu conjugué de la Convention de 1969, en vigueur depuis 1975, et du plan CRISTAL (le Fonds de 1971 n'allait entrer en vigueur qu'en octobre 1978).

(5) Voir Journal de la Marine Marchande, 5 octobre 1978 : l'Etat français réclame 300 millions de dollars, les autres victimes 500 millions de dollars. Le seul coût du nettoyage est estimé à 450 millions de dollars.

(6) Voir Offshore, août 1979, p. 43. Le Monde, 7 août. En septembre, la facture des frais de nettoyage s'élevait déjà à 360 millions de dollars (Le Monde), 27 septembre, p. 28).

dommage : elles tiennent, beaucoup plus qu'à l'aspect pollution, à son ampleur démesurée et aux difficultés de financement du risque qui en résulte. La victime, pour sa part, s'attache davantage au caractère du dommage de pollution des mers ; il lui faut connaitre le type de dommage dont elle pourra réclamer réparation, et comment organiser son action.

Ces deux aspects du problème font l'objet d'un bref examen dans les lignes qui suivent.

I. L'OPTIQUE DU POLLUEUR : L'INDEMNISATION ET LE RISQUE CATASTROPHE

Les accidents de pollution ne représentent qu'une faible part (6%) de la pollution totale des mers (7). Jamais la pollution opérationnelle (8), qui en représente pourtant 37%, n'a suscité la même émotion ni les mêmes efforts. Car la marée noire est avant tout un phénomène catastrophique. Les dommages s'y caractérisent à la fois par leur ampleur et la brutalité avec laquelle ils sont survenus. Or, face à ces deux aspects du phénomène, en droit comme ailleurs, l'organisation défaille. Les mécanismes de responsabilité classiques, supposant qu'un fautif ou un responsable doit régler les dégâts, sont inopérants : aucune personne physique ou morale n'a la surface financière suffisante pour assumer un tel risque. En outre, s'agissant d'un risque industriel, il sera exceptionnel de pouvoir isoler un responsable. L'accident industriel est généralement le produit d'une conjonction de négligences et de hasards malheureux : l'imputer à une personne unique, c'est trouver un bouc émissaire, non pas un responsable. Aussi la collectivisation du risque, entrainant la redistribution de la charge sur les milieux professionnels en cause, apparait-elle comme la panacée.

I.1. Une collectivisation nécessaire

Aussi, les textes destinés à réparer le dommage de pollution font-ils un large usage de Fonds, mutuelles et autres mécanismes d'assurance : la Convention de Bruxelles du 29 novembre 1969, régissant la responsabilité du propriétaire transporteur d'hydrocarbures, contraint celui-ci à souscrire une assurance propre à couvrir sa responsabilité dans la limite prévue par la Convention (9).

_____

(7) Cf. Rapport CNEXO par P. NOUNOU à l'Assemblée Nationale, le 14 juillet 1978 (Journ. Marine Marchande), juillet 1978, p. 1715). Pour des statistiques détaillées, voir Revue de l'I.F.P. 1979, n° 3, p. 483 : les principaux accidents de déversement pétrolier en mer, et la Banque de données de l'Institut Français du Pétrole sur les accidents de navires (1955-1979).

(8) Encore dite "chronique" ou "délibérée" : il s'agit de la pollution liée aux déballastages et vidanges des citernes de pétroliers ou aux opérations normales de forage.

(9) Le propriétaire de navire transportant plus de 2.000 tonnes d'hydrocarbures en vrac en tant que cargaison est tenu de souscrire cette assurance ou une autre garantie financière. L'Etat du pavillon et l'Etat dans les ports duquel le navire fait relâche sont tenus de veiller au respect de cette obligation, attestée par un certificat (art. VII Conv.).

L'obligation des chargeurs, destinataires du brut transporté, de supporter une partie du risque est mise en œuvre au moyen d'un Fonds d'indemnisation : la Convention de Bruxelles du 18 décembre 1971 institue ce Fonds, chargé de compléter l'indemnisation fournie par le navire ou de suppléer ses défaillances financières dans la limite de 54 millions de dollars (10). Le financement en est assuré par une contribution annuelle des industries réceptrices d'hydrocarbures, calculée en fonction de la quantité d'hydrocarbures reçue. En marge de ces mécanismes officiels, l'industrie s'est, de sa propre initiative, collectivement engagée à réparer certains dommages, en attendant l'entrée en vigueur des conventions internationales. Ainsi, le plan TOVALOP, financé au moyen d'une mutuelle d'armateurs, préfigurait-il la Convention de 1969 (11). De même, le plan CRISTAL devait-il assurer une indemnisation intérimaire, à l'image de la Convention de 1971 : on notera qu'il était, en principe, seul applicable lors de l'accident de l'Amoco-Cadiz, la Convention de 1971 n'étant entrée en vigueur que quelques mois plus tard.

Quant aux opérateurs offshore, la structure de leur industrie ne rendait pas utile l'institution d'un Fonds de solidarité. La Convention de Londres du 17 décembre 1976, calquée sur la Convention de 1969, rend l'exploitant de l'installation source de pollution (12). En revanche, la solidarité de l'industrie se manifeste dans le plan OPOL, engagement volontaire d'indemnisation proche du plan CRISTAL (13).

I.2. Les limites de la collectivisation

Il ne faut pas y voir la solution miracle : la collectivisation de la charge représentée par un risque industriel majeur a ses vertus, mais les Fonds d'indemnisation ont leur limites. On notera tout d'abord que les Fonds représentent non seulement la garantie d'une responsabilité, mais également ses limites : qui dit Fonds, dit plafond. Tout Fonds d'indemnisation comporte des limites, au-delà desquelles la victime est abandonnée à elle-même. Or, l'inflation se charge, en quelques années, de rendre ces limites insuffisantes. Sans doute les conventions modernes comportent-elles des palliatifs à la dépréciation monétaire : les montants des Conventions de 1969 et 1971, initialement calculés en francs

---

En france, une loi du 26 mai 1977 répond à cette prescription : les peines encourues sont des amendes de 50.000 à 500.000 francs. (V. Journ. Off., 27 mai).

(10) 30 millions à l'origine ; ce plafond a été relevé en 1979.

(11) Son intervention, sollicitée seize fois de 1970 à 1973, permettait de rembourser aux Etats leurs dépenses de lutte contre les marées noires.

(12) Sa charge est d'autant plus lourde que l'Etat prévoit la faculté pour l'Etat riverain, dit "de contrôle", de relever, voire de supprimer le plafond de responsabilité de 30 millions de D.S.T. qu'il institue (V. DUBAIS, The 1976 Convention on Civil Liability for Oil Pollution Damage from Offshore Operations. Journ. of Mar. Law and Comm., 1977, p. 61).

(13) OPOL. Offshore Pollution Liability. V. Intern. Legal Materials, 1974, vol. XIII. 1049.

Poincaré, ont été convertis en 1976 en D.S.T., unité de compte en principe tirée vers le haut par les monnaies fortes (14). La Convention Offshore prévoit un relèvement automatique de son plafond de responsabilité pour pallier les conséquences de la dépréciation monétaire.

C'est, en outre, le propre d'une catastrophe que d'être unique, de surprendre par l'ampleur imprévue des dommages. Les 15 millions de dollars prévus par la Convention de 1969 correspondaient à peu près au montant des dommages réclamés dans l'affaire du Torrey-Canyon. Des quelques cinquante marées noires de plus de 15.000 tonnes survenues dix années plus tard, aucune n'approchait par l'importance des déversements, et surtout des dégâts, le sinistre du Torrey-Canyon (15). Mais avec l'Amoco-Cadiz, les premières demandes d'indemnisation portent sur 800 millions de dollars. Ce type de phénomène devient dès lors techniquement impossible à intégrer aux données de l'assurance car l'assureur, qu'il gère une mutuelle, un Fonds ou une assurance classique, a besoin de statistiques pour calculer ses primes. Les risques rares et graves sont pour lui les plus difficiles à appréhender.

Rôle inéluctable de la faute. Sur le plan purement juridique même, les Fonds d'indemnisation et autres mécanismes assurant la redistribution d'un risque sur une large assise, présentent enfin deux graves faiblesses, très sensibles en matière de pollution des mers. Quant à l'efficacité du mécanisme d'abord, on peut craindre que la prime ou la cotisation, redistribuée sur une base trop large, perde toute valeur dissuasive ; diluée dans la masse des coûts d'une entreprise, elle cesse d'inciter l'entrepreneur à consentir les investissements propres à améliorer la sécurité du transport ou du forage, portant à réduire le risque de pollution. Aussi, le rôle dévolu à la faute professionnelle de cet entrepreneur est-il un élément essentiel au bon fonctionnement de tels mécanismes ; c'est ainsi que certaines fautes du propriétaire de navire visé par la Convention de 1969 le privent de la prise en charge partielle de ses obligations par le Fonds d'indemnisation de 1969 (16). A l'égard des victimes enfin, le propriétaire de navire à qui une faute personnelle est reprochée se voit privé du bénéfice de sa limitation de responsabilité.

---

(14) Il fallait surtout se défaire de l'unité de compte traditionnelle, le franc Poincaré, lié à l'or (cf. Protocoles Londres, 19 novembre 1976.

(15) Cf. BERTRAND, Etude I.F.P. préc., p. 527.

(16) Cette garantie financière offerte à l'armateur par le Fonds porte sur la partie de sa responsabilité qui excède 1.500 francs par tonneau de jauge, ou 125 millions de francs (francs Poincaré). L'armateur est privé de cette prise en charge en cas de faute intentionnelle ou au cas où, par sa faute personnelle, le navire cause de pollution n'a pas respecté certaines règles de sécurité, provoquant ainsi l'accident. (Conv. 1971, art. 5).

(17) Conv. 1969, art. 5.2 - Conv. 1976.

Il est enfin difficile, et cela aura été une des leçons de l'Amoco-
Cadiz, d'assigner des limites à la collectivisation d'un risque tel
que la pollution des mers. Le but recherché était d'en faire peser la
charge sur les milieux professionnels dont l'activité est génératrice
du risque en cause, armateurs qui tirent profit du transport et armateurs qui tirent profit du produit transporté. Mais ce milieu peut
être envisagé de manière beaucoup plus large et englober sous-traitants,
constructeurs et, de manière générale, tous ceux dont l'activité est
liée au transport d'hydrocarbures ou à la recherche du pétrole en mer.
Or, le système international d'indemnisation permet la mise en cause
de ces personnes. Lacunaire, il ne régit pas toutes les responsabilités
encourues à l'occasion d'une marée noire, mais seulement la responsabilité du propriétaire de navire (Convention de 1969) ou du concession-
naire offshore. Le système diffère ici des textes applicables en
matière de risque nucléaire ; ceux-ci canalisent la responsabilité
encourue pour les dommages nucléaires sur l'exploitant de l'installation à l'exclusion de toute autre personne (18). Aucune disposition du
même ordre n'est inclue dans les Conventions intéressant la pollution
des mers. Rien n'interdit donc la condamnation, sur d'autres fonde-
ments juridiques, d'autres personnes, tels le pilote, le navire assistant venu au secours de l'installation en détresse, voire le sous-
traitant. Or, faute de système pré-établi, la charge pesant sur ces
responsables potentiels reste indéterminée a priori : ne bénéficiant
d'aucune limitation de responsabilité, ils ne peuvent prévoir le
montant maximal des réparations qui peuvent leur être demandées.

Ainsi, l'effort du législateur international apparait-il loin
d'être achevé. C'est maintenant l'ensemble de la question qu'il va
falloir régler, afin que le système global atteigne une cohérence. A
cet égard, le projet de loi actuellement soumis à la discussion du
Congrès des Etats-Unis, envisageant le problème dans son ensemble,
pourrait fournir, sinon un modèle, au moins d'utiles renseignements
(19). Encore faudra-t-il mettre le système en œuvre. Les accidents
récents ont montré qu'il restait encore, là aussi, beaucoup à faire,
notamment quant à l'appréciation des dommages de pollution. A ce
second aspect du problème est consacrée la deuxième étape de notre
réflexion.

―――――――――――――――――

(18) V. Convention de Paris du 29 juillet 1960 sur la responsabilité
     civile dans le domaine de l'énergie nucléaire - art. 3 ; aussi
art. II.2 Convention de Bruxelles du 25 mai 1962 relative à la responsabilité des exploitants de navires nucléaires.

(19)

II. L'OPTIQUE DE LA VICTIME : INDEMNISATION ET DOMMAGE DE POLLUTION

Il est prévu au titre V du plan POLMAR, plan d'urgence destiné à la lutte contre les marées noires et révisé à la suite du naufrage de l'Amoco-Cadiz (20) qu'"une instruction interministérielle préparée par le Ministère du budget traitera des problèmes juridiques et des litiges que peut faire naitre une pollution accidentelle".

On ne saurait mieux dire que les mécanismes applicables à l'heure actuelle ne sont pas au point. De fait, les démons juridiques se sont ici ligués pour rendre la tâche des victimes malaisée. Aux exigences de la responsabilité civile classique, mal adaptées à la nature particulière des dommages liés à l'environnement, se superposent les aléas et les longueurs des procédures internationales.

II.1. Le dommage de pollution

Parce que l'accession de l'environnement au rang des valeurs protégées par le droit est un phénomène récent, les préjudices tenant à sa dégradation sont encore mal perçus et leurs contours mal définis. De plus, ils présentent généralement un caractère collectif dont s'accommode mal la responsabilité civile classique, organisée autour du couple auteur-victime dans une perspective individualiste. Ainsi s'expliquent les difficultés rencontrées par la victime : il lui faut d'abord établir que son dommage existe, et qu'il est certain. En 1973, une juridiction californienne refuse toute indemnité à des plaisanciers à la suite de l'accident de Santa-Barbara parce qu'ils n'avaient été privés que d'un "simple plaisir de pêcheurs du dimanche" (21). En 1965, une Cour d'appel française refuse, faute de preuve, de réparer le préjudice moral subi par la station balnéaire de la Baule, dont les plages avaient été souillées par un pétrolier (22). Notons, d'ailleurs, qu'en dehors des marées noires où elle n'est que trop claire la preuve du préjudice est le plus souvent difficile à rapporter, car il faut prouver, non seulement l'existence du dommage, mais également son lien direct avec le déversement : le dommage doit avoir été causé directement par la pollution incriminée. Ainsi, des hôteliers britaniques se voient-ils débouter de leur demande en réparation d'un préjudice de pollution des mers : leur préjudice est trop "remote", trop lointain (23). Pour les mêmes raisons, l'action des "business men" est rejetée par la Cour de district du Maine en 1973, à la suite de la marée noire causée par le Tamano(24). C'est enfin au même écueil que se heurtent souvent les associations de défense de l'environnement : elles ne peuvent obtenir de

---

(20) V. Instruction ministérielle du 12 octobre 1978 relative à la lutte contre les pollutions marines accidentelles. <u>Journ. Off.</u> 14 octobre

(21) "A simple Sunday piscatorial pleasure". Cf. Oppen V. Aetna Insurance Co. - Cour d'appel 9ème district, 20 septembre 1973.

(22) Le "<u>World Mead</u>" - V. Cour d'appel de Rennes, 3 novembre 1965. Rev. trim. droit commercial. 1967, p. 919, n° 29. Obs. du PONTAVICE.

(23) V. Décisions citées par D.W. ABECASSIS : The Law and Practice of Oil Pollution by Ships. 1978, p. 139.

(24) BURGESS v. M.V. TAMANO (370 F. Supp. 247. sd. Me 1973).

réparation que pour le préjudice subi à titre personnel, en tant que personnes lésées dans leurs intérêts ; ainsi pourront-elles obtenir le remboursement de dépenses raisonnables de nettoyage ou de repeuplement d'une réserve d'oiseaux (25), mais elles ne pourront en aucun cas par ce moyen obtenir la réparation du préjudice subi par leurs membres ou la collectivité (26).

En définitive, une seule catégorie de dommages devrait, en principe, être réparée sans difficulté : les frais de lutte et de nettoyage sont en effet expressément rangés par les conventions internationales au nombre des dommages réparables ; ce sont les "mesures de sauvegarde" qui représentent en pratique la part la plus importante des dommages causés par une marée noire. Toutefois, il ne suffit pas au sauveteur de présenter sa facture : pour en obtenir le remboursement, ces dépenses doivent être jugées raisonnables, ce qui soulève de délicates questions d'évaluation.

Car l'évaluation du préjudice soulève des difficultés redoutables. Ainsi faut-il, à la suite de l'Amoco-Cadiz, évaluer le manque à gagner des pêcheurs tenus de déposer leur rôle, privés de travail ; une circulaire ministérielle des 28 avril et 1er mai 1978 s'y employait (27). Mais comment évaluer le préjudice futur, la durée pendant laquelle il faudra venir en aide aux pêcheurs, les ressources marines ne s'étant pas encore reconstituées ? Encore dispose-t-on ici d'éléments d'appréciation tirés du revenu annuel moyen du travailleur privé d'emploi (28). Mais comment évaluer le préjudice causé aux espèces non commercialisables, à la faune et à la flore marines considérées indépendamment de leur valeur économique ? Une étude menée en Bretagne après l'Amoco-Cadiz, évaluant le dommage subi par les espèces marines à 400 millions de francs, se fonde sur le coût du marché international des espèces commercialisées les plus proches de celles qui ont été détruites (29).

De tels chiffres sont sans doute arbitraires ; ils ne tiennent notamment pas compte de ce que la valeur d'une espèces dépend du rôle qu'elle joue au sein de son écosystème autant que de sa valeur marchande. Pourtant, même arbitraires, ils sont nécessaires : le juge, pour se prononcer, a besoin de chiffres, de même que l'assureur, pour calculer ses primes.

---

(25) De la même manière que les associations de pêcheurs à la ligne obtiennent l'indemnisation de leurs dépenses de réempoissonnement des rivières. V. DESPAX. La pollution des eaux et ses problèmes juridiques. Paris. 1968, p. 117.

(26) V. Trib. administratif Dijon, 23 juin 1969. J.C.P. 1970. 16528. Obs. DESPAX.

(27) V. Rapport de la commission d'enquête de l'Assemblée Nationale n° 665, p. 286.

(28) V. "Les coûts des dommages causés à l'environnement". O.C.D.E. 1974. Spét. les études HAVEMAN, p. 108 et MURATO, p. 148.

(29) V. C. CHASSE. Evaluer pour comprendre, prévenir, réparer et dissuader. Rapport au colloque de l'Union des villes du littoral ouest-européen. Brest, mars 1979.

L'évaluation monétaire est le seul moyen de comparaison précis entre
les sinistres et permet, seule, en outre, de contrôler l'affectation des
sommes versées au titre de la réglementation du milieu marin. Comment,
sinon, répartir l'effort financier entre recherche scientifique et travail
sur le terrain, entre hommes de laboratoires et "chercheurs aux pieds-nus" ?
(30).

II.2. Difficultés liées au caractère international des procédures

L'accident de pollution présente, enfin, un caractère international
marqué : il est exceptionnel, sauf peut-être sur les côtes américaines,
qu'auteur et victimes du dommage soient de la même nationalité. Les procédures visant l'indemnisation des victimes s'en ressentent par des longueurs,
des incertitudes et de sérieuses difficultés d'exécution.

Les procédures, déjà longues en droit interne, sont considérablement
allongées, que le procès se déroule dans l'Etat victime du dommage, comme
le prévoient les Conventions de Bruxelles (31), ou dans l'Etat où se
trouve l'auteur de la pollution, défendeur au procès, suivant le droit
commun. L'enquête sera menée et vérifiée dans les deux Etats : traduction,
compréhension difficile des procédures étrangères, ralentissent encore les
choses.Or, pendant ce temps, les victimes attendent leur indemnisation :
les sommes en jeu portent intérêt et la monnaie se déprécie. Ce sont là
questions de mise en œuvre dont l'importance est en pratique critique.
(Que ce serait-il passé sur les côtes de Bretagne au printemps de 1978
si l'Etat français n'avait organisé de son propre chef un mécanisme d'aide
provisoire aux sinistrés ?) (32).

Les procédures internationales pêchent enfin, surtout, par les difficultés rencontrées lors de l'exécution des décisions judiciaires. L'application d'une décision rendue à l'étranger suppose un contrôle du droit
national, l'exequatur. Or, aux yeux du droit national, la décison étran-
gère, rendue au terme de textes et de procédures qui lui sont peu fami-
lières, parait souvent suspecte et il sera enclin à refuser toute "impor-
tation judiciaire". Ainsi n'est-ce un secret pour personne que l'une des
raisons majeures qui ont poussé les victimes de l'Amoco-Cadiz à porter
leur action devant le juge américain était la crainte, si le procès se
déroulait en France, que l'exécution des décisions ne puisse intervenir
aux Etats-Unis où se trouvaient les principaux défendeurs.

----

(30) A cet égard, on notera l'intérêt du projet de loi en discussion au
    Congrès des Etats-Unis : 10% des sommes destinées à l'indemnisation
des dommages de pollution des mers (le projet de loi prévoit un Fonds de
200 millions de dollars) seraient affectés à la recherche. V. Projet de
loi H.R. 89. Janvier 1979.

(31) Convention, art. 9.

(32) L'insuffisance des structures judiciaires est ici regrettable, et
    l'on songe aux mérites que présenterait ici un plan d'urgence judi-
ciaire mettant en œuvre des moyens à la mesure de l'urgence et de
l'ampleur des dégâts.

De manière générale, les difficultés d'exécution ont toujours fait obstacle à l'effectivité des condamnations prononcées à l'encontre de navires étrangers. On objectera que le droit maritime dispose de voies d'exécution. Les saisies peuvent être pratiquées sur les navires débiteurs pour les contraindre à s'exécuter. Mais cette technique est loin d'être applicable dans tous les ports (33). De plus, nombre de compagnies maritimes ne possèdent pour tout navire qu'un ou deux navires, qui constituent dès lors le seul gage de leurs créanciers et, de facto, la limite des obligations de l'armateur. Les indemnisations de marées noires ne peuvent s'en suffire (34).

Ainsi, les questions de mise en œuvre prennent-elles aujourd'hui, en matière de pollution des mers, le pas sur les principes. Après le Torrey-Canyon, le cadre juridique de l'indemnisation avait été construit. Après l'Amoco-Cadiz, il reste à le remplir.

---

(33) La Convention internationale du 10 mai 1952, qui organise la saisie conservatoire, n'a été ratifiée que par un petit nombre d'Etats.

(34) A cet égard, l'exploitant offshore présente des garanties plus solides : l'Etat riverain, en lui concédant les permis d'exploration ou d'exploitation, s'en est assuré.

SAFETY IN OFFSHORE HYDROCARBON EXPLORATION AND EXPLOITATION

H R George CBE BSc F Inst Pet
Director of Petroleum Engineering, Department of Energy, London
England

INTRODUCTION

The structural safety of offshore installations, the health and
safety of people who work on them and the prevention of pollution
of the environment from oil used or produced on these installations
are all very closely related matters. It is therefore appropriate
that this paper should consider not only safety in offshore
hydrocarbon exploration and exploitation but also the measures
to be taken to minimise the chances of pollution of the oceans,
and the relevant UK legislation governing these matters. Other
papers to be discussed later at this conference discuss the
measures taken to deal with an oil spill, should one occur.
    The UK's attitude has always been that the responsibility
for safety and the prevention of pollution lies primarily with
the oil companies and their employees engaged in offshore
activities. It is the responsibility of Government to set the
standards required and to ensure that the oil companies achieve
the required standards of safety and environmental protection.
Legislation is necessary to provide powers of enforcement although
these rarely have to be used because in such matters the interests
of the oil companies and of Governments are not normally in
conflict. An unsafe operation is unlikely to be an economic one;
there is too large a capital investment at risk and too much oil
is at stake to justify unsafe operations. Nonetheless there
may well be cases where oil companies and Governments have
differing views on the standards that are necessary and it may
not always be possible to resolve these differences by discussion.
In such cases Governments must have the last say as they have
the duty to see that their national resources are developed
properly and safely in the interests of the Nation as a whole
and that its environment and its ecology are properly projected.
Adequate legislation therefore is necessary. The principle
UK legislation is given in Table 1.
    When development of the North Sea began, the risks of
working in such a hostile environment were unknown as were the
areas of particularly high risk. During the early years of
North Sea operations, all accidents and incidents were
investigated and analyzed (and, of course, the process
continues); the most extreme example of the risk was shown by
the collapse of the "Sea Gem" rig in 1965 when 13 lives were
lost. Following this disaster it was decided to issue safety
regulations under the provisions of an Act specifically
designed to cater for offshore safety matters. Accordingly
such an Act was drafted and came into force as the Mineral
Workings (Offshore Installations) Act 1971. It is an enabling
Act. Apart from enabling the Secretary of State to make
regulations for all aspects of safety on or near offshore
installations, this Act also required managers to be appointed

who shall have overall responsibility for all matters affecting
the safety health and welfare of all people employed on the
installation and, where necessary from a safety aspect, for
the maintenance of order and discipline. The Act, and the
regulations made under it, apply to installations concerned
with the offshore exploitation of and exploration for mineral
resources in and surrounding the UK; an installation includes
both mobile and fixed types and so drill-ships, jack-ups and
semi-submersibles are covered in addition to steel and concrete
fixed platforms.

In 1977 the Health and Safety at Work etc Act, 1974, was
applied by Order in Council to the UK Continental Shelf. This
transferred the responsibility for the occupational safety and
health of workers offshore to the Health and Safety Executive
to enforce such regulations as well as existing legislation.

Some of the more important regulations are discussed below
and also some of the research work that was instituted so as to
enable the offshore installations to be safely constructed and
operated.

The Offshore Installations (Construction and Survey) Regulations
1974 These regulations govern the safety of design and
construction of offshore installations to be used on the UK
Continental Shelf. When they were being prepared, an advisory
committee decided that, in many areas, there was insufficient
knowledge to enable detailed regulations to be made and that
technology was evolving very rapidly. Detailed regulations
based on inadequate knowledge would badly inhibit offshore
development without adding to safety. Consequently, the
regulations were broadly framed and were supplemented by
"Guidance Notes" which are an aid to the interpretation of the
Regulations in the light of current knowledge. At the same
time research was instituted whenever needed. As the results of
this research work became available, and in the light of
operational experience on the UK Continental Shelf (and elsewhere in the world), the Guidance Notes have been revised and
re-issued in the light of further improvements in knowledge.

Under these Regulations, no fixed or mobile installation
may be used for exploring for, drilling for or producing minerals
on the UK Continental Shelf unless it has a valid Certificate
of Fitness which says that the design is suitable for the
purpose for which it is intended and that it has been constructed
properly in accordance with this design. These Certificates
are issued on behalf of the Department of Energy by one or other
of the six Certifying Authorities appointed by the Secretary
of State for this purpose. Periodic surveys are carried out
to ensure that the installation continues to be fit for use.

In order to comply with these regulations and guidance
notes it is necessary to ensure that an installation is so
designed and operated as to be safe for use at the location at
which it is intended for it to be used. This involves:-

a) An adequate knowledge of the environmental conditions
likely to be encountered at the location where the
installation is to be used for the duration of its life
there. This includes a knowledge of the winds, waves,
currents, water depth and the sea bed conditions. The air
and sea temperatures may also be important.

b) The installation is properly designed and constructed
so as to maintain its integrity throughout its working life

there when subjected to the forces exerted upon it by the
natural environment and the loads imposed by the operation
it is performing. This in its turn requires a proper
understanding of the forces exerted on a structure by the
winds and waves etc and of the properties of the materials
used and of the methods of construction.

Environmental Data Environmental forces have a significant
effect on almost every aspect of the exploitation of offshore
oil and gas. Design and certification of structures, tow-out
and installation work, pipelining, integrity monitoring - all
depend upon a quantitative knowledge of the environment so that
forces from wind, wave and current can be determined with
confidence.
     The wave climate, the frequency and energy spectra of waves,
are of concern in fatigue calculations made during design, and
in order to design for adequate platform clearance and resistance
to overturning moment, the occurrence of very large waves must
be predicted. The interaction between the assumed wave
conditions and structure design can be illustrated by specific
example. It has been estimated (Ref.2) that a 10% reduction
in the assumed maximum wave height from 30 metres to 27 metres
could result in a reduction of 18-20% on the maximum over-
turning moment for a platform, leading to a decrease in the
design structure weight of 15% with significant cost
implications; a similar saving would be obtained if the
assumed maximum current was halved.
     Knowledge of currents is important for other reasons also,
and data are needed at different depths for :-
     determining forces on fixed and floating structures;
     determining forces on mooring systems;
     assisting pipelaying operations;
     understanding erosion;
     estimating movement of materials, sediments and pollutants;
     making loading and unloading offshore safer; and
     improving navigation and positioning techniques.
With only a limited number of measurement sites covering an
area as large as the UK Continental Shelf, it is essential that
all sources of data are used.
     Figure 1 and Table 2 show the sites at which wind, wave
and/or current data have been, and in many cases still are
being, collected. Observations have been made from weather
ships and light vessels, fixed platforms and data buoys.

Wave Measuring Instruments The majority of wave data are
generated on one of two types of instruments:
     the Datawell Waverider Buoy; and
     the Shipborne Wave Recorder (SBWR)
The Waverider is an anchored, spherical buoy 0.7 metres in
diameter which, as its name suggests, is free to follow the
wave surface. Measurement of the vertical displacement of the
buoy is achieved by double integration of the signals
generated by accelerometers on board.
     The SBWR is readily installed on board vessels maintaining
station such as weatherships or light vessels. Accelerometers
and pressure transducers mounted on either side of the vessel
below the waterline measure the wave height and correct for
the vessel's motion.
     Measurement of wave data from fixed platforms is, in

principle, easier than using Waveriders or SBWR's but there are significant practical and operational problems. The standard instrument is a Wavestaff consisting of two vertical steel wires mounted vertically alongside the platform and with a transducer measuring the length of the wires above the sea surface. Its accuracy can be affected by the proximity of the structure which can modify the wave pattern itself, and it is prone to accidental damage which can be very expensive to repair. Waveriders are used for wave measurements near structures but they add to the complexity and cost of data acquisition and are vulnerable in the heavy shipping traffic in such areas.

Data Buoys   The concept of an unmanned buoy continuously measuring a comprehensive selection of data at a location precisely where such data will be needed, is attractive and manning costs are eliminated. The capital cost is lower but the operational costs depend on the probability of damage and the duration of power supplies and recording systems are the dominant factors in the overall cost.
Two types of databuoy have been used:
    the Marex 3 metre buoy; and
    the 7.6 metre Data Buoy I (DBI).
The former, developed by Marex Limited was used in an attempt to collect data from a position 40 miles West of Foula in the Shetland Isles to replace the "Fitzroy" weathership programme but after a very promising start the long line of communications, the severe weather and a number of teething troubles caused abandonment of the attempt. The delay in getting to the buoy for servicing sometimes stretched to weeks, because of the weather - a classic example of the practical problems out-weighing the theoretical advantages of the data buoy.
    DBI is at present successfully collecting data on the edge of the SW Continental Shelf which is transmitted to shore by an HF link. Amongst its unique features are the surface current meter and a wave direction measuring capability.*

Current Measuring Instruments   Self-recording current meters in which the current drives a rotor have been available for several years and are usually deployed unattended in a string of instruments from a sea-bed anchor to a sub-surface buoy. Meters using acoustic and electromagnetic techniques are available but the former have not been employed (except on DBI) and the latter have only recently started to be used.
    Near-surface current measurements at a depth of 3m are made on databuoy DBI using a large two dimensional ultrasonics transducer array and results to date are excellent. The UK Institute of Oceanographic Sciences (IOS) has already compared measurements from this array with measurements obtained from drift buoys with favourable results.

Other Parameters   The need for data to determine the forces on structures has meant that the measurement of wave data and current strength has been the first priority. Whenever possible, however, other data such as wind strength, atmospheric pressure, sea and air temperatures are recorded. This is usually done when the data are gathered from a ship, but it is also possible when a large buoy is used. Collection of meteorological data at coastal sites has been undertaken on a routine basis for many years, but there is a paucity of such data offshore. For example,

* See Woollen, this volume for further details.

environmental data from the area between Brittany and Eire were
virtually non-existant, other than from ships in passage,
until DBI was deployed there.

Numerical Modelling   The sites at which wave data collection
takes place are few and the data may have to be employed to
provide predictions about a large number of locations in the
UK offshore area.   Even in places where measurements have been
carried out, data are not available for periods much longer
than 3 - 4 years, while design must take into account the worst
conditions expected over a period of decades, usually 50 years
in the case of maximum wave height.   Consequently, a two-fold
extrapolation is required - one in estimating the long-term
conditions from short-term observations; the other in
predicting conditions in locations far from the wave measurement
sites.   This extrapolation can be carried out by means of a
mathematical model which correlates sea conditions with
meteorological data.
    The first Guidance Note 50 year wave map was based on
extreme wave height prediction models developed by IOS which
were simple relationships between the wind and the waves.   A
basic understanding of wave growths and decays was obtained
by the Joint North Sea Wave Project (JONSWAP) and in 1975 work
started on a numerical parametric wave model of the Northern
North Sea (known as NORSWAM Ref.3) which could hindcast waves
from measured winds during the period from 1966 to 1976.   The
work was mainly done at the Hydraulics Research Station, advised
by an international team led by IOS.
    Meteorological data from 42 storms in the period was used
to calculate surface wind strength and direction at points on
a 100 km grid.   The model was used to calculate significant
wave heights at these points and comparison of these results
with actual wave data refined the model, thus improving the
hindcasting back over 10 years.   The results from NORSWAM have
been compared with other independant methods of estimating
maximum wave heights and agreement is generally good but with
significant differences in the maximum wave height contours
around the UK (Fig. 2).   The model also has limitations arising
from the coarse (100km) grid used and the presence of perturbing
influences within this grid such as the Orkneys and Shetlands
which have not been taken into account   There is therefore
still some uncertainty in the available estimates of extreme
wave height which can only be reduced by further refinement
of the model.

Fluid Loading on Installations   In order to design an installation
so as to be able to withstand the winds, waves and currents to
which it is subjected, it is necessary to calculate the forces
which they exert upon it as accurately as possible.   Although
much of the theory of fluid loading on structures has been
known for a long time, the equations used embody empirical
co-efficients and it was considered necessary to institute a
research programme to validate both the theories and the co-
efficients to be used.   In 1973 a series of research projects
to do this was initiated by the Offshore Structures Fluid
Loading Advisory Group (OSFLAG).   The more important parts of
this programme are briefly mentioned below (Ref. 4).
i) Wave loads on large Structures   Wave loads on large off-shore
structures such as concrete gravity oil production platforms,

can only be predicted accurately by taking into account the
diffraction of waves by the structure. The most important parts
of this project involved an experimental validation of the use
of linear diffraction theory to predict the wave loading on
large wide structures and the development of associated
computer programmes for use by designers.
ii) Wave Slam. Each time a wave hits and submerges the upper
horizontal bracings of an offshore installation the members are
subjected to an impulsive loading known as "wave slam". Some
local failures of bracings have been partially attributed to
this type of fluid loading. The object of this programme was
to improve the method of predicting the impulsive loading on
horizontal cylinders set at or near still water levels in wave
flows.
iii) Wave forces on conductor Tubes The major purpose of this
investigation was to measure the drag and transverse forces on
an individual cylinder of an array subject to oscillating flow
and to compare these with the forces on a single isolated
cylinder under the same flow conditions.
iv) Drag co-efficients and surface Roughness The aim of this
project was to provide basic data on the drag force co-efficients
of cylindrical sections roughened by marine growth.
v) Wind Loads Wind loads are important in the construction,
deployment, installation and operation of all types of fixed
and floating offshore installations. In this project models of
typical structures were tested in a wind tunnel with the
objective of improving methods of predicting wind loads.

The Christchurch Bay Tower Experiment The main purpose of this
project was to study wave forces on a tower in a real random
sea and at a larger scale than is possible in wave tanks. An
important secondary objective was to study the behaviour of the
foundation of an offshore gravity platform.
    The main structure of the original tower is shown in detail
in Figure 3; the main steel tower is 13m high, 2.8m in diameter
and was supported on a 10.5m diameter reinformed concrete base
1m thick. The smaller diameter cylinder (the wave staff) is
mounted alongside and is 0.48m in diameter. The main cylinder
is divided into five rings so that wave forces on it can be
measured and the small cylinder has four sleeves for the same
purpose. During storm seas, forces on the main column remain
inertia dominated but the smaller cylinder experiences loadings
which are much more influenced by drag as soon as wave heights
exceed 3m.
    The layout of the other measuring instruments is shown in
the same figure. Considerable redundancy was built into the
instrumentation.
    The tower was placed in 8.4m of water in Christchurch Bay
at a location where the sea-bed has a 3m thickness of fine sand
overlying grey-green stiff fossiliferous clay. Christchurch
Bay was chosen as the site for the experiment for a number of
reasons including: storm conditions arise with a reasonable
frequency due to the long fetch to the south-west which is the
direction of prevailing winds, the shoaling water of the Bay
increases the probability of occurrence of wave heights near to
the design height, the energy in the wave directional spectra
is concentrated in the south and south-west sectors enabling
the platform's measurement stations to be placed so they rarely
shield each other and the tidal range in the Bay is one of the

lowest in the UK waters.

<u>Co-efficients for Morison's Equation</u> The objectives of this project were to :-
    a) appraise the state of the art regarding the fluid loading of the offshore structures from the designers point of view;
    b) to identify those areas requiring further research; and
    c) to suggest any beneficial modifications that might be made to current research projects.
The final report on this project was presented as a working guide to methods of estimating wave and current loads on offshore structures (Ref.5).

    With the exception of the Christchurch Tower experiment mentioned above, all the other experimental work was conducted on relatively small scale models and there is always the uncertainty about the reliability of scale when it comes to applying such results to full size structures in real seas. The Christchurch Bay Tower experiment went some way to bridging the gap, but in order to provide proper validation at full scale instrumentation has been applied to full scale or large scale offshore structures. These include instrumentation on the BP Forties field platform FB, a steel jacket type production platform, and the Shell Brent B structure, a major concrete gravity platform. In additional Exxon have erected an ocean test structure in the Gulf of Mexico which is about one-third the height of typical North Sea production platforms, and there is also a German experimental structure "Nordsee". The overall aim of each of these projects was to collect and analyse basic information about the structures and, using these data, to verify the design principles which have been followed and so improve design techniques.

    Taken together all this work now enables a much greater degree of confidence to be placed upon the calculated forces exerted upon a structure by the environment which have to be taken into account in the design of an installation.

<u>Properties of Materials and Fabrication</u> Even if the environmental conditions were perfectly known and the forces the environment can subject an installation to were perfectly understood, doubts still remain as to how an installation will stand up to these forces over an extended period of time. Oil and gas platforms have to be designed to withstand the frequent storm conditions in the North Sea, while the severity of the almost continuous wave action results in the need to take particular account of fatigue as a possible mode of structural failure. The low temperature of the environment increases the risk of brittle fracture.

    The cyclic loading consideration is especially important for the widely-used steel-jacketed piled structures and the problem is exacerbated by the scale and the complexity of the deeper-water installations. The situation is well recognised and fatigue-life analyses form an integral and significant part of the design and the initial certification procedures. Subsequent detailed checks for damage due to the prolonged cyclic wave loading are routinely made during the regular in-service inspections. The weakness, however, in such fatigue assessment procedures lies in the lack of a gradual evolution of design, based on an extended operational history, which is

commonplace in other engineering disciplines. The extremely
sparse data base which has had to be used represents a
considerable extrapolation of onshore experience of welded
structure.

The UK Offshore Steels Research Project (UKOSRP) In the light
of this situation and following a detailed review, the UKOSRP
was established to provide guidance on the various factors
likely to influence the fatigue and fracture behaviour of the
welded steel jackets of fixed offshore installations.

   The programme of work, which is broad, although by no means
comprehensive, is largely concentrated on the acquisition of
basic engineering data. The substantial body of information
which is being generated will be relevant both to the design
and construction of new installations and to the validation of
existing structures. It will help to quantify the factors of
safety and to identify possible areas of concern.

   However, the project goes further and should provide a
more detailed insight into the fatigue and fracture behaviour
of complicated welded structures in a corrosive environment.
Such information is required :-
   a) to provide a rationale for the inspection of structural
   integrity;
   b) to assist the assessment of the significance of defects
   found during inspection (repair requirements etc); and
   c) to provide an improved basis for action on structural
   cracks or defects resulting from accidental overload (storm
   or collision).
Improved fatigue and fracture data is also likely to be necessary
to provide guidance on the increasingly optimised or novel
strucutres which may be required for oil recovery in deeper
waters.

   The overall programme of fatigue and fracture studies is
made up of a series of related projects involving stress
analysis, basic tests of corrosion fatigue, brittle fracture
studies and full scale fatigue tests on welded joints.

   The results broadly confirm the conservative nature of the
earlier basic design assumptions and, for example, in the case
of stress analysis they have resolved some of the important
discrepancies between different methods of calculation. On the
other hand, the work has also highlighted the very complex and
time-dependant nature of the fatigue process in the sea-water
environment. It has indicated that under certain conditions,
the rates of crack growth may be substantially faster than
expected and it has exposed significant discrepancies between
accepted laboratory methods of endurance and crack-growth
measurement. There are also some preliminary indications of
adverse scaling effects which may be ofparticular import to
the very large steel structures. The fracture programme has
provided some useful results but it has not yet succeeded in
providing straightforward guidance on the requirements for the
heat treatment of thick weldments (Ref.6).

European Offshore Steels Research Mention should also be made
of the related work being carried out under an international
offshore steels research programme supported by the European
Coal and Steel Community. Work under this programme is being
carried out in the UK, France, Germany, the Netherlands and
Italy and close liaison maintained between the programme and

UKOSRP so that the two complement each other.

<u>Concrete in the Oceans</u> In parallel with the UKOSRP a programme
of research was mounted to investigate the behaviour of reinforced concrete in the marine environment. This programme
was designed to investigate such matters as :-
   fundamental mechanism of corrosion of steel reinforcement
      in concrete immersed in sea water;
   corrosion in offshore concrete;
   evaluation of methods for designing against excessive
      cracking and examination of the relationship between
      corrosion and design crack width;
   influence of environment, stress and materials on corrosion
      of reinforcement in concrete;
   the effects of temperature gradients on the walls of concrete
      oil storage structures;
   the performance of existing reinforced concrete marine
      structures;
   strength of large prestressed concrete members in shear;
   fatigue strength of reinforced and prestressed concrete in
      sea water; and
   modes of failure of concrete platforms.
Most of these projects have now been completed and the programme
has recently been extended to cover a new investigation of
implosion strength and a major extension of the work already
started on crack and corrosion criteria.

<u>SAFETY OF DRILLING AND PRODUCTION OPERATIONS</u>

Foremost in the case of drilling operations is the necessity to
ensure that the wells are drilled safely so as to minimise the
chances of an uncontrolled escape of hydrocarbons or a "blowout".
It is essential that each well be programmed with this in mind.
Particularly, attention must be paid to all available information
relating to the sub-surface formations likely to be encountered
in the drilling of the well. The possibility of these formations
being abnormally pressured must be carefully considered and
possible zones of lost circulation must be identified. Casing
points must be selected with these considerations in mind and
with a view to the containment of pressures likely to be
encountered in the next section of hole to be drilled. The
casing, mud and cementation programmes must be designed
accordingly.
   In addition, the drilling rig must be equipped with blowout
preventors capable of containing the maximum shut in pressure
likely to be encountered assuming, in the worst case, the hole
is full of gas. Blowout preventor stacks in common use offshore normally have 3 ram type preventors (one with shear rams
and the other two with different sizes of pipe rams) and an
annular type of preventor through which drill pipe can be
pulled.
   Automatic devices to give warning of unaccountable changes
of mud volumes in the mud circulation system are used as well
as detectors to give warning of gas in solution in the mud.
   If, despite all precautions, a blowout should occur it may
be necessary to drill a relief well and therefore wells must be
surveyed at regular intervals so that at all times the bottom
hole position is accurately known. On a multi-well platform
from which many deviated wells are drilled it is necessary for

the course of all wells to be considered and precautions be
taken to ensure that a new well does not pass dangerously
close to existing wells. Computer models can be helpful in
this respect.
     Wells must be completed with automatic down the hole safety
devices so that in the event of damage to the wellhead from
whatever cause, the well will be automatically closed in.
This is normally achieved by having the tubing-casing annulus
isolated by means of a bottom hole packer and an automatic
down the hole safety valve installed in the tubing string
some 200-300 feet below sea-bed level. This valve is normally
of the "fail-safe" type held open by pressure from the surface.
In the event of damage, the pressure would be released and the
valve would close. Wellheads themselves are also normally
equipped with other fail-safe valves.
     In the production operation, it is necessary that the whole
products system be designed, constructed and tested so as to
safely handle the volumes of fluids to be produced at the
operating temperatures and pressures. Automatic high-low
level indicators and warning devices are employed and
detectors which will automatically shut in a well or unit in
the event of fire, abnormally high or low pressure, smoke, etc.
These devices may also initiate automatic water deluge systems.
These precautions resemble those used in similar installations
on land, but the lack of space on an offshore platform and
the close proximity of various pieces of equipment and pressure
piping systems mean that extra special attention must be paid
to such matters.
     Probably one of the most hazardous operations which take
place on an offshore installation is when a workover is being
performed. Here again, the workover programme must carefully
consider all aspects of what may happen during the course of
operation. A workover is being carried out in order to remedy
something that has gone wrong and frequently by its very nature
one does not always know exactly what the trouble is or what
may be encountered. Therefore extra care has to be taken and
careful consideration be given to any changes that have to be
made to the original programme.
     In the UK the relevant legislation concerning safe drilling
practices etc are the Petroleum (Production) Regulations 1976
(SI No. 1129/1976) which contain Model Clauses attached to the
licences granted to explore for and produce petroleum as
supplemented by various Continental Shelf Operation Notices
which, while they do not themselves have statutory force,
amplify what may be required under the Model Clauses.

HEALTH AND SAFETY OF WORKERS ON OFFSHORE INSTALLATIONS

Apart from the major hazards which could result from damage to
an installation or from a blowout, workers on offshore
installations are at all times subject to normal industrial
type hazards.
     At present on the UK Continental Shelf, the regulations
governing normal working activities on an installation are
the Operational Safety, Health and Welfare Regulations made
under the Mineral Workings Act. However, in 1977 the Health
and Safety at Work etc Act 1974 was extended to the UK
Continental Shelf by Order in Council and in due course, new
regulations will be made under this Act to supplant the former

regulations and as far as is reasonable, to bring offshore and onshore safety into line with each other.

The Offshore Installations (Operational Safety, Health and Welfare) Regulations 1976

These lay down standards of safety in an area which is the source of the great majority of accidents on offshore installations. In particular these Regulations lay duties and obligations on concession owners, owners and managers of offshore installations, employers and employees to ensure that the provisions of the Regulations are complied with; generally the Regulations ensure that operations on and near installations are carried out in a safe manner. They promote the use of safe practices widely recognised as being essential to safety in any industrial environment and, in addition, provisions are laid down for certain procedures unique to offshore installations. Each installation must have at least one trained medical attendant on board and a comprehensive medical store.

The Offshore Installations (Emergency Procedures) Regulations 1976

The Regulations require that there be provided for every installation which is normally manned an emergency procedure manual specifying the actions to be taken in emergencies. They also require the provision of a muster list, the holding of musters and drills and that a vessel is at all times present within 5 nautical miles of any manned installation ready to give assistance in an emergency.

The Offshore Installations (Life-Saving Appliances) Regulations 1977

These require the provision of life buoys, life jackets and means for descending to the water on every offshore installation to which they apply. They also require the provision of survival craft, and general alarm and public address systems on every such installation which is normally manned. There are also provisions requiring the keeping and display of plans of life-saving appliances.

The Offshore Installations (Fire-fighting Equipment) Regulations 1978

These require the provision of fire and flammable gas detection systems, remote control safety devices, fire extinguishers and fireman's equipment on all the installations to which they apply, and the provision of fire alarms, fire mains, hydrants and hoses, water deluge systems or water monitors and automatic sprinkler systems on all such installations which are normally manned. Any equipment provided must be of a type approved by the Secretary of State and must be protected from damage.

The Offshore Installations (Diving Operations) Regulations 1974

These lay duties upon concession owners and other responsible people relating to the provision of necessary equipment and the procedures to be followed in diving operations. They provide for the appointment of diving supervisors to exercise immediate control of diving operations and for the keeping of logbooks. They require that divers shall be certified fit for diving and shall, unless experienced, only dive after instruction and under close supervision by another diver. Requirements are imposed for periodic testing and daily examination of diving equipment and for the provision of rescue services.

ENFORCEMENT OF UK LEGISLATION

Among the first regulations to be issued under the 1971 Act
were the inspectors and casualties regulations which make
provision for the inspection of offshore installations by
inspectors appointed by the Secretary of State and for the
reporting of casualties and other accidents. They detail
the powers given to inspectors and lay duties upon owners of
installations and others to assist the inspectors in the
carrying out of their duties. All installations on the UK
Shelf are inspected at intervals of approximately 3-4 months
and all fatal accidents and serious occurrences are investigated
by inspectors of the Petroleum Engineering Directorate of what
is now the Department of Energy.
    With the extension of the Health and Safety at Work etc
Act to the offshore, the same inspectors are also responsible
for the enforcement there of this Act under an Agency Agreement
between the Health and Safety Executive and the Department of
Energy.

TRAININGJOF OFFSHORE WORKERS

It has long been recognised that inadequate training of workers
is one of the main causes of accidents and training facilities
for offshore workers have been expanded. The Drilling Technology
Training Centre, which was established in 1975 by the Petroleum
Industry Training Board, was moved to Montrose in 1977 and
provides training in all aspects of drilling for new entrants,
experienced drillers and all intermediate grades. Courses in
production technology are available at the Montrose Centre and,
on the same site, the Offshore Fire Training Centre was opened
in January 1978. This provides for full training of offshore
personnel in fire fighting; the equipment at the centre
reproduces the fully enclosed modules found on many installations.
    Facilities for training in survival at sea are also being
expanded and a larger and more diverse fleet of survival vessels
is being built up at Aberdeen and Yarmouth.
    In 1976 the Man Power Services Commission established an
Underwater Training Centre at Loch Linnhe in the West of
Scotland. In addition to providing basic Air and Mixed Gas
Courses, the centre is available for specialist training; for
example - diving supervisors.
    The results of the increased regulation of offshore
installations and of the greater provision of training facilities
is reflected in the overall reduction of the serious accident
rate since the development of the UK Continental Shelf began.
The fatal accident rate is much more variable from year to
year but also shows signs of an overall decrease (Fig. 4)(Ref 7).

DISCHARGE OF OIL FROM OFFSHORE INSTALLATIONS

Apart from oil that may be released into the sea accidentally
as a result of a blow-out, or from some malfunction of equipment
on the platform, offshore installations can also give rise to
two types of discharge that contain small amounts of oil:-
    i) production water present in the natural oil reservoir
    and separated from the stream of produced oil before

storage or transport to shore;
ii) displacement water from the temporary oil storage
system, where oil floats on a layer of sea water and as
more oil is produced and stores before transport to shore,
this sea water is displaced.
The amount of production water is usually small during the
initial life of the oilfield, but increases progressively as
the field is depleted and water invades the rock formation of
the natural reservoir. The maximum rate at which it will be
discharged is governed by the capacity of the plant installed
to separate water from the oil production stream. Oil
production and hence the discharge will presumably cease when
the water content of the stream becomes so high that the
operation is no longer economical.
    The rate at which displacement water will be discharged
is related to the rate of oil production and consequently will
increase during the early life of the field, reach a maximum
and then decrease.
    At one stage it was anticipated that there would be a third
source of oily water discharges from UK oil installations,
originating from oil tanker ballast water. However, almost
without exception, permanent or segregated ballast tankers are
used to transport oil from offshore loading buoys and this
source has been virtually removed.

The Need for Regulation of Discharges
The factors dictating that these discharges should be controlled
on the UK Continental Shelf (UKCS) are that :-
    i) oil is known to be toxic to marine organisms above
    certain concentrations and to cause harmful sub-lethal
    effects at certain lower concentrations;
    ii) the discharge of persistent oils in the UKCS is
    governed by the Prevention of Oil Pollution Act, 1971,
    and exemptions granted to offshore installations by the
    Department of Energy must ensure that the best practical
    means are employed to minimise the amount of oil released;
    iii) neighbouring states and other users of the sea have
    a right to expect that the UK will responsibly control
    discharges from its installations into the North Sea and
    will comply with the provisions of international
    conventions to control such discharges.

Origin of the present provisional standard
The Paris Commission (PARCOM) has adopted a joint proposal
from France and the UK that all new platforms should be
equipped with the best practicable means for separating oil
from discharged water and that these means are corrugated
plate interceptors or gas flotation units, or other equipment
capable of reducing the average oil content of a discharge to
within the range of 30-50 ppm (parts x 10^{-6}). PARCOM has
therefore adopted a provisional target standard of 40 ppm.
However, the acceptance of this proposal was without prejudice
to the view that the control of marine pollution by crude
oil by environmental quality objectives is preferable. The
UK has accepted the provisional target standard of 40 ppm as
a practical equipment performance standard for offshore oil
platforms. All such installations which make discharges of
either production or displacement water are fitted with the
equipment envisaged by PARCOM. All apart from one have, since

production commenced, been required to maintain a monthly
average oil content of less than 40 ppm (large installations)
or 50 ppm (small installations). The consent conditions
require that :-
    i) the oil content should be determined by a specific
    solvent extraction/infrared spectroscopic method;
    ii) samples should be taken at least twice per day and
    analysed by the approved method or a method which is
    demonstrably correlatable with it;
    iii) in any one calendar month not more than 4% of these
    required samples are permitted to have an oil content in
    excess of 100 ppm;
    iv) the volume discharged should not exceed a specified
    limit, based on the treatment capacity of the separation
    plant;
    v) consideration will be given to a requirement that the
    discharge be continuously monitored, once a suitable
    instrument has been identified.
The UK has thus implemented the provisional target standard
for discharges involving more than 95% of the oil discharged in
treated production and displacement water from its installation.
    Discharges from UK installations during 1978 permitted
under the above procedure resulted in the release of about
100 tonnes of oil. Recent predictions indicate that this
could rise to about 3000 tonnes per annum in 10 years time
when the volumes of produced water to be treated reach much
higher levels.
    At the largest single UK discharge, the discharge rate could
rise to a peak of 2500 m^3h^{-1} with a discharge point only 30m
above the sea bed. Under these conditions the maximum bottom
concentration is calculated to be about 30 ppb at 1000m downstream from the source, while the concentration in water column
at the discharge depth does not fall to 50 ppb until 800m
downstream.
    Hence under average water conditions, even at the peak
discharge rate in the largest fields life, we can expect the
proposed environment water quality objective to be met in
bottom waters outside a 1000m zone downstream from the
installation.

INTERNATIONAL COLLABORATION

UK has always valued the benefits of international collaboration
to the safe development of offshore resources of petroleum.
Close liaison has been maintained, particularly with the other
North West European countries bordering the North Sea so as to
achieve as much harmonisation of safety requirements for offshore installations as it is reasonable to expect. In 1973,
the UK convened a conference of these states to discuss the
subject and following this conference working groups were set
up to see how far the objective of harmonisation could be
achieved. A further conference was held in The Hague in 1978
and a large measure of harmonisation has been achieved. It is
intended to hold a further conference at which it is hoped that
the countries involved will give their formal approval to the
recommendations of these working groups. Liaison has been
maintained with the work of other international organisations
in this field, notably the Inter-Governmental Maritime
Consultative Organisation (IMCO) which has recently produced

a code for mobile offshore drilling units.

The collaboration between the UK Offshore Steels Research Programme and the European Coal and Steel Community has already been mentioned. In addition the UK — Norwegian Co-ordinating Committee has discussed many matters of mutual interest, including safety, research and development and anti-pollution measures.

The UK is a contracting party to the agreement for co-operation in dealing with pollution of the North Sea by oil, known as the Bonn agreement. It is also a party to the Convention for the Prevention of Marine Pollution from land based sources, normally referred to as the Paris Convention.

In addition, co-operative arrangements have been made between the UK Offshore Operators Association and their equivalent organisations in Norwegian and Dutch waters, to provide mutual assistance in the event of a blowout or other major disaster, particularly if there is a risk of pollution.

OFFSHORE SAFETY LEGISLATION   Table 1

PETROLEUM PRODUCTION ACT 1918

PETROLEUM PRODUCTION ACT 1934

Petroleum (Production) Regulations 1935 (SI 426),

Petroleum (Production) (Amendment) Regulations 1954 (SI 1378),

Petroleum (Production) (Amendment) Regulations 1957 (SI 1697),

Petroleum (Production) (Continental Shelf and Territorial Sea)
   Regulations 1964 (SI 708),

Petroleum (Production) Regulations 1966 (SI 898),

Petroleum (Production) Regulations 1976 (SI 1129),

Petroleum (Production) (Amendment) Regulations 1978 (SI 929);

CONTINENTAL SHELF ACT 1964

Continental Shelf (Designation of Areas) Order 1964 (SI 697),

Continental Shelf (Designation of Additional Areas) Order 1965
   (SI 1531),

Continental Shelf (Designation of Additional Areas) Order 1968
   (SI 891),

Continental Shelf (Designation of Additional Areas) Order 1971
   (SI 594)

Continental Shelf (Designation of Additional Areas) Order 1974
   (SI 1489),

Continental Shelf (Designation of Additional Areas) Order 1976
   (SI 1153)

Continental Shelf (Designation of Additional Areas) Order 1977
   (SI 1871)

Continental Shelf (Designation of Additional Areas) Order 1978
   (SI 178)

Continental Shelf (Designation of Additional Areas) (No 2)
   Order 1978 (SI 1029),

Continental Shelf (Jurisdiction) Order 1965 (SI 1881),

Continental Shelf (Jurisdiction) Order 1968 (SI 892),

Continental Shelf (Jurisdiction) (Amendment) Order 1971 (SI 721),

Continental Shelf (Jurisdiction) (Amendment) Order 1974 (SI 1490),

Continental Shelf (Jurisdiction) (Amendment) Order 1975 (SI 1708),

Continental Shelf (Jurisdiction) (Amendment) Order 1976 (SI 1517),

Continental Shelf (Jurisdiction) (Amendment) Order 1978 (SI 454),

Continental Shelf (Jurisdiction) (Amendment) (No. 2) Order 1978 (SI 1024),

Continental Shelf (Protection of Installations) Order 1978 (SI260),

Continental Shelf (Protection of Installations) (No. 2) Order 1978 (SI 673),

Continental Shelf (Protection of Installations) (No. 3) Order 1978 (SI 733),

Continental Shelf (Protection of Installations) (No. 4) Order 1978 (SI 890),

Continental Shelf (Protection of Installations) (No. 5) Order 1978 (SI 935),

Continental Shelf (Protection of Installations) (Variation) Order 1978 (SI 1411).

NB Seventeen earlier Orders have been consolidated into SI 260)

THE MINERAL WORKINGS (OFFSHORE INSTALLATIONS) ACT 1971

The Offshore Installations (Registration) Regulations 1972 – (Statutory Instrument 1972 No. 702);

The Offshore Installations (Managers) Regulations 1972 – (Statutory Instrument 1972 No. 703);

The Offshore Installations (Logbooks and Registration of Death) Regulations 1972 – (Statutory Instrument 1972 No. 1542);

The Offshore Installations (Inspectors and Casualties) Regulations 1973 – (Statutory Instrument 1973 No. 1842);

The Offshore Installations (Construction and Survey) Regulations 1974 – (Statutory Instrument 1974 No. 289);

The Offshore Installations (Public Inquiries) Regulations 1974 – (Statutory Instrument 1974 No. 338);

The Offshore Installations (Diving Operations) Regulations 1974 – (Statutory Instrument 1974 No. 1229);

The Offshore Installations (Application of the Employers' Liability (Compulsory Insurance) Act 1969 Regulations 1975 – (Statutory Instrument 1975 No. 1289);

The Employers' Liability (Compulsory Insurance) (Offshore Installations) Regulations 1975 – (Statutory Instrument 1975 No 1443) made under the Employers' Liability (Compulsory Insurance) Act 1969;

The Offshore Installations (Operational Safety, Health and Welfare) Regulations 1976 - (Statutory Instrument 1976 No 1019);

The Offshore Installations (Emergency Procedures) Regulations 1976 - (Statutory Instrument 1976 No. 1542);

The Offshore Installations (Life-Saving Appliances) Regulations 1977 - (Statutory Instrument 1977 No. 486);

The Offshore Installations (Fire-fighting Equipment) Regulations 1978 - (Statutory Instrument 1978 No. 611);

The Offshore Installations (Life-Saving Appliances) (Amendment) Regulations 1978 - (Statutory Instrument 1978 No. 931).

HEALTH AND SAFETY AT WORK ETC ACT 1974

Health and Safety at Work etc Act 1974 (Application outside Great Britain) Order 1977 (SI 1232)

PETROLEUM AND SUBMARINE PIPE-LINES ACT 1975

Submarine Pipe-lines (Diving Operations) Regulations 1976 (SI 923),

Submarine Pipe-lines (Inspectors, etc) Regulations 1977 (SI 835),

Merchant Shipping (Diving Operations) Regulations 1975 (SI 116),

Merchant Shipping (Diving Operations) (Amendment) Regulations 1975 (SI 2062),

Diving Operations Special Regulations 1960 (SI 688).

SUBMARINE TELEGRAPH ACT 1885

MINISTRY OF FUEL AND POWER ACT 1945

COAST PROTECTION ACT 1949

PREVENTION OF OIL POLLUTION ACT 1971

GAS ACT 1972

CONTROL OF OIL POLLUTION ACT 1974

DUMPING AT SEA ACT 1974

OFFSHORE PETROLEUM DEVELOPMENT (SCOTLAND) ACT 1975

FATAL ACCIDENTS AND SUDDEN DEATHS INQUIRY (SCOTLAND) ACT 1976

ENERGY ACT 1976

KEY TO MAP OF WAVE DATA COLLECTION SITES   Table 2

No	Location	Latitude	Longitude	Duration	Respons
1	Fitzroy(Block 205/5)	60°00'N	4°00'W	1974-1976	UKOOA
2	Stevenson (210/21)	61°20'N	0°00'E	1973-1976	UKOOA
3	Boyle (92/13)	50°40'N	7°30'W	1974-1977	UKOOA
4	Brent C (211/29)	61°04'N	1°43'E	1976 continuing	UKOOA
5	Forties (21/10)	57°40'N	0°50'E	1974 continuing	UKOOA
	(22/6)	57°50'N	1°10'E	1974 continuing	UKOOA
6	Frigg (10/1)	59°55'N	2°05'E	1979 continuing	UKOOA
7	Foula	60°08'N	2°59'W	1977-1978	UKOOA
8	Lima	57°00'N	20°00'W	1975 continuing	UKOOA
9	DBI	48°43'N	8°58'W	1978 continuing	UKOOA
10	Casquests	49°55'N	2°56'W	1979 continuing	IOS
11	S.Uist	57°20'N	7°29'W	1976 continuing	IOS
12	Scillies	49°52'N	6°41'W	1977 continuing	IOS
13	Seven Stones	50°04'N	6°04'W	1969 continuing	IOS
14	Kinnards Head	57°56'N	1°54'W	1979 continuing	IOS
15	Berwick on Tweed	55°49'N	1°52'W	1979 continuing	NMI
16	Butt of Lewis	58°43'N	6°09'W	1978 continuing	NMI
17	Famita	57°30'N	3°00'E	1969 continuing	IOS
18	Dowsing	53°34'N	0°50'E	1975 continuing	IOS
19	St Gowan	51°30'N	5°00'W	1975 continuing	IOS
20	Eddystone	50°10'N	4°16'W	1976 continuing	IOS

UKOOA — United Kingdom Offshore Operators Association
IOS   — Institute of Oceanographic Sciences
NMI   — National Maritime Institute

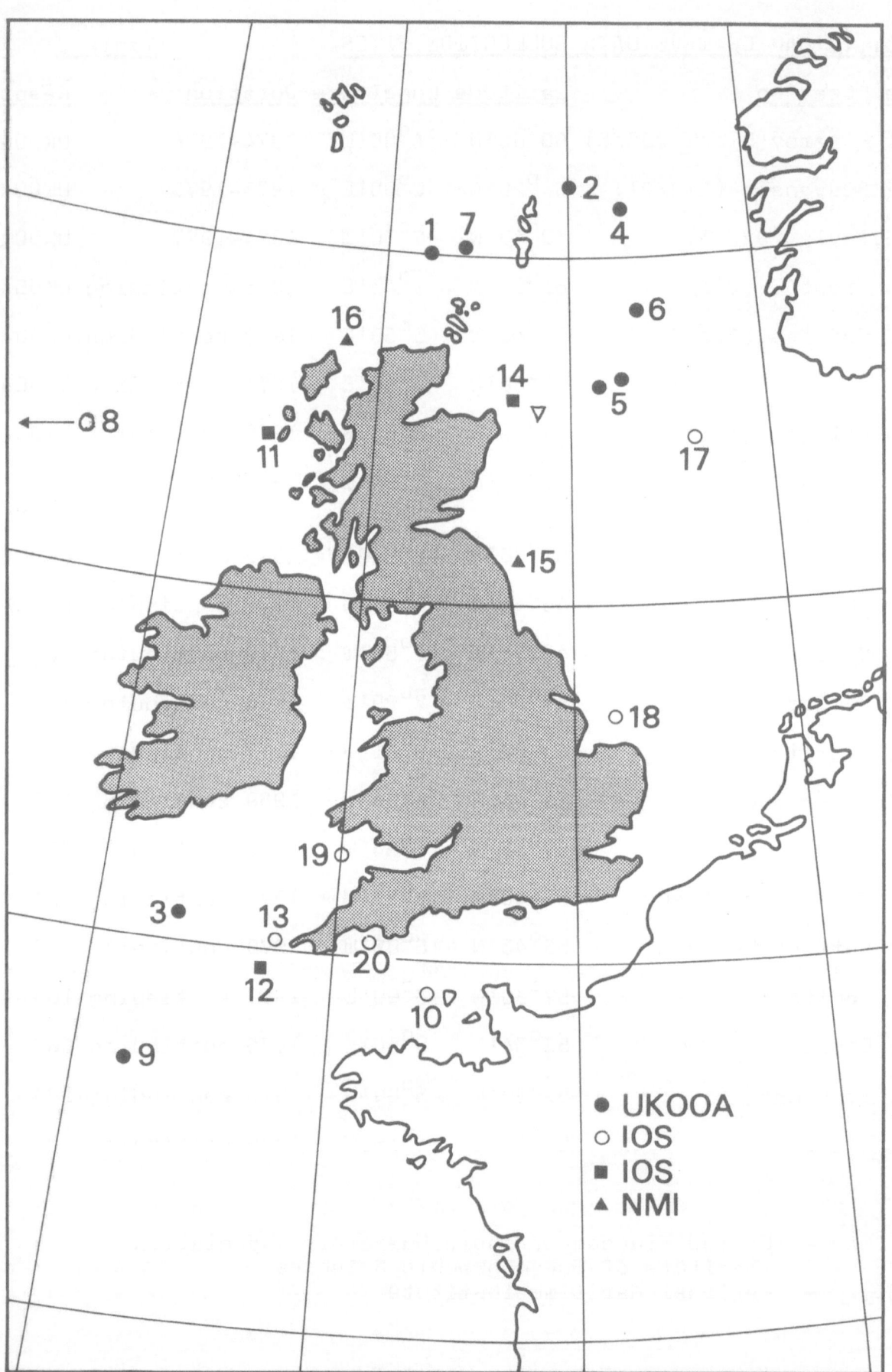

Fig. 1: Wave Data Collection Sites

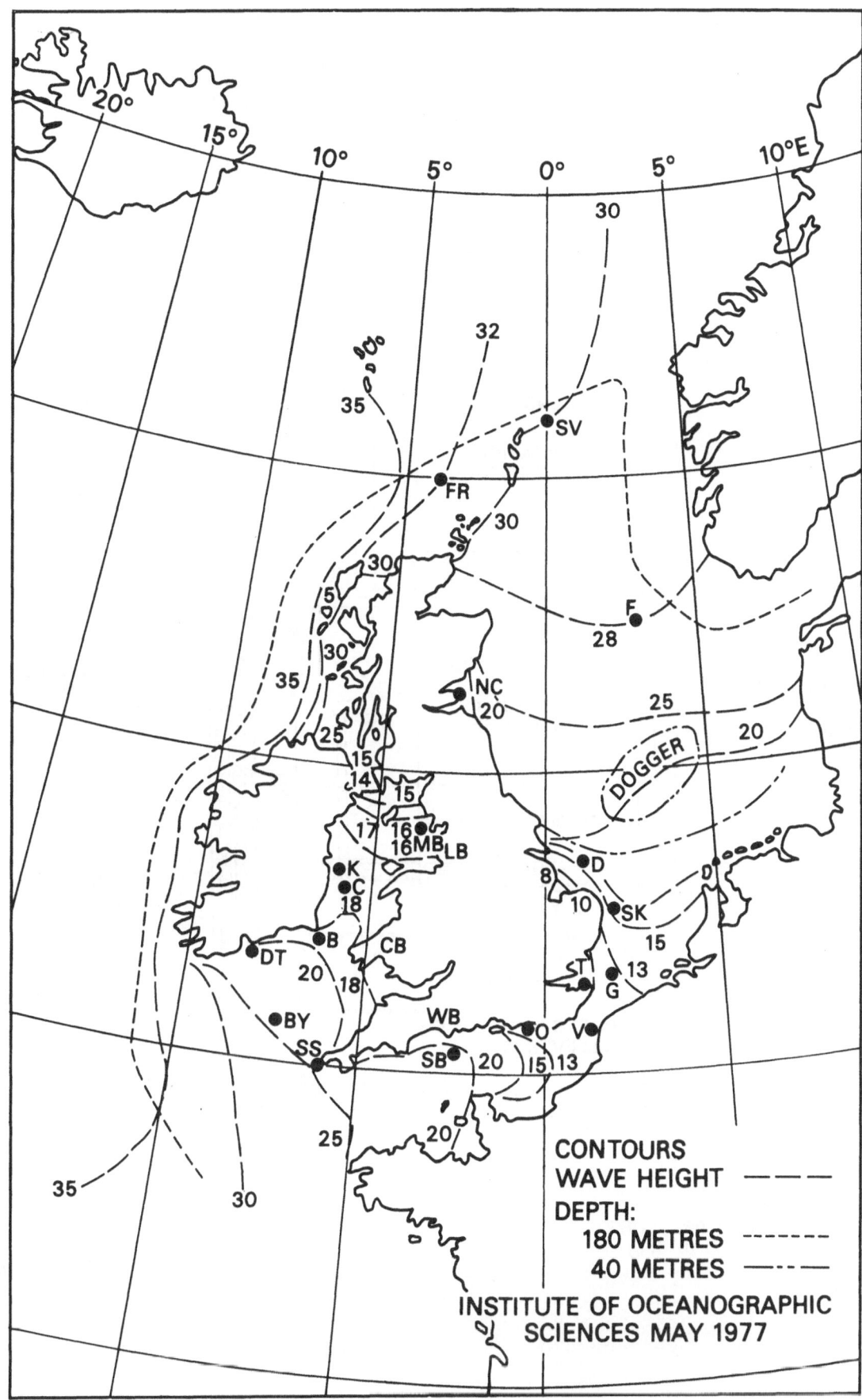

Fig. 2: Fifty year design wave heights in meters.

98

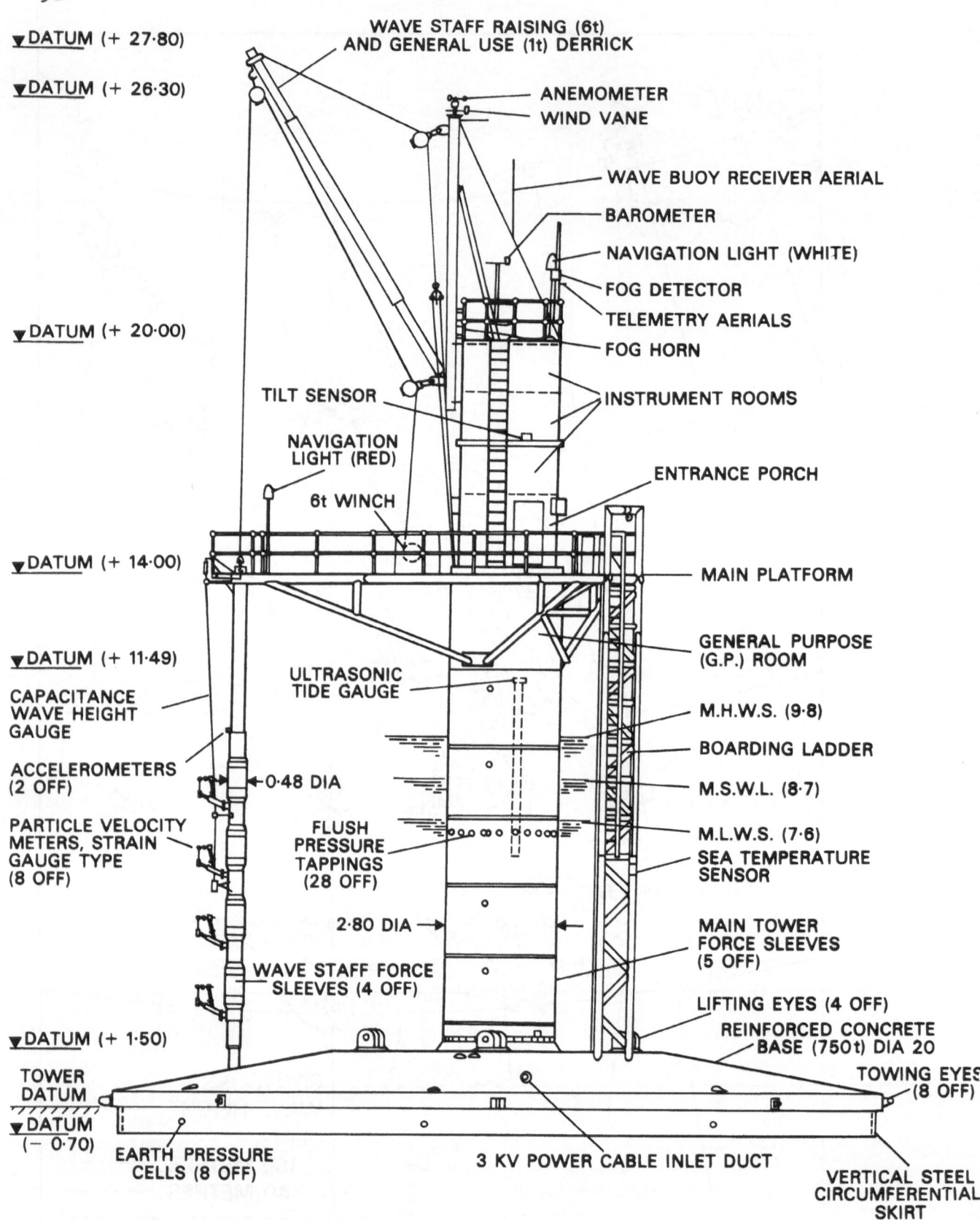

WAVE STAFF RAISING (6t) AND GENERAL USE (1t) DERRICK

▼DATUM (+ 27·80)

▼DATUM (+ 26·30)

ANEMOMETER
WIND VANE

WAVE BUOY RECEIVER AERIAL

BAROMETER

NAVIGATION LIGHT (WHITE)

FOG DETECTOR

▼DATUM (+ 20·00)

TELEMETRY AERIALS

FOG HORN

TILT SENSOR

INSTRUMENT ROOMS

NAVIGATION LIGHT (RED)

6t WINCH

ENTRANCE PORCH

▼DATUM (+ 14·00)

MAIN PLATFORM

▼DATUM (+ 11·49)

CAPACITANCE WAVE HEIGHT GAUGE

ULTRASONIC TIDE GAUGE

GENERAL PURPOSE (G.P.) ROOM

M.H.W.S. (9·8)

BOARDING LADDER

ACCELEROMETERS (2 OFF)

◄0·48 DIA

M.S.W.L. (8·7)

PARTICLE VELOCITY METERS, STRAIN GAUGE TYPE (8 OFF)

FLUSH PRESSURE TAPPINGS (28 OFF)

M.L.W.S. (7·6)

SEA TEMPERATURE SENSOR

2·80 DIA ►

◄ MAIN TOWER FORCE SLEEVES (5 OFF)

WAVE STAFF FORCE SLEEVES (4 OFF)

LIFTING EYES (4 OFF)

▼DATUM (+ 1·50)

REINFORCED CONCRETE BASE (750t) DIA 20

TOWER DATUM

TOWING EYES (8 OFF)

▼DATUM (− 0·70)

EARTH PRESSURE CELLS (8 OFF)

3 KV POWER CABLE INLET DUCT

VERTICAL STEEL CIRCUMFERENTIAL SKIRT

Fig. 3: N.M.I. Christchurch Bay Tower

FATAL ACCIDENT RATE (Approximate rate per 1000 workers)

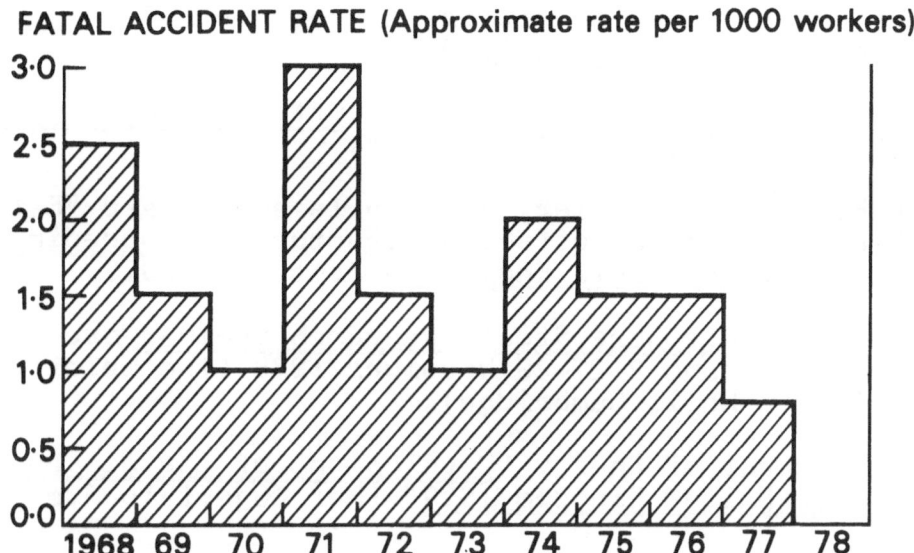

SERIOUS ACCIDENT RATE (Approximate rate per 1000 workers)

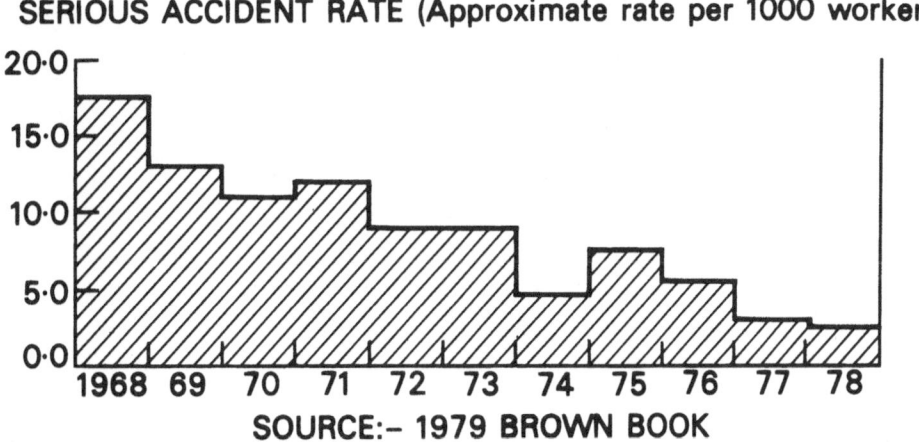

SOURCE:– 1979 BROWN BOOK

Fig. 4: Fatal and serious accident rates from 1968 to 1978

REFERENCES

1. Department of Energy "Offshore Installations: Guidance on design and contruction" (Her Majesty's Stationery Office) 1977.

2. W M McKinney "The effect of variations in design criteria on steel offshore structures" EUROPEC'78 Paper No. EUR 25.

3. J A Ewing, T J Weare and B A Worthington "A hindcast study of extreme wave conditions in the North Sea" Journal of Geophysical Research 1979.

4. Department of Energy "OSFLAG" Offshore Research Focus No 3 (CIRIA September 1977) Page 4.

5. N Hogben, B L Miller, J W Searle and G Ward "Estimation of Fluid loading on offshore structures" Proceedings of the Institute of Civil Engineers September 1977.

6. European Offshore Steels Research Seminar pre-prints Volumes 1 and 2 27-29 November 1978 (The Welding Institute, Abingdon Hall, Cambridge).

7. Department of Energy "Development of the oil and gas resources of the United Kingdom 1979" (Her Majesty's Stationery Office).

MANAGEMENT OF THE ENVIRONMENT

A call for cooperative stewardship

by

F. Hughes

Chairman of the Oil Industry International
Exploration and Production Forum

ABSTRACT

Management of the environment is the management of change. This century
has seen unprecedented development and the rate of change seemingly increases with time. Man's activities no longer affect only his local surroundings, they impact increasingly on a wider scale. Nature has immense
capacity to adapt to accommodate change and is not fragile. There is a
limit however to what the environment can tolerate by way of pollution and
all of us - in Government, in business and as individual citizens - must
contribute to responsible management of the environment.

Society's impact on environment does not respect frontiers whether
these are industrial, local, regional, or national, on or offshore.
Management control requires agreement within society at various levels
and frontiers. There is need to agree on control, on standards to be established for environmental quality, on measuring performance against those
standards, on the deployment of resources and on the real costs and benefits to society. This is Governments' role in management of the environment.

Industry must concern itself with its societal responsibilities, participating responsibly in formulating standards, measuring performance, regulating and controlling its activities and processes in a manner which ensures compliance with national and international standards where these exist,
and with perceptible competence and responsibility where they do not. This
calls for training and maintaining standards of competence among personnel,
establishing and keeping to codes of practice and exercising discipline in
regard to people and resources and their disposition, to ensure adequate
attention to managing industry's impact on the environment.

Examples are given of how the exploration and production section of the
oil industry currently participates in management of the environment.

Private citizens can and should participate in completing the process
of management of the environment by responding responsibly in matters of real
concern in the public interest. Responsible and accurate perception by
properly informed citizens provides an important regulatory mechanism which
can and should be encouraged to contribute to effective management of the
environment.

1. INTRODUCTION

Environmental Management is management of change. Man has changed the environment as long as he lives, cultivating the soil, irrigating arid regions, by hunting attractive species, by urbanization. "Nature" also changes the environment continuously, floods, earthquakes, draughts, climatic changes and ice ages have recorded their impacts not only in written history and folklore but also in the sedimentological profiles of holocene and pleistocene ages. Generally the rate of change was relatively slow.

This century, however, innumerable societies have undergone unprecedented development. Unavoidably their impact on the environment has escalated and the environmental changes have assumed a dramatic pace. Man's activities no longer affect only his local surroundings; often they now impact on the environment on a regional wider scale. A relevant example is the building of the Aswan dam, which over the later years has affected the fishing industry in the eastern Mediterranean so drastically.

The environment, nature, through its natural processes (such as dilution, photo oxidation, biodegradation) has a massive capacity for coping with society's waste.

The sea around us is not a fragile system, it is strong and can easily recuperate and even rejuvenate after "accidents" of our society. But there is a limit to what the environment can take in the way of chronic pollution. Over the last ten years it has increasingly been realised that responsible stewardship over our resources and environment is of paramount importance.

Government, industry and public, each for its specific part, must contribute responsible environmental management.

2. THE ROLE OF GOVERNMENTS

A developed society strongly impacts on the environment. It benefits from the intensive use of the environment: increased food production, recreational pleasures, cooling water, etc. It will suffer from unlimited use of the environment: desertification, smog, dead lakes.

It is therefore the responsibility of Governments to control the use of the environment in society's interest and it has to consider society's needs against the costs. The needs for energy, employment, healthy water and air, land for farming and for recreation have to be weighed against, for instance, use of scarce resources, outlay of capital, pollution of rivers or seas. Economy and ecology should be brought into equilibrium.

Society's impact on environment does not respect frontiers. Air emissions from W. Europe are reported to come down as acidic rain in Scandinavia. The Netherlands suffers now from salination of its potable water resources by the brine effluent of French potassium mines.

Inter-governmental consultation on such matters abounds, and many regional arrangements for regulation of environmental matters have been made or are in the making.

Examples of such arrangements for the marine environment are the 1974 Paris Convention, against pollution of the North Sea, and the 1976 Barcelona Convention against pollution of the Mediterranean.

Such concerted efforts can mean the setting of a framework for the environmental management of the area considered. To be effective, the nations involved must share a well-defined common concern. It may well be that the relatively slow progress in the Mediterranean consultations may, also, be due to the broad sweep of the objective: "a blue Mediterranean". The circulatory patterns in the Mediterranean suggest that restricted "compartmentalized concerns" might draw specific commitments more readily. The Balearic Sea compartment certainly has its own well-defined pollution problems, for instance. Such an approach is already visible in the Adriatic, where Italy and Yugoslavia are active bi-laterally.

Government control on use of the environment can be established by setting standards (for land use, building, industrial development and operations, etc.) and enforced by procedures and inspections. Standard setting immediately raises the question: which standards to set? Like zero risk in every day life, zero pollution is an unattainable goal. And pure distilled water is toxic to quite a few living species. Such unrealistic goals rather illustrate the necessity to use a workable definition of "pollution".

The U.N. definition given by their experts in the GESAMP* in 1974 still appears to provide a workable basis:

"Pollution is the introduction by man, directly or indirectly, of substances or energy into the marine environment (including estuaries) resulting in such deleterious effects as

harm to living resources,
hazard to human health,
hindrance to marine activities including fishing,
impairment of quality for use of sea water, and
reduction of amenities."

It is obvious from this definition that "impact on the environment" per se is not the issue, but the "impact that causes deleterious effects". In other words, it calls for local Environmental Quality Criteria against which human activity should be judged.

Setting the admissible pollution limits at arbitrary fairly low levels may defeat the ultimate government goal: to benefit society in a broad sense. The final "costs" in nett unemployment, wasted energy or capital may be too heavy.

Japanese environmental legislation is a case in point. Their original environmental legislation of 1967 which used "harmonisation with economical development" as a reference, was changed in 1970 to a framework that was geared towards "quality of life".

Admissible ambient SO_2 pollution limits were set at 0.04 ppm in 1973. Industry protested, but attempted to comply, and succeeded through innovative technology at considerable capital investment, to achieve levels below the limit set. Yet, in the same country, admissible ambient NO_x pollution limits

---

*Group of Experts on Scientific Aspects of Marine Pollution

were set at 0.02 ppm in 1973. There were no health data available that
justified so low a limit. In spite of the industry's efforts to comply the
limit turned out to be achievable only against extravagant "costs" in
capital, energy and expected nett unemployment. In view of this failure,
the Japanese Environmental Agency has now announced the relaxation of the
limit from 0.02 to 0.04 ppm. It is expected that this limit can be achieved
in 1985, against at least a capital outlay between 400 and 700 billion yen.
     A similar need to relax too stringent standards comes from the U.S.A.
The EPA relaxed the primary health for ozone from 0.08 to 0.10 ppm, when
research data showed the original standard to be excessively low. Society's
savings, at least in capital investment, are estimated to be around one
billion dollars.
     The basic question for governments, therefore, is: How clean, at
what cost? Or: How clean can we afford to be? It should be answered at a
government level where the views of the Departments of Environment, of Energy,
of Social Affairs, etc., are all taken into account, and compared to the
governments' Environmental Quality Criteria (E.Q.C.).
     For such E.Q.C. both factual baseline data and sound technological
advice are needed. Factual baseline data are essential. They are the only
realistic basis against which to assess at what levels the above deleterious
effects to our ecosystem start under the established local conditions, and
to compare any progress made during time. This, moreover, enables governments to maintain a realistic and flexible environmental policy that sets
standards attuned to local or regional conditions, rather than rigid costly
nation-wide limits. Environmental agencies would be well advised to heed
the warning signal sounded by UNEP's Director, Dr. Mostafa K. Tolba:

     "It is important that environmental issues relating to industry
and industrialization be rationally, pragmatically and efficiently managed...
     ... Information is required as a basis for making correct decisions
on environmentally-sound management in development activities".

3.   THE ROLE OF INDUSTRY

Oil industry, as part of society, also has its impact on the environment.
By nature cost-conscious, the oil industry, in order to protect life, investment and environment, is nowadays most interested in safe and clean operations.
It has become part of its business policy. Although not all of them use the
imaginative language of the Venezuelan oil company MARAVEN in its environmental Creed (see Annex I), all responsible oil companies have formulated environmental policy principles, and have functional and staff divisions to implement them and look after the world-wide adherence to company policy.
For examples see Annex I.

     The operating companies use appropriate technology, company procedures
and training to arrive at an environmental practice adequate to comply with
prevailing standards. In those areas where no environmental legislation
exists, the company standards apply. Good operational housekeeping is promoted by environmental audits and by continued environmental awareness from
the side of top management. Most major oil companies support environmental
base line studies, at Universities for instance, and have their own environmental R and D. Corporate annual environmental research expenditures of more
than a million dollars are no exception.

The result is that, overall, the environmental performance of the oil industry compares well with that of other industries and municipal authorities. Reviews of its impact on various sea or coastal areas bear this out*. It is also supported by the observation that for oil exploration and production operations the greater part of major blow-outs/oil spills is not due to equipment failure but to human failure, poor judgement or just negligence. Where decades of operating experience have perfected the technology and procedures, the human element has become the weakest link, and only continued training and promotion of awareness can bring about further improvement. This therefore, is an area of major concern for the oil industry. It is and must be however, a continuing process. Environmental awareness is an attitude that cannot be legislated into existence. It requires the active co-operation of all members of the work force and the incubation of discipline.

Its vast experience puts the oil industry in a favourable position to be a useful technological adviser to governments, both nationally and internationally. Christian Hambro, early 1979 Deputy Gen. Director of the Norwegian Ministry of Environment underlines this in his observation: "good working relationships between industry and government are of paramount importance."

The Oil Industry's E & P Forum and the Oil Companies International Marine Forum have a record of competent technological advice with IMCO, with the 1974 Paris Commission, etc. This was evidenced, yet again during recent IMCO/MEPC proceedings, where developing nations were advised to make effective use of the experience and expertise of shipping and oil industry to arrange their oil spill contingency planning and their regional spill combat arrangements.

Rather than elaborate generally I would prefer in this audience to highlight two examples of oil industry contributions to responsible management of petroleum in the marine environment. The examples are taken from two major areas of pollution concern, an accidental one and a chronic one.

1.   Oil Spill Combat

Offshore industry has undertaken sophisticated measures in this field. North Sea co-operative spill-combat forces now becoming available in the U.K., Norway and the main producing areas are impressive, and so is the back-up expertise. Low-toxicity chemicals for slick dispersion and beach protection are on stock. Similar provisions are available elsewhere, around the U.S. coast, in the Gulf area, etc. Industry is now moving beyond the fairly uniform contingency measures, towards tailored spill-combat and coastal-protection strategies with appropriate government agencies, in which selective preparations are being made in advance of possible accidents. Basic in such an approach are the estimation of the pollution risk for a specific stretch of coast, and the selective protection measures on the basis of local ecological evaluation.

To estimate this risk, the E & P Forum has developed and made available a forecast model SLIKTRAK, which will provide an estimate of pollution risk, of the arrival times and volumes of oil, and clean-up costs to be expected. It has been tested and updated during and after the Ekofisk-Bravo blow-out and is now widely accepted as a useful tool. A multipurpose version, SLIKFORCAST, will be available soon. This includes a short-term slick-movement forecast capability (based on the Norwegian IKU "OILSIM" model).

---

*See a.o. Ambio, 1977, Nr. 6, p. 312-326

The protection of a certain coastal area has to be adjusted to the existing ecosystem, with its own resilience and repopulation characteristics to natural or man-inflicted disasters. Governments will have to decide on the value of that ecosystem to society, and the economic costs justified to protect it. Priorities should be given to those cases where any harmful effects of beaching crude are most likely, for example confined shallow locations as sheltered bays, estuaries and tidal flats. Again, the absence of relevant data for proper decision-making in local situations, calls for urgent co-operation of all parties involved. Utmost use should be made of the unfortunate occurrence of "spills of opportunity" to collect and analyse such data in a scientific way.

2.   Standards for offshore production water discharges

During production usually reservoir water is co-produced with the crude oil and/or gas, often in large volumes per day. After proper separation, the water must be disposed of by discharge into the sea. It nevertheless still contains oil droplets at very low concentrations, (normally of some 30-33 ppm). To put this in perspective: 30 ppm equals 1 tot whiskey distributed over 1000 bottles of water.

To what levels should we purify produced water before discharging it? Two questions have to be answered before a well-considered decision can be taken:
- what can technologically be achieved, at what cost?
- at what oil concentrations must deleterious ecological effects be expected at the location?

To answer the first question the E & P. Forum in 1977 started two workgroups to evaluate the status of the technology in sampling, analysis and separation of oil from produced water. The conclusions of these workgroups were presented at the June 1979 IMCO/MEPC meeting and the reports are available to any interested party.

The second question calls for formulation of the local Environmental Quality Criteria on the basis of adequate baseline data. Generally such data and even the estimate of orders of magnitude involved are lacking. The growing recognition that responsible environmental management needs such data is now, fortunately, creating more and more projects where such local ecological baseline data are collected. These data, though expensive to collect, may in the end save society sizeable amounts of money. They will allow a flexible Environmental Quality Criteria approach, where for instance, the effluent discharge limits for platforms in a coastal zone or in estuarine conditions can be more stringent than those for platforms offshore in high seas conditions.

4.   PUBLIC, THE THIRD DIMENSION

The press, but above all the mass media currently involve the public at large on an unprecedented scale in environmental issues. First industry, but quite often at a later stage also government agencies are the object of severe criticism. This is not always unfounded.

Industry and government alike will have to adjust to the reality that the public has become a third dimension in environmental management with quite a political impact. It is easy to list large-scale projects which have been seriously affected due to intervention by the public: the Alaskan pipeline, refineries at the eastern U.S. seaboard, the Anglesey terminal and pipeline, the Scottish Mossmorran LPG plant, and so on.

This intervention should not change the basic responsibilities of government and of industry, but it does call for new "societal" qualities in Environmental Management. The public's "perception" is becoming as important to the viability of a project as the "facts". Open project planning, providing for citizens' advisory participation right from the start of large projects, will be necessary.

Examples are becoming available where such planning has been successful, in addition to often being cost-effective. Examples are the Big Cypress oil production development (Florida), the Sullom Voe Oil Terminal development, and the U.S. Steel Corporation steel mill project at Conneaut, Ohio.

In cases where "open planning" led finally to project abandonment, it can well be argued that this would have been the fate of the project at the hands of the public anyway, probably with greater delay and cost.

Through such a participative approach, between industry, Government agencies and the public, the sound technological and scientific data together with well-considered societal values can enable the government to exercise its basic role: to maximize the benefits of today's society without detriment to the opportunities and environmental heritage of to-morrow's.

MARAVEN CONSERVATION CREED (1977)

Maraven shares the genuine concern existing in all spheres for the protection
and improvement of the environment, upon which the survival and prosperity
of mankind depend in great measure. In consequence, it is the policy of
this company to conduct all of its industrial and maritime activities com-
plying with the environmental protection norms known and feasible. All of
this geared to the integral conservation of the natural conditions of the
environment which are vital for life itself and the harmonious development
of the human being. Additionally, Maraven has as its corporate policy, that
of motivating, supporting and cooperating as far as possible with all persons
and institutions which share the concern of the company for the environmental
conditions of the Venezuelan nation; and of participating in such national
and international organisations in fulfilment of this purpose.

BP BRIEFING PAPER ON ENVIRONMENTAL PROTECTION (1979)

Companies which have (and recognise that they have) a major impact on the environment often (but by no means always) have a written policy on environmental protection. The petroleum companies have been in the vanguard here, proably because their operations have been increasingly circumscribed by environmental constraints.

The question remains, however, how do such companies ensure that the pious generalities written into their policies are translated into sound environmental practice lower down the management hierarchy?

BP has written policies on environmental protection and perhaps more importantly, it has its Environmental Control Centre (ECC). The ECC, which was set up in 1970 and is directed by Mr. Geoffrey Larminie (General Manager), co-ordinates the environmental efforts of the various operating departments, operating areas, and of associate and subsidiary companies.

The ECC contributes to training courses for branch managers, divisional managers and general managers. Environmental co-ordinators are scattered throughout the associated and subsidiary companies, but are kept in touch by means of an environmental bulletin, specialist fact sheets and a two-day Environmental Forum held twice a year.

Among the subjects covered at the last Forum, in November 1978, were ecological monitoring; heavy oil recovery; the rate and effects of spilled oil; the long-range transport of sulphur dioxide and other photo-chemical oxidants developments in BP's R & D department and within the ECC; lead pollution; the use of oil dispersants, the Christos Bitas spill, and the role of the ECC.

All major new developments are the subject of an environmental impact assessment (EIA) and ECC endorsement is required for any planned capital investment of more than £5 million. Some of these studies will be reviewed at a later date in the ENDS Report, but it is also interesting to note that BP carries out environmental audits - to evaluate the performance of existing refineries, chemical plants and so on.

SHELL POLICY ON HEALTH, SAFETY AND ENVIRONMENTAL CONSERVATION (1978)

The following statement, which has been endorsed by the Steering Committee for Safety and Environmental Conservation and has the approval of the Committee of Managing Directors, sets out policies to guide individual Shell Companies on Health, Safety and Environmental Conservation matters.

Policy Guidelines

It is the policy of Shell Companies to conduct their activities in such a way that the health and safety of their employees and of other persons are safeguarded and proper regard is paid to the conservation of the environment. In implementing this policy Shell Companies do not merely comply with the requirements of the relevant legislation but promote constructive measures for the protection of health, safety and the environment for all who may be affected by their activities.

In following this policy they:

i)   -   seek to avoid harm to the health of, or injury to employees and others and damage to property from their operations and from those of contractors acting on their behalf
  -   promote safety consciousness in their employees and contractors

ii)   -   use their best endeavours to provide products which, when used for the purpose intended and in accordance with any advice provided will not cause injury or harm to health or the environment

iii)   -   pay appropriate regard to the environment by acting to preserve air, water, soil and plant and animal life from adverse effects of their operations or those of contractors acting on their behalf and to minimize any nuisance which may arise from such operations

iv)   -   work with governments, local authorities, industry and academic and professional bodies as appropriate and take the initiative where necessary to promote workable and improved codes of practice, and timely and practical regulations which relate to the above matters

v)   -   encourage, support or conduct research directed towards the improvement of safety and health at work, towards ensuring the safety of their products and towards the abatement of pollution

vi)   -   facilitate the transfer to others of the know-how they develop in these fields

vii)   -   keep their employees, the relevant authorities and the public well informed on what they are doing to improve the quality and safety of the working environment

viii) - include future requirements and expected developments in these areas in their long-term planning

ix) - recognize the importance of the involvement and commitment of their employees in developing and applying this policy.

COST-BENEFIT ANALYSIS FOR ENVIRONMENTAL MANAGEMENT

Yusuf J. Ahmad
Deputy Assistant Executive Director
United Nations Environment Programme,
Nairobi, Kenya

There is need for an authoritative and objective statement on the application of cost-benefit analysis to environmental protection measures, on the specification of the areas and levels to which such analysis can be applied and on the methodologies that could be appropriately used. Case studies so far undertaken to evaluate the benefits and costs of environmental protection measures have remained tentative, if not eclectic, in nature. There are very few case studies which could be characterized as bona fide cost-benefit analyses; most appear to use the language and technique of such analysis for more abstract purposes, such as general policy evaluations, bolstering decisions affecting the environment which have already been taken on political or other policy grounds, drawing up environmental impact assessment statements of one type or another, research into the structure of existing environmental standards adopted earlier and their secondary and tertiary effects, and so forth.

The reasons for this situation are not far to seek. Environmental goals and constraints are by definition long-term and general in nature while economic policies and planning are largely short-term and specific: methodologies, such as cost-benefit analysis, developed to deal with essentially economic scenarios cannot easily be applied to the more complex environmental parameters.

On the other hand, the current economic situation makes it necessary that a rigorous and economically acceptable analysis of environmental protection measures be attempted in order to carry conviction. In the past decade, the context of environmental policies has changed perceptibly. In the late Sixties and until the recession of 1974-75, the cause of concern was the abatement of the backlog of what may be called the classical pollutants (BOD, COD and SO_2), during a period marked by full employment, cheap energy and sustained economic growth. The economic context in 1979 is not only qualitatively different but environmental concerns have also changed. A host of new pollutants constituting major health hazards have been identified, such as heavy metals and particulates. But most important of all, it has become necessary to concentrate on energy and resource efficient technologies which, at the same time, are less polluting.

Both governments and private industry have become increasingly pre-occupied with questions of productivity and profitability. The central issue has come to be a reconciliation between economic and growth objectives on the one hand and social and environmental aims on the other. Governments are experiencing considerable difficulty in allocating funds to environmental

improvements at a time of growing fiscal restraints and competing demands
for government financial assistance. Increase in energy and raw materials
costs, growing unemployment, and, in certain cases, greater balance of payments difficulties, have all led to significant changes in national
priorities. At the industry level, a natural reluctance to invest in
large fixed cost projects has been aggravated by the anticipated drop in
world economic growth rates and by reduced medium-term profits. The costs
of pollution abatement investment/measures vary little with the level of
output. This combination of factors makes it problematic to adequately
enforce even existing environmental regulations and standards, not to
mention enhancing their effectiveness, even though it is clear that the
nature and structure of future growth, even at its lower rate, is likely
to add substantially to the pollution load in the years to come.

Anti-inflation lobbies and business circles are claiming with growing
insistence that the regulatory burden of environmental measures has grown
so enormously in recent years due to a lack of proper accounting procedures,
confused priorities, unrealistic standards on the basis of dubious evidence
and so forth, that they impose "unreasonable costs" for "questionable
benefits". Statistics compiled during 1978 and 1979, from both academic
and business sources, show daunting figures of what environmental regulations add to the operating expenses of business concerns. Environmentalists counter that the estimates are exaggerated, and that although the
initial costs of regulation are high, they are, in any event, off-set by
the long-term benefits of cleaner air, water, job safety and better
protection of public health.

Although businessmen may be expected, by and large, to over-estimate
the costs of environmental regulations and environmentalists to over-
estimate the benefits, it is necessary to see the situation in its proper
perspective. Such a perspective is perhaps that of options and trade-offs
at the margin. The marginal cost of environmental quality control measures
must be weighed and evaluated against the environmental physical damage
that would otherwise take place if the pollution control investments were
reduced, postponed or abandoned during the period in question. This is
cost-benefit analysis of a sort and involves the balancing of a series
of considerations, one against the other. For instance, what are the
benefits and costs to society of enhancing a given environmental standard,
considering that scarce resources with high opportunity costs are in
question; is the societal cost-benefit ratio higher in the field of environment than in other areas (e.g. employment, inflation); how effective and
comprehensive should a particular environmental policy be as against another
which is less so but more economically feasible and acceptable as a mix with
other policies ?

In order to make a rational choice between alternative options, it is
necessary to have an adequate data base. Such a data base should permit
the assessment of costs, including those of "add-on" and "built-on" production processes, of emissions and effluents, of control efficiency, and
so on. Although environmental costs are largely visible and recorded
in both national income accounts and estimates of economic productivity

changes, certain methodological difficulties still remain in carrying out cost calculations. But they are manageable as compared to the difficulties that are faced in an assessment of benefits. Serious doubts have been expressed, as is known, about any attempt at strict adherence to cost-benefit criteria for environmental policy decisions. It is noted that events have overtaken existing analytical and statistical techniques which have not yet evolved methods for measuring the more outstanding external dis-economies. In the event, quantifiable effects are included and the un-quantifiable ones omitted; conclusions based on the "measurable" effects may, thus, indeed, be unwarranted in many cases.

II.

UNEP's approach to the use of cost-benefit analysis for environmental management has been essentially practical and pragmatic in character. Since the Stockholm Conference in 1972 one of the main pre-occupations of member governments as well as one of the major efforts within UNEP has been to develop the capacity to provide clear and comprehensive advice on the implementation of sound principles of environmental management, including the fashioning of methodological and analytical tools. (Ref. 1). The goal is reflected in several decisions of the Governing Council requesting the Executive Director "to encourage and support an integrated approach to the planning and management of development, including that of natural resources, so as to take account of environmental consequences, to achieve maximum social, economic and environmental benefits." (Ref. 2). In another of its Decisions, the Governing Council instructed that "guide-lines should be elaborated for the integration of the environmental dimension in future development projects, with the aim of ensuring that the inclusion of environmental parameters does not adversely affect development priorities". (Ref. 3). At its Seventh Session in April, 1979, the Governing Council adopted Decision 7/7 which

> "3. Considers that the pursuance of the cost-benefit analysis undertaken by the United Nations Environment Programme is an exercise of major importance for all countries, requiring a sufficient period of prepara-tion and implementation and adequate financial re-sources for its proper conduct.
>
> 5. Urges Governments to provide the Executive Director with available case studies, particularly those which show new ideas and techniques of analysis, and which also cover new areas". (Ref. 4).

Earlier, in accordance with the expressed wish of several governments that methodologies for cost-benefit accounting of environmental measures should be further developed and improved, the Executive Director had proposed certain modalities for dealing with the subject. An inter-governmental consultative meeting was held in Nairobi in May, 1978 to discuss the possibility of carrying out concrete case studies for the evaluation of benefits and costs of environmental protection measures. The inter-governmental meeting agreed on the need to undertake a co-ordinated programme of case studies on the application of cost-benefit analysis to environmental protection measures; to make the existing national and regional case studies available to UNEP and to suggest additional case studies whenever considered appropriate. On the basis of these recommendations and communications received from different member governments since then, the Executive Director proposed to a second inter-governmental meeting in January, 1979, the initiation of a Programme of Action in the use of cost-benefit analysis for environmental management.

The proposed Programme of Action was to consist of two principal phases. Phase I was to cover the collection and classification of case studies already completed and available in different countries; the identification of the methodologies used in those studies; and the harmonization of the methodology base so as to make the case studies comparable and applicable to other countries.

Phase II was defined as the initiation of a selected set of new case studies using cost-benefit analysis undertaken by a few countries and covering different aspects of the environmental concern.

To analyze the results of the Programme of Action and to monitor its progress, an Inter-Governmental Expert Group was established. The Expert Group has held two meetings so far: the first in April and the second in October, 1979.

The general consensus of the meetings (Ref. 5) has been that UNEP is moving in the right direction in its work on the classification and categorization of the case studies. It has been agreed that UNEP should play a stronger catalytic role and help member countries broach environmental problems in an effective manner. The experts have also requested UNEP to prepare a common framework for the presentation of the case studies and to broaden the scope of the exercise to include such problems as those relating to over-exploitation of renewable resources. The experts have suggested the following sequence of activities to develop the Programme of Action:

(a) continuation of the work of classification and categorization of available case studies;

(b) determination of the core of common sensible practices implicit in them;

(c) identification of the remaining gaps in methodology and subject areas; and

(d) development of appropriate guidelines and frameworks for new case studies.

III.

A large number of case studies have since been received by UNEP and
analysed for their scope and the nature of the methodology used. Although
the programme of work is still in progress, it is interesting to note certain common characteristics exhibited by the case studies so far examined.

There does not appear to be one format or one agreed definition of
what constitutes a cost-benefit evaluation of environmental management.
Different countries, and even different agencies within a country, appear
to have different approaches and understanding of the subject matter. UNEP
has received, in addition to a few genuine cost-benefit studies, a large
number of other theoretical material and analyses and evaluations of
national experience which, although interesting in themselves in summarizing
the state of the art and national efforts in different countries in regard
to environmental protection measures, only seem to emphasize the broad
application of cost-benefit analysis in different situations. They do not
contain some of the essential elements which should enter into a cost-
benefit evaluation if such evaluations are to be used as effective tools
for decision-making. They differ "in the concern they express for environmental considerations; in their underlying scale of values; in the depth
and rigour of the research undertaken; in certain specific and implicit
(national) characteristics they display; in the various sectors they
cover; in the scope of the systems they deal with; and in the time
horizons they consider". (Ref. 6). In sum, they vary in both form and
content.

Secondly, despite the variations and substantial differences in content
between them, the case studies have been found to show an analytical
identity of approach. This may appear contradictory at first sight but it
stems from an intriguing aspect of the present state of environmental
economics, namely, the lack of new and innovative techniques of analysis.
The great majority of the case studies attempt to adapt, with different
degrees of success, the traditional economic apparatus, oscillating between
micro-economics and welfare theory, to environmental concerns.

At the same time, and thirdly, it is clear that for most countries and
the concrete case studies they have attempted, cost-benefit analysis for
environmental management is a method of organizing information and data in

respect of a project or a change in the economy. Such an organization of
information shows clearly that although all problems of cost-benefit
analysis have not yet been solved or even broached, a common core of
sensible and good practices does exist.

It is interesting, in this connexion, to note that most case studies
examined by UNEP appear to confirm certain insights and perceptions for
cost-benefit application problems and techniques already established in
the literature on the subject.

It is seen, for instance, that cost-benefit conclusions are very
sensitive to assumptions for key parameters, and even when a more refined
analysis is attempted using "best assumption" parameter values, the results
do not often appear to provide authoritative policy guidance. The Environmental Protection Agency of the United States decided to conduct a cost-
benifit study on the chlorofluorocarbon/ozone problem in 1975. The
decision was based on the expectation that (i) the data and analysis that
had been generated on the effects of stratospheric pollution in relation
to the decision on the development of an SST and (ii) the $1 million
devoted to the study of economic and social impacts of biological and
climatic change under the $20 million Climatic Impact Assessment Programme
of the Department of Transportation, would provide an adequate information
and data base for a successful cost-benefit analysis. In the event, these
expectations were not realized and the cost-benefit analysis was not
successful. Similarly, a report prepared in 1978 for the Canadian Joint
Federal Provincial Committee on South-Western Ontario Dyking about the
costs and benefits associated with dyking and flood control structures
concluded that none of a set of projects as designed up to that time
was viable. (Ref. 7). The implications of the finding were also that
economic rationality or efficiency did not provide the only basis on which
to judge the merits of a publicly funded project: other human, social and
ecological issues were at least as important. A second implication can
also be drawn, namely, that for any analysis undertaken to be explicit
and open so that other interested parties could test, criticize or refute
the assumptions made by the analyst. These aspects call for the existence
of an organizational structure to allow a thorough discussion of environ-
mental/economic matters in which all interested parties could present
their points of view. An example is the Committee on Costs of Environmental
Management established by the Government of Sweden. Its primary responsibility is to examine environmental measures, their economic pre-requisites
and consequences. Issues such as the charting of costs and methods to assess
costs of environmental protection, and the examination of the distribution of
costs between the Government, local authorities, industry and private
persons, are among the current tasks of the Committee. (Ref. 8).

The economic costs of early environmental planning are substantially
lower than the larger costs of subsequent corrective measures. This aspect
of the environmental management problematique has been tested in several
countries, and a broad consensus has been reached on its validity. As
for example, a study on the macro-economics of environmental protection
prepared in 1978 by the National Swedish Environment Protection Board,
has pointed out the necessity of assessing costs and benefits in connection
not only with environmental disturbances but also with conservation

measures, so as to make the best use of finite resources. This conclusion was found valid both for trade-offs within the environmental field and between that field and other societal interests. (Ref. 9).

The conceptual framework of cost-benefit analysis is found to be broad enough to allow its application as much to an existing programme as to a proposed project, since discontinuation is an obvious alternative that warrants consideration. Certain work carried out recently in France illustrates the point. In a study commissioned by the Ministry of Culture and Environment in 1978, an attempt was made to identify links between models of economic growth and the reduction of damages on the one side and emissions of pollutants by source on the other. The purpose of the study was to arrive at a strategic analysis which could consider the problems of selection and orientation of a model of growth at medium and long-term. (Ref. 10).

Many case studies develop and present complex reasoning and arithmetic for adjustments of benefits (to reflect multiplier effects or built-in factors) and costs (to reflect the employment of allegedly unemployed resources). The case studies examined by UNEP make it clear that it is necessary to require a strict and vigorous proof before accepting any departures from prices and discount rates prevailing in the market. This aspect is crucial and many case studies developed in several countries have encountered difficult hurdles in over-coming the problem. For example, in a study on iron and steel and pulp and paper industries requested in 1976 by the Government of Sweden from the Industrial Institute for Economic and Social Research on the Emission Control Costs in Swedish Industry (Ref. 11), the difficulties associated with the calculation of social costs limited the study to an investigation of the direct private costs only.

A conclusion reached in certain case studies (where the quantification of benefits has proved intractable) is that given the state of our analytical tools and technical knowledge, a cost-effectiveness analysis is perhaps a more viable, or at least feasible, alternative than a cost-benefit analysis proper. The fact that a cost-effectiveness analysis can be applied almost directly from an environmental impact assessment has been found especially attractive. Other case studies, however, have concluded that the problem is not one of a choice between the two and that the first constitutes, in reality, only a particular case of the second. It is possible to envisage situations where both tools could be used as complementary to each other.

IV.

A general conclusion which can be drawn is that the use of cost-benefit analysis as a tool for decision-making in environmental management cannot be regarded in isolation or as a unique methodology. Such analyses are part and parcel of a whole range of analytical tools that must be

utilized to improve judgemental decision-making. It is necessary, for
instance, to begin with environmental impact assessments which will
identify the main parameters and considerations to be evaluated in terms
of cost-benefit analysis. Needless to say, such statements will have to
be made as comprehensive and as detailed as the financial and human
resources in a given situation permit. In other words, the data base for
cost-benefit analysis must first be identified and this can be determined
efficiently from an adequate environmental impact assessment. To that end,
the case studies examined by UNEP indicate that several countries have
instituted socio-economic impact analyses for all major governmental
regulations including environmental regulations. One of the techniques
applied to the results of socio-economic impact analyses is a cost-benefit
analysis but other methodologies can also be incorporated to improve the
decision-making process.

The linking of the two exercises helps to overcome the old controversy
between environmentalists and economists as to the time-frame within which
to consider environmental decision-making. The purely economic approach
has been traditionally criticized because of the short time perspective
chosen for decisions. On the other hand, long-term ecological horizons
have been declared unacceptable, or even meaningless, from the analytical
point of view by economists. It is clear, however, that many large development projects have substantial environmental consequences which are sometimes neglected or overlooked by decision-makers. The environmental
impacts could spread to regions or sectors beyond those originally foreseen (or predicted). The economic evaluation of such projects should,
needless to say, be improved so as to take due account of long-term
effects. For this purpose, an environmental impact assessment statement
is an effective framework.

There are in reality two problems. The first is a problem of specification. This problem arises from the difficulties inherent in identifying
the major environmental impacts of development activities (i.e. the need
for hard scientific information to help environmental decisions). The
identification of the primary physical and ecological consequences (to say
nothing of the secondary and tertiary ones) of growth objectives and actions
requires inter-sectoral and multi-disciplinary analysis from ecologists,
economists and physical scientists alike. This is not easy to come by.
Moreover, there will always be dependence in the specification of environmental impacts on the current state of scientific knowledge and available
technology. These are not fixed in time but constantly changing, and bringing new insights and perceptions to the current understanding and appreciation of the seamless web of global interaction. Second, there are the
evaluation problems proper, such as, among others, an inadequate knowledge
of cost-benefit and dose/response relationships, possible irreversible
effects, problems of measurement and projection, and the inability to
quantify intangible factors. It is obvious, however, that many of these
latter problems arise primarily because of the lack of an adequate environmental impact assessment in the first place. For these reasons, cost-
benefit analyses must be supported by the full spectrum of environmental
management tools.

In an abridged form, it is possible to envisage the CBA/EIA relationship in terms of the following sequence (Ref. 12):

Step 1:   Establishment of Environmental Goals and Framework;

Step 2:   Establishment of Environmental Policies and Programmes;

Step 3:   Environmental Actions;

Step 4:   Environmental Impact Assessments;

Step 5:   Economic Appraisal of Environmental Actions (CBA);

Step 6:   Environmental Decision-Making;

Step 7:   Implementation of Environmental Actions;   and

Step 8:   Evaluation of Environmental Actions.

The sequence of steps noted above attempts to clarify the relationship between environmental impact assessment and cost-benefit analysis in the context of an environmental planning process: the implementation of an environmental impact assessment must precede a meaningful economic appraisal through a cost-benefit analysis. It is for the environmental impact assessment statement to provide the correct, and as comprehensive as possible, identification of the nature and consequences of the activities proposed to be undertaken so that the cost-benefit analysis can evaluate the consequences, positive and negative, as an aid to decision-making.

One of the inescapable features of environmental issues is that decisions are required to be taken in the face of significant uncertainties. Another is that the environmental impact of certain types of activities are cumulative and only become noticeable at some time in the future. Two additional elements must thus be kept in view in carrying out cost-benefit analyses. The first is the need to take account of risks and uncertainties in dealing with environmental measures. A study commissioned by the Environmental Protection Agency of the United States in 1978 on stratospheric pollution (Ref. 13) concluded that the only possibility of applying economic efficiency criteria to decisions in the subject area was through the inclusion of uncertainty analysis. At least two substantial uncertainties were detected: knowledge of the transfer functions as regards impact on climatic conditions and the estimate of the economic costs of climatic changes - - both to be evaluated in the context of a long time frame (30 years) and in a system which could be considered as an international global common property, namely, the stratosphere. The study concluded that in the management of such an uncertain natural environment as the stratosphere, a cost-benefit analysis could only be applied to decisions concerning the regulation of its use. With due regard to the quality of the data used, it would appear that in such studies, systemic and uncertainty considerations prevail over purely economic ones.

Identification of the uncertainties involved in activities concerning the environment can substantially affect the decision-maker's choice in dealing with the alternatives set out in a cost-benefit analysis. Uncertainties cannot be measured in money terms. The responsibility for taking

due account of different uncertainties, technical or otherwise, must rest
on the decision-maker but a practical cost-benefit exercise could help to
lighten the burden through sensitivity analysis and other techniques.
Indeed, an important advantage of cost-benefit analysis is that it reduces
a problem with many aspects and dimensions to one with fewer dimensions.
If applied scientifically and sensibly, it can be an efficient tool for
bringing the inherent uncertainty of environmental decisions to manageable
proportions.

The second element that must be kept in view is that environmental
considerations may well require a passage from partial equilibrium analysis
to general equilibrium analysis (as appears to be the case in studies dealing with public health programmes) which immediately brings in the issue of
systems analysis. A misunderstanding of the real physical and/or chemical
changes provoked in nature by human processes can distort a cost-benefit
analysis immensely. It is necessary that issues such as the definition
of a system's limits, the classification of the system according to the
degree of transformation of goods, the flows of goods within and out of
the system (transport), and so forth, be carefully defined. Since cost-
benefit analysis is a tool for decision-making, it is important to know
at which echelon of the organizational structure the final decision will
be made, what are the criteria for the partitioning of the system into
manageable sub-systems, and what are the corresponding integration procedures; cost-benefit analysis would require different approaches when
dealing with different levels of operation, e.g. a national policy or a
plant control device. The issues which arise when considering the
availability of data are also extremely sensitive. An analysis of the
flows and characteristics of information is necessary in order to understand the thresholds at which data requirements become a real problem for
the effective application of a cost-benefit analysis. Environmental
systems are extremely complex and the analysis of the different variables,
parameters, constraints, objective functions and other elements demand
difficult research activity. They may indeed require, in many cases, the
use of fairly sophisticated tools such as econometric modelling, operations research, probability analysis, and risk assessment. The real
problems, however, lie not so much in the degree of sophistication of the
tools utilized, but in the identification of the most effective methodology for a specific system.

A study attempted by the Environment Agency of Japan in 1978 of the
costs of Nitrogen Oxides control measures, introduced an advanced econometric model to predict the economic impact of various NOx control
measures (Ref. 14). An exhaustive classification of emission sources
by districts and types of industries was undertaken in order to reach an
acceptable understanding of the physical limits of the study. The final
objective was to offer basic data for the design of an optimum NOx control
policy. The study, however, came to recognize that the assumptions made
could be affected by various combinations and permutations of impacts
arising from the systems nature of the natural phenomena and eventually
recommended that the conclusions of the report should be considered only
as a useful information base predicting the fields that might be affected
by NOx control measures, and not be regarded as a source of concrete
numerical data on the cost and effectiveness of such measures.

If the different elements, as described, are dealt with in a consequent and consolidated manner the decision-making process in regard to
environmental protection measures could be substantially enhanced. This
does not mean that the complexity of environmental decision-making or the
difficulties in the proper use of the cost-benefit analysis can be easily
brushed aside. For instance, a number of operational problems still
remain to be solved which have major implications for practical policy,
specifically:

- the problem of levels (macro against micro: the
  aggregation problem);

- the methodological problem (neo-classical welfare
  economics, pure and simple, versus systems dynamics);

- the problems arising from institutional approaches
  which criticise both systems dynamics and neo-
  classical economics.

Different non-economic problems also arise in different contexts.
A recent study (Ref. 15) on water quality management in Thailand proposed
control measures to avoid pollution from sugar mills' effluents in the
Mae Klong river. A sudden fall in the price of sugar led to the closing
of the plant in question and the consequent elimination of the source of
the pollution, rendering the cost-benefit analysis partially superfluous,
(the possible re-opening of the plant due to the feasibility of producing
alcohol from sugar has since then made the results topical again).
Another aspect which served to further complicate the problem was that
other sources of pollution of the river were not included in the study.
The explanation for this seeming neglect lies in the fact national water
pollution standards in Thailand are valid for specific industries and not
for the river itself. These two extraneous elements - the changing
economic situation and the characteristics of the national legislation -
became determining factors and led to the non-utilization of an otherwise effective and sound analysis. In another context, the experience of
a country, such as the USSR, presents certain features which reflect
analogous problems (Ref. 16). As may be expected, a large number of cost-
benefit analyses have been carried out in the USSR and the results of some
of them have been negative. Nevertheless environmental protection measures
have been undertaken by the Government on the basis of social or political
considerations. Cost-benefit analysis in the USSR (and perhaps in most of
the centrally planned economies) also faces the problem of attempting to
change the decisions after a national plan is already underway. A yet
another type of difficulty is illustrated by the fate of a study carried
out in the Netherlands in regard to pollution in the Rhine river: the difficulties + problems of international or transfrontier aspects of environmental
concerns. Although the cost-benefit ratio for the Netherlands was positive
(Ref. 17), the other countries involved in the use of the Rhine did not
consider it necessary to carry out similar studies.

One lesson which can be drawn from these cases is that the application
of cost-benefit analysis for environmental management is fraught with high
risks when used as direct support for the decision-making process. These

risks, and consequent failures, result from the isolated application of cost-benefit analysis: they can be minimized if other analytical tools and non-economic considerations are also taken into account.

In so far as the question of establishing a <u>common framework</u> for carrying out cost-benefit analysis is concerned, it is possible to identify in the various case studies considered by UNEP a number of sensible practices and procedures in regard to the estimation of both costs and benefits. Many of them could be used effectively by other countries in order to tackle similar specific problems.

It must be emphasized, however, that in view of the difficulties, both of specification and evaluation that we have identified, emphasis on a common framework of analysis should be at the earlier stages of estimation than the final valuation one. This is seen clearly in the extent to which estimates of benefits have sometimes been pushed into otherwise scientific, rigorous analyses. Some of the credibility problems connected with cost-benefit analysis have arisen precisely because quantification of the appropriate benefits in monetary terms has been carried too far and tended to become academic rather than realistic. It may not be possible, or even necessary, to establish a framework which will aim at presenting a total picture. In many instances it may well be justified, at least in the initial stages, to quantify only the physical effects of pollution. The difficulties of the more advanced stage of collapsing the different data on costs and benefits into a monetary unit need not deter analysts from making gains where such gains are possible.

A common framework for making environmental decisions could, thus, use the existing case studies of environmental protection measures to illustrate how trade-offs between health, economic and energy considerations are made. The case studies could also identify those components of the decision-making process which involve scientific evaluation and economic analysis and those components which involve judgements by the politically responsible leaders on acceptable levels of risk.

References

:

1. Report of the United Nations Conference on the Human Environment:
   U.N. Document A/CONF.48/14/Rev.1.

2. UNEP General Policy Objectives, Governing Council Decision 1(I), 1973.

3. UNEP Governing Council Decision 8(II), 1974.

4. Report of the UNEP Governing Council on the Work of its Seventh
   Session: UNEP/GC.7/19.

5. Reports on the First and Second Sessions of the Inter-Governmental
   Expert Group on Cost-Benefit Evaluation of Environmental Protection
   Measures: UNEP/IG.15/4 of 17 April, 1979 and UNEP/IG.17/18 of
   6 November 1979.

6. Opening Remarks by the Chairman at the Second Inter-Governmental
   Expert Group Meeting on Cost-Benefit Evaluation of Environmental
   Protection Measures convened by UNEP in Paris on 10-12 October,
   1979: UNEP/IG.17/8.

7. Report to the Joint Federal-Provincial Committee on Southwestern
   Ontario Dyking: Costs and Benefits associated with selected dyking
   and flood control structures, 1978.

8. Committee on Costs of Environmental Management: Environmental
   Protection within Industry – A study with the Pulp and Paper Industry
   as a base, 1976.

9. LIDGREN K. and OLSON I.: The Macro-Economics of Environmental
   Protection, 1978; The National Swedish Environmental Protection Board,
   1978.

10. THEYS, J. : Politiques de gestion à long terme des pollutions.
    Incidences de différentes hypothèses de développement socio-
    économique et d'aménagement du territoire sur l'état de l'environne-
    ment à moyen et long terme, 1978.

11. FACHT J. : Emission Control Costs in Swedish Industry: An empirical
    Study of the Iron and Steel and Pulp and Paper Industries, 1976; The
    Industrial Institute for Economic and Social Research, 1976.

12. MUNN R.E.: Environmental Impact Assessment in developing countries:
    principles and procedures; University of Toronto, 1979

13. D'ARGE, R. and SMITH K. : Managing an Uncertain Natural Environment:
    The Stratosphere, 1978; Washington D.C.

14. Environment Agency of Japan: Cost-effectiveness analysis of Nitrogen
    Oxides control measures, 1978.

126 Yusuf J. Ahmad

15. PHANTUMVANIT, D.: Cost-Benefit Analysis as a tool for decision-making – A case study on water quality management in Thailand, 1979.

16. LEMESHEV, M.: Methods of Approach to define Economic Efficiency of Environmental Protection Measures, 1979.

17. HUETING, R. : Some economic aspects of Pollution in the Rhine, 1979.

ENVIRONMENTAL COSTS/BENEFITS : AN INDUSTRY VIEW

Jean Guillaume
Ad hoc rapporteur of the International Chamber of Commerce
Commission on Environment
Paris, France

SUMMARY AND MAIN CONCLUSIONS

Environmental protection measures not only influence the environmental aspects of human activity. They affect many other aspects as well, and first of all the availability and the price of goods and services which are needed and desired. A careful and complete cost/benefit analysis should, therefore, be carried out before any such measure is taken, in order to optimise the protection effort and to avoid measures based on emotion rather than fact.

Industry can only operate properly within a system of well defined rules which make long term planning possible, and will often have to carry in the first instance the cost of these measures ; moreover Industry is a part of the human community and is a specific source of data and expertise. For these reasons Industry supports the cost/benefit analysis concept in the environmental field and is willing to participate in these analyses.

The methodology to be used needs some research and benefits may often be the more difficult to assess, especially in the marine field. It must also be kept in mind that, in many cases, cost/benefit analysis will not do more than to assist the decision-making process which, as such, remains a political process, in which Industry must be given the opportunity to express its opinion.

INTRODUCTION

It may be worthwhile to recall first that the International Chamber of Commerce (ICC) is a world business organization, with members in over 80 countries both developed and developing. It acts to promote business interests at international level, to foster the greater freedom of international trade, and to harmonize and facilitate business and trade practices.

To achieve this, the International Chamber of Commerce commonly operates through specialised commissions or committees of experts, regularly convened at international level. These will review the different aspects of the questions they are in charge of, and will suggest common policies and/or produce recommendations or guidelines for world industry and trade. Environmental issues are thus followed by a Commission on environment which, in particular, produced some environmental guidelines for world industry, (edited as ICC publication n° 282 - september 1974).

The United Nations have granted the International Chamber
of Commerce consultative status I and this has enabled the ICC
to establish very close working relations with ECOSOC and all
the UN agencies and organizations dealing with economic affairs.
Thus the ICC cooperates closely with the United Nations Environment Programme (UNEP).

The latter is, at the moment, investigating the possibility
to derive a common methodology for cost/benefit analyses of environmental protection measures, (i. e. careful and complete analyses of overall costs and overall advantages of such measures
for society at large,) from a number of case studies. It requested the ICC Commission on Environment to help in this task.

This led the Commission, while collecting available case
studies, to envisage expressing its views on the subject and an
ad hoc working party was set up. At the time of writing work is
still in progress, but the main ideas and conclusions of this
working party will be given here. They may be regarded as expressing the views of Industry, as represented in the Chamber membership.

These views have a general character and not all of them are
relevant to the marine environment problems, but they may be
worth to be kept in mind in a field where hasty measures may affect to a large extent the availability and the price of energy
and raw materials.

ENVIRONMENTAL PROTECTION AND COST/BENEFIT ANALYSIS

Some basic concepts
Man's activities are increasingly affecting his environment but
he cannot revert to a state of nature. To face the rapid growth
and urbanization of the world population, he must maintain an
ever higher level of technical development in order to satisfy
his needs, including the most basic ones such as food and shelter. He needs the products of industry just as he needs trade
but a sensible balance must be struck between the need for economic growth and the protection of the environment.

At all times society is in a state of continuous evolution,
all its members - Industry being one of them - influencing each
other to a greater or lesser degree, although the interactions
are not always clearly understood. In this process of evolution,
regulating mechanisms should have a role to play to the benefit
of the environment and its preservation for the future. However,
in considering certain wastes of raw materials, or energy, or the
way in which certain human activities have polluted the natural
environment, it can be concluded that these mechanisms have often failed. There is thus a necessity to intervene deliberately
in the processes of change and, in so doing, avoid adopting easy
short term expedients to the detriment of the long term.

As regards environmental conservation, the measures to be
taken will not only influence the environmental aspects of human
activity but they will affect many other aspects as well. Furthermore the social and economic objectives of a community often
seem to be in conflict with that community's environmental protection objectives and vice-versa.

It is necessary therefore to weigh carefully the total impact of environmental conservation measures before they are adopted and implemented. Each measure must be submitted to studies demonstrating the advantages to be expected and comparing these benefits with the costs (benefits and costs being taken in their broadest sense).

It should also be recognized that the costs of environmental measures, whether taken by the State or by Industry, will ultimately be borne by the general public.

Cost/benefit analysis

Environmental protection follows the law of diminishing returns i. e. at first it is possible to achieve substantial reductions of environmental pollution by relatively low-cost measures such as improved management and supervisory attitudes, good housekeeping and simple control equipment. However, the effects of subsequent increments in pollution reduction will rapidly become less significant whereas the measures required become increasingly expensive. In other words the cost to benefit ratio becomes progressively less favourable.

These remarks emphasize the role that cost/benefit analysis can play in spite of the fact that it is still in its infancy and that there may not be available a technique for every application. The total economic and social impact of proposed environmental control measures must be taken into account without unduly over-emphasizing any one aspect. Examples of the application of cost/benefit analysis may be found in :
- setting priorities for pollutants to be dealt with,
- setting levels for most effective control,
- defining optimum technology.

However it must be recognized that cost/benefit analysis is a technique for assessing the effect of specific environmental control measures, not for assessing the need to apply environmental control per se. Cost/benefit analysis can never be more than one of the tools available to assist in the process of decision-making which is essentially a political process. In that respect cost/benefit analysis, as a concept, is comparable with the environmental impact assessment.

The position and the role of industry

Industry has a role to fulfil in society of which it is an essential part. It has to take into account the interests of others just as they are expected to respect the interests of industry.

A first responsibility of industry is to provide goods and services as needed and desired, and in so doing it employs a great many people, both directly and indirectly. Furthermore modern industry is a result of man's progress. For that progress to be sustained, industry must continue to develop. In the interest of a balanced and harmonious development industry should be in a position to make long-term plans and to set its priorities. Hence it must operate in the context of clear rules, gerally understood and accepted.

Industry recognizes that the future welfare of society is associated with the maintenance of a reasonable environmental balance, and accepts the need for environmental control measures. These will often put constraints on production, and lead to increased costs, both directly and indirectly, for industry itself and for society at large. In view of its responsibilities, industry must seek to contain the effects of restraints on productive activity by minimising the costs of being a good environmental citizen.

Industry wishes to assist in identifying priority objectives for environmental protection in the broadest sense and in assessing the effects of protective measures which may prove necessary. Industry expects to be consulted in good time when measures are being prepared which will eventually affect its activities. Due to its specialised knowledge industry is generally the best and sometimes even the only source of expertise and data required for a proper assessment of such measures.

The environmental protection objectives once identified will, at the national level, almost certainly lead to new or additional regulatory measures. Provided these measures are well conceived and the cost/benefit relationship is favourable, implementation will in general proceed smoothly and lead to rapidly noticeable improvement. On the other hand, ill-conceived measures, insufficiently evaluated, and perhaps based on emotion rather than fact, will create many problems for the enforcing authority and meet with considerable resistance from those who have to comply with them.

Therefore the ICC strongly recommends that :

(i) - industry be involved ab initio in the preparatory process leading up to such regulations if these are relevant to its activities

(ii) - industry be consulted on the implementation of such regulations

(iii) - cost/benefit analysis be carried out, where possible, before final decisions are made on proposed environmental protection measures.

Furthermore, the ICC stresses the importance of ensuring that environmental regulations take into account the concept of liberal trade and free competition.

METHODOLOGY FOR COST/BENEFIT ANALYSIS

Technical considerations

It is doubtful whether a single method will be found for identifying and comparing costs and benefits for each particular environmental measure. Methods will probably be totally different, depending on the level of investigation. Cost/benefit analysis could e.g. be made :

(i) <u>at the level of the individual firm or industrial sector</u>
     (or operations at sea for example)
     - on the cost side one or more of the following items may
       have to be taken into account :
       . cost of treatment plants
       .. cost of modified processes or operations
       . capital tied up and consequences thereof
       . effect on competitive position
       . reduced level of production
       . plant closure
       . impediments to new developments
     - on the benefit side :
       . benefits from recycled material or reduced losses
       . less costly waste management
       . improved community relations
       . reduced civil liability exposure
(ii) <u>at the local community level</u>
     - on the cost side :
       . cost of municipal treatment plants
       . reduced growth of industrial activities
       . loss of employment and relocation costs
       . cost of enforcement
     - on the benefit side :
       . improved health and well-being of people
       . new employment opportunities
       . conservation of amenities
       . attractiveness for tourists
(iii) <u>at the regional/state/national level</u>
     - on the cost side :
       . loss of employment
       . impaired growth possibilities due to capital being
         tied up
       . loss of competitiveness
       . increased energy consumption
       . impediment to the development of new resources (energy
         or raw materials)
       . cost of monitoring, control and enforcement
     - on the benefit side :
       . additional employment opportunities
       . improved well-being and health conditions
       . conservation of resources
       . conservation of amenities
    These three levels could be supplemented by others, in par-
ticular at the international level and in the marine field, as
could be the cost/benefit items listed.
    The examples given already demonstrate four points viz. :
 (i) As stated earlier the relevant cost/benefit items vary
     greatly with the level at which the investigation is being
     carried out.
(ii) The idea that costs are always easily evaluated in monetary
     terms is a misconception. Indeed, some of the social costs
     will be as difficult to evaluate as are many of the benefits.
(iii) Environmental protection measures can involve costs for one
     part of society and benefits for another.

(iv) Costs and benefits do not necessarily occur at the same time. In fact there may be a considerable time delay before the full effects, in one way or the other, of an environmental conservation measure may take place.

To sum up, the difficulties in properly assessing the effects of environmental protection measures should not be underestimated. Costs are not necessarily just those of installing a standard form of treatment, but may involve a manufacturing or product change. The cost additionally involves some knowledge of the control techniques to be applied, and may be significantly at variance with estimates if relatively untried technologies are applied to either treatment or control. Finally the community may also have to bear the costs of alternative sources of energy or raw materials, which, in our rapidly changing world, are not easy to estimate.

The problems of estimating benefits are probably better recognized. Principally these are due to a lack of information on important physical and biological relationships which will vary according to different levels of pollution, as well as to the fact that some benefits will only become apparent in the long term. Benefits also imply the use of realistic value for environmental services, as reflected in market prices which in themselves will be determined by various constraining influences, both in terms of time and location.

The depth of detail of the cost/benefit analysis should obviously be related to the expected impact of the measure under study. Furthermore in some cases it may be considered sufficient to make some random analyses, in others pilot or comprehensive studies might be desirable.

Therefore when considering new environmental measures the following procedures should be envisaged :

(i) A rough estimate of the impact of the measure should be made in consultation with those who will have to play a role in implementing it.

(ii) Depending on the outcome of this estimate, a framework for the cost/benefit analysis should be defined. This should include :
- the level (s) at which the cost/benefit analysis is to be conducted,
- the depth of the study,
- the extent (random or comprehensive),
- options or scenarios.

(iii) Once the effects of the proposed environment measures have been identified it will be necessary to establish possible interactions between them. The resulting model may range from a straightforward addition of plus and minus components on the one hand, to a complicated mathematical model on the other. Here again, the defining of interactions and the selection of data require the input of expertise from all appropriate parties.

Whether it will ever be possible to translate the end result of cost/benefit analyses, especially the more complicated
ones, in terms of monetary value depends on the possibility of
putting price tags on each of the effects entering into the
scenarios. For a number of amenity benefits this possibility
seems rather remote as it is also for some of the social costs.
This means that the result of a cost/benefit analysis will often
have to be expressed in a mixture of different units. Although
at first sight not as attractive as a result expressed in a
single (monetary) unit, the outcome will still be of great value
to the decision-makers.

Rather than concentrating efforts too much on this problem
of monetary values, research into cost/benefit analysis methodology should deal with problems such as :
  (i) Selection of base line criteria in the different sectors
      of environmental protection (air, water, ocean, noise, was-
      te disposal, land-use, conservation of resources, etc).
 (ii) Criteria for quantifying environmental improvements.
(iii) Rules for the presentation of criteria, data and results,
      so as to make them comparable.
 (iv) Criteria for the definition of reference levels for costs
      and benefits since both are often marginal. Depending on
      the base level chosen the results of a cost/benefit ana-
      lysis may vary widely (e.g. a boiler stack of 40 metres
      high will probably be regarded as a "normal" investment ;
      if the clean air requirements result in 70 metres the ad-
      ditional 30 metres are "environmental").

Recommendations
The foremost requirement is to define precisely when and to
what extent cost/benefit analyses are necessary. To this end
the following sequence is suggested :
  (i) First of all, nations should agree what hazards and subs-
      tances need attention and should agree environmental ob-
      jectives for these, based on proven health and environ-
      ment criteria. Opportunity should be provided for challen-
      ging on technical grounds the choices made.
 (ii) Secondly, the initiating authority should carry out a cost/
      benefit analysis of the various options to achieve the ob-
      jectives, in consultation with the interested groups.
(iii) Thirdly, proposals should be put forward, in consequence
      of these preliminary studies and publicized.
 (iv) Finally, where an interested group can demonstrate that, in
      the light of the studies, the proposal is unreasonable, it
      should be able to ask for, and to contribute to a more de-
      tailed analysis. The proposal should be amended/withdrawn/
      left unchanged accordingly and, if substantially altered,
      subject to challenge by other aggrieved interested groups
      having demonstrable evidence.

It should be recognized that replacing ineffective programmes which were intended to benefit the environment can be very costly. In certain cases, therefore, especially where the initial costs are high, research or pilot studies should be carried out in order to ascertain the best possible cost and benefit information, even if this may result in some delay in implementing the proposed environmental improvement.

In cases where the benefits to be expected from a given environmental measure are obvious, it may happen that the measure is readily acceptable by all, and industry in particular. Then the measure may be adopted by governments with no need for a particular cost study. It then remains for industry to carry out a cost-effectiveness analysis to arrive at the lowest implementation cost. Indeed such cost-effectiveness analyses will always be necessary since it is implicit in cost/benefit analysis that the benefits are achieved in the most cost-effective manner.

CONCLUSIONS

Industry recognizes the importance of, and the need for assessing the social effects of its actions, in respect of both the short and the long term. It recognizes also that this is particularly true for activities that affect the quality of life, such as those pertaining to environmental conservation, human health and safety, availability of energy and the conservation of natural resources.

Industry wishes to participate in an active way in any study aimed at assessing the costs and benefits, or otherwise of measures to be taken to protect the environment, for the following reasons :
  (i) It is an integral and essential part of society.
 (ii) It is a unique source of expertise and data.
(iii) It has to carry, in the first instance, the cost of any new regulation.
 (iv) It can only operate properly within a system of well-defined and clearly specified rules.

Industry supports a wider use of cost/benefit assessment techniques as a means of arriving at economically and environmentally sound decisions concerning new additional control regulations. It believes that the most effective cost/benefit assessment procedure is one which provides for dialogue between industry and the authorities so as to arrive at the optimum solution for establishing the quality criteria.

Whilst it is accepted that additional time may be required to arrive at an agreed optimum, industry believes that such a dialogue is essential to achieve economic development and growth alongside environmental protection.

There are facets to the principle of cost/benefit analysis which have not been fully explored. Further research into the technique is advocated, coupled with the study of its application to pollution prevention and control. Industry is prepared to contribute to all discussions on a national, regional or global level.

THE RELATIVE EFFECTIVENESS OF CURRENT TECHNIQUES FOR OIL SPILL CLEARANCE

D Cormack B.Sc Ph.D

Scientific and Technical Adviser, Marine Pollution Control Unit, Department of Trade, London, UK

INTRODUCTION

To assess the effectiveness of any process it is necessary first of all to define clearly the task which it is required to perform. With regard to oil spill clearance it is necessary to consider the range of oil types likely to be encountered, the factors which affect them after spillage and the changes in properties thus introduced which in general render the oils much less amenable to treatment. The salient features of the problem are identified and related to the strengths and limitations of current response techniques with the aim of assessing the chances of success.

It is emphasised that the oil clearance potential of any technique is limited by the rate at which the response unit can encounter the oil to be treated. It is shown how encounter rate limits the current shipborne dispersant treatment rate and how performance can be improved by the use of aircraft. Relative rates of treatment with estimated costs are given for ships and aircraft systems.

Oil recovery is then discussed in terms of encounter rate and it is shown that this being shipborne, must in general be less effective in terms of tonnage per hour than aircraft spraying. Waves on the open sea have in the past made oil recovery extremely difficult in any case but new observations now embodied in newly designed systems may allow the wave problem to be substantially overcome. Encounter rate limitations will however remain. Relative performance limits for aerial dispersant spraying and shipborne oil recovery can be calculated in terms of the respective encounter rates. Costs can be estimated for various recovery options in terms of the numbers of ships required to achieve a common recovery rate.

It is noted that if the spill is big enough or close enough to shore or is of a type which will not respond to dispersants or is not pumpable and therefore not recoverable at sea it will inevitably reach shore in an onshore wind. Consequently although beach cleaning costs are high in relation to sea treatment costs there is no option in such a case but to pay them. It is important to recognise that it is quite impossible in general to avoid beach pollution altogether and so more effort is needed to develop relatively cost effective beach cleaning techniques and equipment. New beach cleaning techniques designed with this idea very much in mind are briefly described. It is however too soon to say what their cost effectiveness will be but a considerable improvement is anticipated.

136 D Cormack B.Sc. Ph.D

THE PROBLEM

Low Viscosity Oils[1]

Oil which is liquid and mobile at sea temperature, spreads rapidly on the sea surface once spilled. Theoretical predictions confirmed by observation indicate that the average thickness for an unconfined oil slick at sea is about 0.1 mm. Consequently the area covered is calculable from the tonnage spilled. Clearly an affected area of the order of several hundred square miles would be quite typical. In addition to spreading, the oil slick will move bodily on wind and tide bringing the pollution to areas other than that of the original spill. Results from experimental oil discharges have shown that the movement is according to the vector sum of the whole current vector and 0.03% of the wind vector. This very simple model has been used successfully to predict movement of Amoco Cadiz, Eleni V and Christos Bitas oil in recent months. In the computations, published tidal data was used in conjunction with meteorological information.

The wind also has the effect, through the generation of Langmuir circulation cells in the upper layers of the sea, of producing windrows of oil. These are long ribbons lying in the direction of the wind in which the oil is thicker than would be the case in the absence of wind. The areas between the ribbons are correspondingly denuded of oil and are essentially clean water surfaces. The affected area therefore has a nonuniform coverage in the presence of wind but the average oil thickness remains about 0.1 mm.

Some evaporative loss of light ends occurs on exposure to the air. Evaporative loss correlates with loss of components having a boiling point < 250°C and is consequently predictable given the distillation properties of the spilt oil. For light crudes such as Ekofisk evaporative loss will account for some 25 - 30% of the spillage volume in 8 - 10 hours. The loss of light ends results in an increase in oil viscosity.

Once exposed to sea water the oil absorbs water to form water-in-oil emulsions. These emulsions consist of small droplets of water distributed throughout the oil phase and water contents can be as high as 80%. Consequently the bulk of the oil is increased by a factor of approximately five. These emulsions differ in many important respects from the original oil. In particular the viscosity is very much higher; so high in some cases that the final pollutant will be unpumpable regardless of the pumpability of the original oil. In addition, the adhesive properties of the water-in-oil emulsion may be very different from the original oil. Water uptake has been roughly correlated with duration of exposure and prevailing seastate. Thus rough seas have given 70% water content in < 2 hours or in calmer conditions only 50% in 10 hours. Once the water content has reached its maximum a second phenomenon has been observed viz that the water droplet size decreases accompanied by further increase in viscosity. The time scales for this second phenomenon are presently the subject of study.

High Viscosity Crudes and Petroleum Products

There is a wide spectrum of properties for both crude oils and oil products. Some are extremely viscous and at ambient temperatures may be solid, while others are very mobile liquids. In the same way petroleum products vary

from petrol to heavy residual fuel oils which are solid or semi-solid at
ambient temperatures. Clearly the general description already given of the
fate of oil once spilled requires modification in detail, in respect of the
properties of specific oils, with regard to the rates and extents to which
the various changes may be expected to take place.

Summary of the Problem Facing Oil Spill Response Units

In general an oil slick observed from the air will be seen to affect an
enormous area of the order of hundreds of square miles at an average thickness of 0.1 mm very soon after the spillage. If windy conditions prevail
the oil coverage will not be uniform but will consist of ribbons of oil
aligned with the wind direction. In the main the area affected will be
covered by a very thin sheen, the bulk of the oil being present in the
windrows. The oil too will have changed from the fresh material spilled
to a water-in-oil emulsion, which depending on sea state and elapsed time
may contain up to 80% water. It will have lost its original black appearance and may be light brown or even orange in colour and may have attained
a viscosity which renders it unpumpable. It may of course be unpumpable at
sea temperature to start with.

Natural Dispersion Rates

Clearly the situation described is not permanent. The pollutant may disperse
into the water column either completely or partially under the action of
natural forces and/or it may come ashore. If the oil is spilled close to
shore there is little opportunity for natural dispersion before the slicks
reach shore. In such circumstances beach pollution seems inevitable. However even in the Amoco Cadiz incident it has been estimated that some 50%
of the spillage dispersed into the water column in the prevailing conditions
of high natural agitation rather than stranding on the beaches.

    With the advent of North Sea oil exploration and production efforts
were made to estimate the quantities likely to reach shore by considering
natural dispersion rates in relation to the speed of travel on wind and
tide[1]. It has been concluded that for Ekofisk oil only very small percentages of the original spillage have any chance of polluting beaches.

    On the other hand viscous crudes such as Boscan and heavy furnace fuel
oils being solid or semi solid at sea temperature and being more resistant
to natural dispersion are much more likely to reach shore. In addition to
differences in initial oil type the differing properties of the water-in-
oil emulsions subsequently formed also control the rates of natural dispersion which can be expected. Thus the viscosity of a 70% water-in-oil
emulsion of Ekofisk oil has a value of 1000 cs whereas that of a similar
water content Kuwait crude oil emulsion is 3500 cs. Kuwait emulsions are
accordingly much more persistent than those of Ekofisk oil.

    Thus although natural dispersion does take place it cannot be relied
upon in general to remove the pollutant from the sea surface before it
reaches shore. The extent to which natural dispersion will assist is a
function of oil type, quantity spilled and distance from shore of the spillage source.

THE RESPONSE

The response to oil spillage may take two forms. One approach is to increase the tendency to natural dispersion by some means so as to decrease the chances of oil reaching sensitive areas. The second is to remove the pollutant altogether by some form of collection process. Either way the problems are daunting and it is well to recognise this at the outset. The main difficulty arises from the spreading rate. In a matter of hours the response units have to cope with an area measured in terms of hundreds of square miles. The rate controlling step in the response is the encounter rate ie the rate at which the response unit comes into contact with the oil. Because of the enormous area covered the oil is spread very thinly (about 0.1 mm on average). Therefore a 1m wide response unit travelling at 1m/sec encounters oil at the rate of 0.36 tons per hour. This is therefore the maximum treatment rate for such a unit, requiring about 30 hours work to deal with 10 tons of oil considering for the moment an unspecified process operating at greater than 90% efficiency.

The two distinct approaches to oil spill response viz dispersants and oil recovery will now be discussed against this background datum.

Increase in Rate of Natural Dispersion

It has been stated that Ekofisk oil disperses naturally at sufficient rate to prevent any substantial quantity reaching shore for certain spill sizes occurring at appropriate distances from shore.[2] If the natural rate were insufficient as for example with large spills occurring close to shore it would in principle be possible to increase the natural dispersion rate by the use of chemical dispersants and in principle this approach could be used for any other oil for which dispersants would work in this way.

It is important to recognise at this point that dispersants are simply a means of increasing the natural dispersion rate so that sufficient dispersion will have occurred to remove the oil from the sea surface before it reaches the shore. Certainly it is necessary to treat an enormous area and the encounter rate will be low but there is a sense in which this is an advantage. Thus the spreading phenomenon itself is a form of dispersion. In fact it is two-dimensional dispersion and all that remains to be done is to add the third dimension ie to disperse the surface oil into the body of the sea.

Because of the very extensive two-dimensional dispersion which has already occurred naturally, the resulting concentrations in the sea are of necessity low. Clearly if a 0.1 mm layer is dispersed uniformly into the top 1 m of the sea the concentration resulting is 100 ppm. Oil concentrations measured under chemically dispersed oil slicks and under the naturally dispersing Ekofisk Bravo oil slicks were much lower than 100 ppm because the oil, of course, disperses throughout the sea water to even greater depths than 1 m. Indeed under experimental slicks the rate of decrease in oil concentration in surface waters is extremely fast showing rapid dilution to greater depths.[3]

The quantities of oil entering the sea by natural means under oil slicks have resulted in certain observable changes in marine life. However at

Ekofisk and in other cases in the open sea these changes have been assigned a relatively low significance[4] and it is generally accepted that it is preferable for the oil to disperse at sea rather than for it to come ashore. The Ekofisk slicks dispersed into the sea during the blowout at a rate of approximately 2500 tons per day.[2] The notion is that if such a rate could be achieved by the action of dispersants on oils which would not naturally disperse at such a rate great advantage would be gained, for no more damage to the environment than attended the natural dispersion of Ekofisk oil at sea.[5] It should be remembered however that it would take a very large fleet of spraying vessels to achieve such a dispersion rate.

It is often argued however that the application of dispersants is unacceptable even if it does achieve the desirable result described above. Against this view it must be emphasised that current UK approved dispersants are less toxic than the oil itself and that their use does not increase the toxicity of the oil. Only those products which fulfil these requirements are approved for use.[6]

The real difficulty in the dispersant approach therefore is not its supposed unacceptable impact on marine life in the open sea but rather the limited extent to which it can cause the oil to disperse there before it reaches more sensitive areas. This limitation arises from the low encounter rate which in turn arises from the very thin layer of oil present on the sea. The thinness of the layer is also of course the factor which prevents the attainment of significant toxic oil/dispersant concentrations in the sea.

The Application of Dispersants to Oil Spills at Sea

If ships equipped with the current Warren Spring Laboratory dispersant spraying equipment[7] are used as oil treatment units the maximum encounter and therefore treatment rate is 15 tons of oil per hour per unit.

This rather disappointing performance together with the development of dispersant concentrates[8] led to consideration of aircraft[9,10] for use in spraying operations, since the encounter rate is proportional to speed of travel of the spraying unit. Thus an aircraft travelling at 200 knots can potentially treat 20 times the amount treated by a ship moving at 10 knots provided that it can carry sufficient dispersant. Obviously there must be a payload limitation.

A further advantage with aircraft arising from their high speed is that they can reach the spill area much faster than can ships. This, together with the higher treatment rate on arrival, enables greater quantities of oil to be treated before water-in-oil emulsion formation sets in and renders the pollutant untreatable by dispersants.[11] Studies are currently in hand to determine the cut off point for dispersant operations as a function of oil type and water-in-oil emulsion formation.

When payload is taken into consideration the question of relative cost-effectiveness of ships and aircraft can be approached in the following way. The larger type of aircraft with a payload of 10 tons could deal with 200 tons of oil per sortie ie 200 tons of oil per hour within a radius of 75-100 miles of the supply base given an operating speed of approximately 200 knots. It should be possible with such an aircraft to treat 400 tons per hour within a radius of 35-50 miles of the operational base. If a retainer of £100,000 and a rate per flying hour of £500 is assumed for the aircraft and a hire charge of £2000 per day for a spray tug the following results are obtained (see Table 1).

It is clear that

1   The operating time required by an aircraft to treat a given size of spill is less than for a tug. This immediately confirms the advantage

of aircraft in treating the oil more quickly before water-in-oil emulsions form.

2   If the annual retainer is ignored operating costs are much less than for a tug. Of course the retained aircraft may be allocated other duties such as oil spill surveillance, traffic routing surveillance or the detection of illegal discharges from ships at sea. These additional uses increase the cost effectiveness of the retainer cost required to ensure availability for spraying purposes.

3   The overall costs are swamped by the dispersant cost, particularly for a large spill. The combination of search, surveillance and treatment which the aircraft provides additionally ensures that the dispersant is applied to best advantage. Ships have great difficulty in finding and treating oil at sea and may over use dispersants as a consequence. There is a clear incentive to restrict dispersant use to the minimum required. This is best achieved by the use of aircraft.

4   If one spill in excess of 5000 tons occurs every year operating costs for an aircraft plus the annual retainer would be less than a ship mounted operation. The aircrafts' high speed increases the chances of involvement in more than one spill per year since it can be readily offered to neighbouring countries on a suitable cost sharing basis in the event of spills.

The figures given cannot be guaranteed but are sufficiently accurate for present comparative purposes. The detailed conclusions can readily be adjusted for any change in the basic assumptions and the basic conclusion will obviously hold viz that aircraft spraying can be activated much more rapidly and can deal with oil spills much more quickly than can ship spraying and that the costs are not prohibitive in relation to ship costs. In addition a spill size can be identified in the present example >5000 tons beyond which it is actually cheaper to use aircraft and this advantage increases with spill size beyond that point.

Oil Recovery At Sea
Before considering oil recovery at sea it is necessary to return to the earlier assessment of the problem. It will be recalled that encounter rate for a 1 m wide unit travelling at 1 m/sec is 0.36 tons per hour.
    It is also a known fact that if a barrier having draught and freeboard, a skimmer or a boom, for example, is moved across a water surface at speeds greater than 1 knot there is set up a flow of water beneath the barrier which is sufficient to carry with it any floating oil which is in front of the barrier. Consequently the encounter rate in the oil recovery context cannot be increased by increase in speed of travel in the manner appropriate to dispersant application. At first sight the only option is to increase encounter rate by increasing the width of the collection unit by the use of booms. Thus a boom array of width 250 m will increase encounter rate to 90 tons per hour. Such an array however is extremely cumbersome in practice and experience has shown that equipment of this type at the Ekofisk and Ixtoc blowouts permitted the collection of oil for only 15% of the total time on scene. If we assume that emulsion viscosity will reduce pump rate from the 90 tons per hour encounter rate to say 50 tons per hour and then apply the 15% factor for effective operation the resulting average performance is 7 tons per hour. It should be borne in mind at this point that such a boom-skimmer combination involves 3 ships, one on each end of the boom and a third

ship for deployment of the skimmer itself.   In the above analysis the problems of operating in waves have not yet been fully accounted for though one assumes that this limitation is to some extent allowed for in the figure of 15%.   One final comment is that some oils and oil products are unpumpable and the water-in-oil emulsions of many others eventually reach an unpumpable state.

Against a clear appreciation of this background, the UK has conducted over the last three years, an extensive investigation of the problems of oil recovery at sea[12].   The reasons for this investigation arose from a desire to remove oil pollution from the marine environment before it could reach sensitive inshore areas and beaches and from a realisation that in spite of claims to the contrary no significant quantities of oil had ever been recovered at sea.   The objectives of the programme were to find out why performance was so low and how to improve it with due regard to cost-benefit concepts.   To this end available equipment was classified in terms of the nature of the skimming element eg adsorption belt, disc, rope; vortex chamber; or weir/direct suction device.   In a similar manner booms were classified in terms of their identifiable design elements.   Examples of each type of skimmer and boom were then acquired for evaluation at sea in the presence of oil.

In the trials of the above equipment results were sought on the overall systems aspects of the problem.   Thus it was not considered sufficient to identify a skimmer in isolation.   The overall effectiveness of the skimmer-boom-ship system was sought in terms of cost-benefit.   Obviously in this connection the number of ships required in a system had to be set against the likely tonnage to be collected per hour.   The cost effectiveness of purpose built skimming vessels had to be set against cheaper options of installing equipment on available ships in the event of a spill.   As a result of these investigations it has been found that recovery equipment fails to perform satisfactorily in waves for the following reasons

1   The inertial mass of the equipment is too great

2   The linear dimensions of rigid elements or of the entire
      system are too great in relation to encountered wave lengths.

This being so purpose built vessels incorporating the skimming function as an integral element in their construction will be inoperative in waves. This conclusion means that great cost savings are immediately possible for such vessels are extremely expensive to build and maintain and would if chosen be required in large numbers in order to overcome the enormous logistics difficulties inherent in this concept.

Another observation was that since in general the oil is present in windrows of emulsion by the time surface ships arrive on scene it might be possible to achieve the optimum encounter rate with single vessel systems. The long boom multiple vessel option offers the chance of a greater encounter rate in that it can collect a number of windrows against the single vessel capability of only one.   However the single vessel is likely to operate for more than 15% of total time on scene and whatever it collects is attributable to one vessel and its cost whereas the multiple ship collection rate which may not be significantly greater is achieved at a cost of three ships.

The Springsweep System[13,14] represents an attempt to produce a low inertial mass/small linear dimension system for deployment from a single ship.   It consists of a separate tension-line boom (for maximum wave following characteristics) equipped into an oil trap from which collected oil is transferred to the ship's cargo tanks by means of an air conveyor system.

The power supplies, air pump etc are all onboard ship so that the specially designed skimming head is of negligible inertial mass.   The elements in the sea are not rigidly connected to the ship which therefore does not contribute to their inertial mass.

The pumping rate on water is 120 tons per hour and because of the airlift principle the system is independent of viscosity provided the oil can flow to the floating head i.e. viscosities up to 5000 cs can be dealt with independent of the value of the viscosity.   This capacity is judged to be adequately in excess of anticipated encounter rates.

Another system which can be reported on favourably is that developed by ORI/Star Offshore, known as the Force 7.[15]   This is also a single ship system and is based on the flexible adsorption rate principle.   Consequently it has excellent wave following characteristics and should recover oil of viscosities allowing adhesion to the rope at rates controlled only by encounter rate.

These two systems meet the identified requirements in different ways. They may differ however in encounter rate potential, the Springsweep system having a collection boom while the force 7 does not.   On the other hand the force 7 may be capable of operating at a greater ship speed.   It is believed that further evaluation of these systems can now only be achieved through operation at real spills.   Decisions can then be taken on a cost-effectiveness footing in the light of the considerations presented here.

The large boom option is not entirely neglected however.   Encouraging results have been obtained with the BP Weirboom which incorporates a flexible weir, and associated pumps which together have good wave following characteristics.   The weirs are positioned close to one of the deployment vessels to which oil is transferred directly from the weir.   The other end of the boom is controlled by a second vessel in order to collect and direct oil to the weirs.   It is thus a two ship system and as such is expected to achieve better results per ship than a three-ship system while being less dependent on sea state.

Final evaluation of this approach also awaits more experience in real spills but encouraging preliminary results have been obtained at the Ixtoc blow out, though once again encounter rate proved to be limiting even close to the blowout itself.

Beaches

Oil spills which are not treated at sea may eventually come ashore.   The extent to which this will happen depends on spill size, oil type and the extent to which it is treatable at sea.   At present oil clearance on beaches is largely a question of scraping away the oily beach material.   It is time consuming and costly.   Some liquid oil or emulsions may be collected free of solids from sheltered water surfaces[16,17,] but this is usually a small percentage of the total.   Overall it is a costly and labour intensive business which has so far justified attempts to deal with the oil at sea.

New techniques are however under development which should improve the cost effectiveness of beach operations.   It is proposed for example to separate the oil or emulsion from the beach material, sand gravel pebbles etc by standard mineral processing equipment in the presence of hot water and to recycle this hot water through an oil-water separator of standard design. In this way the beach material can be returned to the beach in a clean state and only the oil need be transported.   This will reduce transport costs enormously.   Oily beach material in general contains only 5-10% water-oil emulsion.   It will also obviate the need to land fill the oily solids which gives rise to severe problems at present.

Of course cleaning up beaches or recovering oil from inshore water surfaces after it has arrived does not prevent damage to marine organisms in the coastal zone it simply removes the oil after the damage has been done. Only treatment at sea can prevent damage in the coastal zone.   However treatment at sea is as we have seen limited.   It works best if dispersants can be used but they only work on certain oils and the operation must cease when the water-in-oil emulsions have reached a certain stage of development. Oil recovery at sea may at last be made to work but treatment rates are likely to be very much lower than for the aerial dispersant technique and will also decrease to zero for certain oils and water-in-oil emulsions. Beach cleaning must inevitably remain an important area for work designed to improve cost effectiveness of the techniques available or shortly to be available.

CONCLUSIONS

Oil spreads very rapidly over extremely large areas of sea which results in very low oil encounter rates for the response units.

It is possible to increase encounter rate by increasing the speed of travel of the response units. This approach can be fully developed for dispersant application by the use of aircraft. Although retainers are required for aircraft and not for ships there is a spill size, depending on size of retainer beyond which it becomes cheaper to use aircraft. However there are aspects of aircraft use which cannot be duplicated at all by ships which have to be set against aircraft cost. Thus the need to respond before water-in-oil emulsions render the oil untreatable require the rapid response only available with aircraft. In addition aircraft incorporate both the search/surveillance and treatment functions whereas ships if they are to manifest even their more limited capability need to be directed to the oil from the air.

On the otherhand, oil recovery rates which of course are also limited by encounter rate, cannot hope to equal aircraft performance since speed of travel in this case is limited to about 1 knot (cf 200 kts for aircraft). In principle of course it is possible to increase encounter rate, not by speed increase but by greater swathe width using large boom arrays. In anticipation of high encounter rates skimmers have been designed with large pumping capacities but these high pump rates have never been realised for extended periods because of a combination of difficulties. Thus encounter rates were smaller than expected and the effect of waves seriously limited boom and skimmer performance. In addition the independent operations of booming and skimming in 3-ship operations permitted oil recovery to take place for a small fraction of the total time on scene.

It has now been shown however that improvements are possible in overcoming the wave effect and in reducing the complexity of the operation by reducing the number of ships involved by designing low inertial mass skimmers and by concentrating on windrows. Both of these features should extend the fraction of total time on scene in which recovery of oil is possible. Single ship operation in addition offers a better option in terms of cost-effectiveness. A two-ship option is retained however which may show significant advantages over single ship operation if deployable close to oil source in a blow-out.

In general cost-effectiveness will be optimised if uninterrupted operation can be achieved using the minimum number of ships to deploy equipment of maximum seastate tolerance.

In the end however it is necessary to consider encounter rate limitations. Even at Ixtoc close to the blow-out encounter rates were rather less than anticipated. In general for the treatment of fully spread oil slicks at sea, encounter rates are expected to be such that recovery rates per vessel, even with the improvements described, are more likely to resemble the treatment rates of conventional spraying vessels than those of spraying aircraft.

For those oils and water-in-oil emulsions which are not amenable to dispersants therefore a low treatment rate by recovery systems has to be accepted. It must also be accepted that pumping rate will be overall rate determining for certain viscosities thus further reducing the performance below that already limited by encounter rate. Oil will therefore reach shore in quantities proportional to the difficulties encountered in dealing with it at sea.

The means to deal with oil on shore are available but are difficult to operate, slow and costly. Nonetheless they do work. The oil is finally removed. Advances are possible here however and the new beach cleaning systems currently under development will play an increasing role. As these techniques are used in spills of the future it will be possible to assess cost effectiveness. It will also be possible to assess the extent to which treatment at sea is to be pursued and the extent to which it is preferable to treat the beaches. This apportionation will differ for different oil types. It is however already apparent that both sites for oil spill treatment are complimentary, each to be used as the nature of the problem dictates. Beach cleaning must not be viewed only as a failure of the operation at sea which itself is capable of infinite improvement. There are very real limits to seaborne operations and this fact will have to be recognised.

Table 1   Relative rates of oil spill clearance with costs for Ships and Aircraft

Spill Size Kilotons	Cost of Dispersant £K	Aircraft		Ships	
		Operating Time Radius 100 miles hours	Total Cost £K	Operating Time Days	Total Cost £K
2	75	10	180	20	115
4	150	20	260	40	230
5	187.5	25	299.5	50	287.5
6	225.0	30	340.0	60	345.0

REFERENCES

1   Cormack, D., Nichols, J., Lynch, B. Investigation of Factors Affecting
the Fate of North Sea Oils Discharged At Sea Part 1 Ekofisk Crude Oil
July 1975 - February 1978 Warren Spring Laboratory Report LR 273 (OP)
Stevenage 1978.

2   Cormack, D., Nichols, J. Investigation of Factors Affecting the Fate
of North Sea Oils Discharged at Sea. Part 2 The Ekofisk Blowout April/
May 1977 Warren Spring Laboratory Report LR 272 (OP). Stevenage 1978.

3   Cormack, D., Nichols, J. The Concentrations of Oil in Sea Water
Resulting from Natural and Chemically induced Dispersion of Oil Slicks.
Proceedings of Oil Spill Conference, New Orleans. March 1977.

4   Anon. The Bravo Blowout, Fisken og Havet, Series B 1977 No 5.

5   Cormack, D., Nichols, J. A System for the Application of Dispersants
to the Problems of Oil Spill Clearance. ASTM Symposium on Chemical Dispersants Williamsburg, Virginia October 4-5 1977.

6   Norton, M.G., Franklin, F.L., Blackmann R.A.A. Toxicity Testing in the
United Kingdom for the Evaluation of Oil Slick Dispersants. ibid.

7   Cormack, D., Nichols, J., UK Oil Spill Clearance Techniques and Equipment. Petroleum Times Vol. 80, No 2026 April 1976 pp 23-28.

8   Cormack, D., Lynch, B., Smith, J. Dispersants For Oil Spill Clean-Up
Operations at Sea, On Coastal Waters and Beaches. Warren Spring Laboratory
Report LR 316 (OP) April 1979.

9   Cormack, D., Nichols, J. Feasibility Study of Aerial Application of
Oil Dispersant Concentrates for Oil Spill Clearance. Warren Spring Laboratory
Report LR 257 (OP) Stevenage 1977.

10   Cormack, D., Parker, H. Use of Aircraft in Oil Spill Clearance. Proceedings of Oil Spill Conference. Los Angeles, March 1979.

11   Martinelli, F.N., Cormack, D. Investigation of the Effects of Oil
Viscosity and Water-in-Oil Emulsion Formation on Dispersant Efficiency.
Warren Spring Laboratory Report LR 313 (OP) May 1979.

12   Cormack, D. Criteria For The Selection of Oil Spill Containment and
Recovery Equipment For Use At Sea. Warren Spring Laboratory Report LR 318(OP)
May 1979.

13   Cormack, D. Oil Recovery System for Open Sea Use: The Springsweep System. Proceedings of IOPPEC Conference, Hamburg, September 1978.

14   Cormack, D., Dowsett, B.O., Troilboom. An Assessment of Design Concept and General Reliability. Warren Spring Laboratory Report LR 246 (OP) Stevenage 1977.

15   Cormack, D., Dowsett, B.O., Thomas, D.H., Davia N.C. Oil Mop Device for Recovery of Oil on the Open Sea Warren Spring Laboratory Report LR 328 (OP) July 1979.

16   Lynch, B.W.J., Nightingale J., Thomas, D.H. Oil Recovery International: The Barracuda and Piranha Machines Warren Spring Laboratory Report LR 292 (OP) Stevenage 1978.

17   Thomas, D.H., Oil Mop Inc., The Oil Mop Mk 11-9DP Warren Spring Laboratory Report LR 295 (OP) Stevenage 1978.

PREPARING FOR OIL SPILLS : PLANNING FOR OPTIMUM USE OF LIMITED RESOURCES

June Lindstedt-Siva, Ph.D
Society of Petroleum Industry Biologists,and Senior Science
Advisor, Atlantic Richfield Co., Los Angeles, CA.

INTRODUCTION

The Amoco Cadiz spill, the IXTOC I blowout and the numerous
minor oil spills that occur more frequently are evidence that,
in spite of efforts to prevent them, oil spills can occur at
any time. Certainly spill prevention efforts have been
strengthened; ships have more safety equipment, crews on
tankers and offshore rigs are better trained. However, as long
as there is ship traffic and offshore oil development, there
will always be the risk of oil spills.
    Oil spills are costly ecologically, socially, and economi-
cally. Indeed, costs associated with loss of cargo, vessel or
drilling rig may be less than cleanup costs and damage claims.
In the case of the Amoco Cadiz spill, claims for cleanup and
damages to third parties now exceed 2 billion dollars (1).
Ecological costs cannot always be expressed in dollar values
but these can be significant during large spills. Sometimes
these costs are greatly increased through well-meaning "cleanup" efforts. For example, a marsh oiled during the Amoco Cadiz
spill was "restored" by very nearly bulldozing it out of existence.
Long and Vandermuellen (2) concluded that, "massive cleanup
including large-scale removal of oiled sediments has resulted in
radical alteration of the Ile Grande marsh system. Eventual
recovery to pre-oiling conditions ... may take a century". There
are other such examples of spill cleanup which increases, rather
than decreases, the ecological impacts of the spill. Beaches
are bulldozed, rocks steam-cleaned and marsh vegetation cut and
removed. These high ecological costs are unnecessary. Spill
response can be designed to minimize the ecological impacts of
oil spills. With careful planning, responding agencies can
influence the outcome of a spill situation in a positive way and
even prevent impact.

SPILL RESPONSE GOALS

During a spill incident, resources (including time, people and
equipment) will usually be limited. Therefore, spill response
goals and priorities are needed so that optimum use is made of
these limited resources. Since many agencies may participate in
spill response (federal, state, local government agencies,
industry cleanup co-operatives, the spiller company) it makes
sense to reach some agreement beforehand on what, precisely,
spill response should accomplish.
    I propose that, except where life and limb are threatened,
the primary goal of spill response should be to minimize the
ecological impacts of oil spills (3, 4). Although spilled oil

149

is ugly, our primary response goal should not be esthetic, i.e. to remove visible oil or to prevent esthetic impact. There will be times and places when the esthetic goal of removing visible oil may predominate over the ecological goal of minimizing impact (e.g. during cleanup of public amenity beaches), but these should be recognized as exceptions to the rule. In the main, the goal of minimizing ecological impact should take precedence when the two conflict. Ecological impacts can be both longer lasting and harder to repair than esthetic impacts. In addition, it is often those areas most vulnerable to the ecological impacts of spills in the first place that are most difficult to "clean" or "restore" without causing additional impact.

For example, if a choice must be made between booming the entrance to a salt marsh or a marina full of expensive boats, the ecologically sound decision is to boom the marsh. In the best of worlds, both could be boomed, but if resources are limited, the decision to boom the marsh before the marina will already be a part of the response plan. We have technology to clean oiled boats; there is no technology to clean the marsh without increasing ecological impact.

PLANNING SPILL RESPONSE TO MINIMIZE ECOLOGICAL IMPACT

As coastal environments vary both physically and biologically, response to spills in these environments will vary as well. Spill response planning must be done on a site-by-site basis, and response action prioritized to accomplish agreed upon response goals. Six oil cleanup co-operatives on the east and west coasts of the United States have implemented or are in the process of implementing this type of spill response planning. A two-part approach is used: (1) identification of biologically sensitive areas and development of strategies to protect them, and (2) development of guidelines for minimum-impact cleanup of oiled shorelines.

Of course, protection is ideal, it prevents both the initial impacts of the oil and subsequent impacts of cleanup. However, if shorelines do become oiled, often low-impact, rather than high-impact, cleanup methods can be chosen. The no-cleanup alternative should always be considered.

Biologically Sensitive Areas

Ecological impacts of oil spills can be minimized if those sites most vulnerable to such impact could be identified before-hand and protected during a spill event. Biologically sensitive areas have one or more of the following characteristics. (3)

High biological productivity - A highly productive environment is often a source for recruitment into surrounding areas (e.g. marshes are examples of such highly productive environments).

High ecological significance - If an environment has a particular food chain importance, acts as a spawning or nursery area, its disruption would affect not only its residents but its users as well (e.g. mud flats are used as feeding areas by birds).

Unique features or uses - If an area is the habitat of an endangered, threatened or rare species, its disruption could significantly affect this uniqueness.

<u>Vulnerability to oil pollution</u> - Some ecosystems are
particularly vulnerable to impact from spills and/or spill
cleanup activities (e.g. wetlands are vulnerable because
oil collects in complex waterways; cleanup crews add impact
by working oil into sediments, disrupting plant root systems
and drainage patterns).

Identified biologically sensitive areas on the west U.S. coast
include seal and sea lion haul-out and rookery areas, seabird
rookeries, nesting sites for endangered bird species, and
coastal wetlands. Each area was identified, mapped and its
ecological significance documented. The co-operative then
developed specific methods and strategies to protect the
identified site. Marine mammal habitats and coastal wetlands
will be discussed as examples.

<u>Marine Mammal Habitats</u> Seals and sea lions are abundant in southern
California waters and they use the Channel Islands as haul-out
or rest areas. For example, six pinniped species use the west
end of San Miguel Island (Fig. 1.) as a haul-out area. Five of
those species breed there. For two of them (northern fur seal;
Steller's sea lion) San Miguel represents the southernmost
extension of their breeding range. The effects of spilled oil
on marine mammals are not completely known. However, these

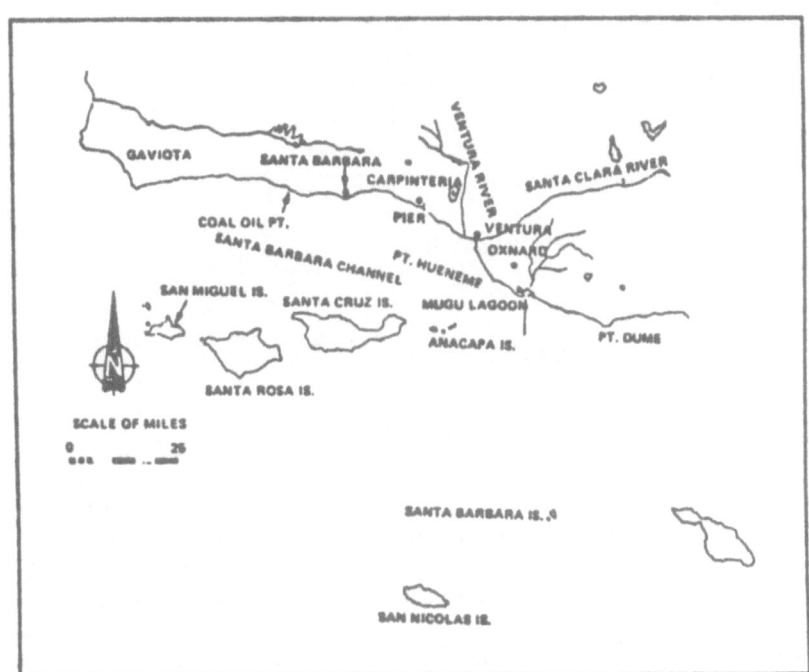

Figure 1. The Santa Barbara Channel.

animals are quite sensitive to disturbance by humans. National
Marine Fisheries Service has conducted tagging and capture
operations on the California Channel Islands. Beach (5) and
DeLong (6) report that animals so disturbed don't reoccupy the
disturbed beach for from one to several days. When the disturbance occurs during breeding season, the whole social structure
of the herd may be changed by a single disturbance(5).
The preferred strategy, then, is to keep oil away from near-

shore waters around the Channel Islands. To accomplish this, the use of chemical dispersants should be considered. Mechanical containment and recovery of spilled oil is not likely to be completely effective, particularly in heavy seas. For example, if a slick is headed toward the west end of San Miguel Island, it could be chemically dispersed in the open sea (at a distance greater than five miles from the island). This will prevent or at least minimize oil impacts on marine mammals and island sea-bird rookeries. If it does contaminate a pinniped rookery or haul-out area, the no cleanup alternative is recommended. Human activities should be kept to a minimum in nearshore waters and on beaches.

<u>Coastal Wetlands</u> Coastal wetlands, including salt marshes, water channels, lagoons, sand and mud flats (Fig. 2.), are productive and diverse environments. In addition, they function as spawning and nursery areas for several fish and invertebrate species,

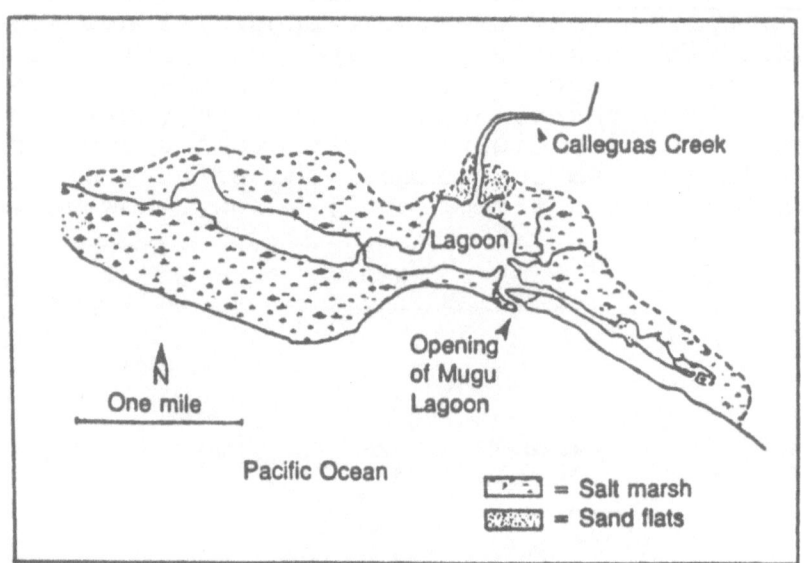

Figure 2. Mugu Lagoon wetlands system.

feeding and resting sites for birds, and vital stopover points for migratory birds. Wetlands systems are particularly vulner-able to impacts from oil spills. If oil enters these systems it may contaminate large portions of them via complex systems of water channels. Organisms known to be sensitive to oil (larval fish and invertebrates, birds) are often concentrated in wet-lands. In addition, most commonly used "cleanup" methods often serve to work oil into the sediments, disrupt plant root systems, soil structure and drainage patterns. The cleanup of the Ile Grande marsh after the Amoco Cadiz spill (2) is a classic, but by no means the only example.

Therefore, protection of wetlands systems should be given high priority in spill response. Openings into the wetlands systems, such as the opening to Mugu Lagoon (Fig. 2) should be identified for each coastal area. Often booming methods can be developed to prevent oil from entering the system with the incoming tide. Oil containment booms can be used to block the opening or to divert oil from it. (7).

During periods of low flow, such as during the summer in Southern California, constructing a sand berm may be an effective

means of preventing oil from entering the system.
    These protective methods must be specifically planned and practiced so that they can be implemented rapidly in an emergency. For example, Clean Seas co-operative (Santa Barbara, CA) stores booms and associated equipment in 40-foot trailers. These are stationed at several locations along the coast, near wetlands sites they are equipped to protect.
    Another strategy was developed by Southern California Petroleum Contingency Organization (Long Beach, CA). Protective booms and associated gear are packaged in helicopter-transportable modules. These are stored in two staging areas in Long Beach and on Catalina Island (8).

Minimum-Impact Cleanup

When oil contaminates shorelines, decisions must be made about whether to attempt cleanup and which methods to use. These decisions should be made to minimize the ecological impacts of oil spills. The different environments within the jurisdiction of the response plan should be identified (e.g. sandy beach, rocky beach, mud flat, etc.) and low-impact cleanup methods recommended for each environment. The no-cleanup alternative should always be considered. These guidelines should be developed by ecologists working with spill cleanup specialists. Minimum impact cleanup has been reviewed in a previous paper (9). Salt marshes, rocky and sandy beaches will be discussed briefly, as examples.

Salt Marshes Most salt marsh ecologists agree that marshes are sensitive to all forms of disturbance and nearly impossible to "clean" or "restore" following an oil spill without causing increased ecological impact (10, 11, 12, 13). Even low-level foot traffic can damage plant root systems, affect soil structure and drainage patterns and work oil into the sediments. Extreme cleanup methods such as bulldozing or raking should be ruled out except where life and limb are threatened. Their effects can ultimately be far more long lasting than those of the oil spill (2). Cutting of marsh vegetation merely to remove visible oil (stained vegetation) is not recommended. If it is absolutely necessary to prevent contamination of unoiled parts of the marsh, it should be done from boats at high tide, using aquatic weed cutters. Westrec (14) recommends use of low-pressure seawater flushing from boats at high tide to direct free oil out of the marsh and into a collection area.
    Under most circumstances, the no cleanup alternative is probably best. Foot or vehicle traffic through a marsh is not recommended (13). Because of site-specific and seasonal variations the advice of salt marsh ecologists is important both during the preparation of the response plan and during a spill event.

Rocky Shores Rocky intertidal environments are best left to self-clean, in most cases. High-impact cleanup methods such as steam-cleaning, hydroblasting or scraping are not recommended. They not only remove visible oil, but also attached plants and animals. Flushing with low pressure seawater can remove lighter oils, while not harming attached organisms.

Sandy Beaches The most common method of sandy beach cleanup has been use of heavy equipment to remove oiled sediments (15). With these methods approximately 10 parts of sediment are removed for

every one part of oil, and in many cases, several feet of
sediment have been removed (16,17).

Hand cleanup methods (hand crews with shovels) are preferred
because they minimize removal of unoiled sediment. A technology
being developed in Canada as part of the Arctic Marine Oilspill
Program (AMOP) bears further study. It may enable removal of
oiled sand with a minimum of sediment disturbance. Russel et
al. (18) have developed a rotating belt with large nails
protruding from it. Nails penetrate the sediments and oiled
sand sticks to the nails and belt, clean sand does not. So far,
only a very small scale version of this equipment has been
tested.

DISCUSSION

To accomplish the primary response goal, to minimize the
ecological impacts of oil spills, the input of ecologists is
important. This input is needed during development of a spill
response plan and during actual spill emergencies. Both industry
and government agencies seem to be moving in this direction. The
National Oceanic and Atmospheric Administration (NOAA) has
appointed Science Support Co-ordinators to advise the Coast
Guard during spill events. In addition, U.S. Fish and Wildlife
Service has appointed regional response personnel to advise
the lead agency (U.S. Coast Guard or Environmental Protection
Agency) on scene co-ordinators during oil spills. Their
presence has been beneficial during the recent spills in the
Gulf of Mexico where spill response included the booming of
passes into the ecologically sensitive lagoons behind the barrier
islands. Available resources (booms, personnel) were concentrated
where they could do most to minimize impact. Even if we know
for two months that oil is headed toward our shores (as in IXTOC
I blowout) it is still not possible to protect an entire coastline. In my opinion, during the IXTOC I spill, available
resources were used wisely.

Likewise, biologists are being consulted to advise industry-
funded oil spill cleanup co-operative during development of their
response plans (19, 20, 21, 22, 23, 24) and are being included
on some industry spill response teams as advisors (7, 25). The
Society of Petroleum Industry Biologists has offered to help
cleanup co-operatives to find biologists to serve on their
response teams. These are certainly steps in the right direction.

Implementation of the spill response planning concepts
described here has been accomplished or is underway by a number
of oil cleanup co-operatives. For older co-operatives like
Clean Seas with much equipment already purchased, very little new
equipment was required. The co-operative merely positioned and
used its equipment differently. Instead of storing everything
in a central yard, specific gear designed to protect wetlands,
for example, was stored in trailers, near those locations. For
new or developing co-operatives, equipment is being purchased
based upon the priorities set forth in their response plans.
For example, the highest priority purchases for the newly
formed Delaware River Co-operative are booms to protect 16
identified wetlands sites (22) and trailers in which to store
the equipment.

Finally, spill response planning is more effective if the
groups who will be responding to the spill (e.g. industry,
federal, state and local government agency representatives, an

industry-funded cleanup co-operative, bird rescue organizations)
have dialog beforehand, not only on what spill response goals
are but on methods to achieve those goals. As much as possible
debates and discussion about why, how and when should take place
in advance of a spill event.

I recommend that input be sought from all responding
agencies during development of the spill response plan so that
the final plan represents a consensus document. This has the
added benefit of building communication between these
responding organizations prior to an event, on a non-crisis
basis.

Once the response plan is written and lines of communication
established, it must be put into practice through regular
drills and training exercises. Planning sets priorities and
reduces the number of decisions that must be made after a spill
occurs. Practice keeps the plan current and operational rather
than only theoretical. Both are essential in order to influence
the outcome of a spill event.

ACKNOWLEDGEMENTS

Helpful comments were received from Bill Ridpath, Dan Gealy,
Bill Kelly and Jeff Pendergraft of Atlantic Richfield Co.

REFERENCES AND NOTES

1. International Environment Reporter. Bureau of National
    Affairs (May 9, 1979) : 661
2. B F Long and J H Vandermuellen, 'Impact of cleanup efforts
    on an oiled salt marsh (Ile Grande) in north Brittany,
    France', Spill Technology Newsletter 4 (4), 218-231 (1979)
3. J Lindstedt-Siva, 'Oil Spill response planning for bio-
    logically sensitive areas', Proc. 1977 Oil Spill Conf.,
    API Publ. 4264, (1977) 111-114
4. J Lindstedt-Siva, 'Why cleanup oil spills - another look',
    Spill Technology Newsletter, 4 (1), 15-16 (1979)
5. D Beach, Marine Mammal Program, National Marine Fisheries.
    Service, Personal Communication (1976)
6. R L DeLong 'San Miguel Island management plan', Marine
    Mammal Community Report, (1975) 38 pp.
7. Clean Seas, Inc., 'Oil Spill Cleanup Manual' Prepared for
    Clean Seas, Inc., by Woodward-Clyde Consultants (1978)
8. C Barker, Manager, Southern California Petroleum Contingency
    Organization, Personal Communication (1978)
9. J Lindstedt-Siva, 'Ecological impacts of oil spill cleanup:
    are they significant?' Proc. 1979 Oil Spill Conf., API
    Publ. 4308, (1979) 521-524
10. J M Teal 'Energy flow in salt marsh ecosystems in Georgia'
    Ecology 43 (4), 614-624 (1962)
11. E B Cowell, M J Baker and G B Crapp 'The biological effects
    of oil pollution and cleaning materials on littoral
    communities, including salt marshes', In M Ruvio (ed.)
    Marine Pollution and Sea Life, Fishing News Ltd, Surrey
    (1972), 359-364
12. D J Ranwell, 'The Ecology of Salt Marshes and Sand Dunes'
    (Chapman and Hall, London, 1972)
13. E B Cowell, 'Ecological effects and minimizing the impacts
    of oil pollution cleanup in intertidal marshes and related
    communities', Working paper presented to Soc. Petroleum

156

Industry Biologists (1978) 19 pp
14. B Westree 'Biological criteria for the selection of cleanup techniques in salt marshes', Proc. 1977 Oil Spill Conf., API Publ. 4248, (1977) 231-235
15. URS 'Evaluation of selected earthmoving equipment for the restoration of oil contaminated beaches', Rept. prepared for Fed. Water Quality Admin., No. 15080EOS 10/70 (1970)
16. E R Gundlach, J Michel, G I Scott, M O Hayes, C D Getter and W P Davis 'Ecological assessment of the Peck Slip (19 December 1978) oil spill in eastern Puerto Rico', Proc. Ecological Damage Assessment Conf., Soc. Petroleum Industry Biologists (1979), 303-317
17. D Soule, Univ. of Southern California Inst. of Marine Studies, Personal Communication (1978)
18. L T Russell, G D M Mackay and W Carlson 'The removal of spilled oil from recreational beaches', Proc. Arctic Marine Oilspill Program Tech. Seminar (March 1979) 24-1 - 24-12
19. J Lindstedt-Siva, 'Oil spill response planning for bio- logically sensitive areas of the Santa Barbara Channel' Rept. prepared for Atlantic Richfield Co. and Clean Seas, Inc. (1976) 41 pp
20. J Lindstedt-Siva 'Oil Spill response planning for biologically sensitive areas in southern California: Pt. Dume to the Mexican border', Rept. prepared for Atlantic Richfield Co. and Southern California Petroleum Contingency Organization (1977) 96 pp
21. J Lindstedt-Siva 'Oil Spill response planning for biologically sensitive areas in the northern Puget Sound region', Rept. prepared for Atlantic Richfield Co. and Clean Sound Co- operative (1978), 138 pp
22. J Lindstedt-Siva 'Oil spill response planning for the Delaware River Estuary I. The upper estuary : Trenton to Delaware City ', Rept. prepared for Atlantic Richfield Co. and Delaware River Co-operative (1979), 104 pp
23. J Lindstedt-Siva 'Oil spill response planning for the Delaware River Estuary II. The lower estuary : Delaware City to the Capes', Rept. prepared for Atlantic Richfield Co. and Delaware Bay Co-operative (L979), 134 pp
24. J Lindstedt-Siva 'Draft outline, sensitivity atlas and mitigation plan' prepared for Alaskan Beaufort Sea Oilspill Response Body (1979), 10 pp
25. SC-PCO, 'Response measures for selected economic and bio- logically sensitive areas', Rept. prepared for Southern California Petroleum Contingency Organization by Ultra- systems, Ltd. (1979), 79 pp.

OIL SPILL PREPAREDNESS AND CONTINGENCY PLANNING FOR THE NORWEGIAN CONTINENTAL SHELF

K. A. Westby, Phillips Petroleum Exploration U.K. Ltd.

INTRODUCTION

The title of this paper refers to the Norwegian operators oil spill contingency planning and preparedness on the Norwegian Continental Shelf. In this connection we are talking about offshore spills associated with production, processing, drilling and work-overs, transportation through pipelines and offshore loading operations.

Increasing public concern over potential pollution of the environment coupled with legislation as well as sincere oil industry concern has made the Norwegian operators to quickly respond to the need of building up a high level of oil spill preparedness with strong emphasis on mechanical equipment.

Today the plan is completed and the state of oil spill preparedness on the Norwegian Continental Shelf now appears to be the highest in the world.

NORWEGIAN PHILOSOPHY AND REQUIREMENTS

The philosophy concerning oil pollution from offshore installations and operations is that the operator is responsible on behalf of the licence holder or holders.

The operator, therefore, has the responsibility and the duties with regard to oil spill preparedness and measures in connection with accidents, which rest upon the licence holder or holders. In accordance with the Royal Decree of July 9, 1976

> "the licence holder or holders shall at all times maintain
> a preparedness in case of accidents or danger situations
> which enable these to be brought quickly under control and
> reduce the damaging effects of such accidents and danger
> situations, as far as possible."

Phillips Petroleum Company Norway therefore has an Emergency Plan Manual which i.e., consists of 18 different contingency plans covering all possible accidents that can occur in all phases of offshore oil and gas operations. The Oil Spill Contingency Plan is one of these 18 contingency plans.

In cases of catastrophe and large-scale oil spills at sea which involve significant danger the Norwegian Government will quickly assemble an Action Command committee consisting of representatives from:
1. The State Pollution Control Authorities
2. Chief of Police
3. The Norwegian Petroleum Directorate
4. The Navy

The Director of the State Pollution Control Authorities, SPCA, will become the Chairman of this Committee and report to the Minister of the Environment.

The operator in trouble has a reporting obligation to the Action Command, which will follow and assess the operators plans and actions, draw

up guidelines for the work, give the necessary instructions, **orders** and
permission to the operator and also take decisions with regard to participation by official authorities. The operator will therefore be present at
all meetings which are not of a purely internal nature and submit reports
regarding the progress of the oil spill clean-up activities.

The Action Command has the power to take over the administrative
leadership of all, or parts, of the operations, if it so deems necessary,
however, such a decision will have to be clearly justified on the basis of
the interest of Norway. It is the author's opinion that this could only
happen if the operator in question is totally incapable of coping with the
organizational, manpower and equipment demands required in blow-out and
other catastrophic situations.

Legislation was put forward during 1976 stating that each operator
south of 62° North was required to have a mechanical oil spill clean-up
capability of 8,000 tons of oil per day.

The environmental criteria for this requirement was 2.5 meter significant wave height and a current of 1.5 knots. However, the operators were
permitted to pool their resources and jointly own their equipment through
their operators committee, North Sea Operators Committee-Norway, NSOC-N.

Since no equipment existed that could fulfil this stringent requirement the operators had to undertake the development of such equipment.

This development programme was initiated in early 1977.

THE NORWEGIAN OPERATORS ROLE

The 14 operators through NSOC-N assigned the responsibility for developing
a programme for design, development and procurement of mechanical and
chemical oil spill clean-up equipment to one of its sub-committees, the
Clean Seas Group. Additionally, the Clean Seas Group was given the task of
putting together an oil spill contingency plan covering the mobilization,
deployment and utilization of this equipment including logistics, mobilization personnel and operational manpower.

However, since the Clean Seas Group was a sub-committee of the NSOC-N
it was an operator organization and therefore could not be held legally
responsible. Therefore, a licencee organization was created in March 1978.
It was called Norwegian Offshore Clean Seas Association for operators or
NOCSA and is completely independent from NSOC-N. NOCSA immediately took
over the responsibility for the Norwegian operators joint oil spill plan
and equipment.

During the design and development period close liaison with SPCA was
maintained since they were empowered to approve the equipment once prototypes
were available and had been tested. This two-way dialogue between NOCSA
and SPCA has been to mutual benefit by making both parties more receptible
to one another's opinions and problems.

THE NOCSA PLAN

A.   General
     Basically the NOCSA Agreement, which today has been signed by 14
     operators representing about 100 licencees, is an agreement of co-
     operation between operators on the Norwegian Continental Shelf south
     of 62°N.

     The purpose of this agreement is:
     1.   To establish a co-operation amongst the operators in order to
          reach a common approach to combating oil pollution offshore.

2. To establish a common administration for future purchasing and
   for already purchased pollution control equipment.
3. To ensure co-ordination regarding the development of contingency
   plans.
4. To ensure the necessary contact with the authorities so that their
   requirements relating to pollution are properly complied with.
5. To plan and carry out equipment maintenance and testing, deployment
   programmes and emergency exercises.
6. To ensure the establishment of an effective response in case of
   emergency and acceptable performance of equipment and personnel.

B. Organization
NOCSA is managed by a 5 member Steering Committee and Phillips currently
holds the chairmanship. Statoil, Mobil, Amoco and Norsk Hydro are also
represented. The Steering Committee reports to a General Assembly, who
approves budgets, plans, etc. Refer to Fig. 1.
       The daily work to maintain and repair the equipment, plan, co-
ordinate and supervise personnel training, equipment deployment programmes and offshore exercises, inspect the equipment at the storage
bases in Bergen and Stavanger, plus re-negotiating associated contracts,
etc. is managed by the NOCSA Technical and Training Co-ordinator, the
T&T Co-ordinator. He is also the head of the NOCSA Secretariate which
is located in Stavanger. The T&T Co-ordinator reports to the NOCSA
Steering Committee.

C. Basic Philosophy
As laid down by the SPCA, the NOCSA plan is based on collection of oil
in open sea conditions.
       A totally integrated mechanical recovery system has been developed
consisting of oil containment booms to contain the oil and stop it from
spreading and thereby provide a means for collection.
       The contained oil is then removed by a skimmer and pumped into
storage tanks on board a special oil spill recovery vessel.
       This dedicated vessel further has a transfer system which is
capable of pumping the recovered oil/water emulsion, which may have a
very high viscosity, into an offshore tanker or other vessel or storage
facility.
       Trained personnel are used in deploying and recovering the booms
and for operating the skimmers and transfer systems. Furthermore, a
trained vessel crew is required to effectively manoeuvre the vessels in
high seas, darkness and close to platforms and other vessels.

D. Vessels
The dedicated vessels previously mentioned function both as oil spill
recovery and shuttling units and due to their oil storage capability
sometimes as oil storage vessels. This flexibility make the dedicated
vessels a key item in the NOCSA plan.
       As per today NOCSA has 8 converted supply vessels each of which has
been modified to achieve a storage capability of 1,000 tons of oil
emulsion and a transfer capability of at least 400 tons/hr at a
viscosity of 5,000 centistokes.
       The dedicated vessels are readily available and used as normal supply
vessels on Statfjord, Frigg and Ekofisk, etc. They are also scheduled
for frequent training programs, exercises, etc.

Conversion from a supply vessel to an oil recovery vessel takes about 4 hours and includes the onloading and installation of skimmers, booms, chemicals and spray equipment.

In addition to the dedicated oil recovery vessels a considerable number of other vessels will be required to operate and manoeuvre the boom sections offshore. Such vessels are ready available in the North Sea today and twice yearly up-dates on available tonnage provides the necessary market overview.

E.   Skimmers

Two different types of skimmers have been selected and developed through NOCSA and the manufacturers.

NOCSA now has 10 Frank Mohn skimmers, Framo ACW-400, and 2 Thune Eureka skimmers, Euroskimmer I, both of which are employing the adhesive disc principle for oil recovery and have a capacity of 40-160 tons/hour each, depending on fluid viscosity. Additionally, a Frank Mohn Artic skimmer is on hand. This skimmer has been designed for more severe weather conditions north of 62°N.

The Framo skimmer is a self-contained unit consisting of a steel base, an operator cabin with controls, an hydraulic arm to which a floating skimmer head is attached and a powerpack. Refer to Fig. 2. The system is controlled by one man who is positioned in the cabin where he can see the skimmer head and the boom during manoeuvring, etc. The automatic load compensation system is always engaged during skimming operations allowing the arm and head to follow the main wave movements in an ideal skimming position.

The Framo skimmer will handle light oil as well as emulsified heavy oil if not too weathered. The Ekofisk operation in April/May 1977 indicated significant reduced performance if weathered oil (7-10 days old) was skimmed. The recovery arrangement can be adjusted from a closed adhesion system for light films to an open flow high volume weir system if thick slicks are encountered, thereby vastly increasing the capacity. Refer to Fig. 3 for technical and performance data. The Framo skimmer is mounted midships close to the rail.

The Euroskimmer is of catamara design, see Fig. 4, which is designed for survival in 5 meter seas. The total system consists of the catamara, which is equipped with a large disc adhesion system, a separate control panel for manual on-board steering and measures 6 x 6 meters, a floating hose with a winch, a deck crane on a frame for hoisting the Euroskimmer in and out of the sea, a diesel power package and a remote control panel.

The 6 ton catamara is preferably operated remotely within a boom system as shown by Fig. 5 and Fig. 6. Both j and u configurations are used, however, the latter requires more vessels but gives better overall efficiency and flexibility. The oil recovery catamara, which is connected to the recovery vessel through the 120 meter long flexible and floating hose bundle is easily manoeuvred by remote control with the help of two 750 kp thrusters, one in each catamara hull. The hose bundle consists of a 18 cm oil conveying hose plus an electrical cable transmitting the remote signals and an hydraulic hose. The hose bundle is stored and operated from a drum with fixed swivel couplings and is connected to the skimmer by a universal joint.

The 8 ton crane is mounted near the rail to facilitate easy deployment and recovery of the skimmer. In order to contain excessive dynamic loads the crane is heave compensated. The Euroskimmer has been extensively tested both in laboratories and in North Sea conditions and its seaworthiness is now confirmed.

Both the Framo skimmer and the Euroskimmer are approved by the SPCA.

F. Booms

Initially two types of booms were designed, however, during the development stage one type was ruled out. The boom now being used is the Nofi boom of which 3,000 meters are on hand, four sections of 500 meters each and four sections of 250 meters each. All booms are used with an hydraulic drum for ease of inspection, repair, deployment and recovery.

The Nofi boom is designed based on the bottom tension principle. Refer to Fig. 7, which also shows the principle behind the booms ability to contain more oil than conventional booms. It consists of a flotation chamber to which a stabilizing rod and a PVC skirt are attached. The design also uses a net section between the bottom tension line and the skirt. This design gives the boom unique dynamic response characteristics, see Fig. 8, since the net section, coupled with the bottom tension line, greatly reduces the counter-acting forces of horizontal towing and vertical movement from big waves when the boom is under tow. The freeboard is 85 cm and the skirt, including the stabilizing net beneath it, is 315 cm. The bottom tension line which is a Kevlar rope, combines high strength with lightness and minimum stretch. A maximum of only 4 per cent stretch at breaking point, which is at 60 tons or so, makes this fibre rope ideal for North Sea conditions.

The Nofi Boom is approved by SPCA.

G. Transfer Systems

A four pump package is provided to achieve the required transfer capability of 400 tons/hr. Two pumps are permanently mounted in the bottom of the vessel and two pumps are located on the deck and serves as booster when required. A pipe manifold allows good flexibility and a diesel power pack is installed to provide power for the pumps.

H. Chemicals and Spray Equipment

Although spilled oil should primarily be recovered by mechanical means chemical dispersants and spray equipment are a part of the NOCSA plan. 6 sets of Ramsfjord spray equipment and 300 bbls of BP 1100 WD dispersant chemical are therefore available.

It is well recognized amongst the operators, and other expertise as well as the SPCA and other government bodies and institutions that an effective oil spill contingency plan will have to subscribe to the philosophy of simultaneous utilization of mechanical equipment, dispersant chemicals and natural evaporation and dispersion.

However, any significant use of dispersant chemical for oil spill combat require prior approval by SPCA.

The following is a summary of the existing regulations for dis-sants:

1. A dispersant shall not contain environmentally poisonous components and shall satisfy toxicity levels in force at any one time.

2. The dispersant shall facilitate bio-degradation of oil.
3. The dispersant properties shall be such that it can be used as low temperatures as - 15 °C.
4. The dispersant shall have a low combustion level.
5. Dispersants shall be capable of being stored for at least 5 years.
6. As a general rule dispersants shall not be used in the following cases:
   a. when the film thickness is above 0.5 cm.
   b. when the film consists of volatile components such as light hydrocarbons.
   c. at low temperatures when the oil slick consists of heavy fractions.
   d. when the slick is strongly emulsified.
   e. small oil slicks in the open sea.
7. All dispersants will have to be approved in advance.
8. Approval from SPCA is required to spray more than 1,000 litres (5 barrels).
9. If a fire or explosion hazard arises in connection with an oil spill around offshore installations or vessels, dispersants may be used irrespective of the restrictions mentioned above.
10. Dispensation may be granted in special cases.

I. Storage Bases
Two storage bases are used, one at Bergen the other at Stavanger. About half the equipment is stored at each base. Refer to Fig. 9.

J. Mobilization of Equipment
The requirement from SPCA is that the equipment shall be on the scene and operating within 48 hours of a major oil spill situation.

NOCSA's area of responsibility is the Norwegian Continental Shelf south of 62º N. The location of the storage bases at Bergen and Stavanger have been chosen with this 48 hour requirement in mind. Even in the "worst possible case" it is believed that the dedicated vessels can be called in, have the equipment installed, have key personnel onboard and be operational in the field within 48 hours of a major spill.

In order to ensure the availability of mobilization personnel and equipment operators, each base has under the supervision of the NOCSA T&T Co-ordinator developed their own mobilization plan which relies on a 24 hour personnel stand-by service.

The following mobilization personnel are available within 2 hours:
1 Manager
1 Foreman
4 Roustabouts

The mobilization force has one sole purpose which is to quickly mobilize and sea-fasten required oil spill clean-up equipment.

Additionally, there are 24 equipment operators on stand-by at each base of whom at least 16 will report within 6 hours. Their duty is to operate the equipment offshore so that the responsible operator can initiate the clean-up work immediately. He will then himself re-place the NOCSA personnel with his own when the time comes. In this connection, it should be emphasized that NOCSA is only responsible for mobilizing the equipment and the key personnel onboard the dedicated vessels. Once the vessels leave the quay the operator in trouble is in charge.

K.   Exercises and Training
Proficiency in the maintenance and operation of the equipment is obtained through courses and exercises.

The basic introductory courses are as a rule conducted by the manufacturer of the equipment. In order to provide further proficiency both in the use of specific pieces of equipment and in co-operation and co-ordination with other participating units, including aircraft and helicopters, both small and large scale exercises are held under realistic conditions.

L.   Research and Development
The Norwegian Operators through North Sea Operators Committee Norway also participate in the financing and conducting of research and development projects directed towards improving the knowledge of the behaviour of spilled oil on water and the improvement of the available equipment.

NORTH OF 62°N

My previous discussion has solely centered around the NOCSA plan which covers the Norwegian Continental Shelf South of 62° N. However, the NOCSA Agreement provides for utilizing the equipment North of 62° N, if required. Statoil, the Norwegian state oil company, has devised a very similar plan for North of 62° N, where a 3 meter significant wave height operating requirement applies. The plan comprises:
1.   4,000 meters of booms of which 3,000 meters are Nofi Boom of a larger type and 1,000 meters are Norsk Oljelense.
2.   10 Frank Mohn Skimmers will be required, however, this type is bigger and is called the Artic Skimmer.
3.   8 dedicated oil recovery vessels will also be required with a storage capability of 1,000 tons each.
The above equipment was made available this year, the reason being that drilling just started North of 62° N.

The same 8,000 ton/day mechanical clean-up capability requirement applies North of 62° N, however, all equipment will have to be approved for 3 meter significant waves.

PHILLIPS PETROLEUM COMPANY NORWAY'S OIL SPILL CONTINGENCY PLAN

This plan or manual covers oil spill reporting, mobilization of equipment and personnel, the organization required to effectively carry out an open sea oil spill recovery operation plus training of personnel.

Figs. 10 and 11 show the concept employed by Phillips in major oil spill clean-up situations. The offshore team is totally responsible for implementation and co-ordination of the clean-up effort. The Offshore Clean-up Co-ordinator is directing and co-ordinating the work and only reports to the Offshore Manager. The offshore team is effectively supported by the onshore team who provide technical support and keeps management and relevant authorities advised of the ongoing operation. This organization was utilized during the Ekofisk clean-up operation in April/May 1977 and proved highly efficient and was able to effectively cope with a difficult situation.

In a major spill situation Phillips will draw on NOCSA equipment and personnel as required, however, for minor spills in the Ekofisk Area a different scheme is in use:

Two dedicated oil spill recovery vessels are available, with one always stationed in the greater Ekofisk Area, so that any platform can be reached within 2 hours. The two vessels, which will be utilized for stand-by and rescue service as well as for normal supply duties, are equipped with 500 meters of Nofi boom and one Framo skimmer. They have also been modified to be able to store 500 tons of oil emulsion and are further equipped with the necessary transfer pumps. The philosophy of having oil spill clean-up equipment permanently installed on a supply vessel in the North Sea is somewhat questionable. However, this is an SPCA requirement. All operators are required to have a minor spill plan during drilling and producing operations.

SUMMARY

1. Experiences during the Ekofisk incident in 1977 and the Ixtoc incident in 1979 plus sea trials, testing and training programs and exercises have shown that the equipment purchased by NOCSA meets expected performance if utilized correctly and operated by trained personnel.

2. It is not known today whether or not the NOCSA scheme is capable of cleaning-up 8,000 tons of oil or oil emulsion per day in 2.5 meter significant waves and a current of 1.5 knots over a longer period of time. However, experience to date coupled with realistic calculations indicate this to be achievable.

3. It is strongly believed that several small, independent, mobile oil spill clean-up units of NOCSA type or equal provides the necessary flexibility required to effectively and safely combat major as well as minor oil spill situations in a North Sea environment.

4. Trained personnel is an absolute must to have a chance of achieving the 8,000 ton/day requirement.

5. The NOCSA plan has been developed by co-operation between Norwegian equipment suppliers, the State Pollution Control Authorities in Norway and the 14 operators on the Norwegian Continental Shelf and seems to represent the ultimate in oil spill preparedness with today's technology.

6. The Statoil plan North of 62° N has been developed by Statoil, but has utilized the same concept originated by the operators south of 62° N.

7. As new technology is made available in the field of oil spill clean-up and existing research and development programs are concluded, the Norwegian operators will have to consider to revise their plan. It is anticipated that it will take 3-4 years before this point is reached.

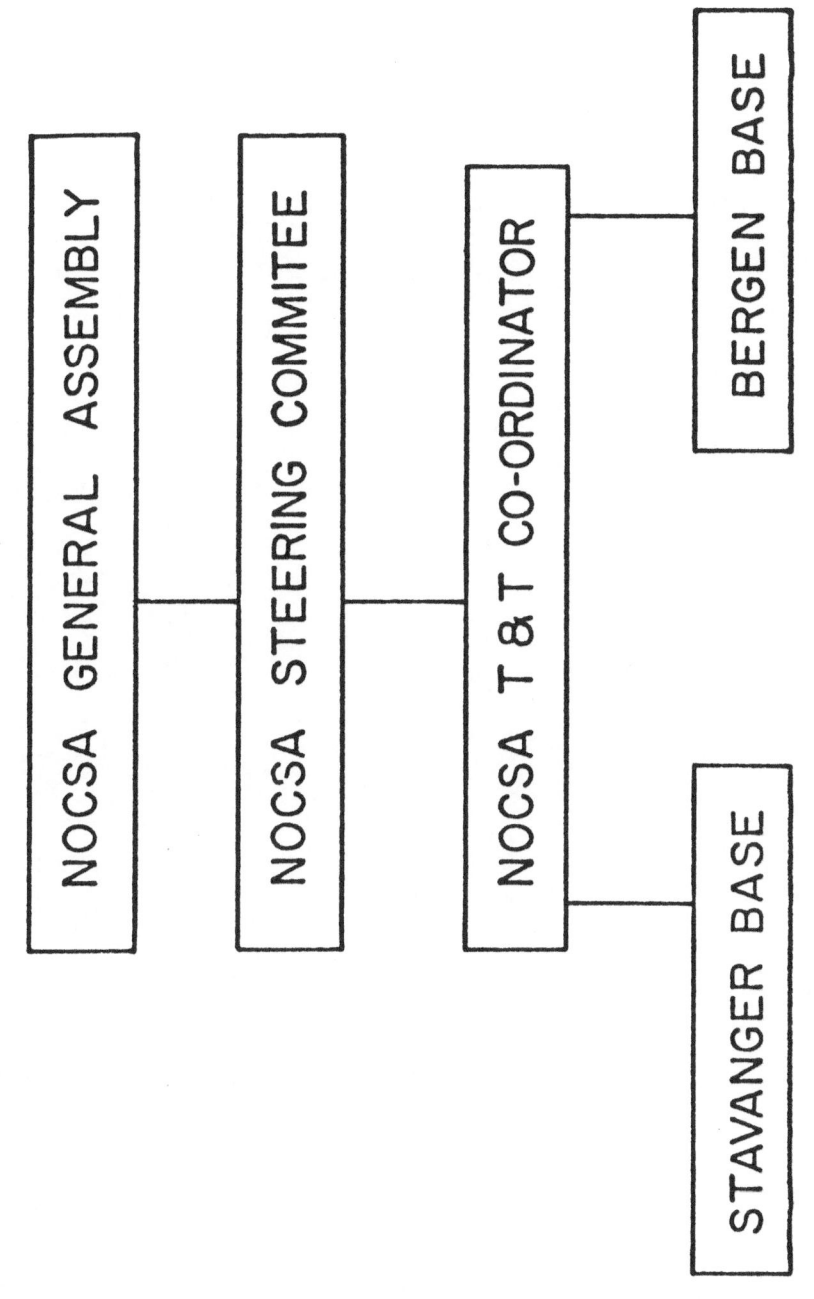

THE NOCSA ORGANIZATION

NOCSA GENERAL ASSEMBLY

NOCSA STEERING COMMITEE

NOCSA T & T CO-ORDINATOR

BERGEN BASE

STAVANGER BASE

FIG. 1

165

166

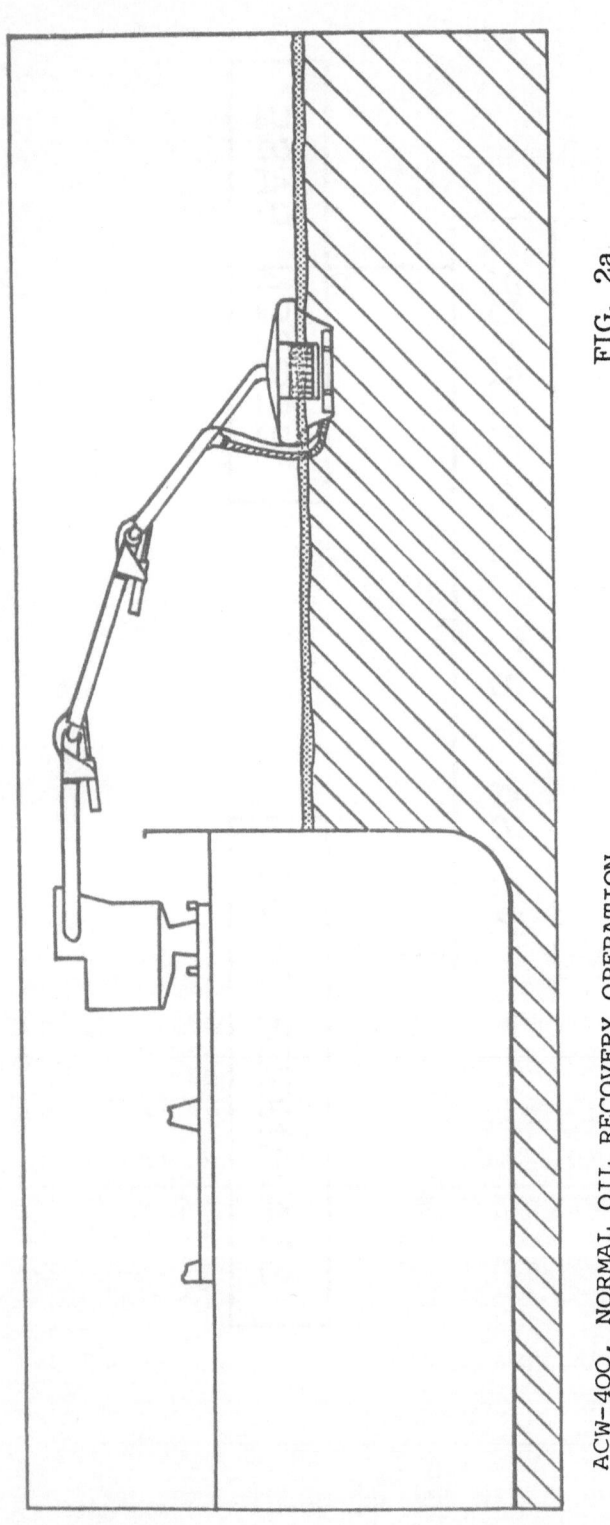

FIG. 2a

ACW-400, NORMAL OIL RECOVERY OPERATION.

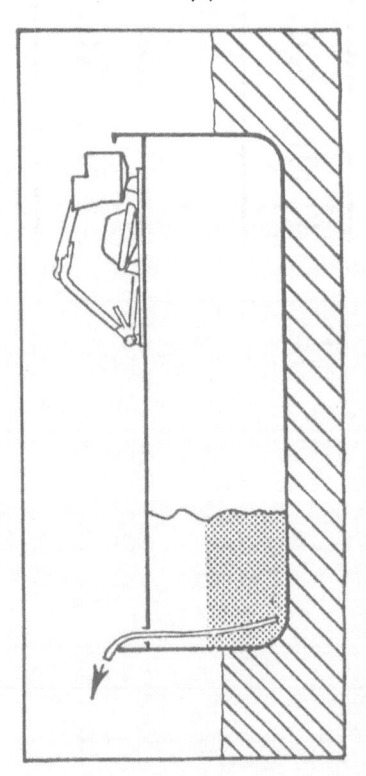

FIG. 2b

RECOVERY UNIT IN PARKED POSITION.

FRAMO ACW - 400 SYSTEM

FIG. 3

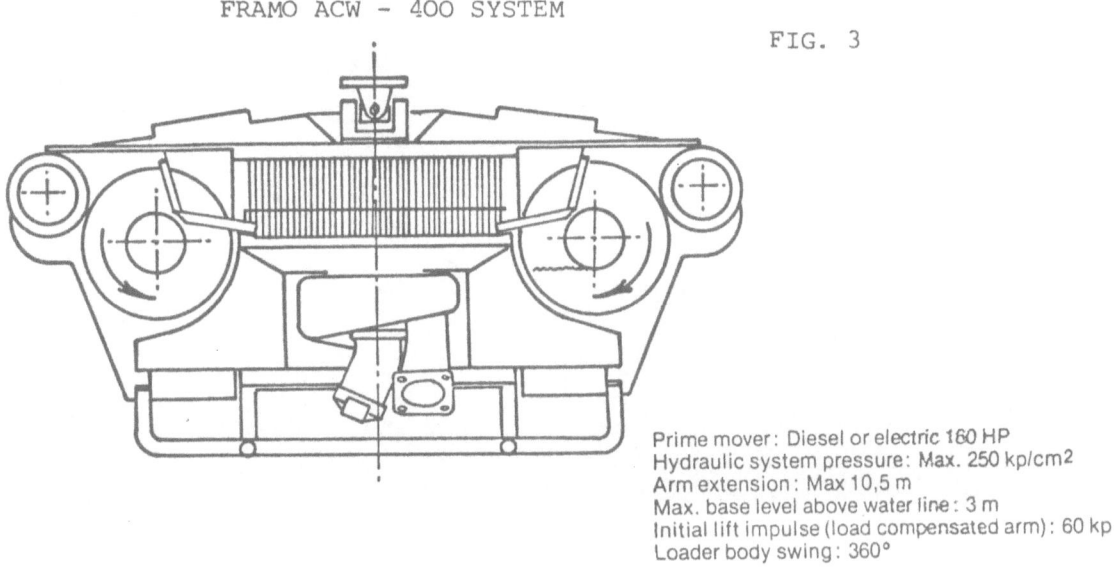

Prime mover: Diesel or electric 160 HP
Hydraulic system pressure: Max. 250 kp/cm2
Arm extension: Max 10,5 m
Max. base level above water line: 3 m
Initial lift impulse (load compensated arm): 60 kp
Loader body swing: 360°

The skimmer head is constructed in SW-resisting aluminium. Four recovery drums with discs are assembled in a square configuration outside the adjustable weir/pumpwell. All functions are hydraulically operated and adjusted from the operator cabin.

Drum speed: 0-20 rpm.
Pump speed: 0-2000 rpm.
Weir level: Water line — 45 m. to + 80 mm.
Material: A57S.

Operational conditions	Capacity/head	Watercontent beyond emulsion
Viscosity 100 C. Stoke Oil layer 20 cm. Oil layer 2,5 cm.	400 m³/h - 35 mwc. 60 m³/h - 40 mwc.	Max. 20% Max. 10%
Viscosity 500 C. Stoke Oil layer 20 cm. Oil layer 2,5 cm.	250 m³/h - 40 mwc. 100 m³/h - 40 mwc.	Max. 15% Max. 10%
Viscosity 1500 C. Stoke Oil layer 20 cm. Oil layer 2,5 cm.	150 m³/h - 35 mwc. 100 m³/h - 40 mwc.	Max 10% Max. 15%

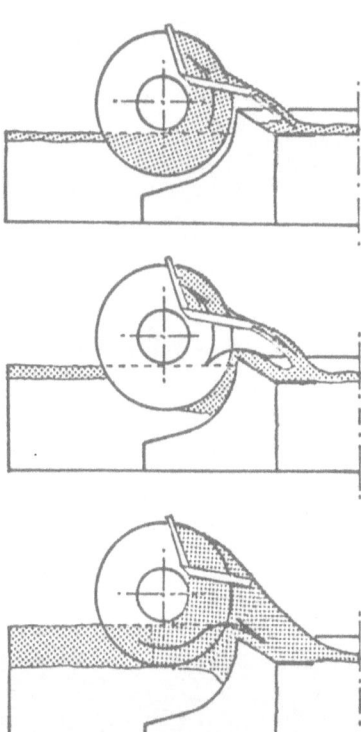

Pumpwell/weir in upper position. All oil being carried over adhesion discs. Maximum capacity abt. 100 m³/h

Pumpwell/weir in middle position. Accumulated oil gravitating over the barrier. Surplus oil under guiding vanes being flushed over the barrier.

Pumpwell/weir in lower position. Surface oil layer gravitating directly into pumpwell. Conveyor effect of recovery drum improving the flow of oil over the barrier.

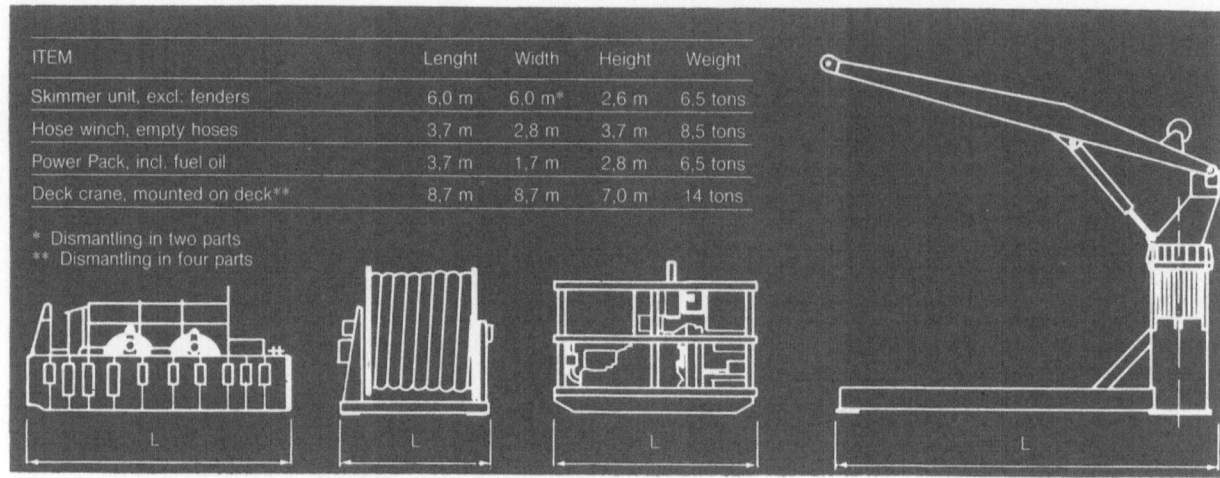

ITEM	Lenght	Width	Height	Weight
Skimmer unit, excl: fenders	6,0 m	6,0 m*	2,6 m	6.5 tons
Hose winch, empty hoses	3,7 m	2,8 m	3,7 m	8.5 tons
Power Pack, incl. fuel oil	3,7 m	1,7 m	2,8 m	6.5 tons
Deck crane, mounted on deck**	8,7 m	8,7 m	7,0 m	14 tons

* Dismantling in two parts
** Dismantling in four parts

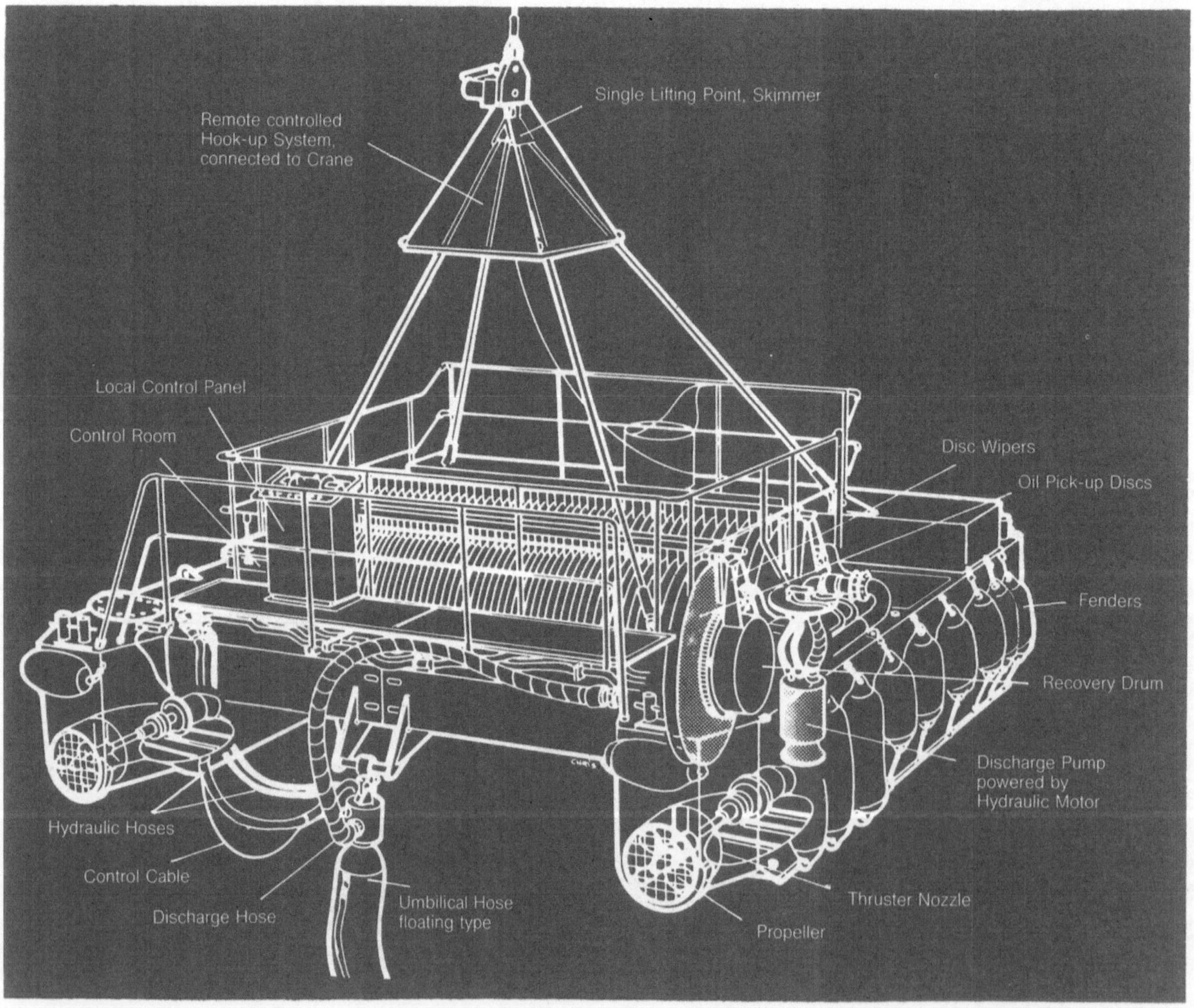

OPERATION

When towing in a
J-formation with
the mother ship
slightly behind the
other tug, only
two vessels are
needed for oil
recovery.

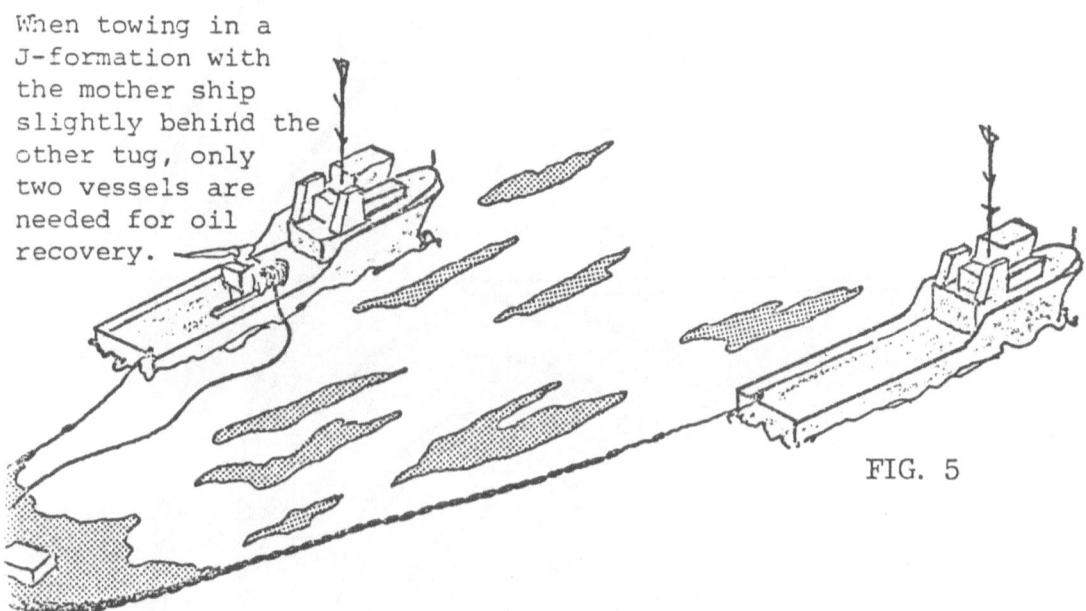

FIG. 5

When towing in U-formation, the mother
ship is best located inside the boom.
If vessels work in two or more
U-formations, they can be serviced by
one skimmer and mother ship, as the
skimmer empties each
boom trap in rotation.

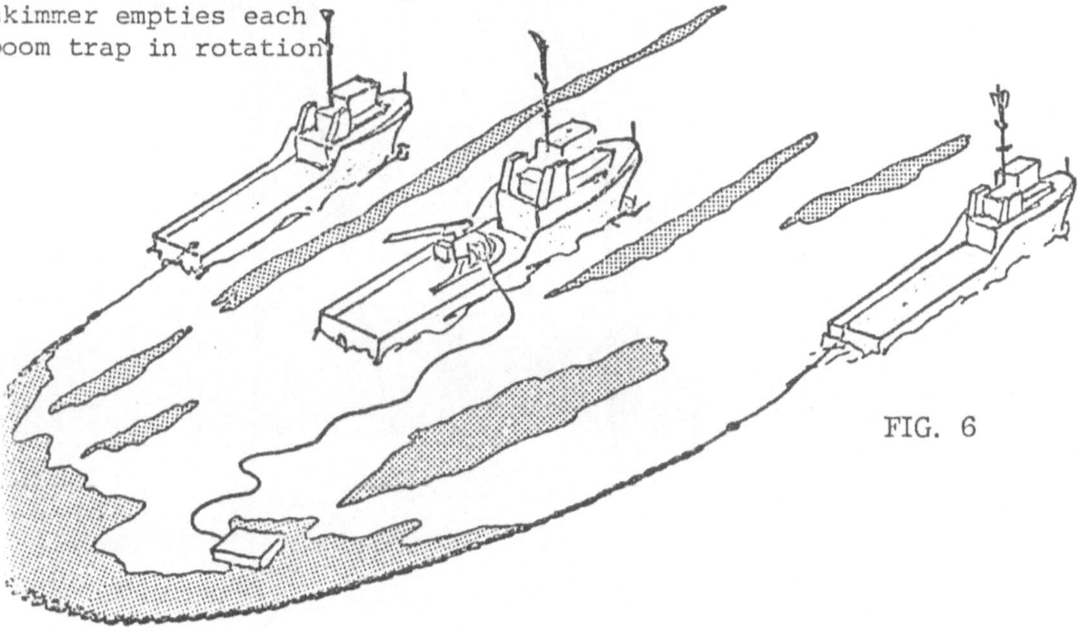

FIG. 6

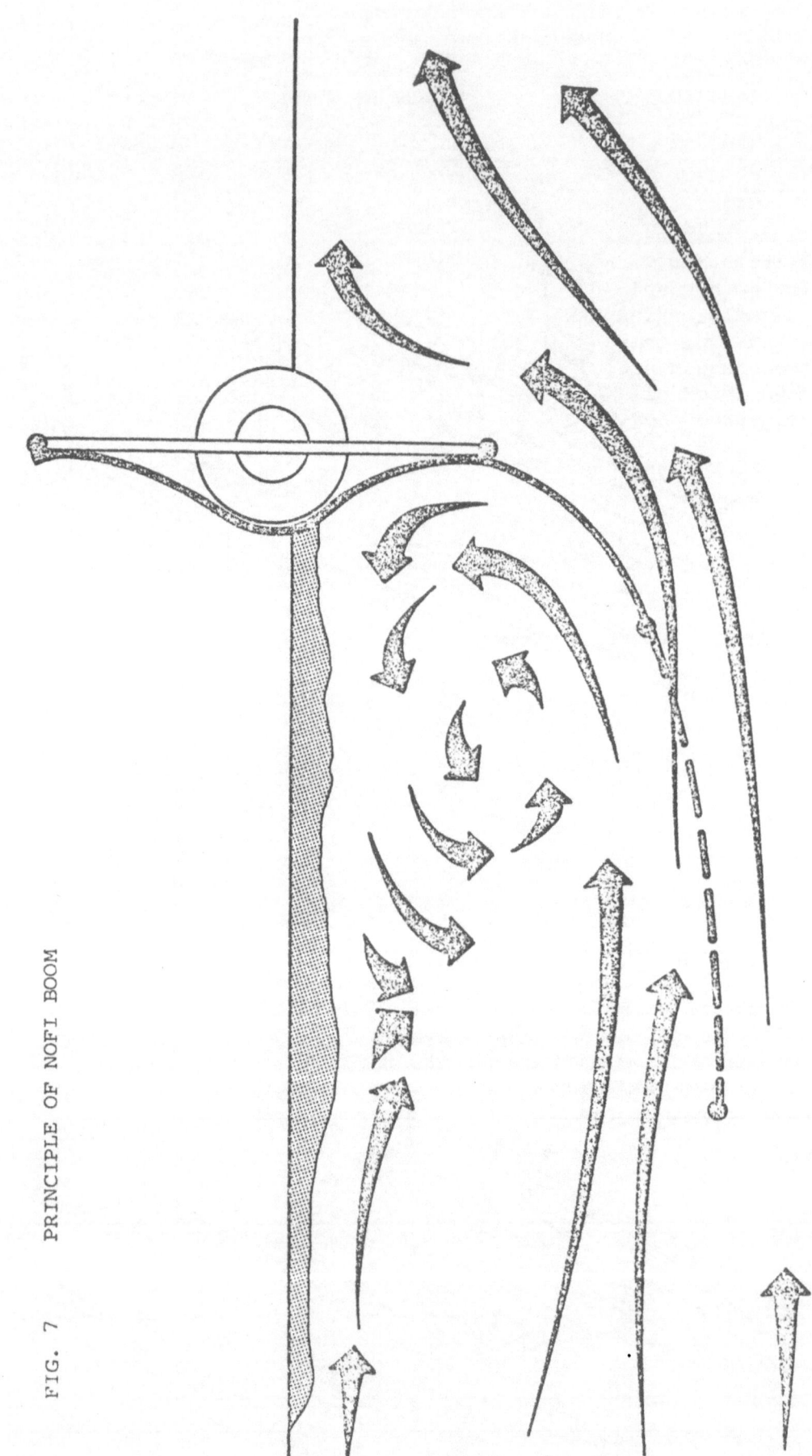

FIG. 7   PRINCIPLE OF NOFI BOOM

NOFI HIGH SEAS
OIL CONTAINMENT BOOM SYSTEM

FIG. 8

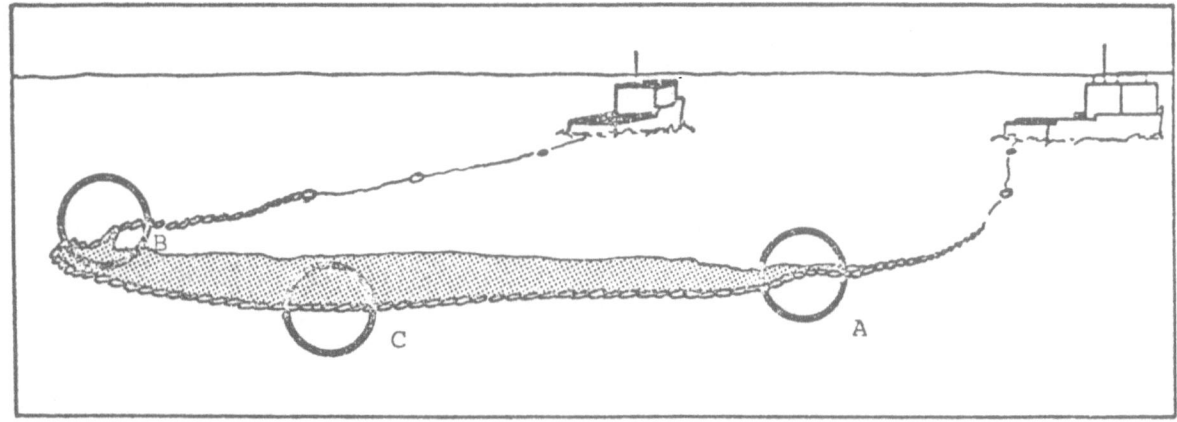

SECTION A

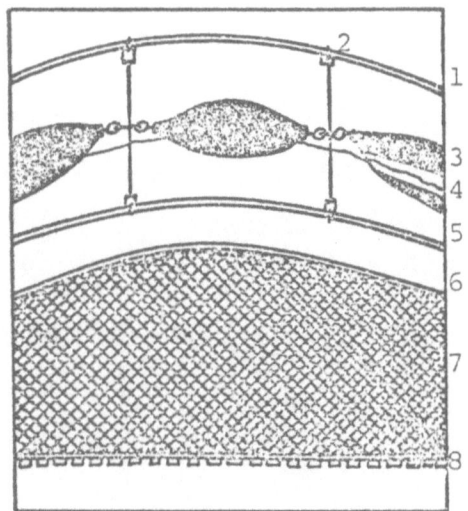

1 Top Line
2 Stabilizing Support Rod
3 Floatation Buoy
4 Water Level
5 Centre Line
6 Lower Line with Lead Weights
7 Stabilizing Net
8 Bottom Tension Line (Kevlar)
        and Lead Weights.

SECTION C.
DYNAMIC RESPONSE CHARACTERISTIC

SECTION B

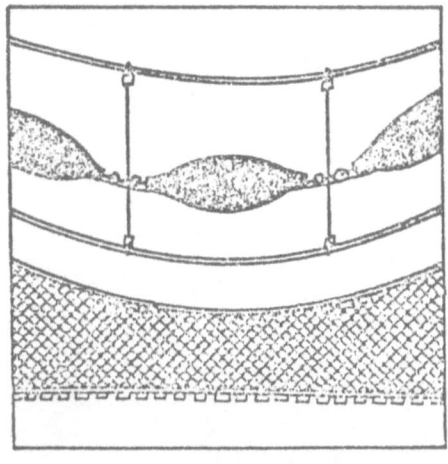

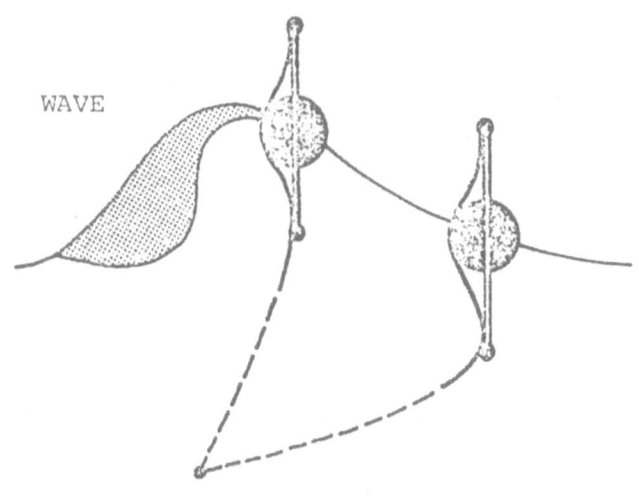

WAVE

172

FIG. 9

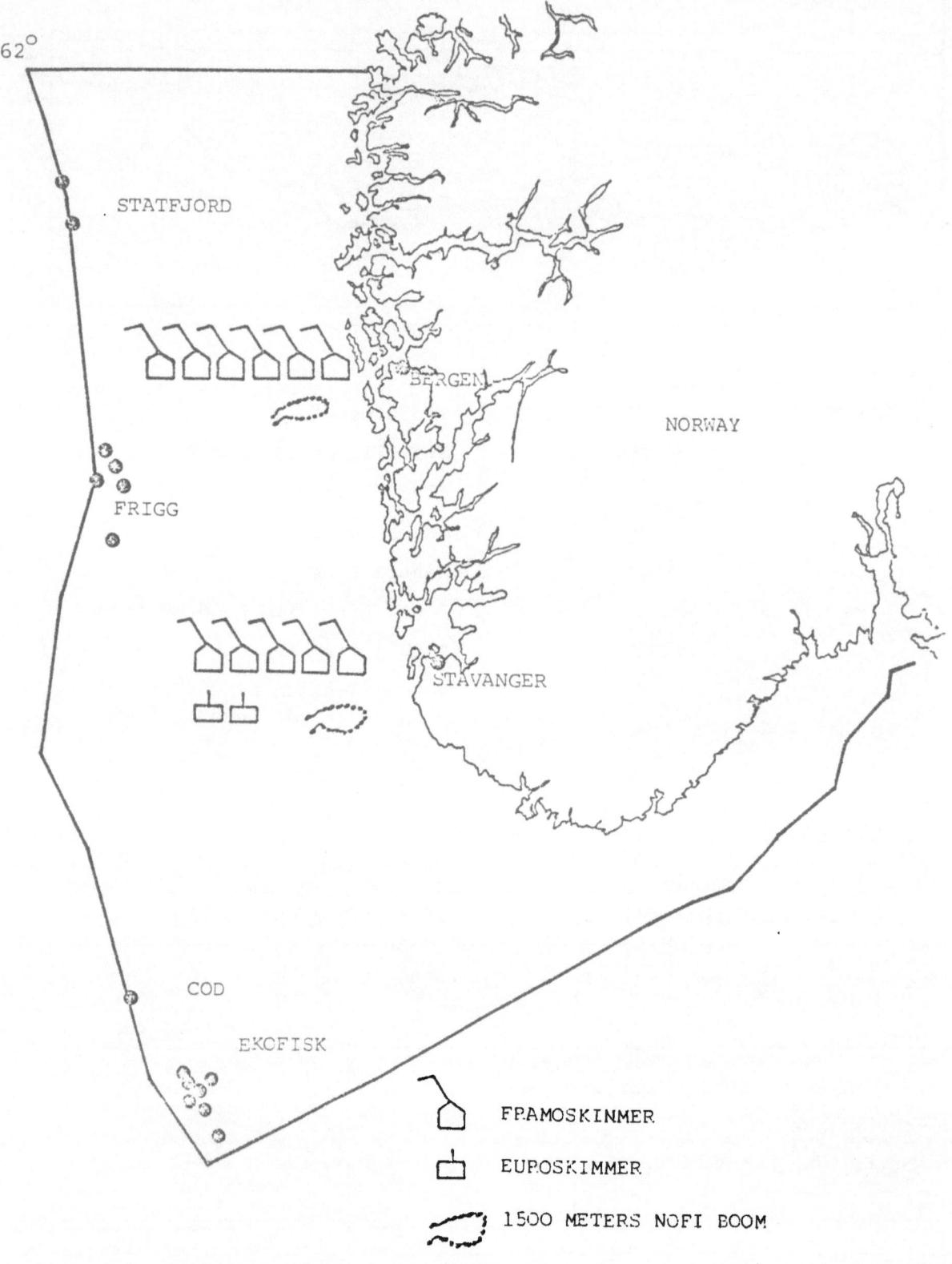

FIG. 10

OFFSHORE CLEAN-UP ORGANIZATION

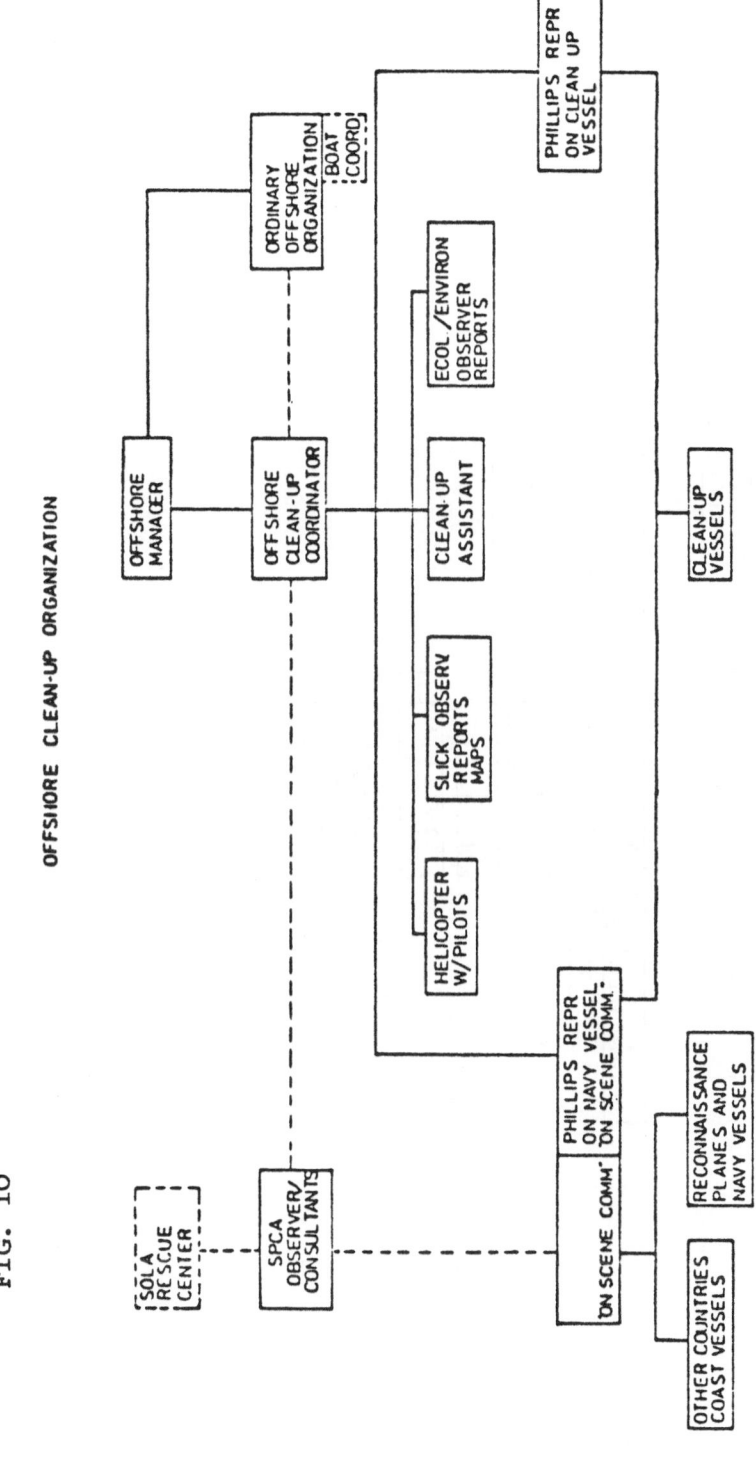

174

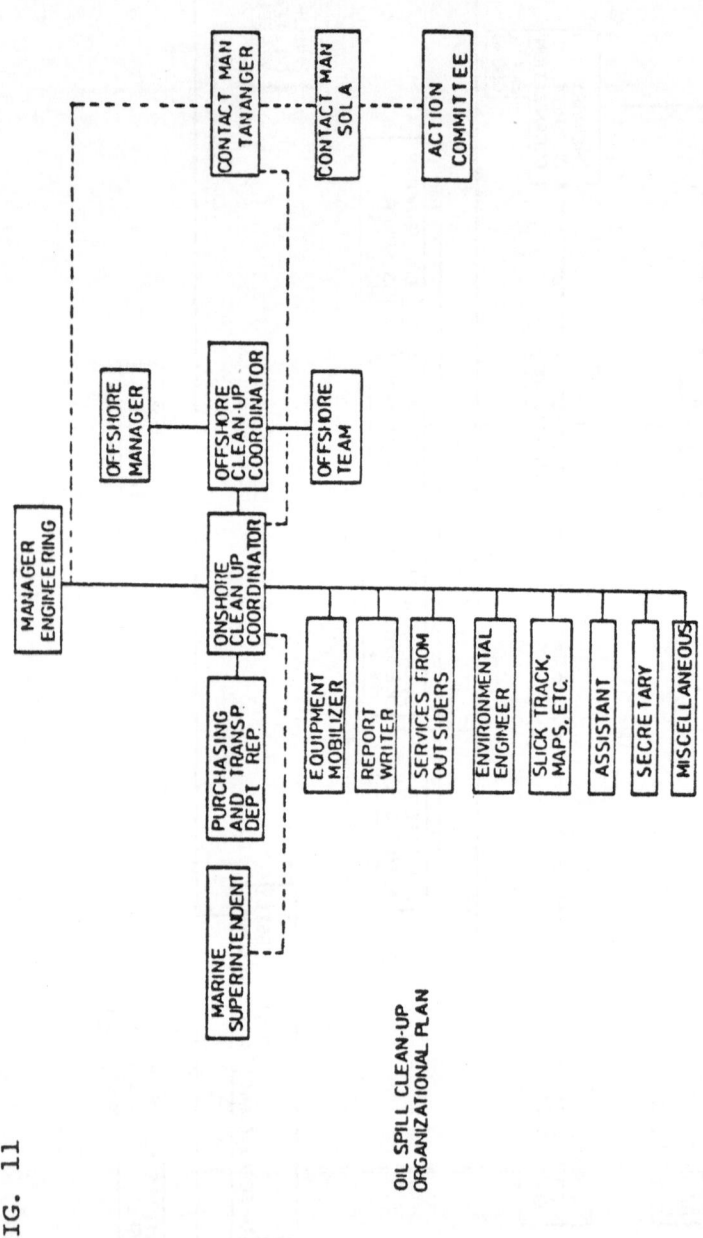

FIG. 11

OIL SPILL CLEAN-UP
ORGANIZATIONAL PLAN

ECONOMIC ANALYSIS IN ENVIRONMENTAL PROTECTION POLICY

By

Douglas B. Seba, Ph.D.
Executive Secretary
Science Advisory Board
U.S. Environmental Protection
Agency, A-101
Washington, D.C. 20460

Barry D. Gold, M.S.
Staff Assistant
Economic Analysis Subcommittee
Science Advisory Board
U.S. Environmental Protection
Agency, A-101
Washington, D.C. 20460

Recognizing that the deterioration of the environment had become a significant national problem, in December 1970 President Nixon signed the law creating the United States Environmental Protection Agency (EPA). The enabling legislation required the EPA to produce a large number of regulations and standards in a relatively short period of time. Almost immediately the nation's environmental goals to clean up U.S. air, water, soil, and workplaces entered into conflict with the nation's goals of full employment, reduced inflation and economic growth. It is no surprise then that industry began to decry the high cost of compliance and that Congress in response requested EPA to produce a series of reports known as the "Cost of Clean..." while the EPA at its own initiative produced macroeconomic analyses of the economic impacts.

The first macroeconomic analyses to assess the overall impact of pollution control costs on the U.S. economy was completed in 1971 by Chase Econometric Associates.

The most recent macroeconomic analyses was conducted for the EPA by Data Resources, Inc., and covered the period from 1970 - 1986. It revealed the following:

- The rate of inflation would be raised by less than 0.5 percent by pollution control expenditures.

- The increase in consumer prices from investment, operating and maintenance costs would average 0.3 percentage points, so if the inflation rate would be 7.3 percent with environmental protection measures, it would be 7.0 percent without.

- Federally mandated pollution control costs will total approximately $360 billion (1977 dollars).

- By 1986 it is estimated that the unemployment rate will be 0.2 percentage points lower than without environmental protection measures.

- An overall 0.1 to 0.2 percent increase in the inflation rate is predicted from the present through 1986.

It should be understood that none of these figures take into account the benefits of the environmental programs in terms of improved public health and welfare simply because such things are not included in standard GNP estimates. It may be that if they were included, the overall impact of pollution control on the economy would be positive. Even in light of such information however, the concern about the costs of environmental regulation is considered to be at an all time high (CRS 1979).

It is this concern about increasing costs due to environmental regulation which has been pointed to as the stimulus for Executive Order 12044 (March 23, 1978) requiring each Executive Agency to adopt procedures to improve existing and future regulations. Specifically, it calls for a regulatory development process which ensures:

1. that the need for and the purpose of the regulation are clearly established;

2. that the heads of agencies and policy officials exercise effective oversight;

3. that the opportunity exists for early participation and comment by other Federal agencies, state and local governments, businesses, organizations and individual members of the public;

4. that meaningful alternatives are considered and analyzed before regulation is issued;

5. and that compliance costs, paperwork and other burdens on the public are minimized.

The immediate results of this Executive Order has been: the establishment of the Interagency Review Liaison Group chaired by Douglas M. Costle, Administrator of the EPA; the rewriting of the Administrative Procedures Act and Title V of the U.S. Code (expected to be passed by Congress in late 1979) with a specific requirement for agencies to conduct benefit cost analyses as part of their "framework for principled decision making"; and the establishment of a number of subcommittees of the EPA's Science Advisory Board to examine the EPA's decision-making process (i.e., Economic Analysis Subcommittee, Decision Analysis Subcommittee).

STATUTORY AUTHORITY

A brief glance at the core of the EPA's regulatory authoritory to deal with health, safety and environmental problems reveals that the following statutes in one or more of their sections currently permits or requires some form of economic anlaysis: The Toxic Substances Control Act, 15 U.S.C., S.S. 2601 et. seq., (TSCA); The Clean Air Act, 42 U.S.C., S.S. 1857, et. seq., (CAA); The Water Pollution Control Act, 33 U.S.C., S.S. 251 et. seq., (WPCA); The Safe Drinking Water Act, 42

U.S.C., S.S. 300f et. seq., (SDWA); and the Federal Insecticide, Fungicide and Rodenticide Act, 7 U.S.C., S.C. 136 et. seq., (FIFRA). And specific regulatory authority for marine activities related to petroleum activities may be found in TSCA via the regulation of both products and pollutants; and the Clean Water Act (CWA) as well as The Marine Protection, Research, and Sanctuaries Act (MPRSA) via the regulation of pollutants (Fig. 1 and 2).

EPA'S DECISION-MAKING FRAMEWORK

In considering the role of economics in the formulation and implementation of environmental policy we need to first consider EPA's decision making process.

EPA's procedures for writing standards and regulations are elaborate, coordinating input from the separate offices for research, media programs, enforcement, and management, and assuring that new rules receive careful review and participation by various EPA components and the public. At the heart of EPA's decision-making, and of increasing importance, is the steering committee of EPA officials (a continuing group representing the seven Assistant Administrators, General Counsel, and certain Office Directors on the Administrator's staff) which reviews a proposed or final regulation before it is sent to the Administrator for his consideration. This procedure governs the development of most of the salient and precedential regulatory policies.

A typical regulation is developed in a four-stage process: 1) defining the need for a regulation, 2) preparing a development plan, 3) preparing a decision package, and 4) conducting a three-part internal review before publication. Each regulation goes through stages three and four twice, first as a proposal and again in final form.

A 1977 National Academy of Sciences study entitled "Decision Making in the Environmental Protection Agency" has reviewed and commented upon EPA's decision-making procedures. NAS stated that the steering committee procedures facilitated the systematic examination of a new policy by all interested parts of EPA, thus providing some safeguards for sound decision-making, without encumbering EPA in a procedural straitjacket. Also, the NAS cited the steering committee procedures for helping to foster openness in the drafting of EPA regulations.

ECONOMIC FOCUS

A review of the Agency's organizational chart provides a perspective on the role of economic analysis within this decision-making framework. The most concentrated economic focus is in the Office for Planning and Management (OPM), Office of Planning and Evaluation (OPE) which has an Economic Analysis Division consisting of three branches: Industrial Analysis; Air

FIGURE 1   FIGURE 2

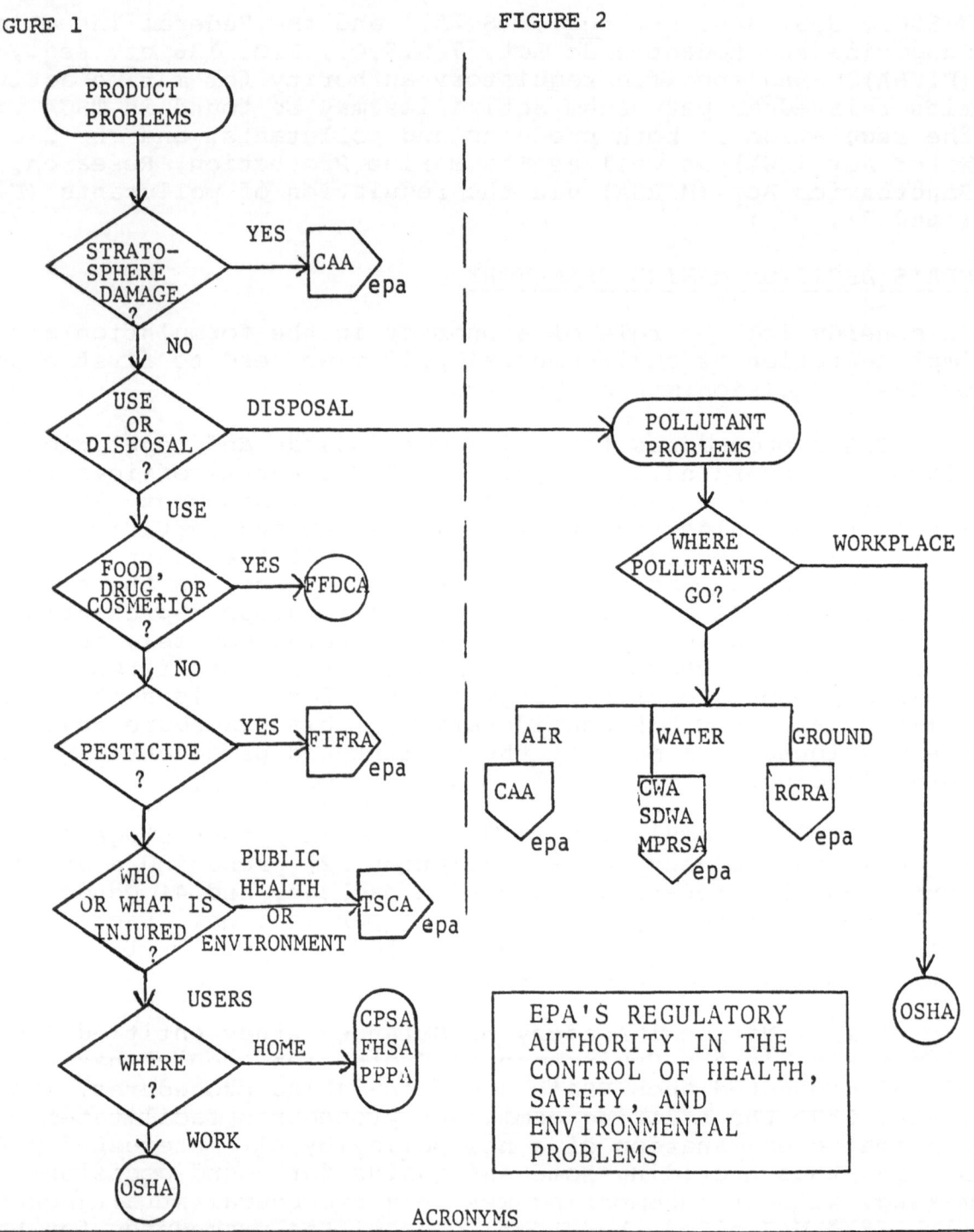

EPA'S REGULATORY
AUTHORITY IN THE
CONTROL OF HEALTH,
SAFETY, AND
ENVIRONMENTAL
PROBLEMS

ACRONYMS

CAA   - Clean Air Act
CPSA - Consumer Product Safety Act
CWA   - Clean Water Act
FFDCA - Federal Food, Drug, and Cosmetic Act
FHSA - Federal Hazardous Substances Act
FIFRA - Federal Insecticide, Fungicide, and Rodenticide Act
MPRSA - Marine Protection Research and Sanctuaries Act
OSHA - Occupational Safety and Health Act
PPPA - Poison Prevention and Packaging Act
RCRA - Resource Conservation and Recovery Act
SDWA - Safe Drinking Water Act
TSCA - Toxic Substances Control Act

Economics & Special Projects; and Water Economics. The Economic Analysis Division of OPE provides the Agency with a means of monitoring the overall impact of the Agency's programs on the economy and functions to insure that all proposed regulations make sense from an economist's point of view.

The Office for Water and Waste Management has a water economics division under the Office of Analysis and Evaluation to conduct its economic analyses. The Office for Air, Noise and Radiation Programs has economic divisions under the Office of Radiation Programs (Economics & Statistical Evaluation Branch), Air Quality Planning & Standards (Economic Analysis Office), and Mobil Source Air Pollution Control (Program Management Office), to conduct its economic analyses. The Office for Research and Development has an Office of Economic Analysis with primary responsibility for methods development and analysis of the benefits of environmental improvement. The Office for Toxic Substances has economic divisions under the Office of Chemical Control (Office of Regulatory Analysis), and the Office of Pesticide Programs (Economic Analysis Branch) to conduct its economic analyses. While the Office for Enforcement lists no specific economics office, the Agency has produced at least three reports dealing with enforcement economics.

Currently, the activities of the Agency in terms of economic analysis may be grouped into the following categories: 1) Macroeconomic analysis, 2) program specific analysis, 3) regulation specific analysis, 4) industry analysis, 5) issue analysis, 6) plant closures, 7) special reports, and 8) methods development. The kinds of analyses are explained as follows:

1) Macroeconomic analysis done for OPE, mainly by outside contractors using econometric models, are designed to assess the overall impact of pollution control costs on the U.S. economy.

2) Program specific analysis done primarily by the various program offices or their contractor's, are attempts to assess the impact of a particular pollution control program (i.e. Water and Hazardous Materials) on the economy.

3) Regulation specific analysis done primarily by the various program offices or contracted out by them, are attempts to assess the economic impacts of specific regulations. Direction to the program offices concerning these studies and a review of their work is conducted by OPE. This is currently the bulk of EPA's economic activities since for many regulations, a formal economic impact is required by law while for others the analysis is required by Executive Order 12044. And, generally even when an economic impact analysis is not required, the Agency conducts one to inform itself of the economic consequences of its activities.

4. Industry analyses are conducted by OPE since most industries are affected by more than one set of environmental regulations. In many cases the costs of the combined set of regu-

lations are much greater than would be included in the economic impact analysis of the individual regulations. An analysis of this type has been completed for the petroleum industry, and one is currently underway for the chemical industry.

5) Issue-specific analyses are economic assessments of special issues which arise from time to time and are done primarily by OPE. Examples of these would be the impact of EPA's programs on jobs, or the development of a mechanism to assess fees so as to remove the economic advantages of avoiding compliance with environmental regulations.

6) Plant closures due to the costs of environmental regulation is a very sensitive issue and prompted EPA to develop an Early Warning System. Every quarter, based on a review of the Early Warning System, OPE sends a comprehensive information report to the U.S. Secretary of Labor detailing possible unemployment problems in affected areas. To date, only twenty plants, fifteen percent of the total one hundred thirty six which have been reviewed, have actually closed due to environmental regulations. In many cases, OPE also investigates whether measures can be taken to allow plants to stay open. U.S. Department of Commerce figures show net job losses to be less than fifty percent for plant closures due to environmental regulations.

7) Special reports of a similar type to those conducted under the heading of issue analysis, i.e. "Cost of Clean..." reports are also done by EPA through OPE.

8) Methods development is the broad area of general responsibility that belongs to the Office of Research and Development's Economic Analysis Division. Much of this work is done by contract with some of it being done at the various regional labs where the physical data is also gathered and analyzed. Currently, ORD is involved with a major program to improve the methodology for determining the benefits of national and regional pollution control programs and make initial applications of such methodologies.

Within these eight areas of economic activity EPA is conducting cost, cost effectiveness, benefit/cost, and benefits analyses (Table 1). Cost analysis or cost impact analysis may be defined as an analysis of all the economic costs, direct, indirect, capital, maintenance and operating, and marginal costs as well as social costs which accrue due to some activity (i.e. environmental regulation in this case). Cost effectiveness analysis may be defined as the ratio of cost in dollars and other tangible values to some effectiveness or performance measure, for example the cost on a per unit basis of pollutant removal. Benefit/cost analysis may be defined as a way of looking at the costs and benefits from a program over time following the principle of maximum social gain. To perform a benefit/cost analysis, one must first know how to measure costs and benefits and second, know how to determine the present value of future benefits (Freeman, 1979). Also, in order to be useful, the

TABLE 1

Types and Locations of U.S. Environmental Protection Agency
Economic Studies to June 1979.

	Cost	Cost Ef-fective-ness	Cost/Benefit	Benefits
Macroeconomics (all acts)	9	1	-	2
Program Specific (a number of related acts)	26	5	3	26
Regulation Specific (section of an act)	133	3	12	11
Industry Specific (impact of a regulation)	116	47	11	-
Methods Development	15	4	5	11
Plant Closures	137	34	-	-
Issues (special reports) (economic incentives) (etc.)	29	4	4	5

Total number of studies is 427.   Column totals may be
greater because some studies included more than one category.

decisionmaker needs to know what constitutes an acceptable relationship between the costs and benefits of a program. Benefit
analysis as conducted by and for the U.S. Environmental Protection Agency is an attempt to determine the values that individuals place on reducing the adverse effects of pollution. For
environmental problems the important questions to be answered
in doing a benefit analysis are, "How does one obtain a dose-
response function and what value does one place on the risks to
life, health, and aesthetic phenomena?" Figure 3 represents
factors as damage costs which should be considered in a benefits
estimate of the value of improving water quality. For a discussion of the state of the art in benefit analysis see Freeman
(1979) or Baram (1979) and for an example of its application
see d'Arge, et. al., (1979).

REVIEW OF ECONOMIC ANALYSIS TO DATE

Table 1 reviews 427 economic analyses which have been done by or
for the U.S. EPA from January 1970 through June 1979. The
table was constructed from a review of abstracts of the various
studies and then checking the characteristics of the study in
the appropriate categories. For example, a single study might
be classified as regulation specific and industry specific, as
well as both a cost and cost-effectiveness analysis. It is
apparent from looking at this table that most of the Agency's
analyses to date have been industry specific or plant closure
studies with the major focus on the cost of regulation. This
has been confirmed to the Science Advisory Board's Economic
Analysis Subcommittee by presentations from Agency economists.
However, the Agency has recognized the value of benefits analysis and is putting more emphasis in this area.

Table 2 similarly considers the various program offices and
whether the types of analysis conducted by or for them were
cost, cost-effectiveness, benefit/cost analyses, or benefit
analysis. It is clear from looking at this table that most of
the Agency's effort has been in the area of cost analysis and
that OPE has played a major role in at least reviewing these
studies. Similarly, we see that most of the benefit/cost
analyses and benefit studies have been done by ORD with primary
responsibility for methods development. In addition, this table
shows that the correlation between the level of regulatory activity of a program office and the level of economic analysis
conducted is a measure of the value of such analysis in developing regulations by the various program offices. Clearly, the
Air and Water programs offices have been the major performers
and users of economic analysis to date and one would expect the
Office of Toxic Substances to become a more active user as they
begin to implement TSCA.

Table 3 examines in greater detail the four categories of
economic analysis that have been done at the industry specific
level and compares that with the actual reported resources
committed by industry to comply with environmental regulations.
It is apparent that of the studies identified, the percentage

Figure 3 Benefits measures as a reduction of damage
as specifically related to water pollution

1) Human Health Costs Related to
   Water Pollution
Cost Impacts of Water Pollution on:
   a) Hospitalization expenses
   b) Mortality and morbidity
   c) Sickness prevention and care
   d) Employee productivity: sick-
      ness, absenteeism
   e) Insurance rates: health and
      life
   f) Pain, suffering and mental
      anguish to diseased and family
   g) Increased welfare costs to
      society
   h) Drinking water decontamination
   i) Chlorinated drinking water and
      correlation with cancer
   j) Providing adequate water supply

2) Animal Health Costs Related to
   Water Pollution
Cost Impacts of Water Pollution on:
   a) Commercial fish supply, quali-
      ty of catch, etc.
   b) Recreational fish supply
   c) Consumer prices for seafood
   d) Maintenance of healthy marine
      environment
   e) Disease transmitted through
      polluted seafood
   f) Marine ecosystem diversity,
      stability
   g) Fish concentrations, movements
   h) Fishing license revenues

3) Damage to Vegetation Related to
   Water Pollution
Cost Impacts of Water Pollution on:
   a) Supply of marine vegetation
   b) Quality of marine vegetation
   c) Ecosystem diversity/stability
   d) Marine food supplies

4) Diminished Real Property Values
   Related to Water Pollution
   a) Property values
      1) Impact of odor
      2) Deterioration of water supply
   b) Tax revenues for State and
      local governments
   c) Habitability
   d) Ability to attract new resi-
      dents or business to area

5) Maintenance Costs and Damage to
   Materials Related to Water
   Pollution
Cost Impacts of Water Pollution on:
   a) Industry's pre-treatment expenses,
      maintenance expenditures
   b) Water-front structure repairs
   c) Vessel maintenance and
      depreciation
   d) Commerical expenditures for clean-
      ing, repair, prevention
   e) Domestic expenditures for clean-
      ing, repair, prevention

6) Diminished Recreational Opportunities Related to Water Pollution
Cost Impacts of Water Pollution on:
   a) Value of existing recreational
      areas
   b) Restricted recreational use
   c) Commercial tourism involving
      water (beaches, streams, lakes,
      etc.)
   d) Business dependent upon tourism
   e) Fishing and boat license revenue

7) Aesthetic Costs Related to Water
   Pollution
Impacts of Water Pollution on:
   a) Amount of odor
   b) Scenery benefits
   c) Turbidity and color

8) Miscellaneous Costs/Benefits
   Related to Water Quality
Impacts of Water Quality on:
   a) Benefits of emerging greenway
      opportunities concomitant with
      water cleanup

9) Avoidance Costs Related to Water
   Pollution
Cost Impacts of Water Pollution on:
   a) Desire to escape polluted areas
      1) Residential movement
      2) Recreational escapes
   b) Preventative measures to avoid
      adverse impacts dirty water
      1) Health costs
      2) Fish and vegetation
      3) Maintenance and prevention

Table 2

Economic Studies by Program Office in the Environmental Protection Agency to June 1979

	Cost	Cost Effectiveness	Cost Benefits	Benefits
Enforcement	3	-	-	-
Planning & Management	358	45	7	12
Research & Development	54	7	17	40
Toxic Substances	22	1	2	1
Radiation	5	-	1	-
Noise	14	-	-	1
Solid Waste	11	-	-	1
Water	61	6	3	10
Air & Hazardous Waste	108	3	1	18

Total number of studies is 427. Column totals may be greater because some studies included more than one category. Benefits include earlier studies identified as harm or damage costs from pollution.

TABLE 3

Review of Industry Specific Economic Studies of U.S. Environmental Protection Agency to June 1978 Compared with New Plant and Equipment Expenditures for Pollution Abatement from 1973 to 1977

	1	2	3	4	5	6	7	8
	Cost	Cost Effective	Cost/ Benefit	Benefit	Total Number of Studies	Percent of Total Studies	Expenditures for Pollution Abatement	Percent of Total Expenditures
All Industry	116	47	11	-	174	100	31,105	100
Manufacturing	83	37	10	-	130	75	20,106	65
Durable Goods	31	14	-	-	45	26	8,297	27
Primary Metals	20	4	-	-	24	14	4,490	14
All Others[a]	11	10	-	-	21	12	3,207	10
Non Durable Goods	53	22	10	-	85	49	11,765	38
Food, Textiles, Paper	6	8	-	-	14	8	3,268	11
Chemicals, Petroleum	44	15	10	-	69	40	8,351	27
All Other Non Durable	2	-	-	-	2	1	136	1
Non-manufacturing	33	10	1	-	44	25	11,000	35
Mining	11	2	-	-	13	7	414	1
Public Utilities	8	7	1	-	16	9	9,197	30
All Others[b]	14	1	-	-	15	9	1,401	5

a. Includes machinery, transportation equipment, stone, clay, glass, etc.

b. Includes transportation, communication, commercial, etc.

c. New plant and equipment expenditures, unit measured in $ x 10^6, taken from Rutledge, G.L., F.J. Dreiling, and B.C. Dunlap, "Capital Expenditures by Business for Pollution Abatement 1973-77 and Planned 1978" Survey of Current Business, June 1978, as cited in The Status of Environmental Economics: An Update, a report by the Congressional Research Service for the Committee on Environment and Public Works, U.S. Senate, Serial No. 96-6, July 1979, pp. 152-153.

distribution closely parallels the percentage distribution of
industrial resources, indicating again that within its budget,
the agency has been sensitive to the economic impact of its
regulatory activities.

Table 4 is a matrix which examines the 48 benefit analyses
which have been done so far in terms of the damage costs considered and the various media programs to which they relate. The
Water Program cell is expanded in Figure 3 to show the kinds of
considerations which need to be valued and entered into a benefits analysis. As the Agency comes under increasing attention
concerning the costs of environmental regulations, and also is
faced with the need to allocate its budget between the various
program offices, this type of study will increase in importance.

MARINE ECONOMICS

Economic analysis directed at the marine environment appears to
be one of the few areas that has not received an appropriate
amount of economic study by the Agency, considering the scope of
the Agency's statuatory authority. While perhaps 20 economic
studies had a marine component, very few studies were identified
that were specifically marine. The most broad based economic
study currently underway is the National Academy of Sciences
study on the development of ocean resources, scheduled to be completed in late 1979. Somewhat more limited in scope is a recent
draft report on the cost impacts of ocean discharge regulations
(Temple, Barker, and Sloane, 1979). Under Section 403(c) of the
Clean Water Act, the Agency has authority to propose regulations
for discharges into the territorial sea, contiguous zone and
ocean. This economic study investigated the feasibility of regulation based on monitoring and treatment in response to identified damage and concluded that this was an advantageous approach
for the communities and industries studied. More industry
specific studies currently underway are the study to investigate
and evaluate manipulations of aquatic ecosystems for the control
of productivity (Peterson, 1979) and the study on the market feasibility of seafood by-products as a method of waste reduction.
(Development Planning and Research Associates, 1979). An example
of an industry site specific study was the study on the economic
impact of effluent regulations on the rum industry in Puerto
Rico and the U.S. Virgin Islands. (Seltzer, Wissman, and Hanra-
han (1979). This study estimated the broad economic effects
which might result from the required application of various control methods and technologies. This study also included an anal-
asis of the economic ramifications of possible production cur-
tailments from plant closures within the rum industry.

Only three benefit studies were identified, two of a similar nature, that investigated the cost impact of marine pollution on recreation and travel patterns. (Binkley and Hanemann,
1976)(Bigler, McCarty, and Shawn, 1979) The studies explored
demand models which are new to recreation analysis and which are
based on site characteristics and individual preferences to estimate benefit measured by consumers surplus. Considerable data

TABLE 4

Media Matrix of Economic Benefit Studies by the U.S. Environmental Protection Agency to June 1979

Benefits	Air and Hazardous Wastes	Water	Solid Waste	Noise	Toxic Substances
Human Health	10	10	1	1	1
Animal Health	1	6	-	-	1
Vegetation	11	3	-	-	1
Real Property	7	9	-	1	-
Maintenance and Materials	8	4	-	-	-
Recreation	2	10	-	-	-
Aesthetic	9	5	-	1	1
Miscellaneous	8	5	-	-	1
Avoidance Costs	7	5	-	1	-

Total number of studies is 48. Column totals may be greater because some studies included more than one category. Benefits includes earlier studies identified as harm or damage costs from pollution. There were no benefit studies identified for the offices of Enforcement or Radiation.

was generated on willingness-to-pay, substitution between sites and activities, water quality perceptions, and general marine recreation behavior. Both of these studies were limited to items 6a,b,c and 7a,b,c as given in Figure 3. The third studied the benefit of water pollution control on property values and was limited to 9a in Figure 3 (Dorbusch and Barracer, 1973). Thus mosts of the components identified in the expanded benefits for the reduction of damage related to water pollution in Figure 3 simply have not been studied by the Agency. Thus, not only has there been a dearth of marine economic studies, almost no attention has been given to determining the benefits of marine pollution control.

This may be due in part to the fact that historically, legislation governing ocean pollution research and development and monitoring has been hazy and fragmented. Additionally, the Agency's authority in the marine environment has been clouded with questions of conflicting jurisdiction with other agencies, particularly which agency should be the lead agency. Thus, Congress passed the Ocean Pollution Research and Development and Monitoring Planning Act (Public Law 95-273) on May 8, 1978. This legislation called for establishment of a comprehensive five year plan for Federal ocean pollution research and development and monitoring, emphasizing coordination among the concerned Federal Agencies. The National Oceanic and Atmospheric Administration in consultation with the Office of Science and Technology Policy was designated lead agency in this endeavor. To accomplish this plan, there was created the Interagency Committee on Ocean Pollution Research and Development and Monitoring under the Federal Coordinating Council on Science, Engineering and Technology. A Subcommittee on National Needs and Priorities was established in July 1978 to help the Committee successfully complete its task. This Subcommittee had strong representation from the EPA. In considering interdisciplinary needs, the Subcommittee concluded that while not everything that needs to be factored in risk analyses reduces to economic terms, there is a strong need to employ methods to relate values measurable in dollars to values measurable in other units, such as increased summer violent crime when urban recreational beaches are closed due to bacterial pollution. Thus, the Subcommittee recommended the inclusion of economists, planners, and other social scientist to a greater extent than heretofore in controlling ocean pollution. The Subcommittee produced a ranking of research needs for marine activities and estimated their socio-economic effects a summary of which is reproduced here. (Gage, 1979)

RESEARCH PRIORITIES FOR MARINE ACTIVITIES

PRIORITY	ACTIVITY	SOCIO-ECONOMIC EFFECTS
High	Industrial Waste Disposal	Important
High	Land Use Practices	Important
High	Municipal Sewage Outfalls	Moderately Important
High	Oil and Gas Development	Important
High	Oil Transportation	Important
High	Steam Electric Power Plants	Important
High	Transportation of Hazardous Material	Important
Medium	Deep Seabed Mining	Of Some Importance
Medium	Fish and Shellfish Processing	Of Some Importance
Medium	Hatcheries and Aquaculture	Of Some Importance
Medium	Ocean Dumping of Dredged Spoil	Moderately Important
Medium	Recreation (Including Small Craft Activities)	Of Some Importance
Medium	Sand, Gravel and Shell Mining	Of Some Importance
Medium	Sewage Sludge Dumping	Moderately Important
Low	Biomass Fueled Systems	Of Some Importance
Low	Brine Producing Activities	Of Some Importance
Low	Kinetic Ocean Systems	Not Very Important
Low	OTEC	Of Some Importance
Low	Salinity Gradient Systems	Not Very Important
Low	Satellite Power Systems	Not Very Important

It would seem that as petroleum exploration and development is extended to include not only the Gulf of Mexico and areas off the Pacific Coast, but the Georges Bank and other areas of the Atlantic Coast. the Agency will have an increasing responsibility to examine the consequences of this development. This economic analysis would probably go beyond looking at effluent from drilling rigs, ships, and outfalls to include a cost/benefit analysis of the effects of potential oil spills both during production and transport and increased seepage resulting from production processes. Perhaps as part of this, a benefits study based on some of the criteria suggested in Figure 3 might be undertaken.

CONCLUSIONS

The information contained in this report was given to the Economic Analysis Subcommittee of the Administrator's Science Advisory Board during a public meeting on November 20, 1979, for their consideration. This is a new Subcommittee of the Board, being chartered in mid-1979, to look at the past and present role of economics in the Environmental Protection Agency and to suggest potential future directions. Some brief, preliminary conclusions of the Subcommittee, available at the deadline time for the writing of this paper are that economic studies are both necessary and useful for the proper functioning of the Agency. Further, both the types and distribution of economic studies

that the Agency had been doing were reasonable and adequate in
view of the historical development of the Agency and its legislative mandates. Thus most of the economic studies done by the
Agency were concerned only with cost aspects, not benefits.
Some tentative recommendations of the Subcommittee are that the
Agency should now move more into the benefits side of economic
analysis so as to enable the Agency to make realistic cost/bene-
fit decisions in a time of increasingly scarce resources. This
reflects the Subcommittee's view that the methodologies for
measuring benefits have improved considerably in the decade
since the inception of the Agency and that, concomitantly, as
the country is entering a period of energy shortfalls and fiscal
restraint, the necessity to measure and demonstrate comensurate
benefits has increased. To implement this the Subcommittee has
suggested that the Agency devote more resources to the development of appropriate benefits data and useful data bases for the
storage and manipulation of the information obtained. This
would require additional efforts to develop benefits methodologies and models and it was felt that this could be obtained by
building some very large data bases and making extensive use of
input/output analysis. The final report of the Economic Analysis Subcommittee with the actual conclusions and recommendations
should be available in mid-1980 and can be obtained upon written
request to the first author.

Baram, Michael. "Regulation of Health, Safety, and Environmental Quality and the Use of Cost Benefit Analysis." Report to the Administrative Conference of the United States, 1979.

Biglar, A.B., H. McCarty and R. Shawn. Cost Impact of Marine Pollution on Recreation Travel Patterns. EPA/600/5-79-003. May, 1979.

Binkley, C.S. and W.M. Hanemann. The Recreation Benefits of Water Quality Improvements: Analysis of Day Trips in Urban Settings. (Contractor's report prepared for the U.S. Environmental Protection Agency). June, 1976.

Congressional Research Service. The Status of Environmental Economics, an Update. 1979.

d'Arge, Ralph C. et. al. Methods Development for Assessing Air Pollution Control Benefits. Executive Summary. EPA/600/5-79/001E. 1979.

Data Resources, Inc. "The Macroeconomic Impact of Federal Pollution Control Programs 1978 Assessment." Report to the U.S. Environmental Protection Agency and the U.S. Council on Environmental Quality. 1979.

Development Planning and Research Associates, Inc. Market Feasibility Study of Seafood Waste Reduction in Alaska. February, 1979. (Draft report prepared for the U.S. Environmental Protection Agency.)

Dornbush, D.M. and S.M. Barrager. Benefit of Water Pollution Control on Property Values. EPA/600/5-73-005. October, 1973.

Freeman, Rick. The Benefits of Environmental Improvement. Resources for the Future, Johns Hopkins Press, Baltimore and London, 1979.

Gage, Stephen J. First Report of the Subcommittee on National Needs and Problems, Committee on Ocean Pollution Research and Development and Monitoring. May, 1979.   EPA 600/8-79-013.

National Academy of Sciences. Decisionmaking In The Environmental Protection Agency. 1977.

Peterson, S.A. Investigate and Evaluate Manipulations of Aquatic Ecosystems for the Control of Productivity. Interim Report to the U.S. Environmental Protection Agency. July, 1979.

Seltzer, Richard E., et. al. Economic Impact of Effluent Regulations on the Rum Industry (Puerto Rico and U.S. Virgin Islands). March, 1979.   EPA-440/2-79-022.

Temple, Barker & Sloane, Inc. Cost Impacts of Ocean Discharge Regulations Summary Report. April, 1979. (Draft Submission prepared for the U.S. Environmental Protection Agency).

A CONTINUOUS MONITORING AND CONTROL SYSTEM FOR HYDROCARBON DISCHARGE
AT SEA

Fréderic GUIGUES
Président d'Honneur, Guigues S.A.
Aix-en-Provence, France

Sea pollution is no longer just a problem of specialists and the
reality of its importance cannot be denied. It becomes more and more
acute, not only because of accidents or related events, but essentially
because of progress of ecological and oceanographical sciences which have
demonstrated how important the sea is to keep the earth in balance.

Among all humankind industrial activities, industry involved with
hydrocarbons, as raw materials or finished products, is a major
preoccupation because of the important quantity used or transported.

More than half the raw petroleum consumption is carried by sea, which
represents a yearly tonnage over 1 500 Mt. Transport gives many problems,
among which those connected to pollution have an important place, due to
the necessity of using sea water for ships' operations.

In fact, a 200.000 tdw empty tanker (i.e. without ballast or loading
charge) offers a lateral surface of 6 200 m^2.

Hence the necessity to "ballast" after unloading its products to sail
to its loading harbour.

The ballast quantity depends on meteorological conditions and may vary
from 35 to 50 % of load capacity in bad weather.

Usually, with clear weather, a little over one third of ballast volume
is placed into special tanks (permanent ballasts) and is therefore
qualified as "clean ballast" since it is not in contact with any petroleum
product. The complement must be placed into the ships' tanks containing
some residue in various quantities, it is therefore a "dirty ballast".

Before loading, the tanker must ballast and discharge its ballast
into the sea, at harbour, or near the sea port. There is no problem with
permanent ballast water, on the other hand widespread pollution would
arise in the sea port or near it if it were to discharge sea-water soiled
by hydrocarbons.

The tanker is therefore brought in, during its return trip, to carry
out a change of ballast, i.e. wash its loading tanks (between 1/4 and
1/3 of the tanks' total volume), and fill them with sea water, thus forming
a clean ballast, But during this operation, it must on one hand
discharge the waters used for washing, and on the other hand, the waters
coming from dirty tanks: "dirty ballast".

Direct discharge of these waters would cause enormous pollution. Even since 1962, three of the main International Petroleum groups have set up a process allowing tankers to limit this operational pollution. It consists of keeping on board the washing residues and dirty ballasts, treating the polluted waters in two settling tanks placed in series and separating the water and hydrocarbon or residues. At the end of the process, the water is removed from those tanks, a new load is then fed over the settled residues, whence the name of Load on Top.

Presently, 80 % of petroleum tankers apply the load on top, which limits the total "operational" discharges to 800 000 - 900 000 t, while they would be, without this process, in the order of 2 000 000 t.

Many meetings and International Conventions have taken place in order to control these operations, the first one dates back to 1926, then 1954, with 1962, 1969 and 1973 amendments.

The International Maritime Consultative Organization, IMCO, voted some conventions (rules) against sea pollution from hydrocarbons. In 1973, a new international convention was adopted by the IMCO Assembly cancelling the convention of 1964 and all amendments. It regulates all hydrocarbon discharges as well as toxic matters, particularly chemicals, waste waters, and garbage.

It concerns all ships of any kind, including submersibles, floating or steady platforms, hovercrafts, excluding, however, pollution resulting from mineral resources work and exploration.

It also covers some accidental ship pollution and prescribes conception and construction standards.

It mentions as well a notion of special areas such as:

-   Baltic Sea
-   Mediterranean Sea
-   Black Sea
-   Gulf Zone,

for which additional conditions are required.

For operational discharge the principal requirements are:

1   Tankers

A tanker can discharge ballast waters if:

-   it does not find itself in a special area
-   it is more than fifty miles away from the nearest coast
-   it is sailing
-   the hydrocarbon instant rate does not exceed 60 litres per nautical mile
-   the total amount of discharged hydrocarbons does not exceed 1/15.000 th for tankers and 1/30.000 th for ships of the total quantity of the particular shipload giving residues

- the tanker uses a permanent discharge control and a clarification network
- this device is used for each discharge and allows an automatic stop when discharge is not in conformity
- outlet piping leads, except in special cases, to uncovered deck or ship's board, or else above floating line with maximum ballast conditions.

2   Other ships (not tankers) and bilge waters of tanker's machinery

a)   Ships with a gross gauge equivalent or exceeding 400 drums are allowed to discharge if:

- they are not in a special area
- they are more than 12 miles away from nearest coast
- they are sailing
- hydrocarbon content is less than 100 ppm
- they are equipped with a permanent discharge control or a hydro-carbon separating system, or else a filtration system appointed by an authority and of such conception that discharges do not contain more than 100 ppm

b)   Ships with a gross gauge equivalent or exceeding 10.000 drums:

Besides the above requirements, they have to be equipped with a permanent discharge control system recording continuous hydrocarbon content of discharge and stating date and hour.

c)   Ships of a gross gauge inferior or equivalent to 400 drums:

For those ships (not tankers) not sailing in special areas, the Authority prescribes, within possible limits, equipment for hydro-carbon residues conservation on board and discharge into reception installations, or at sea, but according to conditions ruling ships with a gross gauge exceeding 400 drums and inferior to 10.000 drums.

To reach such control, and for the application of the convention to take effect during the eighties, it was necessary to immediately start studying a continuous control device which we shall call the "BLACK BOX", able to integrate different parameters and provide the necessary information.

The BLACK BOX presently studied by the GUIGUES, SERES, and BEN companies, with the support of the Merchant Marine Ministry, will allow the Authorities to control all discharges *a posteriori* and check if they conform with the International Convention regulations.

This BLACK BOX will permanently record all hydrocarbon discharges per litre and per nautical mile, and the total amount discharged during a trip.

Information thus obtained, and provided every minute, will be dated and can be recorded on magnetic tape, just like the well-known black box in aircraft.

The BLACK BOX will have to simultaneously control discharge points, and will start the automatic control as soon as a discharge occurs into one of the pipes going from the ship to the sea.

This start-up is set in motion by the observation of a flow in the discharge piping by an ultrasonic flowmeter especially studied for the BLACK BOX, to comply with security problems on board ships. During the further development of operations, this eight-directional flowmetrical setting will provide all flow information required for the BLACK BOX computation.

At the same time as the setting start-up, the flowmetrical device will start samplings and hydrocarbon measures on discharges (oilmeter).

If the main characteristics of this device are observed during discharge, one can see that the entire discharge is clean but, when interface oil/water is approaching, the hydrocarbon content rises suddenly within a few minutes. Therefore, the essential characteristics of the oilmeter have to reside in an extremely short time of response, because if hydrocarbon content rises very fast, the quantity discharged can be great. This happens especially in the case of a strong flow discharge with the tanker's dirty ballast waters.

Other characteristics established by IMCO working teams are, of course, required. They are mainly:

- measuring frequence: sampling and analysis have to be permanent, the range of measures has to cover 0 to 100 ppm and, if possible, 0 to 1 000 ppm
- the time of intrinsic response of the device has to be less than 20 seconds and the total time less than 40 seconds
- measuring accuracy: more or less 20 % even with sediments and iron oxide
- maintenance: the least possible
- black box adjustment possibility in order to allow further control.

There are quite a few devices, some of solvent extraction, others fluorescent, etc ... but corresponding more or less to the above requirements. Most of them are adapted laboratory or industrial devices. We know of about twenty devices in the world, elaborated for control of ballast or bilge waters, but being short of time, we shall only mention the D.R.U.R/SERES, with which we now study our BLACK BOX.

The BLACK BOX computer will execute all hydrocarbon discharge calculations, every minute, per litre and per nautical mile, along with the amount of hydrocarbon discharged during the minute in question.

Results are added in order to know the total amount of hydrocarbon discharged during a ballast operation, just like a kilometrical meter.

If the authorized amount of discharge is exceeded an alarm will signal the excess on board and stop all ballast operations.

It is obvious that such a continuous discharge control system requires some security, and, this is why two recording systems have been provided:

- one on paper tape, allowing port authorities to control the information (very easily),

- and a second one on magnetic tape able to keep the information over a few years.

Moreover, the BLACK BOX records information concerning the day, the hour and the minute of discharge in order to cross-check with the log-book, the ship's exact navigational position at the time of ballast.

Later, another BLACK BOX will be completed for ships other than tankers because, with these too, a continuous dated control will have to be applied for bilge waters.

This survey shows that rules and technical means will be available in the very near future to allow authorities permanently engaged in overcoming sea pollution to exercise positive control, supported by legal and technical means.

SESSION 3

Polar Activities

INDUSTRY, GOVERNMENT AND THE ENVIRONMENT:

HYDROCARBONS IN THE CANADIAN HIGH ARCTIC.

Noel E.J. Boston, Director, Ocean Sciences,
Beak Consultants Limited.

L. Clifford White, Vice-President, Prairies and Northern Division,
Beak Consultants Limited, Canada

DEFINING THE ARCTIC

The Canadian High Arctic could be defined as that area of Canada north of the Arctic Circle to 200 miles north of Ellesmere Island, bounded on the east by Greenland and on the west by Alaska. Geometrically it can be thought of as that part of the globe north of $58^{O}NL$ and south of $85^{O}NL$ between $55^{O}WL$ and $140^{O}WL$ (Fig. 1). In fact the Canadian Arctic cannot be defined by geometrical or political boundaries as those features which are uniquely arctic, low temperatures and distinctive day/night cycles, are determined not only by the geometry of the earth-sun system but also by the geography of the far north. The distribution of oceans and land masses has major effects on the surface air flow which creates what we consider an "arctic" climate.

DESCRIPTION

The single unique feature of arctic regions is the low average intensity of solar radiation coupled with strong seasonal variation in that intensity. The most significant consequence is that water is in solid form much of the time, which in turn results in two effects which lead to even harsher conditions. Much of the energy received goes into changing the phase of water and not temperature. Second, the solid phase (ice and snow) has a high reflectivity sending energy back into the atmosphere and space.

In spite of its harsh climate, the arctic is not just a desolate, snow covered plain, but a region of fantastic variety and beauty both in form and life. Whereas generalization is not possible, a few comments are necessary to indicate the general trends.

In mid-winter there is little or no daylight. Freeze-up begins in September with thaw beginning in May. The permanent arctic ice sheet covers the Arctic Ocean and in some areas reaches the archipelago. Winter storms and heavy snowfall occur in September and October but by December the atmosphere is cold and dry and snowfall is much lighter. Mean average temperatures, for what they are worth, vary between $-26^{O}C$ and $-18^{O}C$. Average annual snowfall is a modest 750 mm. Much of the heat received during May, June and July only melts the ice and snow and does little warming. The soil and surface waters receive radiation for warming in July which is past the peak heating period and decreasing daily. The sea ice cover allows almost no light for photosynthesis and no free exchange of gases. In fact only a portion of the sea surface becomes directly exposed to sunlight as cracks, leads and open patches develop in the ice, and these may close as quickly as they open. Clearly such open areas are crucial to marine biological life and resources.

FIGURE I

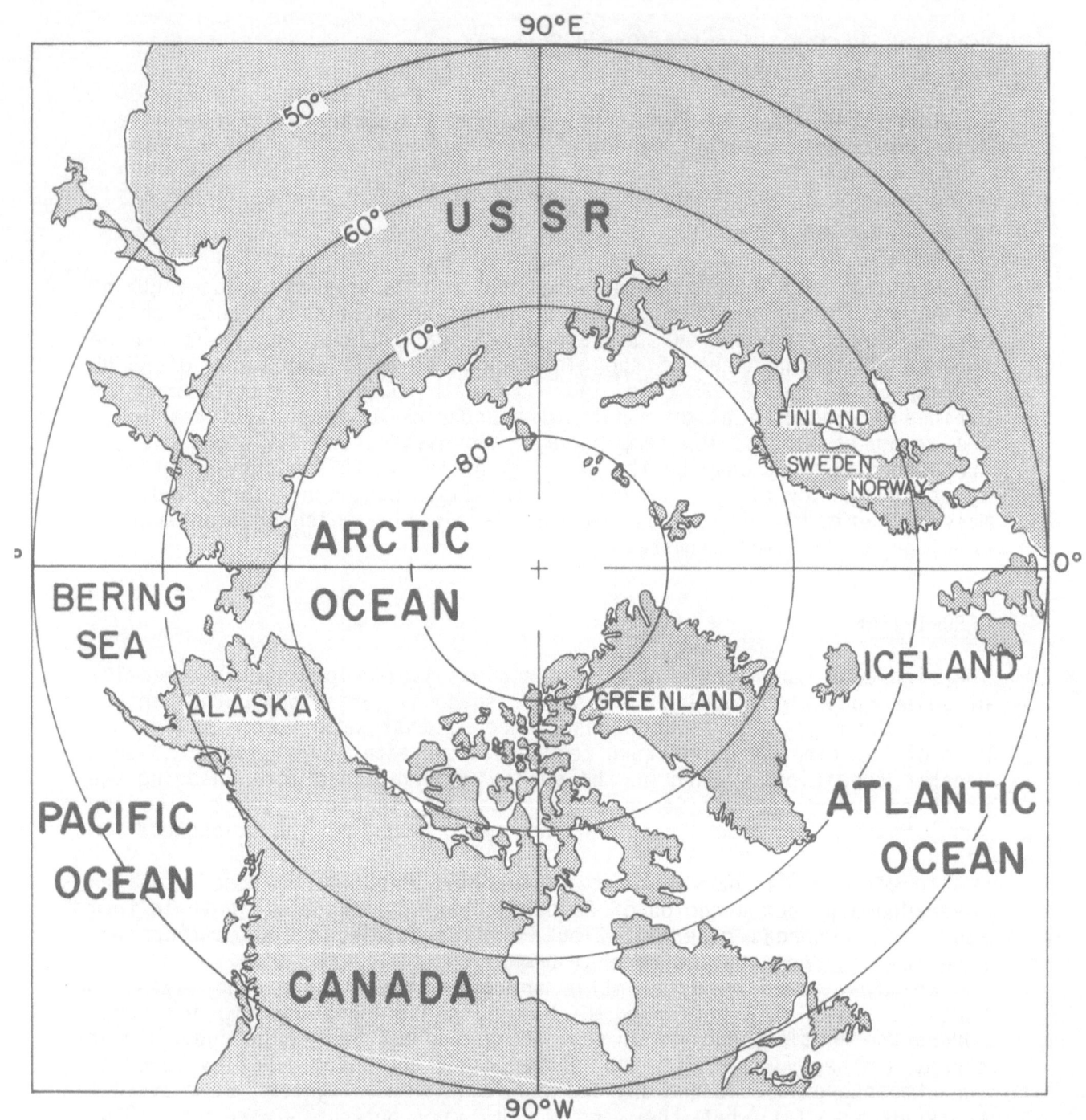

THE ARCTIC REGIONS

FIGURE 2
THE ARCTIC REGIONS

The Arctic is harsh but delicate. Whereas it may impose severe hard-
ship on human activities, it bruises easily.

GEOGRAPHICAL FEATURES

The Canadian Arctic consists of portions of mainland North
America, the islands of the High Arctic and ocean boundaries
(Fig. 2). Administratively, that portion of Canada north
of 60NL is composed of two territories, the Yukon and the
Northwest Territories, and comes under the direction of the
federal government, Department of Indian Affairs and Nor- ..ie
thern Development (DIAND). The Yukon has Alaska on its rs and
western border (141^{U}WL), a mountain divide west of the
Mackenzie River as its eastern boundary and the Beaufort Sea
on the north. East of the Mackenzie and north of 60^{U}NL are the Northwest
Territories bounded by the Beaufort Sea, Arctic Ocean, Nares Strait,
Baffin Bay and Davis Strait. These last three separate the Northwest
Territories from the Danish possession of Greenland. The High Arctic is
usually thought of as consisting primarily of the waterways and islands of
the Canadian Archipelago. The principal islands along and just north of
70°NL from west to east are Banks, Victoria, Prince of Wales, Somerset and
Baffin. Labrador, the mainland part of the province of Newfoundland,
completes the southern geographical extent. Further to the north are
found the Queen Elizabeth Islands and the Sverdrup Islands. Within these
two groups are found major islands such as Melville, Bathurst (location
of north magnetic pole), Devoi, Axel Heiberg and Ellesmere, the most
northerly island. Islands not included above which have received exploratory fame are King Christian Island, Cameron Island of the Govenor
General group, and Ellef Ringnes Island. The waterways comprising the
main sea route from the Beaufort Sea to Baffin Bay are M'Clure Strait,
Viscount Melville Sound, Barrow Strait and Lancaster Sound. The last
three are referred to collectively as Parry Channel. Together they form
the old Northwest Passage.
    The Yukon and Northwest Territories comprise about 4 million square
kilometres, almost 40% of the land area of Canada. Within this vast
region live (as of 1977) less than 65,000 people, including Inuit and
Indians.

GEOLOGICAL FEATURES

There are several geological features in the Canadian High Arctic and
offshore Canada which are of interest to the hydrocarbon industry.
Referred to as the Frontier Basins, these are the Mackenzie-Beaufort,
Sverdrup, Melville-Lancaster, Greenland Offshore, Labrador Shelf,
Newfoundland and Grand Banks (Fig. 3). In the context of oil exploration
in the High Arctic, three general regions involving the above features may
be identified; the Mackenzie-Beaufort, Arctic Islands and Eastern Arctic.

Mackenzie-Beaufort

This region includes the Beaufort Sea and the Mackenzie River delta. It
covers 260,000 km^2 and includes the continental shelf, slope and adjacent
Arctic coastal plains of the mainland (Fig. 4). Primary drill targets are
the overthrust anticlines of the late Cretaceous and Tertiary shales.
Diapiric structures in the Beaufort Sea adjacent to the modern delta front

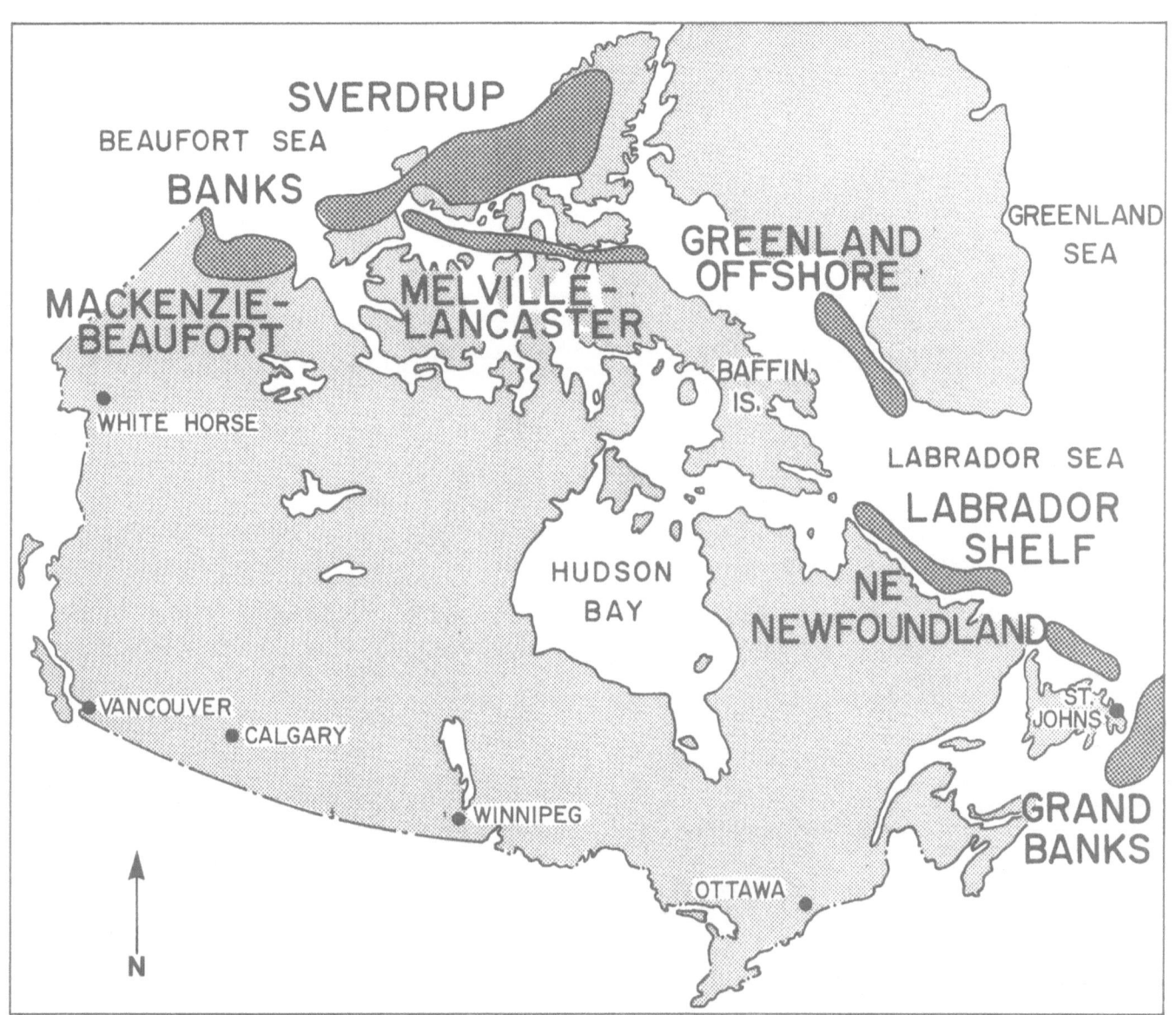

FIGURE 3
THE FRONTIER BASINS

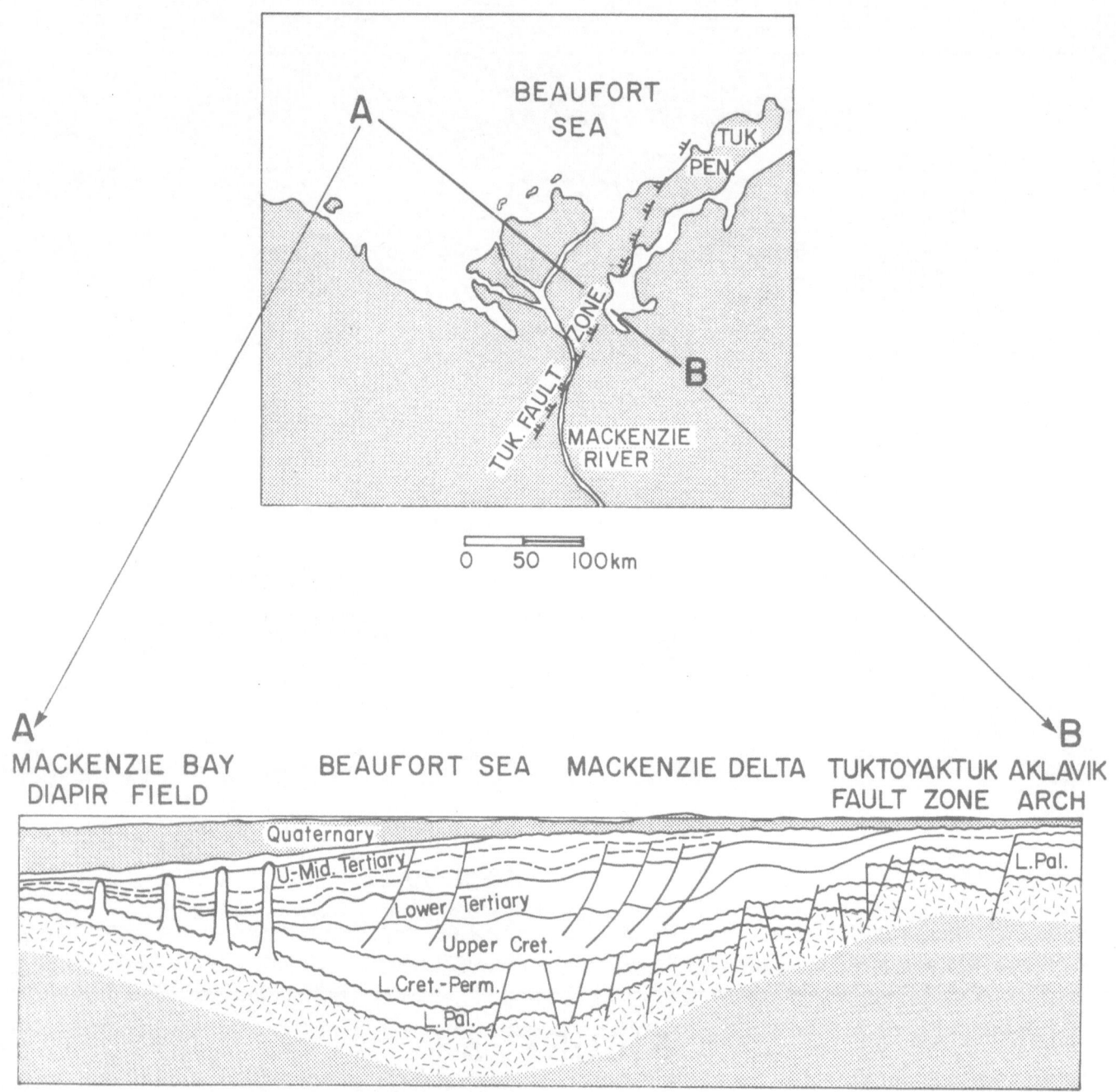

FIGURE 4

THE MACKENZIE BEAUFORT BASIN

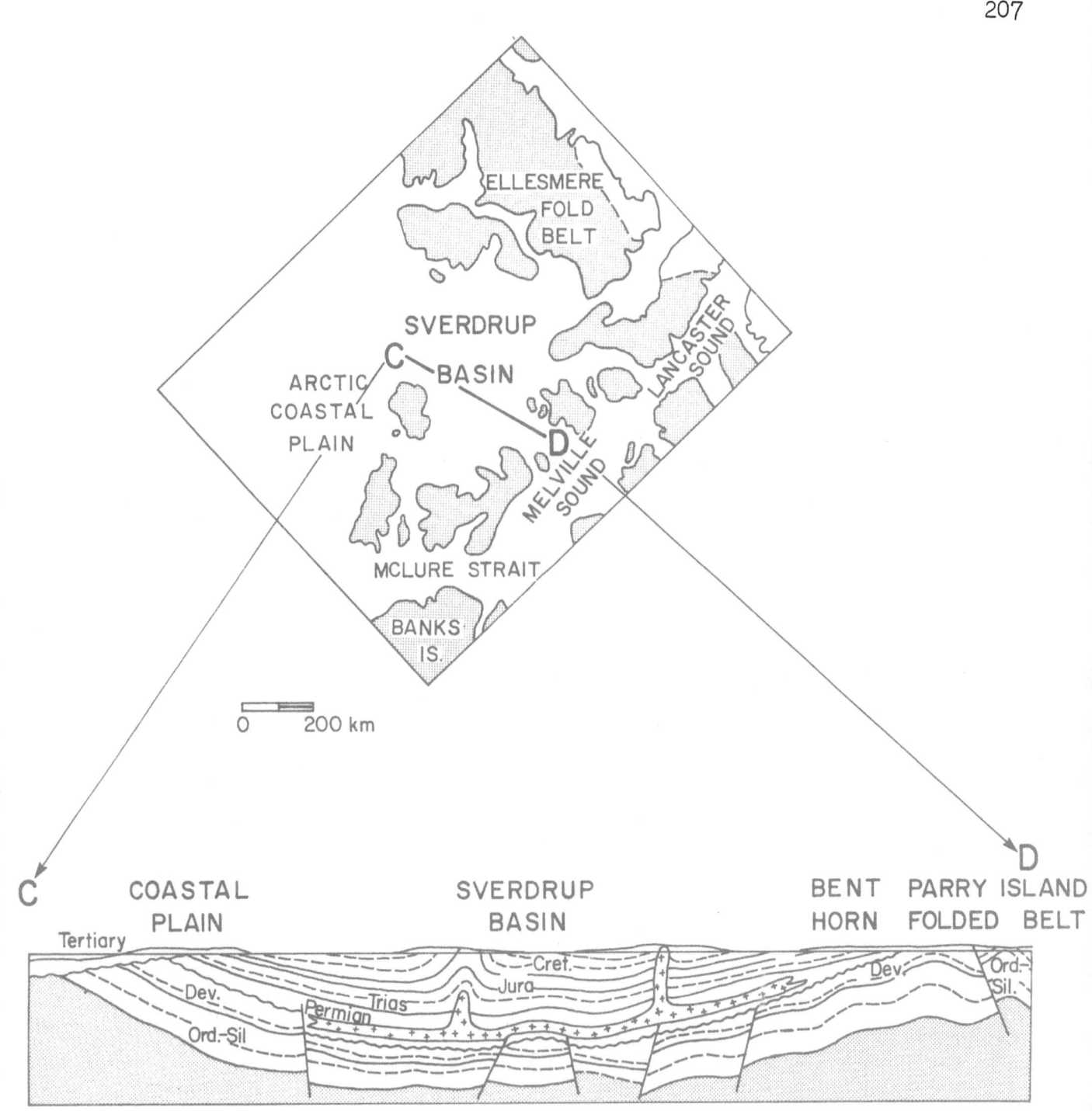

FIGURE 5

THE SVERDRUP BASIN

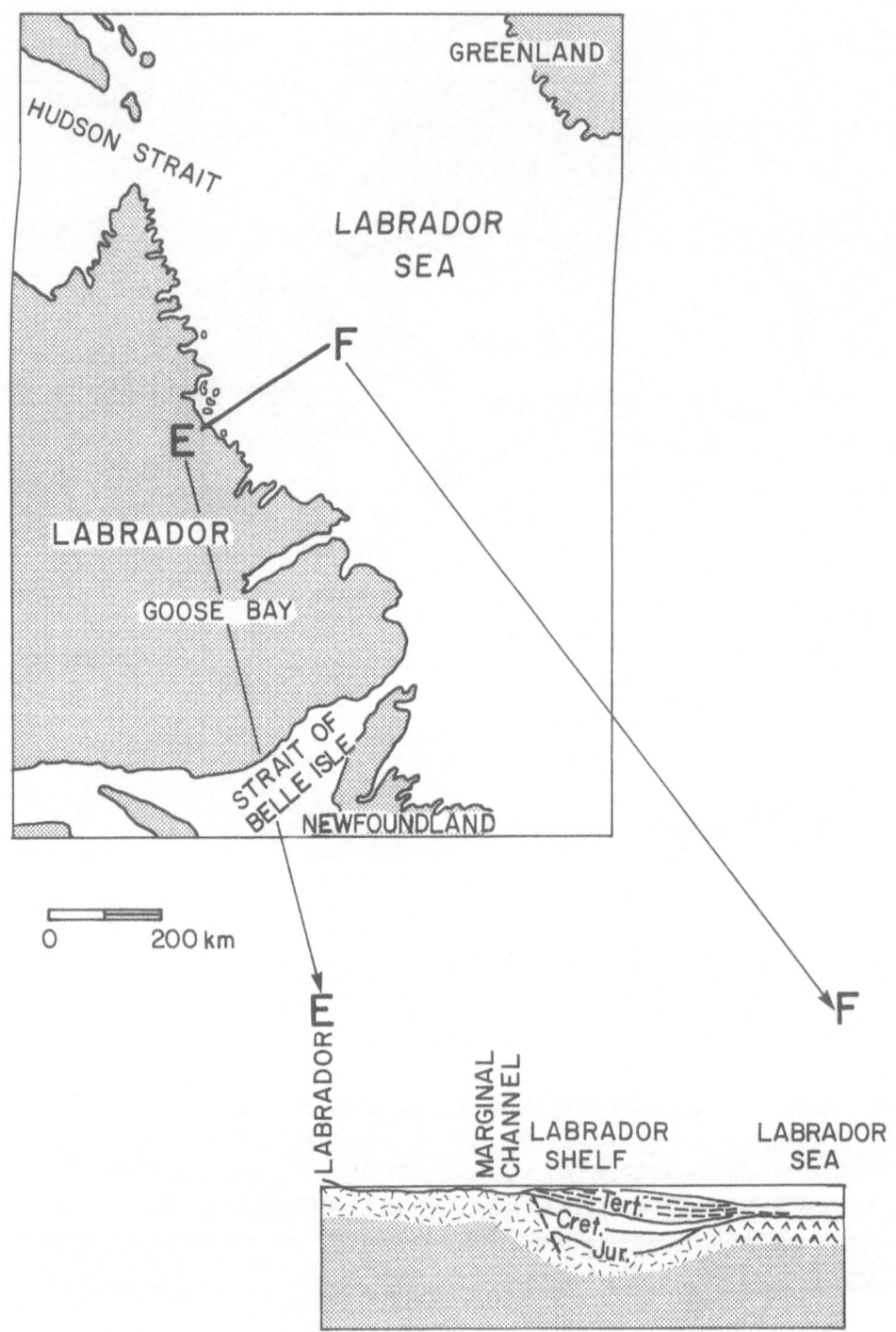

FIGURE 6
THE LABRADOR SHELF

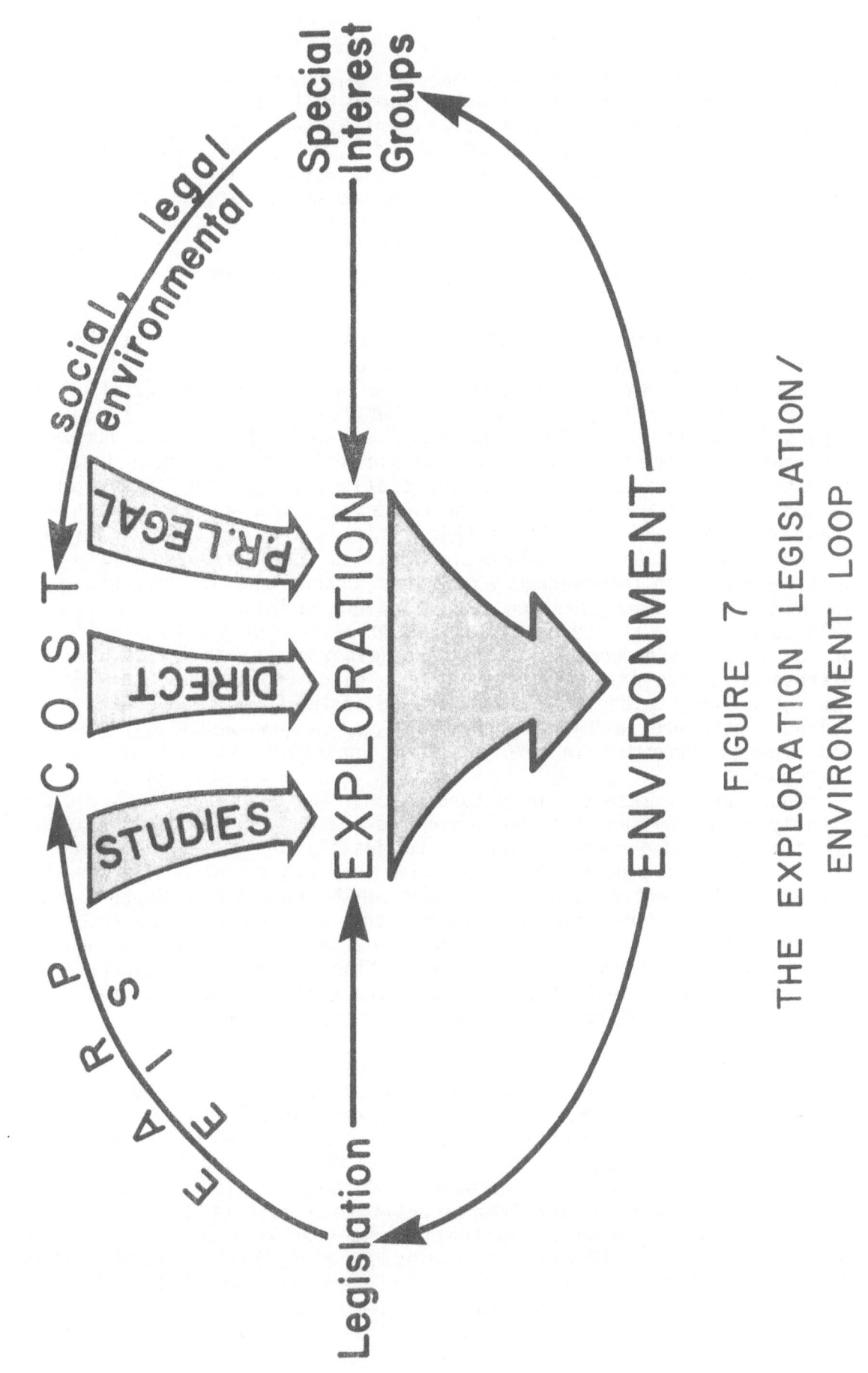

FIGURE 7

THE EXPLORATION LEGISLATION/
ENVIRONMENT LOOP

conditions, transportation problems and a depressed world oil market
followed. In 1969, activities were resumed with vigour when the Prudhoe
Bay, Alaska discovery was made. A Canadian industry-government consortium, Panarctic Oils Limited, was formed, though not specifically to
explore the Beaufort Basin. In fact this area of the Arctic has been
explored primarily by Esso Resources Limited who developed the technique
in 1973 of using artificial islands, and by Dome Petroleum in 1976 by
means of ice strengthened drill ships. Further discussion on this area is
unnecessary as it is discussed elsewhere in this volume (Ref. 4,5).

Arctic Islands

Scientific interest in the Arctic Islands, particularly the Sverdrup
Basin region, was stimulated when the Geological Survey of Canada carried
out a systematic but limited survey in the late 1940's. The first comprehensive geological exploration began in 1955 under a program called
"Operation Franklin". This resulted in a description of a number of
sedimentary basins with the potential for oil and gas reservoirs. The
private sector then began serious exploration and permits for most of the
islands north of M'Clure Strait and Lancaster Sound were issued in 1960.
The first well was drilled in the Arctic Islands in 1961 on Melville
Island at Winter Harbour. Panarctic drilled four more wells during the
60's resulting in the discovery of gas at Drake Point on Melville Island
in the winter of 1969. A second gas discovery at King Christian Island in
1970 ushered in a period of extensive exploration of the high grade area
of the western Sverdrup Basin. To date, eight major gas fields have been
discovered. The first significant oil discovery was made in 1974 by
Panarctic Oils Limited off Cameron Island of the Govenor General group of
islands which are just west of Bathurst Island. A second oil discovery
was made by Panarctic in 1976 off King Christian Island using an ice-
thickened pad.

Total proven reserves of natural gas in the Arctic Islands are constantly being revised, but the current estimate of 450 billion m^3 is near
the threshold of reserves required to justify the proposed 100 cm (42
inch) Polar Gas pipeline project from the Arctic Islands to southern
markets. A recent offshore discovery (Whitefish) near Lougheed Island
may prove to be the find which pushes the reserves past the threshold.
Industry estimates that gas reserves will stand at greater than 800 billion m^3 by the end of the 1981 exploration season. Present reserves,
however, are sufficient to justify the more flexible LNG plant and tanker
system advocated by Petro-Canada in the Arctic Pilot Project.

Eastern Arctic

The Eastern Arctic is the most recent portion of the Arctic to be subjected to intensive exploration. Drilling began in 1979 after earlier ex-
plorations indicated potentially rich oil deposits off the east coast of
Baffin Island. Esso Resources, along with Cities Service, Shell and B.P.,
has carried out drilling from a semisubmersible platform in southern
Davis Strait and Aquataine from a drill ship east of Loks Land. The
government is providing administrative and practical assistance to companies conducting environmental studies prior to drilling in the Eastern
Arctic under EAMES, the Eastern Arctic Marine Environmental Studies program. This will be discussed in more detail later.

METHODS OF EXPLORATION

Exploration for hydrocarbons in the Canadian High Arctic often requires the development of ingenious methods and techniques.   Different approaches have been employed in the three Arctic regions which have been discussed.   In many cases, the solution to a problem results from a blending of old and new methods.   The basic approaches outlined here illustrate methods of exploration in the High Arctic, the difficulties involved and how exploration impinges on the environment.

Beaufort Sea

The two principal means of drilling in the Beaufort Sea have been from artificial islands and ice-strengthened drill ships.   The former have been used by Esso Resources and the latter by Dome Petroleum.   Artificial islands have been utilized primarily in shallow water.   They are constructed of dredged sand and gravel and gravel hauled from onshore deposits over winter roads.

Arctic Islands

Drilling in the Arctic Islands has been accomplished by means of traditional land drilling, using arctic enclosed drilling rigs, and the new technique of drilling from ice-thickened platforms on the landfast ice of offshore areas.   The ice-thickened platforms are the result of extensive research in offshore arctic exploration by Panarctic Oils Limited and deserve further comment.

The technique involves flooding an area of ice approximately 110 m x 90 m in area until a thickness of 4.5 m to 6.5 m is obtained.   Normal ice thickness would be 1.5 m to 2 m.   A conventional land drill rig is then moved onto location on the "ice-platform" and drilling proceeds according to conventional offshore procedures with subsea blow-out-protector (BOP) and a high pressure marine riser to the surface.   The riser is designed to tolerate horizontal ice movement of up to 15% of water depth.   The system has been effective at depths from 60 to 365 m and up to 35 km from shore which means it has wide application in the Arctic Islands.

The completion of the first Arctic subsea producing well in April 1978 by Panarctic was described by Hood et al.(Ref. 6).

A variety of scientific and engineering studies contribute to such successful performance; studies which are beyond the scope of any single company or even of industry alone.   Joint industry associations, such as the Arctic Petroleum Operators Association (Ref. 7,8) in cooperation with government are necessary to provide meaningful progress.   In this case, the successful completion has far reaching significance and benefits all parties since it demonstrates the ability to develop offshore gas reserves in the Canadian Arctic Islands.   Further elaboration on this subject is provided by Kaustinen (Ref. 9) later in this volume.

Eastern Arctic

Procedures used in Eastern Arctic exploration tend to be more conventional in the sense that the same systems are used elsewhere.   Three common systems used are drill ships, jack-up rigs and semisubmersibles.   A major source of worry is provided by icebergs which are found from Baffin

Bay to the Labrador Sea. In fact, little loss of time has been experienced. When icebergs approach a drilling operation they are monitored, tracked and if possible, diverted. If diversion is not possible the remaining option is to secure the well and move out of the way. An efficient crew can carry out this procedure in a few minutes. There are no inefficient crews. Moveable platforms then have distinct advantages, physically and psychologically. In order to re-enter and stay on station, moveable platforms use a dynamic positioning system based on obtaining fixes with respect to acoustic transponders located around the drill site on the sea floor.

The foregoing descriptions have tended to emphasize the "harsh" aspect of the arctic. Now let us look at some of the "delicate" aspects.

GENERAL CHARACTERISTICS OF ARCTIC FAUNA

Accompanying the growth of arctic exploration has been the growth of environmental awareness. In order to appreciate the validity of claims of proponents and intervenors in the numerous hearings held _ad mare usque_ _ad_ _mare_, two levels of environmental knowledge are required: in depth, in order to correctly ascertain consequences of an action, and broad band, in order to maintain perspective. Neither are readily available in a region where relatively few comprehensive field studies have been made. We restrict our review to the arctic fauna of the marine waterways with which we are concerned.

Two major features of the arctic environment determine the nature of the fauna - extreme cold much of the year resulting in a short season of weather amenable to breeding, and, the rich productivity of marine life. These features result in special arctic faunal characteristics.

Most arctic birds and mammals are highly migratory. In the case of marine mammals the movement is in relation to spatially changing ice patterns which determine the availability of food. For example, seal movement is to a large extent governed by ice patterns, and polar bears move relative to food sources such as seals. However, caribou move in relation to the availability of lichens and birds in relation to season. Species which do not migrate, or migrate only in some years, have various physical and physiological adaptions to cold.

The proportion of marine animals in the fauna is high although some terrestrial groups are also present (such as lemmings, hares, ungulates, insects, shorebirds). Since the distribution of marine food (plankton, fish) depends on ice melt and ice movement, this food source is not available in a reliable temporal manner. Species high in the food chain are most affected and respond to food availability by varying their clutch size in relation to food abundance, by not breeding in years of short food supply and by longevity to compensate for non-breeding years. Breeding activities tend to start almost immediately upon the arrival of the migratory species.

Some species have adapted to take advantage of temporary situations in the uncertain arctic. Ivory Gulls have been known to nest on icebergs. In response to fluctuating terrestrial food sources, Snowy Owls and Short eared Owls breed where the shifting rodent populations are found.

Clearly, the balance of the various floral and faunal elements, including native man, is very delicate and therefore easily disrupted. Examples of this fragility are not difficult to find. A few days disturbance at a breeding colony of birds can delay nesting to a point where not enough of the season remains to complete the breeding that year. Because of the high proportion of marine to terrestrial fauna, destruction of plankton and fish by an oil spill could have a severe effect, directly and indirectly, on a major part of animal life. This loss would be augmented by the loss of seabirds which tend to land on oil slicks, presumably because they resemble the calm patches of water with which they associate food. Equipment such as skidoos and rifles enable native hunters to become more efficient which upsets their historic balance with their prey.

The arctic has few niches and therefore a low species diversity. The dimunition of even one species can have profound effects on all the others. The concerns for the environment then are real, and not just academic. The growth of environmental awareness means that decisions on whether to proceed on a project are no longer based on relatively simple economics, but must now take into account social concerns which are far more difficult. How is a value assigned to a park, a non-commercial species, or a wilderness area? However, decisions must be made and we will now examine the Canadian process of making these decisions.

THE GROWTH OF LEGISLATION AND AGENCIES

There is no concise chronological development of the decision making process in the Canadian Arctic. Departments and agencies have come and gone, and even the names of existing departments have changed. An overview can be obtained by examining briefly, events which took place and the legislative methods developed in each of the three arctic regions under discussion.

Beaufort Sea

Whereas environmental concern was never lacking in the Beaufort Sea region, the proposal to use drill ships in ice-infested waters by Canadian Marine Drilling Limited (Canmar), a wholly owned subsidiary of Dome Petroleum, provided a focus for this concern and the federal government was forced to act. In July 1973, the cabinet granted approval-in-principle for exploratory drilling, using drill ships, in the Beaufort Sea. The approval was conditional upon (1) no drilling before the summer of 1976, and (2) constraints determined by a group of studies called the Beaufort Sea Project. This project (Ref. 10), now concluded, provided a basis upon which to formulate subsequent arctic exploratory programs. It established the joint industry, government approach of developing exploratory procedures in concert with environmental concerns. This unique approach saw industry contribute $4.5 million to support the Beaufort Sea Project. The total cost of the project at its completion at the end of 1976 was estimated to be $12 million, including both government and industry. The studies provided ecological baselines and a better understanding of the physical environment. In addition ideas and techniques were developed on methods of preventing and controlling oilspills and their clean-up in the arctic marine environment. The paper by Wadhams (Ref. 5) is an example of research carried out under the Beaufort Sea Project.

Arctic Islands

Unlike the Beaufort Sea experience, there was no clear issue in the high arctic upon which to focus and develop a program or even develop a set of clear guidelines. Yet exploration was proceeding and problems were being encountered which required immediate research in a number of areas such as wildlife, drilling, sea ice, oceanography, scour, sea bottom conditions and transportation. Too large for any one exploratory group, but applicable to all, these problems forced industry in 1969 into creating a non-profit association, the Arctic Petroleum Operator's Association (APOA). Since the role of this association is described elsewhere in this volume (Ref. 8), details on it will not be repeated here. We simply note that it has provided a major role in directing and funding research in the Canadian High Arctic and continues to do so.

As exploration activities and environmental awareness increased, the federal government through the Department of Indian Affairs and Northern Development (DIAND) had to play a larger role. This role is not easy because whereas it needs to encourage exploration for future energy needs (as well as to be active in the High Arctic to establish sovereignty rights), it must provide environmental protection (and a forum by which the concerns of the people can be noted and acted upon). As in any situation without benefit of history, errors occur in the handling of it. Guidelines for exploratory procedures have been, in some cases, too restrictive, unrealistic, too vague or non-existent. A Joint Industry-Government Steering Committee on Hydrocarbon Development has evolved in an effort to clarify guidelines and ensure consistent terms of reference when environmental impact studies are required.

The experience gained in both the Beaufort Sea Project and the High Arctic studies has benefited programs in the Eastern Arctic where a more structured approach was followed. An examination of it illustrates the present Canadian system which has evolved.

Eastern Arctic

The Eastern Arctic Marine Environmental Studies (EAMES) program was created to carry out intensive environmental and physical oceanographic studies in the Eastern Arctic (see Levy (Ref. 12) for example). The history of the establishment of this program was documented by the petroleum industry and presented to the federal government (Ref. 13).

The following exerpt and Table 1 (Chronology of Clearance) is taken from it. Acronyms are given in Table 2.

Clearance Activity Timing - Davis Strait

Esso Resources Canada Ltd. advised the Department of Indian and Northern Affairs in September of 1975, of plans to drill an exploration well on Davis Strait permit acreage in 1978. In response, the Department of Indian Affairs supplied Esso with provisional environmental guidelines in July, 1976. On the basis of these guidelines, Esso initiated environmental field studies in the Davis Strait in the summer of 1976. A more comprehensive environmental program was planned for 1977. At this time, cost sharing arrangements were made with other members of industry, including Aquitaine, Cities Service, Hudson's Bay Oil and Gas, Shell and British Petroleums, who also retained permit acreage in the Davis Strait area. The program was then expanded on the basis of additional discussions with DINA, to provide for an ongoing environmental study probram in 1978. In December 1976 a meeting was held with DINA to update EAMES planning for the Davis Strait program. In April 1977 representatives from DINA and EMR met with all industry-east coast permit holders in Calgary to

Table 1 Federal Environmental Clearance Activity Timing – Davis Strait

Year	DEPARTMENT OF ENVIRONMENT		DEPT. OF INDIAN & NORTHERN AFFAIRS	
	FEARO	DATE	EAMES	DATE
1975			PROJECT INITIATION	SEPT 75
1976			PROVISIONAL GUIDELINES	JULY 76
			FIELD WORK STARTED	JULY 76
			EAMES PLANNING MTG	DEC 76
1977			GOV'T/INDUSTRY MTG	APR 77
			REFERRAL TO EARP	JULY 77
	GUIDELINE TASK FORCE	SEPT 77	EAMES PROGRAM	NOV 77
	EAR PANEL SELECTION	NOV 77	EAMES MANAGEMENT COMMITTEE	DEC 77
1978	OFFICIAL GUIDELINES	JAN 78	EIS SUBMITTED	JAN 78
	EIS REVIEWS – PANEL	APR	EIS DEFICIENCIES	JAN 78
	– OTHER GOV'T	TO	EAMES ADVISORY BOARD	JAN 78
	– PUBLIC	SEPT	EIS REVIEW & BRIEF	SEPT 78
	PUBLIC MEETINGS	SEPT 78	FINAL ADVISORY BOARD REVIEW	NOV 78
	PANEL RECOMMENDATION	NOV 78		
	MINISTERS DECISION	DEC 78		
1979			ADVISORY BOARD REPORT	JAN 79
			MINISTERS DECISION	JAN 79
			SUPPLEMENTAL EIS	FEB 79
			APPROVAL CONDITIONS	MAY 79

Table 2 List of Acronyms Used in Text (or, how to tok Canajun)

AES	Atmospheric Environment Service
DEMR	Department of Energy, Mines and Resources
DIAND	Department of Indian Affairs and Northern Development, sometimes abbreviated to
DIANA	Department of Indian and Northern Affairs, which may also appear as
DINA	
DOT	Department of Transport
EAMES	Eastern Arctic Marine Environmental Studies
EAR	Environmental Assessment and Review
EARP	Environmental Assessment and Review Process
EIS	Environmental Impact Statement
EMR	Energy, Mines and Resources
EMS	Environmental Management Service
EPS	Environmental Protection Service
FEARO	Federal Environmental Assessment and Review Office

review government objectives and responsibilities for east coast environmental study programs. At this time permit holders under DINA were advised that funding for accelerated baseline studies would be funded by industry.

Esso, Cities Service and Aquitaine, as the proponents of the Davis Strait project, received advice from DINA that the project would be referred to EARP in July of 1977. Subsequently, the EAMES program was developed in November, 1977, which was very similar to the provisional DINA guidelines provided for the proponents in July, 1976. An EAMES Management Committee was formed and first met in December, 1977. FEARO assembled a guideline task force in September, 1977 and the Davis Strait EAR Panel was selected in November, 1977. The official guidelines from FEARO were received in January, 1978, which again did not vary significantly from the EAMES guidelines provided earlier.

The proponents' Davis Strait EIS was submitted in January of 1978 and the EAMES Management Committee advised, that if the information in the EIS was in compliance with the guideline requirements, the document would be adequate for an environmental assessment review later in 1978. Deficiencies in the EIS were also identified in Janaury, 1978, and the proponents recognized the need and in fact had already planned, to continue the Davis Strait environmental study program through 1978. The EAMES Advisory Board met for the first time in January, 1978.

EIS reviews were conducted by the panel, other government scientists and the public, during the period from April through August, although a few of the EIS support documents were not distributed to all reviewers until July of 1978. DINA also submitted a review and brief to the EAR Panel. These reviews and briefs were complete and ready for the EARP public meetings scheduled in September of 1978. The Minister of Environment accepted the November, 1978 EARP recommendation, to permit drilling under certain conditions and announced his decision in December, 1978.

Looking back on the other side, we can see that the Minister of Indian Affairs waited on the final EAMES Advisory Board review which took place in November/December, 1978 and the final Advisory Board Report in January, 1979. Subsequently, the proponents were advised that final ongoing environmental study conditions for drilling approval would be completed by May, 1979, after appropriate reviews of the supplemental EIS for Davis Strait, which was submitted in February, 1979.

A period of three years and four months elapsed between project initiation and the final ministerial decision in January of 1979.

With this perspective of legislative development in the three areas of concern, a formal description of the present approval process can now be examined.

THE PRESENT LEGISLATIVE APPROVAL PROCESS FOR OIL AND GAS EXPLORATION IN ARCTIC WATERS

An applicant owning surface rights in territorial Canada who wishes to drill in the offshore areas of the Arctic, applies to the Oil and Gas Division of DIAND for Approval-in-Principal (Diagram 1). If it is received, the applicant then must apply for a Drilling Authority (Diagram 2). The legislative procedures which are followed for major developments where significant environmental impacts could occur are outlined in

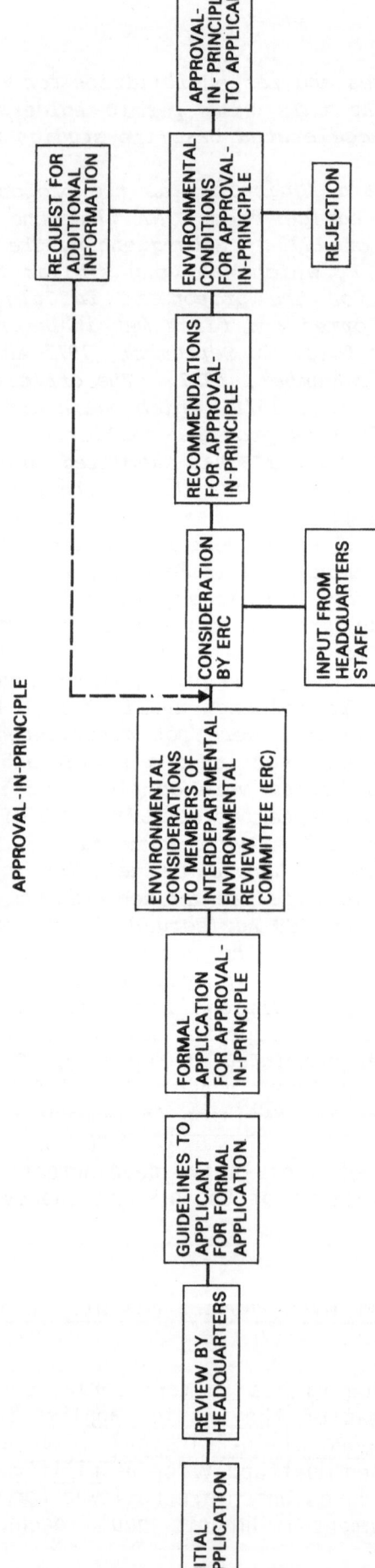

Diagram 1

**APPROVAL PROCEDURE
ARCTIC WATERS
OFFSHORE OIL AND GAS DRILLING
ENVIRONMENTAL CONSIDERATIONS**

APPROVAL-IN-PRINCIPLE

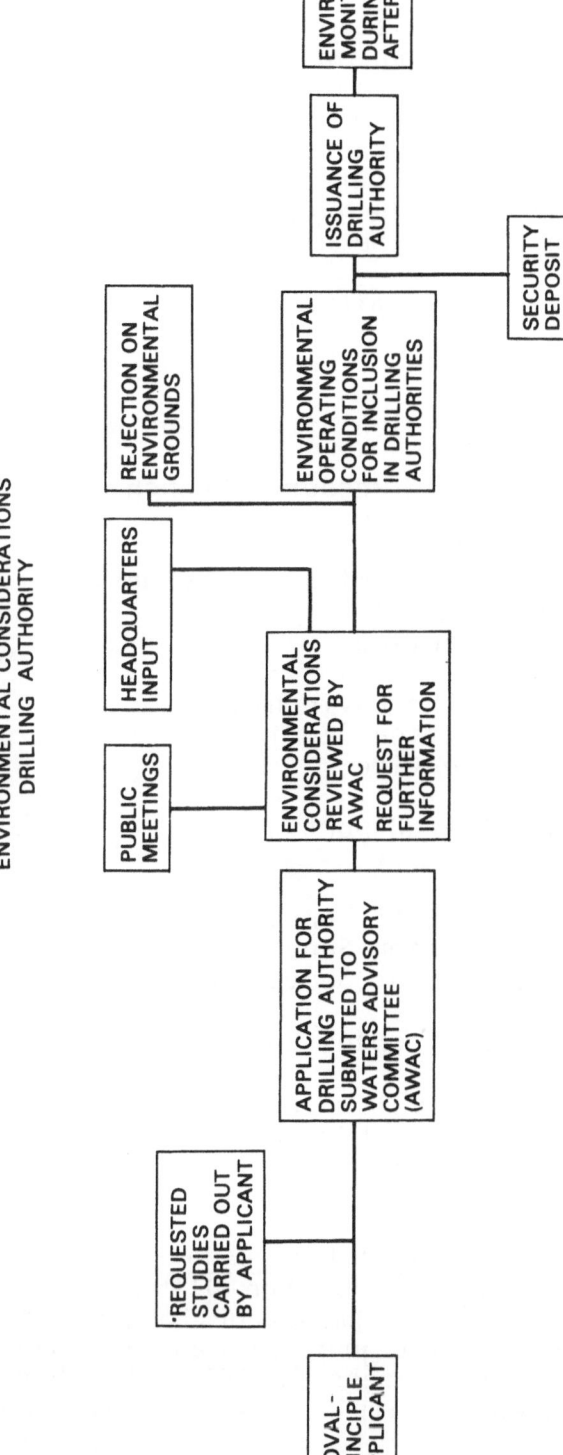

Diagram 2

APPROVAL PROCEDURE
ARCTIC WATERS
OFFSHORE OIL AND GAS DRILLING
ENVIRONMENTAL CONSIDERATIONS
DRILLING AUTHORITY

Diagram 3. If no impacts are identified, the process may be much shorter.
   Three governmental environmental committees are involved in the approval process.   These committees review, at various legislative levels, oil and gas and other developments. They advise government on the potential environmental affect of a project and suggest mitigating measures to reduce a potential environmental impact. Each will be reviewed briefly.

Environmental Review Committee
This committee provides an interface between DIAND and other government departments on matters relating to the environment. The committee consists of members from the Environmental Protection Service, Environmental Management Service and Atmospheric Environment Service, all federal agencies. The Departments of Energy, Mines and Resources and Transport Canada have observer status on the committee. DIAND has committee representatives which cover all of the disciplines of the other departments.

Arctic Waters Advisory Committee
The Arctic Waters Advisory Committee is composed of members who reside in the Northwest Territories. These are:

Federal Department of Indian Affairs and Northern Development:
- Director (Chairman)
- Northern Non-Renewable Resource Branch
- Renewable Resources Branch

Federal Department of Fisheries and Oceans:
- Fisheries Marine Service
- Ocean and Aquatic Service

Federal Department of the Environment:
- Environmental Protection Service
- Canadian Wildlife Service

Federal Department of Transport:
- Canadian Coast Guard

Government of the Northwest Territories:
- Environmental Protection Service
- Environmental Management Service

Committee for Original Peoples Entitlement:
- Representative
   This is the first level whereby industry and public can make formal presentations to federal and local government and affect the way a project is to be developed and/or the environmental safeguards which could be implemented.

The Environmental Assessment Review Process (EARP)
If a project is considered by the DIAND to have the potential to cause considerable environmental impact or to warrant public input, the rejection or acceptance of the project can be referred to the Environmental Assessment and Review process (EARP). In this case, a formal environmental review process is implemented and all those who might be effected

by the proposed project are invited to make submissions at hearings on it.
    The EARP panel is selected by the chairman of the Federal Assessment Review Office (FEARO) and his staff in Ottawa. The panel members are selected for their technical understanding of the project and/or their understanding of the environmental or socio-economic aspects of the proposed project. This process ensures control of hydrocarbon exploration in the High Arctic. It allows for input from both industry and public. If studies are recommended they are required to be paid for by industry. If the studies are major, a co-operative program is carried out. The environmental studies done are reviewed by government and public. If deficiencies are found, further studies are made. When the deficiencies are corrected, or if none were evident, the project proceeds. This is basically the Canadian solution to fostering hydrocarbon exploration in the High Arctic while ensuring its environmental protection.

FUTURE OF HYDROCARBON DEVELOPMENT IN THE HIGH ARCTIC

Marine transportation will become the next major environmental issue in the Canadian Arctic. The Arctic Pilot Project, which involves the shipment of Liquefied Natural Gas (LNG) from the High Arctic to east coast North America, is already in planning by Petro-Canada and it is reasonable to assume that similar plans for the shipment of oil will not be far behind. Environmental studies will have to cover much broader areas, which means the intensive studies started with the Beaufort Sea Project will continue throughout the major seaways of the High Arctic.
    In 1973, a study was released on Canada's Ocean Policy which emphasized that Canada should strive to attain technical excellence in ice-covered waters. The Beaufort Sea Project and the research phase of the Arctic Pilot Project have brought Canada to equal status with any nation in ice knowledge and ice navigation. Continuing studies such as the above and other, which have been carried out or are being planned (Table 3) by the Arctic Marine Oilspill Program, can be expected in this decade to make Canada the world leader in Arctic knowledge, experience and environmental protection.

Table 3   Arctic Marine Oilspill Program, 1980-83

Program	Date	Level of Effort ($ Can)
Oil and gas under Beaufort Sea	Spring 1980	315,000
Oil on cold water	Summer 1979-Summer 1981	470,000
Oil on Arctic nearshore environments	Summer 1979-1983	2,530,000
Oil on Arctic shorelines	Summer 1979-1982	350,000
Oil in east coast pack ice - Phase 1	Feb. - May 1982	} 1,250,000
Oil in east coast pack ice - Phase 2	Feb. - May 1983	

The Canadian High Arctic experience has been largely one of cooperation between government and industry. Certainly differences have occurred and, as in any healthy, evolutionary experience, will likely continue to occur. A working system has emerged which allows and encourages public participation and attempts to incorporate public concerns into the exploration process (Fig. 7) The system has seen the development of intervenor groups, with the ability to present their cases with considerable sophistication, displaying full awareness of the compromise that must exist between environmental risk and energy necessity. Their responsible approach has considerably strengthened their cause and led to improved exploration methods.

Industry will continue to finance environmental studies but with closer cooperation with government and special interest groups. Greater emphasis will be given to early or initial environmental evaluation studies. Further, studies will be closely tied to lease renewals. In general, more long-term environmental planning will be carried out by industry in order to allow efficient phasing of studies and plans. This is in contrast with the past, where the tendency was to do environmental work only when required. Environmental work will be increasingly recognized as being able to contribute to engineering design as well as to assess project feasibility. Companies can thus look forward to more structured programs, with work planned in phases, where abandonment or proceed decisions can be made with some certainty.

ACKNOWLEDGEMENTS

The authors wish to acknowledge the assistance of Suzanne Spohn, BEAK Vancouver, who researched much of this material. Dr. W.J. Stephen contributed to the geological discussion and Dr. M.K. McNicoll, BEAK Calgary provided the information on Arctic fauna. The assistance of Panarctic Oils Limited, Dome Petroleum, and Esso Resources Canada Ltd., is very much appreciated for providing descriptive material on projects and for providing diagrams for figures. A sense of perspective was provided by Mr. Bill Bernard, Dome Petroleum. Editorial comments were made by Stan Mackay, Esso. The authors thank all of the above for their contributions and the Petromar '80 program committee for their encouragement and patience.

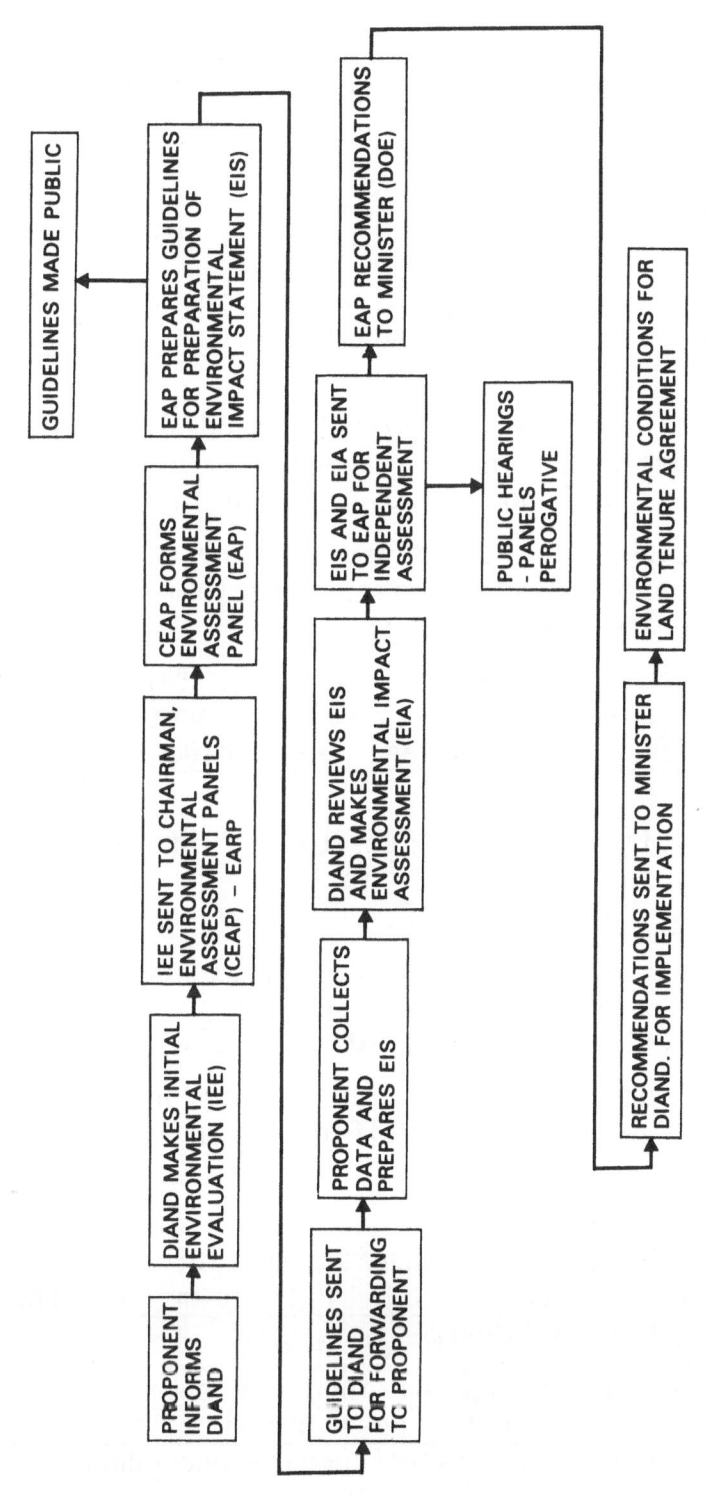

223

Diagram 3

ENVIRONMENTAL ASSESSMENT - MAJOR DEVELOPMENT PROJECTS

GUIDELINES MADE PUBLIC

EAP PREPARES GUIDELINES FOR PREPARATION OF ENVIRONMENTAL IMPACT STATEMENT (EIS)

CEAP FORMS ENVIRONMENTAL ASSESSMENT PANEL (EAP)

IEE SENT TO CHAIRMAN, ENVIRONMENTAL ASSESSMENT PANELS (CEAP) – EARP

DIAND MAKES INITIAL ENVIRONMENTAL EVALUATION (IEE)

PROPONENT INFORMS DIAND

GUIDELINES SENT TO DIAND FOR FORWARDING TC PROPONENT

PROPONENT COLLECTS DATA AND PREPARES EIS

DIAND REVIEWS EIS AND MAKES ENVIRONMENTAL IMPACT ASSESSMENT (EIA)

EIS AND EIA SENT TO EAP FOR INDEPENDENT ASSESSMENT

PUBLIC HEARINGS - PANELS PEROGATIVE

EAP RECOMMENDATIONS TO MINISTER (DOE)

RECOMMENDATIONS SENT TO MINISTER DIAND. FOR IMPLEMENTATION

ENVIRONMENTAL CONDITIONS FOR LAND TENURE AGREEMENT

REFERENCES

1   Grantz, A. and D.A. Dinter, "The constraints of geologic processes on petroleum development in the western Beaufort Sea", in Proceedings: Petromar 80, Petroleum and the Marine Environment,
EUROCEAN, Monaco (1980).

2.   King, R.E., "Canadian arctic promises future gas and oil supply";
World Oil December (1975): 53-59.

3.   Napoleoni, J.G., "Arctic oil exploration" in Proceedings:
Petromar 80, Petroleum and the Marine Environment, EUROCEAN, Monaco
(1980).

5.   Wadhams, P., "Oil and ice in the Beaufort Sea", in Proceedings;
Petromar 80, Petroleum and the Marine Environment, EUROCEAN, Monaco
(1980).

6.   Hood, G.L., D.L. Masterson and J.S. Watts, "Installation of a
subsea completion in the Canadian Arctic Islands", Petroleum Society
of C.I.M. paper no. 79-30-20, reprint (1979).

7.   Hnatiuk, J., "Joint industry research through the Arctic Petroleum Operators Association", Petroleum Society of C.I.M., paper no.
7260, reprint (1972).

8.   Kustan, E.H. and J., Hnatiuk, "The Arctic Petroleum Operators
Association (APOA) and its role in arctic research", in Proceedings;
Petromar 80, Petroleum and the Marine Environment, EUROCEAN, Monaco
(1980).

9.   Kaustinen, O.M., "Feasibility of constructing marine pipeline
crossings in the Arctic", in Proceedings, Petromar 80, Petroleum and
the Marine Environment, EUROCEAN, Monaco (1980).

10.   Milne, A., "Oil, ice and climate change the Beaufort Sea and the
search for oil, Beaufort Sea Project, Fisheries and Environment
Canada, (Ottawa, Ontario, 103 pp, 1978).

11.   Hnatiuk, J., "Results of an environmental research program in the
Canadian Beaufort Sea", Oilfield Technical Conference, paper no. 2445
(1976).

12.   Levy, E.M., "Background levels of petroleum residues in Baffin
Bay and the eastern Canadian Arctic; role of natural seepage", in
Proceedings; Petromar 80, Petroleum and the Marine Environment,
EUROCEAN, Monaco (1980).

13.   Anonymous, Petroleum industry submission to the Honourable John
Fraser, Minister of the Environment with respect to proposed changes
in environmental assessment review procedures for industrial projects,
41 pp, June , unpub MS, 1979.

ARCTIC EXPLORATION : THE LABRADOR CASE - PRODUCTION SCHEMES AND ICEBERG STUDIES

Jean Gérard NAPOLEONI, Marc JOZAN,

Offshore Department

TOTAL-COMPAGNIE FRANCAISE DES PETROLES

1. INTRODUCTION - ABSTRACT

In 1973, the dynamically positioned ship PELICAN drilled its first well on the Labrador shelf for TOTAL-EASTCAN Exploration acting as operator for the Labrador group (*). This was following a first unsuccessful attempt using an anchored type vessel ; indeed the environmental conditions found on this huge arctic acreage (sea figure 1) were, at the time, totally unknown to the oil industry. Prominent features are :

- sea ice six months a year,
- icebergs year round,
- rough seas from late summer until freeze-up.

To cope with these, new techniques were developed and used, while at the same time a research program was conducted to study feasible seasonal production schemes. As some of this work has already been published it will only be briefly summarized here. After this program had ended, it was found that year round production could be feasible either by improving the seasonal scheme, or in specific cases of sea bottom topography.

The presence of icebergs is a very challenging factor of offshore Labrador and extensive work was carried out to improve a very scanty pre-existing knowledge. This, will be presented in more details, including theoretical drift modeling work and 1979 iceberg draft measurement campaign.

2. EXPLORATION DRILLING

Dynamically positioned drill ships proved to be the only viable solution to the iceberg problem. On the ships, ice observers are on duty 24 hours a day, using visual and radar observation to plot the drift of all icebergs within 12 miles of the ship. Icebergs are referenced so that it is later possible to use drift trajectories in conjunction with iceberg measurements and ocean and wind data to study the physics of the drift. When a berg is thought to become too menacing it is towed away using a tug permanently standing by for the purpose. Down time on drilling due to iceberg was thus kept a a very low level (0 to 5 %) during the four months of the season (July to October).

(*) Presently, the Labrador group is composed of :

- AGIP Canada Ltd,
- AMERADA MINERALS CORPORATION OF CANADA Ldt,
- AQUITAINE Company of Canada Ltd,
- GULF Canada Resources Inc.,
- PETRO-CANADA Exploration Inc.,
- SUNCOR Inc.,
- TOTAL PETROLEUM (N.A.) Ltd,
- TOTAL EASTCAN Exploration Ltd.

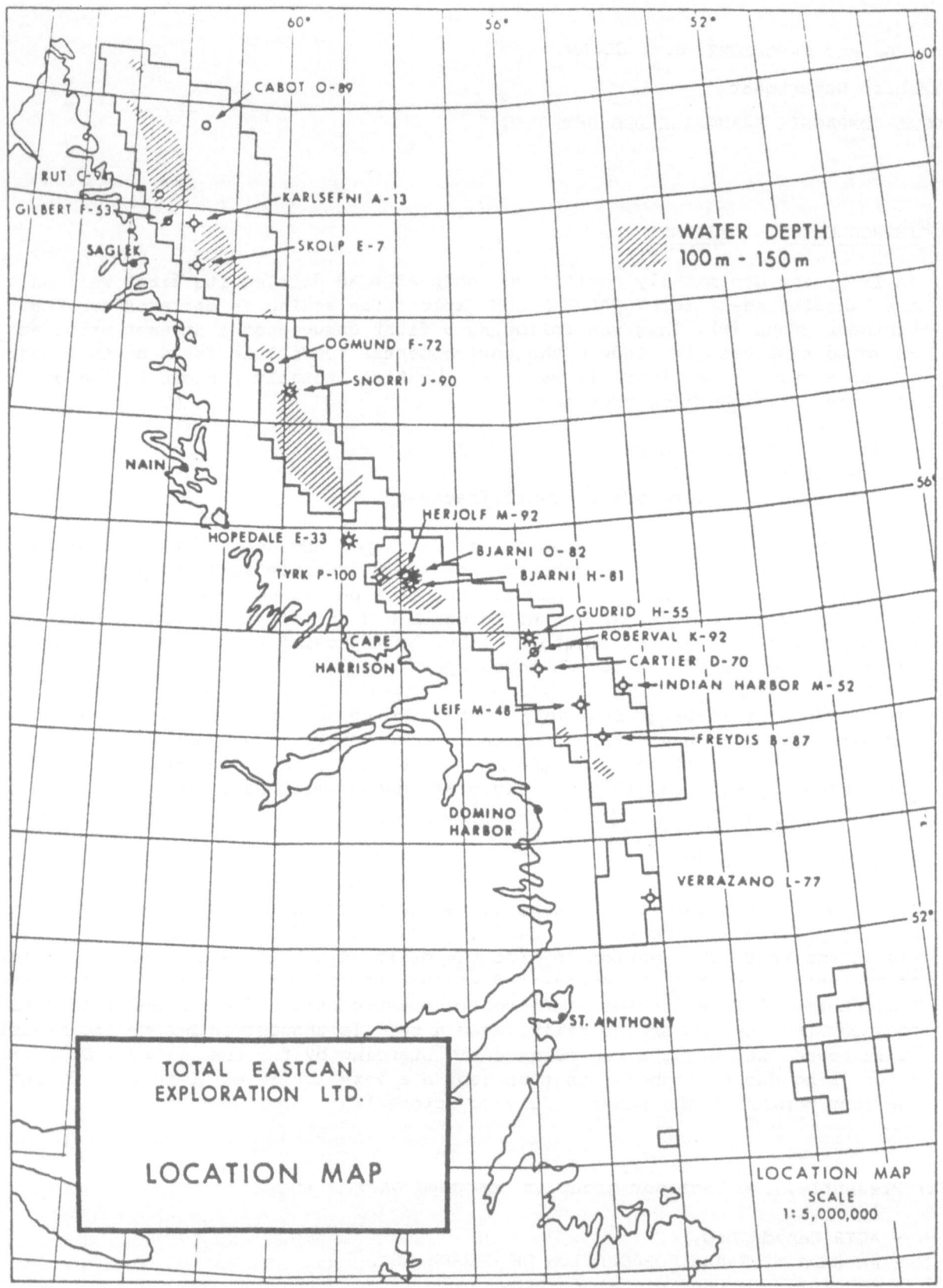

Fig. 1 - Labrador group offshore Labrador acreage.

3. SEASONAL PRODUCTION SCHEME

While exploration went on, a multi-company team led by TOTAL was set up in Paris to study a possible production scheme. Because of the presence of very heavy pack ice during winter and spring the efforts were directed toward the definition of a seasonal scheme. Here again, icebergs were the main worry since they do not only prevent any fixed or anchored structures from being used but also because their keels scour the sea floor and can, in the process, destroy subsea production equipment.

The chosen production scheme is that shown in Figure 2.

- A dynamically positioned platform of the "spar" type (Dypospar) is used for production and storage. Should any iceberg come too close, the platform lifts part of its riser and moves off location.

- Crude oil is exported using shuttle tankers.

- Essential subsea equipment such as well heads and manifold are located in excavations dug deeper than the maximum expected iceberg scour. This means that an excavation technique and a new buried well head had to be developed :

  - the well head is in a caisson, itself buried at the bottom of the excavation. All vital components such as master valve block, etc... are located below a weak point so that if an iceberg keel collides with the top portion of the well head, no damage can be induced to its lower part ;

  - the excavation system shown on Figure 3 was successfully tested offshore Brittany. Using a grab was made necessary since the boulders found in Labrador morainic soil does not allow the use of traditional sand suction techniques.

- As for flowlines, they are made of flexible and are not specially protected.

This program proved that seasonal production was indeed achievable while coping and preserving the very special Labrador environment. First estimates found that development costs would be around 30 % above that of similar but non Arctic (North Sea for example) schemes.

4. ALTERNATIVE SCHEMES FOR YEAR ROUND PRODUCTION

4.1. Floating structures and ice cutting devices

It is not so much the increased investment cost that makes the seasonal scheme very costly but mainly the fact that production is limited to six months a year. It was thought that better economics could be obtained, even though through higher investment costs, if production could be maintained all year round. Such scheme is shown on Figure 4 where basically the same set up is used as for the seasonal scheme but for the following differences :

- the dypospar storage production platform is equipped with ice cutters that will crush drifting sea ice ;

- since dynamically positioned tankers were found unable to remain on location in drifting sea ice for any sufficient amount of time their loading has to be done using dypospar storage platform also equipped with ice cutters. This platform loads from the production platform and then, while drifting along with the tanker, unloads onto it ; finally, this shuttle platform sails

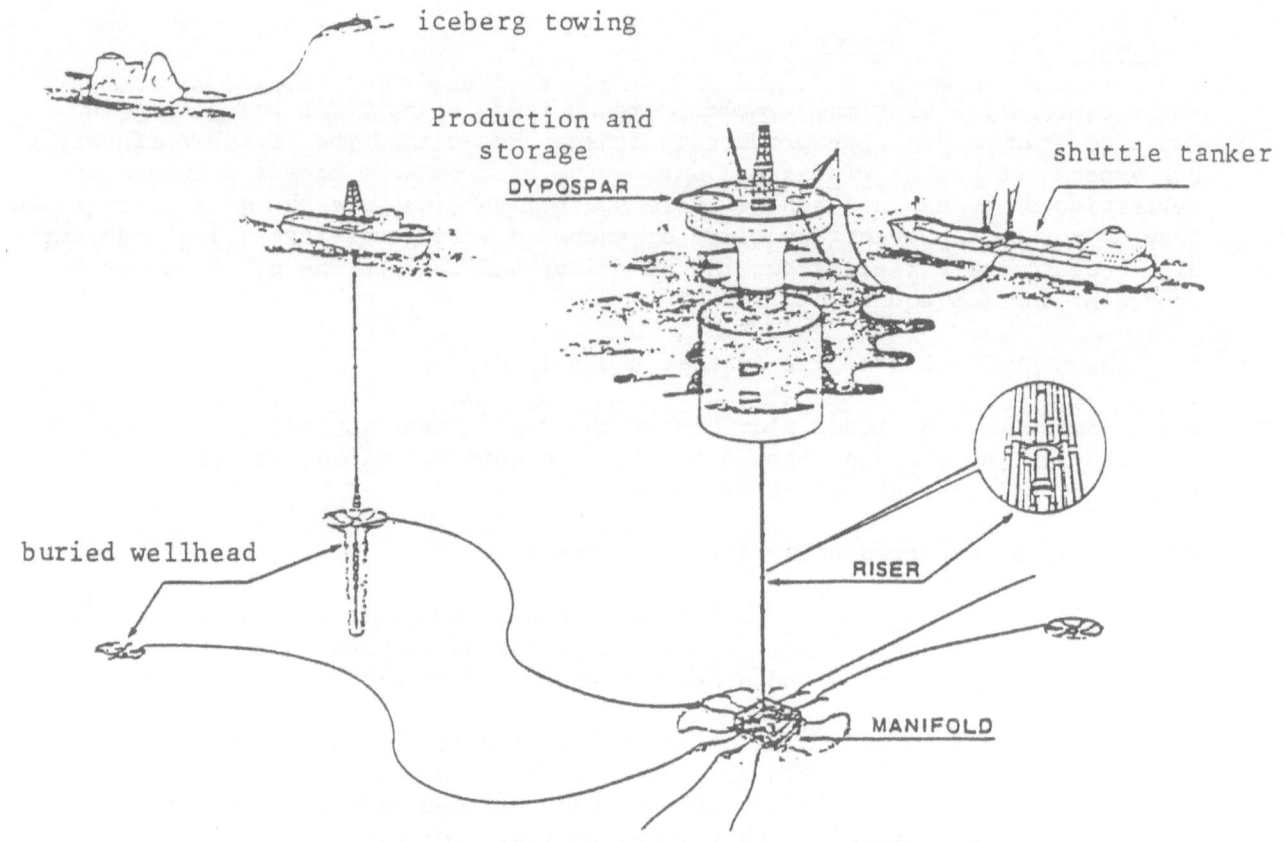

iceberg towing

Production and storage
DYPOSPAR

shuttle tanker

buried wellhead

RISER

MANIFOLD

FIG. 2. SEASONAL PRODUCTION SCHEME

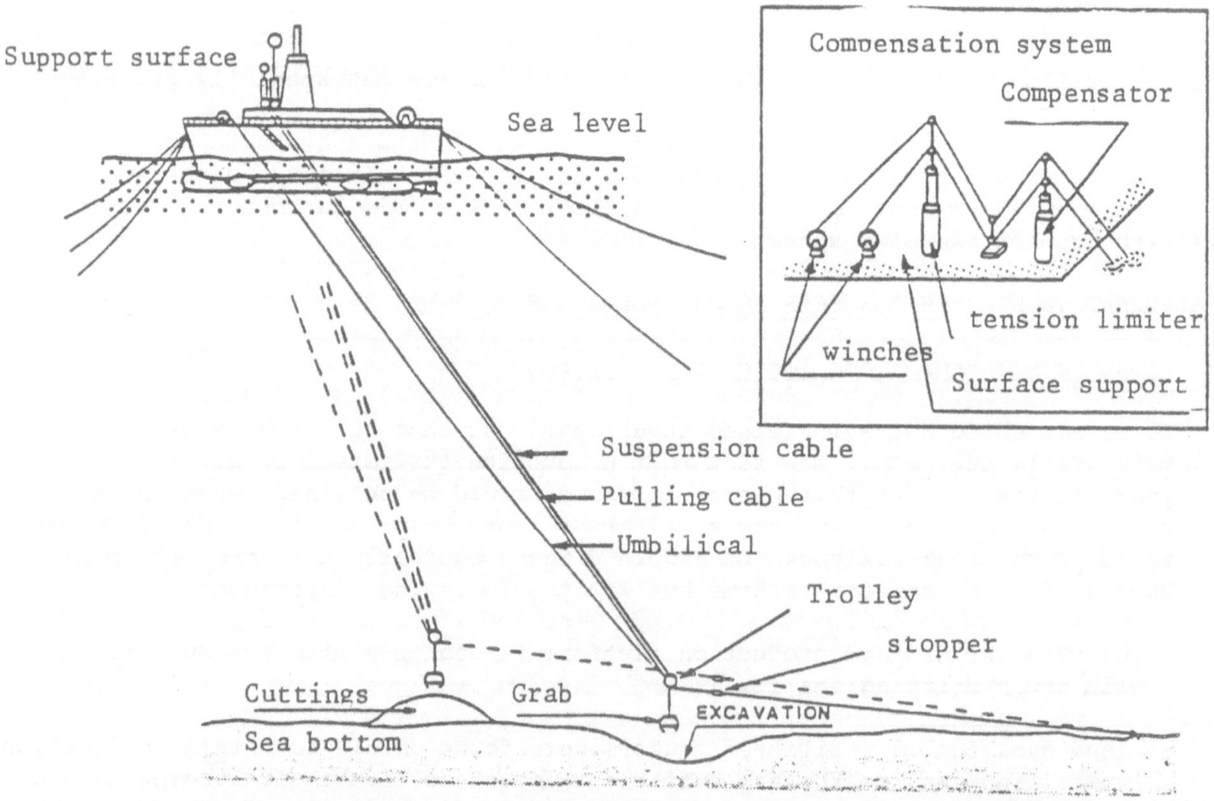

Support surface

Sea level

Compensation system

Compensator

tension limiter

winches

Surface support

Suspension cable

Pulling cable

Umbilical

Trolley

stopper

Cuttings   Grab

EXCAVATION

Sea bottom

FIG. 3. EXCAVATION METHOD FOR ARCTIC MORAINIC SOIL

back to the production platform for a new load.

Although it is believed that this scheme will have an investment cost 70 % higher than a traditional North Sea scheme, because production is extended to twelve months, both pay out and production costs are decreased when compared to the seasonal scheme.

4.2. Site specific all subsea scheme

In many areas of the shelf, the prospects are separated from shore by sea bottom depressions that are partially or totally closed. Any equipment, and in particular, any pipe-line, laid on such depressions, would be perfectly safe from iceberg keels since very deep icebergs cannot enter the area. This led to envision the scheme shown on Figure 5. Well heads, flowlines and manifold are similar to those designed for the other schemes. Production is directly fed into a dual pipe-line without previously flowing to any surface production or process installation. This pipe-line (doubled for safety and maintenance reasons) only needs protection by burying on its way from the field to the protected depression and at the approach of shore. A process and storage facility would be built either on shore or on shallow or rocky islands further out at sea. There also a terminal would be built for year round loading of ice re-inforced tankers. In the case of a gas field this would be replaced by a liquefaction plant and L.N.G. loading terminal. Of course, there are two technological difficulties to overcome if such an otherwise attractive scheme is to be made possible.

● Sending production directly from the wells to the pipe-line : presently this would indeed only be possible in the case of a field producing very dry good quality gas where there would be no problem as far as corrosion or condensates are concerned. But if oil is produced, there will be the need for pumping and this means having an underwater pumping station as well as bringing the power to run it. It would also require mastering two phase flow transport technology i.e. pressure losses prediction, regulating and pumping techniques. Although, thanks to extensive research carried in France by TOTAL, INSTITUT FRANCAIS DU PETROLE and ELF these two phase flow techniques will be soon mastered, this does not provide any easy answer to the question of bringing the energy down, either from shore through electrical cables or from a power plant on board a stand-by vessel.

● Indeed a stand-by vessel will be needed on the field at all time to monitor and control the field. Since a relatively small vessel is sufficient for the job, it should not be difficult to keep it there all year round. One can think of a quick make and break connection system that would allow the vessel even when connected, some freedom of movement. Acoustic systems would be there as back up for times when disconnection cannot be avoided.

The summer season would be time for work-overs and other maintenance and also of seasonal water injection if needed. If pipe-line shore approach turns out to be difficult because of steep or rough and rocky bottoms, recent studies seem to show that tunneling is a viable alternative.

Quite clearly this is an attractive concept that would in some instances provide a much cheaper scheme than any of the other floating schemes. On top of that it may be the only solution available for the development of a gas field.

ICE CUTTING DYPOSAR
EXPORT TANKER
SHUTTLE
RISER
MANIFOLD
WELLHEAD
FLOWLINES
• CABOT
• OG MUND
• BJARNI
• NORTH LEIF
• ROBERVAL

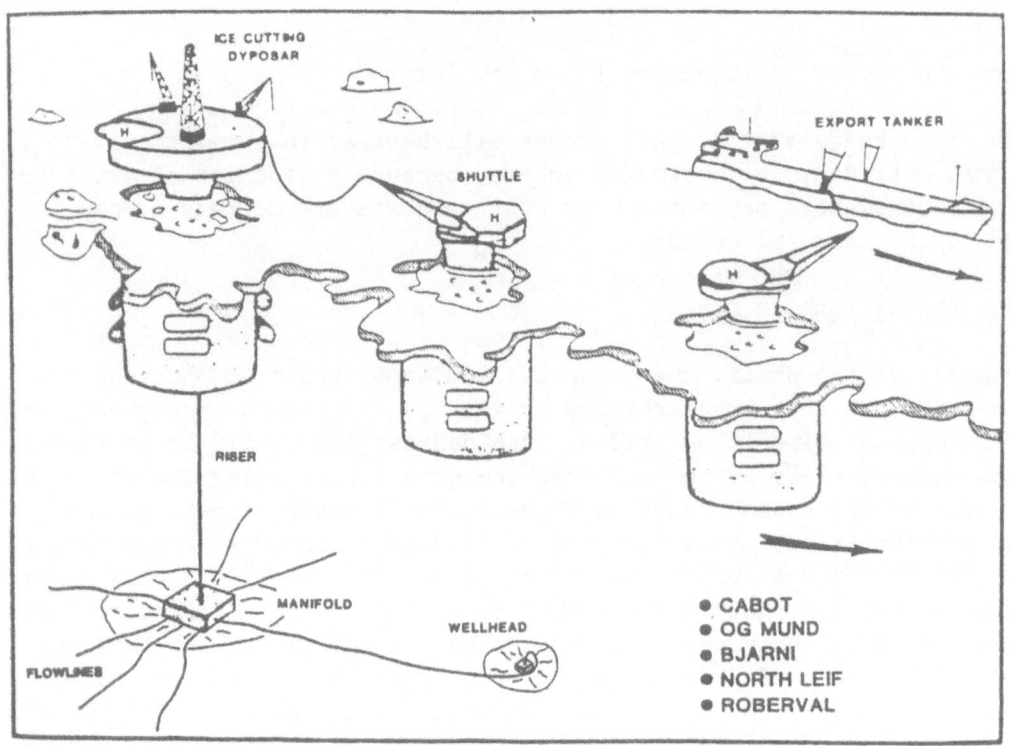

FIG. 4. ALL YEAR ROUND FLOATION PRODUCTION SCHEME

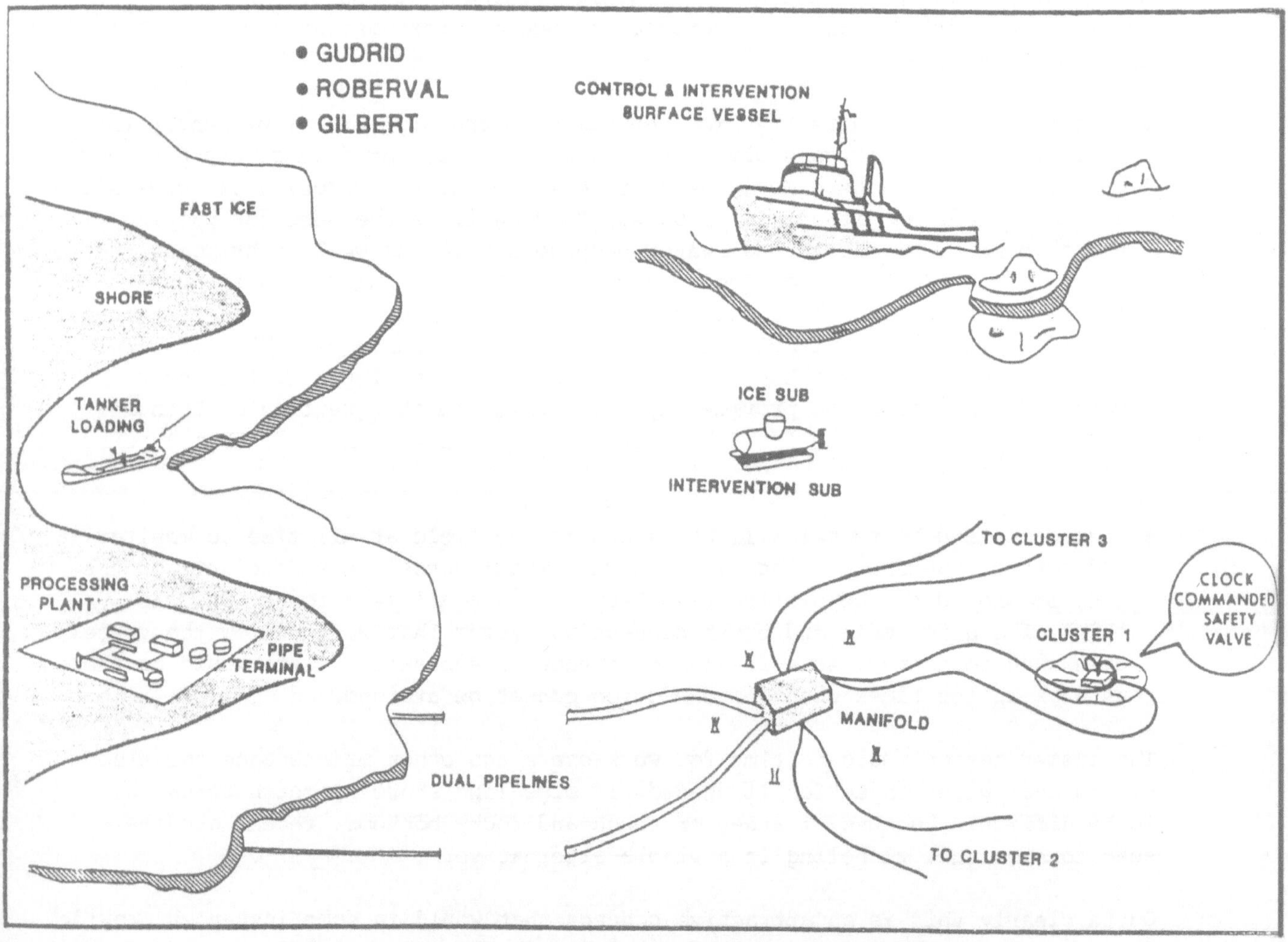

• GUDRID
• ROBERVAL
• GILBERT
CONTROL & INTERVENTION SURFACE VESSEL
FAST ICE
SHORE
TANKER LOADING
ICE SUB
INTERVENTION SUB
TO CLUSTER 3
PROCESSING PLANT
PIPE TERMINAL
CLOCK COMMANDED SAFETY VALVE
CLUSTER 1
MANIFOLD
DUAL PIPELINES
TO CLUSTER 2

FIG. 5. ALL SUBSEA, ALL YEAR ROUND PRODUCTION SCHEME FOR SPECIFIC SITES

5. ICEBERG WORK

5.1. Incentive

Icebergs and sea ice are quite obviously the two most important environment factors of the Labrador area and although little was known at the time on the characteristics of either of them, a lot of work has since then been carried out on sea ice, very often in conjunction with other oil and gas projects in the North American Arctic. Icebergs, however, are only encountered by the oil industry in Labrador and most of the greatly needed knowledge had to be acquired. There are two main topics of concern :

- iceberg drift : collision risk analysis, drift prediction, towing efficiency.
- iceberg draft : advance determination of whether or not a given iceberg is deep enough to destroy sea bottom installations.

5.2. Dynamics of iceberg drift ; numerical model

Understanding the dynamics of iceberg drift allows better evaluation of the collision risk with a fixed or submerged structure ; it also helps in evaluating the scouring risk (water depth where the scour occurs, depth and direction of scour). All these statistical data are essential when deciding on a development policy since if the risk can be proved very small, and as long as no totally unacceptable event remains possible (pollution for example), one may decide not to protect the equipment at all and simply replace it if destroyed.

Drift prediction would allow a better on stream for both drilling and production. Indeed, if, when an iceberg approaches the operation area, its future trajectory can be assessed, one will avoid at least some of the otherwise necessary precautionnary disconnections. Also if the truly dangerous berg can be pinpointed very sonn this will leave more time for efficient towing. This applies to surface as well as to bottom installations, since when a flowline for example is in danger of being destroyed it must preventively be drained of oil to avoid any pollution in case of rupture and this of course induces production down time.

5.2.1. Drift numerical modeling

A numerical model was built to study iceberg drift and this work will be briefly described here. (For further information one should refer to J.G. NAPOLEONI's M.A.Sc Thesis "The dynamics of iceberg drift", Department of Geophysics, the University of British Columbia, Vancouver, Canada).

This model is based on Newton, second law of motion so that the basic motion equation is :

$$\vec{\gamma}_i \ = \ (\vec{F}_{da} + \vec{F}_{dw}) / M_i \ + \ 2\vec{\omega} \times \vec{v}_i \ + \ g \cdot \vec{s} \qquad (1)$$

| Iceberg acceleration | Acceleration due to air and water dynamic drag forces | Coriolis term | Sea slope acceleration |

with : $\vec{F}_{dw} = 1 / 2 \ C_{dw} \ S_w \ \rho_w \ v_{wi} \vec{v}_{wi}$ (2)

232

Where :

C_{dw} : form drag coefficient for portion of iceberg below sea level.

g : acceleration of gravity

M_i : iceberg mass.

S_w : area of the underwater portion of iceberg perpendicular to relative water flow direction.

$\vec{\lambda}$: sea surface slope.

$\vec{v_i}$: iceberg velocity.

$\vec{v_{wi}}$: water velocity with respect to iceberg.

ρ_w : water density.

$\vec{\omega}$: earth angular velocity.

$\vec{F_{da}}$, the dynamic air drag is written in a form similar to $\vec{F_{dw}}$. Quite clearly the product $C_{dw}S_w$ is one major unknown of this equation and furthermore, for a given iceberg it is not a constant but a circular function of the direction from which the water flows around the berg. In the model this was simulated by setting $C_{dw}S_w$ (as well as $C_{da}S_a$) equal to a Fourier function limited to only one non constant term.

Wind and current velocities can usually be measured but unfortunately this is not the case for the sea slope since it usually amounts to no more than a few millimeters per kilometer. However, if as a first order approximation, one neglects wind stress on the sea surface, and supposing an infinite, non viscous and homogeneous ocean, water motion is described by :

$$\vec{\gamma}_w = 2\vec{\omega} \times \vec{v}_w + g \cdot \vec{s} \quad (3)$$

which yields, when replaced in (1) :

$$\vec{\gamma}_i = \vec{\gamma}_w + (\vec{F}_{da} + \vec{F}_{dw}) / M_i - 2\vec{\omega} \times \vec{v}_{wi} \quad (4)$$

With $\vec{\gamma}_w$ = sea water acceleration.

This equation shows in particular that if no wind blows, the iceberg will move in exactly the same manner as does the water. It is therefore inadequate to speak of response time of the berg under a change in current velocity since both current and iceberg will accelerate in phase and it would be a mistake to say that an iceberg passing through an ocean potential field would be accelerated by the flow only after the flow had itself accelerated.

Icebergs being usually highly irregularly shaped they were, in the model, allowed to rotate by introducing "lever arm functions" so that air and water dynamic drag forces may have a spinning effect on the berg. Other features of the model are :

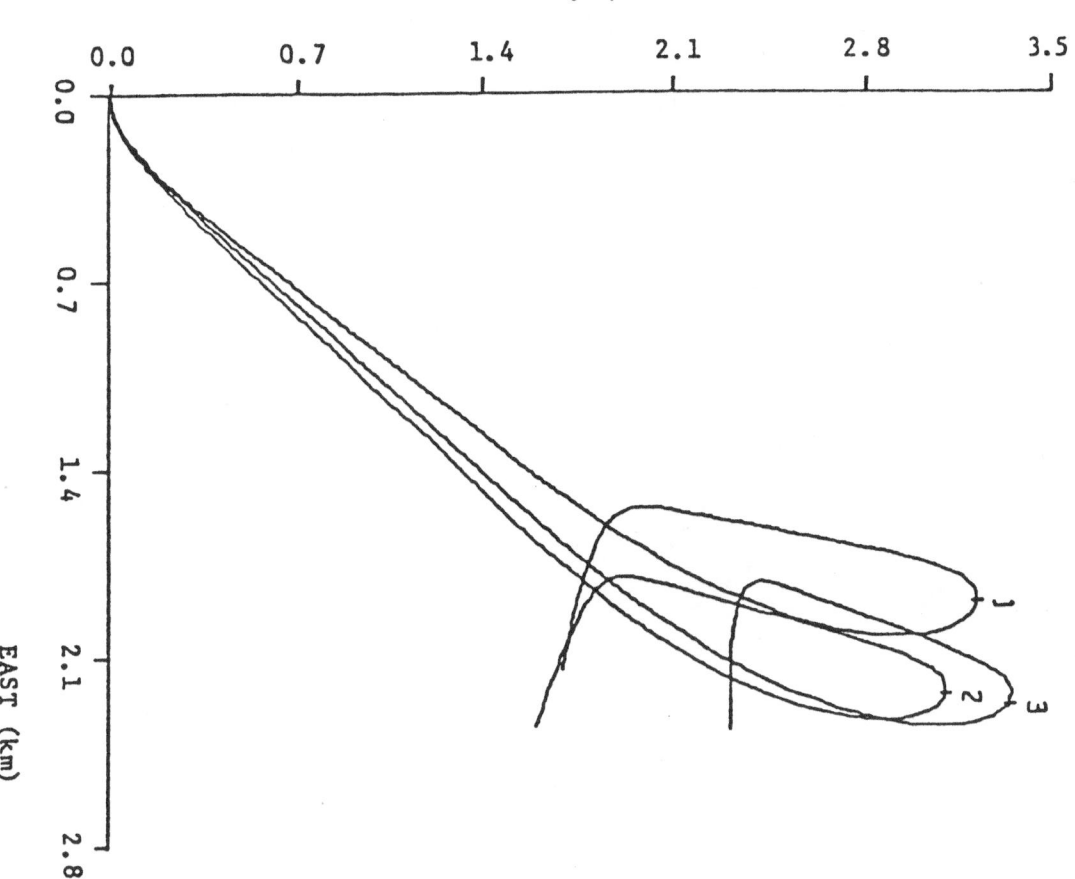

FIGURE 6. Influence of Coriolis term on trajectories. The water velocity is constant eastward at 0.25 m s⁻¹. The wind velocity is given in Figure 6A. The total drift time is 7 hours.

Trajectory 1: no Coriolis term.

Trajectory 2: Coriolis term uses $\vec{v}_i$, iceberg velocity relative to earth's surface. In effect, this amounts to ignoring the sea slope.

Trajectory 3: Coriolis term uses $\vec{v}_{iw}$, iceberg velocity relative to the water. This is the best treatment of the three.

FIGURE 6-A

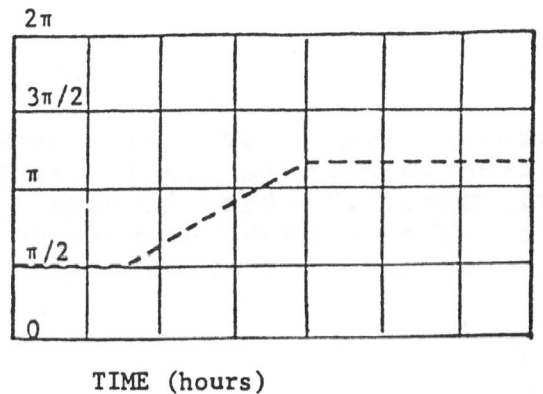

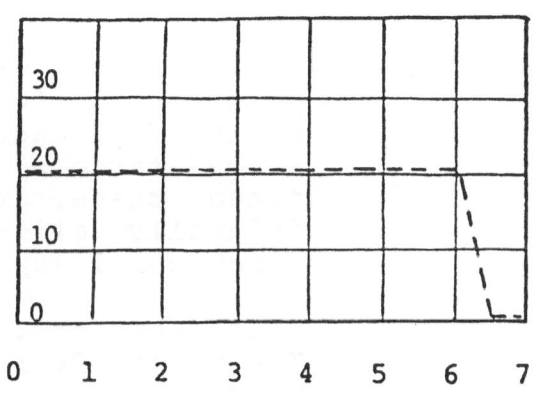

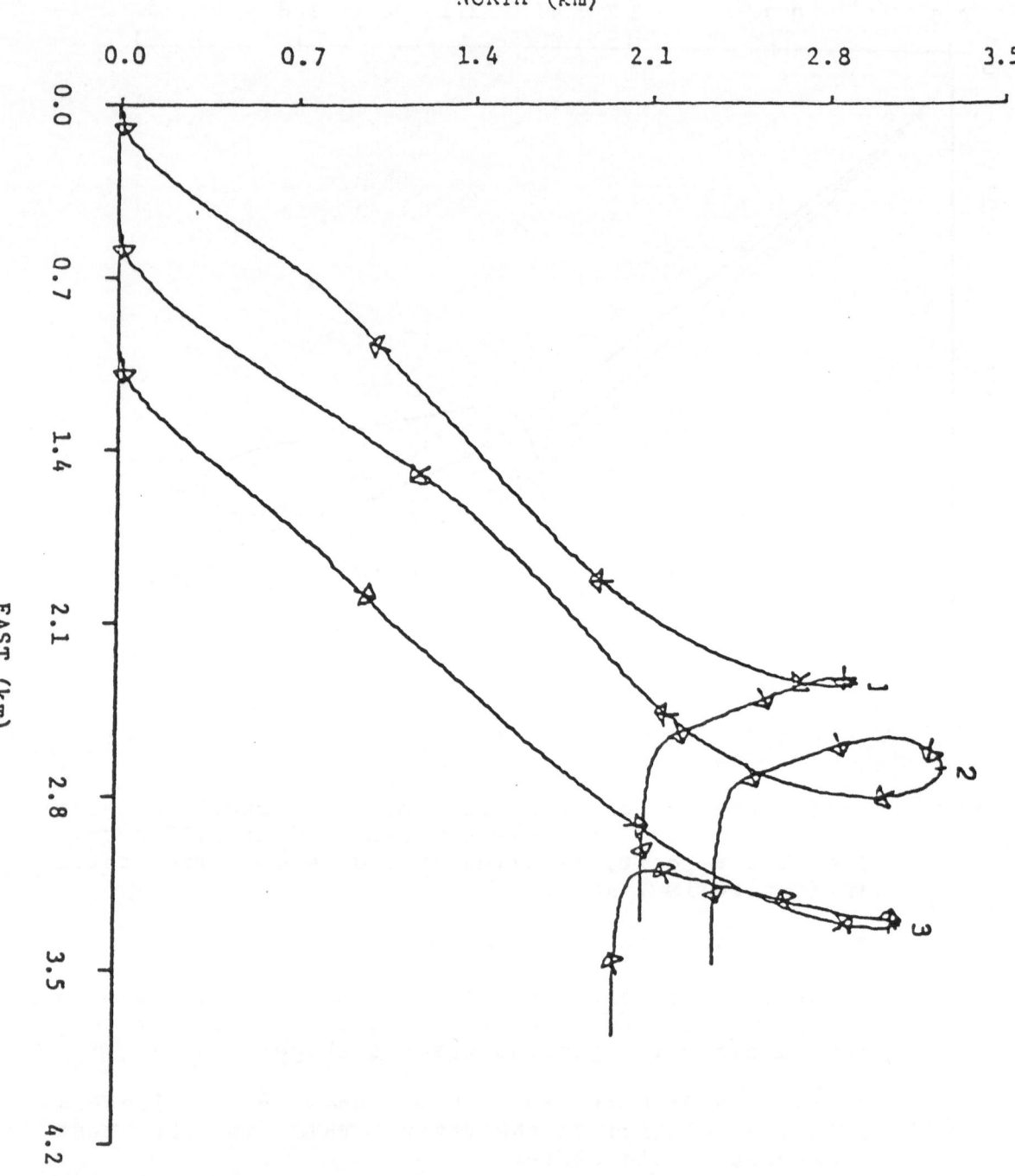

FIGURE 7. Influence of iceberg rotation on trajectories. Note that, for clarity, the origin is shifted eastward at each subsequent simulation. The iceberg shape factors are given in Section 5.3.3. In all three cases, the θ=0 axis of the iceberg is initially aligned east. The below-water moment arm functions l (θ) for the trajectories shown are

Trajectory 1: l (θ) =30sin(θ)
Trajectory 2: l (θ) =10sin(θ)
Trajectory 3: l (θ) =-20sin(θ)

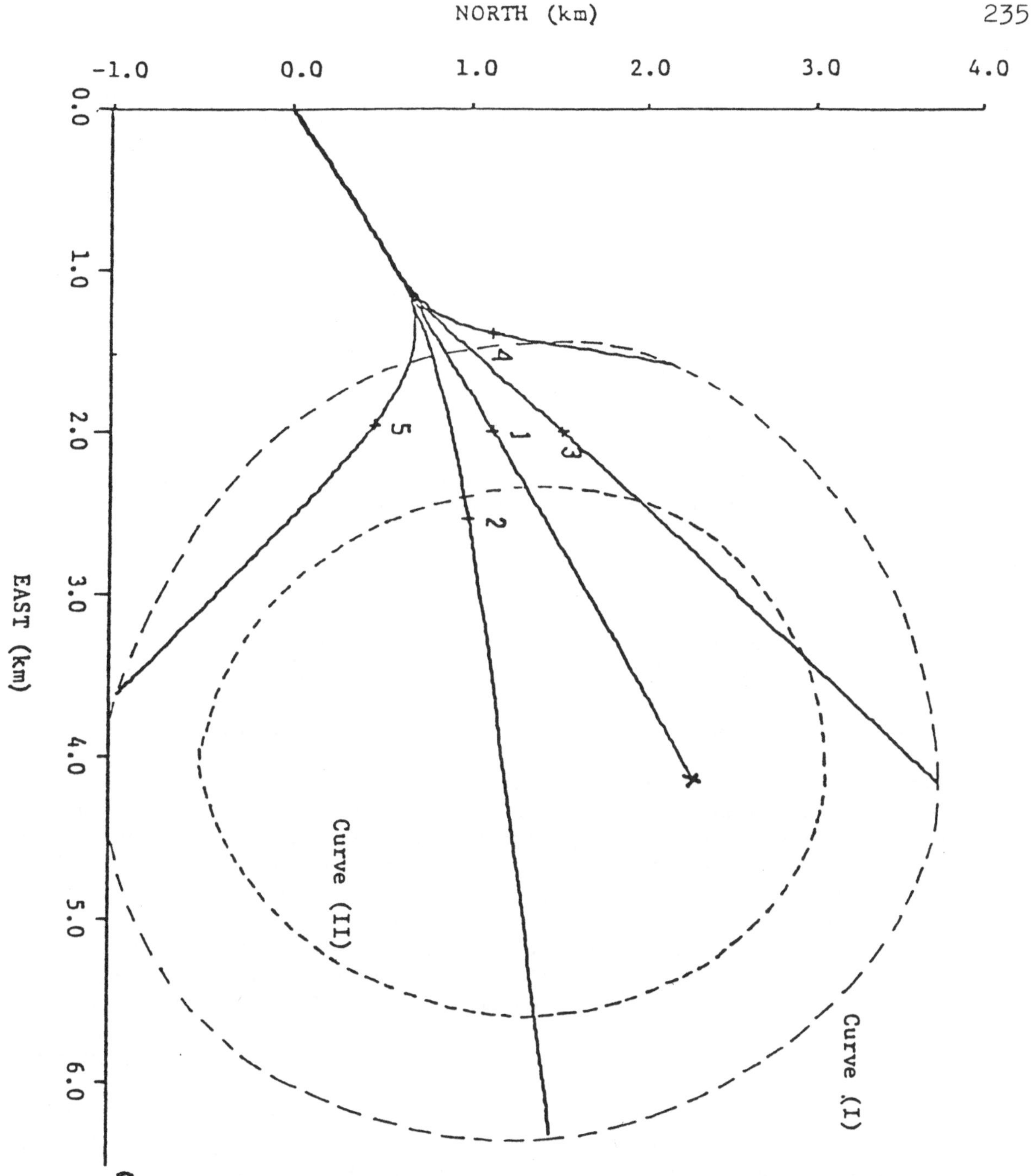

FIGURE 8. Towing efficiency. Effect of constant towing force
F_T applied in direction θ (counterclockwise from
east) from t=1 hour to t=4 hours on a cylindrical
iceberg. Wind is northward
($\theta = \pi/2$) at 10 m s⁻¹ and current is eastward at
0.25 m s⁻¹.

Curve (I) is locus of endpoints for F_T =5 x 10⁵ N
and all θ.
Trajectory 1: no towing
Trajectory 2: θ=0 (east)
Trajectory 3: θ= $\pi/2$ (north)
Trajectory 4: θ= π (west)
Trajectory 5: θ=3$\pi/2$ (south)

Curve (II) is locus of endpoints for F_T =2.5 x 10⁵ N
and all θ.

- water form drag coefficients vary as a function of the Reynolds number account for changes from usual turbulent flow to rarely encountered laminar flow.

- Possibility of simulating a layered ocean where the water velocity varies with depth. In such case, an equation somewhat similar to (3) needs to be written for every layer of water, and this option though providing a more realistic description of the physical reality also introduces more iceberg related unknown so that the accuracy gained in drift prediction is debatable.

- Introduction of an external force to simulate towing, and thus study the best towing strategy in order to keep the berg as far away as possible from a given point to protect.

Examples of model runs are shown on figures **6**, **7** and **8**. The first two show the importance of correct taking into account Coriolis and rotation of berg.

5.2.2. *Drift prediction*

Using current, wind and iceberg data gathered in the field it was possible to compare observed and computed drift trajectories. Good agreement is very often obtained and in most other cases erroneous data is to be blamed, not that current meter readings were false but because the current direction at the berg's location and at the drill-ship were quite different due to a local eddy. It is well known that large eddies (50 to 200 km across) and smaller size ones (a few km across) do exist in the midst of oceanic currents and the Labrador current is no exception in that respect. There may lie the biggest difficulty for achieving accurate iceberg drift prediction. Indeed, ocean current is the most important factor influencing the iceberg drift, this being especially true of the summer season in Labrador when the winds are usually light. Then, ocean current prediction can be attempted based on the knowledge of average values, tidal components, density-temperature profiles, barometric gradients but this will only provide values for the general flow but will not help predicting the onset of eddies. The only solution at the present stage seems to have an array of current meters somewhat upstream (speaking of the prevailing current direction) of the area where prediction is needed, so that eddies carried downstream in the general flow can be spotted in advance ; since they do usually remain active for a few tens of hours current prediction could become reliable. Obviously, one must keep in mind that the on line reading of half a dozen moored current meter strings can easily become an operational nightmare and that it will be necessary to precisely evaluate the gains to be obtained from such a set up before its use is envisionned.

5.2.3. *Determining iceberg parameters from drift data*

Assuming that ocean and wind values are known while an iceberg is tracked it is tempting to feed these data in the model and try an inversion, that is, try to find the shape characteristics of the iceberg, thus allowing more accurate further drift prediction. The numerical model was used to simulate this process and random and systematic errors were deliberately added to perfect theoretical data to test the accuracy of the method. It turns out that one must restrict to simple shaped icebergs for this inversion procedure, and it is also extremely beneficial to set lower and upper limits to some of the iceberg characteristics to be extracted. Such limits concern for example the mass, draft and horizontal dimensions which can at least be estimated visually. It was also found that the improvements gained in the accuracy of a subsequent drift prediction by using this method was probably usually less than that could be gained by improving ocean current prediction. It somehow confirms that the accuracy of oceanographic data and prediction is the key to iceberg drift prediction.

5.3. 1979 Summer field work

Some iceberg work had been routinely carried out every summer ; this included iceberg tracking from onboard the drill-ships and above water dimensions measured by iceberg observers onboard the berg towing tugs. Using these measurements, the iceberg mass was estimated through simple calculation. Draft measurements were also obtained using side scan sonars. Over the years considerable data was gathered which can be extremely valuable for statistical risk analysis. However, the value of this data could be questioned, especially the draft measurements because the accuracy of the side scan sonar was not known. It was therefore decided that in summer of 1979, a vessel would do systematic draft measurements using the traditional side scan sonar and a remote controlled submersible to obtain exact reliable values. This vessel was also used for the deployment and recovery of current meter strings in the vicinity of the drill-ship so that spatial current variations of the current could be checked and that subsequent numerical drift analysis could be performed.

5.3.1. Side scan sonar

The side scan system used was a standard Klein system modified for vertical deployment by relocating the towfish connector to the nose of the housing and adding a tail spinner to effect scanning of the berg. Deployment of the fish took place from the bow. Using a supporting davit and sheave the fish was lowered down the face of the berg until all returns were lost or the fish reached bottom. The fish was then recovered (still scanning the berg) and the procedure repeated up to four times at four different locations.

Sample of results are shown on Figure 9 and 10. Near the top of the plot echoes returning from the vessel hull can be seen. The berg measured on Figure 10 had a draft fairly close to the water depth (103 meters for 129 m water depth) so that returning echoes from the sea bottom begin to appear even before the keel of the berg can be seen.

Side scan sonar measurements are a fairly easy operation especially because the towfish is a fairly light piece of equipment easy to handle. The most difficult part of the operation is keeping the vessel sufficiently close to the berg (around 50 m or less) and this can become quite tricky if not unsafe as soon as they certain waves and wind conditions. It also becomes hazardous with some very unstable bergs or for bergs which have high hanging cliffs or protruding underwater spikes. Usually though, there is always at least one side from which measurement can be obtained but as it turned out not all sides of the underwater portion of the berg provide good sonar echoes.

5.3.2. Remote controlled vehicle (R.C.V.)

The R.C.V. used was the Scorpio of AMETEK, STRAZA Division. Here is a description of the operation procedure.

The R.C.V. was launched with the vessel being downwind around 150 m away from the berg. While umbilical was paid out by the winch, the R.C.V. approached

TYPICAL SIDE SCAN SONAR RECORDINGS.

FIGURE 9 : ICEBERG 12A01/TY-89

Depth (meters)

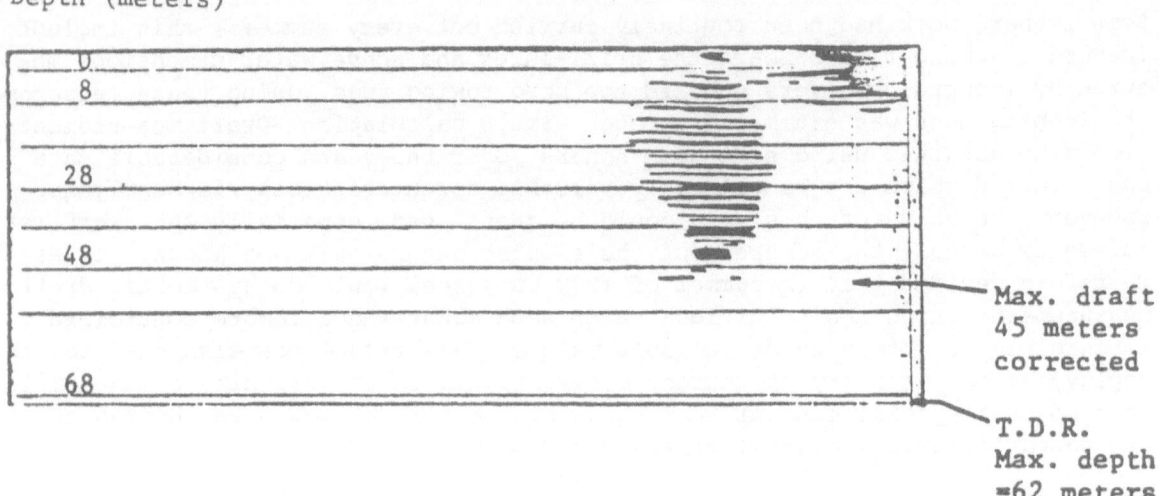

Max. draft
45 meters
corrected

T.D.R.
Max. depth
=62 meters

FIGURE 10 : ICEBERG 2A02

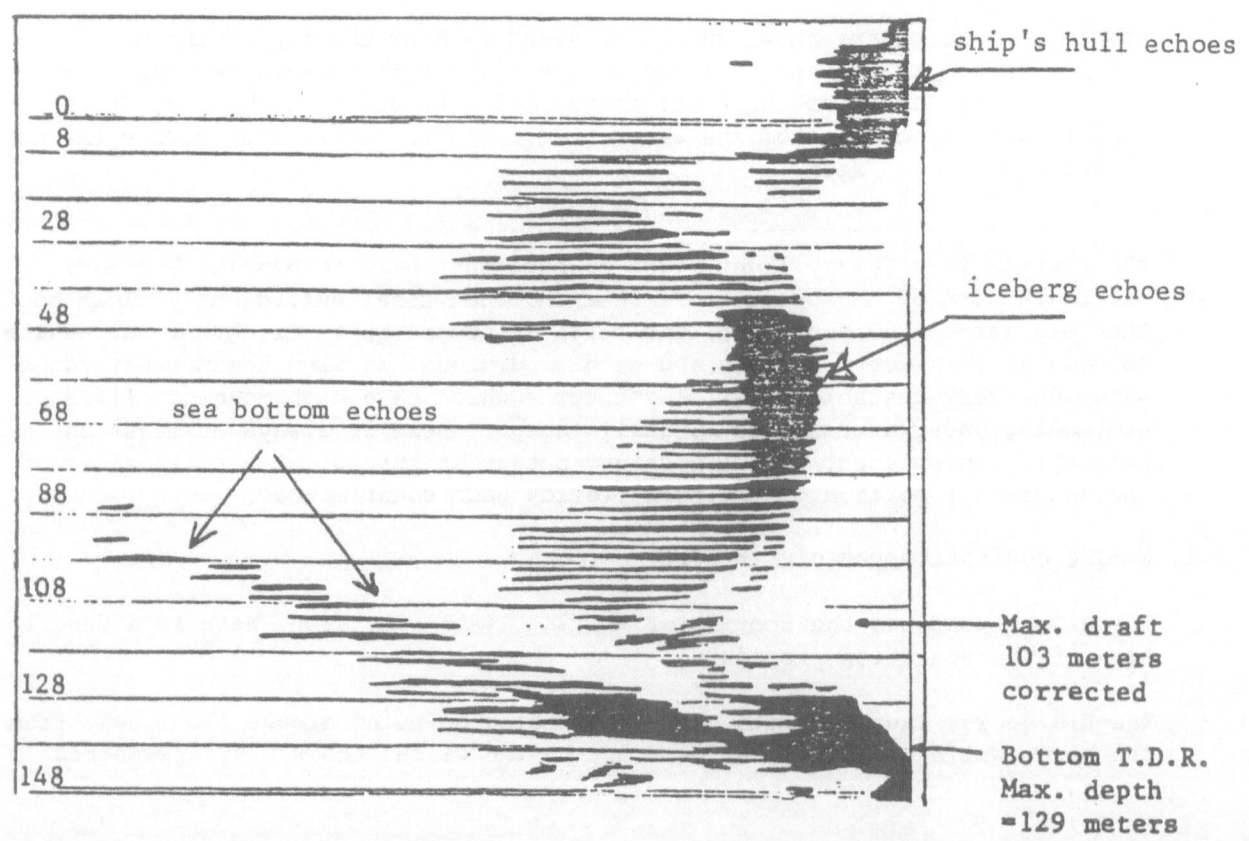

ship's hull echoes

iceberg echoes

sea bottom echoes

Max. draft
103 meters
corrected

Bottom T.D.R.
Max. depth
=129 meters

the berg still swimming at the surface. Sonar scanning was performed in the horizontal mode and the dive started, keeping the submersible 10 to 20 meters away from the iceberg. When below the keel's depth, the sonar was set in the vertical scanning mode and the deepest keel was located, then the submersible positioned right under it. An online mini-computer provided keel depth values by integrating the following : submersible depth, pitch and roll, plus vertical bearings and ranges of all echoes returning during one vertical scan. This set-up provided a draft measurement with absolute accuracy better than one meter. This operation procedure is described on Figure 11 Figure 11 and following show the sort of data obtained on an iceberg using R.C.V., sonar, above water photographs.

5.3.3. Comparison : sonar versus R.C.V.

If accuracy is the main concern, R.C.V. is obviously the best choice. Results showed that the sonar only provided a result good within 10 % to 15 % of the actual draft. It is our belief however, that minor improvements in the system such as better vertical stability of the towfish and real time depth recording (instead of using the cable length later corrected according to plot left by a time depth recorder fastened to the fish) would lead to relative error or around 5 %. This is probably no more than what is needed for operational requirements ; indeed a safety margin of at least 15 meters would probably always be required as clearance between the keel of a berg and any subsea equipment, and this value is well within the possibility of side scan sonars. Such clearance is in particular needed since, among other reasons, one cannot guarantee that an iceberg draft will never increase, when the berg rolls over due to melting and erosion.

Finally, the sonar is a cheap, almost expendable equipment, where the loss of an R.C.V. would be a major failure. If one then wishes to make draft measurement an every day routine (even considering winter use among sea ice) the improved side scan sonar seems the better choice.

CONCLUSION

Coping with icebergs in the Labrador environment is a feasible task. By this it is not meant that no downtime or equipment damage will ever take place but that these can be limited to a minimum and, in the case of equipment damage, to installations deliberately designed to be replaced in case they were destroyed. The optimum set-up is probably one where some drift prediction can be routinely performed and where towing takes a large role. Draft measurement helps in deciding which bergs should be towed when sea bottom installations are at stake, or what preventive measures should be taken when an iceberg is definitely going to enter the production area. In the end, it will be a purely economic decision to weigh the relative protective efforts to be made when compared to production losses. In all cases there is no doubt the environment can always be protected.

FIGURE 11 : ICEBERG 12A01/TY-89

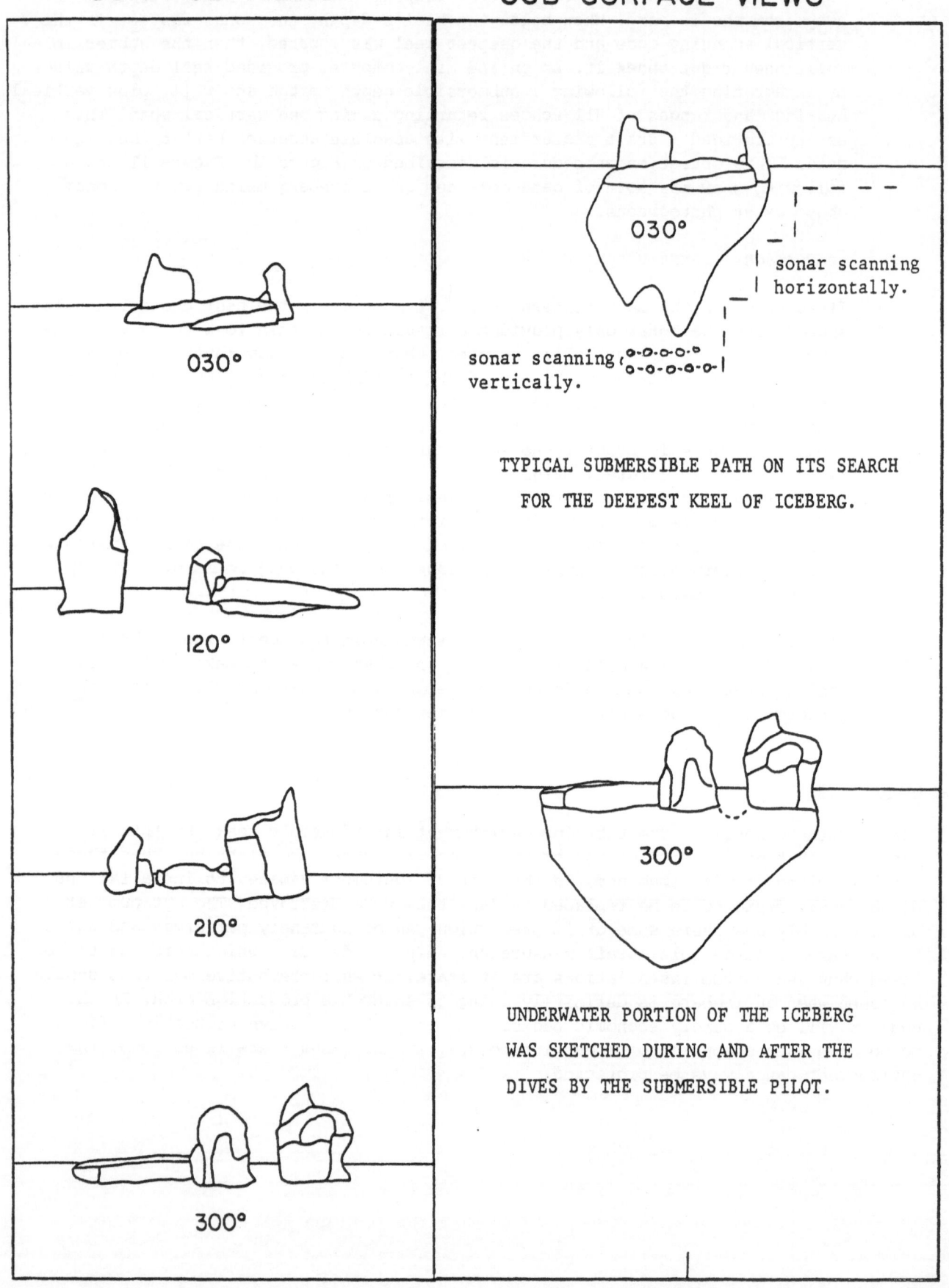

SURFACE VIEWS

030°

120°

210°

300°

SUB-SURFACE VIEWS

030°

sonar scanning
vertically.

sonar scanning
horizontally.

TYPICAL SUBMERSIBLE PATH ON ITS SEARCH
FOR THE DEEPEST KEEL OF ICEBERG.

300°

UNDERWATER PORTION OF THE ICEBERG
WAS SKETCHED DURING AND AFTER THE
DIVES BY THE SUBMERSIBLE PILOT.

0   100   200m

IÇEBERG NUMBER	12 AOI / TY-89
TYPE	DRYDOCK
MAXIMUM LENGTH	137
MAXIMUM WIDTH	94
MAXIMUM HEIGHT	42
MAXIMUM DRAFT (S.S.S.)	46
AT FACE	120°
MAXIMUM DRAFT (SUB.)	53
TOTAL MASS	0.20×10^6 M.TONS
H/D RATIO	1:1.3
H/L RATIO	1:3.3
L/H+D RATIO	1:1.4
REMARKS	

ICEBERG PARAMETERS

FIGURE 12

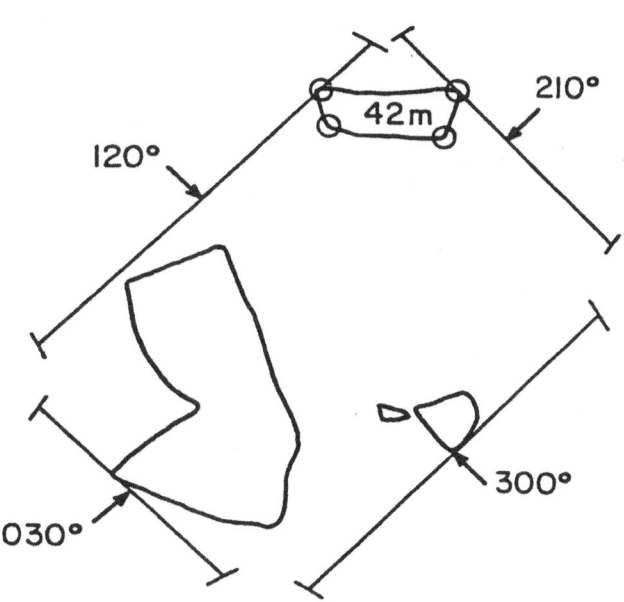

PLAN VIEW

0   100m

REFERENCES

1. "Système de Production Saisonnière d'Huile en Zones Arctiques Marines à Icebergs", J. DUVAL, G. MERCIER, P. MORIN, Congrès Mondial du Pétrole, Panel N° 6, Bucarest 1979.

2. "Offshore Labrador Year Round Production Study", TRI OCEAN, ACRES & NORDCO, April 1979.

3. "A Feasibility Study on Pipe-lines Offshore Labrador", FENCO CONSULTANTS Ldt, September 1979.

4. "Two Phase Flow Pipe-line Study for the Gudrid Field", NEOTECHNOLOGY CONSULTANTS Ltd, 1979.

5. "Iceberg Draft Measurement Labrador Sea", NORDCO Ltd, 1980.

6. "The Dynamics of Iceberg Drift", J.G. NAPOLEONI, Department of Geophysics and Astronomy, University of British Columbia, Vancouver B.C., Canada, 1979.

7. "Potential Schemes for Offshore Labrador Year Round Production", by VERNON F. WETZEL, SUNCOR Inc., Didier M. BERENGER, and Marc M. JOZAN, TOTAL EASTCAN EXPLORATION Ltd, 1980.

THE ARCTIC PETROLEUM OPERATORS' ASSOCIATION AND ITS ROLE IN

ARCTIC RESEARCH

Dr E H Kustan
Manager, Environmental Services, Canadian Superior Oil Ltd.

J Hnatiuk
Manager, Frontier Development, Gulf Canada Resources Inc.,
Calgary, Alberta, Canada

INTRODUCTION

The rapidly increasing demand for oil and natural gas in North
America and the uncertainty of oversees supplies has resulted
in accelerated activity in the Canadian frontier areas.
    The Arctic presents as hostile an environment as any which
will be faced by industry. New technology must be developed
for exploration, production and transportation of hydrocarbons
in the Arctic, particularly in the ice-infested offshore areas.
    Research costs in the Arctic are extremely high due to
remoteness, severe climate, unusual logistics, difficult
operating conditions, the complexity of the problems, the
vastness of the area and often the lack of prior research.

ARCTIC PETROLEUM OPERATORS' ASSOCIATION (Ref 1, 2, 3)

In order to avoid unnecessary duplication of effort and thus
accomplish more research with available funding, joint industry
research in the Arctic has become common. When the need for
such joint effort became apparent, the Arctic Petroleum
Operators' Association (APOA) was formed in January 1970. Most
of the major companies holding permits to search for oil and
gas in the area north of 60°N shown in Fig. 1 belong to the
APOA. Approximately 80% of the 170 million acres under permit
are represented in the Association. The areas encompassed
include the Yukon, Northwest Territories mainland, Beaufort Sea,
Arctic Islands and Davis Strait-Baffin Bay.
    The objectives of this non-profit association are to
encourage joint research, co-operatively develop the necessary
operating technology for the Arctic, serve as a liaison between
industry, governments and universities and collect and
disseminate information related to the Arctic.

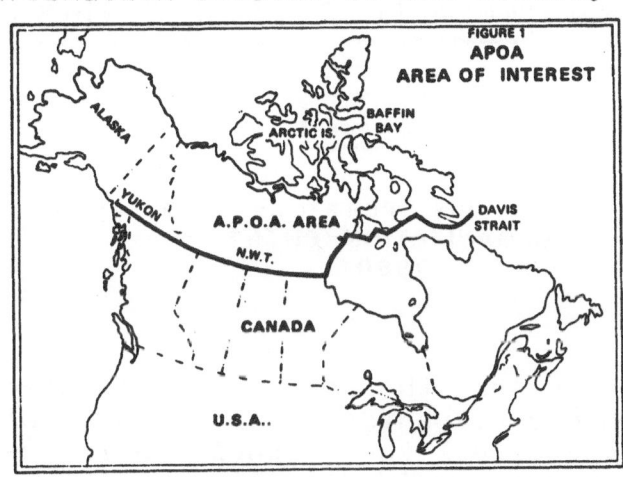

244

APOA has proven very successful in reducing the costs to member companies, has assured a broad look at Arctic problems, and has provided an excellent means for liaison with government and academic institutions.

Membership is currently twenty-three. A total of 150 APOA research projects have been completed, are underway or are proposed at a total cost of 29.4 million. The history of APOA is briefly summarized on Table 1.

TABLE 1

ARCTIC PETROLEUM OPERATORS' ASSOCIATION

DATE INITIATED	JANUARY, 1970
CURRENT MEMBERS	23
PERMITS HELD BY MEMBERS	170 MILLION ACRES OR 78%
PROJECTS COMPLETED OR UNDERWAY	150
COST OF PROJECTS UNDERTAKEN	$ 29·4 MILLION
COMMITTEES	6

The organization of APOA is as shown in Fig. 2 with six active committees; Drilling, Production, Environmental, Oil Spill, Remote Sensing and Public Information. APOA also has representatives on many joint industry-government committees, task forces and working groups.

FIGURE 2

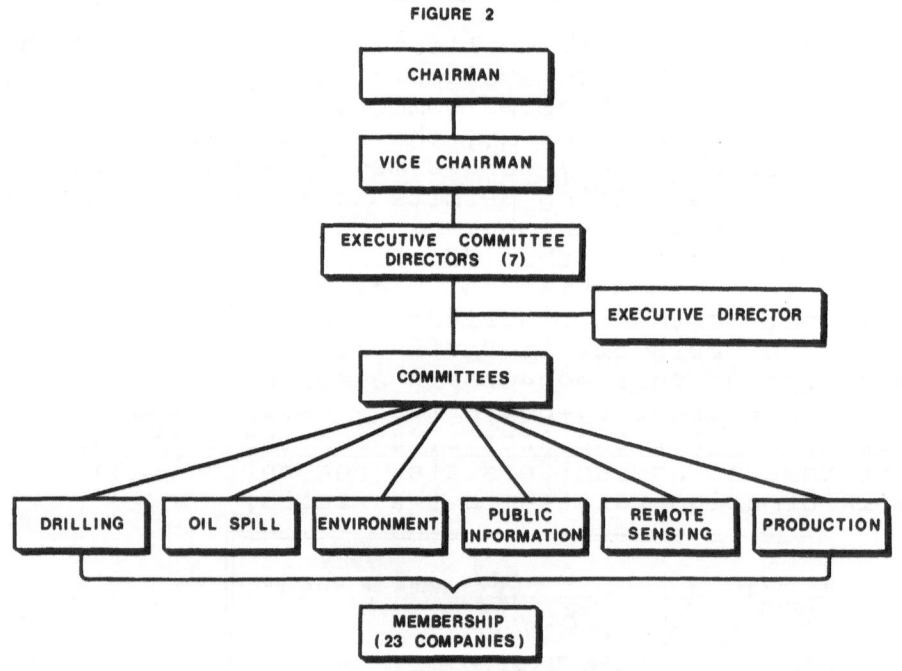

ARCTIC PETROLEUM OPERATORS ASSOCIATION
ORGANIZATION CHART
1980

When a member company sees the need for a specific research project, it may make a presentation to the general APOA membership. Those who feel the expenditure is justified in relation to their needs provide the financial support for the project. A list of the APOA study categories, number of projects and costs are shown in Table 2.

TABLE 2

APOA PROJECTS COMPLETED OR UNDERWAY
DECEMBER 1, 1979

CATEGORY	NUMBER OF PROJECTS	COST MILLION DOLLARS
DRILLING AND PRODUCTION	16	4·0
SEA ICE COVER	11	·8
SEA ICE STRENGTH AND MECHANICS	37	5·7
SEA ICE GEOMETRY	12	·8
SEA ICE MOVEMENT	11	2·4
SEA BOTTOM	10	1·0
OCEANOGRAPHIC AND METEOROLOGICAL	4	3·0
MARINE LIFE	10	4·5
OIL SPILL CLEANUP	15	2·4
ONSHORE ENVIRONMENTAL	12	2·6
OTHERS	12	2·2
TOTAL	150	29·4

Although the environments in all areas are very harsh, the most difficult problems are those encountered due to sea ice and icebergs. It may be noted from Table 2 that 71 APOA projects representing one-third of the expenditures have been for research related to sea ice.

As study results become available to the public, the information is widely distributed through an APOA newspaper, microfiche and hard copy reports. A consultant is retained for this public information program.

ENVIRONMENTAL CONDITIONS (Ref. 4)

Beaufort Sea

The primary environmental constraint in the Beaufort Sea is sea ice, since the area is usually ice-covered for at least nine months of the year. With the onset of winter in early October, freeze-up commences and progresses seaward with the growth of landfast ice. As shown in Fig. 3, this ice may be subdivided into two zones, a smooth area extending 15 to 20 miles offshore to the 30 foot water depth contour, and beyond that, a rough pressure-ridged area growing to the 60 foot contour where the landfast ice stabilizes in early January. A recurring open water lead is found along the outer edge of the landfast ice, which opens and closes under the influence of offshore and on-shore winds respectively. Between this and the polar pack is the so-called transition zone in which deforming, sporadically moving, pack ice is present. The polar pack generally blows well offshore during the summer months, but its nearshore boundary in winter usually lies about 100 miles from the coast, within the 1500 foot isobath near the edge of the Continental shelf.

Figure 3

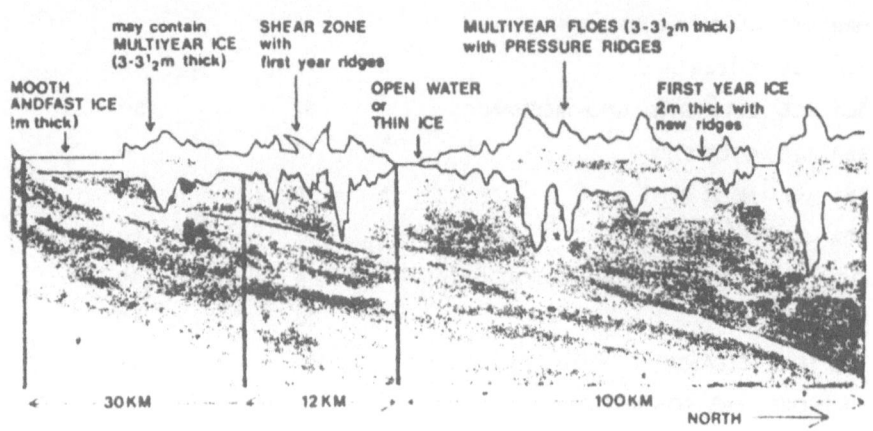

LATE WINTER
SCHEMATIC CROSS SECTION OF COASTAL SEA ICE
SOUTHERN BEAUFORT SEA

By late July, the coastal fast ice usually disintegrates and moves offshore with the polar pack. However, if westerly or northwesterly winds prevail throughout most of the summer, a heavy ice year will result in which very little open water is experienced. Alternately, if winds are predominantly from the east or south, an extensive open water season will be experienced.

The ice type which predominates in the landfast zone is first-year ice, which generally forms to thicknesses between 6 and 7 feet and reaches its maximum thickness in May. Multi-year ice floes in varying concentrations reach equilibrium thicknesses of between 10 and 15 feet in total thickness. The pack ice in the southeastern Beaufort Sea is in continual motion.

Although infrequent, ice islands (large tabular icebergs) and ice island fragments are sometimes found in the Beaufort Sea region. Most of the ice island fragments sighted in this area are in the order of 75-100 feet thick.

Sea bottom gouging or scouring by ice features is common in the water depths out to 150 ft. Many of these may be ancient, formed when the sea level was lower. In addition, sea bottom permafrost is common, reaching depths of nearly 2,000 feet below sea bottom at some locations.

In the Beaufort Sea, temperatures vary from lows of -50°F in the winter to highs of 80°F during the summer, while mean annual temperatures are roughly 10°F. Annual wind speeds in the region average 10 knots, while extremes of 75 knots may be expected. Annual precipitation in the area is low and usually averages 6 inches per year. The most obvious restriction to visibility is the long polar night in winter. Visibility in the area is also reduced by fog, whiteouts and blowing snow.

Waves in the Beaufort Sea are quite small since the presence of pack ice limits the wind fetch, and thus wave heights. Generally, waves in the area are in the order of several feet, although extremes during storms may be as great as 25 feet.

Arctic Islands Offshore

In the northwestern portion of the Arctic Islands, or the Sverdrup Basin, the dominant environmental factor in the off-shore region is sea ice. Here, the waters of the Archipelago are almost completely ice covered for 10 or 11 months, while during the remainder of the year, drifting pack ice is prevalent. Generally, the ice type which predominates in the Sverdrup Basin area is multi-year ice with average thicknesses of about 20 feet. First-year ice, which consolidates the ice cover, reaches thicknesses of 6 or 7 feet. Multi-year ridges 30 - 40 feet in thickness are not uncommon. The ice in the inter-island area is landfast for most of the year, and as a result small, sporadic movements occur throughout the winter and spring periods. Ice islands, ice island fragments and icebergs which calve from the most northerly islands may also drift southward into this area. However, their frequency of occurrence is extremely low. Sea bottom gouging is also present in nearshore areas and some offshore permafrost is present. Weather and wave conditions in the Arctic Islands are similar to those described for the Beaufort Sea.

Davis Strait

The ice cover in Davis Strait is characterized by very rough first-year ice and an abundant supply of icebergs and multi-year ice. In the northern reaches of this area, freeze-up begins in late November with the formation of first-year pack ice which, together with multi-year floes and icebergs that are already present, extends 100 to 150 miles offshore. As winter continues, this pack ice is transported southward by winds and currents. Because of the dynamic nature of the pack in this area, widely varying ice concentrations and ice types may be expected.

The ice type which predominates in the Davis Strait pack is first-year ice which generally grows to undeformed thicknesses of 3 feet by late April. Multi-year floes, which are scattered throughout the pack, are about 10 feet thick and can contain pressure ridges of significantly greater thicknesses. Icebergs are the most extreme features in the pack and may have thick-nesses ranging from tens to hundreds of feet.

The majority of the icebergs found in Davis Strait are calved from the glaciers of West Greenland with smaller populations originating from East Greenland and the Canadian Archipelago. Generally, the density or number of bergs per given area is lowest in the last summer and fall and highest in the spring and early summer months. These bergs vary in size from very small growlers, or pieces calved from larger bergs at sea, to extremely large ice forms a thousand or more feet in diameter with heights of several hundred feet above water and keel depths up to 1000 feet. Extreme iceberg masses may be as high as 30 million tons. Typical velocities are in the order of 0.5 knots while extremes may be several knots.

Temperatures during the winter range from $-35^\circ F$ to $20^\circ F$ and are quite temperate during the spring, summer and fall. This area lies in the path of the majority of the storms that sweep across North America. When the area is free of sea ice, high winds, combined with the large fetches available for wave generation, result in a severe wave climate. Visibility is restricted during the winter and spring by frequent snow storms, blowing snow and whiteouts. Extreme fog also reduces

the visibility in this area, particularly during the late
spring and early summer months.

Mainland
The most important physical environmental condition on the
mainland is permafrost which appears as frozen soil, ice lenses
and ice wedges. In those tundra areas covered by vegetation,
severe erosion can take place when the insulating layer of
vegetation is removed or damaged. Thawing of the ice rich
soil or ice lenses causes new drainage courses which quickly
enlarge. Careful choice of operating vehicles and operating
seasons is important.

Temperatures range from -60 to 90°F. Poor visibility due
to fog, blowing snow and whiteouts is a problem. Precipitation
is about 10 inches. Freeze-up usually occurs in October and
winter roads can be used from then through most of April in the
more northerly areas. Much of the APOA work on the mainland
has been in the Mackenzie River Delta area.

Arctic Islands
The Arctic Islands are less vegetated than the mainland, with
less precipitation and longer periods of cold and darkness.
Permafrost is also present here and erosion will occur after
disturbance. In general, the terrain is less sensitive than
the Mackenzie Delta.

ICE RESEARCH

Arctic Offshore Drilling Concepts (Ref. 5)
As offshore permits to explore for oil and gas were acquired,
the petroleum industry began to address the technological
requirements for exploitation of the Arctic's potential
hydrocarbon resource. In addition to the normal constraints
imposed by an offshore environment, unique problems were
introduced by the presence of various types of ice. Initially,
feasibility studies involving the design of systems for
seasonal or year-round explatory drilling were conducted for
various areas in the Canadian Arctic. These resulted in a
number of concepts, some of which are shown in Fig. 4. Each
of these systems is designed for specific environmental
constraints, reflecting the variability in offshore conditions
from the Beaufort Sea to the Davis Strait. However, these
exploration approaches are most sensitive to ice conditions and
water depth and may be subdivided into three basic alternatives.
These include (1) the use of a bottom founded strucutre or
vessel capable of resisting lateral ice forces, (2) the use of
a floating vessel during the ice free season and (3) drilling
on the ice cover itself.

Figure 4

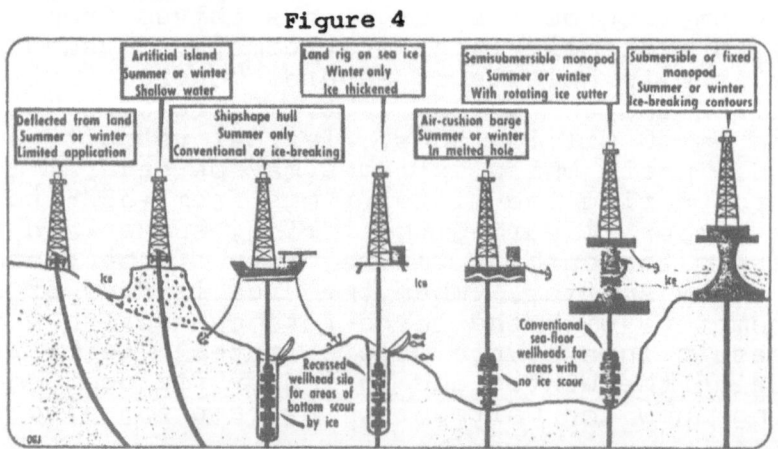

Arctic Drilling Concepts \ND GAS JOURNAL—MAY 6, 1974

To date, three of the offshore drilling concepts shown in Fig. 4 have been employed. In the shallow water areas of the Beaufort Sea, artificial islands have been constructed for use as exploratory drilling platforms in water depths to 65 feet. The majority of these islands have been built from locally dredged material during the open water season and are designed to withstand the forces imposed by the winter ice cover (Fig. 5). In the deeper water areas, much greater quantities of fill material are required to construct artificial islands and, as the water depth increases, this concept becomes uneconomical for exploration.

Figure 5   Beaufort Sea Artifical Island During Winter

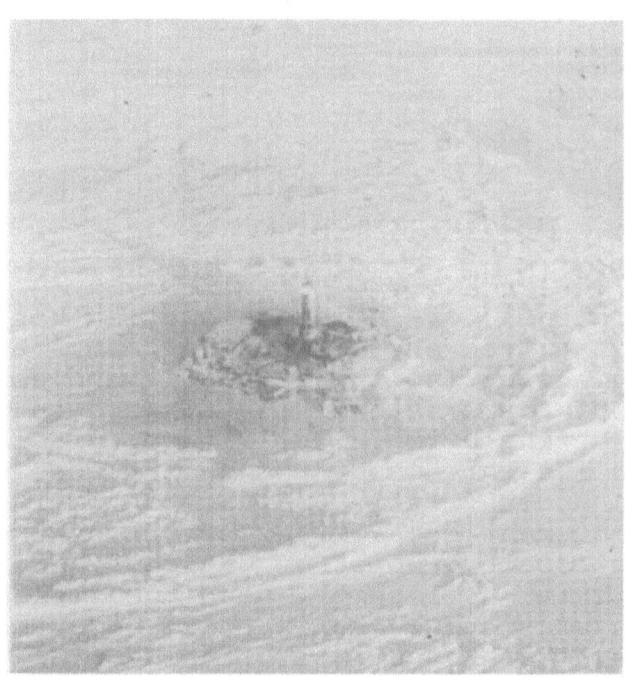

As a result, ice strengthened drill ships have been used for exploratory drilling during the open water/thin ice season in the 75 to 250 foot water depth range. Although the drill-ships are supported by icebreaking workboats, they cannot with-stand significant ice forces and must move off location when thick ice intrusions occur (Fig. 6a).

Figure 6

A

Drillship in the   a) Beaufort Sea b) Eastcoast

In the Arctic Islands, offshore exploration has been conducted
using the ice itself as a drilling platform. Construction
usually takes place during the late fall and early winter
periods by flooding the natural ice surface to a thickness
capable of supporting a rig. Drilling occurs later in the
winter when the ice throughout the region does not move
significantly. However, this approach is limited to areas
where the ice movement does not exceed 5% of the water depth.

Off the northeastern coast of Canada, exploratory drilling
in the Davis Strait has recently been undertaken during the
lengthy open water season. Here, drillships and semi-
submersibles have been used in accordance with the demands of
the wave climate. Due to the constant threat of icebergs
(Fig. 6b) these vessels are dynamically positioned and are
capable of quickly moving off location to avoid the iceberg
hazard.

Clearly, the design and successful use of the exploratory
drilling systems has required considerable research, particularly
related to the ice environment. As noted earlier, 71 or the
150 APOA projects carried out to date have been directed
towards a better understanding of the problems associated
with sea ice and icebergs. Because of the variability in ice
conditions across the Canadian Arctic, this APOA research has
generally been regional in nature with the results being used
for a variety of applications.

However, the ice research can be broadly classed into the
three areas of ice cover, ice movement and ice mechanics.
Some of the research that has been undertaken in these three
subject areas and is currently being used for exploration,
production and transportation considerations is highlighted
below.

Ice Cover (Ref. 6)
In order to preperly design offshore exploration and production
systems, a detailed knowledge of the ice cover, its composition
and its variability is essential. If offshore drilling is to
be conducted from conventional floating vessels, the length
of the available open water season must be adequately defined.

Alternatively, the design of ice resistant structures requires a thorough understanding of the normal ice conditions and extreme features that will be encountered in a given area, so that ice forces can be assessed. In order to supplement the ice cover data accumulated by government agencies, the APOA has conducted aerial reconnaissance and marine vessel surveys in various areas. Other field programs have been undertaken to obtain more information on extreme features such as pressure ridges, ice island fragments, and icebergs.

In the Beaufort Sea, some of the early APOA work was directed towards a better definition of the composition of the ice cover, its thickness and properties such as temperature, salinity and crystal structure which affect its strength. APOA Project No. 2 was carried out during the winter of 1970 and measured these ice characteristics along with small scale ice strenghts at a number of locations in the landfast ice zone. The following summer, APOA Project No. 14 obtained similar information on multi-year pack ice that had been driven into the seasonal open water area during a severe storm. Aerial reconnaissance conducted in conjunction with this survey provided supplementary information on the extent of the polar pack intrusion along with ice cover parameters such as concentration and floe size. The results of these projects showed first year landfast ice thicknesses to be in the order of 5 to 6 feet in April, while the measured multi-year ice floe thicknesses were all greater than 13 feet. Ice temperatures reflected the ambient air temperatures, being low in the cold first year ice and near freezing in the summer's multi-year ice. Salinities and small scale strengths were in the expected range of values although ice salinity was found to increase from the nearshore waters influenced by the Mackenzie River towards the outer edge of the landfast ice. The results of these two projects tended to confirm preliminary information on the Beaufort Sea's ice regime, but general observations on factors such as the extent of the landfast and pack ice zones and the distribution of pressure ridges emphasized the spatial and temporal variability exhibited by the ice cover.

After this initial exposure, several APOA projects were conducted to more systematically document and quantify various characteristics of the Beaufort Sea ice cover. These studies involved a number of aerial photographic overflights throughout the winter and summer periods and tended to concentrate on the shallow water or landfast ice areas. In APOA Project Nos. 31, 46 and 54, information on the growth and extent of the winter's landfast ice zone was collected along with data on the ice concentrations and floe sizes which characterized break-up and summer ice intrusions during the 1971-73 period. The growth and extend of the landfast ice was found to follow a similar pattern each year. Although fall temperatures and wind patterns did influence its growth rate and final extent, the outer edge of the landfast ice typically stabilized in the vicinity of the 60 foot contour in January, remaining there until break-up. Ice concentrations shortly after break-up were high while floe sizes in the order of hundreds to thousands of feet were most common. As break-up progressed, increasing temperatures and offshore winds were found to quickly reduce ice concentrations to an open water condition. A corresponding decrease in floe size was also observed as the ice abated and experienced wind action. During the summer of 1973, additional information on the

thickness, temperature and salinity of both first and multi-
year floes was obtained in APOA Project No. 60. The results
of this project generally corroborated earlier work but
indicated the mean thickness of multi-year ice to be in the
order of 15 to 25 feet.

Stereo photographs collected during the winter overflights were used to obtain the height and frequency distribution
of first-year pressure ridges in the area. Typically, one to
twenty ridges per mile were found in the landfast ice zone
while near its outer edge and in the pack ice adjacent to it,
up to eighty ridges per mile were observed. The highest ridge
sail measured was 22 feet, although ridges with 10 foot sails
were common in all but the nearshore ice areas. An analytical
function used to describe the statistical height distribution
of these ridges indicated a mean sail height to keel depth
ratio of 1 to 3.8 and suggested that from 1 to 4% of the ridge
keels were grounded, depending upon location.

During these projects, the polar pack was generally
located about 150 miles offshore and few ice island fragments
or multi-year pressure ridges were seen. However, these
features are recognised as the largest and most hazardous ice
formations in the Beaufort Sea (Fig. 7) and a knowledge of
their size, geometry and frequency of occurrence was needed
for design purposes. As a result, APOA Project Nos. 53 and 99
were undertaken to count the number of ice island fragments
trapped within the coastal fast ice during the 1972-76 period.
Photographic overflights established typical counts of several
hundred fragments per year, but the vast majority of these
features were located in Alaskan waters. Additionally, those
observed were generally quite small with average diameters of
150 - 200 feet and estimated thicknesses of 75 - 100 feet.
An earlier study, APOA Project No. 36, had established the
feasibility of ice island destruction with explosives and this
approach was considered as a feasible method of dealing with
small ice island fragments expected in the Canadian Beaufort
Sea.

Figure 7

a) Grounded ice island fragment, Beaufort Sea

b) Largest observed multi-year pressure ridge sail, Beaufort Sea

Since multi-year pressure ridges were more commonly observed in the area, they were felt to pose a more serious design problem for fixed offshore structures and APOA Project Nos. 17, 89 and 91 were undertaken to gain a better understanding of their size and internal properties. In these projects, the thickness, geometry and structural integrity of twenty multi-year ridges was established during on ice investigations. Standard elevation surveys were carried out on the ridge sails while a submarine sonar profiling technique was used to obtain their keel contours. Supplementary information on the structural integrity of the ridges was also collected by drilling and some small scale ice strength testing. The results of these studies showed that the multi-year ridges had completely consolidated from their rather porous first-year form and were comprised of low salinity, solid ice. As shown in Fig. 8, typical ridge thicknesses were in the order of 50 to 70 feet, but one extreme multi-year ridge with a total thickness of 137 feet was profiled. A relatively constant sail to keel ratio of 1 to 3.2 was established for the features investigated and it was found that their bowl-shaped keel geometry was adequately described by the Kovac's multi-year ridge model.

FIGURE 8

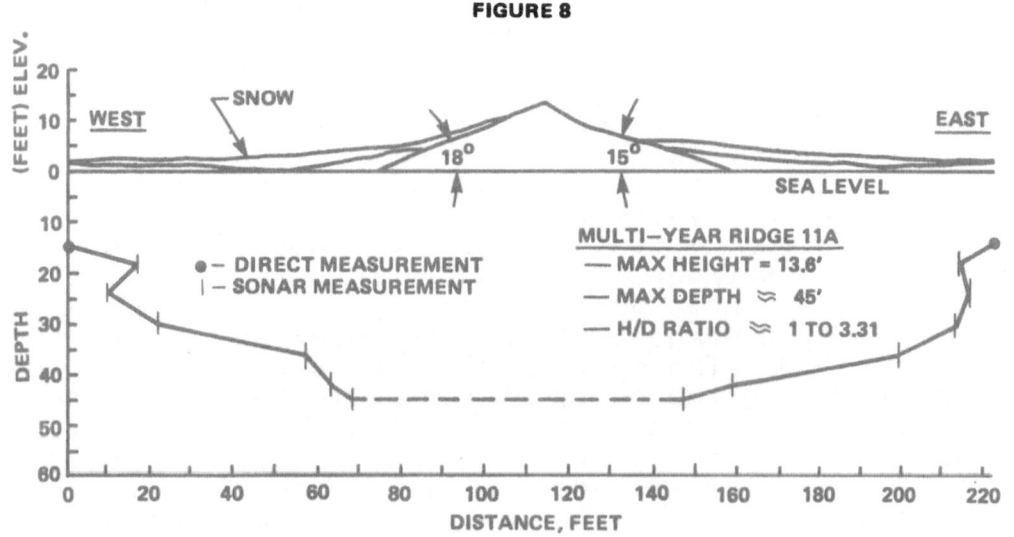

CROSS SECTION OF MULTI—YEAR PRESSURE RIDGE

In the Arctic Islands and Davis Strait areas, similar APOA work has been undertaken to better define the normal and extreme characteristics of the ice cover. In APOA Project No. 92, historical information from various government agencies was supplemented by satellite imagery and used to produce maps showing the spatial and temporal distribution of various ice forms in the Arctic Islands. The study results showed relatively high concentrations of multi-year ice characterizing the landfast ice in the western portion of the Sverdrup Basin with first-year ice predominating towards the east. Although there were some areas that commonly displayed severe pressure ridging, the surface relief of the landfast ice cover was generally much smoother than that in the Beaufort Sea. The short summer break-up period with its high concentrations of pack ice was found to vary according to air temperatures, wind patterns and the regional ice distribution over previous winters. In some years, break-up did not occur at all. Records of ice thickness obtained from seismic shot holes in the Sverdrup Basin during the winters of 1971 to 1976 were systematically analyzed in APOA Project Nos. 96, 117 and 142. The resultant ice thickness distributions confirmed the presence of high concentrations of multi-year ice in the area with typical thicknesses in the 20 to 30 foot range. Additional information on the extreme thicknesses of multi-year ridges was required for the ice cutter design studies, and APOA Project No. 102 investigated the geometry of fiteen of these features in the Arctic Islands.

ICE MOVEMENT

In addition to information on the ice cover, a knowledge of ice movement is necessary for the proper design of offshore exploration and production systems. In landfast ice zones, the magnitude of ice movement determines the feasibility of ice platform drilling while its rate influences the forces that may be experienced by ice resistant structures. Alternatively, the velocity distribution of icebergs must be known to assess potential impact forces and set operational procedures for hazard avoidance while pack ice motions are required for ice/ structure interaction considerations.
     During the 1970-75 period, landfast ice movements were determined at a number of Beaufort Sea locations in APOA Project Nos. 2, 33, 51, 67 and 83. The earlier studies tended to concentrate on the shallow water areas within the nearshore "barrier" islands while subsequent studies established ice movements throughout the landfast ice zone. In each of these projects, ten to fifteen stations consisting of a spring tensioned reel/wireline system which connected the ice with the seafloor were deployed. Potentiometric readings of ice move-ment were remotely interrogated at hourly intervals and recorded at a master station onshore via a telemetry system. At the nearshore locations, average landfast ice movements of roughly 10 feet were observed. The major portion of these displacements usually occurred during the early winter period when the ice cover was thin, with the remainder occurring in the spring, due to thermal expansion of the ice. Stations deployed beyond the protection of the barrier islands typically showed larger net movements which increased towards the edge of the landfast ice. These ranged from roughly 50 to 100 feet and showed rates as

high as 10 feet per hour. Although the presence of pressure ridges and large scale floes caused variations in movement between stations, the general ice movement patters was characterized by spatial continuity. Some oscillatory motions were observed in the movement records and were associated with pack ice stresses on the boundary of the land fast ice coupled with an elastic response of the ice sheet. No interpretive motion models were developed in these APOA projects but wind stress, therm expansions and pack ice pressures were identified as the major causes of landfast ice movement in the Beaufort Sea.

In these projects, attempts were also made to measure winter movements in the seasonal pack beyond the landfast ice edge. However, significant displacements which exceeded the wireline system capabilities usually occurred within several days and limited information was obtained. As a result, several more recent studies have used different techniques to determine pack ice movement in the Beaufort Sea. In APOA Project No. 72, summer and winter pack ice movements for 1973 - 75 period obtained by tracking identifiable ice floes on sequential satellite imagery as shown in Fig. 9. In another component of this major project, radio beacons were deployed on the ice during the summer of 1974 and subsequent tracking from aircraft provided movement information for isolated floes and groups of floes.

The results of these studies showed average pack ice motions of several miles per day with extremes as high as thirty miles per day. The longer term average drift direction was from east to west, but on shorter time scales, movements from all directions were observed. Wind stresses were found to be the primary driving force for pack ice motions in the Beaufort Sea. During the winter period, major relative movements within the continuous pack ice cover appeared to be equalized by large scale momentum transfers between its component floes while the more variable drift patterns of isolated floes or groups of floes observed during the summer period were explained in terms of local winds and currents. More recent pack ice movement studies have involved the deployment and tracking of on ice "RAMS Transmitters" for sequential ice drift information. These buoys send doppler shifted frequency transmissions to an orbiting satellite that, after each pass, relays this information to a ground station where it is processed for ice buoy position. Due to accuracy limitations, this system and the techniques of the earlier studies have only provided one day to one week of averages of pack ice movement. In OPOA Project No. 139, a new on ice satellite reporting buoy system is being developed to provide more accurate ice motion information on shorter time scales.

During the winters of 1975-77, landfast ice movements in the western portion of the Sverdrup Basin were also determined in APOA Project Nos. 79, 95 and 118. Three to four movement stations, comprised of a seafloor pinger and four hydrophones suspended beneath the ice, were deployed in each of these projects. Time differences in signal transmission from the pinger to the hydrophones were recorded at hourly intervals and analyzed for ice movement information at each location. Due to the harsh environment, this system experienced a number of component failures which limited the amount of data collected. In APOA Project No. 118, position fixes were also

256

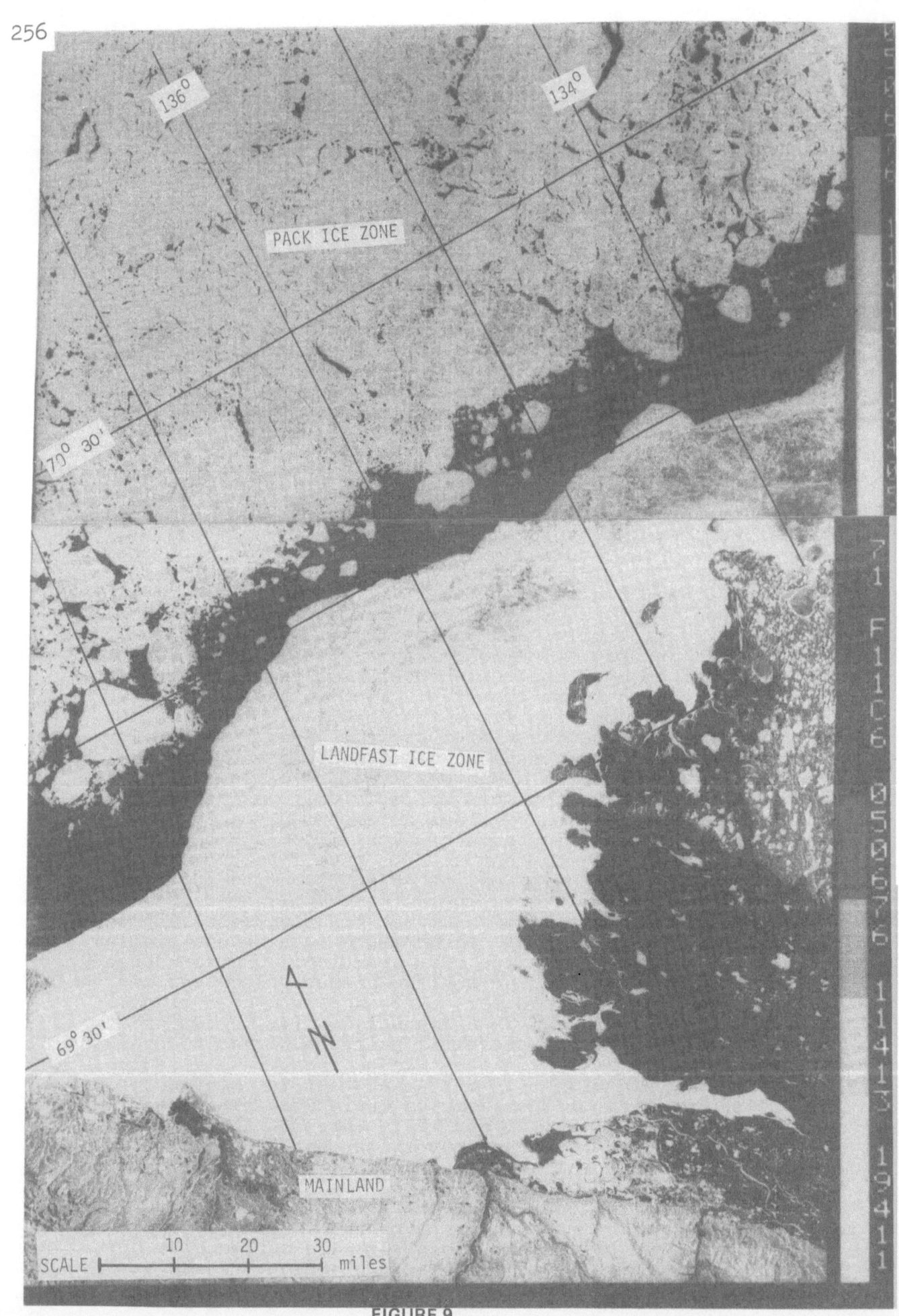

PACK ICE ZONE

136°

134°

70° 30'

LANDFAST ICE ZONE

69° 30'

N

MAINLAND

SCALE |——|——|——|——| miles
10   20   30

FIGURE 9

LANDSAT IMAGERY OF THE SOUTHERN BEAUFORT SEA, JUNE 5, 1976

obtained at two week intervals with on ice Doppler Satellite
Survey techniques which ensured a complete ice movement record
throughout the winter. As evidenced by recent ice platform
drilling, winter landfast ice movements in the western part of
the Arctic Islands were found to be small. Attempts to
correlate the ice movement records with wind, current and
temperature data measured at several of the locations was
largely unsuccessful.

Ice Mechanics

A major portion of the APOA's research has been directed towards
a better understanding of the basic mechanical behaviour of sea
ice and the forces that it will exert on fixed offshore
structures. Clearly, a thorough understanding of the ice
structure interaction problem is essential for the safe and
economic design of year-round exploration and production
systems.

In order to establish design criteria for offshore structures,
a relationship between ice forces and the strength properties
of the interacting ice cover is required. These forces are not
an intrinsic ice property since they depend not only on the
strength of the ice itself, but also on the size and shape of
the structure and the particular conditions that produce the
ice failure. For a given structural concept, the failure mode
of the ice must first be defined and the resultant forces then
computed on the basis of relevant strength parameters. Crushing,
for example, involves the compressive strength of ice and may
be expected when ice moves against a vertically sided structure,
while bending failures involving flexural strength values are
expected for an inclined structure. Since ice is a highly
complex, anisotropic material whose mechanical behaviour and
strength depends on factors such as crystal structure, sample
size, strain rate, confining pressure and temperature, the
development of adequate ice design criteria has been difficult.

During the winter of 1970, the first APOA project was
undertaken to obtain more information on the strength of Arctic
sea ice. In this project, large scale ice crushing strengths
were determined in situ by freezing four rather unique "Nutcracker"
test devices into the nearshore fast ice on the Beaufort Sea.
As shown in Fig. 10 these test devices were constructed in the
form of large inverted nutcrackers with their long tubular leg
members hinged at the bottom. To crush the ice, the legs were
forced apart by hydraulic cylinders mounted between them. The
pressure within the hydraulic system at the time of ice failure
provided a measure of the ice crushing strength. A large semi-
cylindrical attachment was added to one of the units to
investigate the effect of changes in the indentor width to ice
thickness ratio (aspect ratio), while an angular wedge was
attached to another to assess the influence of indentor shape.
Two test series comprising seven tests were carried out in two
different ice thicknesses. During the winter of 1971 only three
in situ nutcracker tests were repeated in APOA Project No. 9
due to equipment problems. The measure crushing strengths
ranged between 600 and 1050 psi which indicated that the ice
was much stronger than expected. The influence of the
indentor shape, loading rate and aspect ration on the ice
crushing strength was not quantitatively determined due to the
small number of tests.

Figure 10

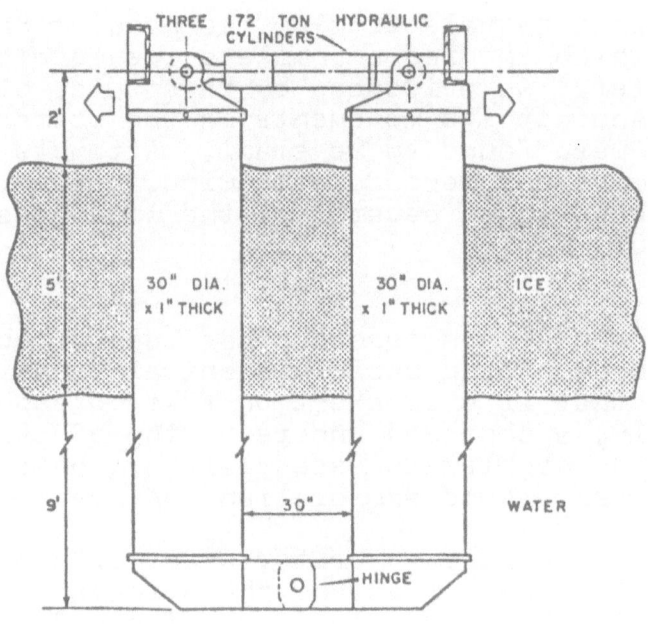

Nutcracker ice strength tester - general configuration

As a result of the experimental difficulties associated with the
Arctic tests, a portable device was designed and used to conduct
a large number of more cost effective crushing strength tests
on a fresh water lake near Calgary in APOA Project Nos. 9, 52
and 66. This device consisted of two parallel flat plates
with four hydraulic rams mounted between them. To perform a
test, the device was lowered into a pre-cut rectangular hole,
centered on the ice and hydraulic pressure then applied to the
rams, which forced the load plates apart. The purpose of these
test series was to examine the effects of temperature, strain
rate, aspect ratio, shape and ice/indentor bond on the crushing
strength of ice. The resultant strength values ranged from
roughly 300 to 1700 psi with an average of about 1000 psi.
However, these strengths were found to be quite sensitive to
the variables mentioned above. Generally, the measured crushing
strengths showed a tendency to increase with decreasing
temperature. Strengths measure at strain rates near the ductile
to brittle failure transition (5×10^{-3}/sec) were typically larger
than those at higher or lower rates. A very pronounced size
effect was also observed with effective crushing pressures
decreasing to an asympotic value of 300-400 psi as the aspect
ratio increased. The influence of indentor-shape and the ice/
indentor bond could not be assessed due to scatter in the
experimental results, but the variation of crystall orientation
in the ice was found to significantly affect the strengths
obtained. In order to relate the strenght results from the
relatively thin lake ice to the thicker, colder ice of the
Beaufort Sea, additional theoretical and laboratory studies
were carried out in conjunction with these projects. More
recently, APOA Project Nos. 93, 123 and 148 have involved
laboratory studies of the crushing strength of ice sheets that
are in irregular contact with very wide structures and the

non-simultaneous aspects of the associated ice failures.

In APOA Project Nos. 104, 105 and 122, full scale in situ ice pressures have also been measured around artificial islands in the Beaufort Sea. These measurements were obtained with an array of large area (4 ft x 6 ft) ice pressure sensors frozen into the landfast ice cover around each island and in turn related to full scale ice forces on offshore structures. Other full scale field tests on the load-bearing capacity of sea ice have been conducted to provide design data for Arctic Islands ice platform drilling. In APOA Project Nos. 64 and 81, large storage tanks placed on natural and artificially thickened sea ice were filled with sea water until the supporting ice sheet collapsed. Factors such as ice deflections, stresses and strains were monitored as the loading magnitudes and the rates were varied and the results used to evaluate safe loading conditions.

The APOA has also used scale modelling techniques to obtain information on the ice failure modes, forces and movements that characterize various ice/structure interactions. Although the models and scaling relationships must be calibrated with larger scale data, this approach has been a cost effective means of systematically investigating the effects of ice property variations on ice forces and behaviour. Several modelling techniques were evaluated on specific structural concepts in APOA Project Nos. 25, 40, 41 and 68 with synthetic and natural ice. The majority of the work has involved the conical structure with model scale factors ranging between 10 and 50 (Fig. 11). In APOA Project Nos. 65 and 77, the interaction between cones and level ice was investigated while pressure ridge interactions were studied in APOA Project Nos. 82 and 86. The results of these projects have been used to confirm and refind the analytic models for ice/cone interactions developed in APOA Project Nos. 57 and 87. More recent physical modelling studies have involved the ice forces and the ice clearance problems that occur when pack ice moves against different types of artificial islands.

Figure 11

Model Test of Natural Ice/Conical
Structure Interaction

260

CONCLUSIONS

The Arctic Petroleum Operators' Association has provided an
effective vehicle through which the industry can conduct
jointly-sponsored, high-cost Arctic research. To date, 150
different projects relating to petroleum activities have been
proposed or undertaken in the areas of Drilling and Production,
Sea Ice, Sea Bottom, Oceanography and Meteorology, Marine Life,
Oil Spill Cleanup, Onshore Environmental and others. In
these projects, which total nearly thirty million dollars,
duplication of effort has been avoided as the projects are
presented to the membership and jointly undertaken by
interested participants. A broad look at Arctic problems and
more research has been possible with the joint funding
available. The quality of the joint Arctic research has also
been improved because of greater funds, more comprehensive
study definition and the availability of more expertise. A
large portion of the research effort has focused on better
defining ice cover, its movement and its mechanics since this
information is needed for the proper design of offshore
exploration and production systems. Through APOA research
activities, excellent liaison has been established with
government agencies, university institutions and the interested
public.

ACKNOWLEDGEMENTS

The authors wish to express their appreciation to the members
of the Arctic Petroleum Operators' Association for permission
to present this paper.

REFERENCES

1. J Hnatiuk, Joint Industry Research through the Arctic
    Petroleum Operators' Association, the Journal of Canadian
    Petroleum Technology, July – September 1972.

2. Arctic Petroleum Operators' Association "Description of
    Research Projects, May 1979".

3. Arctic Petroleum Operators' Association Information Service,
    "APOA Review", Vol. 1 Nos. 1-4, 1978 Vol. 2 Nos 1-3 1979.

4. B Wright and J Hnatiuk, "Satellite Applications to the Oil
    and Gas Industry in the Canadian Arctic, proceeding from
    Satellite Applications to Marine Application", New Orleans,
    1977.

5. K R Croasdale, "Ice Engineering for Offshore Petroleum
    Exploration in Canada", Proceedings of the Fourth Inter-
    national Conference on Port and Ocean Engineering under
    Arctic Conditions, Memorial University, St John's,
    Newfoundland, Sept. 1977, 1.

6. B D Wright, J Hnatiuk, A Kovacs, "Sea Ice Pressure Ridges
    in the Beaufort Sea", Proceedings of IAHR International
    Association for Hydraulic Research, Luten, Sweden, August
    1978.

FEASIBILITY OF CONSTRUCTING MARINE

PIPELINE CROSSINGS IN THE ARCTIC

by O. M. Kaustinen

INTRODUCTION

The Polar Gas Project was formed in 1972 to find the best means of transporting the natural gas discovered in Canada's Arctic to southern markets, and to do so in a manner that would be economically, environmentally and socially acceptable to Canadians. Polar Gas is a joint public/private sector project sponsored by TransCanada PipeLines (Project Manager), Tenneco Oil of Canada Ltd., Panarctic Oils Ltd., Ontario Energy Corporation and Petro-Canada.

Natural gas was first discovered in the Arctic in 1962 and since that time, close to $650 million has been spent on exploration. The substantial potential for natural gas in this region has been known since the early fifties as a result of work undertaken by the Geological Survey of Canada and others, but it wasn't until recent years that it was considered to be technically and economically feasible to bring these frontier hydrocarbons to southern markets. Other gas reserves on Canada's Arctic frontier have been established in the Mackenzie Delta/Beaufort Sea area. The first discovery in this area was made in 1965. The gas reserve areas of the Arctic Islands and the Mackenzie Delta/Beaufort Sea areas are shown on Figure 1.

The initial thrust of the Polar Gas studies was directed towards determining the best method of moving gas from the hostile environment of the Arctic. After consideration of a variety of modes of transportation including LNG tankers, railway and airships, it was concluded that the preferred method was a large diameter pipeline. The energy efficiencies of some of the alternative modes studied are shown in Figure 2.

TECHNICAL CHALLENGES

One of the major technical challenges for Polar Gas was to establish feasible methods for crossing the marine channels in the Arctic Islands. Very little technical data was available on such aspects as bathymetry, currents, ice technology and sea bed conditions at the marine crossings locations. Also, appropriate environmental and social data for northern regions was not available. The first step of the study program began in early 1973 and the study work centred on ice research in the Byam and Austin Channels. These channels were considered to be representative of conditions that might be encountered in the three other crossings contemplated on the initially selected route. Water depths in these crossings range up to 275 metres and ice conditions include both new and multi-year ice. (Figure 3)

As it was considered advantageous to use the ice as a working platform for pipeline construction, it was necessary to gather information on the strength of the ice, its stability and the time frame in which the ice could be depended upon. It was also necessary to gather information concerning the sea bottom profiles and soils information at cross-

ing locations to determine the suitability of installing a pipeline on
the bottom of the channels.

A potential threat to the submarine pipeline was chunks of ice islands
or ice ridges whose keels might traverse the pipeline routing. In order
to ensure that the pipeline would not be damaged by ice scour, it was
necessary to determine the source and migratory routes of ice island fragments and their maximum keel depths.

MARINE CROSSINGS RESEARCH

In the spring of 1973, the initial field ice research program began and a
base camp was established at Rea Point, Melville Island. Polar Gas modified a conventional pipeline ditcher and transported it to Byam Channel off
Melville Island. This modified ditcher cut 48 km of trench in one channel
at rates of up to 3.5 metres per minute, in ice up to 4.5 metres in thick-
nais. In order to determine the strength of the ice, large tanks were con-
stercted on the ice surface, filled with water and the deflection of the
ice measured. Tests were performed on various ice thicknesses to determine
the force that would cause initial cracking and the force that would result
in ultimate failure of the ice surface. Deflections were also measured over
time to determine the creep strength of the ice. This is shown on Figure 4.

In the summer of 1973, data collection was undertaken from a chartered
vessel, the "Percy M. Crosbie". The ship travelled some 27,000 km during
this period. Two smaller launches, designed and constructed specifically
for Polar Gas, were used to gather information in difficult areas and near
the shore line of the marine crossings. The vessels gathered data on channel
bottom conditions, currents and sub-bottom soil conditions.

In 1974, a more comprehensive field program was undertaken to obtain
channel bathymetry from the ice. Bottom profiles were taken initially at
300 metre intervals, and once a generally acceptable route had been established, tracked vehicles equipped with electronic and sonar devices with
suitable recording equipment followed the routing selected. A continuous
profile was established along the corridor. Additional information was
gathered on ice thickness, salinity and strength. As a method of strengthening the ice to serve as a working platform, areas were covered with sea
water and then allowed to freeze, thereby creating an artificially thickened ice platform. The strength of the ice was determined by loading methods
similar to those used for natural ice surfaces.

Cantilever tests were also undertaken to determine the short term and
long term strength of ice which would simulate equipment moving or standing
on the ice surface. The short term cantilever test is shown on Figure 5.

Polar Gas was able to determine early in its research that icebergs
would not cross the marine crossings of the proposed route, since glaciers
were north and east of the crossings and currents are generally in a south-
easterly direction. However, large pieces of ice or ice islands calved
from the main Polar Ice Shelf may move through the marine channels crossed
by the pipeline, and to assess the extent of this challenge, measurements
were taken to determine the maximum depth at which scour could occur. An
assessment was also made as to the frequency of calving of ice islands. Two
factors primarily influence the water depth at which ice scour may occur,
namely:

    (i)   the thickness of the parent ice shelf
   (ii)   the rate of ablation (loss of thickness)

It was determined from measurements that ice islands calve from the ice shelves with thicknesses up to 55 metres. Based on the relative densities of fresh water ice and sea water in the Archipelago, such an ice fragment would have a keel depth of some 48 metres immediately prior to calving. Allowing for a small amount of ablation to take place during its passage, it was concluded that the maximum probable water depth of scour from ice island fragments is 45 metres.

Calving from the ice shelf is not a common occurrence, as in the past forty years, there have been probably only two such events, but protection would have to be provided should ice islands pass the marine crossing locations in the future. It was decided, therefore, that the pipeline would require protection to 45 metre water depths to prevent potential damage from ice scour on the sea bed.

Two methods of protecting the pipe in the foreshore areas were considered:
    (i)   trenching and burying the pipe to the required depth
    (ii)   installing the pipeline in a tunnel below the sea bed
         in the shore approach zone.

It was concluded that the tunnel scheme would be more economical and assure protection to the pipeline in the shore approach area. The method proposed is shown in Figure 6.

During the period of field engineering investigations, environmental scientists were on site to observe the research activities and to report on the environmental factors that should be considered in the design and construction of marine crossings, and also to make recommendations for future study programs that would be required.

MARINE CROSSING CONSTRUCTION TECHNIQUES

Continuing research involved the study of a variety of techniques and specialized equipment designs that would be required to construct the pipeline across these marine channels, working from an ice surface or from the water in the short open water season.

The equipment to lay pipe in the winter from the ice was not available, however, the techniques of marine pipelaying from the sea surface by laybarge or pullships, or shore-to-shore bottom pull systems for river crossings and relatively short marine crossings are well known and proven. (Figures 7 and 8)

Ice Island Bottom Pull
For the longer crossings where open water cannot be counted on for pipeline installation, one of the first proposals made by Polar Gas' consultants was the ice island bottom pull method, which utilizes the principles of bottom pull but from an artificially thickened ice island. This method shown in Figure 9, entails the cutting of a trench in the ice through which the cable would be lowered. From a make-up yard on shore, strings of 914 mm diameter pipe, approximately 10 km long, would be pulled by a steel cable 8 cm in diameter along the channel bottom. At the pulling site, the ice would be artificially thickened to form an ice island strong enough to support the equipment and resist the anchoring forces. A pull sled, running along the trench on skids would be equipped with a buoyant fin extended into the sea water to reduce the load on the ice surface. This sled would be connected by cable to the pipe string and pulled

along the ice by the main cables attached to the gripper jacks and main
winches. This configuration would reduce the downward force on the ice
to lower values, and consequently allow the sled to move over weaker ice
surfaces that may be encountered along the route.

Other Construction Methods

Several other methods were proposed and studied, such as utilizing an
ice-strengthened pull ship or pull barge in those areas where open water
was available during the summer months. For the channels requiring off-
the-ice construction techniques, several variations of the ice-island
bottom pull were studied, such as an under-ice pull and lowering technique
through ice holes, which eliminates the need for a continuous ice trench.

MARINE CROSSINGS ON THE ORIGINAL ROUTE

There are six marine crossings on this route from the Arctic Islands with
channel widths and depths as follows:

East Barrow Strait
58.4 km with a maximum water depth of 180 metres.

Pullen Strait
4.2 km in width with a maximum water depth of 50 metres.

Crozier Strait East
6.4 km in width, and this crossing contains the maximum water depth of
the Polar Gas marine crossings at 322 metres.

Crozier Strait West
3.3 km in width with a maximum water depth of 62 metres.

Austin Channel
39.1 km in width with a maximum depth of 275 metres.

Byam Channel
32 km in width with a maximum depth of 171 metres.

East Barrow Strait is the channel which experiences the most open water
and where winter ice cover cannot be depended upon. For these reasons,
it was decided to build this crossing during the summer season, using a
water-borne construction method. Third generation laybarges are available
with the design capability for laying 914 mm pipes and larger at depths
in excess of the deepest sections of this channel.
    Pullen Strait is ideally suited for the construction of a full-length
tunnel by conventional tunnelling techniques. Since shore approach tunnels
are proposed on all channel crossings, these tunnels would have extended
across 80 per cent of Pullen Strait; and therefore, to complete the crossing it was necessary to merely plan on extending the tunnel lengths for
the remaining 20 per cent of the crossing.
    Crozier Strait East was designed to utilize a shore-to-shore construc-
tion technique. As the name implies, shore-to-shore bottom pull consists
of pulling one length of pipe along the bottom of the channel with the
pulling device on the opposite shore.

The relatively very short channel of Crozier Strait West will be
crossed with a full-length tunnel for the reasons described for Pullen
Strait.

The open water season in Austin Channel is generally too short to
enable pipeline installation to be undertaken from the water; however, a
stable ice cover does exist in the channel every year. As a result, Polar
Gas engineers decided to capitalize on this thick stable ice platform by
developing a "pull from the ice" technique, in which winches are mounted
on the ice surface at suitable locations across the channel. This technique is illustrated in Figure 9.

Byam Channel has similar ice conditions to those in Austin Channel and
the "pull from the ice" technique is also proposed for this crossing.

Land Routes

A variety of land routes were considered for this Polar Gas pipeline, but
late in 1977 application was made to the National Energy Board of Canada and
the Department of Indian Affairs and Northern Development to construct a
pipeline down the west coast of Hudson Bay, as illustrated in Figure 10.

MARINE CROSSINGS FOR A COMBINED SYSTEM

While the single line down the west side of Hudson Bay remains a feasible
option, Polar Gas has indicated that its current plan is to file a new combined system application. This would allow for the connection of two sources
of natural gas, the Sverdrup Basin in the Arctic Islands and the Mackenzie
Delta in a combined pipeline system referred to as the "Y" Line. Two
examples of a "Y" Line are shown in Figure 11.

The "Y" Line concept for alternative routes has been considered feasible
in the light of recent advances in the technique of laying pipelines in
deeper waters, which would make possible the crossing of M'Clure Strait,
which is 120 km wide and 500 m deep. One other crossing is involved on the
"Y" Line, Dolphin and Union Strait, which is more typical of those crossings
previously surveyed, being 30 km wide with a maximum depth of 110 metres.

M'CLURE STRAIT STUDIES

In early 1979 a major field investigation was carried out on M'Clure Strait
in order to establish the physical characteristics of the Strait in sufficient detail to confirm that it was technically feasible to construct a submarine pipeline in this channel. The previous studies had already shown that
it was possible to use the ice surface as a working platform.

The field work indicated that the bottom profile and soil conditions
presented no problems, having a comparatively gentle contour from shore to
shore, and that microtopography of the sea bottom is relatively smooth.
The soil conditions vary from mud overlying laminated clay on the south side
to a thick silt and sand combination on the north.

Currents, reversing with tidal influences, are relatively low, the
maximum being in the range of .7 km per hour.

Severe weather conditions were encountered during the surveys, with
whiteouts, fog and temperatures down to minus 35 degress Celsius and wind
forces up to 120 km per hour. Any on-ice construction operations would
therefore have to be fully enclosed and be provided with complete life

support systems capable of providing a safe environment for working personnel.

The ice conditions between January and May can provide a competent
working base. The ice thickness varies from 2 m to 12 m with the average
thickness being about 3 metres. Because of the severe surface roughness,
together with the unusual thickness, an alternative construction method to
ice island bottom pull was devised.

The new system proposed is the ice hole method, as illustrated in
Figures 12 and 13. Through a series of holes cut in the ice along the route,
spaced about 2 km apart, a cable system would be attached to the pipe, which
would be pulled from the shore in six strings each about 19 km long, by
specially designed gripper winches located over an opening in the ice.
These pull units, operating in tandem are equipped with 150 ton pull capability each, providing a total of 300 ton force. The final tie-in of the
six strings would be welded either on the surface of the ice sheet or on the
sea bed. Welding on the channel bottom could utilize either the one-
atmosphere pressure or the hyperbaric method. A selection of the method
will be determined following further research and tests.

The surface weld tie-in method has been successfully undertaken in the
North Sea, and studies by Polar Gas' consultants indicate that it would be
technically feasible from the ice surfaces in the Arctic channels. Overlapping strings would be raised vertically to the ice surface by winches,
the pipe ends welded and then lowered to the sea bed in a curving configuration, having a radius equal to the water depth from the underside of the
ice. (Figure 14)

The one-atmosphere welding technique is undertaken inside a chamber on
the sea bottom (Figure 15), where the pipe ends are enclosed by a clam
shell mechanism, de-watered, and a welder is lowered in a small submersible
to perform the weld and complete the tie-in operation.

The hyperbaric technique has been used for many years, but because of
the depth of M'Clure Strait (500 metres), the decompression time required
for welders would be almost two weeks, which lengthens the construction
period required considerably. This method is therefore considered less
appropriate for the deep water tie-ins.

DOLPHIN AND UNION STRAIT STUDIES

In the summer of 1979, field studies were undertaken on the Dolphin and
Union Strait, which is 30 km wide and 110 m deep. This channel is normally
ice free in August and September every year, and favours the lay barge or
pull method of pipe laying.

The survey consisted of obtaining bottom profiles, current readings,
bottom core samples and all of the additional information outlined for
the other channels.

The survey vessel was taken by barge from the Mackenzie Delta area
and all other equipment was trans-shipped by air to Lady Franklin Point,
Victoria Island on the north side of the channel.

LAND PIPELINE

The majority of the land pipeline will be constructed by conventional
methods; however, about one-third to one-half of the line traverses continuous permafrost areas. Here special precautions are necessary and
the gas will be cooled to prevent thawing and possible settlement of the
pipe. In the discontinuous permafrost zone, the gas flowing in the pipeline will be kept above zero Celsius, and the settlement will be controlled
where necessary to prevent damage to the pipeline.

The land portion of the Polar Gas pipeline as proposed is a 1 067 mm
pipe operated at 11 583.2 kPa. The right-of-way required for construction
of this pipeline will be approximately 40 metres in width.

Studies undertaken pertaining to construction include computing
thickness of construction pads necessary to insulate permafrost terrain;
location of compressor station, airstrips, access roads, construction
sites and borrow pits from which needed sand or rock can safely be removed;
design and construction planning for drainage and erosion control. Where
necessary, the land will be recontoured to prevent erosion and drainage
will be assisted by installation of appropriate systems.

LOGISTICS

The logistics required for a project of this magnitude are considerable.
Approximately two million tonnes of steel pipe will be transported to
staging sites. The project will also require one million tonnes of other
materials and one million tonnes of fuel.

Staging sites and transfer points for movement of materials and equipment will be established before construction commences. Most of the
materials will be moved by water transport or rail with smaller tonnages
being transported by air when necessary.

OPERATIONS AND MAINTENANCE

During the life of the pipeline, the primary elements to be operated and
maintained are: the pipeline, the compression, refrigeration and metering
systems, gas control centre and the satellite-based communications network.
Backup support will be provided by trained personnel at divisional headquarters.

Specialized equipment for the inspection, maintenance and, if necessary, repair of the pipeline will be available. Regular aerial patrols
of the line will be part of the inspection routine. In addition, inspection of marine crossings will be a regular operational requirement.

The compressor stations and gas refrigeration facilities will be
fueled by natural gas. These stations will provide storage for materials
as well as fuel and accommodation for personnel; they will be self-sufficient
in terms of electric power, water supply and waste disposal. Heavy equipment for maintenance and repairs will be kept at selected compressor stations, and all stations will have year-round airstrips.

PEOPLE

The Polar Gas route passes through some of the most sparsely populated
areas in Canada; however, the areas in the Northwest Territories are
undergoing varying degrees of cultural, social and economic change.
These changes are not entirely of recent origin nor the result of modern,
industrial development. They are a continuation of changes which began
with the arrival of whalers, fur traders and missionaries and have accelerated in the past few decades with the systematic provision by government
of various services, including health care, education, housing and social
assistance.

Polar Gas recognizes that the construction and operation of the pipeline could affect this process of change in a number of ways and has tried
to develop a flexible approach in meeting the concerns of local people.
The Project is attempting to work with local people, communities, and
government to ensure that changes likely to arise from its construction
and operation are understood well in advance, and that appropriate programs and policies are developed prior to construction which will mitigate
the impact of such change.

ENVIRONMENT

Throughout its planning and design, Polar Gas has been concerned about the
environment through which the pipeline passes. Environmental scientists
have been studying the environment along the routes since the earliest
planning stages. They have identified locations along the route as areas
of special concern because of important environmental or land use features
and, as a result of their work, the proposed location of the original route
and its associated facilities was re-routed to reduce or eliminate the
environmental impact of the pipeline.

In addition to relocation of the pipeline route and facilities, a
number of other measures have been developed to minimize environmental
impacts. These include specific restrictions and controls to reduce disruption of terrain and vegetation, to minimize disturbance of mammals, fish
and birds in their habitats, to reduce pollution, and to minimize effects
on existing uses of land and water. Monitoring programs are planned to
determine the effects of the project on the environment. Most of the
environmental impacts expected will occur during the construction of the
pipeline and are short-term and minor.

SUMMARY

Construction of a Polar Gas pipeline will offer significant benefits to
Canadians and to the Canadian economy. It will transport the large volumes of natural gas discovered in the Arctic and will ensure continued
exploration activity to tap the large potential reserves of these Canadian
Arctic frontier areas. The channel crossings in the Arctic Islands are
a significant part of the challenge to be faced. However, the methods
proposed by Polar Gas in constructing a pipeline from these frontier areas,
and the research and investigations undertaken by Polar Gas in developing
construction techniques for these marine crossings has reduced the challenge to one of manageable proportions.

The Project has been able to adapt the high level of existing technology developed in other parts of the world to the unique conditions of the Canadian Arctic, and, in so doing, ensure the technical feasibility of the pipeline. The Project will make it possible to connect a major Canadian energy source in the Arctic to meet a forecast deficiency in the 1980's and will reduce Canada's dependence upon foreign energy sources.

270

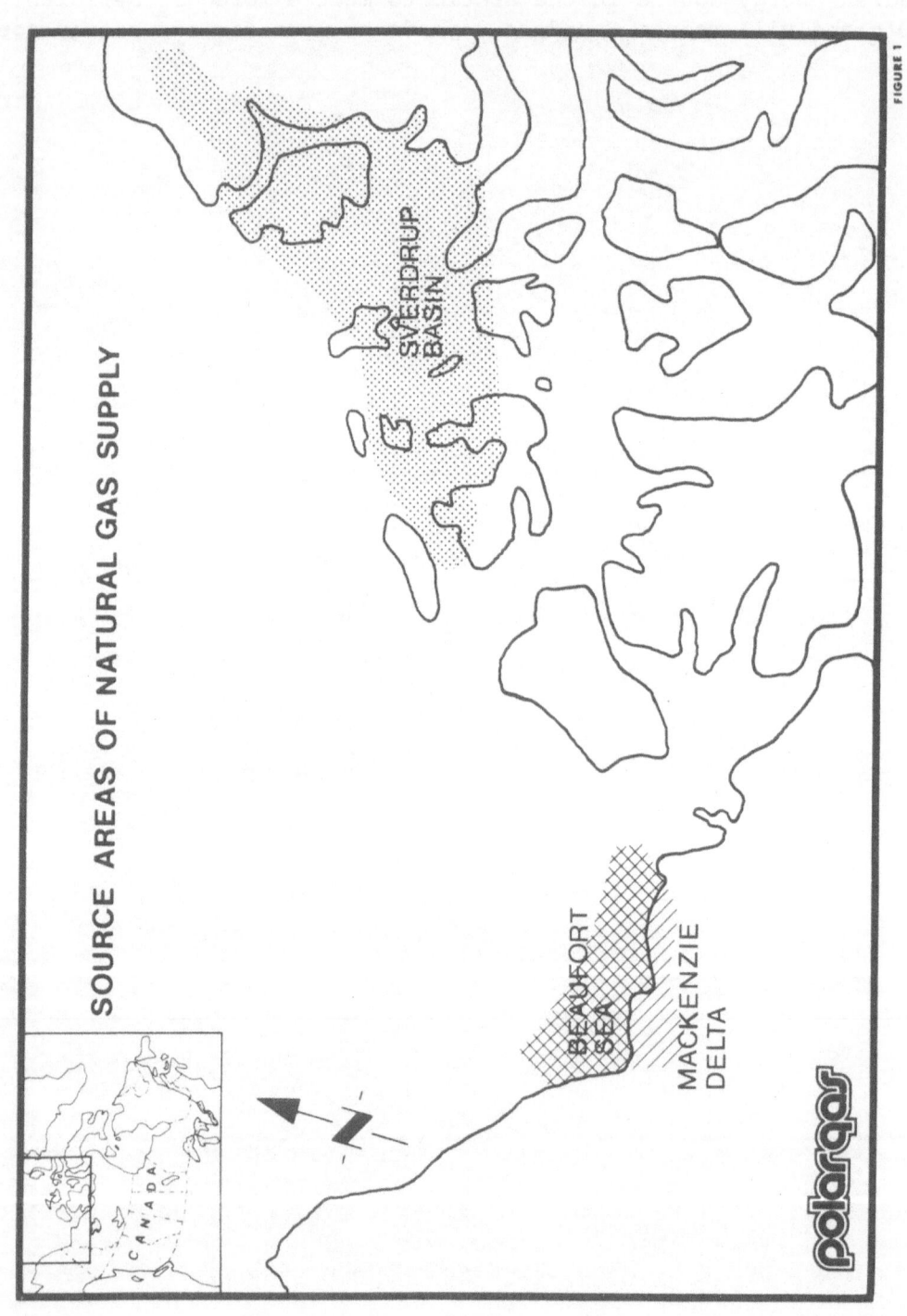

SOURCE AREAS OF NATURAL GAS SUPPLY

SVERDRUP BASIN

BEAUFORT SEA

MACKENZIE DELTA

CANADA

polargas

FIGURE 1

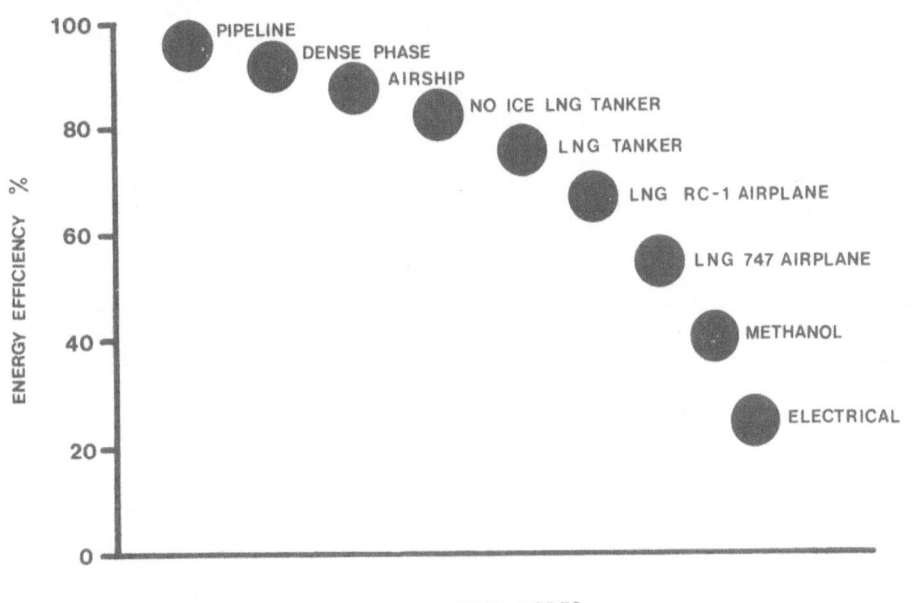

ENERGY EFFICIENCY OF ALTERNATE MODES
OF TRANSPORTING ARCTIC ISLANDS GAS

Figure 2

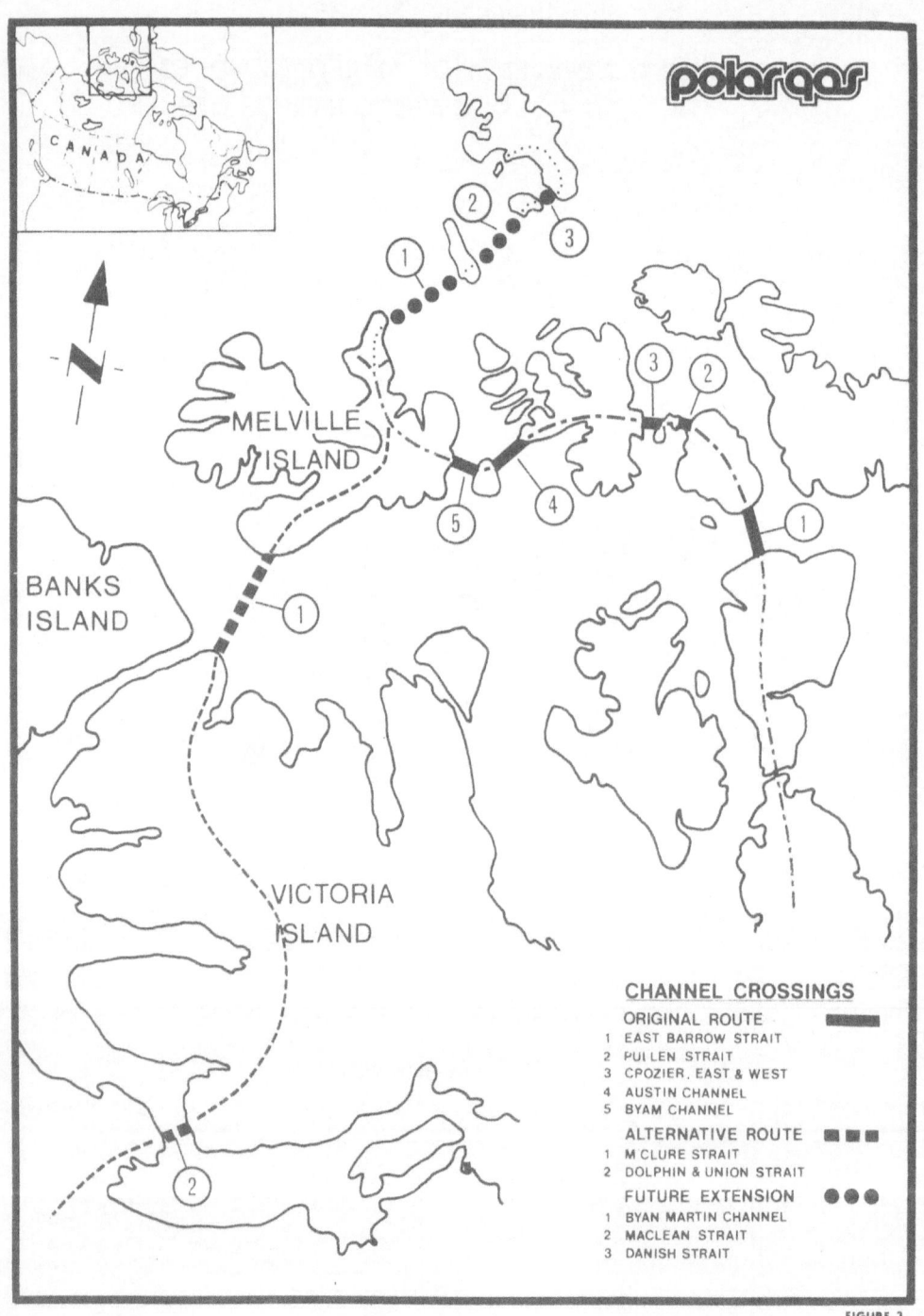

CHANNEL CROSSINGS

ORIGINAL ROUTE
1 EAST BARROW STRAIT
2 PULLEN STRAIT
3 CPOZIER, EAST & WEST
4 AUSTIN CHANNEL
5 BYAM CHANNEL

ALTERNATIVE ROUTE
1 M CLURE STRAIT
2 DOLPHIN & UNION STRAIT

FUTURE EXTENSION
1 BYAN MARTIN CHANNEL
2 MACLEAN STRAIT
3 DANISH STRAIT

FIGURE 3

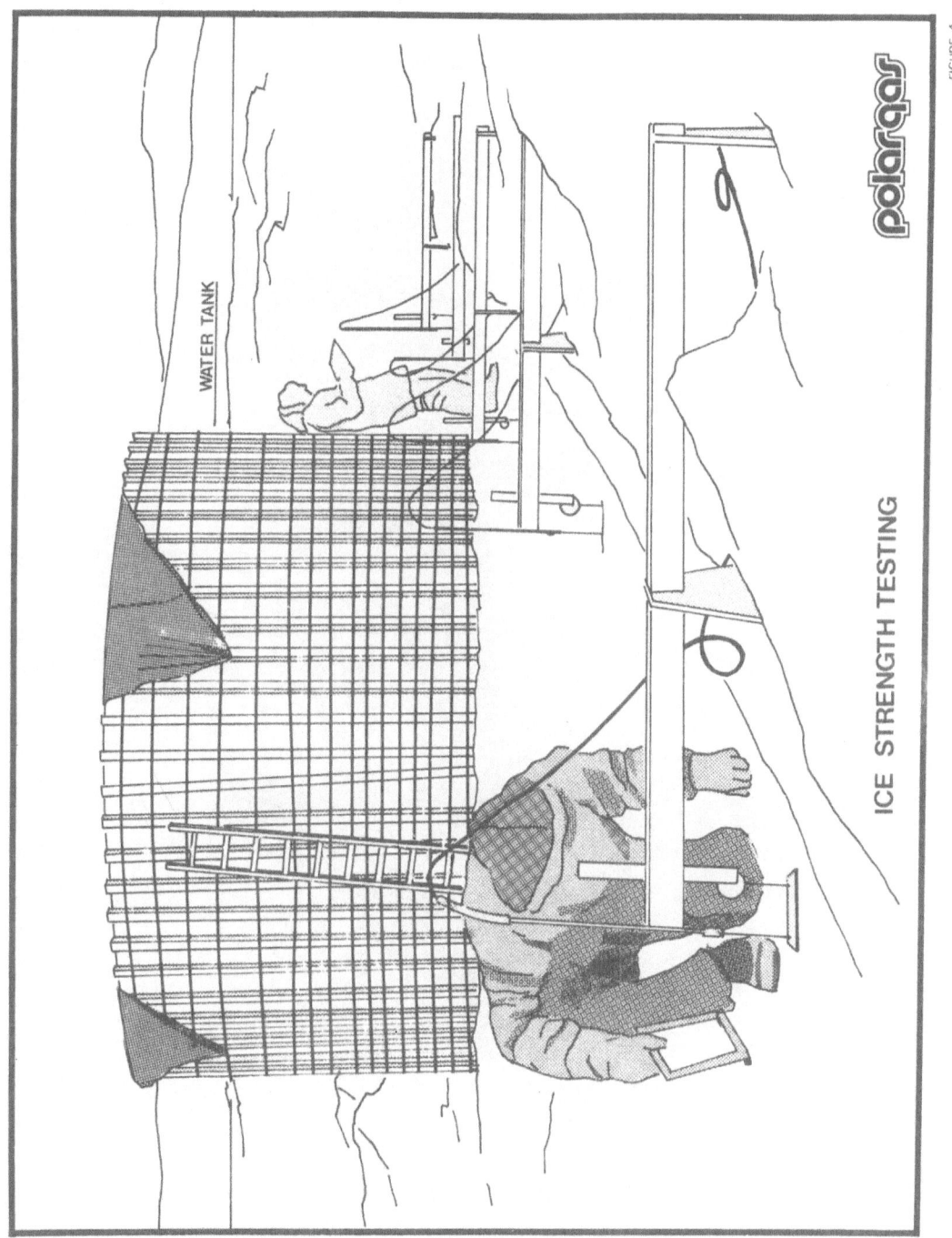

WATER TANK

ICE STRENGTH TESTING

polargas

FIGURE 4

CANTILEVER TESTING

polargas

FIGURE 5

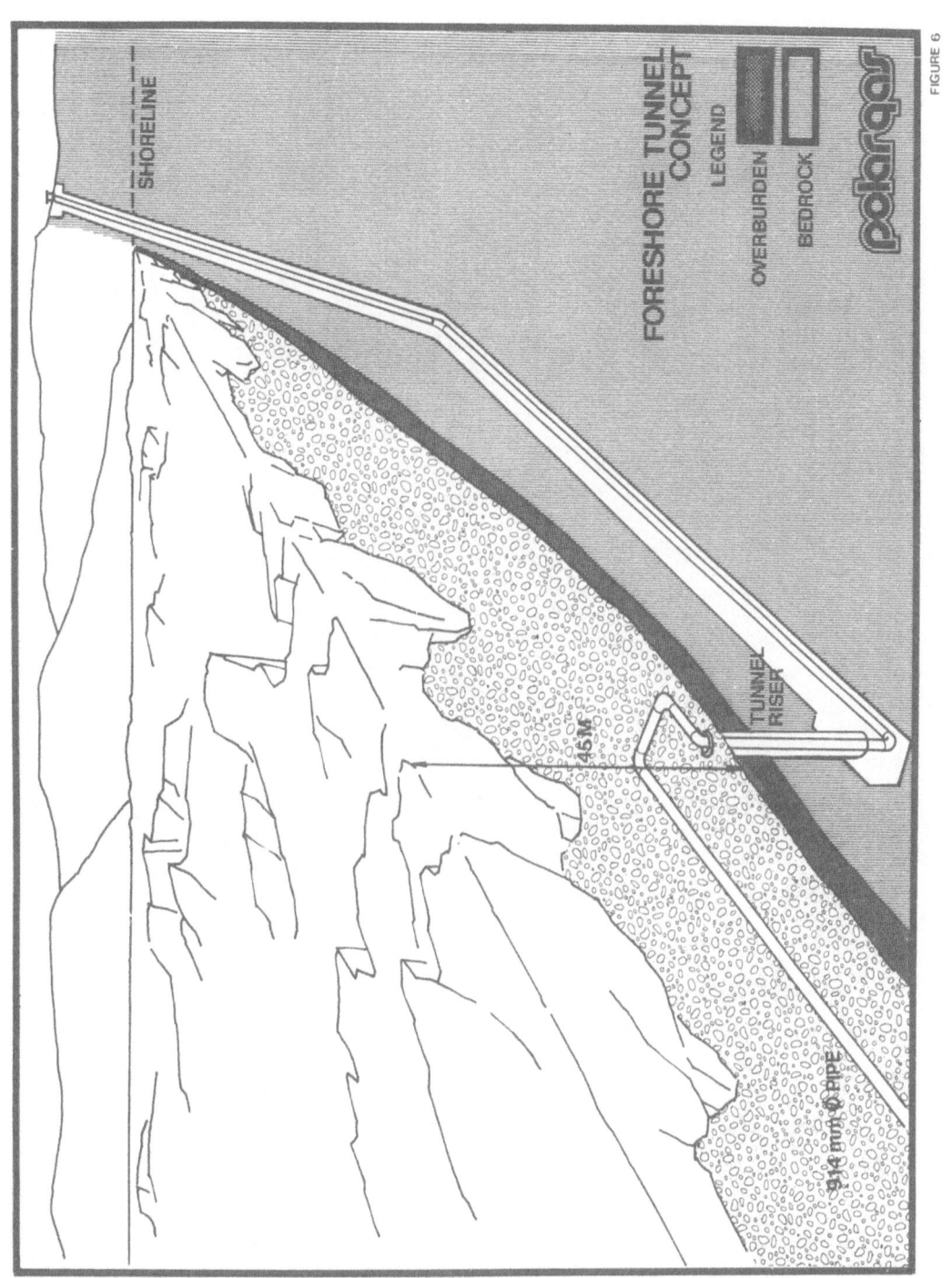

FORESHORE TUNNEL CONCEPT

polargas

LEGEND

OVERBURDEN

BEDROCK

SHORELINE

TUNNEL RISER

45 M

914 mm Ø PIPE

FIGURE 6

276

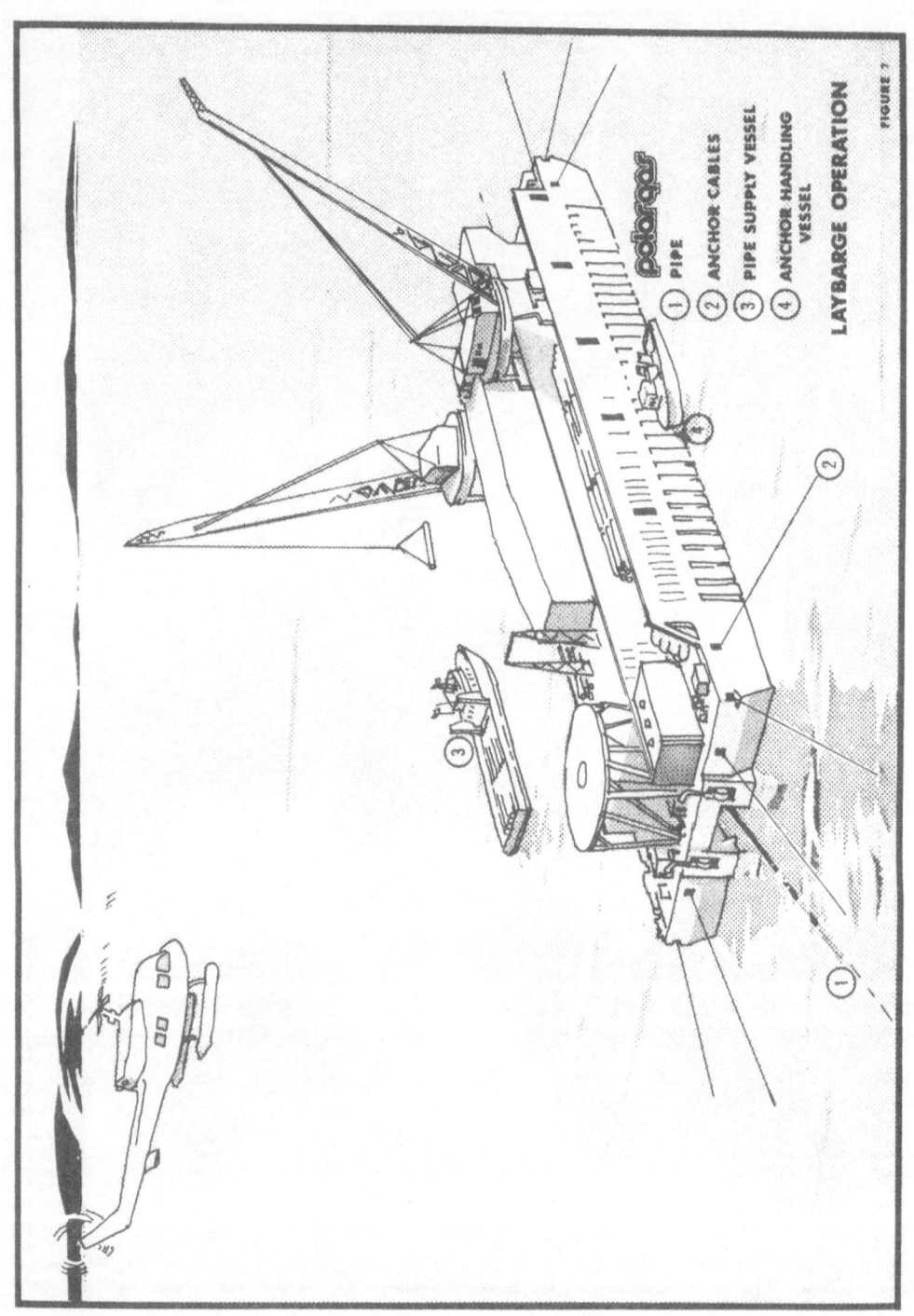

1 PIPE
2 ANCHOR CABLES
3 PIPE SUPPLY VESSEL
4 ANCHOR HANDLING
     VESSEL

LAYBARGE OPERATION

FIGURE 7

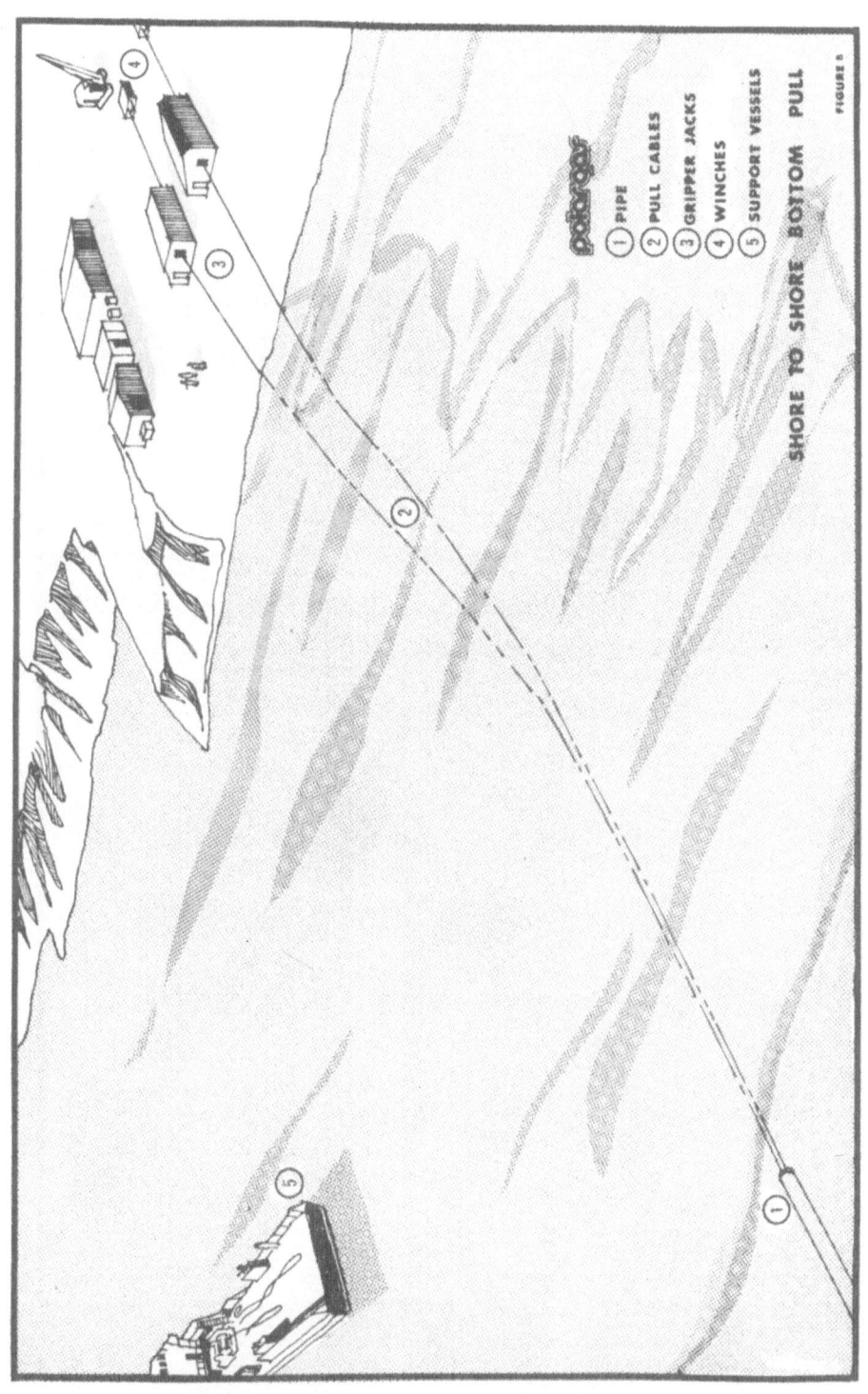

pipelays
1. PIPE
2. PULL CABLES
3. GRIPPER JACKS
4. WINCHES
5. SUPPORT VESSELS

SHORE TO SHORE BOTTOM PULL

FIGURE 5

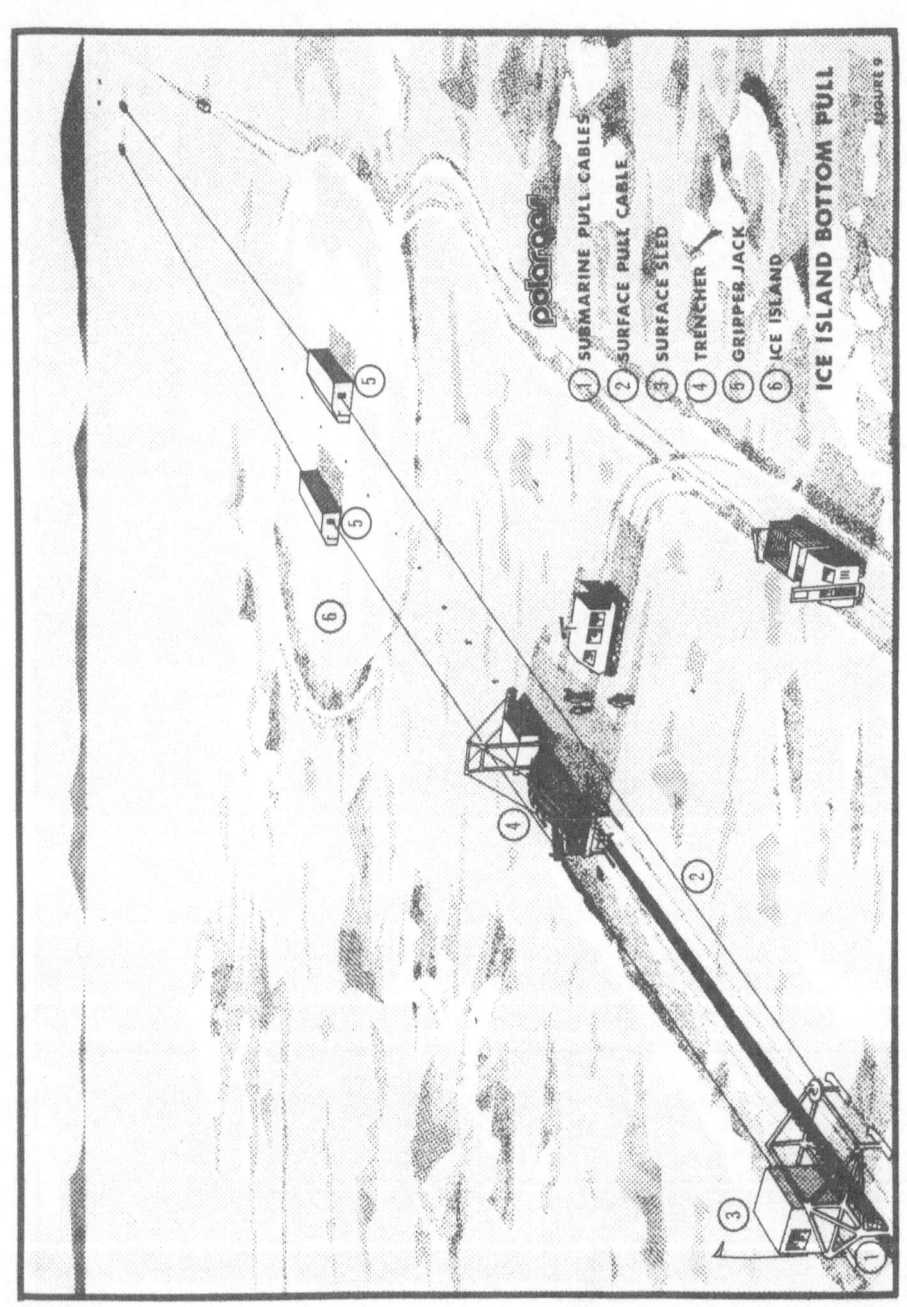

polaroof

1 SUBMARINE PULL CABLES
2 SURFACE PULL CABLE
3 SURFACE SLED
4 TRENCHER
5 GRIPPER JACK
6 ICE ISLAND

ICE ISLAND BOTTOM PULL

FIGURE 8

279

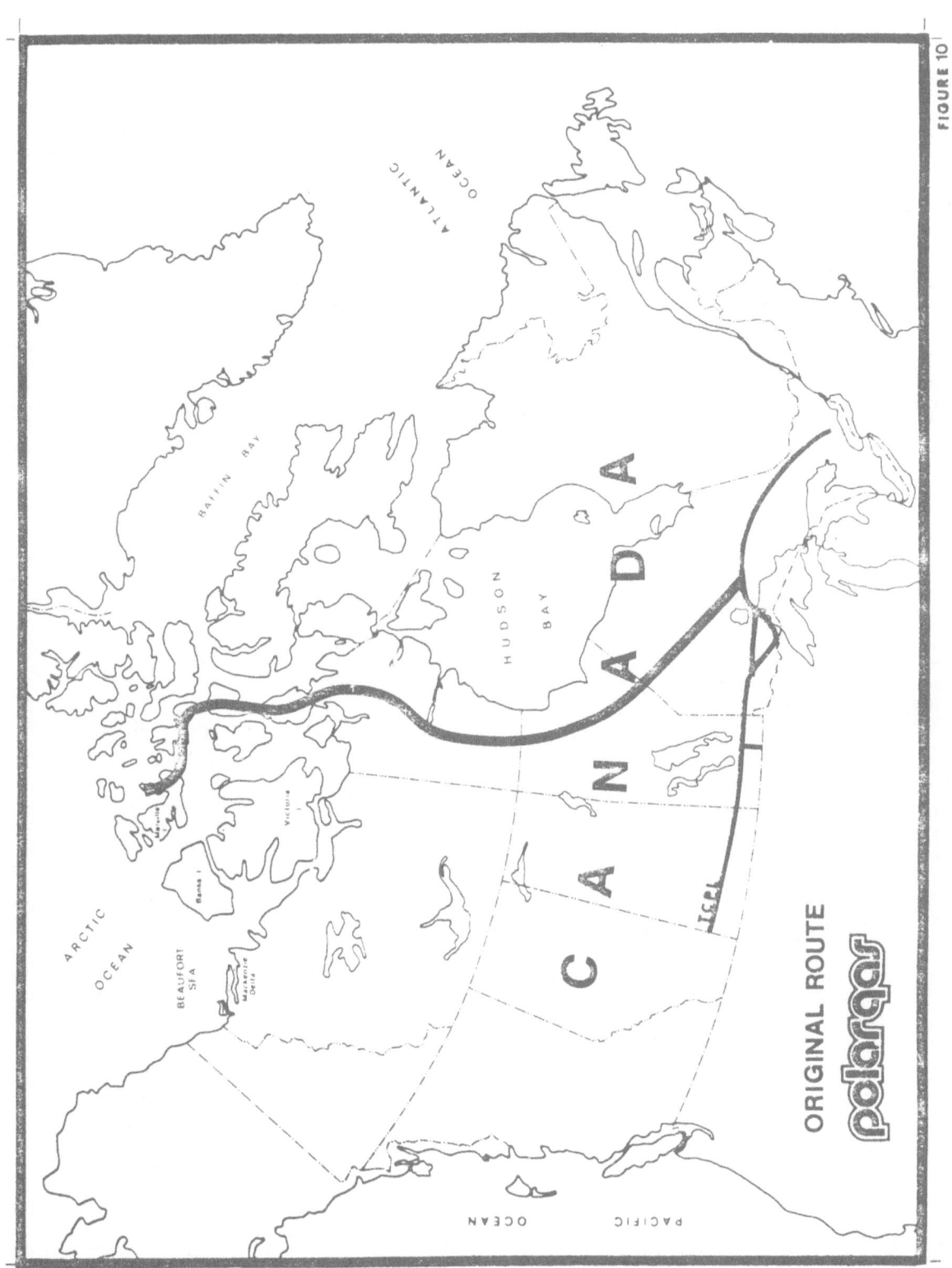

ORIGINAL ROUTE

polargas

FIGURE 10

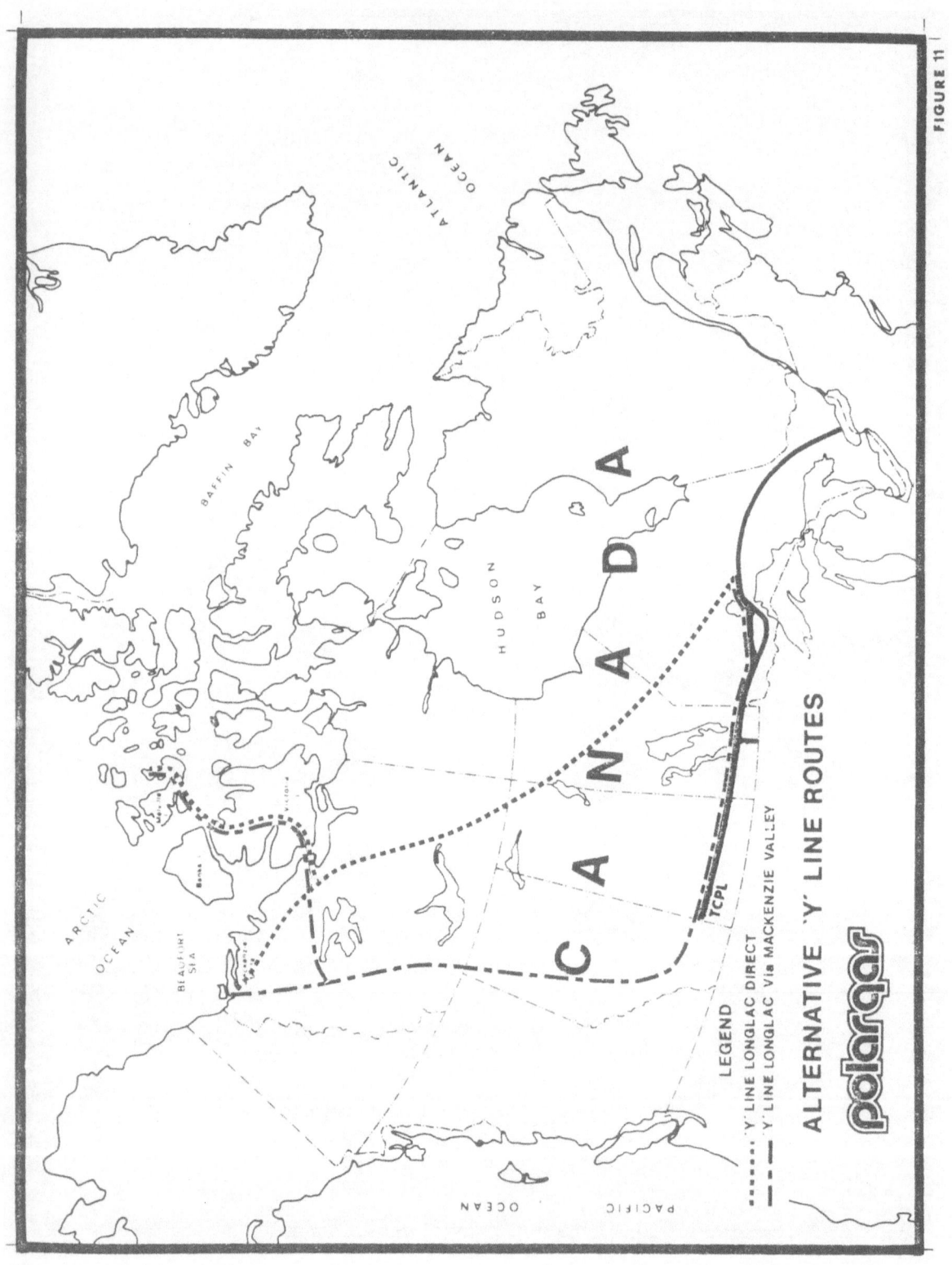

LEGEND

'Y' LINE LONGLAC DIRECT
'Y' LINE LONGLAC via MACKENZIE VALLEY

ALTERNATIVE 'Y' LINE ROUTES

polargas

FIGURE 11

ICE HOLE BOTTOM PULL CONCEPT

polargas

FIGURE 12

ICE HOLE

914 mm Ø PIPE
ON SEA BOTTOM

282

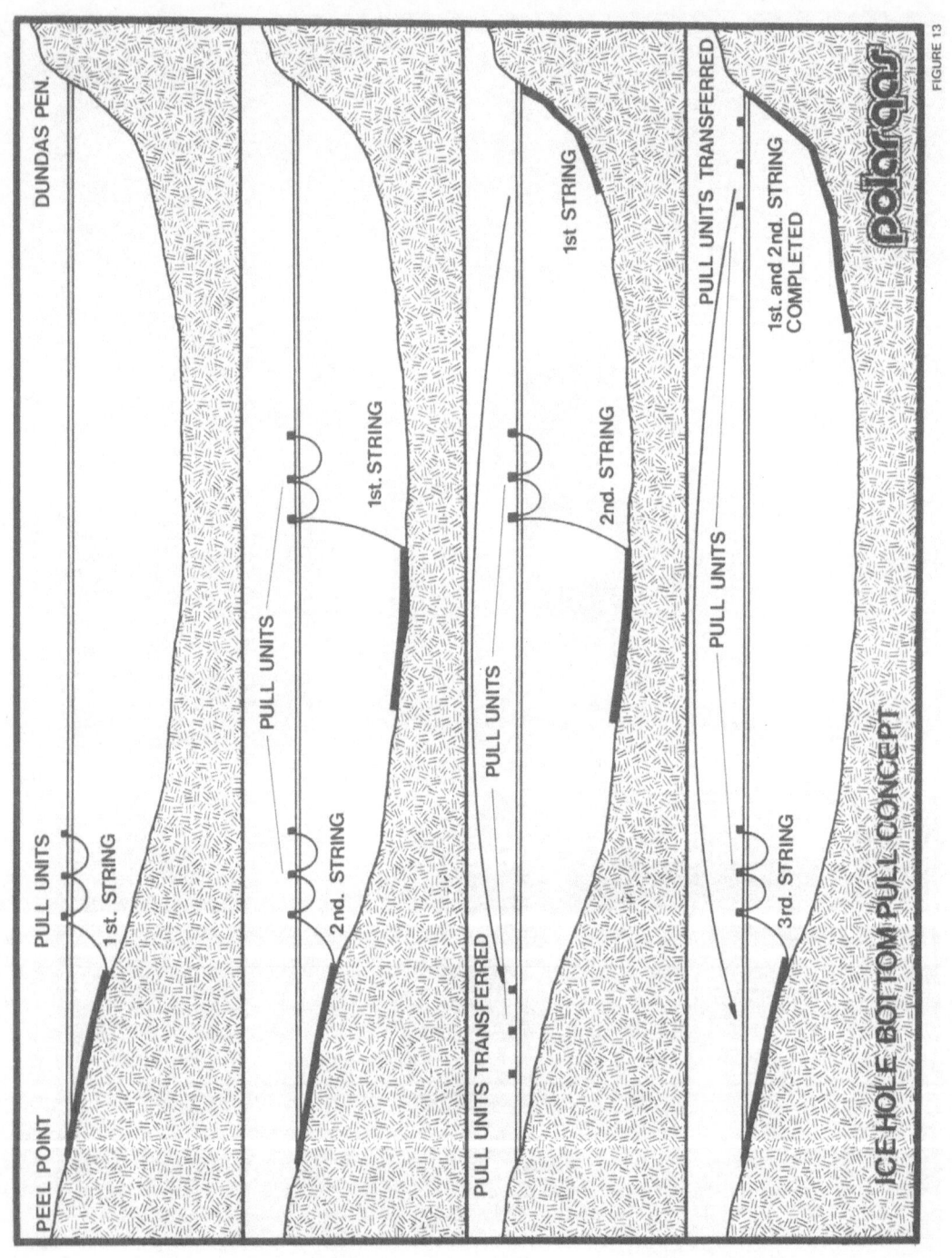

PEEL POINT   PULL UNITS   DUNDAS PEN.

1st. STRING

PULL UNITS

1st. STRING

2nd. STRING

PULL UNITS TRANSFERRED

PULL UNITS

2nd. STRING

1st STRING

PULL UNITS TRANSFERRED

PULL UNITS

3rd. STRING

1st. and 2nd. STRING COMPLETED

ICE HOLE BOTTOM PULL CONCEPT

polargas

FIGURE 13

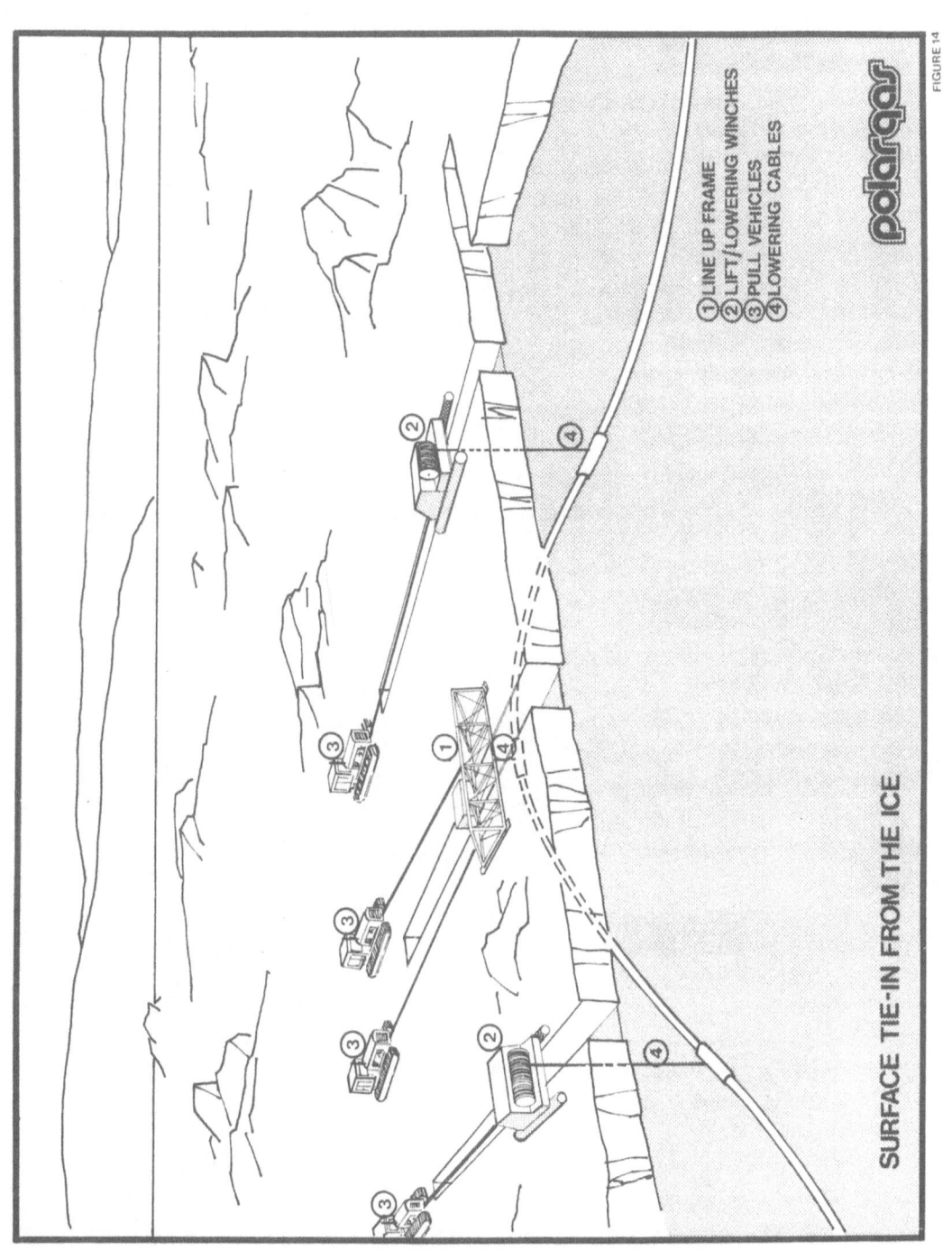

SURFACE TIE-IN FROM THE ICE

① LINE UP FRAME
② LIFT/LOWERING WINCHES
③ PULL VEHICLES
④ LOWERING CABLES

polargas

FIGURE 14

284

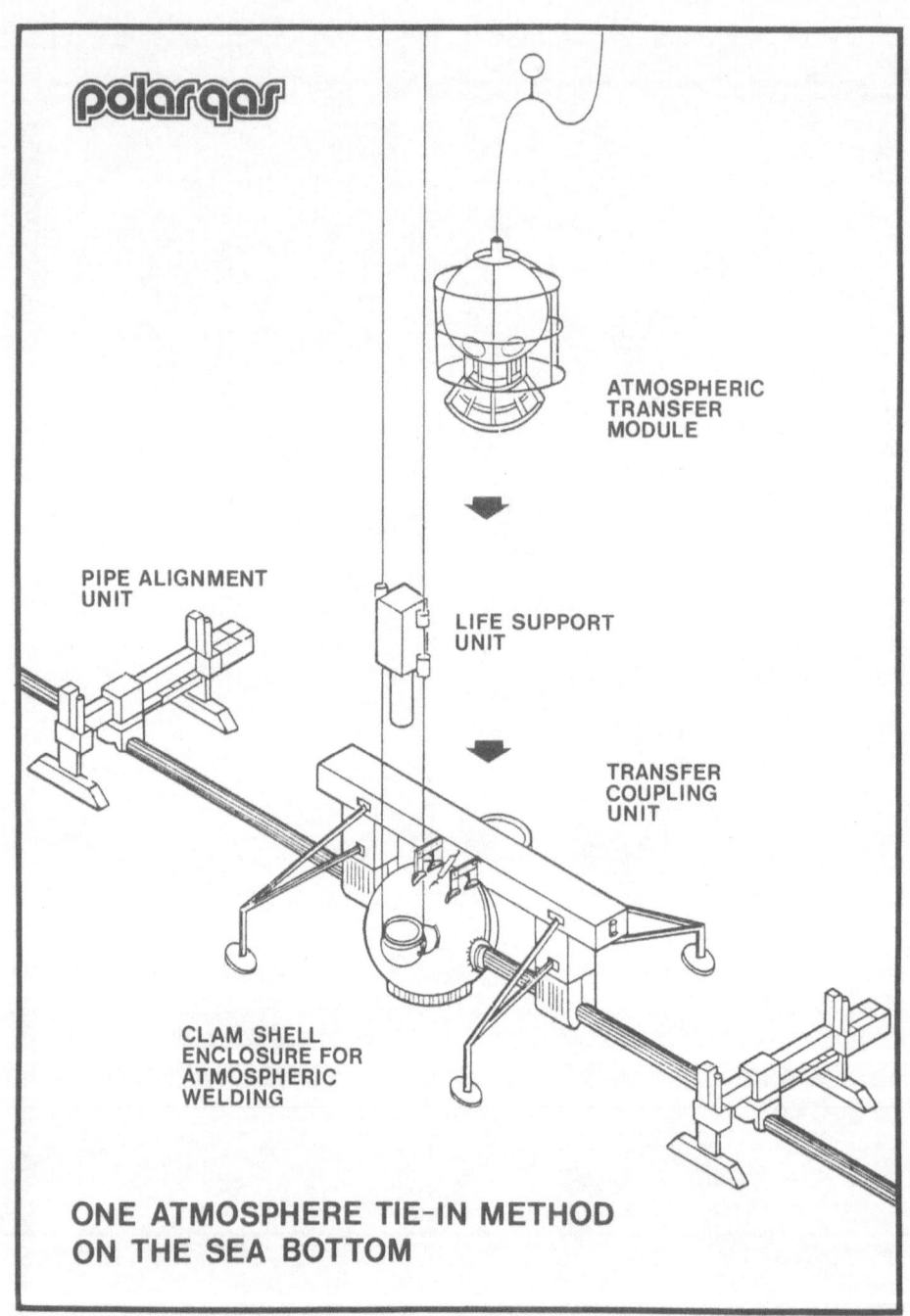

polarqas

ATMOSPHERIC
TRANSFER
MODULE

PIPE ALIGNMENT
UNIT

LIFE SUPPORT
UNIT

TRANSFER
COUPLING
UNIT

CLAM SHELL
ENCLOSURE FOR
ATMOSPHERIC
WELDING

ONE ATMOSPHERE TIE-IN METHOD
ON THE SEA BOTTOM

FIGURE 15

CONSTRAINTS OF GEOLOGIC PROCESSES ON PETROLEUM DEVELOPMENT IN THE WESTERN
BEAUFORT SEA

Arthur Grantz and David A. Dinter
United States Geological Survey
Menlo Park, California 94025

ABSTRACT

Active progradation, Holocene tectonism, and frigid temperatures have created
geologic conditions that uniquely complicate petroleum development in the
Beaufort Sea. Although most of the western Beaufort shelf is ice free and
accessible to drill ships for up to two months during favorable years, drilling will still be hampered by troublesome subsea conditions in large areas.
Natural gas hydrate with a zone of free gas at its base 300 to 700 m beneath
the seafloor is widespread where water is deeper than 400 to 600 m. Shale
diapirs, presumably overpressured, disrupt the continental slope east of 147°
W long. Shallow, low-angle bedding-plane slides as wide as 38 km underlie
most of the outer shelf. The slides are sufficiently young to preserve open
crevasses as deep as 17 m (exceptionally 37 m). Subsea permafrost, commonly
containing gas pockets, extends from near the sea-bottom to depths of several
hundred meters in sediments near the coast and may present thawing and com-
paction hazards to drillholes on the inner shelf.

The westward-drifting polar ice pack, which lies seaward of a zone of
shorefast and bottom-fast ice inside the 10- to 20-m isobath, will be a
major obstacle to production activities. In addition, the seabed between the
coastline and at least the 60-m isobath is gouged by keels of drifting ice.
Subbottom completions in this zone, and pipelines crossing it, will have to
be buried below the gouges, which locally exceed 5 m in depth. East of 146°
W long. production facilities may be subject to earthquakes as large as magnitude 6 and in places to active faulting, uplift, and subsidence.

LOCATION, PETROLEUM DEVELOPMENT AND ENVIRONMENTAL CONCERNS

The continental shelf and upper slope of the western Beaufort Sea north of
Alaska (Fig. 1) is probably the most promising untested terrane for petroleum
in the United States, but environmentally one of the most difficult to deal
with. The area is underlain by a minimum of 1 km and a maximum of more than
8 km of sedimentary strata that are prospective for petroleum (Fig. 2). Representatives of every geological system from the Mississippian to the Tertiary
are present, although pre-Jurassic beds occur only near shore and are probably
absent beneath the eastern third of the shelf. Ten of the thirteen major geologic units into which the stratigraphic section in adjacent northern Alaska
is divided contain strong shows or commercial pools of oil and gas. As of the
end of 1979, a giant and a supergiant oil field, a small producing gas field,
six undeveloped oil discoveries of unknown or unannounced economic potential,
and several oil and gas seeps are known to lie within 10 km of the coast.
Petroleum deposits of probable commercial size have been discovered in the
Canadian sector of the Beaufort Sea in strata that trend toward the western
Beaufort shelf. Seismic reflection profiles demonstrate the presence of many
structural and stratigraphic configurations within the sedimentary section that
are potentially favorable for trapping oil and gas. These factors suggest that
the Beaufort shelf has excellent prospects for large petroleum deposits.

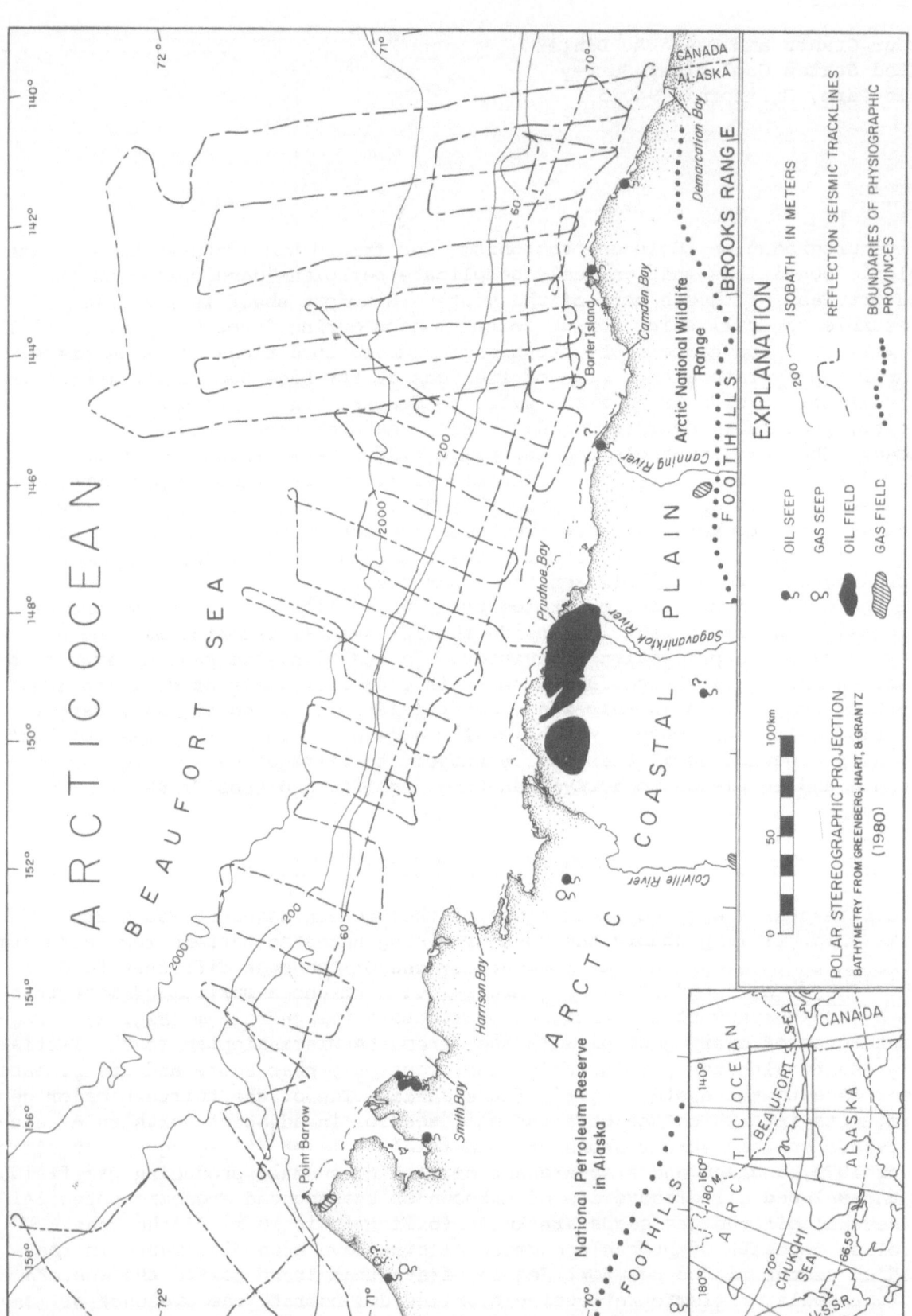

Figure 1. Bathymetric map of the continental shelf, slope, and rise and seismic trackline coverage of the Beaufort Sea north of Alaska.

Oil and gas leases for the nearshore zone of the central part of the western Beaufort shelf were sold in December, 1979. Another possible lease sale, initially including the entire Beaufort shelf out to the 200-m isobath, is tentatively set for February, 1983. Petroleum exploration of this area can be accomplished using existing methods during the summer months of years with favorable ice conditions. Year-round operations to develop discoveries, however, will be constrained by the remoteness and frigid climate of the region, the polar icepack, and   various geologic factors. The successful development of large reserves of oil and gas in northern Alaska has demonstrated that the petroleum industry can cope with Arctic cold and remoteness, and innovative designs are being developed for production platforms suitable for seas usually covered by polar pack ice. Preoccupation with these obvious, major impediments, however, has perhaps detracted from the attention that eventually must be paid to a number of geologic processes and conditions which will also constrain development. The present report, based on a U.S. Geological Survey open-file report (Ref. 1), summarizes our current knowledge of these processes and conditions.

Active geologic and oceanographic processes, some indigenous to the western Beaufort Sea, and others to all polar seas, will make exploratory drilling and the establishment and maintenance of production facilities more difficult there than in more southerly seas. West of Camden Bay (Fig. 2) the structure of the Beaufort shelf is dominated by gently seaward dipping strata broken by down-to-the-north gravity faults, some of which are growth faults that have been active in Quaternary time. Similar faults also occur in the outer shelf east of Camden Bay. The inner shelf east of Camden Bay is underlain by large folds, in places actively growing and seismogenic, developed in Neogene and Quaternary beds. Surficial hazards include seabed faulting and earthquakes, submarine sliding, unconsolidated sediments, storm surges, and the drifting polar ice-pack. Shallow gas, permafrost, and over-pressured shales pose additional drilling and development problems.

The environmental impact of oil spills may be especially acute in the western Beaufort Sea. Low temperatures and the usual presence of pack ice will retard degradation of spilled oil and render clean-up difficult, and often impractical. Innovative structures with large safety factors will be necessary to recover oil and gas from this area without undue risk to the work force or the environment.

GEOHAZARDS CONSTRAINING PETROLEUM DEVELOPMENT

Polar ice pack and gouging of the seabed

The western Beaufort Sea is almost completely covered by ice for about nine months of the year, and even during the nominal open season (mid-July to early October) scattered to solid pack ice is commonly present. "Freeze-up" in shallow water along the coast generally begins in late September or early October, and by May the seasonal shorefast ice is about 2 m thick. Winter ice conditions and areas of seabed ice gouging in the western Beaufort Sea are shown in Fig. 3. According to Reimnitz and Barnes in Ref. 1, pages 22-23:

"The winter ice canopy overlying the shelf can be divided into three broad categories: 1) seasonal floating and bottom-fast ice of the inner shelf, 2) a brecciated shear (stamukhi) zone containing grounded ice ridges that marks the zone of interaction between the stationary fast ice and the moving polar pack, and 3) the polar pack of new and multi-year floes on the average 2 m to 4 m thick, pressure ridges, and ice-island fragments that are

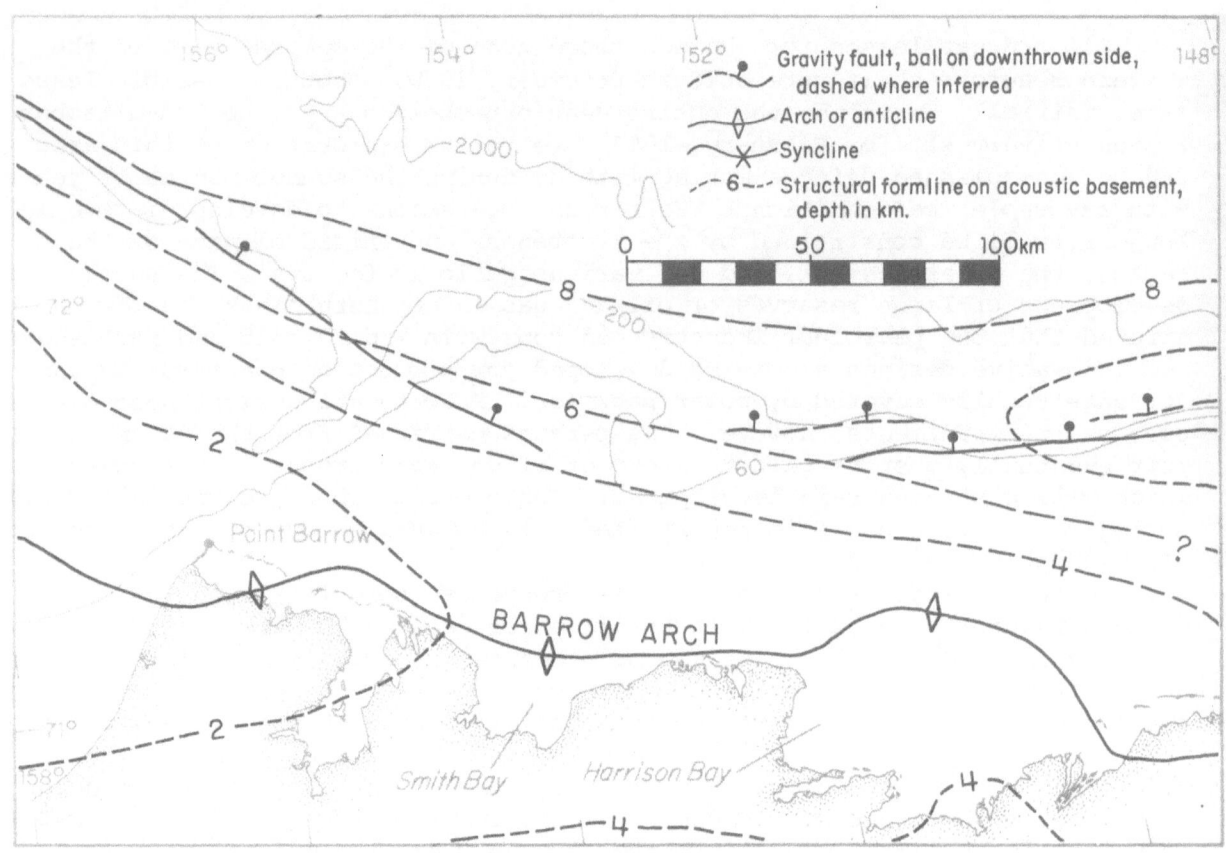

Figure 2:   Generalized geology beneath the western Beaufort Sea.

in almost constant motion. The deepest pressure-ridge keel that has been
measured had a draft of 47 m. The general drift of the pack on the Beaufort
shelf is westerly under the influence of the clockwise rotating Pacific Gyre.
     "The fast ice zone is composed mostly of seasonal first-year ice, which,
depending on the coastal configuration and shelf morphology, extends out to
the 10- to 20-m isobath. By the end of winter, ice inside the 2-m isobath
rests on the bottom over extensive areas. In early winter the location of
the boundary between undeformed fast ice and the westward-drifting polar pack
is controlled predominantly by the location of major coastal promontories and
submerged shoals. Pronounced linear pressure and shear ridges form along
this boundary and are stabilized by grounding. Slippage along this boundary
occurs intermittently during the winter forming new grounded ridges in a
widening zone (the stamukhi zone)....
     "Grounded pressure-ridge keels in the stamukhi zone will exert tremen-
dous stresses on the seabottom and on any structures present in a band of
varying width between the 10- and 40-m isobaths. In some places the extent
of shorefast ice may be deflected seaward by artificial structures.
     "Ice moving in response to wind, current, and pack ice pressures often
plows through and disrupts the shelf sediments, forming seabed gouges from
nearshore out to water depths of at least 60 m. Gouges are generally ori-
ented parallel to shore and commonly range from 0.5 to 1 m deep. However,
gouges cut to a depth of 5.5 m have been measured on the outer shelf. When
first formed, the gouges may be considerably deeper. Regions of high gouge-
density are common within the stamukhi zone and along the steep seaward
flanks of topographic highs.... Rates of gouging inshore of the protective
stamukhi zone have been measured at 1 to 2 percent of the seafloor per year."

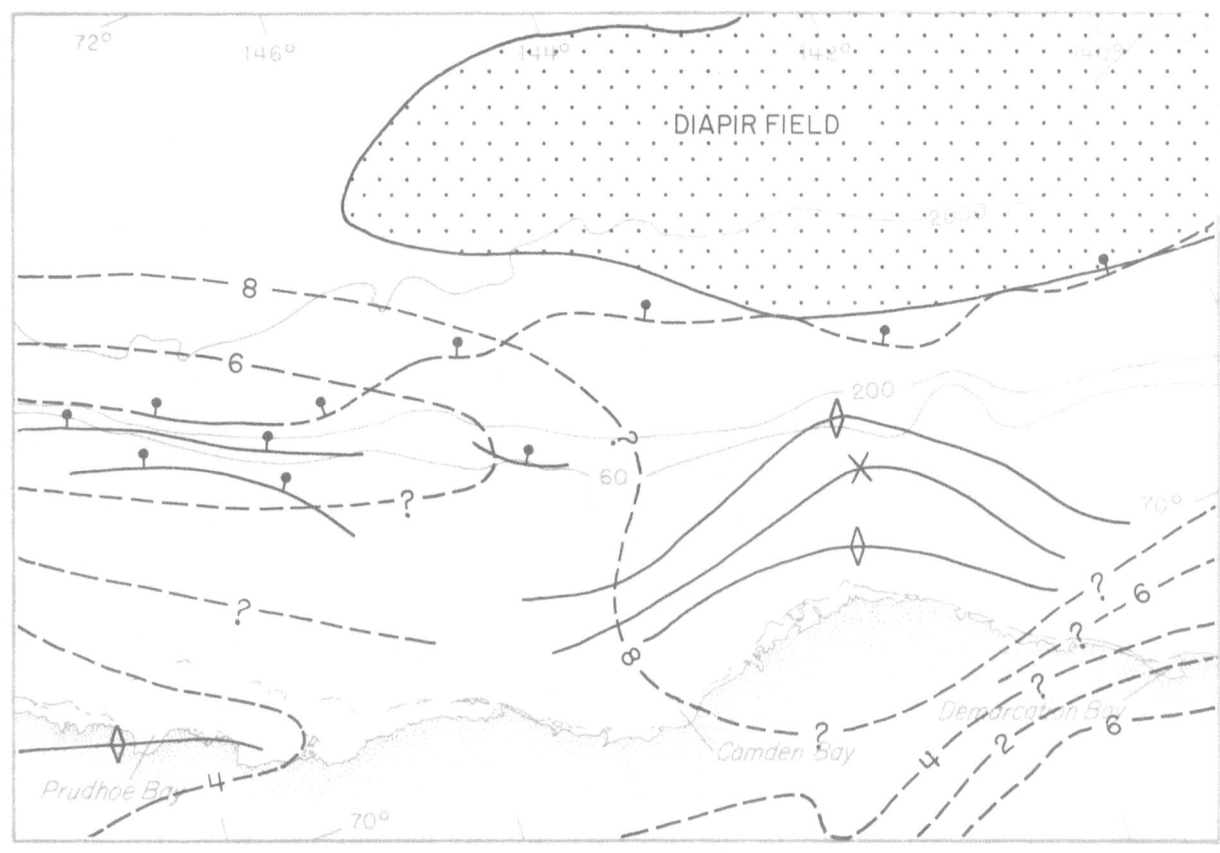

Holocene sediments

The most poorly consolidated sediments on the Beaufort shelf are the Holocene
marine muds and silts whose thickness is contoured in Fig. 4. They have been
identified and measured on high-resolution seismic profiles on the basis of
acoustic properties and geometric considerations. Inshore, drilling information corroborates the acoustic data where they overlap. The Holocene
deposits apparently form a wedge that thickens offshore to a maximum of more
than 45 m near the shelf-break, where the wedge overlaps the shoreward boundary of the chaotic slump terrane shown in Fig. 5. The wedge is apparently
thinner on the western half of the shelf than in the east, but records in
the west are much less definitive. Except in paleovalleys, Holocene sediments are notably absent along the coast. Little sediment appears to be ac-
cumulating in deltas and offshore from river mouths.

The Holocene sediments and probably a substantial thickness of the
underlying Pleistocene sediments in many areas of the middle and outer shelf
have low shear strength, as indicated by the development of very low-angle
bedding-plane slides, and they may be susceptible to liquefaction as well.
The instability of these deposits poses a hazard to pipelines, platforms,
and artificial islands where the deposits are thick and where they lie near
the active seismic zone northeast of Camden Bay. A band of unknown width
shoreward from the bedding-plane slide terranes of the outer shelf, subject
to failure through headward extension of the slides during earthquakes, is
also hazardous.

Holocene tectonic deformation

Several areas of the Beaufort shelf are disrupted by faults that displace

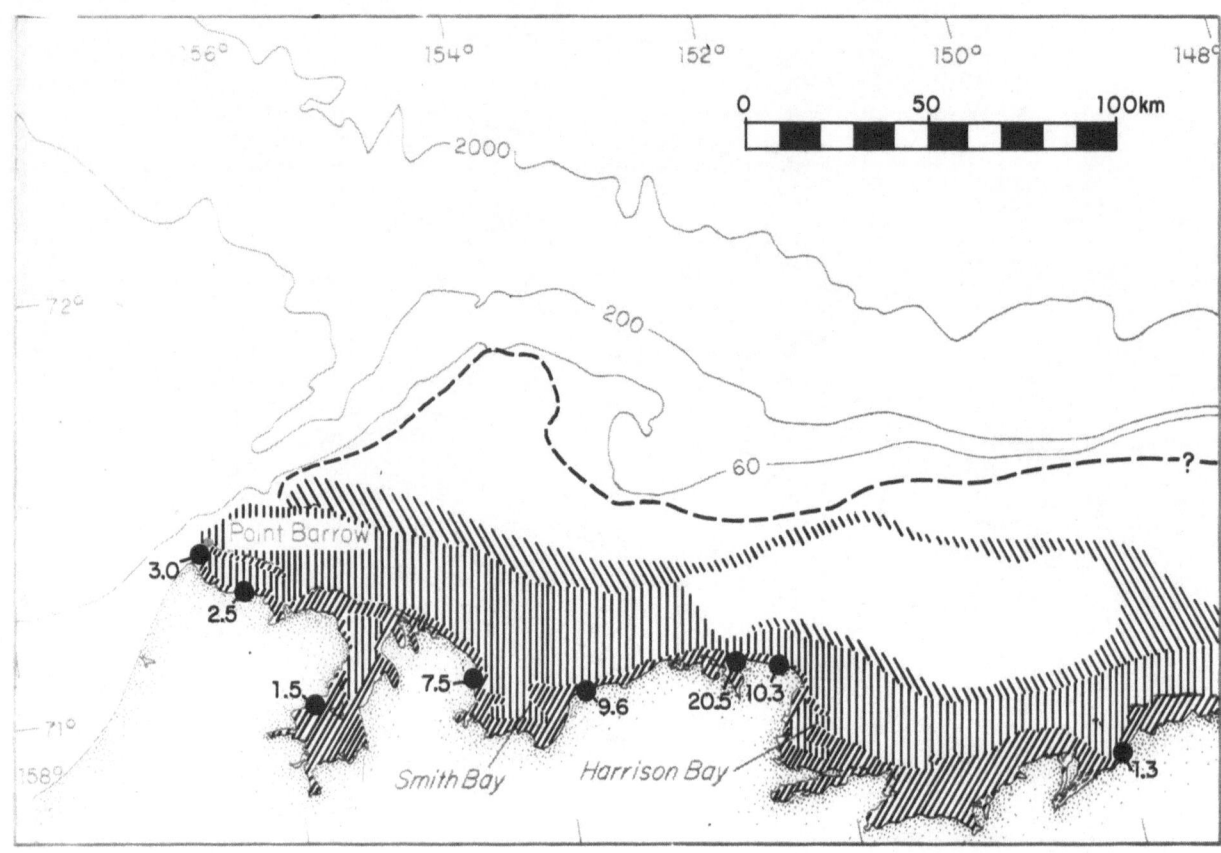

Figure 3:   Ice zonation, ice-gouging, and coastal erosion rates in the western
            Beaufort Sea from Reimnitz and Barnes in Grantz et. al. 1980.

Pleistocene and, in places, Holocene sediments and the seabed, or by deeper
faults that underlie monoclinal warps in the younger sediments (Figs. 2 and
4). Off Camden Bay, an east-northeast-trending zone about 30 km wide and at
least 60 km long, marked by localized seismicity and negligible Holocene sediment cover, coincides with a terrane of normal faults, thrust faults, and
monoclines. This zone probably represents the active front of the tectonic
system that produced the large folds in Tertiary rocks that characterize the
shelf and coastal plain to the south and east. The greatest earthquake risk
to petroleum exploration and production on the Beaufort shelf is within this
zone.

Earthquakes. The only part of the western Beaufort shelf from which earthquakes have been recorded is the area of Holocene uplift and faulting off
Camden Bay (Fig. 4). The earthquake history of this area consists of data
from the worldwide seismographic network (Ref. 2), supplemented by regional
and local network data beginning in 1968 (Ref. 3). The short instrumental
record is dominated by the largest recorded event, a magnitude 5.3 earthquake
on January 22, 1968, and its aftershocks. Biswas and Gedney (Ref. 3, p. 1)
state that one of the design criteria for structures built for the exploration and development of hydrocarbons in this area of the Beaufort shelf
should be the ability "...to withstand ground vibrations corresponding to
those from a shallow earthquake (less than 20 km) of at least magnitude 6.0."
     The Beaufort shelf east of the active zone off Camden Bay may also be
seismically active because late Cenozoic folds and local Holocene warping,
which are spatially associated with that zone, extend eastward to the Canadian
border. However, only two earthquakes have been reported from the Alaskan

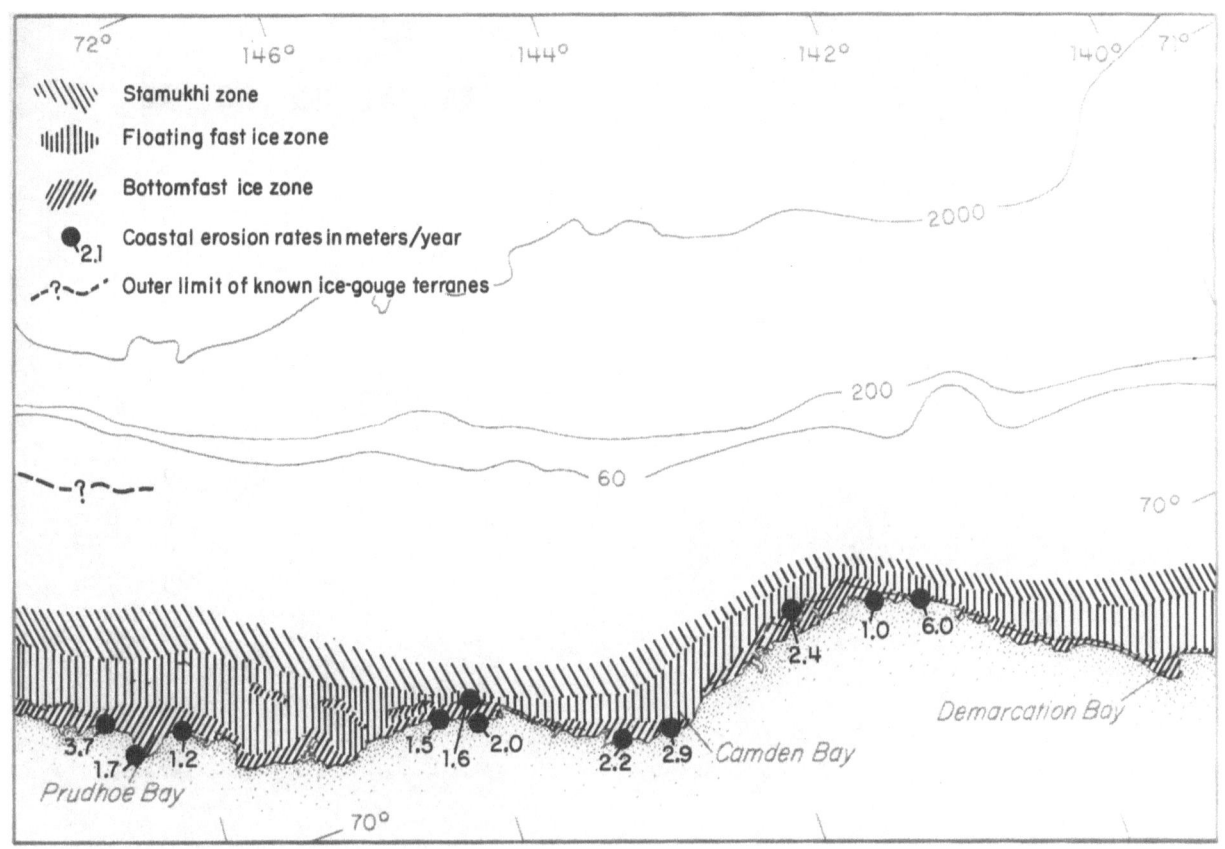

shelf east of the active zone and the Holocene warping east of that zone is smaller in amplitude and areal extent than it is off Camden Bay. If the area to the east is seismic, it may be less so than the active area off Camden Bay.

Warping.   Holocene warping in the zone of localized seismicity and negligible Holocene sediment cover off Camden Bay has uplifted the contoured base of Holocene sediment (Fig. 4) on the order of 10 m to 25 m above its mean level in adjacent areas.   This structural high is not delineated by the bathymetry because the related troughs at its perimeter have been mostly filled by Holocene sediment.   The epicenters of recent earthquakes on the western Beaufort shelf plot·almost exclusively within this area of rapid uplift, and this evidence implies that deformation here is presently active.

North of Demarcation Bay a small, rapidly subsiding basin has accumulated more than 20 m of sediment along its west-northwest-trending axis during the 10,000 or so years since the latest glacial sea level lowstand. The rate of subsidence here is slightly greater than that of sedimentation, and as a result the seabed is slightly concave above the basinal axis. Just northwest of this basin a bathymetric high covered by anomalously thin Holocene sediment coincides with a broad regional arch developed in Tertiary strata. The coincidence north of Demarcation Bay of an area of thin Holocene sediment with a bathymetric high and arched Tertiary strata and of an area of relatively thick Holocene sediment with a bathymetric low suggests that the warping here is also tectonically controlled and active.

Tectonic faulting.   Normal and reverse faults, many of which offset Holocene deposits and the seabed, and monoclinal warps overlying such faults are abundant off Camden Bay (Fig. 4).   The youthfulness of these features and their

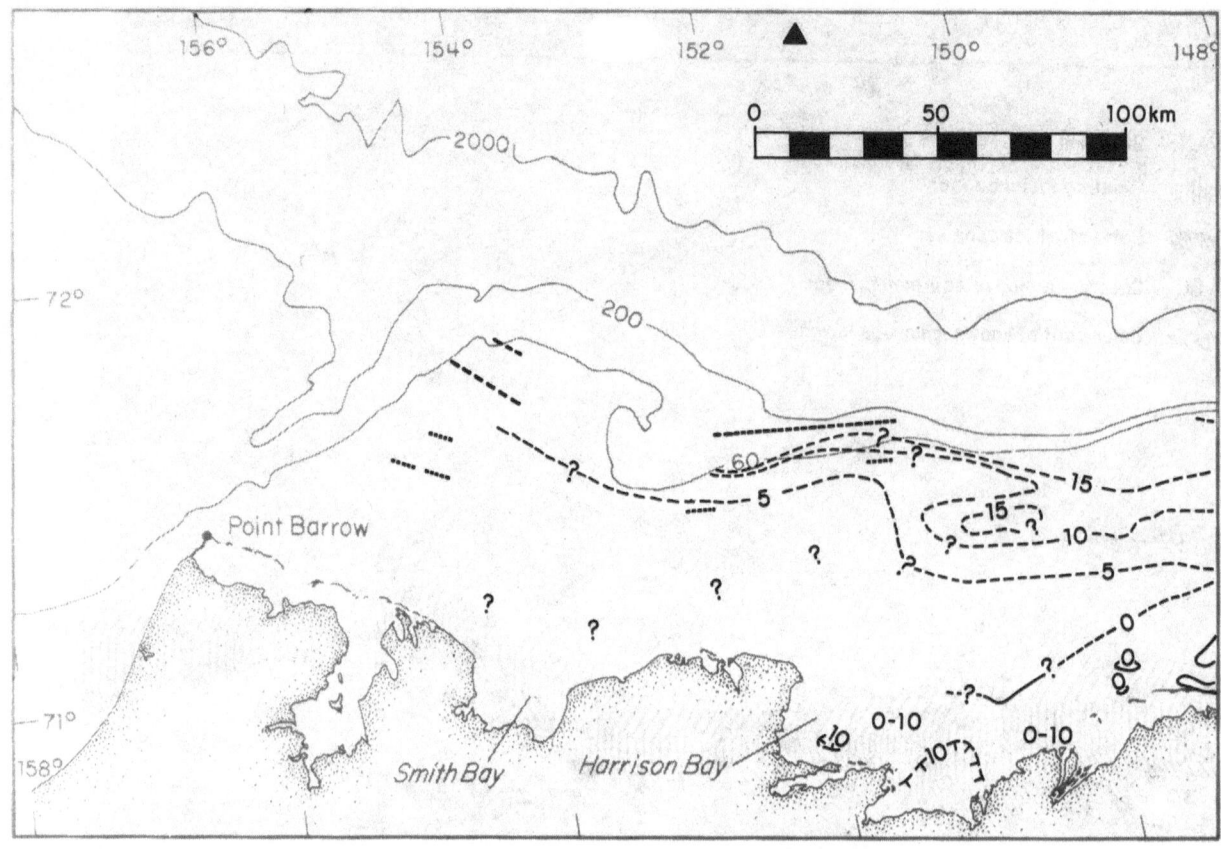

Figure 4: Holocene sediment thickness, tectonic faulting, and earthquakes
           beneath the western Beaufort Sea.

local concentration in the area of Holocene arching and shallow earthquakes
suggest that at least some of them are active. The maximum displacement of
the seabed that we recorded, 6 m, was found at two faults.
     There are insufficient data from which to calculate a definitive recur-
rence interval for displacement events on the faults off Camden Bay. A
rough estimate can be made, however, if the area is capable of generating
shallow magnitude 6 earthquakes, as suggested by Biswas and Gedney (Ref. 3),
and if we assume that fault displacement events here are episodic and produce earthquakes. Surface fault offsets accompanying shallow historic world-
wide earthquakes of magnitude 5 to 6, according to a compilation by Slemmons
(Ref. 4), have ranged from 0.015 to 0.61 m. A curve relating the Holocene
rise in sea level off northern and western Alaska (Ref. 5) to time suggests
that inundation of the seafloor offset by the faults occurred about 13,000 to
16,000 years ago,if allowance is made for the 10 to 25 m of Holocene tectonic
uplift that affected the area. As the fault-related seabed morphology does
not appear to have been modified by the passage of a strand-line, the offset
seabed is probably younger than 13,000-16,000 years. From these observations
and assumptions we calculate that the faults which produced the 6-m seabed
offsets have a maximum recurrence interval of between 30 and 1,600 years for
displacement events and related earthquakes. The longer intervals would per-
tain if typical earthquakes are closer to magnitude 6, and the expected offsets would be in the upper part of the range of offsets compiled by Slemmons.
In view of the short instrumental seismic record, the magnitude 5.3 earthquake
of 1968 should be considered the minimum, and not necessarily the typical or
maximum magnitude of earthquakes in this area. We suggest that a recurrence
interval of a few hundred years may be the most realistic estimate that can

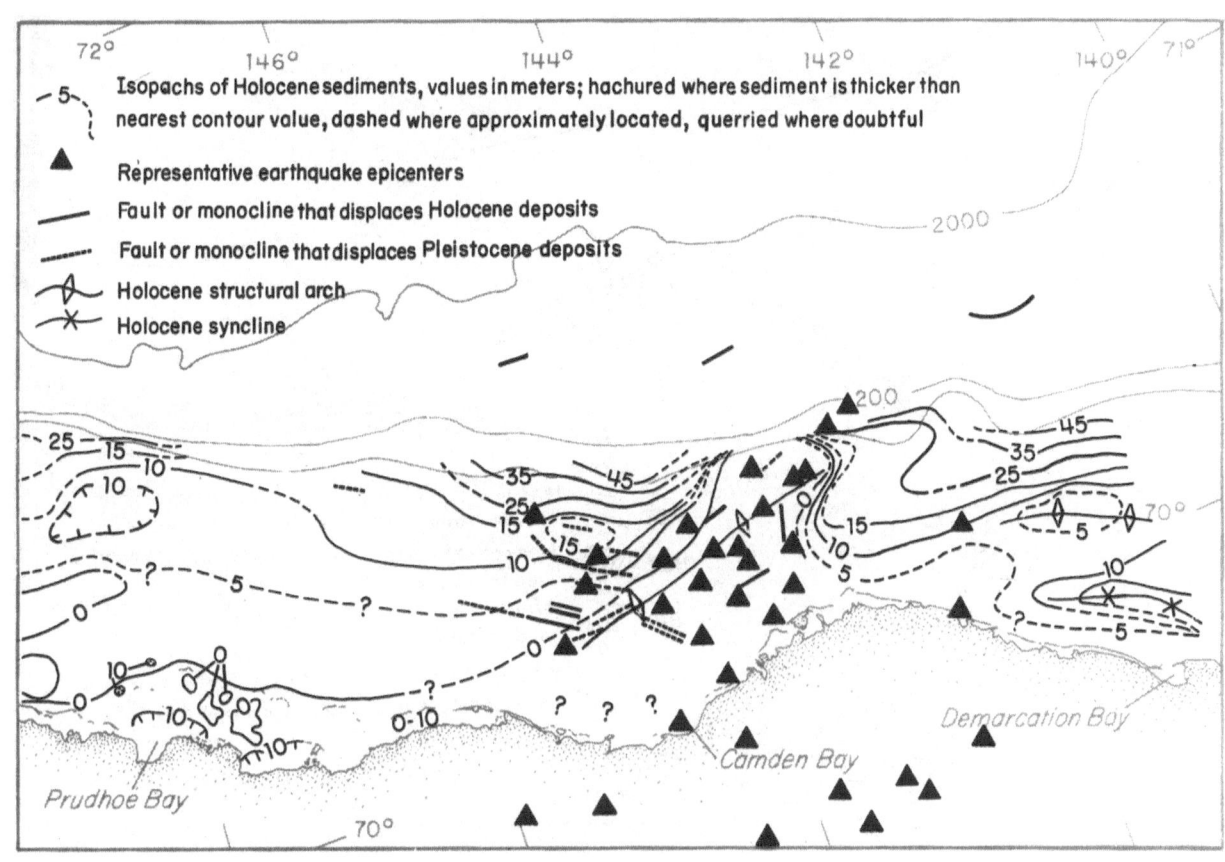

Isopachs of Holocene sediments, values in meters; hachured where sediment is thicker than nearest contour value, dashed where approximately located, querried where doubtful

▲   Representative earthquake epicenters

___   Fault or monocline that displaces Holocene deposits

-----   Fault or monocline that displaces Pleistocene deposits

Holocene structural arch

Holocene syncline

be made at present for earthquake-generating displacement events on the largest seabed-offsetting faults off Camden Bay.

Gravity faults. Two types of down-to-the-basin, north-dipping gravity faults, along which the sedimentary prism of the shelf failed and moved toward the free face of the continental slope, underlie the western Beaufort shelf (Figs. 2 and 4). These faults include all those shown outside of the northeast-trending zone of seismicity and Holocene uplift off Camden Bay. The first type of gravity fault, which is restricted to the outermost shelf and upper slope, has total displacements as great as 1,055 m and bounds shallow structural blocks that are akin to large rotational slumps. Most of the offset along these faults may have occurred in one or a few large displacement events. Additional features of this type are likely to disrupt the adjacent outermost shelf in the future.

The second set of gravity faults, which occurs beneath the middle and outer shelf, is characterized by much smaller offsets of Quaternary deposits and the seabed than the first set and includes many growth faults with a long history of activity. The outer shelf faults of this set displace Holocene deposits and the seabed as much as 15 to 20 m, and in one area possibly as much as 70 m or more. Those on the mid-shelf displace sediments no younger than Pleistocene or early Holocene.

The gravity faults are active in the sense that they formed in the present tectonic environment and displace Pleistocene or Holocene sediments. However, they have not generated earthquakes of sufficient magnitude to be detected by the regional and local seismograph networks in place since 1968. The absence of seismicity may be due to the fact that low stress drops are characteristic of movement along gravity faults. In the absence of earthquakes or

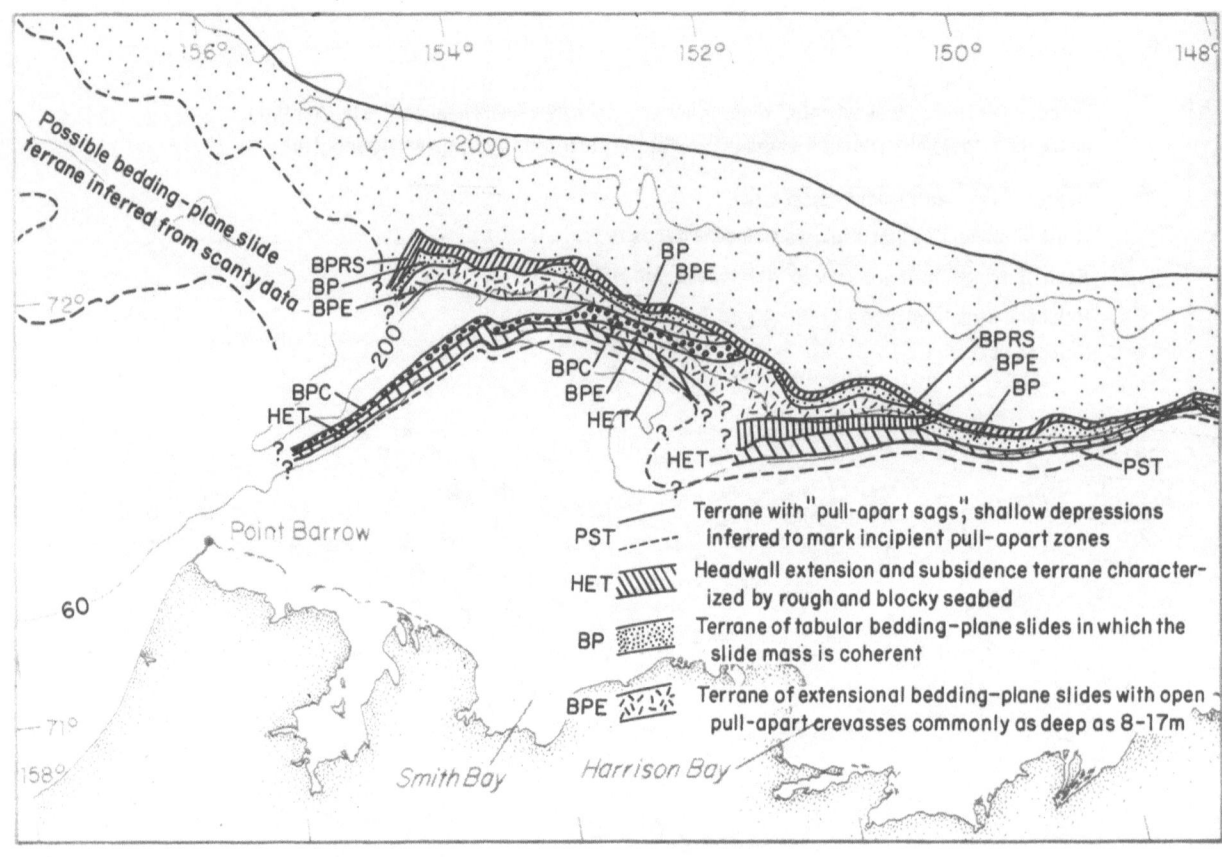

Figure 5:   Slumping and sliding beneath the western Beaufort Sea.

detailed physical stratigraphy, the recurrence interval of displacement events
along the gravity faults is impossible to calculate. We estimate, however,
that the faults beneath the outer shelf, which show large Holocene offsets,
may have recurrence intervals in the range of a few hundred to several thousand years. Those on the mid-shelf, which show only Pleistocene or early
Holocene offsets, may be quiescent or have very long recurrence intervals.
Some displacement events on the gravity faults of the outer shelf and upper
slope off northeast Alaska may be triggered by the larger earthquakes originating in the seismic zone off Camden Bay.

Slumping and sliding
A large part of the Beaufort outer shelf and upper slope seaward of the 50-m,
locally the 60-m, isobath is pervasively disrupted by young bedding-plane
slides in unconsolidated or poorly consolidated Holocene and Pleistocene sediments.   Several distinct instability terranes (Fig. 5) have been delineated
on the basis of high-resolution seismic records (Fig. 1).   These include sag
terranes at the heads of extensional terranes, coherent bedding-plane slide
terranes in which large tabular blocks moving seaward are separated by open
crevasses as deep as 17 m (exceptionally 37 m), rotational slump terranes in
which huge slump masses break along listric (concave upwards) surfaces and
slide downslope at high angles, and hummocky rubble piles at the bases of
slopes where these slump masses accumulate.

    The bedding-plane slide masses are tabular sheets as much as 38 km long
and typically 20 to 230 m thick that move seaward along slip planes which,
since they dip only 0.5° to 1.5°, must include materials of very low shear
strength. Locally, as many as three generations of slide masses are superimposed, and reactivation of sliding is apparently common. No samples of the

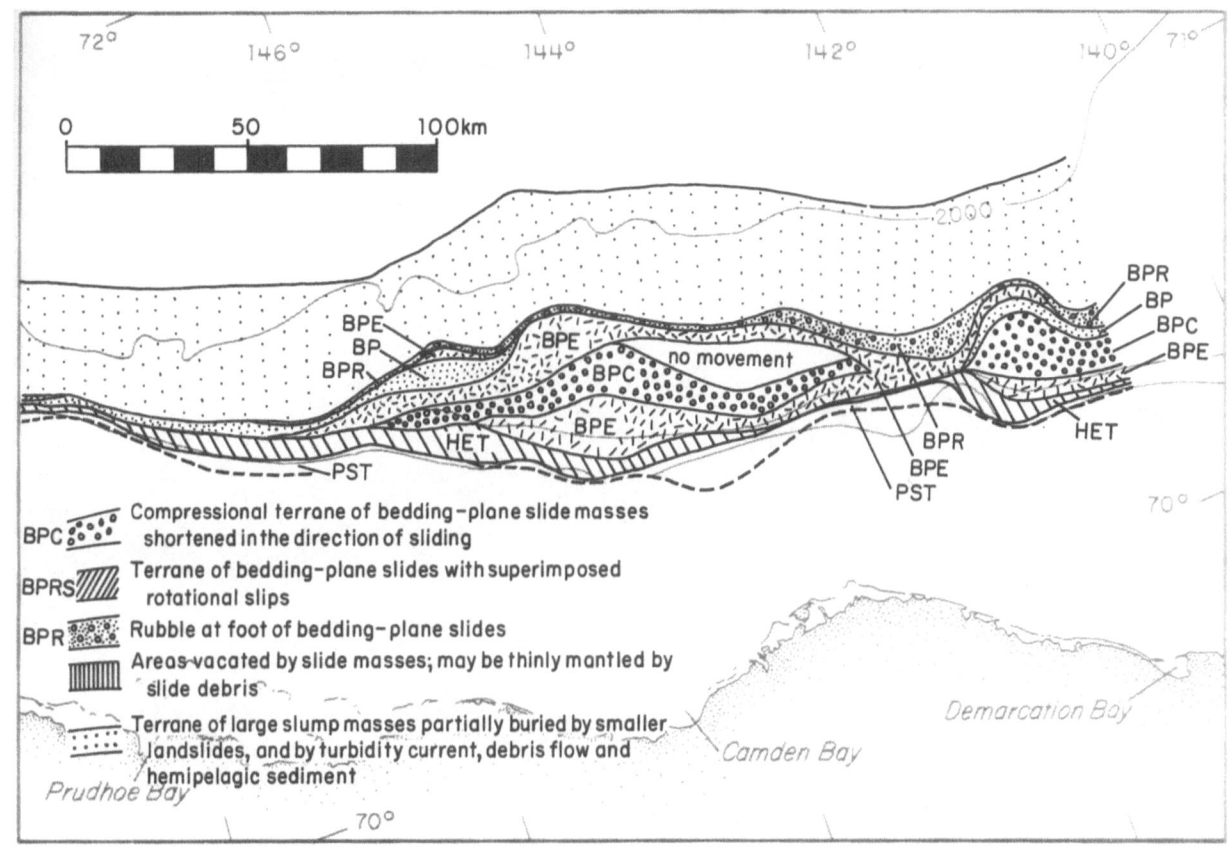

sediments involved in sliding have been dated; however, the thinner slides and the uppermost parts of the thicker ones must be mostly Holocene since Holocene deposits near the head of the slide terrane are typically 30 to 50 m thick (Fig. 4). In view of the evidence for repeated failure of broad, thick masses of unconsolidated and poorly consolidated sediment at several levels in the stratigraphic column, much of the western Beaufort shelf and slope seaward of the 50-m, locally the 60-m, isobath must be considered hazardous or potentially hazardous to oil and gas development structures.

The continental slope seaward of the bedding-plane slide terrane is underlain by massive slumps and chaotic slump terrane. East of 150° W long. the slumps are bounded upslope by gravity faults (Fig. 5) with large vertical displacements. The top of the continental slope is relatively steep at these faults, and the adjacent outermost continental shelf may be vulnerable to additional slumping.

Diapirism and overpressured shale

Diapirs disturb the Tertiary sediments beneath the continental slope and rise east of the longitude of Camden Bay (Fig. 6) and in places arch the seabed, creating dams that pond clastic sediments and trap submarine landslide debris. The diapirs apparently consist of overpressured shale which may have originated beneath the continental shelf and slope in beds observed to have relatively low seismic interval velocities (∿2.7 km/sec.). These possible source beds lie about 3 km below the seabed beneath beds with somewhat higher interval velocities. They are inferred to extend shoreward beneath the outer shelf, where they may create problems for deep drilling.

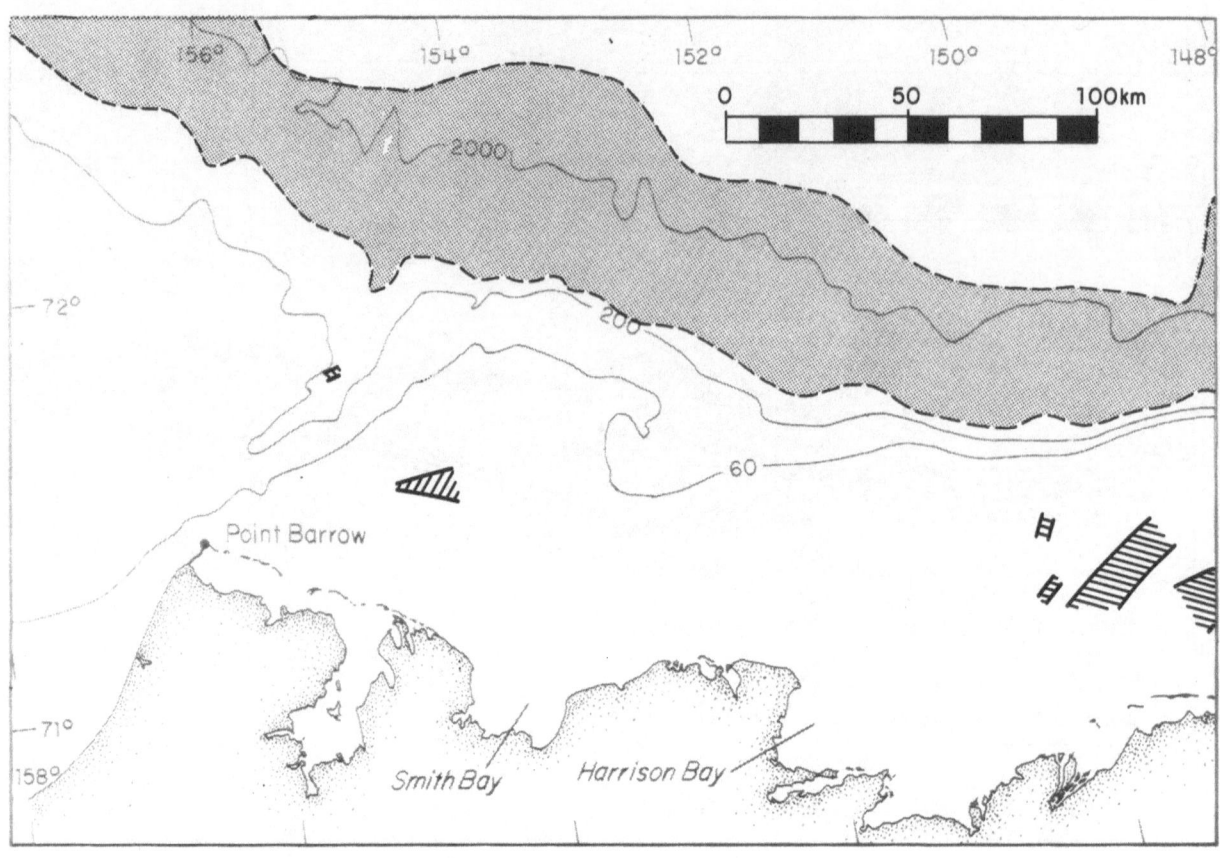

Figure 6: Shallow gas, natural gas hydrate, diapirs, and overpressured shale
          beneath the western Beaufort Sea.

Shallow gas in Quaternary sediment, and gas hydrate

Shallow gas may accumulate in three distinct geologic environments in the
western Beaufort Sea. It forms isolated pockets within and beneath permafrost (see below), is inferred to underlie solid gas hydrate beneath the
slope and deepest parts of the outer shelf, and forms scattered, isolated
concentrations in loosely consolidated Quaternary sediment (Fig. 6). Seismic
reflection profiles delineate the base of gas hydrate fairly well but yield
little direct information pertaining to permafrost distribution. The existence and extent of shallow free gas are only nominally established, owing in
part to its rather erratic distribution and in part to intermittent degradation of our high-resolution data during storms.

Gas hydrates (solids composed of light gases caged in the interstices of
an expanded ice lattice) are stable under the conditions of low temperature
and relatively high pressure prevailing within the uppermost 300 m to 700 m or
so of sediment beneath the continental slope of the Beaufort Sea. Hydrates
decompose and release large volumes of gas during drilling operations unless
care is taken to cool the drilling mud. Such precautions are mandatory because the gas is commonly methane, and hence inflammable, and because decompo-
sition can cause caving in drill holes. In addition, extensive, possibly
overpressured pockets of free gas probably occur beneath the sheets of gas
hydrate.

"Turbid" zones, in which strong, continuous reflectors are "wiped out"
for short distances on high-resolution seismic profiles, have been ascribed
to the presence of shallow gas, probably mostly biogenic methane in Quaternary
sediments. The "turbid" intervals are apparently restricted to the inner
shelf and are most abundant in a zone about 50 km north and northwest of
Prudhoe Bay (Fig. 6).

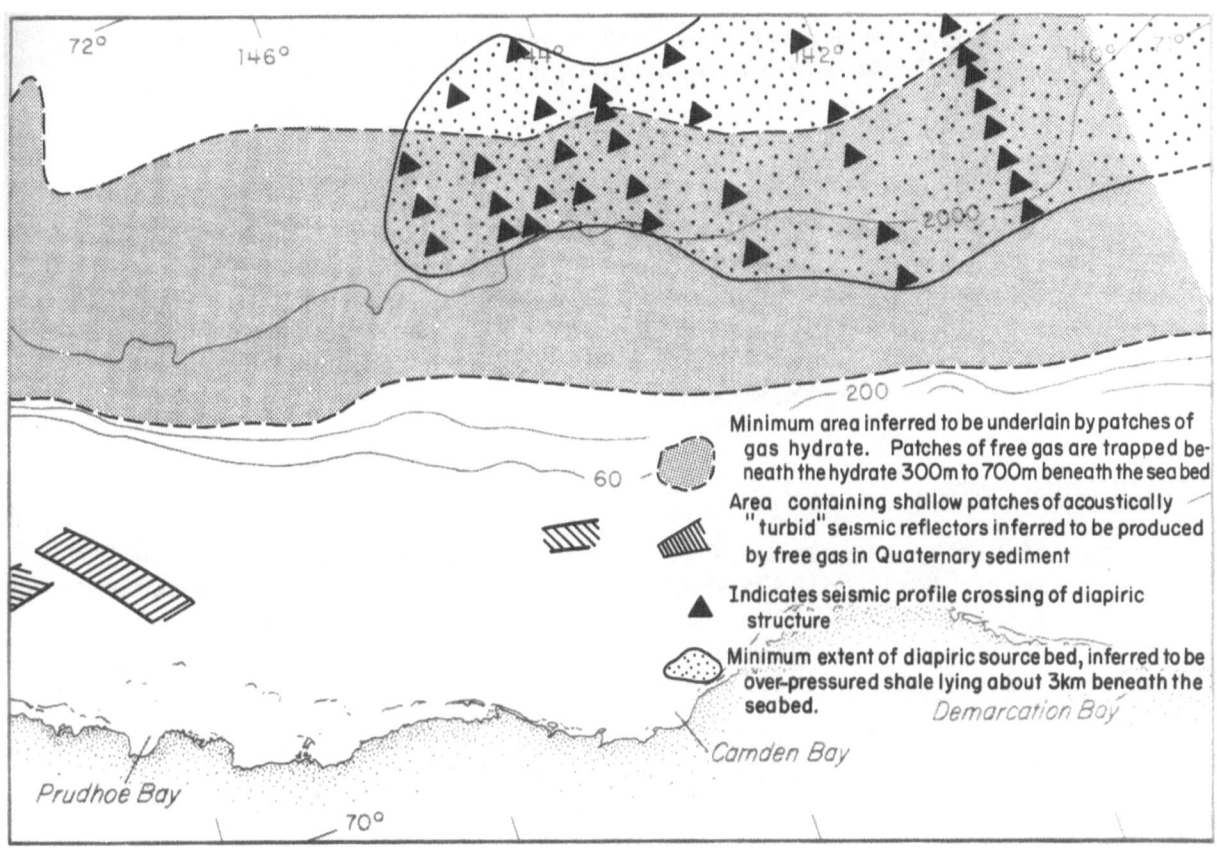

Permafrost

Prior to about 10,000 years ago, during the last glacial sea level lowstand, the Beaufort shelf was exposed subaerially to frigid temperatures and ice-bonded permafrost probably aggraded downward in the sediments to depths exceeding 300 m. Reflooding of the shelf exposed these sediments to saline water at temperatures above the freezing point and much of the permafrost terrane has probably warmed and melted. To date, the distribution of permafrost has been determined only in the vicinity of Prudhoe Bay.

Sellman and Chamberlain (Ref. 6) found that in this area three obvious groups of seismic velocities are apparently related to the degree of ice-bonding in the sediments. Fully ice-bonded permafrost with ice-saturated pores and velocities greater than 4.0 km/sec. crops out at the surface onshore and on the barrier islands, but lies at highly variable depths as great as several hundred meters beneath the seafloor in between. It is overlain in this area mostly by materials with velocities of about 2.7 km/sec. which are taken to represent partially ice-bonded sediments containing varying proportions of unfrozen pore water. Materials with velocities less than 2.2 km/sec are sparse and assumed to be unbonded.

Although the extent of relict permafrost seaward of the barrier islands has not been determined, the base of Holocene marine sediments contoured in Fig. 4 provides a probable minimum depth to the top of permafrost. It is unlikely that permafrost aggraded upward into the saline marine muds deposited on the shelf during the Holocene rise in sea level. By analogy with the conditions described nearshore, relict permafrost in the uppermost sediments beneath this Holocene "wedge" is probably melted or partially melted down to unknown depths. If fully ice-bonded permafrost is drilled offshore, care must be taken to avoid melting-induced failures beneath pipelines and drill platforms, and within frozen intervals penetrated in drilling. Pockets of methane gas, which commonly occur within and beneath permafrost bodies may, in places, constitute an additional hazard.

CONCLUSIONS

The promise of large discoveries has already brought oil and gas leasing to large nearshore areas of the Beaufort Sea, and the decade of the 1980's will undoubtedly see exploration of selected offshore areas of the shelf. It is clear that geologic processes and conditions, as well as remoteness, frigid temperatures, and the polar ice pack will constrain petroleum exploration and development in this region.

REFERENCES CITED

1.   Arthur Grantz, P.W. Barnes, D.A. Dinter, M.B. Lynch, Erk Reimnitz, and E.W. Scott, "Geologic framework, hydrocarbon potential, environmental conditions, and anticipated technology for exploration and development of the Beaufort Shelf north of Alaska, a summary report", U.S. Geological Survey Open-File Report 80-94 (1980).

2.   H. Meyers, "A historical summary of earthquake epicenters in and near laska", National Oceanic and Atmospheric Administration Technical Memo EDS NGSDG-1 (1976), 57.

3.   N.N. Biswas and L. Gedney, "Seismotectonic studies of northeast and western Alaska", University of Alaska, Geophysical Institute (1978), 45.

4.   D.B. Slemmons, "Faults and earthquake magnitude, Report 6 of State-of-the-art for assessing earthquake hazards in the United States", U.S. Army Engineer Waterways Experiment Station, Vicksburg (May 1977).

5.   D.M. Hopkins, "Landscape and climate of Beringia during Late Pleistocene and Holocene time" in W.S. Laughlin and A.B. Harper, Eds., "The first Americans: origins, affinities, and adaptations", Gustav Fischer, New York (1979), 15-41.

6.   P.V. Sellman and E. Chamberlain, "Delineation and engineering characteristics of permafrost beneath the Beaufort Sea", National Oceanic and Atmospheric Administration, Arctic Projects Office, Fairbanks, Alaska, Quarterly Report to U.S. Department of Commerce (1979).

7.   Jonathan Greenberg, P.E. Hart, and Arthur Grantz, Bathymetric map of the continental shelf, slope, and rise of the Beaufort Sea north of Alaska, U.S. Geological Survey Miscellaneous Geologic Investigation Map I-1182, scale 1:500,000. (in press).

OIL AND ICE IN THE BEAUFORT SEA - THE PHYSICAL EFFECTS OF A HYPOTHETICAL

BLOWOUT

Peter Wadhams

Scott Polar Research Institute

University of Cambridge

Cambridge CB2 1ER, England

1. INTRODUCTION

In summer 1976 a drillship of Canadian Marine Drilling Ltd. (Canmar), of
Calgary, began the first exploratory well in the southern Beaufort Sea. This
began a new era for the Arctic Ocean, bringing the possibility of oil
pollution to a sea that was formerly immune from Man's activities. Oil
spills and blowouts in open water are a familiar experience, but the
présence of an ice cover lends a unique character to Arctic oil pollution
incidents. During Canmar's exploratory programme the worst accident that
could occur would be a blowout from the seabed, e.g. by uncontrollable
pressure or by the drillship being pushed off station by ice, at the end of
the open water drilling season (October). A relief well could not then be
drilled until the following summer (July) since the present Canmar fleet is
unable to operate in winter ice. Therefore a year's continuous release
would occur, involving up to a million barrels of oil.

In this paper we describe the ice conditions in the southern Beaufort
Sea and review the studies that have been carried out of oil-ice interactions.
Applying the results to a Beaufort Sea blowout, we estimate the likely
physical effects and the disposition of the oil after a winter. Space does
not permit us to deal with the biological effects of a blowout or the results
of the large body of research into clean-up techniques except in passing.

2. ICE CONDITIONS IN THE BEAUFORT SEA

The Beaufort Sea proper is part of the Arctic Ocean and participates in the
ice dynamics of the ocean as a whole. The mean long-term ice motion in the
Çanada Basin consists of a clockwise circulation known as the Beaufort Gyre,
with a centre at about 80°N 140°W. The rate of ice drift is very variable,
with the ice being driven by a mixture of geostrophic forces, direct wind
stress and internal stress transmitted across great distances through the
ice cover. The mean distance made good in the circulation has been estimated
as 2.2-2.6 km/day (Ref. 1), 2.4 km/day (Ref. 2) and 2.7 km/day (Ref. 3), but
the random motion of the ice results in a given piece of ice tracing out a
path which is between 2 and 4 times this value. During the AIDJEX main
experiment, for instance, the ice in the region 73-75°N, 140-150°W had a
mean velocity during a year (April 1975 - April 1976) of 6.0 km/day, which
rose to 9.2 km/day during the summer. On occasions during the winter the
pack stopped moving altogether for days at a time (Ref. 4).

In the southern Beaufort Sea, the area of greatest interest to oil
companies is the wide continental shelf which reaches 80-120 km from the
mainland coast and up to 150 km from the coast of Banks Island. Within this
area it is the mainland shelf in the region of Mackenzie Bay which has seen
most of the exploratory activity to date. Here the ice cover disappears for
up to three months of the year (July-September), leaving open water which may

extend up to 200 km from the coast, although both the duration and area of
the open water are highly variable from year to year. In winter the ice
conditions are heavily affected by the presence of the coast and shallow
water. Three zones can be distinguished.

(i) Landfast ice. Ice begins to form in sheltered bays along the coast at
the beginning of October. A stable sheet forms initially against the shore,
mainly because the shallow water reduces the depth of convection necessary
to cool the water column to the freezing point; other contributing factors
are the quieter sea surface conditions and the lower salinity caused by
Mackenzie river discharge and summer melt of floes. The sheet spreads
seaward during October until it encounters the moving offshore pack, which is
driven periodically against the coast by early winter storms. If a shoreward
drive is of limited duration, the pack simply buckles and rafts the outer part
of the fast ice sheet, forming an irregular relief of rafted blocks. If the
pressure is greater and more prolonged, the fast ice sheet is deformed into
ridges which may grow large enough to become grounded. Other grounded
ridges originate from the pack itself but become incorporated in the fast ice
together with occasional ice island fragments. The limit of fast ice extent
is the depth limit at which pressure ridges are found with a high enough
frequency to stabilise the fast ice sheet and is about 18-20 m (Ref. 5)
although areas of heavy ridging can become temporarily attached to the fast
ice out to a water depth of 27 m. Thus the landfast ice consists of a smooth
sheet near shore, changing to deformed ice near the outer edge. Small
motions still occur within the landfast ice; usually the displacements are
of the order of 10 m (Ref. 6) but off the Alaskan coast motions of up to 160 m
have been observed (Ref. 7). The small displacements are due to thermal
expansion of the ice, and the larger to the compressive effects of offshore
pack moving against the fast ice; the ice returns to approximately its former
position after a large excursion.

The fast ice breaks up in July. Usually in May a large polynya opens in
Amundsen Gulf from which leads radiate into the Beaufort Sea, especially a
large lead along the edge of the fast ice; aided by surface melt due to
insolation and bottom melt due to Mackenzie flow the weakened fast ice now
breaks up and moves away with the pack.

(ii) The transition zone. This is the region extending from the fast ice edge
to the point at which the ice characteristics merge with those of the central
Arctic pack. This point is ill-defined, but corresponds roughly to the edge
of the continental shelf or to the mean limit of summer retreat of the Polar
pack. For this reason the transition zone is sometimes called the seasonal
sea ice zone. A transect of the Beaufort Sea by submarine at 144°W in April
(Ref. 8,9) showed that heavy ridging existed out to a distance of 160 km from
shore, but that beyond this point ice characteristics were very uniform.
Laser profiles (Ref. 10) show the same effect. It is possible that north of
Mackenzie Bay the edge of the transition zone lies further out to sea, because
of the more gently sloping shelf; undoubtedly all possible drillship activity
can be said to lie within the transition zone. The section of the transition
zone that abuts the fast ice edge is known as the shear zone, and possesses
especially heavy ridging. Here almost all of the slip between the circulating
pack and the fast ice is taken up within a zone only 50 km or less across
(Ref. 11). In fact the slip seems to take place in a small number of major
lead systems; in each of these the massive and prolonged shear generates huge
fields of continuous ridging, which have been termed "stamukhi" (Ref. 12).
Areas of stamukhi can become temporarily attached to the fast ice, because
of the generation of very deep ridge keels which run firmly aground. At

other times, when the pack is not pressing down against the coast, the lead system which is the source of the stamukhi becomes a perfect slip plane for the moving pack. Further out to sea the ridging intensity is still greater than in the central pack; typically (Ref. 8) there is a mean ice draft of 4-5 m and 3-5 ridges per km of draft exceeding 9 m. There is a greater proportion of first-year ice than in the central pack.

(iii) The Polar pack. This extends onward into the central Canada Basin. The mean ice drift is similar in the Polar pack and outer transition zone, but the Polar pack contains a greater proportion of multi-year ice and a lesser degree of ridging. Typically (Ref. 8) the mean ice draft is about 3.7 m, with only 1-2 keels per km with draft exceeding 9 m. The mechanisms of ice drift and deformation in the Polar pack are becoming increasingly well understood thanks to the AIDJEX model (Ref. 13) and other numerical models (Ref. 14).

3. A SUMMARY OF OIL-IN-ICE STUDIES

Interest in the Beaufort Sea as a source of oil began in the 1960's, when a successful exploratory well was drilled at Atkinson Point on the Tuktoyaktuk Peninsula. From 1963 to 1969 the Canadian government issued leases for offshore exploration covering most of the mainland shelf, and to exploit these a consortium of ten (later 34) companies founded the Arctic Petroleum Operators' Association (APOA) in 1970. APOA funded environmental and engineering research in the Beaufort Sea, including field and laboratory studies on oil-ice interaction, but the results were kept confidential until 1974. Meanwhile offshore drilling had begun in 1972 from artificial gravel islands in very shallow water (less than 4 m), and an application was made to the government for approval of offshore exploration by drillship. It was clear that such drilling would involve the possibility of an under-ice blowout. A second chain of events which stimulated scientific concern over oil spills in ice was the voyage of the "Manhattan" through the Northwest Passage in 1968-9 (Ref. 15) in an attempt to establish a route for the export of Prudhoe Bay crude oil. This was closely followed in February 1970 by the wreck of the "Arrow" in Chedabucto Bay, Nova Scotia, which spilled Bunker C oil into a nearshore zone which included a winter ice cover. The intensive studies of this spill by the Bedford Institute and others (Refs. 16-18) led to further laboratory and field investigations of oil under and on ice, especially by the U.S. Coast Guard (Ref. 19) and the University of Toronto (Ref. 20). By 1973 the data base, although small, was sufficient to arouse considerable concern about long-term effects of oil spills in ice, particularly the possible climatic effect due to albedo reduction and increased ice melt rate induced by oil on the ice surface (Ref. 3). Such possibilities were discounted by oil industry researchers (Ref. 21).

In 1974 the Canadian government and APOA began a comprehensive joint programme of research into the environmental problems associated with offshore drilling. The study, known as the Beaufort Sea Project (BSP), cost over $12 m and lasted until the end of 1975. The project co-ordinator was A.R. Milne, Institute of Ocean Sciences, Patricia Bay, B.C., and the aim was to determine the conditions under which offshore drilling could be allowed to proceed. The studies in BSP are summarised in (Ref. 22) and (Ref. 23), the latter reference containing a list of the 39 Beaufort Sea Reports and six overview reports which were produced; these are available

from the Institute of Ocean Sciences. Two of the overviews (Refs.24,25),
seven of the reports (Refs. 26-32) and a later paper (Ref. 33) dealt directly
with oil-ice interaction and cleanup techniques. Major studies in this
area during BSP included a year-long field programme in Balaena Bay, Cape
Parry, during which oil was released at intervals under smooth fast ice
(Ref. 26); a small oil spill under moving pack ice (Ref. 26); theoretical
and field simulations of the bubble plume associated with an oil-gas
blowout (Ref. 27); laboratory studies of oil entrapment under ice (Ref. 31);
remote sensing studies of ice conditions and their likely effect on oil
containment and trajectories (Ref. 28,29); and studies on cleanup (Ref. 32)
and possible climatic effects (Ref. 30).

Offshore drilling began in the summer of 1976 at two sites in 30 m and
56 m of water (A, B respectively in Fig. 1; A is called Tingmiark and B
Kopanoar by Canmar). To date Canmar's programme of exploratory drilling,
involving three ice-strengthened drillships, has investigated 9 sites. A
loss of well control at Kopanoar in 1976 led to a high-pressure water
blowout, but to date there have been no oil spills although oil and gas
have been found. The BSP had shown, however, that in the event of a blowout
no adequate countermeasures were available for oil cleanup or containment.
To remedy this, the Canadian government began a $7 m, five-year study known
as the Arctic Marine Oilspill Program (AMOP) in December 1976 (Ref. 34).
AMOP is managed by S.L. Ross, Environmental Emergency Branch, Environmental
Protection Service, Ottawa, and its aim is to develop countermeasures
techniques and equipment suitable for use in the Arctic environment. This

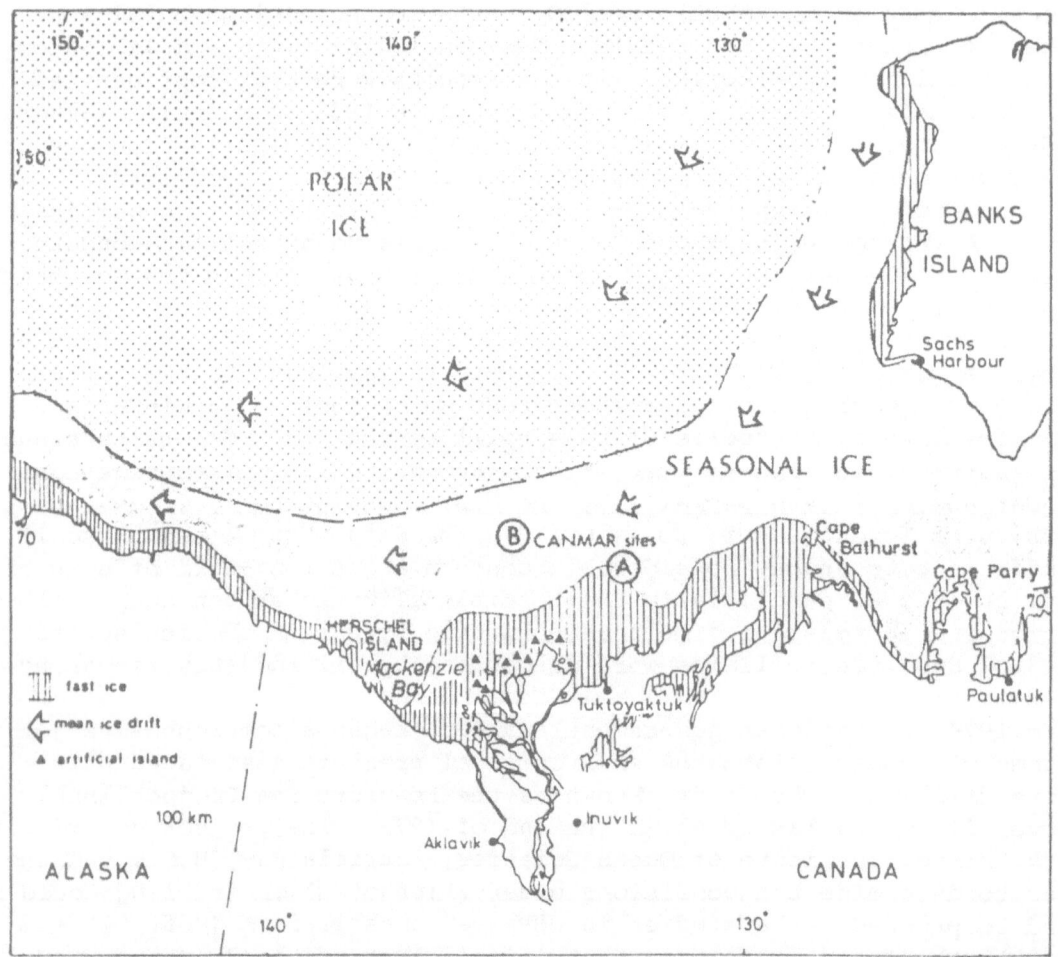

Fig. 1.   The Beaufort Sea, showing ice regimes in winter and the positions
          of artificial islands and of the first two Canmar drilling sites.

requires extensive work on the fate and effects of oil in ice, dealing not only with the Beaufort Sea but also with the Labrador Sea, Hudson Bay and Lancaster Sound, where drilling is in progress or proposed. Its largest field project, planned for spring 1980, is a series of experimental oil spills, including some under pack ice in the Beaufort Sea done in co-operation with Canmar. These will be made from depth using an oil-gas mixture to simulate the mechanisms of a true blowout (Ref. 35).

A forthcoming lease sale off the north coast of Alaska makes it likely that drilling will commence in the US sector of the Beaufort Sea, probably involving artificial islands. The US government has therefore increased its interest in oil-ice studies, which now form part of the Outer Continental Shelf Environmental Assessment Program (OCSEAP), funded by the Bureau of Land Management (Ref. 36,37). Its scope involves the whole Alaskan shelf, but a major component is the heavily ridged shear zone off the North Slope.

4. PATHWAYS OF OIL FROM AN UNDER-ICE BLOWOUT

We shall now consider the effects of a blowout that occurs at the end of the summer drilling season and that continues until a relief well can be drilled the following summer. According to oil industry sources, the likely flow rate from such a blowout would be 2500 bbl/day (400 m^3/day) at first, reducing after a month to a steady rate of 1000 bbl/day (160 m^3/day) by partial closure of the hole, and accompanied by 23 m^3 of gas at atmospheric pressure per barrel of oil. Whether correct or not, this estimate was used throughout the BSP as the basis for a standard blowout scenario and we shall continue to use it. The probability of a major blowout of this kind has been estimated (Ref. 38) as $2-6.10^{-6}$ per drillship-season, with the probability of any kind of blowout being 5.10^{-3} per drillship-season.

4.1. The rising plume

The oil and gas from the blowout rise together in a bubble plume, which deposits the oil at the surface over an area which depends on the water depth. The problem was investigated by Topham (Ref. 27), both theoretically and in field and laboratory experiments with bubble plumes. In two shallow water tests (23 m and 60 m) the gas rose in a turbulent convective plume (Fig. 2), entraining water from near the bottom. In the first stage the plume expands in a cone, since the bubbles emerge as a jet and then decelerate. In stage II it becomes almost cylindrical although with some spreading (Ref. 42) and with a constant rate of rise of some 0.6 m/s generated by buoyancy force alone. Finally the oil and gas reach the surface in a region of expanding vortices similar to smoke rings hitting a ceiling (stage III). The entrained water flows out radially at the surface until at a critical radius it sinks in order to complete a convective cell, the surface effect of this convergence being seen in open water as a wave ring. The oil emerging with the gas coats the outside of each gas bubble, and as the bubble expands on rising the oil is broken up into particles with diameters of about 0.5-1 mm which continue to be carried up within the plume. The gas bubbles themselves become unstable after reaching a diameter of 3 cm and break up into smaller bubbles. It was found (Ref. 39) that the wave ring cannot contain to any extent the oil which reaches the surface except in very calm conditions; the sinking water carries oil particles down with it and hence allows the oil to

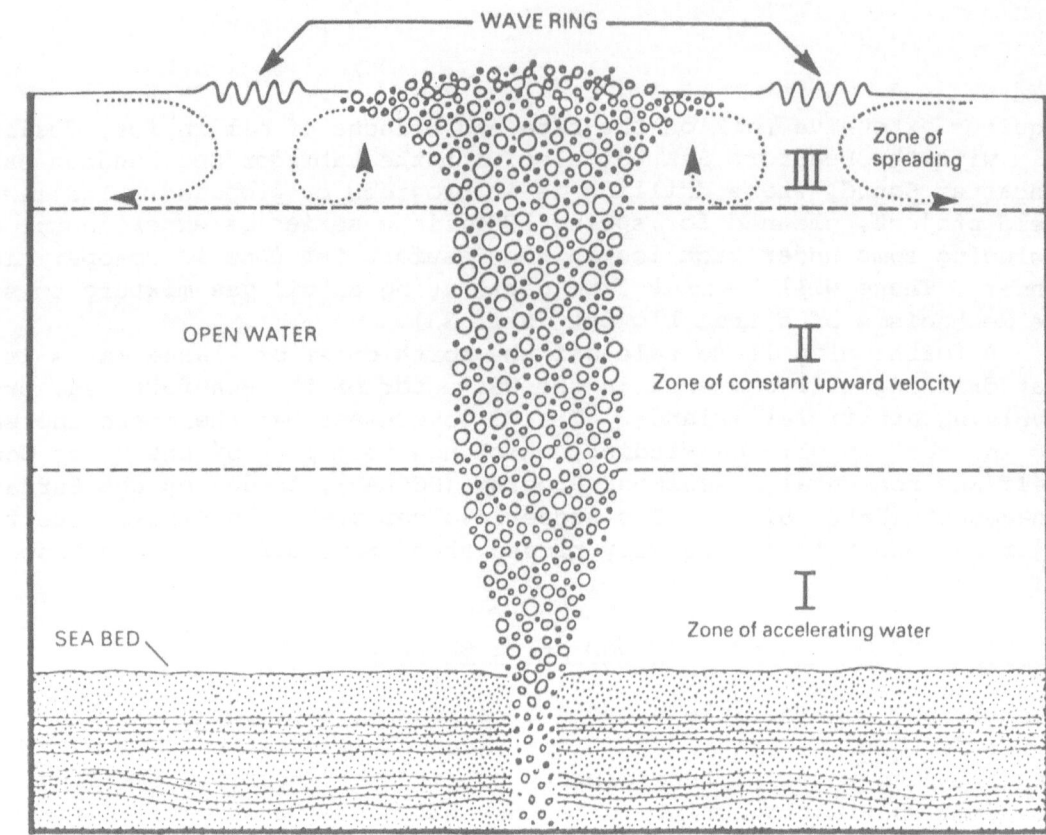

Fig. 2. The plume from a shallow oil-gas blowout in open water (after Ref. 25).

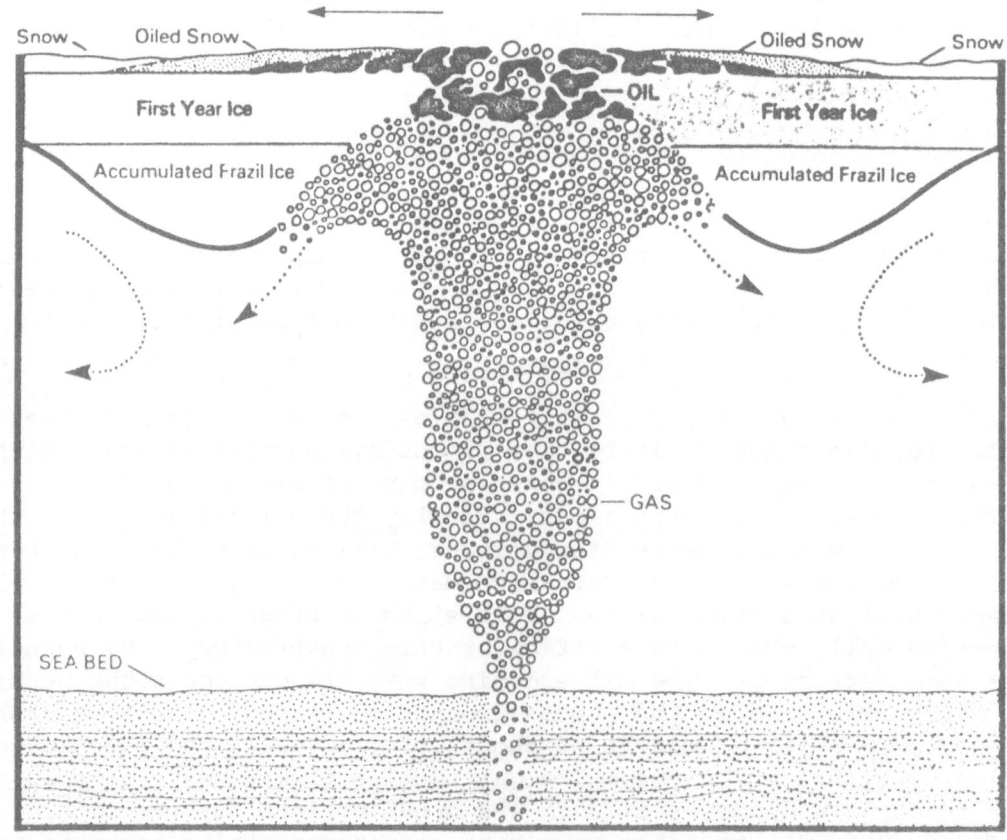

Fig. 3. The plume from an oil-gas blowout under smooth fast ice (after Ref. 25).

spread beyond the ring.

In deeper water three differences occur. Firstly the oil and gas are more likely to leave the seabed outlet in "slug" flow - with alternating slugs of oil and gas - rather than in "annular" flow, where the oil surrounds a core of gas rising through the pipe. This does not affect the shape of the plume, since mixing occurs soon after emergence. Secondly, there is the possibility of gas hydrate formation. Gas hydrates are waxy solids with specific gravities of 0.92-0.96, which form under large hydrostatic pressures. Topham (Ref. 40) saw hydrates forming around natural gas bubbles released at 325-640 m depth from a Pisces submersible and observed from the viewing port. In laboratory tests with a high pressure vertical water tunnel (Ref. 41) total hydrate formation occurred at pressures above 1130 p.s.i. (800 m of water). Thus in a very deep blowout there will be no bubble plume; instead the oil will rise slowly by itself in large drops, although it is possible that nearer the surface the hydrates will break down again. Thirdly, in a stratified water column it is possible for fluid to be shed from the edge of the plume, allowing the oil to spread further before reaching the surface (Ref. 42). The water depths for current drilling operations are too small for these phenomena to occur in the Beaufort Sea.

In open water what is observed at the surface is a turbulent circle of gas bubbles escaping through oil. The oil has partly cohered as a surface slick, is partly emulsified (depending on oil type) and is partly still in suspension as small droplets. The droplets which remain in the water column are swept outwards and under the wave ring, finally surfacing some distance away. The ring diameter is 70-80 m for 60 m water depth, and 40 m for 23 m depth, with the expanding surface vortex rings extending down to a depth of about 10 m. Fig. 2 assumes absence of currents, but in the presence of a surface current the oil droplets in the vortex will be swept downstream before reaching the surface. For instance, a small 0.05 mm diameter droplet (a lower limit for droplet size) initially at 10 m depth will take 5½ hours to reach the surface (Ref. 33), and a local current of 10 cm/sec will transport it 2 km downstream before surfacing.

Therefore our picture of a blowout in the open water season is that of oil initially contained within the ring of bubbles - where part of it could be burned off if the gas were kept continuously ignited - but quickly spreading beyond, where it comes under the influence of wind and surface currents. The normal phenomena of weathering then occur, more slowly in cold water, and the oil may menace the low-lying coastline and the feeding grounds of migratory birds (Ref. 46). If it drifts north, however, it will run into the loose floes at the edge of the summer pack ice. Laboratory experiments (Ref. 43,44) have shown that in loose pack ice the collisions and grinding between floes tend to pump oil and slush up into the floe edges which thus acquire an oiled rim. The oiled floes then continue to drift as part of the pack. This was also observed in the field during the 1979 "Kurdistan" oil spill off Cape Breton Island.

Under a smooth fast ice cover - corresponding to a blowout in the nearshore zone, possibly from a seabed fracture near an artificial island - the plume itself has the same shape as Fig. 2. However, its surface effect is different (Fig. 3). The entrained water causes a vertical heat flow which exceeds by at least an order of magnitude the heat content of the warm oil itself, and which is sufficient to melt ice above the plume at an estimated rate of 2 cm/day (Ref. 33). This means that if the blowout begins in the open water season, the vertical heat flux will be sufficient to keep a hole open above the plume throughout the winter. The gas will vent through this (Fig. 3). Tens of cu.m. of oil are contained within the hole, some of it slopping over the edge of the hole and oiling the surrounding snow cover. The energy balance with the atmosphere is maintained by the production of frazil ice at the surface, which is swept under the edge of the hole to accumulate under the surrounding ice as a rim which helps to contain the oil. In the coldest part of winter a continuous ice sheet may try to form over the hole, but this would soon be fractured by the gas from the blowout; the rim of frazil may then include sheets of rafted nilas. The hole itself, of course, can hold only a few hours' emission of oil, but this concentration does provide an opportunity for in situ burning or pumping out.

If a blowout under fast ice begins when there is already an overlying ice cover, the heat from the rising plume will begin to erode away the ice. However, what is likely to happen first is that the buoyancy of the gas entrapped under the ice will fracture the ice sheet by upward bending, provided there are no existing cracks or leads to enable the gas to vent away. The natural undulations in a smooth ice sheet are sufficient to trap a layer of gas of average thickness about 10 cm; it can be shown that a 70 m diameter zone of such gas will have sufficient upward thrust to rupture a 1 m thick ice sheet (100 m diameter for a 2 m sheet) (Ref. 33). This requires 30 minutes of emission. In deformed fast ice the area is much less because the average gas pocket thickness is greater. Therefore a pool will be created over the blowout, as in Fig. 3, but much more violently and the result will be an irregular hole or series of cracks, filled with oiled ice blocks.

Under moving pack ice in the transition zone the effect of the plume on the ice depends on the speed of the ice and the frequency of leads. If the ice pack remains stationary for a period, the situation of Fig. 3 will have a chance to develop, but as soon as the ice starts to move again the entire pool and contents will be carried off downstream. Whenever a lead or crack crosses the blowout site gas will be able to vent through it and oil will collect in it. If the ice cover is continuous and fast-moving it will simply collect gas in the troughs of its topography (Ref. 45) together with oil, as described in section 4.2. A slow-moving continuous ice cover, however, may rupture since the gas collects fast enough to exceed the buoyancy criterion for failure. If we convert the critical diameters given above to swath widths, we find that a 2 m ice cover will fracture if the ice is moving over the site at less than 6 cm/s, since the emitted gas will fill a swath of depth 10 cm and width 100 m. During the AIDJEX experiment such velocities occurred 54% of the time (Ref. 4). Therefore the trajectory of ice moving downstream from the site will reveal a trail composed partly of oiled leads, partly of discrete heavily oiled holes from periods of no motion, partly of ruptured floes displaying oil-coated rubble, and partly of unbroken ice with gas and oil hidden beneath it.

4.2. Entrapment of oil under ice

So far we have considered the plume and how it impinges on the ice cover. Next we consider how the oil is deposited and how it gets trapped in the ice. The simplest case is a perfectly smooth ice surface. Several experiments have been done on the behaviour of oil droplets at a smooth ice/water interface (Refs. 31, 47-49). The droplets behave like mercury on a table top, forming lenses or slicks whose thickness depends on the balance of surface tension versus buoyancy forces around the rim of the lens. A lens of large diameter has a thickness H given by

$$H^2 \, \delta\rho \, g = 2 \, T \, (1 - \cos \theta) \qquad\qquad (1)$$

where

$\delta\rho$ = density difference between oil and sea water;

T = surface tension at oil/ice interface;

θ = oil/ice contact angle measured within the oil.

Norman Wells crude, the properties of which are thought to resemble those of Beaufort Sea oil, has ρ = 0.847, T = 1.5 - 2. 10^{-2} N m^{-1}, θ = 140°-170°. This gives H = 0.55-0.68 cm, in good agreement with the thickness of slicks observed by diver when Norman Wells crude was spilled under moving pack ice (Ref. 26), which were in the range (0.56±0.08) cm. In laboratory tests most other oil types had H in the range 0.5-1 cm, although a few light crudes actually wetted the ice surface, i.e. spread out into a lens of infinitesimal thickness. We note that in one paper (Ref. 49) a theoretical value of 0.7 cm was derived for H which, although of the correct magnitude, was based on faulty reasoning employing heat flow arguments rather than surface tension.

H as given by (1) is a minimum value because any irregularity in the ice topography can allow the oil to pool in a pocket of greater thickness. An important question is that of the stability of such a slick or pocket when exposed to a shear current. Most evidence points to thin slicks being very stable. For instance, during the 1975 spill (Ref. 26) which took place in a surface shear current of 10 cm/s relative to the ice, the oil emitted from a point source spread downstream so long as fresh oil was being added, but the motion ceased as soon as the slick had reached an equilibrium position where its thickness was everywhere H or else defined by surface topography (some of the oil ran uphill into a depression alongside a pressure ridge, forming a pocket of thickness 10 cm). We conclude that once the oil droplets have hit the ice and cohered to form a continuous or discontinuous slick, they will not be stripped off again by currents to re-enter the water column. Their only subsequent motion will be a spreading along the ice either uphill under gravity or else in a generally downstream direction under the stress of new oil being added. It is estimated (Ref. 33) that a 2° slope is required for uphill motion and a 30 cm/s shear current (seldom attained in nature) for downstream motion in the absence of forcing by new oil addition. There is some evidence from laboratory studies (Ref. 50) that a thick pocket of oil, for instance held in a depression beside a pressure ridge keel, may be stripped away by a lower shear current through the generation and breaking of internal waves at the oil/water interface, but this requires the pocket to be both thick and long for the instability to develop, and so is probably rare in nature. Further, within a few hours of deposition, a slick is locked in position even more firmly because a raised lip of ice forms around its edge. This is because the low thermal conductivity of oil (0.05 to 0.07 that of ice) reduces the vertical heat flux through the ice above the oil slick, and as a result more heat flows to the surface from the area of the ice/water interface immediately around the edge of the slick, causing enhanced ice growth (Refs. 26, 33). It is therefore reasonably safe to assume that an oil slick which reaches an equilibrium position will then stay in place and will be carried passively by the ice wherever it goes.

We can now estimate the maximum area affected by a blowout under fast ice by assuming that the plume described in section 4.1 is a point source emitting a coherent slick of thickness H to the surrounding ice. In the absence of surface current the shape will be a circle; otherwise it will be more like an ellipse pointing downstream with the source as one focus. A year's emission (4.10^5 bbl) from a "standard blowout" will then occupy 12.7 km^2 if H = 5 mm and 6.4 km^2 if H = 1 cm. These are small figures, and under deformed fast ice the area will be even less. Kovacs (Ref. 45), for instance, found by impulse radar profiling that the containment factor (i.e. average oil layer depth) for a first-year floe was 2.7 cm and for a multi-year floe was 29.3 cm; 10 cm for deformed fast ice seems a reasonable figure, which will reduce the area of the spill to less than 1 km^2. The effects of a blowout in fast ice are therefore essentially local. One factor which may spread the oil over a greater area, however, is the possibility that gas pockets will fill all the irregularities under the ice; this will reduce the containment factor to below H, since the oil now sits at the water/gas interface in a thin film. Such a possibility will be investigated during the 1980 AMOP/Canmar discharge experiments (Ref. 51).

In a moving ice cover we have already mentioned ways in which oil can reach the upper surface of the ice - by discharge into leads, by gas-induced rupture of the ice sheet, or by melting of a hole at times of no motion - but if we ignore these and assume that the ice cover is continuous and moving fast enough to avoid rupture, we find that the oil slick deposited underneath it is very diffuse and discontinuous. If V is the ice velocity over the site and R is the rate of oil emission, then if the blowout is considered to be a point source the width W of a continuous oil swath traced out on the ice bottom by the blowout is

$$W = R / V H \qquad\qquad (2)$$

Using two typical values for V (4.8 and 9.6 km/day) and two for H (0.56 cm for smooth ice and 2 cm for rough ice) we find (Table I) that W always comes out much lower than the diameter of the bubble plume, which is typically 40-80 m for the 20-60 m drilling depths in the Beaufort Sea.

TABLE I

R bbl/day	V km/day	H cm	Swath width W m
1000	4.8	0.56	5.9
1000	4.8	2.0	1.7
1000	9.6	0.56	3.0
1000	9.6	2.0	0.8
2500	4.8	0.56	14.8
2500	4.8	2.0	4.1
2500	9.6	0.56	7.4
2500	9.6	2.0	2.1

Since oil is in fact deposited over the entire area of the plume surface, which can be thought of as a wide paintbrush, this result means that the ice moves too fast to receive a complete "painting" with oil. In fact the figures in Table I suggest that the coverage will be 10% or less, a very diffuse spattering of oil over the ice bottom. This spattering will take place in isolated lenses and drops each of minimum thickness 0.56 cm, since surface tension does not allow a thinner but continuous slick to exist. In section 4.4 we shall consider the horizontal trajectories of oiled floes, but it is important to recognise that an "oiled floe" implies, usually, a floe possessing a small percentage cover of oil in discrete lenses and drops.

4.3 Vertical motion of oil through ice

We now consider what happens to the oil after it has been deposited as a
stable lens or pocket on the ice underside in winter. The best field data
on its incorporation into the ice cover come from the 1974-5 controlled
discharge experiment in Balaena Bay for the BSP (Refs. 26, 52), and agree
well with the results of laboratory experiments (Ref. 53). The immediate
results of the deposition, apart from the growth of the ice rim mentioned
in section 4.2, is that the oil percolates some 1-2 cm up into the very
permeable skeletal layer where ice is actively growing in long dendritic
platelets. The oil also penetrates into the lower parts (about 7-8 cm
according to Ref. 52) of brine drainage channels. Eide and Martin (Ref.
54) have shown that discharge from brine drainage channels in the lower
part of an ice sheet occurs in regular pulses, the vertical tube of the
drainage channel acting as a "salt oscillator" which cycles between two
positions of hydrostatic equilibrium. Normally the pulse which expels
dense brine allows seawater to enter on the return stroke, but with a slick
beneath the skeletal layer it is oil that will enter instead; this has
been shown to occur in laboratory experiments. Thus within a short time
vertical filaments of oil have begun to extend into the ice sheet.

Meanwhile, within a week of deposition, the first ice has begun to
grow underneath the oil lens, so that the oil becomes completely encapsulated. The new ice (Ref. 52) is crystallographically isolated from the
ice sheet above, and as it grows it passes through the same stages of fabric
development as an ice sheet growing on an open water surface:- first a
polycrystalline region consisting of small crystals with random orientations
but with a preponderance of vertical c-axes; then a transition region with
an increasing proportion of horizontal c-axes; then, after 15-20 cm, a
columnar region with long vertical candle-shaped crystals with horizontal
c-axes. This crystallographic isolation may cause measurable changes in the
mechanical properties of the ice sheet. Initially the ice under the lens
grows more slowly than surrounding oil-free ice, partly because the crystal
structure does not offer such energetically favourable sites as the skeletal
layer of the oil-free ice, but mainly because the low thermal conductivity
of the oil gives the oil layer an "equivalent thermal thickness" of from
14 to 20 times its true thickness. Thermistor chains inserted through the
ice show a sharp increase in temperature gradient in the oil layer. After
weeks or months, however, this effect fades and may even reverse (Ref. 26).
One possibility is that the oil layer is being bridged by ice which has
formed from descending brine, but a more likely explanation is that vertical
convection cells have been set up within the oil lens, increasing its
effective conductivity. Theoretical studies (Ref. 33) show that such cells
can occur in oil layers as thin as 2-3 cm.

It is not until spring that vertical migration of the oil begins in
earnest. The increased solar radiation penetrates first-year ice down to
the level of the tips of the oil filaments, where energy absorption by the
dark oil increases its temperature and enables it to melt its way upwards,
initially following the line of the drainage channel, until it eventually
reaches the ice surface. The process is assisted by the fact that the
brine drainage channels (which by late winter may contain frozen sea water)
themselves melt in the spring, partly because of the high salinity of the
ice within them and partly because of the waveguide effect of the channel
walls in concentrating solar radiation. In the Balaena Bay experiment the
first oil appeared at the ice surface on May 10th in isolated patches, each
at the head of a brine drainage channel, and by July almost all of the
encapsulated oil had risen in this way. An indication of the ease with

which oil can rise through an open drainage channel was provided by an
experiment in Balaena Bay on May 15th (Ref. 52) in which oil was injected
under 2 m thick ice which had had its snow cover shovelled away. Within
45 minutes the oil had begun to reach the surface through drainage channels
spaced 0.1-0.5 m apart. A small proportion of the oil, instead of reaching
the surface, spread out to fill horizontal layers within the ice that were
possibly produced by meltwater percolation. In fast ice the process of oil
migration may be quicker than in first-year moving ice and may also leave a
greater proportion of the oil sitting in the open drainage channels. This
is because the channels are wider and closer together (Ref. 52), since the
fabric of the fast ice possesses pronounced directional·orientation of the
horizontal c-axes (Ref. 55).

Oil migration through multi-year ice is much more difficult. Such ice
is a sandwich of crystal fabrics, with annual layers of refrozen meltwater
near the surface and annual layers of winter growth near the bottom. Brine
drainage channels in winter exist only in the lower part of the ice sheet, in
the region of the present winter's ice growth (which may be 0.5 m or less),
and since the ice is thick (3 m or more) they receive a lesser amount of
insolation in the following spring. The oil finds it much more difficult
to melt its way upward, and probably will not do so in a single summer. The
worst possible case would be if the oil did not migrate at all; it would then
emerge on the surface only as a result of the natural annual cycles of
surface melting in summer followed by bottom freezing in winter. Since the
surface melt is about 0.5 m such a process would take 6 years, long enough for
the floe to have made almost a complete circuit of the Beaufort Gyre (which
take 7 - 10 years). This is a pessimistic view, and the likely time required
is 2-3 years although equivocal results from a small field test under multi-
year ice (Ref. 56) suggest that migration may have occurred in a single
summer.* It is very important that a large-scale field experiment be done.

Once on the ice surface the oil begins to weather in the same way as an
open water oil spill, partly by evaporation of lighter fractions and partly
by microbial degradation. While encapsulated in the ice the oil was
protected from such weathering, and sampling of such oil showed that its
composition was essentially unchanged from that of fresh oil. Therefore it
should be possible to burn oil which has just arrived on the surface, and
about 90% of the oil in the Balaena Bay experiment was successfully burned
in July (Ref. 26). This is the most promising, and perhaps the only
feasible, means of cleaning up oil spilled under moving pack. The oil will be
easily flammable for only a few days, and repeated burns may be necessary
since the oil does not all come up at once. Some method such as air
droppable igniters must be developed to repeatedly burn a large number of
small oil pools spread over a vast area.

The isolated patches of oil on the ice surface cause accelerated
melting in their vicinity and eventually float on melt pools of their own
making. These melt pools join to form larger pools and, if a thaw hole or
similar route to the underlying water develops, a number of pools will drain
towards this common centre, forming the typical spider-like surface drainage
pattern seen from the air in summer. At the thaw hole some of the oil carried
by the melt pool drainage may be driven back down below the ice surface to be
deposited on the nearby underside (Ref. 26), but most will remain at the
centre of the pattern. This concentrating mechanism does not facilitate
burning, however, since by now the oil is probably partly emulsified and too
heavily weathered to burn; an air-portable skimmer may be of use at this
stage. At the end of summer the weathered oil that remains on the ice
surface becomes incorporated in the refrozen melt pools that form the upper

* Field tests on the permeability of multi-year ice to sea water (Ref. 61) also suggest that
migration may occur in one year.

layer of what is now second - or multi-year ice, and continues to drift
for another winter until the next summer's surface melt frees it again.
    Oil that was deposited in leads during the winter forms a layer under
the thin ice that is encapsulated in the same way as for a thicker ice sheet,
except more quickly. If the ice is very thin or non-existent (a newly-opened
lead) the oil quickly forms a frozen sludge with the new ice that is forming
and the snow that is falling or blowing into the lead (Ref. 52). When the
lead is crushed to form a new pressure ridge the oiled blocks are incorporated in the structure of the ridge. Such a process has not yet been observed,
but it is probable that an oiled ridge would have a weaker structure than a
normal ridge and that the oil in the ridge sail would leach out early in the
spring and run out over the surrounding floes. The fate of oil in the ridge
keel is uncertain; it may rise vertically in spring due to the open permeable
structure of the ridge, or it may leach out of the keel laterally to form a
layer under the neighbouring smooth ice. Deposition in a lead followed by
leaching from a ridge sail is a mechanism for rapid vertical transport
through the ice cover. It is not a mechanism for horizontal transport, as
originally suggested by Campbell and Martin (Ref. 3) who postulated a "lead-
matrix pumping" effect whereby oil is pumped longitudinally from one lead
system to another as a lead closes. This cannot happen, since the lead edges
come together initially at a number of points of contact along the lead's
length; these isolate the closing lead into a number of independent segments.

4.4. Horizontal transport of oil around Beaufort Sea

Finally we examine the travels of an oiled ice floe during its winter drift
in the Beaufort Sea. Deposition under fast ice, of course, does not involve
any horizontal transport except from early July onwards, when the fast ice
breaks up and the fragments join the moving pack. By then the oil has already
appeared on the ice surface and should have been disposed of during the
seven-week period between its appearance and the fast ice breakup. Otherwise
large-scale pollution of the open water region can occur, of especial peril
to fish spawning areas and to migrating birds (Ref. 46).
    In moving pack ice, the chief problem is that every oiled floe which
passes over the blowout site will have a different subsequent trajectory, so
that the total downstream "trail" can only be modelled stochastically using
reasonable assumptions concerning ice dynamics at small and large scales.
In the author's opinion such modelling has not yet been done satisfactorily.*
One model (Ref. 57) has attempted to estimate the broad division of oil
quantities among geographical regions, given a "standard blowout" occurring
on October 5th at Canmar site A or B. The disposition of the oil is
calculated for the following May 5th, just before the first oil appears
on the ice surface and before the breakup of fast ice. The results are as
follows:-

TABLE II

		Site A (m^3)	Site B (m^3)
(i)	Oil coating shoreline of Tuktoyaktuk Peninsula in vicinity of Kugmallit Bay	3500	3500
(ii)	Oil blown northeastward towards Banks Island during autumn	5000	5000
(iii)	Oil encapsulated under landfast ice	22000	18000
(iv)	Oil encapsulated under moving pack ice and distributed westward in Beaufort Gyre	23000	27000

* because the models (Refs. 57, 62, 63) have followed the trajectory of the leading floe from the
blowout site without studying the breakup and subsequent diffusion of the trail of oiled floes
behind it.

The justification for (i) is that violent storms occur in the southern Beaufort Sea during October; a sequence of storms may break up the young fast ice, drive it out to sea, and then drive back the oil from the blowout to pollute the exposed shoreline. Storm surges (Ref. 58) may then drive this oil over the beaches, into a few of the shallow fringing lakes, and into bays and lagoons. Other storms would drive some oil to the northeast (type ii) where, after a winter under the ice of Amundsen Gulf, it would perhaps appear in the large polynya which opens there in early spring, and thence spread into the wide lead which opens across the edge of the fast ice zone. Sites A and B, being in different water depths, differ in the distribution of oil between fast and moving ice. Site A is located in the shear zone close to the fast ice, and so some of the oil spilled in early winter will be incorporated in the outer deformed fast ice, while much that is spilled in spring will go straight into the wide lead. Site B is further out in the moving pack, and it is difficult to agree with the reasoning which predicts large oil deposits under fast ice. In the author's opinion almost all of the oil listed as of types (iii) and (iv) will actually be of type (iii), i.e. deposited under the moving pack in which it will move generally westward.

The question then arises of how far the oil will travel in the pack. Using the drift data given in section 2, we can assume a mean distance made good to the west of some 2.4 km/day. By May, then, the leading edge of the oil will have reached some 500 km (400 km according to Ref. 57) to the west of the blowout site, putting it north of Point Barrow. To achieve this distance the leading oiled floe will have travelled some 6 km/day (according to AIDJEX data), i.e. 1250 km in all, with a meander coefficient of 2.5 (ratio between distance travelled and distance made good in mean direction of drift). The oil is then distributed between offshore Point Barrow and the blowout site in a ragged, broken trail, whose discontinuous and sinuous nature is determined by two sets of factors:-

(1) As described in sections 4.1 and 4.2, the ice drifting across the blowout site does not do so with constant velocity. At times of no motion (19% of the time according to Ref. 4) a fully- or partly-developed fast ice blowout scheme (Fig. 3) evolves, in which the hole and contents move off bodily when drift restarts. When the motion is slow, the ice "blows up" to leave a trail of ruptured, oiled blocks and rubble, while at times of fast motion a very diffuse trail is generated, consisting of a spattering of drops and lenses covering only a few percent of the ice underside over a swath width of 40-80 m. Such a trail is invisible except where a lead has passed over the site and become oiled. A reliable remote sensing technique for detecting oil under ice has not yet been developed (Ref. 59) so in winter there is no way of telling whether a floe is oiled or not except by tagging it as it passes over the blowout site.

(2) Each oiled floe follows its own route across the Beaufort Sea, although there is some coherence between the motions of adjacent trail segments as determined by general ice dynamics. The path of each trail segment cannot be modelled as an independent Markov process with a meander coefficient of 2.5, nor can the trail be thought of as a determinate, connected line. Instead the trail segments break up, rotate relative to one another and diverge so that the envelope of the oiled trail grows wider with increasing downstream distance. It is this process which has not yet been satisfactorily modelled, either to predict the envelope of the downstream trail or the disposition of oil concentrations within it.

The problems of modelling are manifest. The trail would have to be modelled in a number of segments with a time increment of δt. Each segment is of length $\underline{V}δt$ where $\underline{V}(t)$ is chosen realistically. The width of the segment is a constant value $\overline{S}$ (the width of the plume corresponding to the water depth at the blowout site) and the mean oil thickness within the segment is (R/VS), this being the actual thickness H of the lenses multiplied by the fraction of the ice "painted". The locus $\underline{x}(t)$ of each segment must then be calculated using realistic data or else a numerical model which takes account of small-scale deformation as well as bulk motion. No such model exists at present, nor are there adequate ice velocity data available, even from the AIDJEX experiment, since a synoptic net of ice velocities must be known for an entire winter.

The modelling of the downstream trail from a blowout is an important task, since the results define the magnitude and nature of the clean-up operation that must be carried out over a few weeks from mid-May onwards. All that can be said at present is that the oiled area will be about 500 km long, though of uncertain width, and that the oil will be randomly scattered within this area in a vast number of small patches.

5. CONCLUSIONS

Research into the consequences of an under-ice blowout in the Beaufort Sea has laid to rest some of the more extravagant fears that were expressed in the early stages of scientific concern. Modelling of the thermodynamics of oiled ice (Ref. 30), combined with an increased knowledge of oil behaviour on the upper ice surface, have shown that the climatic effect of a blowout is negligible, causing fluctuations which would certainly be far smaller than natural year-to-year variations in ice extent. The "lead-matrix pumping" mechanism of oil spread has also been discounted, and it is clear that the oil's motion is limited by, and in fact determined by, the normal motion of the ice which carries the oil. It is unlikely that any surprising new physical mechanisms of oil-ice interaction will be discovered to add to those described in this review, which therefore define the ways in which clean-up and containment might be attempted.

On the other hand, it is also clear that the extent of the oil spread from the blowout, the persistence of the oil in a toxic state in the environment, and, above all, the difficulty of clean-up, are all considerably greater than for a blowout in the open sea. The ice acts as a diffusing mechanism, or "blotting paper", which takes up the oil, preserves it in an unweathered state and spreads it in very low concentrations over a vast area. In much of that area there is no way of detecting the presence of the oil for several months, until it reappears in May possessing its full toxicity. Some of the oil will not appear for two or more years, during which time it will have travelled thousands of km across the Arctic Ocean. The remoteness and harsh climate of the Arctic Ocean make any cleanup attempt difficult, expensive, and dangerous to the lives of the operatives. Burning appears the only feasible option, but it must be carried out within a short time of the oil's appearance and requires air-droppable igniters because of the huge number of small oil patches that will appear. It is questionable how much oil can be disposed of in this way. Skimmers may have a role in certain well-defined locations, such as the wide lead at the edge of the fast ice, and sensitive coastal areas can be protected by booms in summer. Various schemes (e.g. trenching, subsea containment domes) have been devised to cope with the oil at the blowout site, although all of these are quite

314

impracticable except for a blowout under fast ice.

There is no space here to describe the biological effects of a blowout, except to say that, in relation to the standing crop of biota they would be severe. The chief damage would be to the several million migratory birds which rest in leads in spring and summer, to inshore fish spawning and nursery areas, and to the coastline. Fish would be tainted and the environmental stress on larger animals (seals, polar bears) would also be considerable. It is true, of course, that the commercial value of the "catch" from the Beaufort Sea (including birds later killed by "sportsmen" in lower latitudes) is small, probably less than $1 m per annum (Ref. 60), which is considerably less than the present annual cost of scientific research into blowout effects. Even a total kill of Beaufort Sea wildlife, which in any case is unlikely, would probably be less costly in commercial terms than the most modest attempt at clean-up. Why worry about oil spills at all, then, or bother to clean them up? One answer is that the social cost, through the destruction of the way of life of the few small Eskimo communities that still depend on hunting, would be immeasurable. Another is that in advanced industrial countries the unnecessary pollution of the environment is now regarded as unacceptable, especially a hitherto unspoiled environment such as the Arctic Ocean. The few remaining areas of absolute wilderness in the world are perceived as possessing an incalculable value because of their irreplaceable nature, so that we have a duty to minimise our disruption of such areas and, regardless of cost, to clean up whatever mess we create.

ACKNOWLEDGEMENTS

I wish to acknowledge the support of the British Petroleum Co. Ltd., and of the Natural Environment Research Council of Great Britain and the Office of Naval Research. The opinions expressed in the paper are, however, my own, as are any errors of fact.

REFERENCES

1. L.K. Coachman and C.A. Barnes, 'The contribution of the Bering Sea water to the Arctic Ocean'. Arctic, 14, (1961), 146-161.

2. L.K. Coachman, 'Physical oceanography in the Arctic Ocean: 1968'. Arctic, 22, (1969), 214-224.

3. W.J. Campbell and S. Martin, 'Oil and ice in the Arctic Ocean: possible large-scale interactions'. Science, 181, (1973), 56-58.

4. A.S. Thorndike and R. Colony, 'Large-scale ice motion in the Beaufort Sea during AIDJEX, April 1975 - April 1976'. In Sea Ice Processes and Models (R.S. Pritchard, ed.) Univ. Washington Press, Seattle, in press, 1979.

5. W.J. Stringer, 'Sea ice morphology of the Beaufort Sea shorefast ice'. In The Coast and Shelf of the Beaufort Sea (J.C. Reed, J.F. Sater, eds). Arctic Inst. N. Amer, Arlington, (1974), 165-172.

6 P.F. Cooper, 'Landfast ice in the southeastern part of the Beaufort Sea'. In The Coast and Shelf of the Beaufort Sea (J.C. Reed, J.F. Sater, eds). Arctic Inst. N. Amer., Arlington, (1974), 235-242.

7. W.B. Tucker, W.F. Weeks, A. Kovacs and A.J. Gow, 'Nearshore ice motion at Prudhoe Bay, Alaska'. In Sea Ice Processes and Models, (R.S. Pritchard, ed.) Univ. Washington Press, Seattle, in press, (1979)

8. P. Wadhams and R.J. Horne, 'An analysis of ice profiles obtained by submarine sonar in the AIDJEX area of the Beaufort Sea'. Scott Polar Res. Inst., Cambridge, Tech. Rept. 78-1, (1978) 65pp. Also J. Glaciol., in press.

9. P. Wadhams, 'Ice characteristics in the seasonal sea ice zone'. Workshop on problems of the seasonal sea ice zone, Monterey, Feb-Mar 1979. Proc. in press, Office of Naval Research.

10. W.B. Tucker, W.F. Weeks and M. Frank, 'Sea ice ridging over the Alaskan continental shelf'. J. Geophys. Res., 84 (C8),(1979), 4885-4897.

11. W.D. Hibler, S.F. Ackley, W.K. Crowder, H.L. McKim and D.M. Anderson, 'Analysis of shear zone deformation in the Beaufort Sea using satellite imagery'. In The Coast and Shelf of the Beaufort Sea, (J.C. Reed, J.F. Sater, eds). Arctic Inst. N. Amer., Arlington, (1974), 285-296.

12. E. Reimnitz, L. Toimil and P. Barnes, 'Arctic continental shelf processes and morphology related to sea ice zonation, Beaufort Sea, Alaska'. AIDJEX Bull., 36, (1977), 15-64.

13. M.D. Coon, 'A review of AIDJEX modelling'. In Sea Ice Processes and Models, (R.S. Pritchard, ed). Univ. Washington Press, Seattle, in press, (1979).

14. W.D. Hibler and W.B. Tucker, 'Some results from a linear-viscous model of the Arctic ice cover'. J. Glaciol., 22(87), (1979), 293-304.

15. C.W.M. Swithinbank, 'Northwest Passage trade route by supertanker'. Geographical Mag., 42(7), (1970), 478-490.

16. Report of the Task Force, Operation Oil (Clean up of the ARROW oil spill in Chedabucto Bay). (1970). Ministry of Transport, Ottawa. 3 vols.

17. F.G. Barber, 'Oil spilled with ice: some qualitative aspects'. Proc. Joint Conf. on Prevention and Control of Oil Spills, Washington D.C., June 15-17 1971, 133-137.

18. W.D. Forrester, 'Distribution of suspended oil particles following the grounding of the tanker "Arrow"'. J.Mar. Res., 29(2), (1971) 151-171.

19. J.L. Glaeser, 'A discussion of the future oil spill problem in the Arctic'. Proc. Joint Conf. on Prevention and Control of Oil Spills, Washington D.C., June 15-17 1971, 479-484.

20. G.D. Green, P.J. Leinonen and D. Mackay, 'An exploratory study of the behaviour of crude oil spills under ice'. Can. J. Chem. Engng, 55(6), (1977), 696-700.

21. R.C. Ayers, H.O. Jahns and J.L. Glaeser, 'Oil spills in the Arctic Ocean: extent of spreading and possibility of large-scale thermal effects', Science, 186 (1974), 843-846.

22. J. Hnatiuk, 'Results of an environmental research program in the Canadian Beaufort Sea'. Proc. 8th Offshore Technology Conf., Houston, May 1976.

23. P.Wadhams, 'Oil and ice in the Beaufort Sea'. Polar Record, 18(114), (1976), 237-250.

24. S.L. Ross, W.J. Logan and W. Rowland, 'Oil spill countermeasures'. BSP Overview Rept. M6, Dept. of Fisheries and Envt., Ottawa, (1977).

25. A.R. Milne, 'Oil, ice and climate change'. BSP Overview Rept. M5, Environment Canada, Victoria, (1978).

26. D. Dickins, J. Overall and R. Brown, 'The interaction of crude oil with Arctic Sea ice'. BSP Tech. Rept. 27, Inst. of Ocean Sciences, Patricia Bay, B.C., (1975).

27. D.R. Topham, 'Hydrodynamics of an oilwell blowout'. BSP Tech. Rept. 33, Inst. of Ocean Sciences, Patricia Bay, B.C., (1975).

28. J.R. Marko, 'Satellite observations of the Beaufort Sea ice cover'. BSP Tech. Rept. 34, Inst. of Ocean Sciences, Patricia Bay, B.C., (1975).

29. P. Wadhams, 'Sea ice morphology in the Beaufort Sea'. BSP Tech. Rept. 36, Inst.of Ocean Sciences, Patricia Bay, B.C., (1975). Reprinted as "Sea ice topography in the Beaufort Sea and its effect on oil containment". AIDJEX Bull., 33, 1-52 (1976).

30. E.R. Walker, 'Oil, ice and climate in the Beaufort Sea'. BSP Tech Rept. 35, Inst. of Ocean Sciences, Patricia Bay, B.C., (1975).

31. L.W. Rosenegger, 'Movement of oil under sea ice'. BSP Tech. Rept. 28, Inst. of Ocean Sciences, Patricia Bay, B.C., (1975).

32. A.R. Milne and B. Smiley, 'Offshore drilling for oil in the Beaufort Sea: a preliminary environmental assessment'. BSP Tech. Rept. 39, Inst. of Ocean Sciences, Patricia Bay, B.C., (1975).

33. E.L. Lewis, 'Oil in sea ice'. Pacific Marine Sci. Rept. 76-12, Inst. Ocean Sciences, Patricia Bay, B.C., (1976), 26pp.

34. S.L. Ross, 'Arctic Marine Oilspill Program (AMOP)'. Spill Technology Newsletter, 2(1), (1977), 25-32. Envtl. Protection Service, Ottawa.

35. Experimental Oilspills General Plan. AMOP, Envtl. Protection Service, Ottawa, May 1979.

36. 'Environmental assessment of the Alaskan continental shelf: interim synthesis in Beaufort/Chukchi'. OCSEAP Office, Envtl. Res. Labs, Natl. Oceans & Atmospheres Admin., Boulder, Co. 362pp. (1978).

37. G. Weller, 'Oil pollution in ice-covered Arctic waters'. Proc. 5th Intl. Conf. on Port & Ocean Engng Under Arctic Condns., Trondheim, (1979), 393-406. Univ. Trondheim/Norwegian Inst. Tech.

38. F.G. Bercha, 'Probabilities of blowouts in Canadian Arctic waters'. Rept. EPS-3-EC-78-12, Envtl. Protection Service, Ottawa. 139pp (1978).

39. A.S. Telford and M. Metge, 'Preliminary study of the fate of oil from a subsea blowout on the east coast'. Rept. IPRT-2ME-77, Imperial Oil Product Res. Div., Calgary. (1977).

40. D.R. Topham, 'Observations of the formation of hydrocarbon gas hydrates at depth in seawater'. Spill Technology Newsletter, 2(4), (1977), 185-187. Envtl. Protection Service, Ottawa.

41. P.R. Bishnoi and B.B. Maini, 'Laboratory study of behaviour of oil
     and gas particles in salt water relating to deep oil well blowouts'.
     <u>Spill Technology Newsletter</u>, <u>4</u>(1), (1979), 24-36. Envtl. Protection
     Service, Ottawa.

42. D.E. Thornton, 'The flow structure of an underwater oil blowout'. <u>Spill
     Technology Newsletter</u>, <u>3</u>(1), (1978), 46-59. Envtl. Protection
     Service, Ottawa.

43. M. Metge and A.S. Telford, 'Oil in moving pack ice - laboratory study'.
     Proc. 5th·Intl'. Conf. oh Port & Ocean Engng under Arctic Condns,
     Trondheim, (1979). Univ. Trondheim/Norwegian Inst. Tech.

44. S. Martin, P. Kauffman and P.E. Welander, 'A laboratory study of the
     dispersion of crude oil within sea ice grown in a wave field'.
     <u>Science in Alaska</u>, Vol. II, Proc. 27th Alaska Sci. Conf. Amer.
     Assoc. Advancement of Sci., Fairbanks. (1976).

45. A.Kovacs, 'Sea ice thickness profiling and under-ice oil entrapment'.
     Proc. 9th Offshore Technology Conf., Houston, (1977), <u>3</u>, 547-554.

46. E.L. Lewis, 'Some possible effects of Arctic industrial developments
     on the marine environment'. Proc. 5th Intl. Conf. on Port & Ocean
     Engng Under Arctic Condns., Trondheim, <u>1</u>, (1979), 369-392. Univ.
     Trondheim/Norwegian Inst. Tech.

47. L.W. Rosenegger, 'Oil-in-ice studies'. Lab. Rept. L-12075, Imperial
     Oil Co. Ltd., Res. Prodn. & Tech. Services Lab., Calgary, (1975).

48. D. Mackay, M. Medir and D.E. Thornton, 'Interfacial behaviour of oil
     under ice'. <u>Can. J. Chem. Engng</u>, <u>54</u>, (1975), 72-74.

49. L. S. Wolfe and D.P. Hoult, 'Effects of oil under sea ice'. <u>J. Glaciol.</u>,
     <u>13</u>(69), (1974), 473-488.

50. J.R. Moir and Y.L. Lau, 'Some observations of oil slick containment by
     simulated ice ridge keels'. Unpubl. Rept., Canada Centre for
     Inland Waters, Burlington, Ont., (1975).

51. W.M. Pistruzak, 'A proposed study of oil and gas under ice'. <u>Spill
     Technology Newsletter</u>, <u>4</u>(5), (1979), 304-313. Envtl. Protection
     Service, Ottawa.

52. S. Martin, 'A field study of brine drainage and oil entrapment in
     first-year sea ice'. <u>J. Glaciol.</u>,in press, (1979).

53. B.E. Keevil and R.O. Ramseier, 'Behaviour of oil spilled under floating
     ice'. Proc. Conf. on Prevention & Control of Oil Pollution, San
     Francisco, March 25-27 1975, 497-501.

54. L. I. Eide and S. Martin, 'The formation of brine drainage features in
     young sea ice'. <u>J. Glaciol.</u>, <u>I4</u>(70), (1975), 137-154.

55. W.F. Weeks and A. Gow, 'Preferred crystal orientation in the fast ice
     along the margins of the Arctic Ocean'. <u>J. Geophys. Res.</u>,
     <u>83</u>(ClO), (1978), 5105-5122.

56. D.E. Thornton, 'Oil-ice interaction'. <u>Spill Technology Newsletter</u>,
     <u>4</u>(3), (1979), 160-161. Envtl. Protection Service, Ottawa.

57. Environmental Protection Service, Ottawa, 'Oil-spill countermeasure
     study for the southern Beaufort Sea'. BSP Tech. Rept. 31a.
     Inst. Ocean Sciences, Patricia Bay, B.C., (1975).

318

58. R.F. Henry, 'Storm surges in the southern Beaufort Sea'. BSP Tech Rept. <u>19</u>, Inst. Ocean Sciences, Patricia Bay, B.C., (1975).

59. C-CORE. 'Investigations of the Use of Microwave Systems in detecting and monitoring oil slicks over ice and ice-infested waters'. Contract Rept. 78-18 (1978), Centre for Cold Oceans Resources Engng., Memorial Univ., St. John's, Nfld.

60. D. Mackay, 'Why clean up oil spills?' <u>Spill Technology Newsletter</u>, <u>3</u>(4), (1978), 215-220. Envtl. Protection Service, Ottawa.

61. A.R. Milne, R.H. Herlinveaux and G. Wilton, 'A field study on the permeability of multiyear ice to sea water with implications on its permeability to oil'. Technology Development Rept. EPS-4-EC-77-11, Envtl. Impact Control Directorate, Fisheries & Envt. Canada, Ottawa. (1977)

62. NORCOR Engng & Res. Ltd, 'Probable behaviour and fate of a winter oil spill in the Beaufort Sea'. Technology Development Rept. EPS-4-EC-77-5, Envtl. Impact Control Directorate, Fisheries & Envt. Canada, Ottawa (1977).

63. D.R. Thomas and R.S. Pritchard, 'Oil movement in the ice covered Beaufort and Chukchi Seas'. Res. Rept. 138, Flow Res. Co., 21414-68th Ave S, Kent, Wash. (1979).

THE CHEMICAL AND BIOLOGICAL DEGRADATION OF PETROLEUM: A FOREMOST CHALLENGE FOR THE ANALYTICAL CHEMIST

Donald C. Malins, PhD, DSc
Director, Environmental Conservation Division
Northwest and Alaska Fisheries Center
National Marine Fisheries Service
Seattle, Washington, U.S.A.

INTRODUCTION

Petroleum consists of complex mixtures of aliphatic, alicyclic, and aromatic hydrocarbons, as well as polar organic compounds and high molecular weight polymers [1]. These compounds may be transformed into other structures on entering the marine environment. The predominant type of molecular change is oxidation. This occurs either through abiotic chemical reactions (usually catalyzed by light) or chemical and enzymatic reactions in microorganisms and higher forms of life. Thus, after petroleum enters the marine environment, the proportions of oxygenated compounds increase substantially. This results in a dramatic change in the composition and probably the toxicity of "petroleum".

Attempts to analyze for the oxygenated petroleum products are hampered by limitations in analytical techniques. For example, present methodology (e.g., glass capillary-gas chromatography) is largely confined to the determination of hydrocarbons and certain other weakly polar structures, notably those containing up to 5 or 6 fused rings.

The study of petroleum pollution necessitates both an understanding of the chemical and biological conversions of petroleum compounds in marine environments and the development of methods to analyze samples for oxidation products. This paper gives an overview of (a) what is known about the oxidative conversions of petroleum hydrocarbons in marine environments, and (b) steps being taken in our laboratories to analyze for oxygen-containing petroleum products in laboratory and field samples.

POLAR COMPONENTS OF PETROLEUM

The proportions of polar compounds in different types of petroleum vary considerably [1]. Prudhoe Bay (Alaska) crude oil contains about 5% polar and insoluble compounds, whereas No. 6 fuel oil (Bunker C) contains over 40% [1] (Fig. 1). The polar oxygenated compounds consist primarily of diols, phenols and carboxylic acids; however, ketones, esters, lactones, ethers and anhydrides are also present. The polar functional groups are associated with both straight-chain and closed-ring compounds. Gas chromatography is routinely used to analyze for aliphatic, alicyclic, and aromatic hydrocarbons and certain weakly polar compounds, such as phenols and alkylated benzothiophenes [2]. However, high boiling point compounds are generally not analyzed by gas chromatography. In No. 6 fuel oil, for example, about 5% of the total hydrocarbon fraction (the oil is composed of 55% hydrocarbons) is routinely separated, identified and quantitated by gas chromatographic methods employed by environmental chemists [1]. As degradative changes produce high-boiling point conversion products in the marine environment, analyses for hydrocarbons alone become progressively less reflective of petroleum pollution.

320

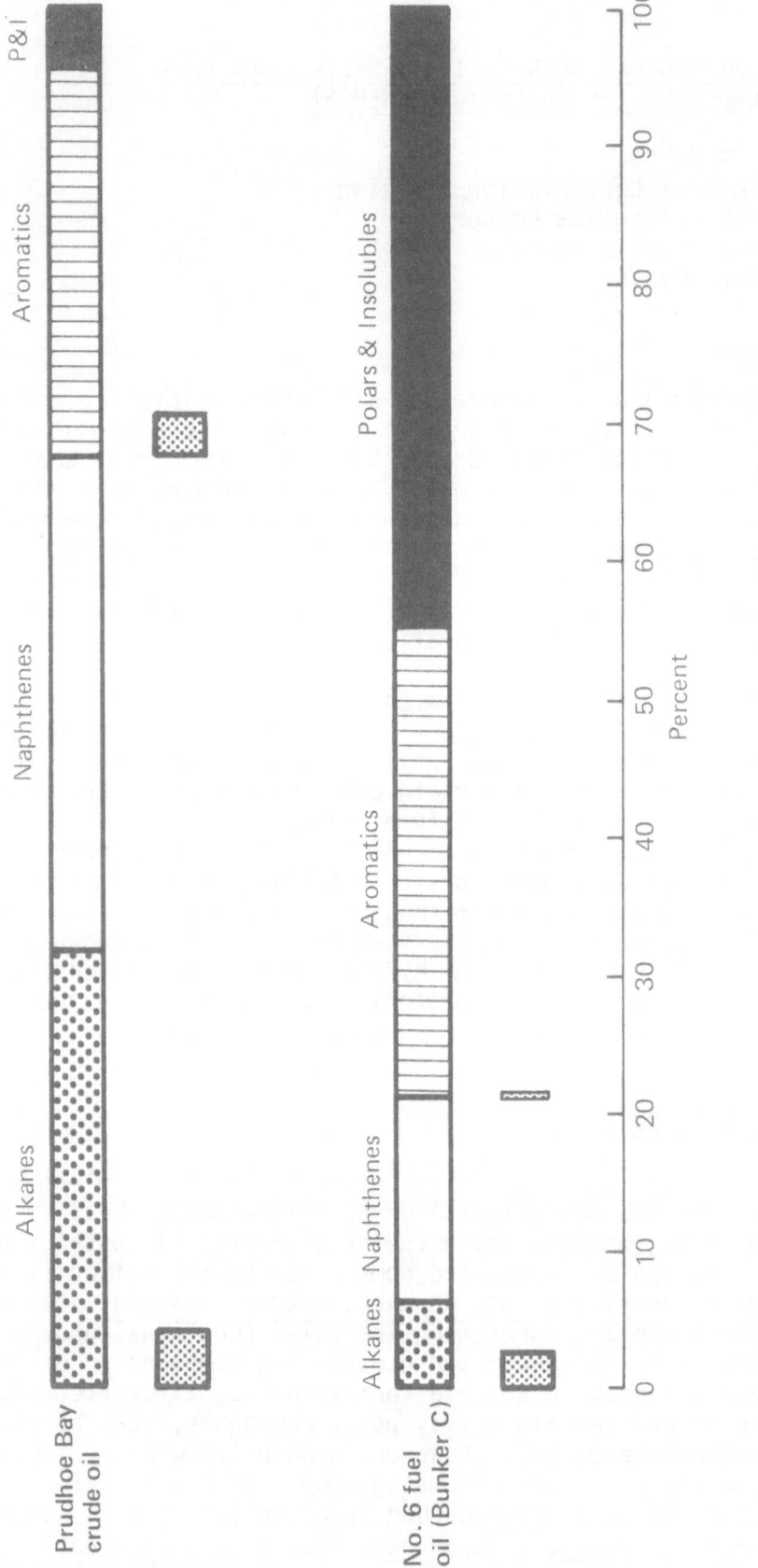

FIG. 1. The limited amount of hydrocarbon data routinely reported by environmental chemists for oil pollution studies (Clark, R.C., Jr., D.W. Brown, and W.D. MacLeod, Jr., unpublished data [1]).

CHEMICAL OXIDATIONS

Free radical chain reactions catalyzed by light (photooxidation) lead to the formation of a variety of oxygen-containing derivatives. Burwood and Speers [3] demonstrated that the components of saline-soluble extracts of Middle East crude oil undergo extensive oxidation when aerated (Fig. 2). Partial gas chromatograms revealed relatively large "envelopes" of oxidized products under both subdued and direct sunlight. These envelopes were substantially greater in magnitude than those obtained from partial gas chromatograms of non-aerated samples. Chemical reactions responsible for the profound changes observed by Burwood and Speers [3] involve the following conversions: chain initiation (free-radical formation), oxygen addition, chain propagation, and chain termination (Fig. 3). Some of the products are alcohols, ethers, dialkyl peroxides, and carbonyl compounds. Polymers are also formed when more than one free radical attack occurs on a single molecule.

Burwood and Speers [3] suggest that the thialkanes in sulfur-containing crude petroleums (e.g., Kuwait, Alaskan, and North Sea crude oils) [4] undergo oxidation to thiacyclane oxides (Fig. 4). Thus, thiacyclopentane, thiahydrindane and perhydrothiaxanthene undergo oxidative conversions to yield thiacyclane oxides and phenols. Further oxidations result in the formation of a variety of carbonyl compounds. These reactions lead to pollutants of increased boiling point and seawater solubility. Under natural environmental conditions a number of factors regulate the type and extent of oxidation. Examples are oxygen concentrations, irradiation intensity and time, and the thickness of oil films.

Patel et al. [5], for example, have provided data on light catalyzed oxidations of aromatic hydrocarbons. They showed that phenanthrene is converted to a variety of oxygen-containing compounds under simulated environmental conditions. Patel et al. [6] have also shown in laboratory and field studies that various alkylated dibenzothiophenes are extensively converted to oxides via photooxidation.

OXIDATIONS PERFORMED BY MICROORGANISMS

The statement was made [7] that "microbial degradation of oil is undoubtedly the most important process involved in weathering and eventual disappearance of petroleum from the marine environment." Microorganisms have a general ability to degrade petroleum compounds. ZoBell [8] showed that even microorganisms from the arctic are able to degrade petroleum. Moreover, marine sediments exposed to petroleum contain high numbers of hydrocarbon-utilizing microorganisms [9]. The psychrophilic pseudomonads are ubiquitous, generally the dominant species in marine environments. Hence the hydrocarbon conversions they carry out are of particular interest.

Leadbetter and Foster [10] proposed that the oxidation of alkanes by pseudomonads proceeds via free radical formation. The reactions yield hydroperoxides, secondary alcohols, ketones, and fatty acids. Oxidations of this type proceed readily with alkane substrates varying from C_6 through C_{18} chain-lengths [11].

Trudgill [12] reviewed the microbial conversions of alicyclic hydrocarbons. He stressed the importance of hydroxylation reactions in initial stages of cycloalkane degradation in both pure- and mixed-cultures. The initial formation of hydroxy derivatives allows subsequent oxidative reactions to occur that ultimately lead to ring cleavage. Pathways have been

322

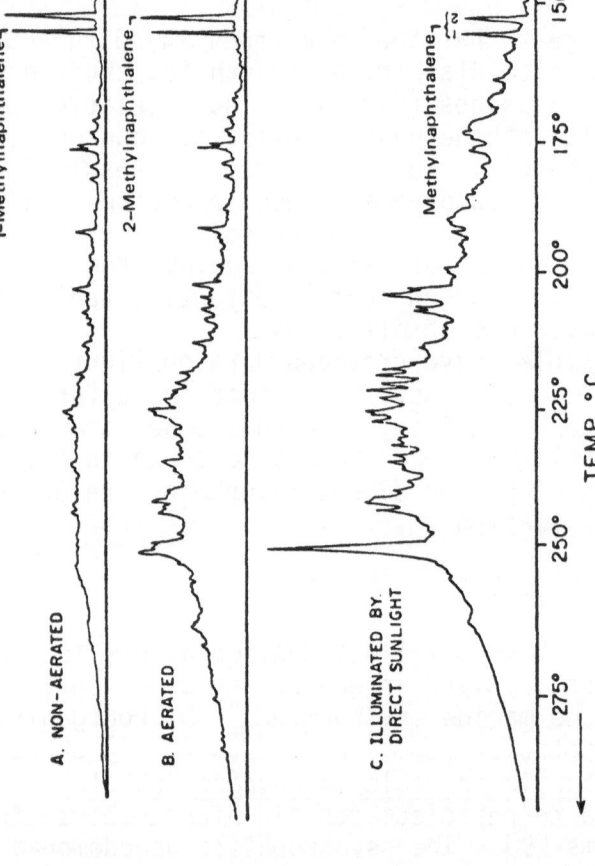

FIG. 2. Partial gas chromatograms of concentrated total extracts obtained from: (A) Non-aerated crude oil--saline equilibration; (B) aerated crude oil--saline equilibration; (C) solar illuminated crude oil--saline equilibration (Burwood and Speers [3]).

Chain initiation: $R_2 \rightarrow 2 R\cdot$

Oxygen addition: $R\cdot + O_2 \rightarrow ROO\cdot$

Chain propagation: $R'H + R\cdot \rightarrow R'\cdot + RH$

$RH + ROO\cdot \rightarrow R\cdot + ROOH$

$ROOH \rightarrow RO\cdot + \cdot OH$

$AH + RO\cdot \rightarrow A\cdot + ROH$

$2 ROO\cdot \rightarrow O_2 + 2 RO\cdot$

Potential chain terminations: $R\cdot + R'\cdot \rightarrow RR'$

$RO\cdot + R\cdot \rightarrow ROR$

$2 RO\cdot \rightarrow ROH + R'=0$

$2 RO\cdot \rightarrow ROOR$

FIG. 3. Steps in the chemical oxidation of petroleum hydrocarbons (K.U. Ingold, Pure Appl. Chem. 15 (1967) 49-67).

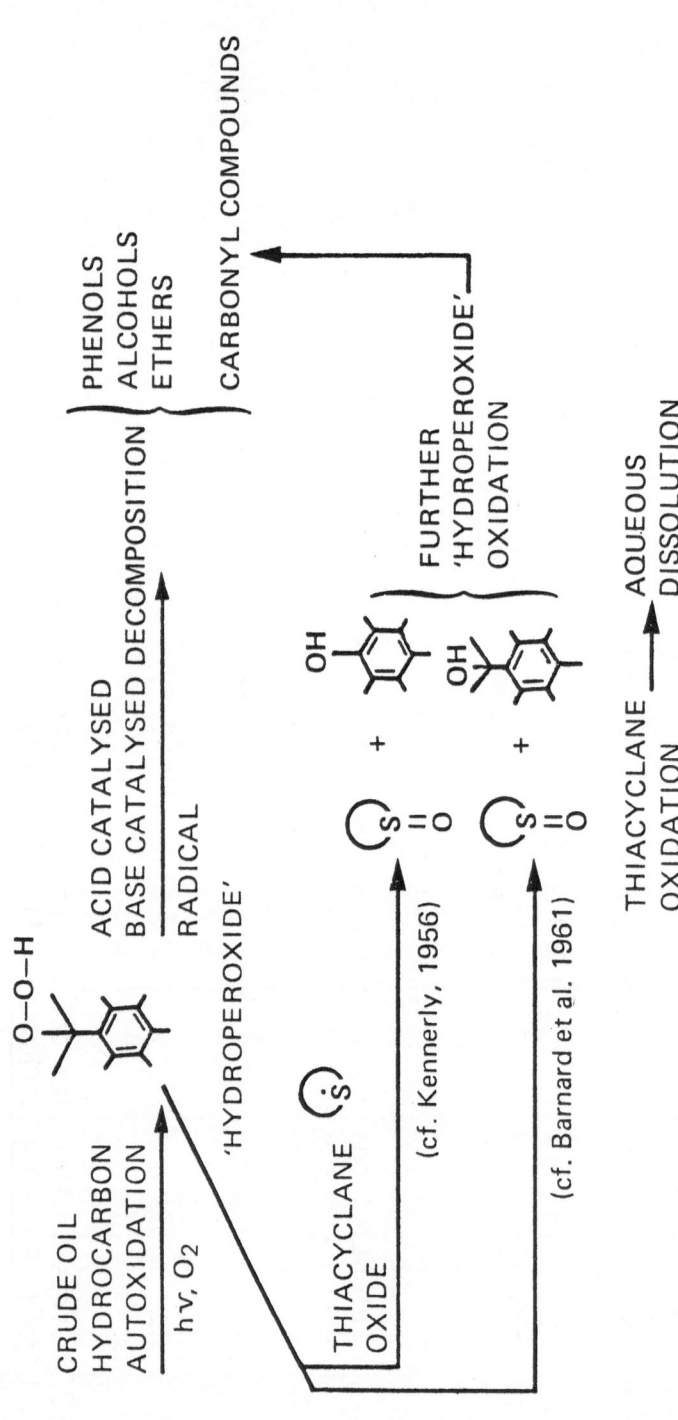

FIG. 4. Suggested route for crude oil hydrocarbon and thiacyclane autoxidative processes (Burwood and Speers [3]).

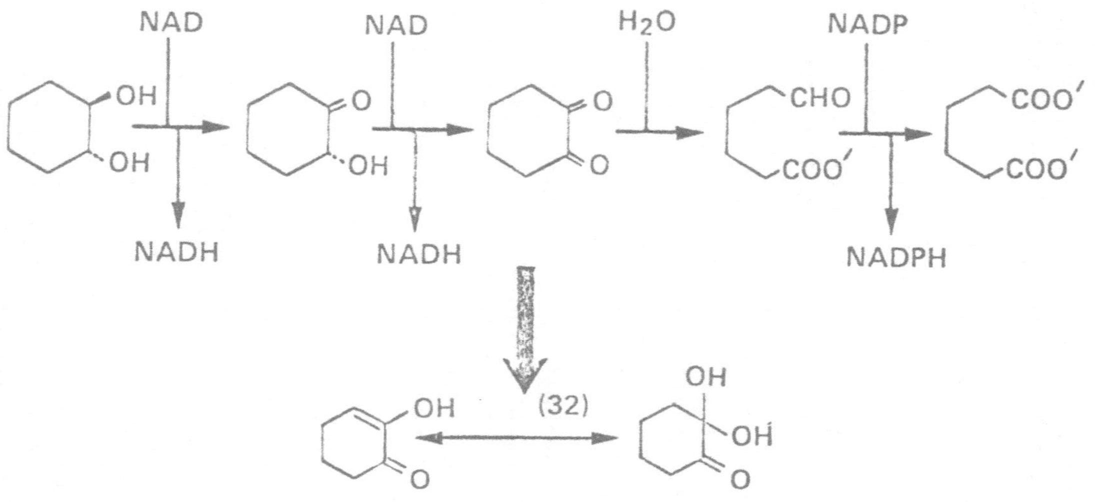

FIG. 5. The pathway proposed for the degradation of trans-cyclohexane-1,2-diol by Pseudomonas C-4. (Taken from the data of Yugari [11] as modified by Trudgill [12]).

proposed for the degradation of trans-cyclohexane-1,2-diol by a Pseudo-monas sp. (Fig. 5) [13]. Formation of cyclohexane-1,2-dione (an unstable intermediate) occurs in the presence of oxidized pyridine nucleotide (NAD). Hydration of this transitory molecular species results in ring cleavage or the formation of the mono-enol in equilibrium with the monohydrate. The conversion of cyclohexane demonstrates how microorganisms produce a number of relatively stable oxidation products.

The microbial oxidation of aromatic hydrocarbons was reviewed by Gibson [14], Karrick [7], and Hopper [15]. Non-substituted and alkyl-sub-stituted aromatic hydrocarbons are converted to dihydrodiols by microoga-nisms, such as Pseudomonas sp. and Alcaligenes sp. [14]. The dihydrodiols undergo dehydrogenation to form ortho-dihydroxy derivatives (catechols). Catechols may then undergo enzymatic ring cleavage in the presence of oxy-gen to form carboxyclic acids. For example, catechol forms cis, cis-muconic acid via ortho-cleavage and -hydroxymuconic semialdehyde via meta-cleavage.

Hopper [15] described the oxidative degradation of the methyl-substi-tuted aromatic hydrocarbon, toluene, by a Pseudomonas sp. (Fig. 6). Both ring and alkyl-group oxidations occur. Ring oxidations yield cis-2,3-dihydro-2,3-dihydroxytoluene which is converted to 3-methyl catechol. Alkyl-group oxidations produce methanol derivatives, such as benzyl alco-hol which is converted to benzaldehyde. The aldehyde is then further oxidized to benzoic acid which is converted to catechol. The catechol [14] is then further oxidized through ring cleaving reactions.

Atlas et al [16] have shown in studies of the Amoco Cadiz oil spill that the microbial degradation of alkanes and aromatic hydrocarbons was extensive, and that substantial amounts of oxidized compounds derived from petroleum were present in the water column. Moreover, Ballerini et al. [17] demonstrated in laboratory studies with a mixed culture of bacteria obtained from littoral sediments polluted by Amoco Cadiz oil that over 50 percent of the saturated hydrocarbons and about 40 percent of the aromatic hydrocarbons of crude oil were degraded. Significant amounts of water soluble oxidized products were found.

In summary, it is evident that aliphatic, alicyclic, and aromatic hydrocarbons in petroleum are highly susceptible to oxidative degradation by microorganisms [18]. The products are complex mixtures of polar compounds of largely unknown toxicity that are not presently being determined in the search for petroleum contamination in marine environments.

OXIDATIONS PERFORMED BY MARINE ANIMALS

Emphasis is placed on conversions of aromatic hydrocarbons because these compounds have been most extensively studied and are considered to be the most toxic to marine organisms. In 1962 Brodie and Maickel [19] sug-gested that fish do not have the capacity to convert aromatic hydrocarbons to metabolites; however, a substantial body of evidence now exists indi-cating that such a capacity is widespread among marine life. The conver-sion of aromatic hydrocarbons to oxygenated derivatives takes place either via the mixed function oxygenase (MFO) systems or through nonenzy-matic chemical reactions. Studies from our laboratory [20-29] and else-where [30-32] have shown that these transformed products are complex mix-tures of both conjugated and nonconjugated derivatives. An example of the extent of the oxidative changes taking place in marine fish exposed to naphthalene is given in Fig. 7. Starry flounder (Platichthys stellatus)

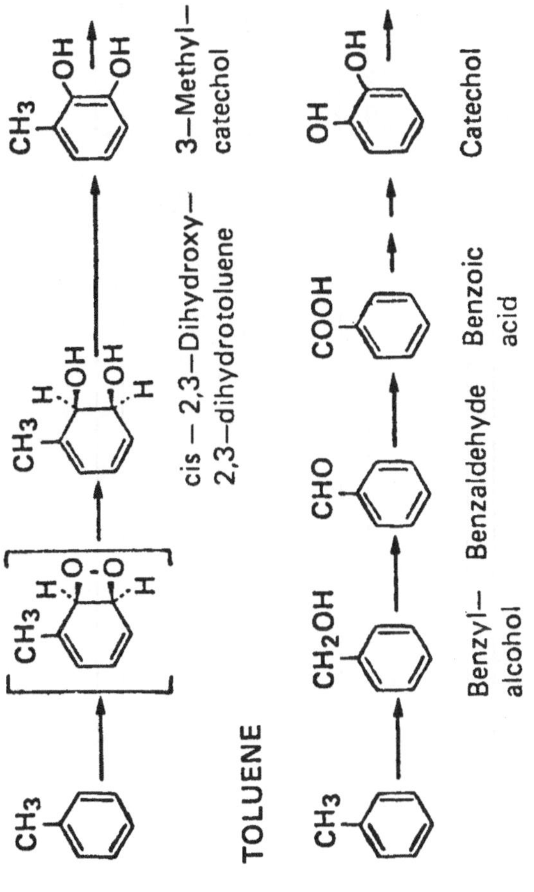

FIG. 6. Pathways for toluene metabolism (Hopper [15]).

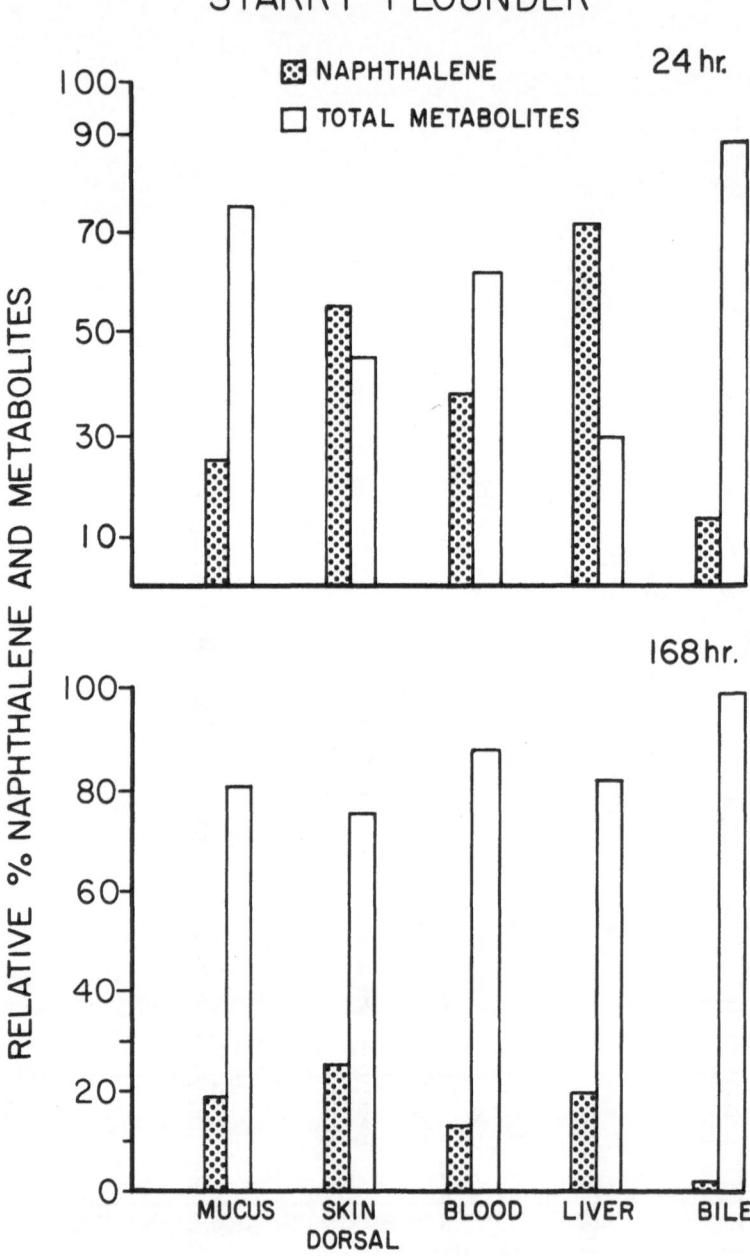

FIG. 7. Relative percent of naphthalene and its metabolic products (expressed) as naphthol) in tissues of P. stellatus exposed to ^{3}H-naphthalene via force-feeding (adapted from Varanasi et al. [24]).

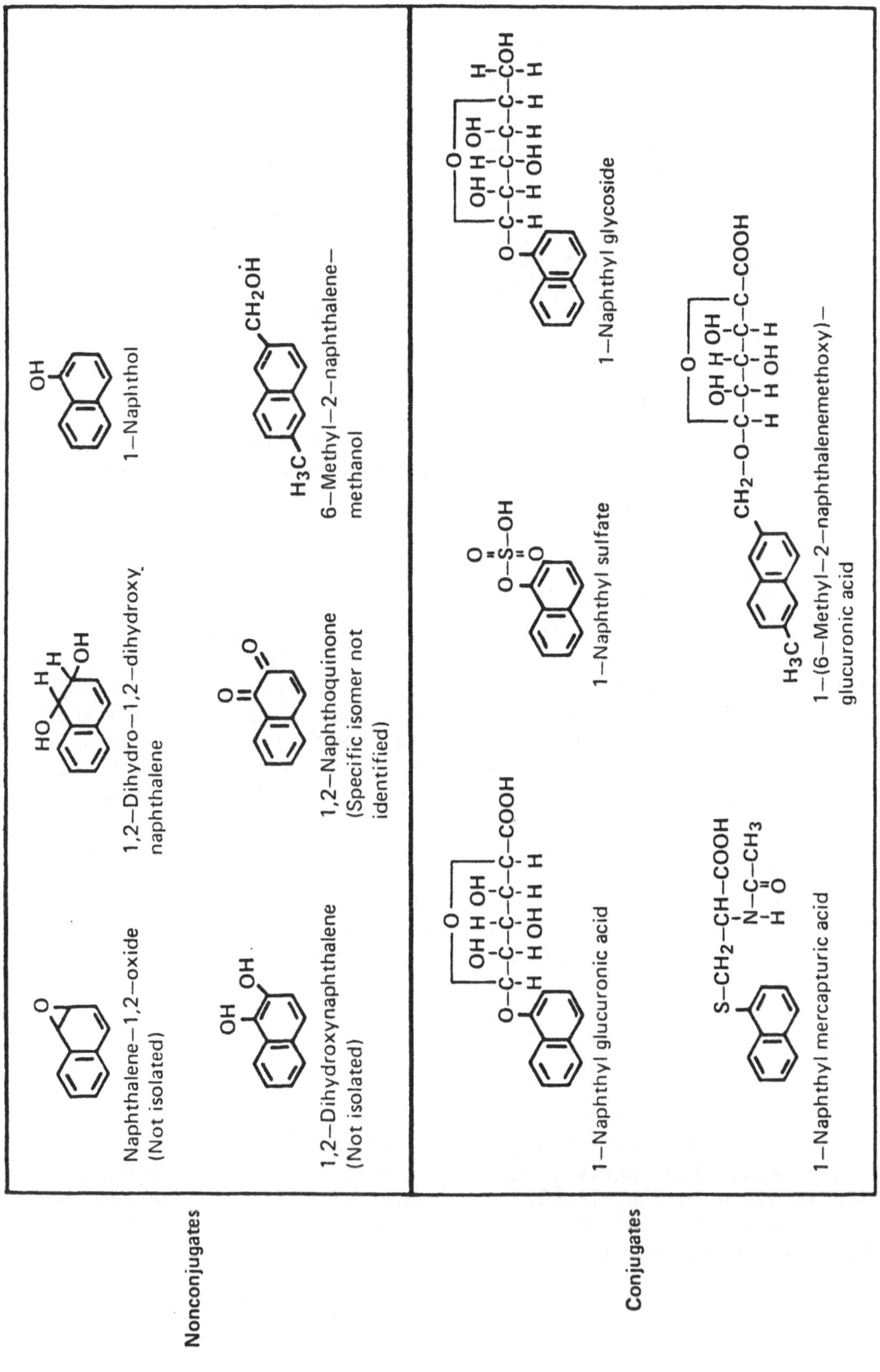

Nonconjugates

Naphthalene—1,2–oxide
(Not isolated)

1,2–Dihydro–1,2–dihydroxy,
naphthalene

1–Naphthol

1,2–Dihydroxynaphthalene
(Not isolated)

1,2–Naphthoquinone
(Specific isomer not
identified)

6–Methyl–2–naphthalene–
methanol

Conjugates

1–Naphthyl glucuronic acid

1–Naphthyl mercapturic acid

1–Naphthyl sulfate

1–Naphthyl glycoside

1–(6–Methyl–2–naphthalenemethoxy)–
glucuronic acid

FIG. 8. Example of nonconjugated and conjugated metabolites found in marine
organisms exposed to naphthalene and 2,6-dimethylnaphthalene.

receiving radiolabeled naphthalene through the diet accumulated large proportions of total metabolites in mucus, dorsal skin, blood, liver, and bile [24]. After 168 hr, the total metabolite fraction of blood, for example, was six times greater than the naphthalene fraction and metabolites predominated in each site examined. Several studies [22,33,34] have indicated that after termination of exposures, tissues and body fluids are readily depurated of hydrocarbons; however, the metabolites tend to increase or remain constant for extended periods of time.

Individual conversion products of aromatic hydrocarbons from tissues and body fluids can be analyzed by thin-layer chromatography (TLC) or high-performance liquid chromatography (HPLC). Examples of the structures identified by one or both of these chromatographic methods are presented (Fig. 8). As with microbial conversions, reactions occur on the aromatic ring and on the alkyl side chains. Methanol derivatives are formed in the latter case. Conjugating reactions also take place. These conversions, involving both the aromatic ring and alkyl-substituents, lead to the formation of glucuronides, sulfates, glycosides, and mercapturic acid derivatives [28,35]. High molecular weight compounds, such as benzo[a]pyrene, are converted in marine organisms [23] to compounds (e.g., diol epoxides) that are considered to be ultimate carcinogens in mammals. Varanasi and Gmur [23] have shown that--consistent with this hypothesis--benzo[a]pyrene is converted in marine fish to structures that bind with DNA (see Fig. 9). Interactions of this type may contribute to the formation of tumors that have been identified in marine fish [36].

Virtually no information is available on the ability of marine organisms to accumulate and metabolize the myriad oxidation products present in crude oils or arising from chemical or microbial oxidations.

ANALYSES OF OXIDIZED PRODUCTS

Most studies on the disposition and metabolism of hydrocarbons in marine organisms have employed radiolabeled hydrocarbons. The radiolabel allows for the detection and quantitation of metabolites isolated by chromatography. TLC is useful for the resolution and quantitation of metabolites isolated from marine organisms exposed to radiolabeled aromatic hydrocarbons [2]. For example, the identification of a number of individual metabolites in the skin of exposed rock sole (Lepidopsetta bilineata) was achieved by Varanasi et al. (Fig. 10) [24]. As with certain other tissues of fish exposed to this hydrocarbon, 1,2-dihydro-1,2-dihydroxynaphthalene was the main nonconjugated metabolite identified. The 1-naphthyl-β-glucuronic acid, found in the skin of rock sole, was the principal conjugated metabolite present in the liver of salmonids exposed to various aromatic hydrocarbons [21]. Varanasi and Gmur [23] have used TLC to analyze nonconjugated metabolites of benzo[a]pyrene isolated from livers of exposed starry flounder and coho salmon (Oncorhynchus kisutch) (Fig. 11). In this study, phenols, diols, and quinones were isolated and quantitated. The results were compared with metabolites formed in benzo[a]pyrene-exposed rats.

HPLC is an effective means of analyzing for aromatic hydrocarbon metabolites isolated from marine organisms. An example of results obtained with naphthalene derivatives from livers of naphthalene-exposed coho salmon is given in Fig. 12 [37]. A number of distinct fractions were isolated and quantitated. Five derivatives--the glucuronide, sulfate, dihydrodiol, glycoside, and 1-naphthol were identified.

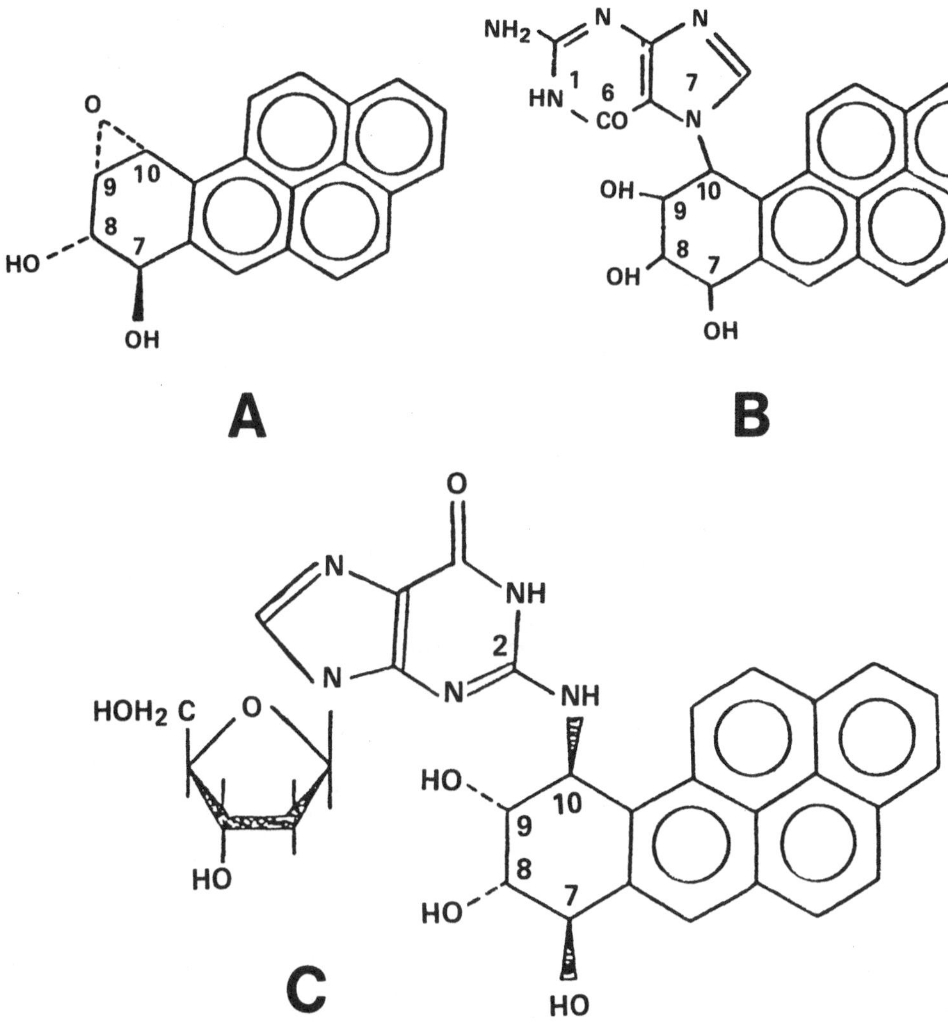

FIG. 9. (A) Benzo[a]pyrene-diol-epoxide (anti-isomer)(+)-7α,8β-dihydroxy-9β,10β-epoxy-7,8,9,10-tetrahydrobenzo[a]pyrene, a metabolite of benzo[a]pyrene that reacts with DNA in vivo; (B) the product of reaction of this metabolite with guanine in DNA at the N^7 position of guanine; (C) the product of reaction of this metabolite with deoxyguanosine in DNA at the extranuclear N^2 position of guanine (P.D. Lawley, IN: Chemical Carcinogens and DNA, P.L. Grover, ed., Vol. 1, CRC Press, Inc., 1979, 1-36).

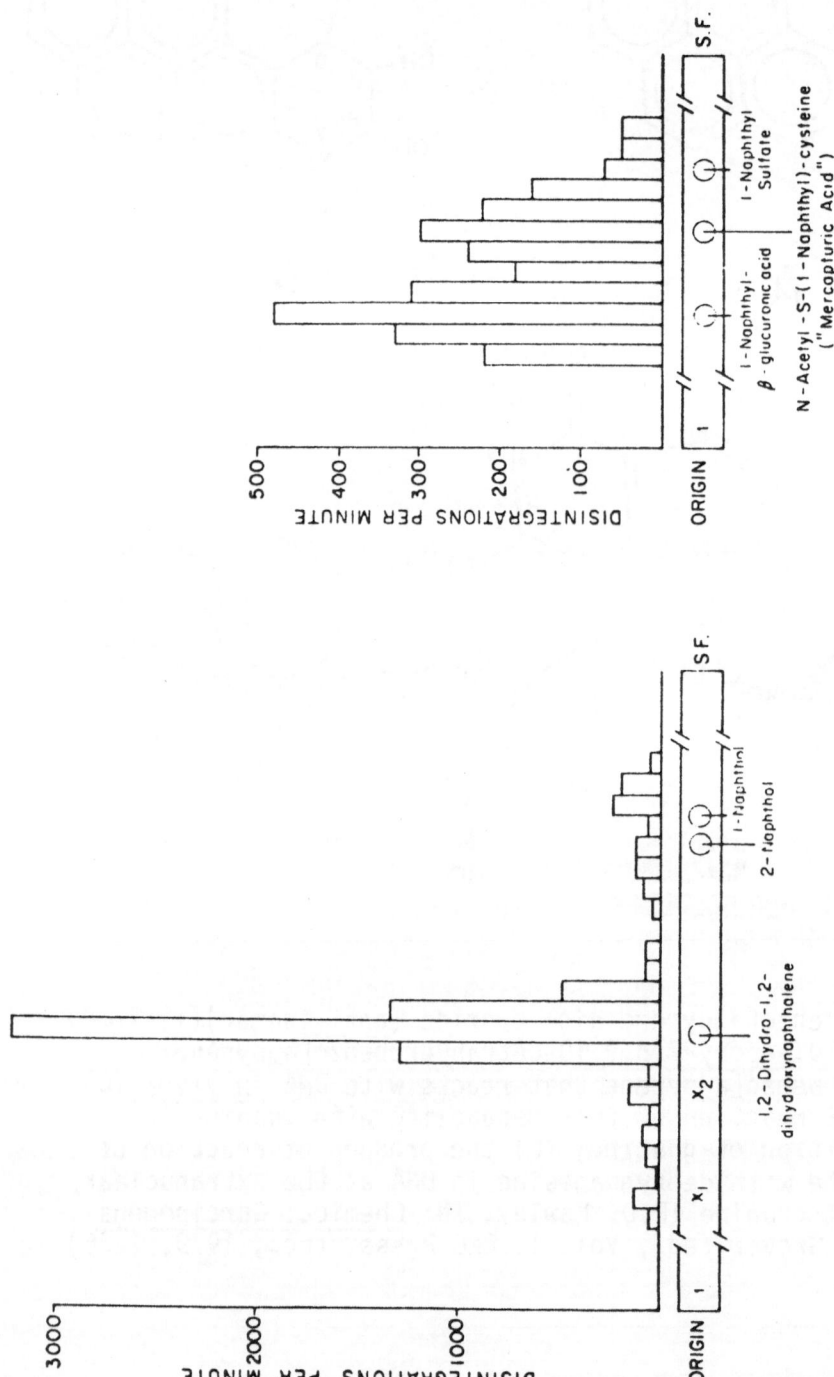

FIG. 10. TLC profiles of naphthalene metabolites in skin of exposed rock sole (adapted from Varanasi et al. [24]).

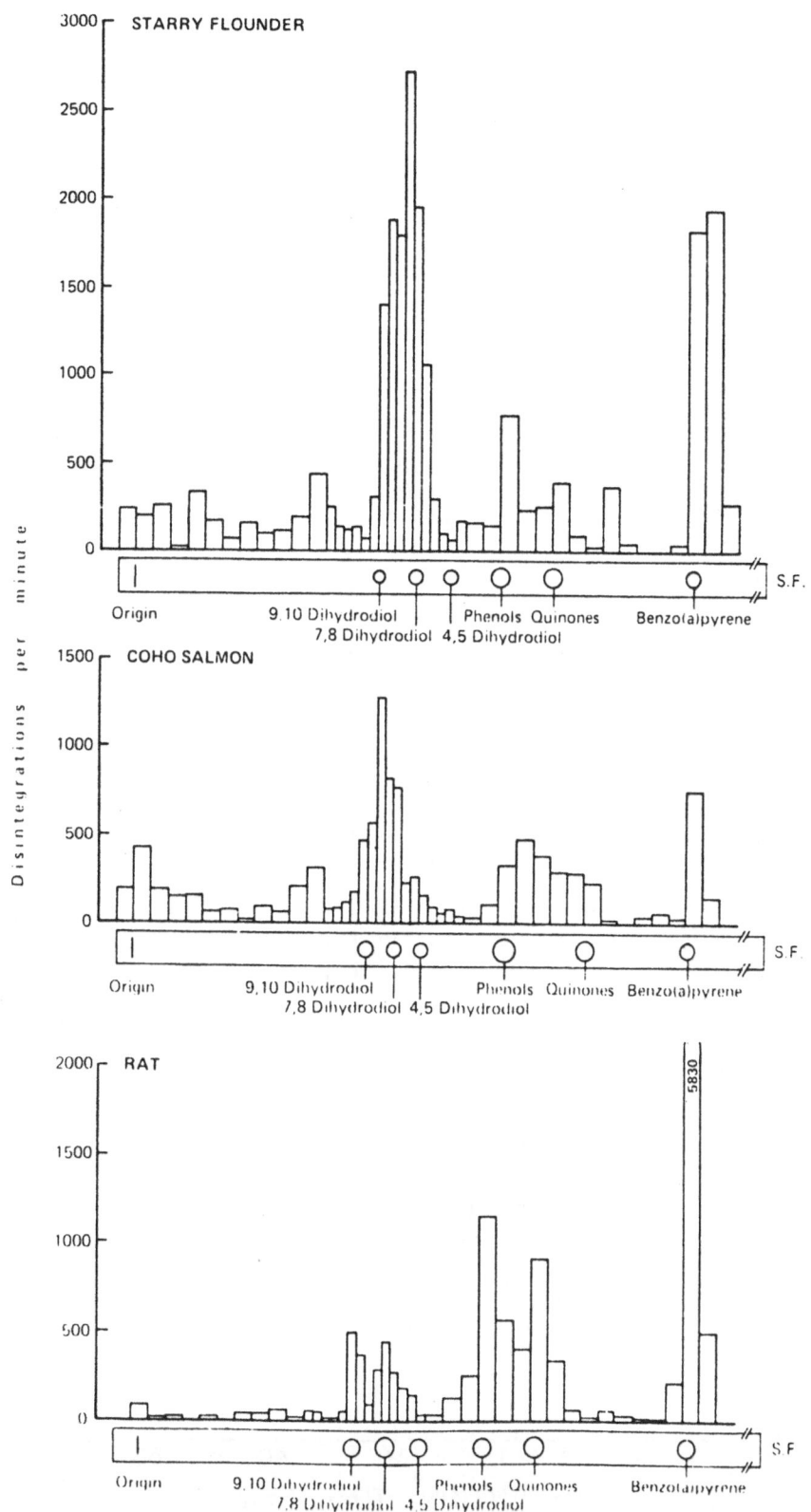

FIG. 11. Ethyl-acetate extractable metabolites of radiolabeled benzo[a]pyrene produced by liver supernatants from methylchol-anthrene-treated animals (adapted from Varanasi and Gmur [23]).

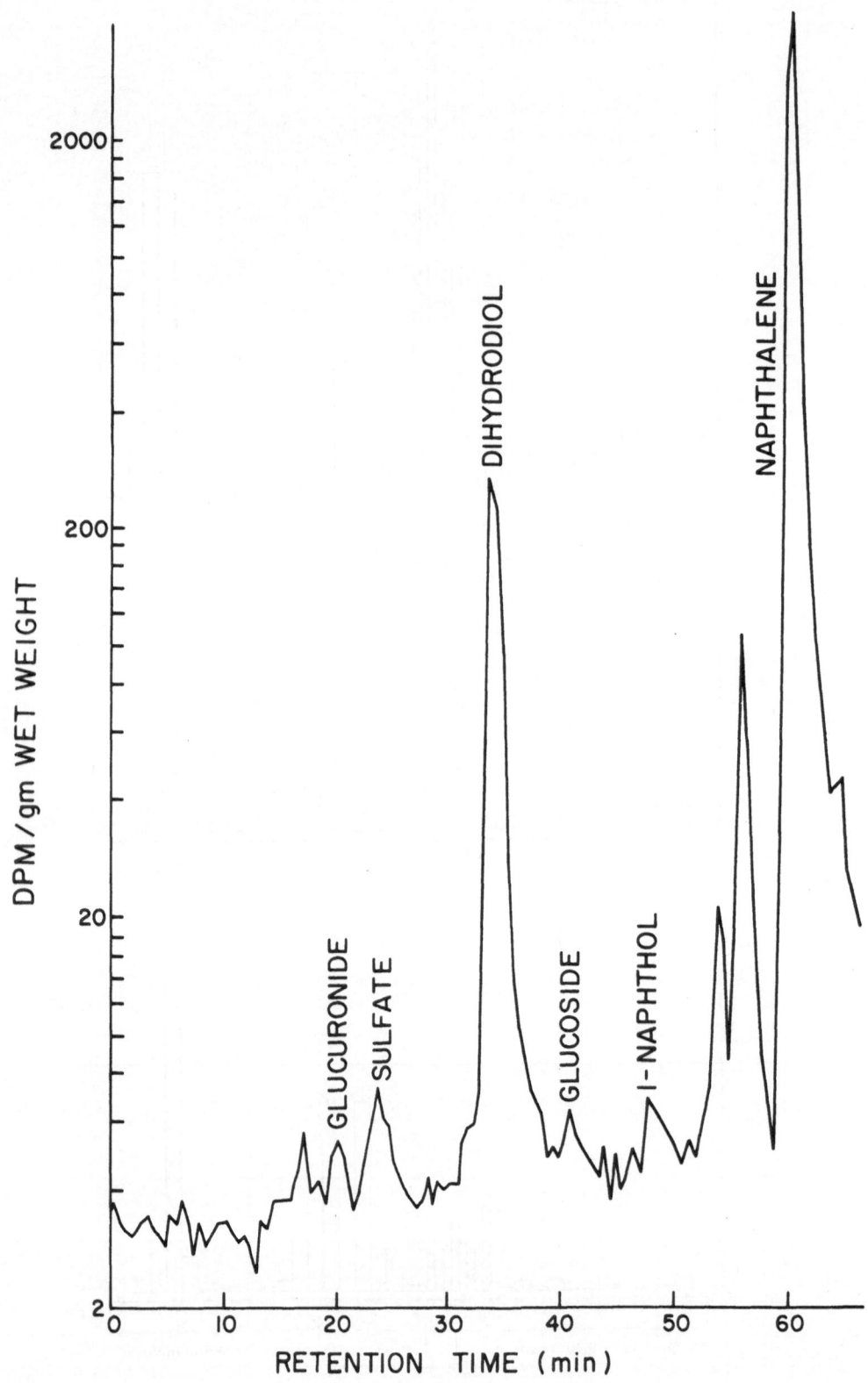

FIG. 12. HPLC profile of derivatives of naphthalene in liver of coho salmon 16 hr after force-feeding of ^{3}H-naphthalene (Collier et al. [37].

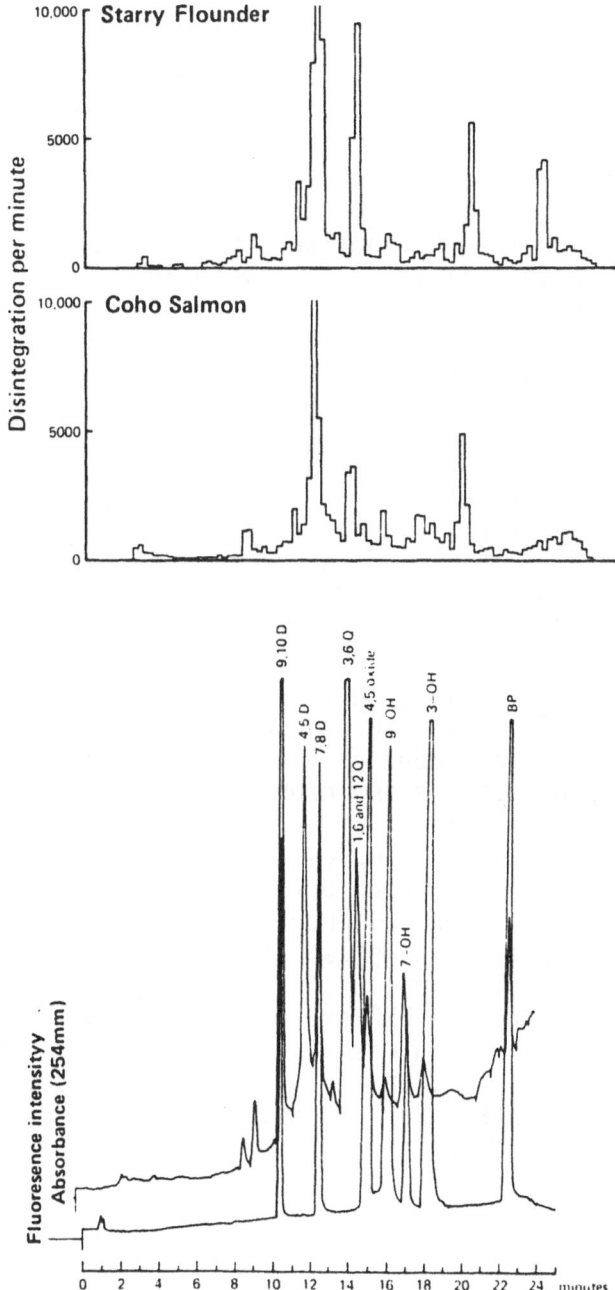

FIG. 13. HPLC of reference benzo[a]pyrene metabolites and ethyl acetate-extractable metabolites of ^{3}H-benzo[a]pyrene by liver supernatants of methylcholanthrene-treated starry flounder and coho salmon. Reference compounds and metabolites were detected by UV and fluorescence spectrometry (adapted from Varanasi et al. [23]).

Separations of benzo[a]pyrene metabolites from the liver supernatant of exposed starry flounder and coho salmon further illustrate the potential of HPLC for the resolution of radiolabeled metabolites in exposed marine organisms (Fig. 13) [29].

There are inherent limitations to the use of radiolabeled aromatic hydrocarbons in laboratory exposure studies. For example, radiolabeled compounds are often expensive and only a limited number of compounds are available. Accordingly, methods were developed to separate and quantify nonradiolabeled hydrocarbon derivatives isolated from tissues and body fluids.

Krahn et al. [38] reported a technique for the rapid analysis of naphthalene metabolites in marine and other animals. HPLC is employed in conjunction with "on-line" ultraviolet-fluorescence detection. Structural characterizations are made by plasma desorption chemical ionization mass spectrometry. This method allows for the separation and quantitation of nonradiolabeled metabolites.

An example is given (Fig. 14) of an HPLC/ultraviolet fluorescence chromatogram obtained from metabolites in the bile of rainbow trout (Salmo gairdneri) that received 11 mg/kg of naphthalene in a force-feeding study. Structures, such as the naphthyl glucuronide (the major component) were confirmed by mass spectrometry using a conventional direct insertion probe with a specially modified tip for use with thermally labile compounds.

The technique described by Krahn et al. [38] has been applied to the analysis of field samples. Significant differences were found when the HPLC/ultraviolet-fluorescence technique was used to analyze for aromatic polar compounds present in mussels (Mytilus edulis) held directly in the path of the Amoco Cadiz oil spill, out of the direct path, or taken from a reference area (Fig. 15) [2]. The obvious increases in polar aromatic compounds found in mussels from impacted sites was associated with observed cellular damage (e.g., increases in lipid and lysosomal granules) [39].

Some of the weakly polar, more volatile, oxidation products of petroleum can be analyzed by gas chromatography coupled with mass spectrometry (GC/MS). For example, a number of alkylated and non-alkylated phenols (e.g., derivatives of benzene and naphthalene) in a water-soluble fraction of No. 2 fuel oil were analyzed by GC/MS (Fig. 16) [40]. These approaches have application to seawater samples from the laboratory and field.

Free radicals form in the oxidation of aromatic hydrocarbons; however, virtually no attention has been given to the analysis of these structures in environmental samples. Yet free radicals are potentially useful indices of petroleum pollution and are known to readily interact with macromolecules, such as DNA and proteins, to alter their structures [41,42]. Using electron spin resonance spectrometry (ESR), Roubal et al. [43] have made progress in identifying free radicals in English sole (Parophrys vetulus) exposed to petroleum and other pollutants in marine environments; however, more work has to be carried out before reproducible, reliable methods are developed.

CONCLUSIONS

The evidence suggests that petroleum is extensively converted to polar products which accounts, in part, for its apparent "disappearance" in the marine environment. Yet little is known about the steady state concentrations of the polar fractions in relation to total hydrocarbons or about

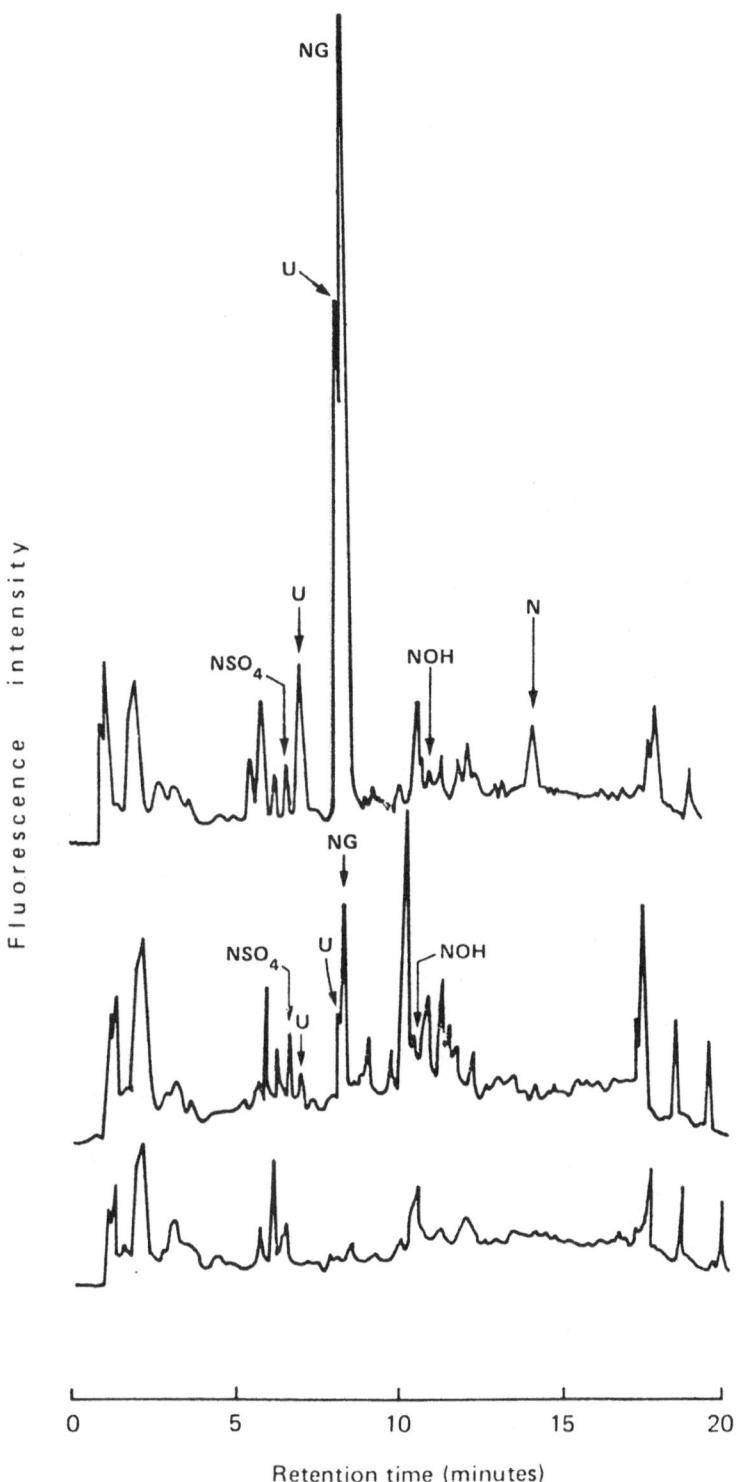

FIG. 14. HPLC profiles of naphthalene and its metabolites in the bile of rainbow trout. Fish received two force-feedings 24 hr apart. Samples of bile were taken 24 hr after the last feeding. Five µl raw bile were injected. Upper: Chromatogram of trout receiving 8.0 mg of naphthalene (73 mg/kg, higher level). Middle: Chromatogram for trout receiving 0.8 mg naphthalene (7.3 mg/kg, lower level). Lower: Chromatogram of the control bile. Detection of compounds at UVF wavelengths λ_{ex} 305 nm and λ_{em} 340 nm (Krahn et al. [38]).

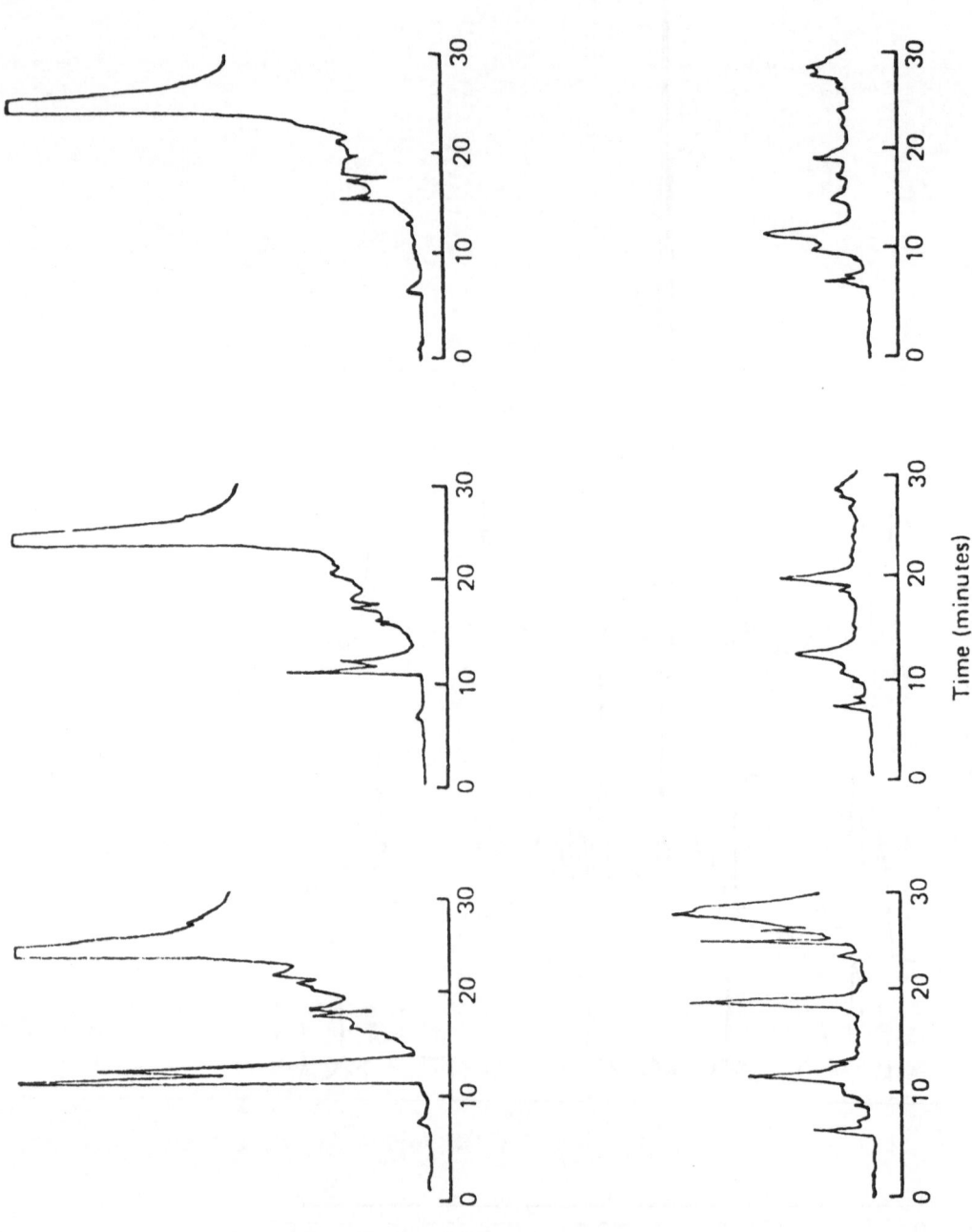

Fluorescence intensity

Time (minutes)

FIG. 15. HPLC profiles of methanol extracts of mussel samples: (A) directly in the path of the AMOCO CADIZ spilled oil; (B) out of the direct path of the AMOCO CADIZ spilled oil; and (C) mussels before exposure to AMOCO CADIZ oil. The profiles on top were detected at λ_{ex} 270 nm and λ_{em} 305 nm (characteristic of phenol); the profiles on the bottom were detected at λ_{ex} 305 nm and λ_{em} 340 nm characteristic of naphthalene metabolites) (Malins et al. [2]).

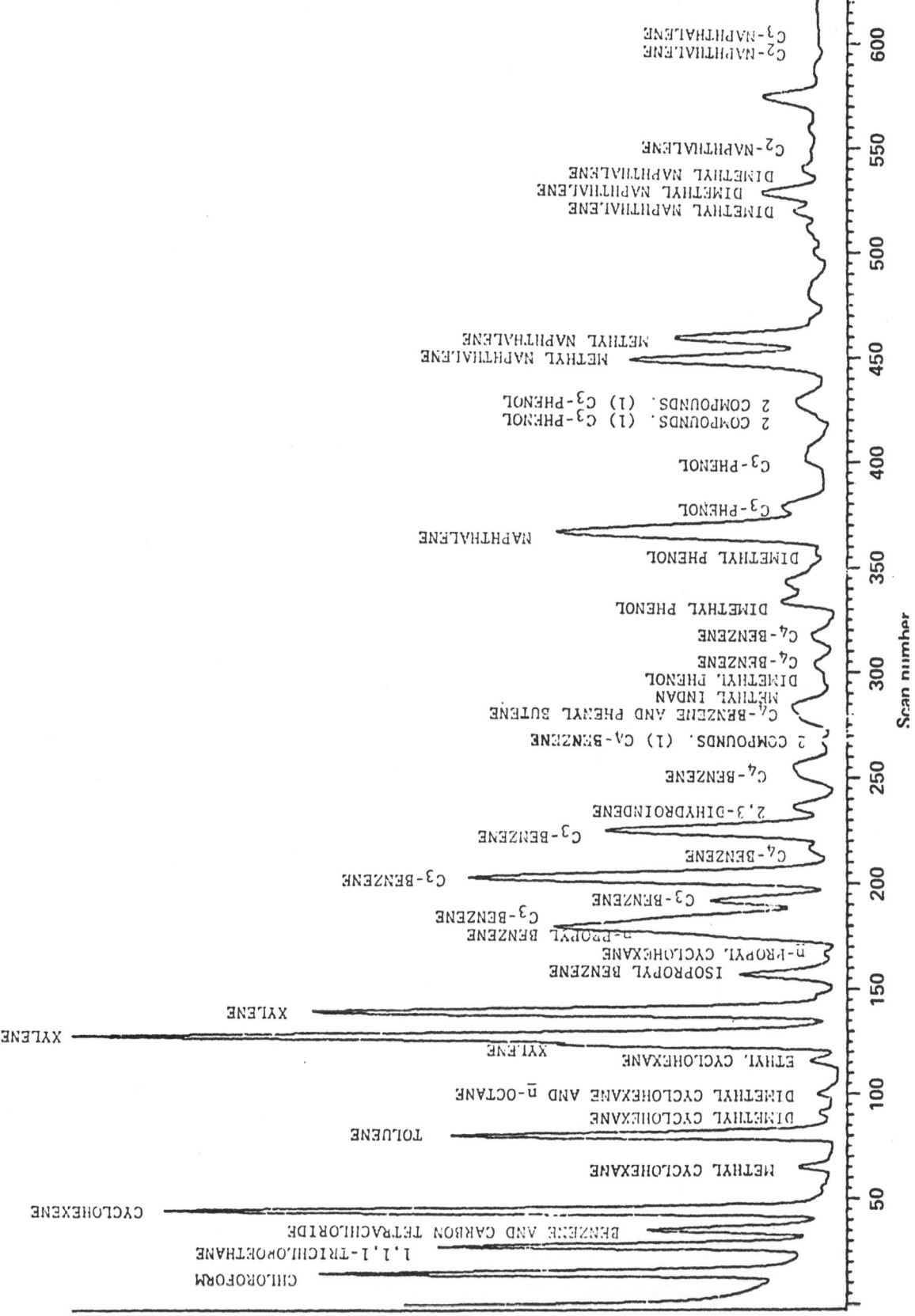

FIG. 16. GC/MS chromatogram of a seawater-soluble fraction of No. 2 fuel oil (Rice et al. [40]).

the quantitative significance of individual petroleum pollutants in marine environments. Problems arising from the transformations of petroleum in the marine environment create a formidable challenge for the analytical chemist. Progress in the analytical determination of the largely over-looked polar degradation products would be rewarding in itself. But an additional advantage may be that progress in this area may stimulate an interest in research on the presently unresolved problem of the toxicity of the oxidation products to various forms of marine life.

REFERENCES

1. R.C. Clark, Jr., D.W. Brown, and W.D. MacLeod, Jr., Unpublished data (1979).
2. D.C. Malins, M.M. Krahn, D.W. Brown, W.D. MacLeod, Jr., and T.K. Collier, "The analysis of petroleum products in marine environments," Helgol. wiss. Merresunters. (1979) In press.
3. R. Burwood and G.C. Speers, "Photo-oxidation as a factor in the environmental dispersal of crude oil," Estuarine Coastal Mar. Sci. 2 (1974) 117-35.
4. I.W. Morrin, Unpublished data (1979).
5. J.R. Patel, I.R. Politzer, G.W. Griffin and J.R. Laseter, "Mass spectra of the oxygenated products generated from phenanthrene under simulated environmental conditions, Biomed. Mass Spectrom, 5(1978) 664-670.
6. J.R. Patel, E.B. Overton and J.L. Laseter, "Environmental photoxidation of dibenzothiphenes following the Amoco Cadiz oil spill," Chemosphere No. 8 (1979) 557-561.
7. N.L. Karrick, "Alterations in petroleum resulting from physico-chemical and microbiological factors." IN Effects of Petroleum on Arctic and Subarctic Marine Environments, D.C. Malins, ed., Vol. I (Academic Press, New York, 1977) 225-99.
8. C.E. ZoBell, "Bacterial degradation of mineral oils at low temperatures." IN The Microbial Degradation of Oil Pollutants, D.G. Ahearn and S.P. Meyers, eds., Publ. No. LSU-SG-73-01 (Center for Wetland Resources, Louisiana State University, Baton Rouge, Louisiana, 1973) 153-61.
9. J.D. Walker and R.R. Colwell, "Microbial degradation of model petroleum at low temperatures," Microb. Ecol. 1 (1974) 63-95.
10. E.R. Leadbetter and J.W. Foster, "Bacterial oxidation of gaseous alkanes," Arch. Mikrobiol. 35 (1960) 92-104.
11. A.C. Van Der Linden and G.J.E. Thijsse, "The mechanisms of microbial oxidations of petroleum hydrocarbons," Adv. Enzymol. 27 (1965) 469.
12. P.W. Trudgill, "Microbial degradation of alicyclic hydrocarbons." IN Developments in Biodegradation of Hydrocarbons-1, R.J. Watkinson, ed. (Applied Science Publishers Ltd., London, 1978) 47-84.
13. Y. Yugari, "Metabolism of cyclohexane-diol-(1,2)-trans by a soil bacterium (I)," Biken's J. 4 (1961) 197-207.
14. D.T. Gibson, "Biodegradation of aromatic petroleum hydrocarbons." IN Fate and Effects of Petroleum Hydrocarbons in Marine Organisms and Ecosystems, D.A. Wolfe, ed. (Pergamon Press, New York, 1977) 36-46.
15. D.J. Hopper, "Microbial degradation of aromatic hydrocarbons." IN Developments in Biodegradation of Hydrocarbons-1, R.J. Watkinson, ed. (Applied Science Publishers Ltd., London, 1978) 85-112.
16. R.M. Atlas, P.D. Boehm and J.A. Calder, "Chemical and biological weathering of oil, from the Amoco Cadiz spillage, within the littoral zone." IN International Symposium: The Amoco Cadiz: Fates and Effects of the Oil Spill (Centre Oceanologique de Bretagne, Brest, France, 1979). In press.
17. Ballerini, J.R. and J.P. Vandecasteele, "Simulation en laboratoire de la degradation microbienne du petrole en milieu marin." IN International Symposium: The Amoco Cadiz: Fates and Effects of the Oil Spill (Centre Oceanologique de Bretagne, Brest, France, 1979). In Press.
18. The oil Industry and Microbial Ecosystems (Eds. K.W.A. Chates and H.J. Somerville) Heyden Press, London (1979)

19. B.B. Brodie and R.P. Maickel, "Comparative biochemistry of drug meta-
    bolism. IN Proceedings of the First International Pharmacology
    Meeting, B.B. Brodie and E.G. Erdos, eds. (MacMillan, New York, 1962)
    299.
20. D.C. Malins, "Metabolism of aromatic hydrocarbons in marine organ-
    isms," Ann. N.Y. Acad. Sci. 298 (1977) 482-96.

21. W.T. Roubal, T.K. Collier, and D.C. Malins, "Accumulation and metabo-
    lism of carbon-14 labeled benzene, naphthalene and anthracene by
    young coho salmon (Oncorhynchus kisutch)," Arch. Environ. Contam.
    Toxicol. 5 (1977) 513-29.
22. U. Varanasi, M.I. Uhler, and S.I. Stranahan, "Uptake and release of
    naphthalene and its metabolites in skin and epidermal mucus of sal-
    monids," Toxicol. Appl. Pharmacol. 44 (1978) 277-89.
23. U. Varanasi and D.J. Gmur, "Metabolic activation and binding of
    benzo[a]pyrene to DNA catalyzed by liver extracts of marine fish,"
    Biochem. Pharmacol. (1979) In press.
24. U. Varanasi, D.J. Gmur, and P.A. Treseler, "Influence of time and
    mode of exposure on biotransformation of naphthalene by juvenile
    starry flounder (Platichthys stellatus) and rock sole (Lepidopsetta
    bilineata)," Arch. Environ. Contam. Toxicol. 8 (1979) In press.
25. L. Thomas, W.D. MacLeod, Jr., and D.C. Malins, "Identification and
    quantitation of aromatic hydrocarbon metabolites in marine biota."
    IN Trace Organic Analysis: A New Frontier in Analytical Chemistry,
    H.S. Hertz and S.N. Chesler, eds., Spec. Publ. 519 (National Bureau
    of Standards, Washington, D.C., 1979) 79-86.
26. H.R. Sanborn and D.C. Malins, "The disposition of aromatic hydrocar-
    bons in adult spot shrimp (Pandalus platyceros) and the formation of
    metabolites of naphthalenes in adult and larval spot shrimp,"
    Xenobiotica (1979) In press.
27. D.C. Malins, T.K. Collier, and H.R. Sanborn, "Disposition and metabo-
    lism of aromatic hydrocarbons in marine organisms." IN Pesticide and
    Xenobiotic Metabolism in Aquatic Organisms, M.A.Q. Khan, J.J. Lech,
    and J.J. Menn, eds., ACS Symp. Series 99 (American Chemical Society,
    Washington, D.C., 1979) 57-75.
28. D.C. Malins, T.K. Collier, L.C. Thomas, and W.T. Roubal, "Metabolic
    fate of aromatic hydrocarbons in aquatic organisms: Analysis of
    metabolites by thin-layer chromatography and high-pressure liquid
    chromatography," Intern. J. Environ. Anal. Chem. 6 (1979) 55-66.
29. U. Varanasi, D.J. Gmur, and M.M. Krahn, "Metabolism and subsequent
    binding of benzo[a]pyrene to DNA in pleuronectid and salmonid fish."
    IN Carcinogenesis: A Comprehensive Survey. Vol. 5. Polynuclear
    Aromatic Hydrocarbons, A. Bjorseth, ed. (Raven Press, New York, 1979)
    In press.
30. R.F. Lee, R. Sauerheber, and G.H. Dobbs, "Uptake, metabolism and dis-
    charge of polycyclic aromatic hydrocarbons by marine fish," Mar.
    Biol. 17 (1972) 201-8.
31. R.F. Lee, "Fate of petroleum hydrocarbons in marine zooplankton."
    IN Proceedings 1975 Conference on Prevention and Control of Oil
    Pollution (American Petroleum Institute, Washington, D.C., 1975)
    549-53.

32. E.D.S. Corner, C.C. Kilvington, and S.C.M. O'Hara, "Qualitative studies on the metabolism of naphthalene in Maia squinado (Herbst)," J. Mar. Biol. Assoc. U.K. 53 (1973) 819-32.

33. H.R. Sanborn and D.C. Malins, "Toxicity and metabolism of naphthalene: A study with marine larval invertebrates," Proc. Soc. Exp. Biol. Med. 154 (1977) 51-5.

34. W.T. Roubal, S.I. Stranahan, and D.C. Malins, "The accumulation of low molecular aromatic hydrocarbons of crude oil by coho salmon (Oncorhynchus kisutch) and starry flounder (Platichthys stellatus)," Arch. Environ. Contam. Toxicol. 7 (1978) 237-49.

35. U. Varanasi and D.C. Malins, "Metabolism of petroleum hydrocarbons: Accumulation and biotransformations in marine organisms." IN Effects of Petroleum on Arctic and Subarctic Marine Environments and Organisms, D.C. Malins, ed., Vol. II (Academic Press, New York, 1977) 175-270.

36. B.B. McCain, W.D. Gronlund, M.S. Myers, and S.R. Wellings, "Tumours and microbial diseases of marine fishes in Alaskan waters," J. Fish. Dis. 2 (1979) 111-30.

37. T.K. Collier, L.C. Thomas, and D.C. Malins, "Influence of environmental temperature on disposition of dietary naphthalene in coho salmon (Oncorhynchus kisutch): Isolation and identification of individual metabolites," Comp. Biochem. Physiol. 61C (1978) 23-8.

38. M.M. Krahn, D.W. Brown, T.K. Collier, A.J. Friedman, R.G. Jenkins, and D.C. Malins, "Rapid analysis of naphthalene and its metabolites in biological systems: Determination by high-performance liquid chromatography/fluorescence detection and by plasma desorption/chemical ionziation mass spectrometry," J. Biochem. Biophys. Methods (1979) In press.

39. D.A. Wolfe, R.C. Clark, Jr., C.A. Foster, J.W. Hawkes, and W.D. MacLeod, Jr., "Hydrocarbon accumulation and histopathology in bivalve molluscs transplanted to the Rade de Morlaix and the Rade de Brest." IN International Symposium: The AMOCO CADIZ: Fates and Effects of the Oil Spill (Centre Oceanologique de Bretagne, Brest, France, 1979) In press.

40. S.D. Rice, S. Korn, and L. Cheatam, "The toxicity, uptake and availability of phenol, substituted phenols, and naphthols from crude oil water-soluble fractions," (1979) In preparation.

41. W.T. Roubal and A.L. Tappel, "Damage to proteins, enzymes, and amino acids by peroxidizing lipids," Arch. Biochem. Biophys. 113 (1966) 5-8.

42. W.T. Roubal and A.L. Tappel, "Polymerization of proteins induced by free-radical lipid peroxidiation," Arch. Biochem. Biophys. 113 (1966) 150-55.

43. W.T. Roubal, K. Gleim, and D.C. Malins (1979) Unpublished data.

BACKGROUND LEVELS OF PETROLEUM RESIDUES IN BAFFIN BAY AND THE EASTERN
CANADIAN ARCTIC: ROLE OF NATURAL SEEPAGE

by E.M. Levy
Research Scientist, Department of Fisheries and Oceans
Ocean and Aquatic Sciences, Atlantic Oceanographic Laboratory
Bedford Institute of Oceanography, Dartmouth, Nova Scotia,   B2Y 4A2

ABSTRACT

        Concentrations of petroleum residues in the waters and surficial
bottom sediments of Baffin Bay and the Eastern Canadian Arctic were
measured during Bedford Institute of Oceanography surveys in 1977 and 1978.
While floating particulate residues were only rarely encountered,
concentrations of extractable residues ranged from 3-1726 µg/L in the
surface microlayer, 0-87.5 µg/L in the water column and 1-41 µg/g in the
sediments. The geographical distribution of the residues in Baffin Bay
suggests that they are derived predominantly from atmospheric fall-out of
combustion products, although there are significant inputs from natural
seepage at Scott Inlet and elsewhere along the northeast coast of Baffin
Island

INTRODUCTION

        On several occasions slicks and gas bubbles erupting at the sea
surface were observed off Scott Inlet on the east coast of Baffin Island
(Fig. 1) (Ref. 1,2). Gas chromatographic and other chemical analyses
indicated that the slick-forming material was probably a partially
weathered crude oil and that unexpectedly high concentrations of dissolved/
dispersed petroleum residues were present in the water column (Ref. 2).
Interestingly, this location was one of only four in the region north of
the Gulf Stream/North Atlantic Current system where floating petroleum
residues were found during an extensive study of the distribution of
particulate petroleum residues in the North Atlantic (Ref. 3). Lacking
any other plausible explanation, it was postulated that these materials
were the consequence of natural seepage (Ref. 2). To explore this
possibility, a cursory chemical survey of the Scott Inlet area was carried
out in 1977 as part of an investigation of the background levels of
petroleum residues in the water column and surficial bottom sediments of
Baffin Bay and the Eastern Canadian Arctic. On this occasion not only
were gas bubbles and slicks present at Scott Inlet, but slicks were also
observed farther north (Fig. 2) (Ref. 2,4). Because surface currents in
this region flow southward, additional sources to the north were indicated.

        The northeast coast of Baffin Island is characterized by fjords
that penetrate into the mountainous interior and by submarine troughs that
extend from the mouths of the fjords seaward across the continental shelf.
At Scott Inlet, the trough has been cut some 700 m into the underlying
rocks. Seismic studies (Ref. 5) suggested that a geological environment
suitable for natural seepage in this area was afforded by a structural
high near the seaward end of the trough and by the truncated beds forming
the walls of the trough and the folded and faulted strata of its floor.
Subsequent geological studies (Ref. 6) indicate that migration of
hydrocarbons updip in the strata flanking the basement high or along the

Fig. 1. Map of Arctic, showing study area

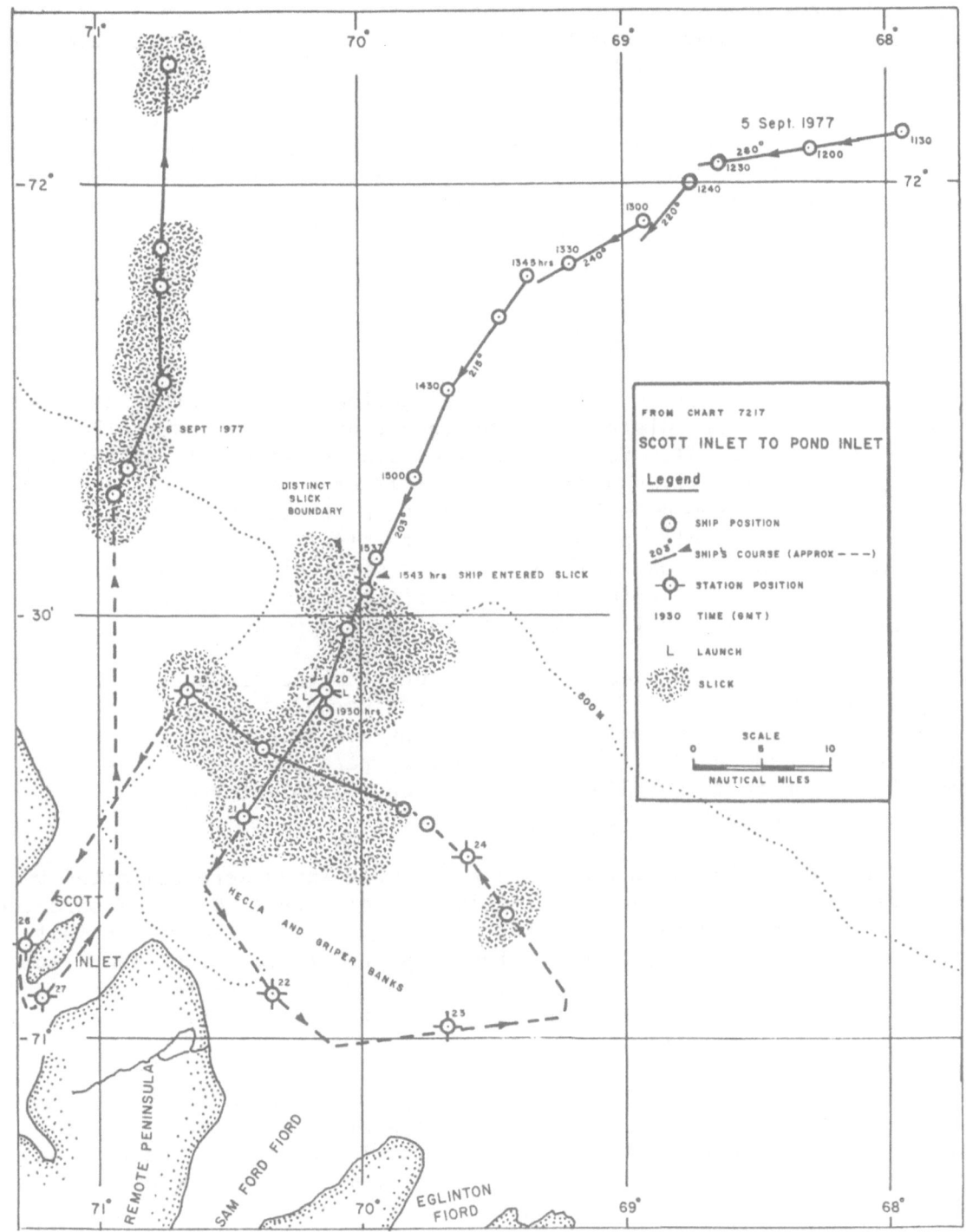

Fig. 2.   Surface slicks at Scott Inlet - Sept. 1977

contact with the basement rocks is probably the mechanism of seepage at this site. Seepage from a similar trough at Buchan Gulf was suspected as being the source of the slicks observed to the north of Scott Inlet.

Because of the potential economic and environmental implications of natural seepage of petroleum in the Arctic, a more detailed chemical survey of Scott Inlet, Buchan Gulf, and the entrance to Lancaster Sound was carried out in 1978. This paper addresses the existing data concerning background levels of petroleum residues in Baffin Bay and explores the role played by natural seepage.

SAMPLING AND ANALYTICAL PROCEDURES

Sampling locations for the investigation of background levels of petroleum residues were distributed throughout Baffin Bay and the Eastern Canadian Arctic (Fig. 3). For the detailed studies at Scott Inlet and Buchan Gulf, sampling sites were on 19 km x 19 km grids (Fig. 4) located over the submarine troughs while at the entrance to Lancaster Sound stations were more widely spaced. In both the 1977 and 1978 surveys, the methods used to collect and analyze floating particulate petroleum residues, samples of sea water from the surface microlayer and throughout the water column, and surficial bottom sediments were as follows:

Floating particulate residues were collected from the sea surface by towing a neuston net for 1.9 km. Any tar or congealed oil was recovered and kept frozen until quantified by a solvent extraction/ gravimetric procedure (Ref. 3). Only on very rare occasions did samples from this region contain solidified tar or anything except marine organisms and fragments of seabird feathers.

Water samples from the *sea surface microlayer* were collected by repeatedly lowering a triangular stainless steel screen (0.12 m^2) onto the surface of the sea and recovering the water retained by the meshes (Ref. 7). Sampling was sometimes precluded by the presence of sea ice or hampered by the water freezing on the sampler. The water was extracted with carbon tetrachloride and subsequently analyzed quantitatively by fluorescence spectrophotometry (Ref. 8).

Samples for the determination of *dissolved/dispersed petroleum residues* were collected at standard oceanographic depths throughout the water column using a CTD/Rosette with 5-L Niskin samplers. The rosette was lowered to within a few metres of the bottom and the samplers were closed at the desired depths during retrieval. This procedure provided a thorough flushing of the samplers and avoided contamination from shallower waters. On recovery of the device, 1-L subsamples were drawn into teflon bottles for immediate extraction with carbon tetrachloride. Subsequently, the concentration of petroleum residues was determined by fluorescence spectrophotometry (Ref. 8).

Surficial bottom sediments were collected with a Shipek grab sampler and immediately frozen in glass jars. The sediment was subsequently extracted with hexane for analysis by fluorescence spectrophotometry. The grain size distribution of the sediments was also determined.

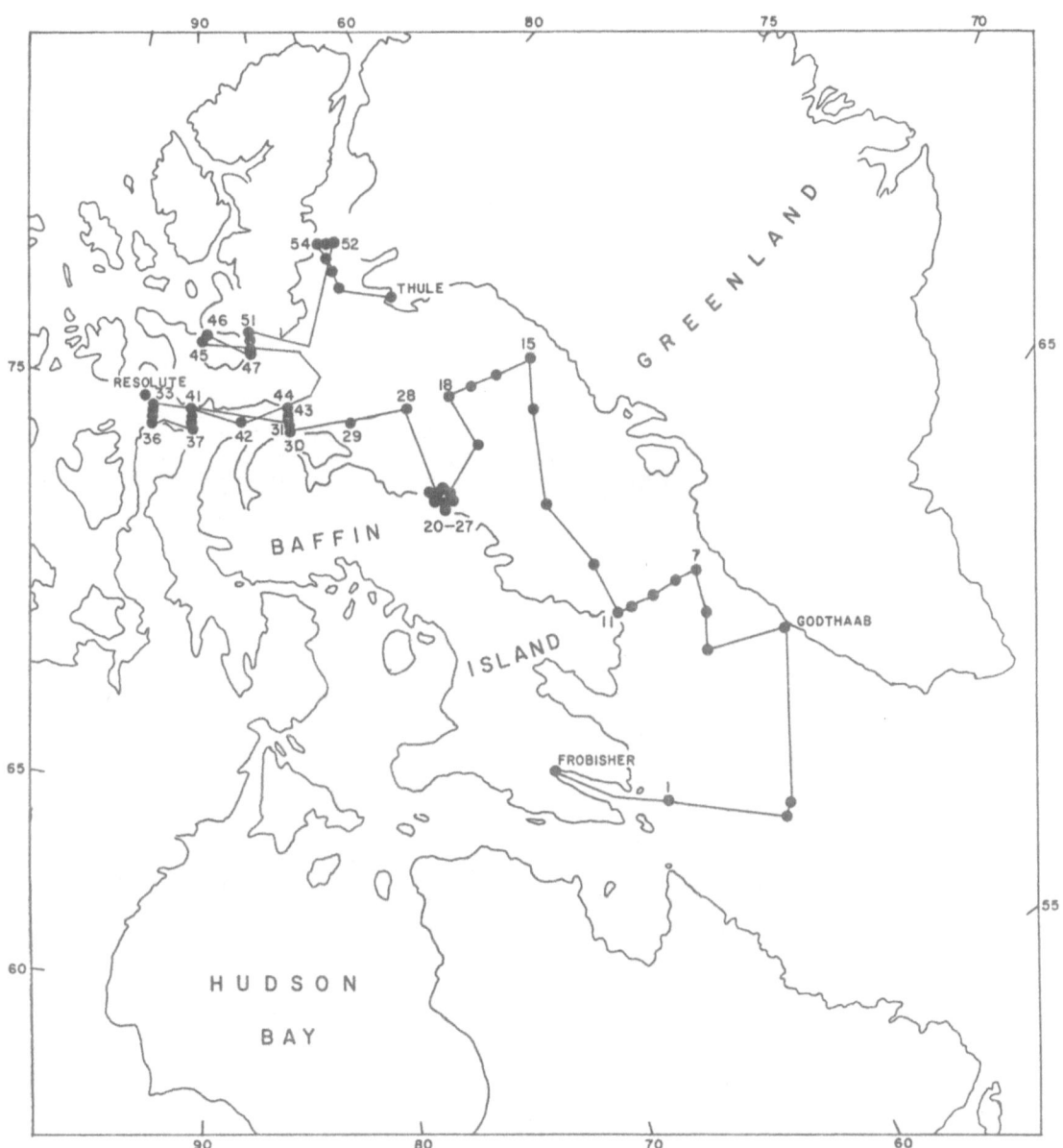

Fig. 3. Cruise track and sampling sites in Baffin Bay
and adjoining areas Aug./Sept., 1977

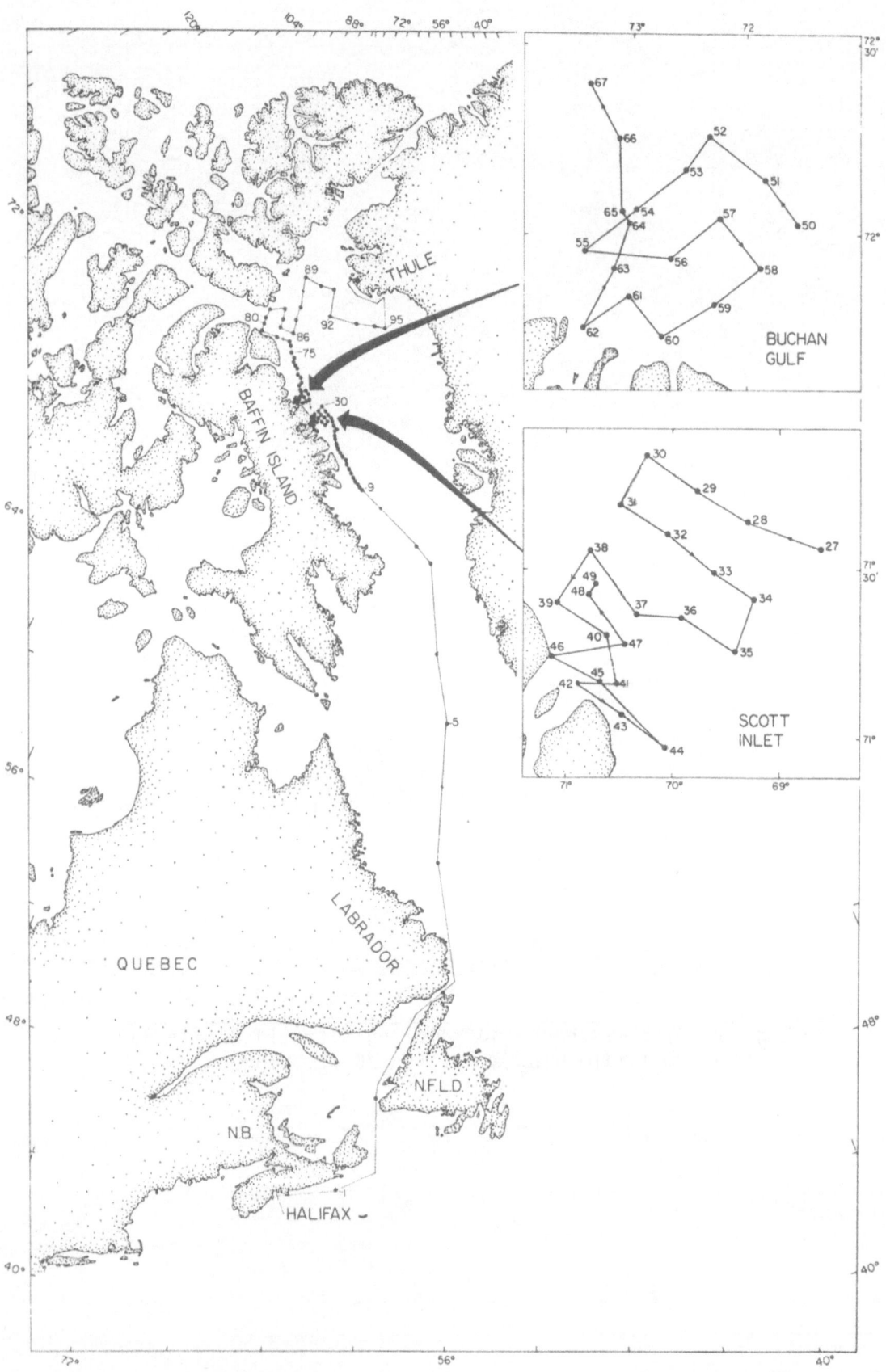

Fig. 4.   Cruise track and sampling sites Aug./Sept., 1978

RESULTS AND DISCUSSION

Floating particulate petroleum residues - During the 1977 survey of Baffin
Bay and the Eastern Canadian Arctic, floating particulate petroleum residues
were encountered at only four of the 41 locations sampled. One of these,
off Scott Inlet (Station 22), was in the area where slicks had been observed
on several previous occasions while the others were situated along a section
across the entrance to Lancaster Sound (Stations 31, 32, 43) (Ref. 9), an
area which is recognized as having a high potential for petroleum reserves.
The concentrations in these areas (0.02 - 0.08 mg/m^2) were, however,
considerably lower than those commonly found farther south in the North
Atlantic (Ref. 3). Except for this comparatively small area off the
northeast coast of Baffin Island, floating particulate petroleum residues
were absent in the extensive region surveyed. It may be concluded, therefore,
that the existing background level of this form of petroleum pollution in
Baffin Bay is zero.

During the detailed study of Scott Inlet, Buchan Gulf and the
entrance to Lancaster Sound in 1978, floating residues were encountered at
only one (Station 32) of the 44 sites sampled in these areas. It is perhaps
surprising that floating residues should be encountered so infrequently in
areas where natural seepage is believed to occur. When the seep material
arrives at the sea surface it is in a fluid form which, under suitable
conditions, may be present as a sufficiently compressed film that a
measurable amount of it adheres to the net. However, as the film spreads and
is carried away by surface currents, it becomes sufficiently diffuse that it
is no longer possible to collect a measurable amount of oil by this procedure.
This process apparently occurs at a rate such that the formation of large
numbers of tar balls is precluded. It may be concluded, therefore, that
while low concentrations of floating petroleum residues are sometimes present
in the immediate vicinity of the seeps, the effects of seepage are very
localized and have not given rise to a detectable background level of tar
in Baffin Bay as a whole.

Surface microlayer - Petroleum-derived and other hydrophobic substances in
seawater tend to accumulate at the sea surface and, under favorable
meteorological and oceanographic conditions, there may be a marked
enrichment of the sea surface microlayer relative to the water below. If
this enrichment advances to the stage that capillary waves are damped, the
light-reflecting properties of the sea surface are modified and the
phenomenon may be visible as a surface slick. Although extensive slicks
were present at Scott Inlet during 1977 (Fig. 2) and in 1976 (Ref. 1),
adverse conditions prevented their formation during the 1978 cruise and
only one sighting was reported (71-23N, 70-66W). Nevertheless, the repeated
observations of slicks at Scott Inlet and those off Buchan Gulf provide
strong evidence of natural seepage from the continental shelf along the
northeast coast of Baffin Island. Petroleum-like slicks were not observed
elsewhere in Baffin Bay.

The background concentrations of dissolved/dispersed petroleum
residues in the surface microlayer of Baffin Bay during the summer of 1977
(Fig. 5) were 3-10 µg/L on the Greenland side but were considerably higher
on the Canadian. The highest concentrations, 20-40 µg/L, were present over
the northeast portion of the Baffin Island shelf in the Scott Inlet/Buchan

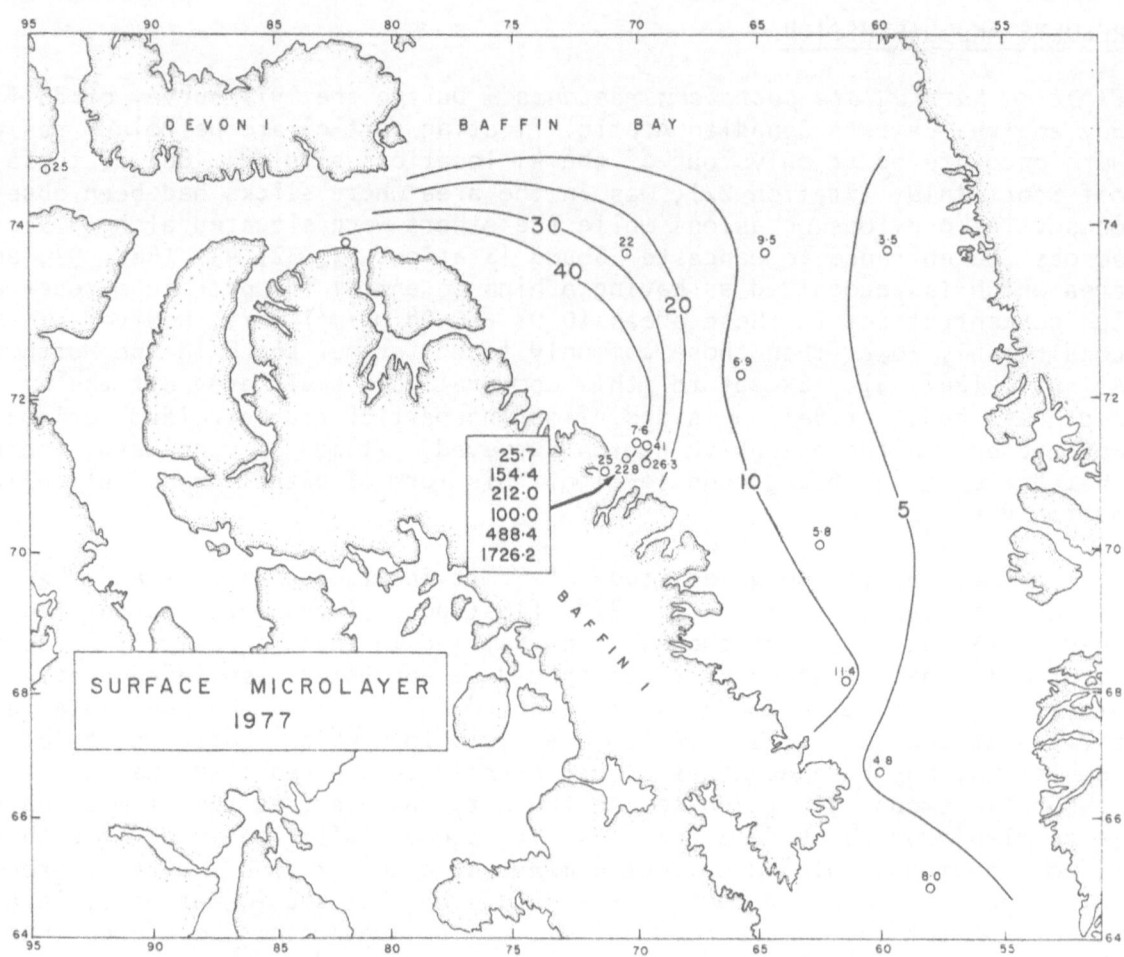

Fig. 5. Concentrations (µg/L) of petroleum residues in the surface microlayer in Baffin Bay, 1977

Gulf/Bylot Island region.   Particularly high concentrations were encountered in the slick which was present off Scott Inlet (Station 20).   Concentrations in excess of 1700 µg/L were present where the surface film was compressed into a small patch of brown mousse-like material.

Although conditions at Scott Inlet during the 1978 cruise were unsuitable for a pronounced enrichment of the surface microlayer, concentrations of petroleum residues at the sea surface were highest over the south wall of the trough and its southern extension (Fig. 6A).   At Buchan Gulf (Fig. 6B) conditions were more favorable and a well-developed surface microlayer was present at Station 61 where the concentration was 1200 µg/L.   Higher than background concentrations were also observed to the southeast.   This is probably a consequence of an input from seepage somewhere in the Buchan Gulf trough followed by southeastward movement of the oil with the surface waters.   At the entrance to Lancaster Sound (Fig. 6C) concentrations were generally higher than the background for Baffin Bay, but no "hot spots" were identified by the sampling grid used.

Surface slicks and enriched surface microlayer concentrations are almost certainly an indication of a local source of the slick-forming material.   The aromatic compounds present in this material indicate a petroleum rather than a biological source, and the absence of any major anthropogenic input in the area suggests natural seepage.   However, the development of a well-defined surface microlayer is so highly dependent on environmental conditions and so transient in nature that failure to observe these surface manifestations does not preclude the presence of submarine seepage.   The persistence in both time and space of visible slicks and enriched surface layers at Scott Inlet, Buchan Gulf, and at the entrance to Lancaster Sound provides a strong indication of seepage in these areas and also that these sites have been active for some time.   The impact on the level of "pollution" in the surface waters of Baffin Bay as a whole is, however, not detectable.

Water column - Concentrations of dissolved/dispersed petroleum residues in the waters of Baffin Bay and adjoining sounds during 1977 ranged from 0 to 87.5 µg/L.   Highly skewed distributions such as this are best treated by lognormal statistics and the geometric mean, the most appropriate measure of central tendency, for this set of data is 0.47 µg/L with a 95% confidence interval of 0.43-0.50 µg/L (Ref. 9).   It may be concluded, therefore, that the general background level of dissolved/dispersed petroleum residues in the region during the summer of 1977 was somewhat less than 0.5 µg/L with the most probable value being 0.47 µg/L.   Closer examination of the data failed to reveal a clear-cut relationship between concentration and depth, nor could a significant difference in concentration be found among the various sub-regions.   However, all the samples having concentrations greater than 10 µg/L were clustered at Scott Inlet (Stations 22, 25) and the entrance to Lancaster Sound (Stations 32, 33).   Since there is no reason to suspect that these samples were inadvertently contaminated during collection or analysis, these values reflect genuine nonhomogeneities in the distribution of petroleum residues in the water column.   That these nonhomogeneities are more frequently encountered at some locations than others implies that there are local inputs.   Further, the data at Scott Inlet were not lognormally distributed (χ^2-test) and therefore did not belong to the general Baffin Bay population.

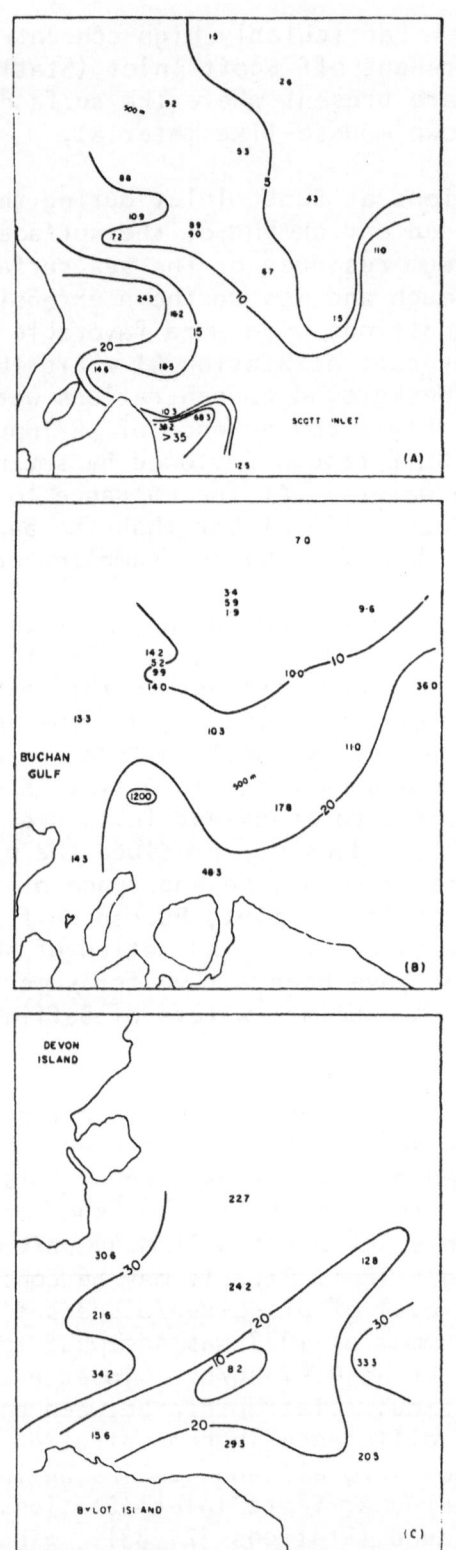

Fig. 6. Concentrations (µg/L) of petroleum residues in the
surface microlayer during 1978 at Scott Inlet (A),
Buchan Gulf (B), and the entrance to Lancaster Sound (C)

The frequency distribution plots (Fig. 7) for the detailed data
collected at Scott Inlet, Buchan Gulf and at the entrance to Lancaster
Sound in 1978 exhibited a bimodal character indicating the presence of two
different populations. It is tempting to surmise that one population
pertains to the background while the other reflects an additional input
of dissolved/dispersed petroleum residues from seepage. However, the
overall concentration levels in these three areas were shown by t-tests
to be significantly lower than the 1977 background level in Baffin Bay.
It would seem, therefore, that local variations in background levels
exist on a scale that is not resolved by widely separated sampling sites as
were employed in 1977. In contrast with the data from 1976 and 1977, those
collected in 1978 did not demonstrate abnormally high concentrations of
dissolved/dispersed petroleum residues in the water column at the sites
sampled at Scott Inlet and elsewhere along the Baffin Island coast where
seepage is believed to occur. This suggests that the three-dimensional
sampling network used in this instance failed to sample the "plumes"
emanating from the sea bed. On the other hand, the choice of sampling
locations during 1977 was guided by the presence of oil slicks which were
visible on the sea surface, and the observed concentration anomalies in
the water column suggest that the plumes had, indeed, been sampled in
several instances.

Although the 1978 data did not demonstrate any notable
concentration anomalies, geographical plots (Fig. 8) of the percentage of
samples having concentrations greater than 0.52 µg/l (i.e., the upper
limit for the 99.9% confidence range of the geometric mean of the 1977
data) clearly show that higher than average concentrations were most
frequently encountered over the southern wall of the trough at Scott
Inlet and its extension between Hecla and Griper Bank and Baffin Island.
In addition, there was a high percentage of higher concentrations at the
seaward extremity of the sampling grid. At Buchan Gulf the higher
concentrations were associated with the trough but there was a pronounced
offshore component. At the entrance to Lancaster Sound concentrations
higher than the 1977 mean were only observed occasionally and were absent
in the northern portion of Lancaster Sound and the region southeast of
Devon Island. It may be concluded, therefore, that, whereas seepage has
had a detectable impact on the background level of dissolved/dispersed
petroleum residues in the water column in the northwestern portion of
Baffin Bay, it is not a major contributor elsewhere. The relatively
uniform, low background observed throughout the water column in Baffin Bay
must originate from a diffuse, non-localized source rather than a number
of point sources, and atmospheric fall-out of combustion products would
seem to be a major contributor. The observed concentrations of petroleum
residues in the water column at Scott Inlet suggest that the seepage
consists of small discrete releases from the walls and bottom of the submarine
trough and that the seepage is neither continuous nor localized but occurs
sporadically at a number of sites distributed over a considerable area of
the seafloor. One of the largest and most persistent sites seems to be
associated with the structural high near the seaward end of the trough.
Immediately upon escaping from the seabed, the oil droplets begin to rise
under the force of their own buoyancy and are swept along by water currents.
The result of this is the presence of "plumes" of oil droplets which are
nonhomogeneously distributed in both space and time in the water body.
Within a plume particles of various sizes are separated by various distances

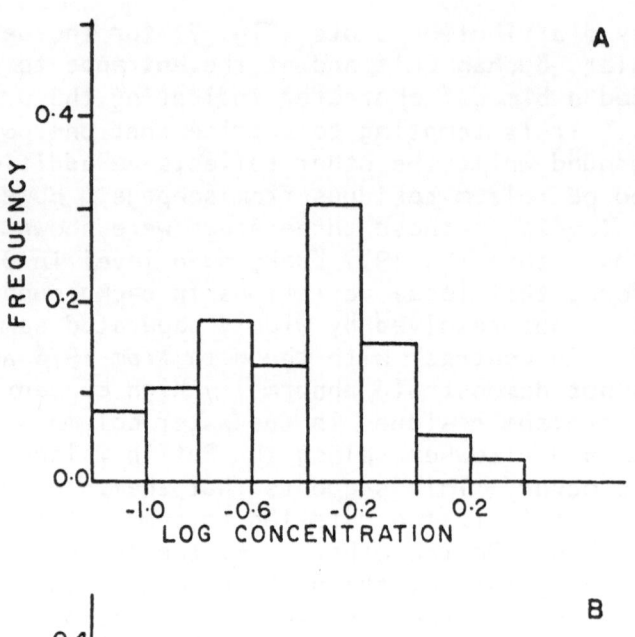

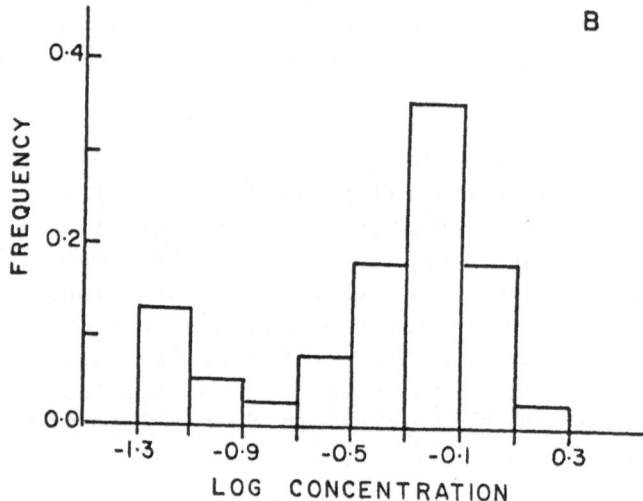

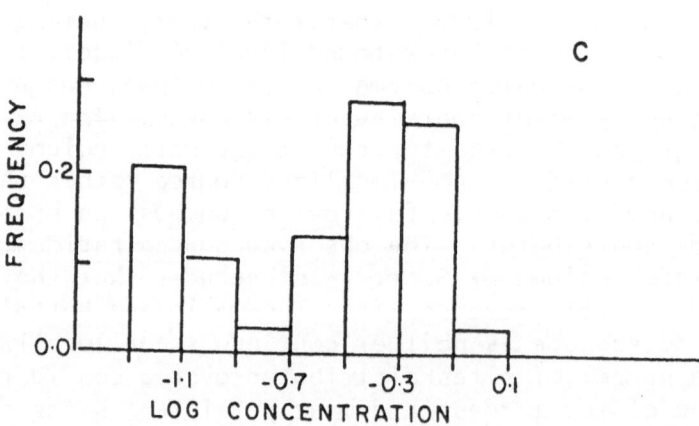

Fig. 7. Frequency distribution histograms of 1978 data for dissolved/dispersed petroleum residues in the water column at Scott Inlet (A), Buchan Gulf (B), and the entrance to Lancaster Sound

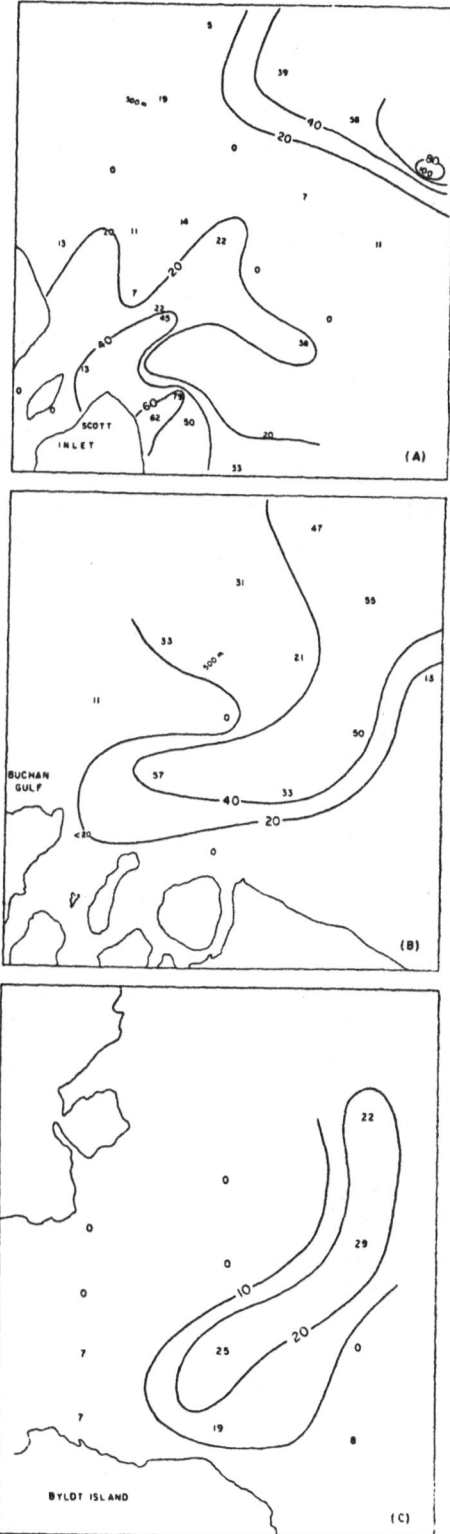

Fig. 8. % of concentrations higher than background in the
water column at Scott Inlet A, Buchan Gulf (B), and
the entrance to Lancaster Sound in 1978

which may be large in relation to the size of the samplers. As a result, the probability of catching one or more of them in a sampler at a given time and place in the water column accounts for some of the variability in the concentration data and the fact that not all samples from a given station, even in the seep areas, have concentrations above the background level.

Surficial bottom sediments - The concentrations of petroleum residues in the surficial bottom sediments of Baffin Bay and the Eastern Canadian Arctic (Fig. 9) in 1977 ranged from 1 to 41 µg/g. Concentrations were less than 5 µg/g in Davis Strait and along the Greenland side of Baffin Bay, 5-10 µg/g in the central region, and 10-20 µg/g along the northeast coast of Baffin Island. In Lancaster Sound concentrations increased in a westward direction from the entrance. One sample collected at the northern extremity of Baffin Bay at the entrance to Smith's Sound contained 26 µg/g. However, the samples taken at Scott Inlet in 1977 did not demonstrate the presence of elevated concentrations of petroleum residues in the sediments from the vicinity of the seep.

The 1978 survey demonstrated a close relationship amongst the geographical distribution of petroleum residue concentrations, grain size and bathymetry (Fig. 10). In general, the grain size distribution closely paralleled the bathymetry; i.e. the coarser sediments were present over the shallow shelf areas (<200 m) while the predominantly fine grained sediments were present in the deep troughs. Similarly, petroleum residue concentrations were lowest in the coarse sediments on the shelf areas and highest in the fine grained sediments in the deep areas. This general trend is commonly observed since the finer grained sediments with their larger surface areas and greater capacity to adsorb organic compounds tend to be transported from the shelf and accumulate in the deep areas. What is remarkable, however, is that anomalously high concentrations of petroleum residues are present in the surficial bottom sediments at both Scott Inlet and at Buchan Gulf in areas where surrounding water depths and sediment grain sizes are uniform. This indicates a source of petroleum within the seabed. At Scott Inlet, this anomaly is in the general area where gas bubbles and slicks have been observed on the surface of the sea on several occasions and where elevated concentrations were observed in the surface microlayer and water column. Most importantly, all of these occur in the general vicinity of the basement high and together provide strong evidence that this structure is a source of the seepage.

CONCLUDING REMARKS

Visible slicks and petroleum concentration anomalies in the sea surface microlayer, water column and surficial bottom sediments demonstrate that natural seepage of petroleum is occurring from the seabed at several locations on the continental shelf of northeastern Baffin Island. Sea surface phenomena are ephemeral since they are highly dependent upon environmental conditions, while concentration anomalies in the water column are not always observed even in the vicinity of the seep because of the remote chance of encountering an oil droplet in a water sample at any specific place and time. Accordingly, these properties provide only a general indication of the locations of the seeps. By comparison, surficial

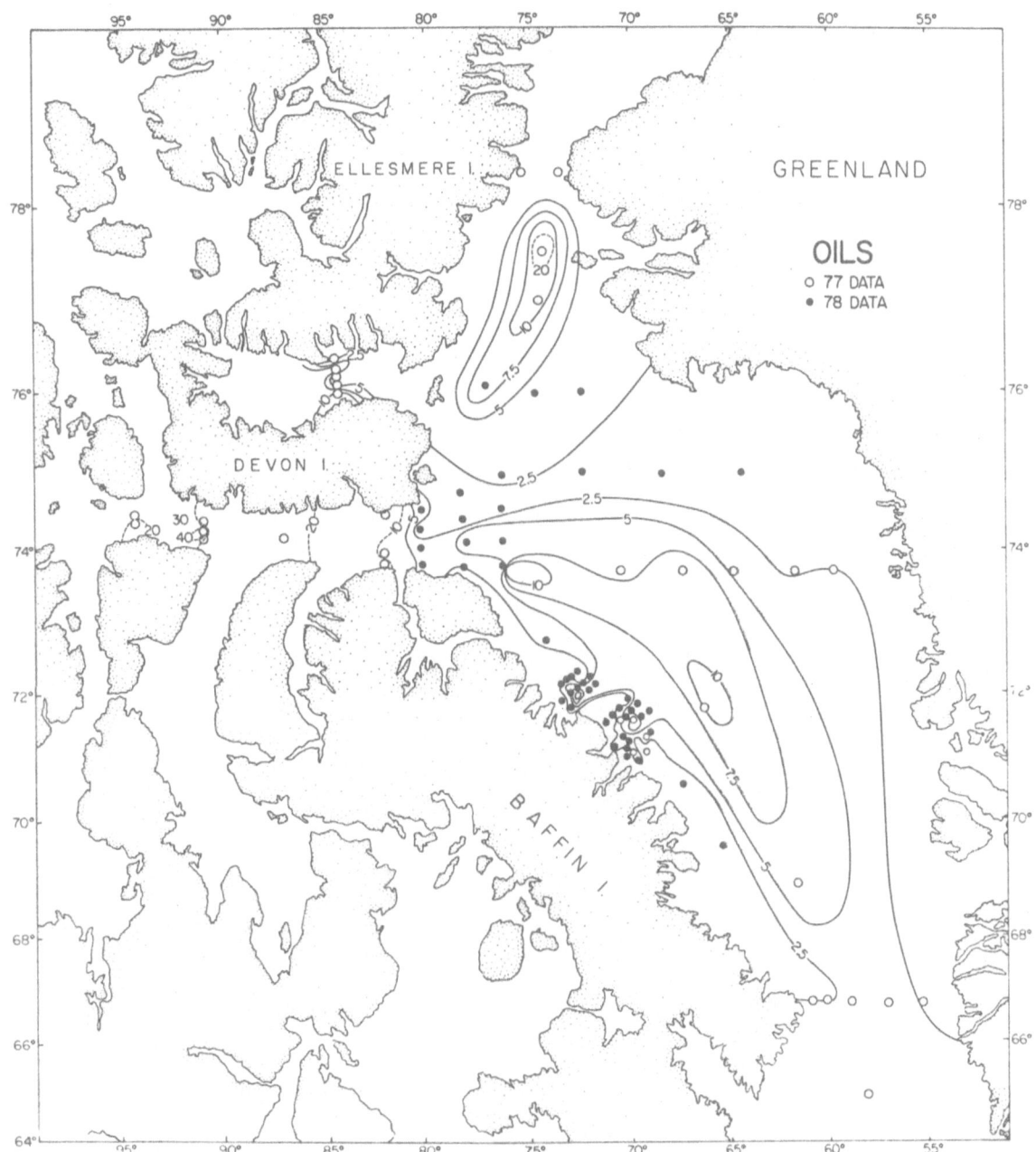

Fig. 9.   Concentration (μg/g) of petroleum residues in the
surficial bottom sediments of Baffin Bay 1977/1978

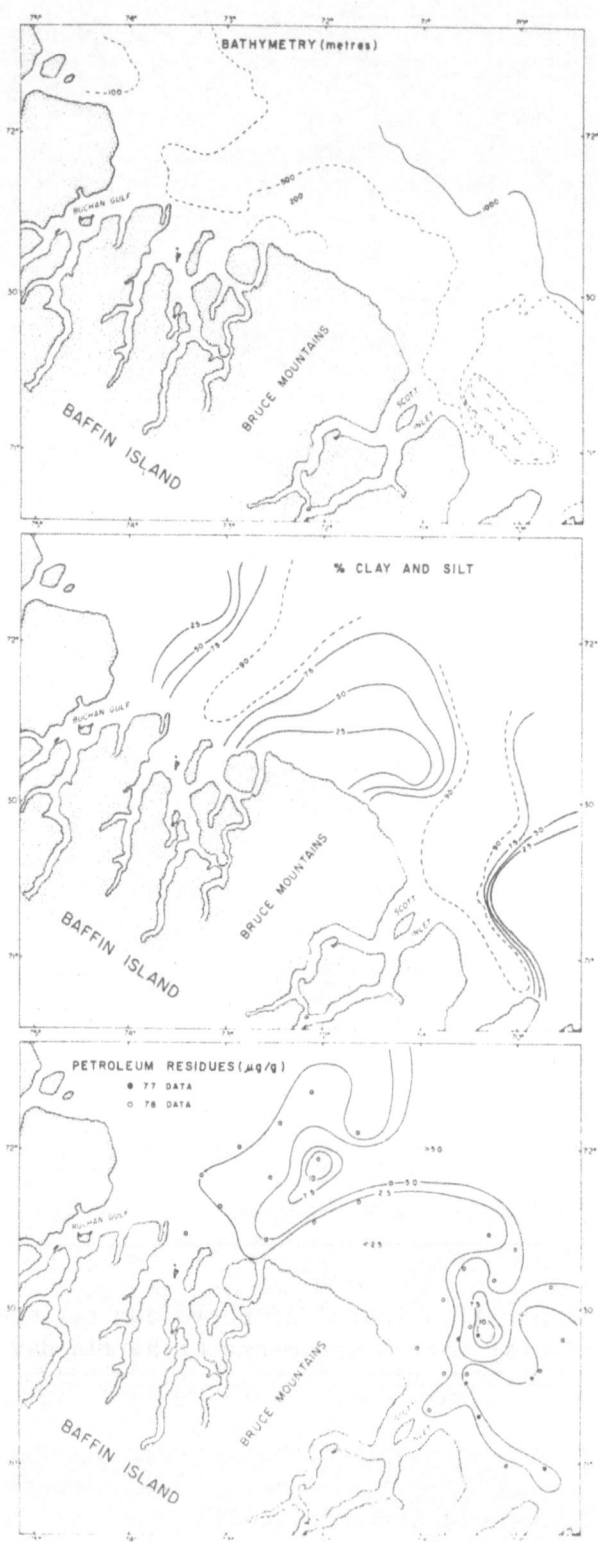

Fig. 10. Bathymetry (m), grain size distribution (% clay and silt), and petroleum residue concentration (µg/g) in surficial bottom sediments at Scott Inlet and Buchan Gulf in 1978

bottom sediments are relatively immobile and permanent and thereby
provide a considerably more direct indication of the seep sites. All
parameters, however, suggest that the seepage is associated with the
submarine troughs that cut across the Baffin Island continental shelf at
Scott Inlet and Buchan Gulf. A structural high near the seaward portion
of the Scott Inlet trough provides a geological environment favorable for
seepage from the underlying rocks and is probably the dominant site of
seepage in that area.

Whereas seepage has had a demonstrable impact on the background
level of petroleum residues in the northwestern portion of Baffin Bay,
it has had no noticeable impact on the region as a whole, and the low
background level observed appears to be the result of atmospheric fall-out
of combustion products. This supposition has been supported by recent
analyses by gas chromatography/mass spectrometry of the hydrocarbons
present in the sediments from Scott Inlet and Buchan Gulf (Ref. 11) and
will be addressed more fully in future investigations in the Arctic.

ACKNOWLEDGEMENTS

The co-operation and assistance of the officers and crew of
C.S.S. HUDSON and the technical staff of the Chemical Oceanography
Division in collecting and analyzing the samples and of Dr. J.M. Bewers
and Dr. R. Pocklington in critically reviewing the manuscript are
gratefully acknowledged.

REFERENCES

1. B.D. Loncarevic and R.K. Falconer, "An oil slick off Baffin Island", in Report of Activities, Part A, Geol. Surv. Can., Paper 77-1A, (1977), 523-524.

2. E.M. Levy, "Scott Inlet slick: an Arctic oil seep?", Spill Technology Newsletter, November/December 1977.

3. E.M. Levy and A. Walton, "High seas oil pollution: particulate petroleum residues in the North Atlantic", J. Fish. Res. Bd. Can., 33, (1976), 2781-2781.

4. E.M. Levy, "Visual and chemical evidence f~r a natural seep at Scott Inlet, Baffin Island, District of Franklin"; in Current Research, Part B, Geological Survey of Canada, Paper 78-1B, (1978), 21-26.

5. B. MacLean, "Marine geological-geophysical investigations of the Scott Inlet and Cape Dyer-Frobisher Bay areas of the Baffin Island Continental shelf in 1977". Current Research, Part B, Geological Survey of Canada, Paper 78-1B, (1978), 13-20.

6. B. MacLean and R. Falconer, "Geological/geophysical studies in Baffin Bay and Scott Inlet - Buchan Gulf and Cape Dyer-Cumberland Sound areas of the Baffin Island shelf", Current Research Part B, Geological Survey of Canada, Paper 79-1B, (1979), 231-244.

7. W.D. Garrett, "Collection of slick-forming materials from the sea surface", Limnol. Oceanogr., 10, (1965), 602-605.

8. E.M. Levy, "Fluorescence spectrophotometry: principles and practice as related to the determination of dissolved/dispersed petroleum residues in sea water", Bedford Institute of Oceanography, Report Series/BI-R-77-7, July 1977.

9. E.M. Levy, "Concentration of petroleum residues in the waters and sediments of Baffin Bay and the Eastern Canadian Arctic, 1977", Bedford Institute of Oceanography Report Series/BI-R-79-3, January, 1979.

10. E.M. Levy, "Further chemical evidence for natural seepage on the Baffin Island shelf", in Current Research, Part B, Geological Survey of Canada, Paper 79-1B, (1979), 379-383.

11. E.M. Levy and M. Ehrhardt, "Natural seepage at Buchan Gulf, Baffin Island", (unpublished results).

STUDIES AFTER THE METULA OIL SPILL IN THE STRAITS OF MAGELLAN, CHILE

Leonardo Guzman
Marine Biologist, Head of the Department of Hydrobiology,
Instituto de la Patagonia, Punta Arenas, Chile.

Italo Campodonico
Marine Biologist, Head of the Laboratory of Marine Biology,
Department of Hydrobiology, Instituto de la Patagonia, Punta
Arenas, Chile.

INTRODUCTION

At present, the oil spill from the supertanker Metula occurred
in August 1974 in the First Narrows of the Straits of Magellan
is the largest registered for the coasts of South America.
Information about the official causes of the grounding and
details on the spillage, refloating and unloading of the cargo
is found in Opazo (1) and Hann & Young (2), respectively.

As a consequence of the spill a large area of the coast of
Tierra del Fuego and the continent was oiled. The intertidal
biota, certain saltmarshes and bird populations in this area
were impacted.

Data on the amount of oil which landed on the beaches, its
fate, behaviour and immediate biological impact are given by
Hann (3), Baker et al (4) and Straughan (5), among others. A
certain amount of pollutant is still present in many localities
and   it will remain there *for many years.*

The magnitude of the spillage, the fact that the oil was
not treated with dispersants nor removed by other means and
that it took place in a cold temperate area, attracted the
attention of people associated with these types of accidents
and of investigators in this field, with the object of gaining
experience which could be applied on future accidents in areas
of similar characteristics.

As a result, numerous articles have been published dealing
with studies that go from coastal geomorphology (6) to microbial ecology (7).

In September 1976 and after being involved in the initial
surveys, the Instituto de la Patagonia started a series of
studies in the area of Puerto Espora in order to investigate
the nature and degree of environmental recuperation.

Research has been conducted in the fields of intertidal
ecology, saltmarshes vegetation, and insect and bird populations.
Some results have already been published (8, 9, 10, 11, 12)
but other studies are still continuing.

An indirect result of this spillage was the initiation, by
the same Institute of baseline studies in the intertidal area
of the eastern straits, in order to obtain information for the
assessment of the environmental impact produced by eventual
oil spillings due to the activities of the Chilean Oil Company
or from other sources. A further aim of these studies is the

monitoring of the community dynamics in sites already polluted.

This article presents an account of the most outstanding published results and gives information about the findings of the studies carried out by the Instituto de la Patagonia. An evaluation of the impact of this oil spill on the *macrobiota of* the eastern area of the Straits is intended.

AFFECTED AREA AND FATE OF THE OIL

As pointed out by several authors, no efforts were made to contain or disperse the oil released by the Metula, or to remove or stabilize the oil which landed on the beaches. Detailed information about this matter may be found in (3, 13, 4).

The oil spilled was rapidly dispersed by the currents and by the normal gravity spreading. Oil slicks were moving back and forth in the Straits for over two months after the spill.

A non quantified portion of the oil was suspended and dissolved into the water column while another part was lost by evaporation. Evaporative loss was estimated as low as 15-20% and as high as 40% (4, 2).

However, because of the meteorological and hydrographical conditions during the spill, most of the oil formed a *water-in-oil* emulsion (mousse) which, in a few hours, stranded on the coast of Isla Grande de Tierra del Fuego, particularly on the beaches located on First Narrows and some kilometers to the west of it. Significant amounts of mousse were deposited in two inlets and adjacent saltmarshes existing in the area of Puerto Espora.

As a consequence of new spills and/or redistribution of the mousse, by late August the affected area in the coast of Tierra del Fuego extended as far as Bahia Felipe on the west, and mousse deposits were observed in Banco Lomas, west of Punta Catalina; on the continental side deposits were observed between Posesion and Punta Dungeness (Fig. 1). By early September the total area affected by these deposits was in the order of 120

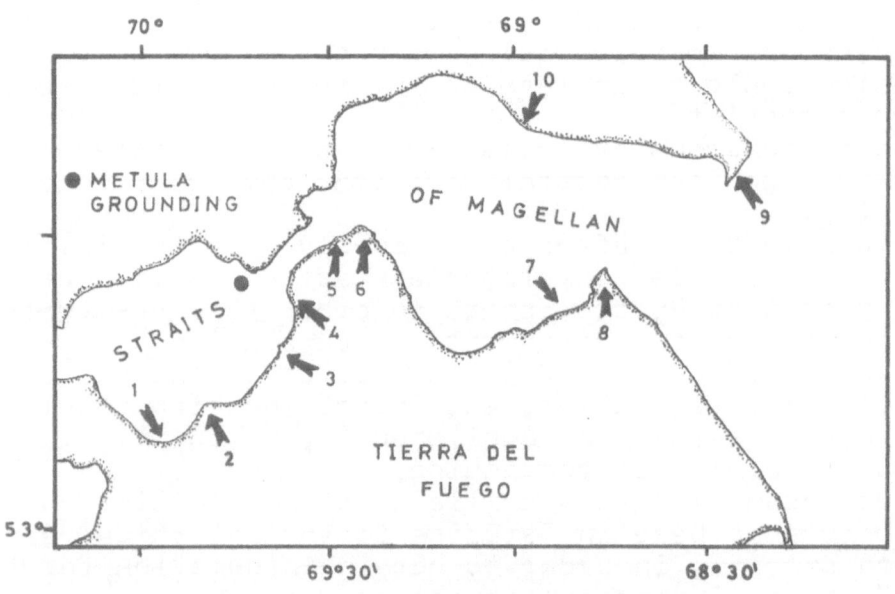

Fig. 1.   Eastern area of the Straits of Magellan.

1, Bahia Felipe; 2, Punta Piedra; 3, Punta Remo; 4, Punta Baxa; 5, Puerto Espora; 6, Punta Anegada; 7, Banco Lomas 8, Punta Catalina; 9, Punta Dungeness; 10, Posesion

to 130 Km., and it was estimated that between Punta Piedra and Punta Anegada, the most contaminated sector, there were about 60,000 cubic meters of mousse (3). Calculations by Baker et al (4) and accepted later by Guzman (15) determined that in this same area there were around 45,000 cubic meters of emulsion, not including the mousse that had made its way into the inlets of Puerto Espora. (Fig. 2).

Surveys practised a year after the spill by Hayes & Gundlach (6) and others effected in February 1976 (see 14) showed that the contamination spread over approximately 250 Km. of coast, the beaches of First Narrows and those between Punta Remo and Punta Baxa being the most heavily affected.

At first it was estimated that the oil volume deposited in the more contaminated beaches of Tierra del Fuego was of approximately 40,000 tons (3) but in a more recent report Hann & Young (2) point out that some 56% of the residual oil (assuming 40% evaporation) might have washed ashore. This would mean between 17,500 to 18,000 tons.

The initial deposits of mousse were highly unstables and had a variable width and thickness. Authors such as Baker et al (4) and Hann (16) point out that the thickness of these layers varied, according to the different localities, between less than 1 cm. to 15 cm., and that their width fluctuated between 2 and 65 meters.

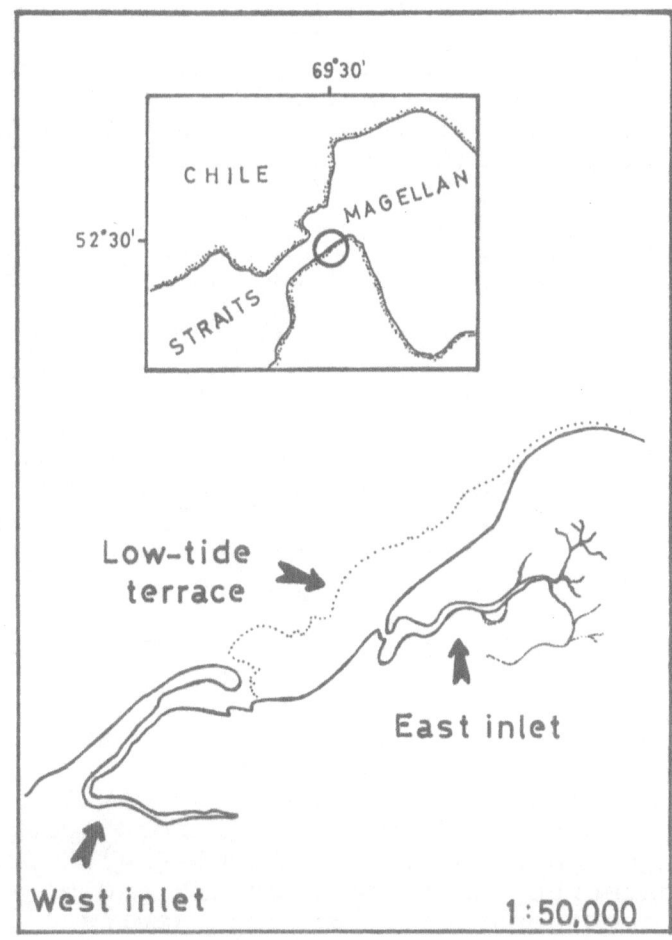

Fig. 2. Area of Puerto Espora, First Narrows of the Straits of Magellan; East and West Inlet

In the East inlet of Puerto Espora the mousse had a thickness
of up to 20 cm. Part of the stranded mousse was carried over
the high spring tide line and to the upper levels of the beaches.
With time, the deposits that remained on the surface weathered
and hardened forming crusts that in some places were covered by
sand and gravel. In the upper reaches of some beaches oil was
also incorporated into the sediments and was likewise covered
by sand.

However, the major part of this emulsion stranded on the
low-tide terraces (excluding its lower levels which were not
reached by this oil), and the subsequent fate of the mousse
varied according to the characteristics of each site. In some
places the mousse remained on the sediments as an almost fresh
mixture being periodically removed and redeposited.

In other sites the mousse penetrated into the sediments,
mixed with them, was trapped beneath large boulders or was
covered by sand layers of up to 15 cm. in thickness.

Most of the oil accumulated in the inner channels of the
inlets of Puerto Espora was later removed by high tides and
deposited on the flats between the channels. In some sectors
this oil hardened but in other the mousse is still soft.

The most noticeable fact resulting from the deposition of
mousse in the coasts was the formation of an asphalt pavement
in the intertidal zone of certain localities, specially that
of Puerto Espora. This site has a low-tide terrace of about
300 to 400 meters width and a very gentle slope constituted by
coarse sand, gravel and pebbles. Very soon after the spill
this terrace was completely covered by a thick layer of mousse
which in some sectors was later naturally removed. In other
sectors however this mousse intensively mixed with the sediments
and weathered forming a relatively stable and artificial sub-
strata, 15 to 20 cm. thick, which at present extends discontin-
uously for several hundreds of meters.

In other localities this oil/sand/gravel pavement was less
extensive and less thick and with time was eroded or covered
by clean sediments.

The oil content in the sediments varied largely according
to the localities and also within the same locality. Samples
collected in January 1975 showed that in the most polluted sites
(First Narrows) the oil content on the surface sediments were
from 30 to 40% (17). Analysis carried out by Blount (14) in
samples collected during 1976 indicated that the volumetric oil
percentage varied from approximately 0.03 to 51.5% and according
to Langley and Lembeye (11) the surface sediments of the East
and West inlet of Puerto Espora contain an average of 3.27 and
0.59% of oil, respectively.

Samples collected by the authors in several sites of the
Strait during January 1977 had an oil content ranging from 0.03
to 6.26%, while others collected in August along two transects
on the asphaltic pavement at Puerto Espora contained from less
than 0.05 to 3.5%.

Detailed information about the behaviour of the oil, its
deposition, retention and erosion in different localities of
the Straits of Magellan is given by Blount (14). Based upon
the information gathered during her surveys and on the previous
data, the author concluded that the most heavily oiled areas
were those of low energy closest to the spill site and that
the oil was commonly deposited on the supratidal areas and low-
tide terraces.

Blount estimates that the deposition of oil on the coasts of the
Straits had depended on the distance from the spill site, the
local tidal range, the tidal phase at the time of oil landfall,
the prevailing winds and the path and velocity of tidal currents,
the former being the most important factor. Likewise, she
concluded that the retention of the oil depended on the extent
of initial contamination, the accretion of clean sediments over
the oiled; the wave action; the water circulation; the existence
of biota on the oiled substrate and the character of the sediments.

Although in certain exposed and high energy places the oil
has been gradually eroded and/or removed, in many other sites
there is still a high level of contamination and authors such
as Hann & Young (2) and Blount (14) have pointed out that in
protected areas (.i.e. the inlets and saltmarshes of Puerto
Espora) or where the pollutant has deeply penetrated into sediments
or formed asphalt pavements, the complete removal of the oil will
take many years and even decades.

According to Blount, the erosion of the oil is being controlled
by the wave energy, the wind velocity and direction, the tidal
currents and the biological activity.

In her opinion, the occurrence of storms might be an important
factor for a rapid removal of the pollutant.

Observations done by the authors from late 1976 to the
beginning of 1980 indicate that the pavement existing in the
intertidal zone of Puerto Espora is being gradually but continuously
eroded on its land and seaward edges as well as in its sides,
mainly by the action of waves and the tidal currents.

EFFECTS OF THE SPILL ON THE BIOTA

Saltmarshes and littoral vegetation

The observations done by Baker et al (4) and Straughan (5) between
September 1974 and January 1975 demonstrated that different
species of phanerogamic plants (among them Salicornia ambigua,
Leymus arenarius, Lepidophyllum cuppressiforme) had been covered
by mousse. The root systems of these plants had not been affected
and new shoots were observed in sites where the mousse layer was
not too thick.

Studies conducted by Pisano (8) in the area of Puero Espora,
between July 1975 and the end of 1976 showed that the fringe of
littoral vegetation, mainly constituted by sand grass L. arenarius,
was affected since mousse penetrated the base of the clumps, and
was also found in the lower sections of stalks and leaves. This
author did not observe the occurrence of seedlings.

Species such as Phacelia magellanica and Senecio candidans
which occurred together with this sand grass, were also impacted.
Among the saltmarsh plants, Salicornia ambigua was the most
damaged species and 80 to 90% of the specimens were estimated
to be dead. The magnitude of the damage depended on the location
of the plants within the saltmarsh and on the type of substrate
they were growing on. In some places the substrate was completely
unfit for recolonization.

After comparing the growth of S. ambigua at Puerto Espora in
January 1975 and the same month of 1977, Straughan et al (18)
suggested that in areas of cold water the full impact of a
massive spillage on saltmarsh plants might occur at least two
years after the oiling.

Recent studies carried out in the same area (9, 10) indicate
that the amount of oil, the location of the plants within the

saltmarsh, the composition and compactness of the soil, the
location of the renewal buds, and the depth of root systems
were all important factors in determining the magnitude of the
effects. In some sectors of the saltmarshes the original
substrate, according to the author, was considered unfit for
recolonization for several years. Out of the two communities
occuring in the flood plains of the saltmarshes: Salicornia-
Suaeda and Frankenia-Atriplex, the first of these was the most
seriously affected, because although the conforming species
are very resistant and have a high recovery potential, the large
amount of accumulated oil covered most of the plants causing
their deaths.

The communities of L. arenarius and L. cuppressiforme
located on the non-flooded plains and at the border of the dunes
did not show noticeable damage and only the marginal specimens
reached by the spring tides were affected.

Bird Populations

The very first surveys after the spill (3, 4) indicated that
about 17 species of birds had been affected by the oil. Until
February 1975, 3,000 to 4,000 individuals were estimated to
have been killed. Cormorants (Phalacrocorax atriceps and P.
magellanicus), penguins (Spheniscus magellanicus, Eudyptes
crestatus) and the kelp-gull Larus dominicanus were the most
impacted birds.

Most of the dead individuals were found near or at the
area of First Narrows. Several breeding colonies of penguins
and cormorants were visited or observed between September 1974
and February 1975, but only a few oiled specimens were found.

No population counts were available at the time the spill
occurred and therefore the actual impact of the spill on birds
could not be properly quantified. However it was estimated
that the oil did not substantially affect the total populations
mainly because the accident happened before the main migration
time of the penguins and before the annual return of certain
shore birds. Nevertheless great mortality of young terns was
observed in a nesting site located in the area of Puerto Espora,
but the causes of this were not specified (13).

A 3-year study done by Venegas et al (12) in many sectors
of the West inlet and saltmarsh at Puerto Espora, showed that
the number of species visiting the area has gradually increased
with time. This area was completely uninhabited after the spill.
Higher numbers of species were recorded in the sites located
in the Straits and in the outer sectors of the inlet than in the
inner ones.

Thirty-six bird species were recorded between September 1976
and September 1979, and one third of them were frequent in the
area. Charadriiformes birds were the most important not only
because of its numerical predominance and wide spatial distribution. Carnivorous species such as the kelp-gull Larus dominicanus, the sandpipers Calidris fuscicollis and C. bairdii
the plover Charadrius falklandicus, the oyster-catcher Haematopus
leucopodus and the dotterel Zonibyx modestus, were highly
frequent, numerous and widely distributed in the studied area.
Herbivorous species were poorly represented and in the opinion
of these authors, this might be due to the heavy contamination
of plant communities in some sectors.

Insect populations

The available information concerning these organisms is limited to studies and observations carried out by the Instituto de la Patagonia from December 1976 to January 1978 in the saltmarsh adjacent to the East inlet of Puerto Espora and refers only to soil insects (19). No insects were recorded in those sites heavily polluted and dominated by the community of Salicornia-Suaeda. The only exception was the occasional appearance of Collembola.

In the sites where the community of L. cuppressiforme is established, the soil-surface insect fauna was represented by 66 families and 111 species and groups such as Tenebrionidae, Heliomyzidae, Curculionidae, Sciaridae and Cicadellidae were the dominant in terms of number of species and individuals.

The more abundant species were Praocis bicarinatus, Plathestes depressa, Emmallodera multipunctata and Cylydrorhinus angulatus.

Captures were much higher in the inner sectors of this community than in the marginal areas of it, adjacent to the community of Salicornia-Suaeda, but the species diversity was lower. Sixty-five percent of the species were common to the inner and marginal areas but the latter had a greater number of exclusive forms. In both types of areas, the qualitative and quantitative composition of the insect fauna showed seasonal variations and 2 groups, corresponding to spring-summer and fall-winter, were distinguished. Both in the inner and marginal areas, the total density increased gradually in the spring, reaching a maximum in summer and decreasing in the fall to its winter minimum. Diversity, as determined by the Shannon-Wiener index, followed this same tendency with the lowest values during the winter and the highest during the summer.

Lanfranco concluded that the absence of insect fauna in the sites heavily oiled was directly related to the strong contamination, but stated that the differences detected between the marginal and inner sectors in the community of L. cuppressiforme might be due to other factors.

Microbial populations

Studies conducted by Colwell et al (7) with samples collected between May and August 1976 and January 1977, in different localities of the Straits of Magellan, showed that in the oiled sites there were significant and persistent increases in the numbers of aerobic heterotrophic micro-organisms as well as in the number of petroleum degraders. The oil had a selective stimulatory effect which modified the structure of microbial populations.

The experiments carried out by these authors demonstrated that the Metula oil had little if any effect on glucose metabolism micro-organisms and that low temperatures seemed not to be a limiting factor for petroleum degradation.

However, natural in situ conditions along the Straits were not ideal for a rapid and complete degradation of oil, and the relatively low concentrations of nitrogen and phosphorus, together with a restricted accessibility to degradable compounds in many cases, were limiting the rate of this process.

Marine macrobiota

The first data on marine macrobiota was supplied by Hann (3) who carried out surveys of the intertidal zone of several sites on

the coast of Tierra del Fuego in August and September, 1974.
In his opinion the largest impact of this spill would be on the
organisms that inhabit the intertidal zone of the affected areas.
Mousse samples analyzed by the author showed a great abundance
of dead worms and other organisms that had been imprisoned in
the mousse. Hann specified that he did not find fish or other
pelagic organisms dead due to the spill and concluded that the
recolonization of the intertidal area would be difficult because
of the extreme climatic conditions.

In January 1975, the same author made broader and more careful observations of the intertidal zone (13). In the contaminated
sites he observed discoloured algae, mussels in bad conditions
and the absence or substantial disappearance of juvenile mussels,
limpets of all sizes and crustaceans. He concluded that the
impact on intertidal biota was much greater than he had
estimated in September 1974.

Between September to October of 1974 and January to February
of 1975, Baker et al (4) surveyed the affected intertidal zone.
The upper levels presented scarce macrobiota because of the
instability of the sediments, while the lower levels supported
a rich and abundant community, dominated by the mussel Mytilus
chilensis and the limpet Nacella magellanica. A variety of
algae was associated with these species, among them Porphyra
columbina, Ulva lactuca, Adenocystis utricularis, Enteromorpha
intestinalis and Iridaea sp., as well as some chitons, small
bivalves and anemones. In September the mussel beds seemed
relatively unaffected, and most specimens were still alive,
although in several sites there was mousse trapped among them.
In February 1975, a greater mortality was obvious and the beds
were still covered by petroleum as a consequence of the continued
run-off. In the First Narrows district there were ill and
dead limpets which had dropped off the rocks although the
animals were still numerous. In several places P. columbina,
U. lactuca and other algae were in bad conditions or dead.

Upon observations of the fronds and holdfasts of the brown
algae Macrocystis pyrifera, it was seen that there was apparently
no damage to the alga itself nor to the organisms inhabiting
the holdfasts. No dead or ill marine mammals were found as a
result of the spill. Commercial fisheries were not affected,
as the principal fishing grounds are located outside of the
contaminated area.

Straughan's studies (5) of samples collected in contaminated
and non-contaminated intertidal sites five months after the spill
showed that marine biota were still negatively affected. In
the sampled areas the author recorded 46 species of marine
invertebrates of which 28 were polychaetes. Analysis of mussel
tissues showed 1,000 to 5,000 ug/g of hydrocarbons. The
physical presence of oil had not only affected the distribution
of intertidal organisms, reducing their abundance and variety,
but had also altered the characteristics of the sediments,
which would influence the intertidal resettlement process.

The author concluded that the beaches of the Straits of
Magellan were comparatively more contaminated than the British
coasts affected by the Torrey Canyon spill.

In January 1977 the author and her collaborators carried
out a new survey of the areas visited in 1975. Preliminary
analysis of the data obtained (18) showed that, in general terms,
there had been an increase in species and specimens in comparison
with 1975, although this tendency could not be correlated with

the oil pollution. The authors suggested that it may have been due to the higher temperatures and salinities in 1977.

In some places where oil was still present there was no evidence of recuperation and at one site there was a decrease in the number of species. In those localities where the pollutant was being gradually removed the number of species and specimens increased. This the authors interpreted as an apparent recovery.

They added that intertidal resettlement is originating from the subtidal zone.

In September 1976 the Instituto de la Patagonia initiated a one-year study on the macrobiota of the inlets of Puerto Espora and their potential repopulation. Special emphasis was given to Eastern inlet, where no epifauna was observed. The scarce infauna consisted of occasional specimens of polychaetes, olygochaetes and nematodes. Of all the algae the only abundant species was Ulothrix flacca; P. columbina, E. intestinalis and U. lactuca were seldom found. Laboratory analysis showed a high hydrocarbon content in surface deposits, but in the opinion of the authors it is likely that the scarcity of benthic marcrobiota in this inlet is due not to the presence of oil, but rather to other factors such as the degreee of exposure to air at low tide (11).

From September 1976 to the present we have conducted studies along a transect in the intertidal zone of Puerto Espora in order to evaluate the recolonization process which is occurring on an asphaltic pavement located in the low-tide terrace of this locality. The data presented below comprise the period of September 1976 to mid-1979.

During this time the surface temperature of the water varied from 5 to 15°C, and the salinity from 29.52 to 30.81 p.p.t. Both these parameters showed clear seasonal variations, with maximum values in the summer.

Twenty-eight months after the spill the algae, represented by 11 species, occupied almost the entire intertidal zone of Puerto Espora, while animals were just beginning to inhabit it.

Among the algae the species E. intestinalis, U. lactuca, P. columbina and U.flacca were dominant in terms of coverage. The abundance of algae along the transect has shown clear seasonal fluctuations; being highest in spring/summer and lowest in fall/winter. If one compares the coverage of the flora for the same period over the years it becomes clear that in 1978 it was lower than in the preceeding year and in 1979 it was higher than in 1978.

Experimental quadrants established in the spring of 1976 at three levels in the intertidal zone in which all the epibenthic macrobiota was removed, showed that recolonization was very rapid. Two months after the complete removal of macrobiota, the area was re-inhabited by E. intestinalis, U. lactuca, P. columbina, U. flacca and A. utricularis.

The abundance of algae along the transect present local variations and, in general terms, greater coverage occurs at mid and upper levels. A significant decrease in abundance is noted towards the lower levels of the transect.

Up to the present no zonation pattern of algae is observed in the intertidal area of Puerto Espora, but along the transect some species are found almost exclusively in certain levels, (i.e. A. utricularis, Chaetangium fastigiatum, Iridaea sp.) while others such as U. lactuca and P. columbina are more widely distributed but nevertheless demonstrate a preference for certain levels.

Of all the species recorded along the transect only E. intestinalis is widely distributed and has a relatively uniform abundance at most levels.

During the same period the fauna was represented by 22 species, 14 of which belonged to the epifauna and 8 to the infauna. The mussel M. chilensis, the gastropods Siphonaria lateralis and Nacella magellanica, the isopod Edotea chilensis and a unidentified species of anemone were among the former, and were present throughout the entire study period. The polychaetes Onuphis dorsalis, Notocirrus cf. lorum, Cirratulus cirratus and the nematode Thoracostoma setosum were infauna which appeared in all the surveys.

Thus far, the colonization process has been characterized by a constant increase in the total number of specimens with the exception of the winters of 1977 and 1979, in which there was a slight decrease. A greater density is evident moving from the upper levels of the transect to its lower intertidal levels. No increase in the number of species through time has been observed.

Among epifauna, M. chilensis and S. lateralis accounted for 93.7 and 99.9% of the total of animals counted and both species have gradually colonized the intertidal zone starting with its lower levels. Settlement by these species of the low-tide terrace along the entire transect took place approximately 3 years after the oil spill.

The density in number and in biomass of the infauna was low, increasing from the upper to the lower levels of the intertidal zone. No clear tendency in the density was apparent through time. The infauna was numerically dominated by the nematode T. setosum, but in terms of biomass, the polychaete O. dorsalis was dominant, with 81.8 to 99.9% of the total.

Recent studies by Guzman (20) in various locations on the Straits of Magellan, including sites contaminated by the Metula spill, have demonstrated that the oil has affected the degree of aggregation of N. magellanica as well as the age structure of the species in the most contaminated areas.

DISCUSSION

Although this article summarizes almost all the available inform- ation on the Metula oil spill, the uneveness in depth and continuity of the studies and the fact that information is lacking on many subjects, does not permit an extensive, inte- grated evaluation of the long-term effects of this spill. This is even more difficult if one considers that the pre-spill conditions are unknown.

Because of the nature of this paper and the objectives of the Conference itself, the following discussion refers only to general aspects of the impact of the spill on the biota and its recuperation.

The magnitude and extent of the ecological alterations depend mainly on the community under consideration and on its location along the eastern area of the Straits.

Thus there were communities which were not affected by this spillage. Swartz & Gallardo (21) have informed on a subtidal survey carried out in April 1976 in the eastern area of the Straits of Magellan. The results of this study have not yet been published, but according to a personal communication of

Dr. V. Gallardo (Universidad de Concepcion, Chile), the subtidal community did not show signs of having been altered, and
the sediments were free of oil.

Although no specific studies are available on the impact of
this spill on the kelp beds (Macrocystis pyrifera) and the
organisms associated with them, we consider that this community
may have been only slightly affected, if at all. This assumption
is based upon early observations (4) and the virtual absence of
oil in the subtidal and lower intertidal sediments, as well as
on the apparently healthy conditions of the biota inhabiting
these two sectors.

Certainly the intertidal communities were the most hard hit
by this spill. The large amount of oil which went ashore (mainly
on the Isla Grande of Tierra del Fuego) and the continued
presence of the pollutant in many sectors are the factors which
have been principally responsible for this situation. However
the magnitude and significance of the disturbances have been
poorly studied. In heavily oiled sites such as Punta Remo and
Punta Baxa, where the mousse was deposited on the upper beach
face and later covered by clean sediments, the age structure
and the degree of aggregation of the limpet Nacella magellanica
were affected (20). This was evident at least until October
of 1979. It is worthwhile pointing out that in these two areas
a high percentage of the specimens of this limpet is parasitized
by larvae of trematodes. Although the incidence of the infest-
ation in those localities contaminated less is much lower or
inexistent, it is not yet known whether the high degree of
parasitism constitutes another long-term effect of this spill.

The formation of an asphaltic pavement on the low-tide
terrace of Puerto Espora annihilated all the intertidal biota,
but the recolonization occurring on it is a process of particular
interest. The gradual, though slow, erosion of this artificial
substrate must be affecting the repopulation. The replacement
of this pavement by the original substrate will bring about the
transformation from a stable to a mobile substrate community.
As suggested by Straughan et al (18) the animal resettlement
was originated from the lower intertidal levels. At present,
the established community is relatively simple in its composition.
The number of species has not increased through time but animal
specimens, five years after the spill, are still growing in
number. The algae have reoccupied the space much more rapidly
than have animals. In a short time opportunistic and fast growth
species of algae have become established; while animals took
approximately three years to repopulate all the intertidal zone.

The study of the resettlement process which is occurring
in the asphaltic pavement could provide valuable information
regarding the elasticity of intertidal communities of cold
temperate ecosystems. A more detailed analysis of these aspects
will be published elsewhere.

Observations carried out in October 1979 at the East inlet
of Puerto Espora, demonstrated that the benthic macrobiota of
this area is still very scarce (G Lembeye, pers. com.). This
could be interpreted as a lack of recovery, but it is also
possible that this scarcity is a natural condition due to factors
other than oil contamination, as mentioned by Langley & Lembeye
(11).

The surface temperature and salinity records obtained by the
authors from September 1976 to the present show clear seasonal
fluctuations in these parameters, and that maximum and minimum

values may shift slightly through time. However there are no indications of a generalized increase or decrease in temperature and salinity. Therefore it is probable that the growth in the number of species and specimens detected by Straughan et al (18) between January 1975 and January 1977 constitutes a sign of actual recovery.

The diverse and abundant biota present in the low-tide terraces of most sectors which were contaminated seems to indicate that intertidal communities are in a healthy condition. However, the changes detected in the populations of N. magellanica suggest that at least in the areas of Punta Remo and Punta Baxa, the intertidal community is disturbed. Only further and more detailed studies would permit the elucidation of this problem and the evaluation of its magnitude and significance.

Heavy contamination and the subsequent death of numerous plants in the community of Salicornia-Suaeda, as well as the unsuitability of the substrate to a natural repopulation constitute a serious setback to the saltmarshes of Puerto Espora, although it is very restricted in extent. Considering the unfavourable conditions existing in these areas for the erosion and removal of the oil, it is estimated that an experiment in artificial repopulation could be advantageous and even desirable. The impermeability of the substrate which has retained the oil in its uppermost levels, the capacity of the plants for vigorous sprouting as long as at least one bud remains undamaged, and the deep and branched root system of the species conforming the community of Salicornia-Suaeda would support the feasibility of such experiment. This not only would help in breaking apart the oil, making it susceptible to further erosion, but would also contribute to improving the deteriorated physical appearance of these saltmarshes.

The absence of soil-surface insect fauna in the most contaminated sector of the east saltmarsh, which was evident at least until January 1978, is also a long-term and serious effect, again highly restricted in its extent. It is considered that this effect will last as long as the vegetation is not re-established.

As pointed out by Venegas et al (12) the notable increase through time in the number of bird species visiting the west inlet and saltmarsh would indicate a recuperation of this area. However, it has not been confirmed that this is actually the case. On the other hand, it would be difficult to prove or disprove whether the currently existing populations are the same which inhabited the area before the spill occurred. Probably the most severe effect of this spill on birds was the destruction of a tern breeding colony (Sterna hirundinacea) located on a spit facing of the mouth of the west inlet of Puerto Espora. Although Hann (13) mentions that the great mortality of young observed in this site in January 1975 was a result of a variety of causes, it was certainly a consequence of the spill. The oil not only caused the death of many young but completely eradicated the colony. Thus far there are no indications of re-establishment and the species has not been observed in the area of Puerto Espora since.

Although there have been other large spillages since the Metula, this one has provided the scientific community with a great natural laboratory, which has not been utilized to its fullest by the academic community of Chile. The public's lack of a conservationist attitude, the small size of Chilean academia, the relative remoteness of the spill site and economic considerations are some of the reasons which might account for this situation.

ACKNOWLEDGEMENTS

The authors appreciate the opportunity extended to the Instituto de la Patagonia to participate in this Conference. We are grateful to our colleagues, O Dollenz, D Lanfranco, G Lembeye, E Pisano, C Venegas and E Zamora for their suggestions and information; to Laurie Nock and Ricardo Gibbons for revising the English text; and to Arlette Pantoja for typing the manuscript. Finally, our thanks to Shell Marine Limited, London, and the United Nations Environmental Program (UNEP) for covering our travel and living expenses.

REFERENCES

1. A Opazo, 'Los casos Napier y Metula: Los antecedentes de hecho'. In: Preservacion del Medio Ambiente Marinto (F. Orrego-Vicuna, Ed., Editorial Universidad Tecnica del Estado, Santiago, 1976), 153-76.

2. R Hann & H Young, 'Fate of oil spilled from the supertanker Metula'. Final Report from the Texax A & M Research Foundation, College Station, Texas (1979), 148 p.

3. R Hann, 'Oil pollution from the tanker Metula'. Report to the U.S. Coast Guard Research and Development Program (1974), 61 p.

4. J Baker, I Campodonico, L Guzman, J Jory Texera, B Texera, C Venegas & A Sanhueza, 'An oil spill in the Straits of Magellan'. In: Marine Ecology and Oil Pollution (J M Baker, Ed., Applied Science Publishers Ltd., Essex 1976) 441-71.

5. D Straughan, 'Biological Survey of intertidal areas in the Straits of Magellan in January 1975, five months after the Metula oil spill'. In: Fate and Effects of Petroleum Hydrocarbons in Marine Organisms and Ecosystems. (D A Wolfe, Ed., Pergamon Press, New York, 1977) 247-60.

6. M Hayes and E Gundlach, 'Coastal Geomorphology and Sedimentation of the Metula Oil Spill Site in the Straits of Magellan'. Final Report for National Science Foundation, Washington (1975), 103 p.

7. R Colwell, A Mills, J Walker, P Garcia-Tello & V Campos, 'Microbial Ecology Studies of the Metula Spill in the Straits of Magellan'. J. Fish. Res. Board. Can. 35, (1978), 573-80.

8. E Pisano, 'Contaminacion por petroleo del B/T Metula en vegetacion fanerogamica litoral. Observaciones preliminares'. ANS. INST. PAT., Punta Arenas (Chile), 7 (1976), 139-53.

9. O Dollenz, 'Estado de la flora vascular en Puerto Espora, Tierra del Fugego, contaminada por el petroleo del B/T Metula. I. Reconocimiento de la entrada de mar noreste'. ANS. INST. PAT., Punta Arenas (Chile) 8 (1977), 251-61.

10. O Dollenz, 'Estado de la flora vascular en Puerto Espora, Tierra del Fugego, contaminada por le petroleo del B/T Metula. II. Reconocimiento de la entrada de mar suroeste'. ANS.INST.PAT., Punta Arenas (Chile), 9 (1978), 133-39.

11. S Langley & G Lembeye, 'Algunos antecedentes sobre el macrobentos, granulometria y contenidos de petroleo en los

sedimentos de dos entradas de mar en Puerto Espora (Tierra del Fuego) contaminados por el derrame del B/T Metula'. ANS.INST.PAT., Punta Arenas (Chile), 8 (1977) 375-88.

12. C Venegas, W Sielfeld & A Atalah, 'Dinamica espacio-temporal de la ornitofauna asociada a una marisma conta-minada por petroleo'. Presented at the ler Encuentro Iberoamericano de Ornitologia y Mundial sobre Ecologia y comportamiento de las Aves, (Buenos Aires, 1979).

13. R Hann, 'Follow-up Field Study of the oil pollution from the tanker Metula'. Report to the U.S. Coast Research and Development Program (1975), 57 p.

14. A Blount, 'Two years after the Metula oil spill, Strait of Magellan, Chile: Oil interaction with coastal environ-ments'. Tech. Rept. No. 16-CRD, University of South Carolina (1978), 214 p.

15. L Guzman, 'Algunas consideraciones ecologicas en torno a la contaminacion producida por el B/T Metula en el estrecho de Magallanes'. In: Preservacion del Medio Ambiente Marino (F. Orrego-Vicuna, Ed., Editorial Univer-sidad Tecnica del Estado, Santiago, 1976), 178-98.

16. R Hann, 'Fate of oil from the supertanker Metula'. Report from the Texas A & M Research Foundation, College Station, Texas (1977), 13 p.

17. J Warner, 'Determination of Petroleum Components in samples from the Metula Oil Spill'. Report to Marine Ecosystems Analysis Program National Oceanic and Atmospheric Admini-stration (1975), 15 p.

18. D Straughan, T Licari & F Piltz, 'Intertidal Biological Studies of the Metula Oil Spill in the Straits of Magellan, January 1977'. Presented at the Symposium Long-Term Recovery Potential of Cold Water Marine Environments after Oil Spill (Dartmouth, Nova Scotia, 1977), 44p.

19. D Lanfranco, 'Estado de la entomofauna en Puerto Espora, despues de la contaminacion provocada por el petroleo del B/T Metula'. Informe al Instituto de la Patagonia, Punta Arenas (1979), 4 p.

20. L Guzman, 'Estudios de lineamiento basico en el intermareal del estrecho de Magallanes: Estado actual y perspectivas'. Medio Ambiente (Chile), 4 (1980), in press.

21. R R Swartz & V Gallardo, 'Subtidal survey of the Strait of Magellan in the vicinity of the Metula oil spill'. Antarctic Journal. 11, (1976), 186-87.

THE DEVELOPMENT OF COUNTERMEASURES FOR OIL SPILLS IN CANADIAN

ARCTIC WATERS

S L Ross, Ph.D., P.Eng.
Manager, Arctic Marine Oilspill Program, Environmental Protection
Service, Department of the Environment, Ottawa, Ontario K1A IC8,
Canada.

BACKGROUND

As the world's petroleum resources in easily accessible areas
become depleted, increasingly more activity is being directed
toward promising, but more difficult, offshore areas. Within
the last decade, the oil industry has been developing a technical
capability to explore and exploit hydrocarbon resources in
hostile and previously unfamiliar ocean environments. Nowhere in
the world is this technological challenge more acute than in
Canada where exploration has moved into Arctic offshore areas
with the use of ice resistant drill ships, reinforced ice
platforms and artificial islands.

The advances in drilling technology that have made these
programs possible are remarkable, considering the harsh Arctic
environment - the remoteness, the low winter temperatures, the
long periods of complete darkness and the presence of ice for
much of the year. These realities create similar challenges in
dealing with another problem that is closely associated with
petroleum resource development, namely oil spills.

The technological challenge in developing effective countermeasures for Arctic oil spills is particularly formidable. The
developer of Arctic oil spill countermeasures, unlike the developer
of drilling systems for the same area, cannot draw upon the sound
and proven technologies applied in more moderate climates. It
is universally recognized that the general state-of-the-art for
dealing with major oil spills is inadequate and unsatisfactory.
The low level of countermeasures capability and the environmental
damage and social dislocation that can result from major spills
have been demonstrated time and time again. This has led to a
world-wide concern over oil pollution, particularly major oil
spills from accidents involving large tankers as well as blowouts
which occur in the marine environment.

It was not surprising then, when in 1973 the Canadian oil
industry proposed to explore for oil in the waters of the
southern Beaufort Sea, that much concern developed over the
probabilities of oil blowouts and the possibilities of being
able to control them and their effects. It was this concern,
primarily, that led to the development of a $12 million
environmental impact assessment study to determine the fate and
effects of oil blowouts in the area (Ref. 1,2). As part of
this study, an investigation was undertaken in 1974 and 1975
to review and analyse the feasibilities of controlling and
cleaning up large oil spills in the southern Beaufort Sea. The
study found, not unexpectedly, that there were few proven
countermeasures available to deal with an oil well blowout or

any large oil spill in this offshore area (Ref. 2,3). This was
estimated to be particularly true in the pack ice where, because
of the moving ice, logistical problems and a lack of countermeasures technology, it was possible for a blowout to run
uncontrolled for at least a year. In view of these pessimistic
findings and the early-1976 government decision to allow the
drilling to proceed, the Canadian Department of the Environment
received approval for a five-year $7 million technology program
to develop oil spill countermeasures for these and other ice-
infested waters.

ARCTIC MARINE OILSPILL PROGRAM (AMOP)

This technology program, started in early 1977 and now referred
to as the Arctic Marine Oilspill Program (AMOP), is being
managed by a small group of technical specialists in the
government's Environmental Protection Service with the assistance
of an interdepartmental committee of other scientists, each of
whom is either an Arctic or oil spill expert. In addition, the
program interfaces very closely with researchers in the oil
industry and in those organizations, such as the Canadian Coast
Guard, which have operational responsibilities for actually
fighting spills in Canada.
    The program consists primarily of engineering projects in
such areas as oil spill recovery and containment devices,
combustion and incineration systems, dispersant equipment and
strategies, remote sensing equipment and shoreline cleanup
methodologies. Research is also proceeding on the fate and
physical effects of oil blowouts and the movement and behaviour
of spills in ice-infested waters. In addition, some work is
being done in the area of biological assessment, particularly
as it relates to the use of chemical dispersants. At the present
time, AMOP is also heavily involved in a series of expensive and
complex experiments involving the controlled discharge of oil
in selected Arctic environments.
    Since the program was initiated, dozens of R & D projects
have been conducted under contract to private contractors or
consultants, university researchers and government scientists,
with the bulk of the work being performed by private companies.
Before summarizing the technical results of the program to date,
most of which are available in other literature sources (Ref.
4,5), it might be of value to review the risks and areas of
particular concern in the Arctic.
    At the present time, the most active area for oil exploration
is in the southern Beaufort Sea where a large and expensive
operation using sophisticated ice-reinforced drillships is
underway. Drilling is also taking place from reinforced ice
platforms in the deep waters of the Sverdrup Basin, and from
drill ships and semi-submersible rigs in the Labrador Sea and
off the coast of Newfoundland, Figure 1. Exploration programs
have taken place in the eastern Arctic in the Davis Strait and
further efforts are being proposed north of these waters in
Lancaster Sound and Baffin Bay.
    Although the immediate concern has been that of a major un-
controlled oil blowout, tanker and pipeline accidents leading
to large spills are considered to be far more probable (Ref. 6).
The transportation of oil out of the Arctic is not expected to
begin for several years and this will provide some lead time for
coming to grips with the anticipated spill problem. It is

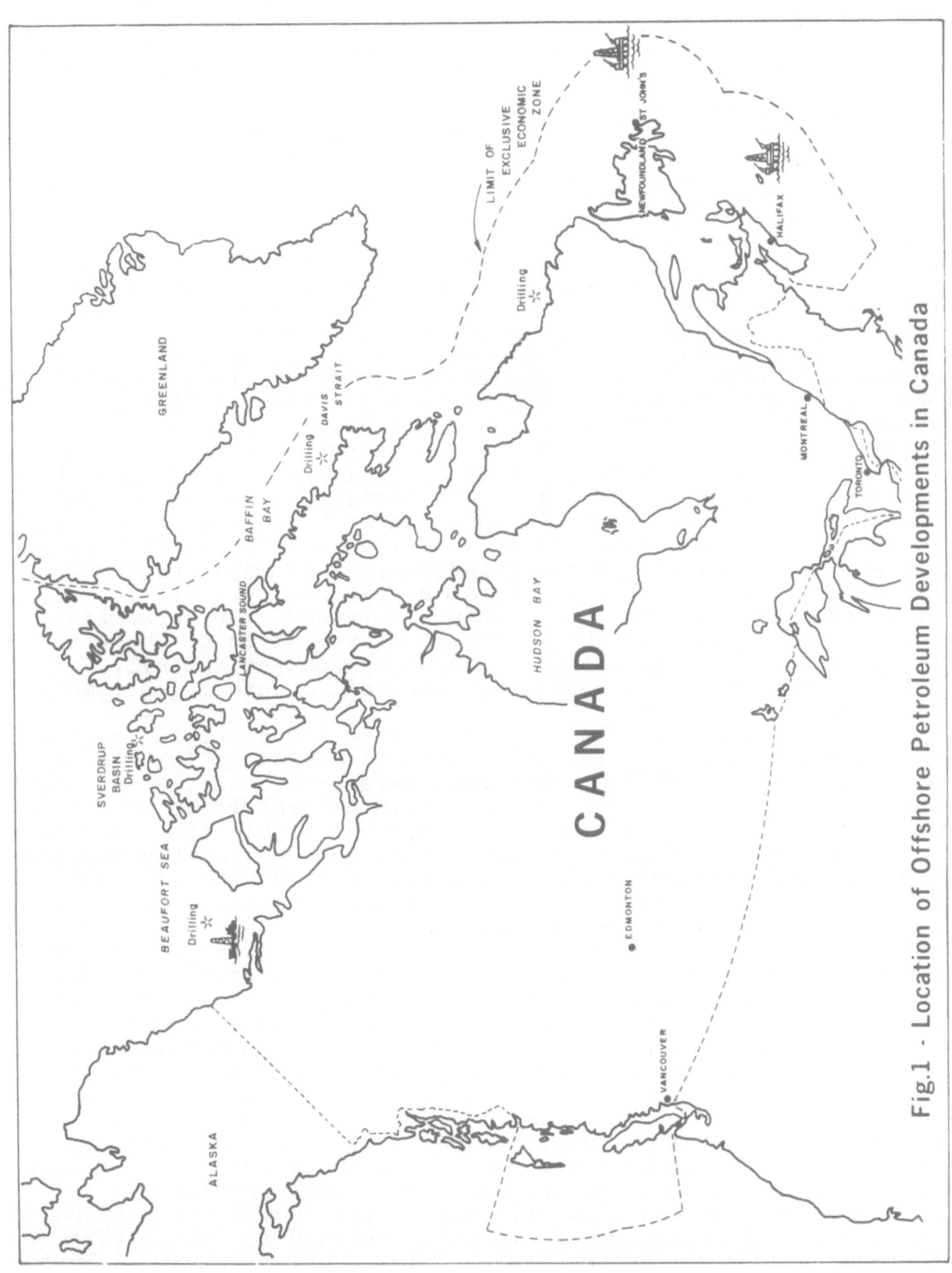

Fig.1 - Location of Offshore Petroleum Developments in Canada

particularly important to avoid complacency, however, in view
of recent major oil discoveries in the southern Beaufort Sea
and off Newfoundland. The proposal to ship Beaufort oil through
the Northwest Passage by 1985 is also worrisome when considering
the risks of oil spills.

SOUTHERN BEAUFORT SEA

Of all Arctic regions, the southern Beaufort Sea has been the
most studied insofar as oil spill contingency planning and
countermeasures development are concerned. There are two ice
regimes in the southern Beaufort Sea, zones of stationary and
zones of moving ice, each quite different from the viewpoint
of oil spill countermeasures. In the landfast (or stationary)
ice zone near the coastline, where drilling from artificial
islands has taken place, it is thought that the oil spill
problem is relatively controllable. An oil blowout at the sea
bottom will either lead to the discharge of oil to the under
surface of the stationary ice, or to the opening in the ice
sheet caused by and directly above the blowout plume. It is
felt in either case that this discharged oil can be burned in
situ in the winter by direct flaring or in the spring when the
oil migrates through the ice to the surface. The latter method
was demonstrated successfully in a research study carried out
in 1975 (Ref. 2). Unlike an oil spill on open water that spreads
out and thins quickly to cover large areas of water surface,
(rendering containment and control operations very difficult),
oil under ice tends to spread very slowly and remain relatively
thick. The stationary ice in the landfast-ice zone, therefore,
tends to act as a natural containment system for the oil,
allowing time to either burn the thick oil or otherwise remove
it by mechanical means.
        This unfortunately is not the case in the other ice regime
in the Beaufort Sea, where oil exploration from drill ships is
now taking place. This is the zone referred to as the seasonal
ice pack zone or transition zone and is characterized in winter
by sporadically moving, heavily ridged and highly irregular ice.
If an oil blowout were to occur in this zone during the two-to-
three month, relatively ice-free summer period, conventional
marine oil spill countermeasures equipment and techniques could
be employed; however the general state-of-the-technology, as
noted earlier, is not perfect and it might be expected that a
removal efficiency of perhaps 50% could be realized (Ref. 3).
        With the onset of the winter, when the ice moves into the
blowout area and forces a halt to all countermeasures operations
with mechanical equipment, the situation changes considerably.
As pack ice moves across the blowout site, with a velocity in
the range of 2-7 km per day (Ref.7), a film of oil is deposited
on the under surface of the moving ice floes. Due to the
dynamic nature of this zone, accessibility would be a major
problem and surface operations would be hazardous. The oil would
either be frozen into the growing ice or would collect in
pockets under the ice where it would remain until the melting
and breakup processes would gradually release it to the surface.
There is a possibility that small volumes of oil might be burned
in situ if sufficiently thick layers of oil were to collect on
the surface of the ice in the spring. However, due to the
movement of the ice, the oil which surfaces might only collect
in very thin oil slicks, making burning techniques inefficient

or even impossible. There also is the problem of actually
locating the oil prior to attempting removal procedures.
Conventional remote sensing techniques, useful in more
southerly ice-free environments, would be less useful in
locating and tracking oil in the ice-infested waters and total
darkness of the Arctic winter. Another problem is the lack
of knowledge concerning the migration of oil through the pack
ice. Despite some studies of the upward movement of oil
through an ice cover, there are still serious unknowns
regarding mechanisms and rate processes. The migration time
can range from one season for first-year ice to several seasons
for multi-year ice, (i.e., ice that has survived several
Arctic seasons) (Ref.8). All of these unknowns place severe
restraints on effective oil spill contingency planning for the
area.

EASTERN ARCTIC

Although most of the exploration play in the Canadian Arctic is
in the southern Beaufort Sea, it is possible that the waters
of the eastern Arctic (Lancaster Sound, Baffin Bay and Davis
Strait) will shortly be the scene of major drilling efforts
(Ref.9). From an oil spill control viewpoint, the major problem
in this area, aside from the Arctic constants of ice, cold and
remoteness, relates to water depth. These waters are sufficiently
deep (over 500 metres) to cause free gas at the sea floor to
solidify and form gas hydrates having a density close to that
of ice (Ref. 11). The upflowing mixture of oil and solid hydrate
from a deep-water blowout will therefore rise slowly and tend
to diffuse greatly in the water column, causing widespread
contamination (Ref. 10). If this happens, only very large areas
of thin oil slicks will eventually appear on the water surface.
This situation would not be amenable to cleanup using conventional
countermeasures techniques, except perhaps by chemical dispersion
methods. Unfortunately, however, it is known that even the most
promising dispersants on the commercial market have limited
effectiveness in dealing with aged oil spills in cold waters.

LABRADOR SEA

Although not geographically an Arctic body of water, the Labrador
Sea, commonly referred to as Iceberg Alley, has as many oil
spill control problems as any area in the Arctic, particularly
in view of the fact that it is not only an ice-infested
environment but also an open-ocean one. The pack ice formed in
these turbulent waters tends to be less consolidated than the
ice found in more northerly and quiescent environments. Ice
fields are created which have highly irregular surface profiles
and move rapidly. The turbulent conditions of the region also
create a condition atypical of Arctic marine systems, namely,
extremely high biological productivity. Preliminary research
indicates that it may be the most difficult area in Canada for
which to develop reasonably effective oil spill countermeasures
(Ref. 12). In view of the on-going exploration in the area and
its biological sensitivity, this is disquieting, to say the
least.

DEVELOPING COUNTERMEASURES

As with oil spill control systems for other environments, no
unique and comprehensive system can be developed for all regions
of the Arctic environment. In general, countermeasures packages
have to be custom-designed for selected typical regimes and
specific regions. For example, Arctic oil spills in stationary
ice situations are obviously different than those in the summer
ice-free environment, and hence will require the development
of different control techniques. Similarly, the complete
countermeasures system for blowouts in the relatively shallow
waters of the southern Beaufort Sea with its areas of stationary
and moving ice, will be very different than that required for
deep water blowouts in the eastern Arctic, as discussed earlier.

   Therefore the task of developing oil spill countermeasures
for the entire Arctic involves several approaches and priorities.
In order to be realistic in defining the objectives of AMOP with
its limited budget, it has been necessary to place certain bounds
on the program's activities. First, on a geographical basis,
certain highly vulnerable regions have been selected for
particularly concentrated study. These include the southern
Beaufort Sea, Lancaster Sound and the Labrador Sea. Second, it
was decided to take the R & D work to the design and prototype
development stage only; the manufacturing of finished or
operational products has been left as a responsibility for others
in the private sector. Third, it was decided to place more of
an emphasis on developing countermeasures on a generic rather
than on a site-specific basis. That is, the program is
developing technologies which should have at least some application
in all Arctic regions. Again we are essentially leaving the
site-specific countermeasures design and development to the private
sector, viz., the oil industry and its various co-operatives.
It is obvious, however, that the R & D and the site-specific
work must be closely integrated. Fortunately, that happens
to be the case in Canada where government and industry scientists
and engineers work closely together.

   Finally, although AMOP is essentially an engineering or
technology program, it was realized at the outset that it is
impossible to develop effective oil spill control equipment and
techniques without having a good understanding of what is being
"controlled". One must put some effort into developing and
consolidating information on the relevant Arctic environment
and into scientific projects related to the fate, behaviour
and effects of Arctic oil spills. For example, a knowledge of
the properties of Arctic ice, the interaction of oil and ice,
and the movement of oil in ice-infested waters is a prerequisite
to the design of appropriate technologies to deal with northern
spills. Field work, such as coastal surveys, and studies on the
effect of dispersed oil on marine life, are important for the
same reason; namely, one must understand the geomorphology of
typical Arctic coastlines, and the biological effects of
chemically dispersed oil, before designing sensible shoreline
cleanup and dispersant-use strategies. Notwithstanding the
above reasoning, efforts are being made in AMOP to minimize
baseline work and to concentrate on the direct activities
associated with countermeasures.

OVERVIEW AND SCENARIO STUDIES

AMOP Arctic Atlas
When the program started, it was decided that a number of overview studies were required in order to identify the exact nature of the oil spill problem in the Arctic. First, it was felt necessary to have available a summary of the salient features of all Arctic areas of interest which have a bearing on the program objective. Hence an Arctic Atlas was developed, containing a synthesis of multi-disciplinary information which provides a common and easily-accessible reference source for personnel involved in the program (Ref.13). The information categories in the Atlas are:

A. Petroleum Development
   1. Coastal bathymetry and topography
   2. Major geological basins
   3. Seismic information
   4. Lease and permit blocks
   5. Oil and gas finds, current and proposed drilling sites
   6. Marine areas suited to particular drilling platforms and associated seasonal timing
   7. Potential petroleum transportation routes
   8. Exploration and development area priorities

B. Physical Environment
   1. Ice (extent, thickness, type, polynia location, iceberg data)
   2. Meteorology (air temperatures, precipitation, wind direction and speed, hours of daylight, probability of low visibility and storms)
   3. Oceanography (surface and deepwater currents, sea surface temperature)
   4. Coastal information (geomorphology, longshore currents and drifts)

C. Biota and Socio-economic
   1. Seasonal concentrations of mammals (seal, walrus, beluga, narwhal, bowhead, foxes, bears)
   2. Wildlife sanctuaries and reserves
   3. Season concentrations and migratory routes for seabirds
   4. Social information (settlements, airports, hunting and trapping areas)
   5. Fisheries and productivity information.

Only 400 copies of the atlas were printed and these were distributed to a very carefully selected group who have a role to play in improving the Canadian capability to respond to Arctic oil spills. Periodic updates will be carried out to incorporate newly acquired baseline information.

Regional oil spill scenarios
On a more regional basis, detailed oil spill scenarios were developed for two areas, Lancaster Sound and the Labrador Sea. Previous studies had already accomplished the same objective for the southern Beaufort Sea and the Davis Strait (Ref.1,9,12).

Labrador Sea
The scenario for the Labrador Sea describes a hypothetical blowout of an exploratory well in 200 metres of water (Ref.12). It was envisaged that the blowout occurred late in the drilling

season near the end of October and was not contained until nine
months later.

Twelve study topics are presented to provide input to the
scenario narrative. These fall into the following categories :
the physical environment (with emphasis on ice); resource
utilization practices (in terms of potential, immediate and
direct impact); the behaviour of oil as it rises from the sea
bed and as it interacts with ice; the environmental prediction
and logistics capability of the Labrador area (to support a
major countermeasurers effort); and finally an appraisal of the
Canadian state of preparedness to respond to the spill event.

The hypothetical spill releases a total of 490,000 cubic
metres of oil, one half of which is dispersed as small droplets
in the water column. Of the remaining oil, which surfaces as a
slick, 14% is handled by countermeasures efforts, 9% is burned,
4% recovered, and 1% chemically dispersed. The ice cover
distributes the oil in low concentrations (of the order of 0.1%)
over an area of 10,000 km^2 along the northeast coast of
Newfoundland and Labrador. Seven percent of the oil comes
ashore. Immediate and major impacts are felt by seabirds,
spawning capelin, whelping harp seals and offshore fishing.

Lancaster Sound

Because of a very poor understanding of currents in Lancaster
Sound, and hence of the fate of oil spills in these waters, AMOP
supported some work in 1977 to monitor ice movement in these
waters (Ref.10). Using shore-based radar, ten satellite-tracked
buoys were released to follow surface currents, and four sub-
surface current meter strings were deployed to measure sub-
surface currents. Following this field work, a study was
conducted to consider the potential fate of four hypothetical
deepwater blowouts in Lancaster Sound during the open water
season. The blowouts were in depths ranging from 280m to 770m,
and the assumed daily flowrates were 950 m^3 of oil and 2850 m^3
of gas.

With the vertical distribution of currents typical of
Lancaster Sound, the patch of oil resulting from a subsea blowout
in waters deeper than 400 m would extend approximately five km
downstream and one km laterally. The highest surface concentration
containing about 85% of the oil over one km^2, (equivalent to a
0.15 mm thick film) occcurred within one km of the patch's
trailing edge. The surface extent of the patch was essentially
determined by the difference in the rise times of the smallest
and largest droplets in the assumed oil droplet distribution.
Local circulation patterns tended to hold oil in the area.

FATE AND BEHAVIOUR STUDIES

Extensive scientific work has taken place in three areas related
to the fate and behaviour of oil spills in Arctic waters :

Fate of oil on open water

The general objective here was to undertake experimental and
theoretical studies to understand more fully the various
processes which occur when oil is spilled on water, particularly
the colder waters of the Canadian Arctic. Much of the work
described below was performed at the University of Toronto.

A literature review has demonstrated the inadequacy of
existing spreading models especially in their ability to

quantify oil spreading into thick and thin patches. A new set
of spreading equations has been assembled which satisfactorily
describes the behaviour of a number of actual spills in this
respect.

A mathematical model describing evaporation losses and
property changes of oil spills has also been developed (Ref.14).
A given oil is assigned a simulated ten-component composition
which yields the same (computed) distillation curve, thus
enabling one to compute the evaporative characteristics of a
spill as a function of time, temperature, spill area and wind
conditions. In addition, the oil property changes with time
can be calculated including viscosity, solubility and density.

Regarding the problem of water-in-oil emulsions, an equation
has been derived for predicting the rate of "mousse" formation
and for characterizing its stability. In addition, an expression
has been developed, and fitted to available data, to yield
viscosity as a function of water content. Some experimental
work has been undertaken in a small-scale variable turbulence
apparatus and in a six metre-long windwave tank in an attempt
to verify the theoretical approach. Finally, work is proceeding
on improving the ability to predict the dispersion by natural
processes of oil spills.

By observing the mechanisms of naturally and chemically
induced dispersion in a small-scale test apparatus and in a
windwave tank, an equation has been developed in which dispersions
are quantified in terms of oil droplet particle size. There
appear to be two primary mechanisms of dispersion. One occurs
under non-breaking wave conditions, and the other when breaking
waves are present. The latter dispersing mechanism is
approximately two orders of magnitude more effective than the
former. In non-wave breaking conditions a slick stretching/
compression process, which is fairly insensitivie to oil thickness,
forces oil into the water column. In the presence of breaking
waves, oil is driven downwards by water cascading from the wave
crest. Empirical expressions are postulated to attempt to
quantify the processes. Some exploratory analysis has also been
performed on the diffusion of dispersed oil in the water column.

The individual processes summarized above have been combined
into an oil spill behaviour model similar in character to a less
refined one developed earlier (Ref. 5).

Oil-Ice Interactions

In view of the paucity of knowledge on the interaction of oil
in ice, and the importance of this knowledge in designing
effective countermeasures, much work has had to be performed in
AMOP in this research area. For an oil blowout in relatively
shallow water under sea ice the basic unknowns to be determined,
from an oil spill point of view, are the mechanisms and rates
of migration of the oil through the ice and the properties of
the oil once it reaches the surface. One laboratory experiment
that was performed in this regard involved injecting crude oil
and natural gas under 35 cm of saline ice and allowing 20 cm
of additional ice to grow under the oil and gas (Ref. 15). The
oil was observed to spread in a thin layer at the under-surface
of the gas bubble. When the sheet was thawed, the gas was
observed to escape when the minimum temperature in the ice
sheet rose to $-3.6^{\circ}C$. The bulk of the oil emerged at the same
time as an oil spill in a control experiment without natural
gas. The conclusion is that the presence of gas greatly increases

the area over which spilled oil surfaces, but does not affect
the timing of its appearance.

Further laboratory experiments using several oil types
provided additional new knowledge of the mechanisms and rates
of the migration of oil and gas through ice. In addition,
two major field studies were performed; one involved a small
controlled experiment in multi-year ice, and the other involved
an investigation of a bunker oil spill which became intermixed
with pack ice off the east coast of Canada after the tanker
KURDISTAN broke in half in April 1979.

All of these studies have advanced our knowledge of the
behaviour of oil spills in ice. Several theories regarding this
behaviour have been advanced, and some of these are currently
being tested in a large controlled oil spill experiment taking
place in the southern Beaufort Sea (Ref. 16).

Deep Water Blowouts
In order to prepare realistic contingency plans for oilspill
countermeasures following a subsea oil well blowout, it is
necessary to develop an understanding of the probable state and
configuration of oil which will surface and be accessible for
cleanup operations. Studies in recent years on the consequences
of a blowout in shallow waters have predicted the existance of
large-scale, vertical circulation in the sea above the blowout,
driven by the released gas, which will tend to confine the
lateral dispersion of rising oil droplets and quickly transport
them to the surface (Ref. 17). A subsea blowout in deep water
(greater than about 250 m) will not conform to the above pattern
because of the removal of gas from the rising plume by the
formation of gas hydrates under the high pressure at depth,
allowing the slowly rising oil droplets to be dispersed more
easily by currents in the water column.

In 1977, government scientists undertook an experiment
involving the release of simulated natural gas at sea depths
up to 650 m near Vancouver Island using the Pisces IV submersible. Hydrates were observed to form on the surface of gas
bubbles within 2 m of the gas release nozzle (Ref. 18). A
theoretical analysis on the rate of extraction of dissolved gas
from a rising droplet indicated that near the exit from the
seabed most of the oil becomes saturated with the gas. This
was speculated to provide a source of energy that might cause
emulsification during the oil's journey to the surface.

To further examine these processes, a high pressure vertical
water tunnel was built in which oil drops and gas bubbles
could be held in view under conditions of programmed pressure
release (Ref. 11). The preliminary experiments involved the
injection of simulated natural gas bubbles about 1 cm in diameter
into $3^{o}C$ sea water held at constant pressure and under reducing
pressures.

The work to date indicates that hydrates are indeed produced
and because of this, a relatively localized blowout plume driven
by rising gas bubbles will not form throughout the water column
in cold water deeper than about 450 m. Moreover, preliminary
experiments suggest that gas dissolved in oil will emerge during
the buoyant ascent and form a hydrate shell around the oil
droplets or dissolve immediately into the water.

IN-SITU BURNING

Crude oil spilled beneath growing sea ice is encapsulated within
the ice sheet and eventually migrates through the ice and collects
in melt pools in an unweathered state on the surface during the
spring and summer months. Oil spilled on the ice surface in the
winter also eventually collects in melt pools in a fairly unweathered state, since during the winter months evaporation is
inhibited by the protective snow blanket and the low temperatures.
In either case, therefore, in situ combustion of oil in
contaminated melt pools is a promising cleanup technique since
the oil should not be significantly weathered and would be
naturally contained. In situ combustion may also be a possible
cleanup option in leads or in broken pack ice, since the ice
edges tend to limit the spreading tendency of the oil. Winds
or currents may even act to concentrate thin slicks to thicknesses
greater than that required for self-sustained combustion.

In the event of a large oil spill in a zone of moving ice,
the area of contaminated ice could be quite extensive. Rather
than to place cleanup crews on the ice, it would be desirable
logistically to deploy incendiary aids from aircraft into the
multitude of contaminated melt pools (and possibly into oiled
leads) which might develop.

With this in mind, AMOP performed several studies to evaluate
a variety of devices and chemicals for igniting confined oil
slicks on water (Ref. 19). The candidates selected were gasoline
with sodium; Kontax (sodium and calcium carbide); Kontax with
gasoline; gelled Kerosene; and solid propellant (ammonium
perchlorate oxidiser, aluminium or magnesium fuel plus poly-
urethane binder). The testing included static experiments where
the devices were simply placed into the oil, and simulated air
deployments directed into an oiled pool from a tower 11.5 m high.
In these tests, the devices were activated by a variety of
delayed starter mechanisms so that ignition occurred after
landing. During the static test runs, the igniters were tested
with two slick thicknesses (3 to 10 mm) under both calm and
windy (at least 10 knot wind speed) conditions, using fresh and
weathered (two weeks outdoors) Normal Wells crude oil. The
preheating, ignition and burning times were determined for each
of the 42 static experiments. In addition, burn efficiencies
and rates, ignition probabilities, and qualitative, observational
information on igniter performance were also obtained. The
ignition probabilities of the various igniter types and fuse
mechanisms evaluated in a total of 48 tests ranged between 60
and 80 percent. Some of the factors important for successful
ignition systems were: a high flame temperature (to maximize
radiation levels and minimise preheating times); a large flame
size (to maximise the volume of oil preheated); and a flame
directed close to the oil surface (to minimize radiation losses
with distance), but not too close so as to push oil away from
the preheating and ignition zone.

The testing indicated that all of the various prototype
air-deployable incendiary devices showed some promise for igniting
confined oil slicks. In each case, with further work, any of
the devices could probably be developed to provide a high
probability of oil-ignition. However, the solid propellant
fuel was considered to offer the most attractive system for
further development. This combination was relatively simple
and safe, and operated particularly well in wind conditions.

In general, the solid propellant fuel is particularly suitable since it contains its own oxygen supply and should be effective even if submerged totally in the oil. Also, rubber-based solid propellants may be mised to provide optimum combinations of burn times and flame temperatures, and be molded to generate suitable flame configurations.

Following this work, in 1978, AMOP began the development of a mass-producible incendiary device using the services of specialists in the Department of National Defence who have considerable expertise regarding the preparation of pyrotechnic compositions such as solid propellent fuels. This work is well underway and several concepts have been developed and tested. Particular attention is being paid to ensuring that the device is inexpensive and safe for deployment from helicopters.

Aside from the above studies and developments, several other AMOP investigations were carried out with regard to improving the in-situ burning operation. These included the development of a floating burner which re-ignites on a periodic basis; the study and evaluation of several combustion promotors to improve the efficiency and chances of combustion; and a basic study of the limits of the combustion process itself (Ref. 5). This latter study is particularly important because there are many factors, such as slick thickness and "age", which greatly affect the combustion potential.

INCINERATORS

Two incinerators have been developed by AMOP for Arctic use but most of the effort has been devoted to the design, development and testing of an air-portable, forced-air, open pit incinerator with a nominal 1 tonner-per-hour capacity for oiled-combustibles debris such as used sorbents and oiled vegetation (Ref. 5). Such a device is seen as a requirement in remote locations where excavation is precluded and the only access is by helicopter.

The combustion chamber consists of six L-shaped sections that form the sides and floor and four end sections, one of which has an access cleanout door. Most of the forced air, supplied by a diesel-driven blower, is directed through a bank of nozzles at the top edge of one side of the combustion chamber and causes a recirculation of the combustion gases to essentially eliminate smoke providing the feed rate remains within design limits; a small percentage of the forced air is injected through under-fire nozzles near the bottom of the chamber.

The complete incinerator can be moved in 14 separate lifts by a medium lift helicopter, since no one section or sub-assembly exceeds 900 kg (2000 lbs). It has been successfully assembled on a prepared (levelled) site using a Bell 212 helicopter to lift each piece into position. A series of field tests using a variety of oiled materials have demonstrated that the design specification has been met.

The second incinerator is a derivative of an improvised reciprocating kiln design developed in co-operation with the petroleum industry. The incinerator can be assembled from used oil drums, car wheels, angle-iron, sheet metal, and similar commonly available materials. A basic welding skill is required but otherwise construction is easily within the normal capability of most rural Canadian communities. The device was designed primarily for restoring oiled beach sand to an acceptable condition, but one was used with good success during

the recent KURDISTAN cleanup operation to dispose of oil and
debris collected on an offshore island. The design has since
been modified to make the incinerator helicopter transportable
and one is being built for use at the base camp on Baffin Island
during the forthcoming Arctic experimental oil spills in that
vicinity.

DISPERSANTS

In certain offshore and Arctic areas in Canada, serious
consideration is being given to the use of chemical dispersants
as a prime oil spill countermeasures technique. In fact, in
some specific areas no other approaches to oil spill control
seem viable.
    There are two major problems related to dispersant use: the
first is that commercially available products and application
techniques are not very effective in dispersing oil spills
rendered viscous by means of aging processes or cold temperatures.
This problem can be solved, hopefully, through technology
development. The second problem is far more difficult and relates
to understanding the environmental advantages and disadvantages
of using the technique and making the most ecologically-sound
decision.
    Within the Arctic Marine Oilspill Program, both the
technological and the decision-making problems are being
addressed.
    The effective use of dispersants requires not only an
effective product but an efficient application technique. In
many situations, such as spills in remote areas or ice-infested
waters, use of vessel-mounted spray apparatus is considered to be
inappropriate. For these cases, which are very common in Canada,
AMOP has been investigating the use of aircraft as the application
platform. This program started with a logistic study for the
Beaufort Sea (Ref. 20) and has been followed by two dry land
trials at Riviere-du-Loup, Quebec and Abbotsford, British
Columbia (Ref. 21). These trials were designed to measure drop
size, swath width and maximum dose rates. The studies revealed
that there is significant loss of small droplets of dispersant
due to wind drift and that the high dose rates (<10gal/acre)
which are needed for highly weathered, emulsified or cold viscous
oils may not be possible with the high airspeeds of aircraft
such as the DC-6 used in the trials. Two studies were then
initiated to solve these problems. The first study investigated
the effect of dispersant drop size and dosage on the percent
dispersion of various thicknesses of oil. The second study
investigated the various parameters which affect the formation
of drops at the wings, and the effects of wind shear and air
turbulence on the size and distribution of droplets. The DC-6
and CL-215 aircraft installations have been modified as a result
of these studies, and an evaluation trial over land is scheduled
for May 1980.
    The effectiveness of dispersants on spills on cold water will
be scientifically determined in a full scale field test
scheduled for October 1980 in the offshore waters of Newfoundland.
    In parallel with this very active program to improve
techniques for applying dispersants, AMOP and the Environmental
Protection Service are also making headway in the more difficult
and controversial area of developing scientifically-based
methodologies for deciding when and where to use these treating

agents in Canada.

It is generally accepted that dispersants should be allowed to be used to treat oil spills if their use results in minimization of overall environmental damage. The most environmentally acceptable countermeasures solution, of course, is to remove the oil from the water environment by mechanical or other means (e.g. burning). When these techniques are not practicable, the only choices left are to leave the oil alone and hope that natural forces break up the slick or to chemically disperse the oil. The appropriate decision, then, should be based on an assessment and comparison of the environmental effects associated with each of these choices.

Before one can truly estimate the biological effects of chemically dispersed oil, the fate of that oil must be appreciated. As indicated earlier, AMOP in conjunction with the University of Toronto is developing a three-dimensional oil spill model for predicting the concentration and distribution of oil after chemical treatment, and the model is based on newly developed experimental laboratory and field data (Ref. 5).

With regard to biological effect studies, AMOP has performed a number of experimental programs to evaluate the acute toxicity of oil and oil/dispersant mixtures on arctic invertebrates (Ref. 22). From this work it has become apparent that naturally and chemically dispersed oil have about the same toxicity. This may have been expected in view of the low toxicities associated with the new dispersants on the market. It should be remembered however, that when comparing the toxic effects of chemically dispersed oil to that of oil on the water surface, the dispersed oil is far more available to aquatic biota because dispersing greatly increases the amount of oil in the underlying water column, at least in the short term.

The acute toxicity is only part of the problem. The longer term sublethal effects may be the most significant problem related to oil spills. This aspect has been studied by investigating the concentrations of oil that affect the growth, respiration, assimilation and feeding rate of fish and several benthic invertebrates. Another "sublethal" study is aimed at investigating the consequences of oil and dispersants on the harvest and reproduction of clams.

Because laboratory studies on the fate and effects of oil are often difficult to extrapolate to field conditions, AMOP is planning a major field spill to investigate the fate and effects of a dispersed oil spill in an Arctic nearshore environment. This is discussed later.

In summary, decisions on the use of dispersants are based upon many considerations such as effectiveness of the dispersants, the fate of the oil and biological considerations, to name a few. Although there will always be some intuition and guessing involved in the decision-making process, AMOP is attempting to fill in some of the most important information and technological gaps. The ultimate aim is to improve the methods of dispersant application and to produce dispersant-use guidelines which will realistically and scientifically provide guidance to dispersant users.

SKIMMERS AND BOOMS

Although numerous skimmers and booms incorporating various principles are commercially available, AMOP engineers have had to develop an improved breed of equipment to contend with the harsh nature and distinctive characteristics of the Canadian Arctic (Ref. 23,24).

Skimmers

At the beginning of the program, it was thought that the clean-up of an Arctic offshore oil spill would involve the use of large skimming units which could withstand the forces of moving floes and masses of sea ice. However, as the problem was researched in depth, it became clear that small skimming packages offered many advantages over larger ones. Immediately obvious was that because of shifting ice and variable sea and weather conditions, the skimmers would have to be light enough to be quickly deployable and retrievable. Logistics also demanded that they be readily and easily transportable. Because smaller systems can be transported to and from the spill site while stowed on vessels of opportunity, the expense of buying large hulls and propulsion systems as integral features of the skimmers is avoided. A mother ship designed to operate in the Arctic, such as a supply vessel or other form of working craft, would be capable of withstanding the extremes of the open ocean. Many such vessels possess an ice-breaking capability and in almost all cases, have a limited lifting capacity adequate for skimmer deployment. Helicopters are also widely used in the Arctic for transfers of personnel and goods and could be utilized for small skimmer transportation. Suitable working platforms and carriers (both air and sea) for smaller skimming units thus appear to be readily available and their use offers many advantages.

Ultimately, the decision was made to pursue the development of three relatively small devices which incorporate proven collection principles. These were the disc, rope mop and oleophilic submersion belt. Each functions by effecting an initial oil/water separation and all have a capability to process a certain amount of debris including ice. Additionally, each craft has been hydrodynamically conceived to allow for oil retrieval at relative velocities of up to one to two knots and do so at lower ambient temperatures.

The three oil skimmers include a number of items common to all. Because they are deployed from a working platform (either ship or dock), a control panel remote from the skimmer is used to operate them. A discharge line feeds collected product to the mother ship via a positive-displacement type pump so that emulsification is minimized. The three skimmers all have catamaran style hulls with the oil collection component housed between the pontoons. They incorporate as much off-the-shelf hardware as possible.

There are other features which distinguish one machine from the other. The disc and rope mop skimmers have been designed primarily to retrieve oil within a containment boom. As such they include small twin propulsion units for added manoeuverability. A diesel/hydraulic system on board each craft drives all functions related to oil pickup and propulsion. The rope mop is operated via radio control while the disc skimmer works through a remote console connected by an umbilical bundle. Both skimmers are approximately five metres in length, weigh less than two metric tonnes each and could be worked equally well from a dock or a vessel. The disc device has been constructed to permit the flow-through of water while the discs rotate in a quiescent, concentrated pool of oil. The rope mop is driven in a zero-relative-velocity fashion.

The third machine, the submersion belt unit, is also a zero-relative-velocity skimmer. At eight metres and approximately six tonnes it is the largest of the three systems. A diesel/hydraulic power pack has been developed to operate and control skimming from either a vessel of opportunity or the skimmer itself. A remote console has also been constructed which allows for connection to the hydraulics of the mother ship should this be required and/or feasible. The belt skimmer was developed primarily for application in a current; in the event of a blowout it would be positioned downstream from the release point. Either conventional oil barriers or, preferably, a water enhancement or herding system (already in place) affixed to either side of the mouth of the skimmer would serve to sweep a larger area and, at the same time, direct a more concentrated product back into the device.

In summary, the task of designing recovery devices for the Canadian Arctic called for a re-evaluation of conventional thinking on countermeasures hardware. It is believed that the resulting concept of small and inexpensive skimming units in association with an appropriate mothership may be a generally useful approach for other geographical areas as well.

Booms

Early in the program it was realized that all of the available oil containment booms suffered from one or more drawbacks that would seriously limit their usefulness in the event of an Arctic oil spill. Some were so large and heavy that they could only be deployed and recovered by ships; some required support pumps and/or air compressors; many were only designed for use in harbours and similar sheltered waters. Accordingly, a project was undertaken to develop a boom that would be effective in exposed offshore applications, withstand extended storage under Arctic winter conditions, be as small and light as possible, and function without the need of spring mechanisms, compressors and similar potential mechanical breakdowns.

One of the first decisions made was that the Arctic oil boom would not be expected to oppose any significant amount of ice; it should be readily removable from the path of intruding ice and be able to withstand ice abrasion if allowed to stream freely in the presence of scattered ice floes.

Because of the almost total lack of road access to Canada's Arctic coastline, the boom would have to be air portable. Because of its intended use offshore it must be able to withstand the loads to be expected when a boom is deployed, towed and recovered by ships in a seaway. On the other hand, it must also be deployable by a small work-force and the small watercraft typical of remote northern communities. In addition to selecting materials with adequate cold temperature properties and developing the structural design features, the optimum diameter of the flotation chamber had to be determined to prevent splash-over in a six foot wave.

Accordingly, two 30-metre lengths each of two sizes were manufactured and moored side-by-side in an exposed area off Vancouver, B.C. for eight days during which time winds of 20 to 25 knots were experienced; waves of up to 0.9 m were super-imposed on a 1.5 m swell. One of the booms had a buoyancy chamber diameter of 460 mm and the other was 600 mm in diameter.

The former had a skirt of 460 mm and the latter had a 600 mm skirt.

Performance was monitored daily and both booms met all requirements; no splashover was observed with either one. In the interests of economy, weight and bulk, the 460 mm by 460 mm size was chosen. Since then, an additional 120 m have been manufactured for offshore testing in the cold waters off St. John's, Newfoundland in February 1980.

Concurrent with the development of the boom, a multi-model transportation container was also developed. A box large enough to accommodate at least 300 m of boom was incorporated into a 7.8 m fibreglass boat hull and fitted with a remotely-releasable door in the stern. The module has been towed by boat at speeds up to 9 knots and the remote boom deployment and towing capability has been demonstrated using 300 m of boom. The loaded module has also been successfully carried at a speed of 85 knots, launched, towed and recovered by a Sikorsky S-61 helicopter. A trailer for road transportation completes the multi-model capability desired.

Other AMOP boom projects have included some laboratory and model testing of air bubbler devices, water sprays, and fire-resistant oil boom with a view to development of a capability to contacin oil for in situ burning on the water. Lastly, AMOP contributed funds towards the development of an oil-ice deflector boom for use in ice-infested fivers to divert ice but allow oil to pass through for collection in the ice-free area created. A prototype of that device was successfully demonstrated in the spring of 1978 and an improved design is now commercially available.

REMOTE SENSING

The Canada Centre for Remote Sensing of the Canadian Department of Energy, Mines and Resources undertook the development of a remote sensing package as an AMOP project. The primary objective was to develop the specifications for an airborne multi-sensor system for detecting oil in ice-infested waters. Emphasis was placed on the development of the capability for the long-range, all-weather detection and monitoring of oil in situations where ice is present.

The two-year project was initiated in September 1977. The first portion of the work involved an examination of available sensors to determine which ones would prove useful in an Arctic situation. Several sub-studies were initiated : one to investigate the feasibility of microwave sensors, another to examine radar systems and a third to optimize fluorosensing techniques. Experiments were performed to evaluate the following sensors: multi-spectral scanner, low-light-level television, visible spectrometer, laser fluorosensor, ultra-violet camera, synthetic aperture radar, microwave scatterometer and an infrared/ultra-violet line scanner (Ref. 25).

In 1978 two experimental missions were flown with the use of two aircraft. The first mission was over the natural oil seep at Scott Inlet on Baffin Island (September 1970) and the second, the American Petroleum Institute test spill (November 1978) off the Eastern Coast of USA (Wallops Island). These missions allowed the evaluation of the available sensors in real situations. The preliminary results of these tests indicate the following :

<u>Low-Light-Level Television</u> : This sensor produced high-contrast imagery of oil-on-water over both sites. The sensor generally worked well and in view of its low cost is considered to be a highly promising instrument.

<u>Laser Fluorosensor</u> : Due to instrumentation problems, the fluorosensor did not function well over the Scott Inlet seep; however, after modification, the unit worked very well during the API test spill. The latter test revealed that the sensor was not only able to provide distinct indication of oil but also provided unique responses to various types and mixtures of oil, dispersant and dye.

<u>Multi-Spectral Scanner</u> : This instrument provided good thermal data over both test sites. It was found that the longer wave lengths in the visible spectrum were the most useful.

<u>Ultra-Violet Photography</u> : Good imagery was obtained at both test site locations.

<u>Synthetic Aperture Radar</u> : Results from the Arctic trial showed that L-band radar was very sensitive to fresh-water and wind "slicks". Results from both trials demonstrated that X-band radar was not as susceptable to these interferences and also provided indication of the presence of oil.

<u>Microwave Scatterometer</u> : Results of tests using this instrument indicated that vertically-polarized X-band radar is the optimal radar for the detection of oil on water.

<u>Infrared/Ultraviolet Line Scanner</u> : This sensor provided good imagery of oil at the Wallops test site; it was not flown at the Scott Inlet seep.

In 1979, some of the sensors were tested in a real situation over the KURDISTAN spill. This test proved to be very interesting since in this particular spill oil was present in the environment in a number of different locations: on water, on ice and with ice. The overflights were conducted with two aircraft: a DC-3 with low-light-level television, multi-spectral scanner and dual-channel line scanner; and a Falcon Jet with dual channel line scanner and photographic equipment.

The large amount of data collected in these three field trials are currently being analysed to assess the performance of each sensor and to determine how each sensor can be optimized. The data analysis portion is scheduled for completion shortly. Following this analysis, the entire project will be documented and recommendations will be made for developing the optimal remote sensing package.

SHORELINE SURVEYS, PROTECTION AND CLEANUP

Many coastline sections of the Arctic have not been well studied, and knowledge of shoreline types and geological sensitivity is minimal. This information is obviously needed in order to give direction to shoreline cleanup research projects and to aid in establishing action plans for the areas. Two particularly sensitive and unfamiliar areas, the coasts of Labrador and the Lancaster Sound were therefore the subject of intense and expensive shoreline examinations in the first two year's of the Arctic Marine Oilspill Program.

In addition, two projects were sponsored which addressed new countermeasures technology: one involved the engineering design and selection of landfill disposal sites along the permafrost-rich Beaufort Sea coast (Ref. 26) and the other involved an investigation of in situ combustion techniques for the disposal of oil deposited on shorelines. The latter study indicated that the burning of oil on certain shorelines is a feasible technology. The success of this preliminary study was encouraging enough to suggest a need for continued research in this area. The combustion technique is particularly attractive for the Arctic because of the lack of manpower there for a massive manual cleanup effort.

Another project, funded by the oil industry with a small amount of AMOP monies, involved the development of a shoreline cleanup and protection manual for the Beaufort Sea (Ref. 27). The manual establishes the relative importance of all sensitive coastal regions along the Beaufort Sea coastline and recommends oil spill and protection cleanup strategies for these areas. The 2000 km of coastline were included on 24 maps at a scale of 1:250,000. Each map is provided with a supporting text which includes information on recommended countermeasures, coastal sensitivity and coastal access. Coastal sensitivity was established with the use of a ranking system which is based on the cumulative assessment of human, biological and geological sensitivity. The whole coast was assessed and 45 important areas were identified as requiring some attention by the On-Scene-Commander in the event of an oil spill. The manual is a major asset to those faced with oil spill control decisions in the Beaufort Sea, and, in fact, is an excellent example of what can be done for other geographical areas as well.

EXPERIMENTAL OILSPILLS

As stated, the primary objective of AMOP is to develop oilspill countermeasure techniques and equipment suitable for use in the Arctic marine environment. Hence, during the first two years, the program has included work on the development of cleanup equipment such as booms, skimmers, incinerators, combustion aids, dispersants, etc. Work has also been undertaken on the fate and effects of oil in a cold marine environment. This has been necessary because the successful development of countermeasure techniques depends on understanding the behaviour and fate of the spilled oil. Also, the acceptability of oilspill counter-measures must be based not only on their effectiveness but also on their biological effects.

Most of the AMOP work has been conducted in the Canadian South and much of it on a laboratory scale. We have now reached a stage at which it is essential that newly developed counter-measure techniques and equipment be tried in the North on actual oil spills. It is intolerable to develop countermeasures and place trust in their effectiveness without significant trials under realistic conditions. Only by attempting to clean up actual oilspills, will a true picture of the countermeasures effectiveness emerge and our capability be improved.

With this in mind, AMOP officials in 1978 formed a committee of scientists to identify specific research needs and to develop an integrated experimental oil spill program for the Arctic.

After over one year's effort and receiving advice and recommendations from over 100 scientists in Canada and abroad, five experiments were designed and described in a publication entitled "Experimental Oilspills: General Plan" (Ref. 28).
   These experiments are summarized below:

1. Oil and Gas Under Beaufort Sea Ice
    This study involves the discharge of 2 - 8 m^3 oil and gas, in a realistic volume ratio for a blowout, beneath landfast ice a few miles offshore in the Beaufort Sea. The oil will be discharged to create an average thickness of about 1 mm in the area of contamination. The main objectives are to verify laboratory results on the behaviour of oil and gas discharged under saline ice and, after the oil has surfaced, to examine possible oil concentrating mechanisms such as wind or current herding, and to deploy and evaluate oil spill incendiary devices under realistic conditions.

2. Oil on Cold Water
    This study involves a controlled evaluation of the performance of selected mechanical devices for cold water and cold weather conditions. The equipment includes a variety of skimmers and booms currently under development by AMOP and the associated support systems. The studies also include an evaluation of the effectiveness of dispersants selected after the completion of ongoing laboratory testing. A series of discharges, each involving about 2 - 10 m^3 of oil, are required.

3. Oil in Arctic Nearshore Environments
    An attempt will be made to select three similar bays, which have a fairly low energy wave environment, and which exhibit a typical biological productivity. Prior to any oil discharges one year of baseline information will be obtained for each bay. One bay will remain uncontaminated and serve as a control, and the other two will each be contaminated by up to 5 - 15 m^3 (about 30 - 95 bbl.) of oil. Dispersants will be used in one of the two contaminated bays. A comprehensive chemical and biological sampling program will be followed for at least the subsequent three years. The objectives of the study are to determine whether dispersants can be used on nearshore oilspills in the Arctic to reduce the biological damage and to quantify the environmental effects of oil alone.

4. Oil on Arctic Shoreline
    It is propsed to that 8 - 16 m^3 of oil be discharged in a test bay with a variety of shoreline types present (high priority will be given to sand, gravel and cobble beaches, mudflats and rocky shorelines). In addition to monitoring the fate of oil on the shoreline, promising cleanup and protection techniques requiring minimal logistical support will be evaluated. The techniques will be selected from a variety being examined in a quasi-laboratory study which is currently underway.

5. Oil in East Coast Pack Ice
    Phase 1 (February-May 1982): the first phase is largely exploratory, and involves the discharge of a total of 2 - 6 m^3 of oil in low concentrations (average thickness

of oil 0.1 - 0.5 mm) under a variety of conditions in the
pack ice in Davis Strait or the Labrador Sea. A vessel
will be used as an operating base in the pack ice. The
primary objective is to examine small-scale, short-term
processes such as oil entrainment into ice edges, emulsion
formation, migration of oil through porous ice, etc.

Phase 2 (February-May 1983): Contingent upon the results
obtained during the first phase, a larger volume of oil
will be released (perhaps 25 - 65 m^3) to examine relatively
long-term, large-scale processes such as spreading amongst
the pack and release of oil from melting ice near the pack
edge. Countermeasures techniques currently under development (mechanical equipment, aerial application of dispersants,
in situ burning) would also be tested on parts of the oil
discharge.

The status of each of these experiments is as follows :
Experiment No. 1 began in December 1979 at a cost of approximately
$1 million. The oil industry is funding and managing the entire
effort. The testing of AMOP skimmers and booms took place off
St. John's, Newfoundland in February and March 1980 and involved
AMOP engineers, contractors and staff of the Canadian Coast Guard.
Dispersant effectiveness tests are planned to take place off St.
John's in the fall of 1980. An industry/government committee
of scientists have been responsible for designing and developing
this experiment with industry funding approximately 80% of the
experiment. Experiments 3 and 4 have been combined into one
experiment. The project will likely take place in selected bays
on the northern tip of Baffin Island. The $3 million four-year
program is an international effort involving the governments of
Canada, U.S.A. and Norway and the Canadian oil industry. No
plans have been developed for Experiment No. 5 because of its
estimated high costs and because research on the oil spill from
the KURDISTAN tanker in early 1979 provided excellent information
on the macro-scale movements of oil in pack ice.
When the above studies are completed, they will significantly
enhance Canada's capability for responding effectively to Arctic
marine oil spills. In the long term the spills will be environmentally beneficial by assuring that Canada has tested, and has
in place, more effective countermeasures equipment to treat
actual spills, some of which could, of course, be thousands of
times larger than any experimental spill. Also, they will provide
a considerably more accurate appreciation of the impact of an oil
spill in the Arctic marine environment to Canadians in general and
to northern residents in particular.

SUMMARY

From the western Arctic to the Labrador Sea, we are faced with
serious problems associated with the development of reasonably
effective oil spill countermeasures. It is unlikely that at the
completion of the Arctic Marine Oilspill Program that all problems
will be solved. Certainly, however, we will be in a better
position than we are today to deal with a large oil spill in an
ice-infested Arctic environment. Hopefully, we will never have
to demonstrate this capability.

398

REFERENCES

1. A.R. Milne and B.D. Smiley, Offshore Drilling for Oil in the Beaufort Sea, (Beaufort Sea project, Victoria, British Columbia, 1976)

2. S.L. Ross, W.J. Logan and W. Rowland, Oil Spill Countermeasures: The Beaufort Sea and the Search for Oil, (Beaufort Sea project, Victoria, British Columbia, 1977)

3. W.J. Logan, D.E. Thornton and S.L. Ross, Oil Spill Countermeasures for the Southern Beaufort Sea, (Environmental Protection Service, Ottawa, Ontario, 1975)

4. Anonymous, "Year-End Review: April 1978 – March 1979", Spill Technology Newsletter, 4 (1979), 149-201

5. Anonymous, Proceedings of the Arctic Marine Oilspill Program Technical Seminar, (Environmental Protection Service, Ottawa, Ontario, 1979)

6. F.G. Bercha, Probabilities of Blowouts in Canadian Arctic Waters, (Environmental Protection Service, Ottawa, Ontario, 1978)

7. Anonymous, Probable Behaviour and Fate of a Winter Oil Spill in the Beaufort Sea, (Environmental Protection Service, Ottawa, Ontario, 1977)

8. A.R. Milne, R.H. Herlinveaux and G. Wilton, A Field Study on the Permeability of Multiyear Ice to Sea Water with Implications on its Permeability to Oil, (Environmental Protection Service, Ottawa, Ontario, 1977)

9. A.R. Milne and B.D. Smiley, Offshore Drilling in Lancaster Sound, (Institute of Ocean Sciences, Sidney, British Columbia, 1978)

10. J.R. Marko, Deep-Water Blowout Trajectory Models for the Lancaster Sound Region, (Environmental Protection Service, Ottawa, Ontario, 1980)

11. P.R. Bishnoi and B.B. Maini, "Laboratory Study of Behaviour of Oil and Gas Particles in Salt Water Relating to Deep Oil Well Blowouts", Spill Technology Newsletter, 4 (1979),24-36

12. B.R. LeDrew and K.A. Gustajtis, Oil Spill Scenario for the Labrador Sea, (Environmental Protection Service, Ottawa, Ontario, 1979)

13. Anonymous, An Arctic Atlas: Background Information for Developing Marine Oilspill Countermeasures, (Environmental Protection Service, Ottawa, Ontario, 1978)

14. D. Mackay and P.J. Leinonen, Mathematical Model of the Behaviour of Oil Spills on Water with Natural and Chemical Dispersion, (Environmental Protection Service, Ottawa, Ontario, 1977)

15. W.F. Purves, <u>The Interaction of Crude Oil and Natural Gas with Laboratory-Grown Saline Ice</u>, (Environmental Protection Service, Ottawa, Ontario, 1978)

16. D.E. Thornton, "Experimental Oil Spill: Oil in Arctic Coastal Environments", <u>Spill Technology Newsletter</u>, 4 (1979) 325-329

17. D.R. Topham, <u>Hydrodynamics of an Oilwell Blowout</u>, (Beaufort Sea Project, Victoria, British Columbia, 1975)

18. D.E. Thornton, "The Flow Structure of an Underwater Oil Blowout", <u>Spill Technology Newsletter</u>, 3 (1978) 44-59

19. Anonymous, <u>Testing of Air-Deployable Incendiary Devices for Igniting Oil on Water</u>, (Environmental Protection Service, Ottawa, Ontario, 1978)

20. P.B. Hildebrand and A.A. Allen, <u>The Feasibility of Oil Spill Dispersant Application in the Southern Beaufort Sea</u>, (Environmental Protection Service, Ottawa, Ontario, 1977)

21. R.W. Dennis and B.L. Steelman, <u>Oil Spill Dispersant Tests Held at Abbotsford, B.C.</u>, (Exxon Research and Engineering Company, Linden, New Jersey, 1979)

22. A.kSekerah and M. Foy, "Acute Lethal Toxicity of Corexit 9527/Prudhoe Bay Crude Oil to Selected Arctic Species", <u>Spill Technology Newsletter</u>, 3 (1978), 37-41

23. L.B. Solsberg, "A Strategy for Skimmer Development", <u>Spill Technology Newsletter</u>, 4 (1979), 345-348

24. K.M. Meikle, "Equipment Development for Arctic Oilspill Countermeasures", <u>Spill Technology Newsletter</u>, 3(1978),35-41

25. R.A. Neville, V. Thomson, R.A. O'Neill, et.al., "Remote Sensing of Oil Spills", <u>Spill Technology Newsletter</u>, 4 (1979) 111-146

26. Anonymous, <u>Oiled Debris Disposal and Storage Sites: Beaufort Sea Coast</u>, (Environmental Protection Service, Ottawa, Ontario, 1979)

27. B.W. Worbetts, <u>Shoreline Oil Spill Protection and Cleanup Strategies: Southern Beaufort Sea</u>, (Dome Petroleum Limited, Calgary, Alberta, 1979)

28. Anonymous, <u>Experimental Oilspills General Plan</u>, (Environmental Protection Service, Ottawa, Ontario, 1979)

SESSION 4

Temperate Zone

UNITED KINGDOM MARINE POLLUTION CONTINGENCY PLANNING

Rear Admiral Michael L Stacey, CB
Director, Marine Pollution Control Unit
Department of Trade, London, UK

INTRODUCTION

First, let me introduce myself. I am a simple naval officer with a traditional naval officer's background, recently retired, who took up this
appointment nine months ago as Director of the United Kingdom Government's
Marine Pollution Control Unit, which I will describe later.

The fact that the Director of this new Unit is a naval officer is a
coincidence of appointment and not one of specific policy. This point is
made deliberately because people are inclined to draw the wrong conclusion,
i.e. that the Royal Navy in the United Kingdom now plays a larger part in
our marine pollution control than it played in the past. As some of you
are probably aware, in some other countries the Navy has the entire
responsibility in this matter but not so in the UK and my appointment is
one of coincidence only.

UNITED KINDOM GOVERNMENT REVIEW

After the Amoco Cadiz, the United Kingdom Government in step with many
other governments reviewed the arrangements they had for marine pollution
control. The first aspect of their investigations was to study the pros
and cons of using dedicated forces to counter pollution and the alternative
of expanding the existing organisation. There are attractions of course
for both courses of action. In general, ships are used to combat spills at
sea, although we are looking for the future into them being augmented by
the use of aircraft. By their nature, ships are relatively slow to deploy -
especially in the bad weather of winter around our coastline which totals
more than 3,000 miles, and this fact mitigates against any system of over-
centralised control. Also, since a casualty occurs relatively infrequently,
we have to draw a distinction between dedicated forces allocated full-time
to pollution clearance - which clearly involves heavy financial expense and
imposes morale problems on personnel - and making the best use of the substantial facilities which already exist within the United Kingdom.
Accordingly, the system which we have adopted is based on the existing
facilities provided by the Department of Trade Marine Survey Service, to
which has been added a dedicated Marine Pollution Control Unit.

CURRENT ORGANISATION

The United Kingdom is divided into nine districts which together cover
the entire coastline each of which is headed by a highly experienced
Principal Officer. Whilst recognising that the main task of the Marine
Survey Officers is to survey ships visiting British ports and to ensure
their sea-worthiness, safe operation and adequate manning - the very great
experience of these Principal Officers, and their intimate knowledge of the
sea condition in their own areas and the relationships which they have
established with local authorities plus all other elements concerned with
this problem, make them ideally suited to handling oil pollution incidents.

These districts also provide convenient positions for the storing of
the UK Government Warren Spring Laboratory designed spraying gear and

dispersants which constitute our present main response capability. While recognising that this is not an ideal technique, which can perhaps in the future be complemented by the use of dispersants from aircraft and mechanical recovery devices, it is at present the only proven deep-sea method which is generally effective in the often turbulent waters around our coasts. Our aim is, therefore, to have sufficient equipment and dispersants in each district - positioned at strategic points - to meet the maximum requirements for the first 48 hours of an operation; this period having shown by experience to be sufficient to arrange the critical re-supply of detergents from the manufacturers. It is surely realised that once one gets into the business one uses a very great deal of dispersant and the re-supply becomes a logistic problem that requires very careful attention.

Each Principal Officer has drawn up a contingency plan for his district and in general this will include the identification of the areas where he considers the risk of an oil spill to be greatest - perhaps on a focal point of shipping or in the approaches to an oil discharge port. Also, he will identify similarly the areas where oil pollution is particularly damaging taking account for example of holiday beaches, amenity facilities, bird sanctuaries and fishing activities; and within this to identify the areas where dispersant spraying should be avoided if at all possible for fish preservation - such as oyster beds and fish breeding grounds. Perhaps most important of all, to identify within his area all agents who are able to provide facilities to be called on when required. This generally includes the identification of tugs and similar vessels suitable for dispersant spraying, and establishing relationships with personnel able to respond speedily when required - and indeed in most regions certain tugs have been identified as being particularly suitable and are permanently fitted with spray gear and equipment and their ships tanks filled with dispersant. He will have identified those places best suited to providing logistic support for the spraying force. Also, he needs of course to identify suitable operational headquarters for use during an incident. The question of command and control of any incident is of paramount importance. It is absolutely essential that everybody knows of their precise responsibilities in this matter and knows where the command lies and where they lie within the command structure.

The operational headquarters will very often be the HM Coastguard Rescue Coordination Centres, of which there are six around the coastline ideally suited for this task with first-class communications and supported by a dedicated aerial surveillance team. But of course an incident may well be remote from such a coordination centre and *ad hoc* facilities and command vehicles will be used. Finally, since these incidents nearly always seem to happen at a weekend, he must have a list of people's home telephone numbers so that they can be quickly called to play their part in such an incident. These plans also, of course, must include the establishment of relationships with local authorities against the time when the oil comes ashore - local fishery departments, Nature Conservancy Council on wild-life interests, and of course the UK Government's Warren Spring Laboratory where lies the scientific and technical advice to assist in classifying the oil and optimising the manner in which it is dealt.

In order to enable reserves to be called in from adjacent regions, plans must be coordinated with the Principal Officers in their neighbouring areas. Finally, there is the involvement of the Ministry of Defence, who almost certainly will be called upon to provide aircraft or helicopters for air reconnaissance and guard ships for incidents which require such assistance and other ships which they may provide for spraying duties.   The Ministry of Defence maintains two maritime headquarters - one in Scotland and another in Plymouth - and those are highly developed command centres with communications networks which have their part to play in any incident in which we are involved.   Sometimes they may geographically be rather too far away to become the on-scene command centre but as a back-up and rear-link facility, and as an agent for requesting and controlling aircraft and helicopters, they have a very real part to play.

REACTION TO AN INCIDENT

The action to be taken - once information has been received of a potential oil spill incident - is of interest and of course of very great importance.   We encourage all agencies, ships and aircraft to report anything that is seen in the way of oil pollution at sea and in general these reports are received from service aircraft via maritime headquarters; from ships via coastguards; and from aircraft via Civil Aviation Authorities.   Once such a report is received - and people are very good at making such reports - the Principal Officer has to assess and to analyse the detail.   Many reports turn out in fact to be false alarms and so he firstly needs to assess (using all the local information and facilities available in his region) the degree and scale of the emergency.   Once this has been established, he can put into operation immediately the various elements in his contingency plan and call expert assistance available to him.   At the same time, he will notify the Director of the Marine Pollution Control Unit in London of the incident in order that they may be kept informed and Ministers may be kept in the picture.   In order to assist in this process, the Department of Trade maintains a Marine Emergency Information Room in London, which serves as a focal point for collecting information for briefing Ministers and keeping other Whitehall Departments informed.   It also provides a channel for liaison with other governments who may be able to provide assistance.

This organisation which I have described lies within Government and involves three separate Departments.   The Marine Division of the Department of Trade - within which the Marine Pollution Control Unit operates - is responsible for marine pollution at sea from a line one mile off the coast. Inside that line, and ashore, is the responsibility of the Department of the Environment who coordinate the many local authorities concerned.   In the offshore oil industry, the Department of Energy works in close accord with the operating oil companies.

THE MARINE POLLUTION CONTROL UNIT

And now for the new Marine Pollution Control Unit.   This is essentially a full-time, dedicated command unit and is based in London.   It consists of the Director supported by:

a)   A senior Scientific Officer who recently worked at the UK Government Warren Spring Laboratory where he gained a great deal of practical experience in oil pollution matters.

b)   A senior Coastguard Officer who recently served at Dover and so is well versed in problems associated with the English Channel.

c)   A small supporting staff.

d)   Last but by no means least, a Director of BP Tanker Co Ltd - Captain Ralph Maybourn - who has been made available as Consultant to the Unit.

This last appointment to the Unit is seen as particularly significant for two reasons. First, it is a tangible demonstration of the responsibility and concern felt by the oil industry for pollution matters, and secondly it is an indication of Government's willingness to go outside into industry to seek professional advice, of which in matters concerning oil pollution there is a very great deal available.

The Unit's responsibilities are:

a) To establish a national contingency plan and to make the best use of the resources available.

b) To establish necessary international relationships so as to provide as much mutual support as possible.

c) To review developments of oil pollution equipment.

d) To investigate particular problems arising from the carriage of chemical and noxious cargoes.

e) To take charge of any marine pollution incident, reporting direct to the Minister.

As has been said, our present policy on dealing with oil at sea, assuming that circumstances do not permit nature to do it for us, is to use dispersants. We are satisfied that great improvements have been made regarding the toxicity of dispersants but nevertheless deploy them only after careful consideration and consultation with the Nature Conservancy Council and the Ministry of Agriculture and Fisheries to ensure minimum damage to the environment. Ideally, one would wish for an all-weather mechanical recovery system for use at sea and there are systems available which show promise and on which the Government will take a decision to purchase very shortly.

TRANSFER OF OIL

If it is possible, the best way to remove the risk of oil pollution from a damaged ship is to transfer the oil into the sound bottom of another tanker and we would always aim to do this. There are at present seven commercial tankers specially modified for cargo transfer around the UK coasts for carrying out routine lightening operations - but to ensure the absolute availability of such an important option, the UK Government are establishing two caches of salvage equipment.

One cache is based at Pembroke Dock in Wales and the other is at Rosyth in Scotland, and each will consist of transfer pumps, inert gas equipment and associated gear - all of which is helicopter transportable - so that in the event of a specialist ship not being available any tanker can be given the necessary fuel transfer capability.

INTERNATIONAL RELATIONSHIPS

In the area of international relationships, all countries bordering on the North Sea are drawn together under the Bonn Agreement which ensures that they keep each other informed of their contingency plans and are ready to assist each other as appropriate. Particularly and obviously for the English Channel area there is a need for a specially developed relationship with the

French and this we have in the MANCHEPLAN which is well progressed and is exercised each year. On the other side, we are in the process of coming to arrangements with the Irish in order that we can work together in the Irish Sea where the pattern of tanker traffic is increasing now that the Sullom Voe terminal is on stream.

SALVAGE OPERATIONS

In the salvage operation which inevitably follows a tanker casualty, it is the Government's desire - and also it is believed of the salvor and the ship-owner - that the salvage contract remains a straightforward commercial relationship between the parties concerned. However, the Government does have a real and positive interest where pollution arising from the incident may reach the shores of the UK.

INTERVENTION

The UK Act of 1971, which arises directly from the 1969 Convention, permits the Government to give directions to those in charge or, if of the opinion that the giving of such directions will not suffice, may take direct action either of its own or through some agent.

This right, which permits intervention, means that during a salvage operation the Director of the Marine Pollution Control Unit, on behalf of the British Government, works very closely with the salvors and encourages them to pay full and proper respect to any pollution problems which might arise. If for any reason this respect is not paid, he has recourse to advise the Minister that the Government should intervene. Such recourse in general is deemed to be highly undesirable since it can represent a break-down in the relationship with the salvor which puts both parties into an extremely difficult professional position and could put the whole salvage operation at risk. To avoid such breakdowns, an absolute close relationship between Government and salvor in incidents involving a pollution risk is of absolute essence.

SAFE HAVENS

One important problem arising from such salvage operations is the frequent necessity for the casualty to be moved to calm and sheltered waters - namely a safe haven - in order that the cargo transfer operation can be more safely carried out. The recognition of this need and the identification of such safe havens is of great importance and is something that must be addressed. For understandable reasons, nobody likes a tanker casualty being brought to their doorstep but there is no doubt that for the common good of all the community, and to ensure that the risk of heavy pollution of our beaches is minimised, such actions will have to be taken in the future.

THE FUTURE

And finally to the future. There can be no doubt that tanker casualties will continue to occur and that risk of heavy oil pollution will persist. There is much to be done in their prevention but in the matter of cleaning-up, which is our topic today, the following pattern of actions has been established:

a) Transfer of fuel from the casualty into a sound ship whenever possible.

b) The dealing with oil on the sea to be vested in as large a variety of
    systems as possible in order to account for all possible circumstances.
    Dispersants will continue to be sprayed from surface ships but we have
    already made substantial progress in delivery of dispersants by air-
    craft. This method has been used in Bantry Bay and also in the Gulf
    of Mexico, and the UK Government will shortly be making arrangements
    to have this facility available for itself. Mechanical recovery has
    shown substantial progress in some areas and similarly the Government
    will be taking early decisions in this area.

c) It is our belief that in order to deal satisfactorily in the future,
    one's contingency plan must be flexible so as to enable a quick response
    to any situation. The command organisation must be ready to improvise
    and to draw on any relevant experts for particular advice. On the
    equipment side, we must have a sensible investment in as broad a cross
    section of proven systems as are available, both in the area of
    dispersants and of mechanical recovery.

d) Finally, plans must be well prepared and well rehearsed. Speed of
    reaction to an oil pollution threat is essential.

U.S. NATIONAL CONTINGENCY PLAN TO CONTROL OIL AND HAZARDOUS SUBSTANCES DISCHARGES *

Kenneth E Biglane
Director, Oil and Special Materials Control Division, Office
of Water Program Operations, U.S. Environmental Protection
Agency, Washington, D.C. 20460. Also, Chairman, U.S. National
Response Team (NRT).

The TORREY CANYON tanker casualty off the Cornwall Coast of
England in 1967, found most nations quite unprepared to respond
effectively to large oil discharges to the oceans. The United
States was most certainly unprepared to deal with such a large
discharge. Those of us from the U.S. who were privileged to
observe the governmental responses of both the United Kingdom
and France in their efforts to organize response forces, bring
to bear effective cleanup techniques and technologies, assign
priorities of cleanup, find disposal sites for recovered oil
and oily debris, restore damaged environments, and assessing
the damage to resources and commerce, were likewise overwhelmed
at the thought of preparing our country to deal with such an
insult. Many other countries were also shocked into the
realities of planning a response mechanism for large and costly
spills of environmentally polluting materials.

In planning a national system to respond to spill events
in the U.S., a task force began with the concept of how to
bring the "best in government" together with the "best in
industry" so as to assure both an efficient and effective
response program.

The "best in government" ranges from involving those departments with appropriate statutory authorities (including funding)
to those agencies that can furnish or provide immediate information on human health hazards; environmental information such
as fisheries spawning seasons; endangered species and critical
habitates; ship configuration and operations; climatic data;
physical and chemical oceanographic data; limnological or
estuarine data; disposal sites; land usage; hazardous chemical
data; worker safety; international and/or judicial considerations;
aerial emissions; and an awareness of the state-of-the-art to
deal with any given episode.

The "best in industry" approach starts with the identification
of available spill control techniques and equipment. How one
goes about identifying, evaluating, cataloging and then utilizing
existing technology has become part of this country's national
planning.

And so it was, in 1968, the U.S. proposed its first National
Oil and Hazardous Substances Pollution Contingency Plan (here-
inafter referred to as the Plan). Although the Plan has undergone five revisions since 1968, the basic concepts, as described
above, have remained the same. The Plan has the following major
components.

* A presentation before the Petroleum and the Marine Environment
International Conference & Exhbition, Monaco, May 27-30, 1980.

I. Federal Policy
The primary thrust of the Plan is to provide a co-ordinated
Federal response capability at the scene of an unplanned or
sudden, and usually accidental discharge of oil or hazardous
substance that poses a threat to the public health or welfare.

II. Implementing Units

A. The National Response Team (NRT)

The National Response Team, Washington, D.C., consists of
respresentatives from the appropriate agencies. It serves as
the National body for planning and preparedness actions prior
to a pollution discharge and for co-ordination and advice
during a pollution emergency.   Except for periods of activation
because of a pollution incident, the representative of the
Environmental Protection Agency (EPA) shall be the chairman
and the representative of the Department of Transportation (DOT)
(represented by the U.S. Coast Guard) (USCG) shall be vice-
chairman of NRT. The vice-chairman shall maintain records of
NRT activities along with National, regional and local plans
for pollution response. When the NRT is activated for a
pollution incident, the Chairman shall be the representative
of EPA or DOT, depending upon the area in which the response is
taking place.

B. The Regional Response Team (RRT)

The RRT serves as the regional body for planning and preparedness actions before a pollution discharge and for co-ordination
and advice during a pollution discharge. The RRT consists of
regional representatives of the participating agencies, State,
and local government representatives as appropriate. The full
participation of high level representation from States and
local governments with major ports and waterways is desired.
     Except when the RRT is activated for a pollution incident,
the representatives of EPA and DOT shall act as co-chairmen.
When the RRT is activated for a pollution incident, the chairman
shall be the representative of EPA or DOT, depending upon the
area of the spill and response.
     Participating States and local governments should also
designate one member and at least one alternatve member to the
team. Agencies may also provide additional representatives as
observers to meetings of the RRT. Persons presenting Federal
and State agencies shall be specified in each regional contingency plan.
     RRT members shall designate representatives from their agencies
to work with the On-Scene Co-ordinators (OSC) in developing local
contingency plans, providing for the use of agency resources,
and in responding to pollution incidents.
     Each of the States lying within a region is invited to
furnish liaison to work with the RRT and OSCs in developing
regional and local plans, plan for and make available State
resources, and serve as the contact point for co-ordination
with local government in responding to pollution incidents.
When the RRT is activated for a pollution emergency affected
States are invited to participate in all RRT deliberations.
Any State or local government representative who participates
in the RRT has the same status as any Federal member of the RRT.

During a pollution emergency, the members of the RRT shall insure that the resources of their agencies are made available to the OSC as specified in the regional and local contingency plans.

When not activated for a pollution incident, the RRT serves as a standing committee to recommend needed policy changes in the regional response organization, to revise the plan, and to evaluate the preparedness of the agencies and the effectiveness of local plans for the Federal response to pollution incidents. The RRT shall :

(1) make a continuing review of regional and local responses to pollution incidents, considering equipment readiness and co-ordination among responsible public agencies and private organizations;

(2) recommend revisions to the National Contingency Plan to the NRT, on the basis of observations of response operations;

(3) consider and recommend necessary changes in policy on the basis of the continuing review of regional responses to pollution incidents;

(4) develop procedures to insure the co-ordination of Federal, State, local government and private responses to pollution incidents;

(5) review the functioning of OSC's to insure that local contingency plans are developed satisfactorily;

(6) be prepared to respond to a major discharge of oil or hazardous substances outside its region;

(7) monitor incoming from all OSC's and activate the RRT when appropriate;

(8) meeting quarterly to review response actions carried out during the preceding period and consider changes in both regional and local contingency plans. In those regions having both coastal and inland RRT's, RRT meetings held in alternating quarters (inland in March, coastal in June, etc.) would meet this requirement; and

(9) RRT's shall provide letter reports on their activities to the NRT twice a year, no later than January 31st and July 31st. The reports will help to identify techniques and procedures that have worked well and subjects requiring improvement and should be circulated to other RRTs. At a minimum, reports will contain paragraphs covering :
(i) Summary of Activities, containing highlights of routine meetings and activation during the reporting period;
(ii) Organizational Matters, outlining improvements made since the last report. Organizational matters requiring NRT actions should be included. RRT's are encouraged to add detailed accounts of successful procedures;
(iii) Operations, including recommendations, comments or observations on response methods, equipment, training or other operational matters which have not been addressed in the review of OSC reports.

Each coastal RRT is required to conduct an annual training exercise in which response equipment is actually deployed. These exercises should use all existing capabilities in the local port area. Any funding required to support the exercise should

be requested through the normal agency budget process. The RRT
shall co-operate to the fullest extent possible in field exercises
of member agencies.

RRT's for inland regions are strongly encouraged to conduct
an annual training exercise in which response equipment is
actually deployed. RRT's for inland regions shall co-operate
to the fullest extent possible in field exercises of member
agencies.

The RRT shall be activated as an emergency response team
when a discharge :
(1) exceeds the reponse capability available to the OSC in
the place where it occurs;
(2) transects regional boundaries; or
(3) poses a substantial threat to the public health and
welfare or to regionally significant amounts of property.
Regional Contingency Plans shall specify detailed
criteria for activation of RRTs.

The RRT shall be activated automatically in the event of a major
or potential discharge. The RRT may be activated during any
other pollution emergency by an oral request from any RRT
representative to the chairman of the team. Requests for team
activation shall later by confirmed in writing. Each representative, or an appropriate alternative, shall be notified immediately by telephone when the RRT is activated.

When activated for a pollution incident, agency representatives
shall meet at the call of the chairman and shall :
(1) request other Federal, State or local government, or
private agencies to consider providing resources under
their existing authorities to combat a discharge or
monitor response operations, and
(2) help the OSC prepare information releases to the public
and for communication with the NRT.

C. The On-Scene Co-Ordinator (OSC)

Co-ordination and direction of Federal pollution control effects
at the scene of a discharge or potential discharge shall be
accomplished through the OSC, predesignated by regional plan to
co-ordinate and direct such pollution control activities in each
area of the region.

In the event of a discharge of oil or hazardous substance,
the first official on the site from an agency having responsibility
under this Plan shall assume co-ordination of activities under
the Plan until the arrival of the predesignated OSC.

The OSC shall determine pertinent facts about a particular
discharge, such as its potential impact on human health and
welfare; the nature, amount and location of material discharged;
the probable direction and time of travel of the material; the
resources and installations which may be affected and the priorities for protecting them.

The OSC shall initiate and direct as required Phase II,
Phase III and Phase IV operations. Advice provided by the EPA
representative on the RRT on use of chemicals in Phase III and
Phase IV operations in response to discharges of oil or hazardous
substances shall be binding on the OSC, except as otherwise
provided by the Plan.

Advice provided by the Fish and Wildlife Service (DOI) or
by the National Oceanic and Atmospheric Administration (Commerce)
on cleanup actions that may affect endangered and threatened

species or their habitats shall be considered at all times and shall be binding on the OSC unless in his judgment actions contrary to this advice must be taken to protect human life.

The OSC shall call upon and direct the deployment of needed resources in accordance with the regional plan to evaluate the magnitude of the discharge and to initiate and continue removal operations.

The OSC shall provide necessary support activities and documentation for Phave V activities. The Phase activities will be described elsewhere in this paper.

The EPA shall furnish or provide for OSC's on inland waters.

The USCG shall furnish or provide for OSC's for the coastal waters, and for Great Lakes waters, ports and harbors.

III. Membership of the NRT and RRT
        Department of Agriculture
        Department of Commerce
        Department of Defense
        Department of Energy
        Evnironmental Protection Agency
        Federal Emergency Management Agency
        Department of Health, Education and Welfare
        Department of the Interior
        Department of Justice
        Department of Labor
        Department of State
        Department of Transportation
In addition, the State Governor of each State has been requested to nominate his State's representative to the RRT.

As a point of clarification, the States do not have represent- ation on the NRT. The Federal agency membership on the NRT is the same for the RRT.

IV. Action Phases
The actions taken to respond to a pollution discharge can be separated into five relatively distinct classes or phases. For descriptive purposes, these are : Phase I - Discovery and Notification; Phase II - Evaluation and Initiation of Action; Phase III - Containment and Countermeasures; Phase IV - Removal, Mitigation and Disposal; and Phase V - Documentation and Cost Recovery. It must be recognized that elements of any one phase may take place concurrently with one or more other phases.

Phase I - Discovery and Notification
A discharge may be discovered through : (1) a report submitted by a discharger in accordance with statutory requirements; (2) through deliberate search by vessel patrols and aircraft; and (3) through random or incidental observations by government agencies or general public.

In the event of a deliberate discovery, the discharge will be reported directly to the NRC. Reports of random discovery may be provided by fishing or pleasure boats, police department, telephone operators, port authorities, news media, or others. Reports generated by random discovery should be submitted to the nearest USCG or EPA office. Regional plans shall provide for such reports to be channeled to the RRC as promptly as possible to facilitate effective response action. Reports of major and medium discharges received by either EPA or USCG

shall be expeditiously relayed by telephone to the other agency.
Reports of minor discharges shall be exchanged between EPA and
USCG as agreed to by the two agencies.

The Agency furnishing the OSC for a particular area is
assigned responsibility for implementing the appropriate
Phase activities in that area.

Phase II - Evaluation and Initiation of Action

The OSC shall insure that a report of a discharge is immediately
investigated. Based on all available information, the OSC
shall : (1) evaluate the magnitude and severity of the discharge;
(2) determine the feasibility of removal; and (3) assess the
effectiveness of removal actions.

The OSC shall, when appropriate and as soon as possible
after receipt of a report, advise the RRT of the need to
initiate further governmental response actions. This may be
limited to activation of the RRT or a request for additional
resources to conduct further surveillance or initiation of
Phase III or Phase IV removal operations.

The OSC shall insure that adequate surveillance is maintained
to determine that removal actions are being properly carried out.
If removal is not being done properly, the OSC shall so advise
the responsible party. If, after the responsible party has
been advised and does not initiate proper removal action, the
OSC shall take necessary action to remove the pollutant.
Funding for these actions will be expended utilizing appropriate
Federal funding authorities.

If the discharger is unknown or otherwise unavailable, the
OSC shall proceed with removal actions.

Phase III - Containment and Countermeasure

These are defensive actions to be initiated as soon as possible
after discovery and notification of a discharge. These actions
may include public health and welfare protection activities,
source control procedures, salvage operations, placement of
physical barriers to halt or slow the spread of a pollutant,
emplacement or activation of booms or barriers to protect
specific installations or areas, control of the water discharge
from upstream impoundments and the employment of chemicals and
other materials to restrain the pollutant and its effects on
water related resources.

Phase IV - Cleanup, Mitigation and Disposal

This includes actions taken to recover the pollutant from the
water and affected public and private shoreline areas, and
monitoring activities to determine the scope and effectiveness
of removal actions. Actions that could be taken include the
use of sorbers, skimmers and other collection devices for
floating pollutants, the use of vacuum dredges or other devices
for sunken pollutants; the use of reaeration or other methods
to minimize or mitigate damage resulting from dissolved,
suspended or emulsified pollutants; or special treatment
techniques to protect public water supplies or wildlife resources
from continuing damage.

Phase V - Documentation and Cost Recovery

This includes a variety of activities, depending on the
location of and circumstances surrounding a particular discharge. Recovery of Federal removal costs and recovery for

damage done to Federal, State or local government property is
included; however, third party damages are not dealt with in
this plan. The collection of scientific and technical information of value to the scientific community as a basis for
research and development activities and for the enhancement
of understanding of the environment may also be considered in
this phase. It must be recognized that the collection of samples
and necessary data must be performed at the proper times during
the case to fix liability and for other purposes.

V. Annexes to the Plan

There are Annexes to the Plan which include :
  1. designating a National Response Center in Washington,
      D.C. to receive and disseminate spill reports;
  2. providing for public information;
  3. describing an oil dispersant use schedule to be followed
      by an OSC; and
  4. providing useful technical information and describing
      geographical boundaries of the various U.S. regional
      and district offices.

The U.S. experiences about 10,000 reported discharges of oil
to its surface waters each year. Additionally, several thousand
discharges involving hazardous substances and other materials
are also reported. Obviously, the response to all reported
discharge incidents by an Agency On-Scene Co-ordinator is
impossible. Likewise, the convening of the various response
team to each incident is not only impossible, but also
undesirable. Major spill incidents, however, trigger the full
response mechanism. Such incidents are defined by the Plan :
(a) according to geographical areas for oil, and (b) according
to real or potential hazardous substances discharges.

The Plan encourages routine meetings of the Regional
Response Teams for planning purposes and to engage in hypothetical
spill problems for training purposes. We have found that the
relationship developed between agency participants during these
kinds of meetings is valuable in program response effectiveness
during actual discharge events. At this writing, RRTs have
been activated for 15 spill incidents, the largest being the
Texas Gulf Coast Team because of the Mexican wild well in the
Bay of Campeche.

The National Response Team (NRT) convenes once each month
to consider such business as amendments to the Plan, displaying
case histories to the membership, and providing through the
Department of State, assistance for treaty negotiation on joint
contingency planning with neighbouring countries. NRT can be
convened during an emergency situation to aid the various RRTs
in their mission.

In the nearly 12 years of its operation, the Plan has served
the Nation in responding to such discharges that occurred during
the Santa Barbara, California, oil well blowout, three wild wells
off the Louisiana Gulf Coast, the Bay of Campeche wild well,
many tanker casualties such as the ARGO MERCHANT off Nantucket
and the ZOE COLOCOTRONIS off Puerto Rico, and more recently,
the hazardous substance spills from waste dump sites containing
chemicals.

Literature Cited :

The United States Oil and Hazardous Substances Pollution
Contingency Plan as amended, March 1980.

Albert H. Lasday, B.Sc., M.A., D.Sc.
Coordinator, Division of Environmental Affairs
Texaco Inc.
Beacon, New York, U.S.A.

INTRODUCTION

The decade of the '70's can be identified with a worldwide upsurge of environmental concern. In the United States the American Petroleum Institute (API) has recognized the importance of environmental protection by planning and funding research on the fate and effects of oil entering the marine environment. Starting in 1970, when the amount of information in the scientific literature dealing with fate and effects of oil was quite small by today's measures, the API undertook pioneering work to elucidate the physical and biological roles of oil in the temperate zone marine environment. Many other researchers have followed and benefitted from API's lead. Because of the upsurge of research interest in the fate and effects of oil that emerged during the past decade, the portion of projects sponsored by the API is smaller today than formerly. However, the API continues its research sponsorship and focuses on fate and effects of oil topics of greatest interest and merit. The API has acted to sponsor the research reported in the following, through its Committee on the Fate and Effects of Oil in the Marine Environment.

The API initiated its research sponsorship on Fate and Effects of Oil in 1970 with a laboratory study. It soon became apparent to all concerned that the artificiality of laboratory conditions, at best, simply was not representative of the conditions and environment found in actual field ecosystems and marine waters. For example, in the field many influences act to degrade and to disperse spilled hydrocarbons as a function of time. In the laboratory, on the other hand, experimental oil dosages usually are maintained at fixed levels for convenience, or to elicit a worst case response, or to establish a dose versus response relationship. Accordingly, the API began emphasizing field-based research. Exceptions were made for those projects where complexity of the desired experiments, logistics and cost, and/or need to control precisely the experimental conditions made the laboratory approach or a combined field and laboratory approach the most feasible one.

A number of varied research efforts has been included within the scope of the API sponsored program. Such disciplines as marine biology, microbiology, oceanography, analytical chemistry, and various of their subdivisions, and others have been represented among the research scientists. The following discussion will highlight many of the more important researches sponsored by the API. Projects are grouped under five headings:
1. General
2. Fate
3. Laboratory Studies of Effects
4. Field Studies of Effects
5. Laboratory and Field Studies of Effects.

This categorization of the research projects is intended to indicate their dominant or most important emphasis.

GENERAL PROJECTS

Standard Reference Test Oils

Starting with its first sponsored research project in 1970, the API Fate and Effects of Oil Committee supplied to the contractor a set of oils which were to be used exclusively in that research effort. They were:

      (1)   South Louisiana crude
      (2)   Kuwait crude
      (3)   No. 2 fuel oil
      (4)   Venezuelan Bunker C fuel oil

Thus there were two crudes and two refined oils. They were intended to be representative of typical crudes and refined oils most commonly transported and hence most likely to be spilled. In retrospect, the particular No. 2 fuel oil chosen was not typical with respect to its high content of aromatic hydrocarbons (38%). Due to the toxicity to marine biota of these aromatics, this particular choice of a No. 2 fuel oil tended to lead to more severe effects on test animals than would have resulted had an oil of lower aromatics content been used.

It was obvious that subsequently sponsored API research projects would benefit from comparison of results with the initial project if these same test oils were used. Thus, all API researchers were supplied with oil samples for their experimental use from the same stock of test oils. Analyses of the chemical and physical properties of these "standard reference test oils" were performed by an API contractor -- Exxon Research and Engineering Company (Ref. 1) -- and the data supplied to all researchers.

Starting in 1974, API provided samples of the reference test oils to any researcher who requested them from a storage and distribution center established at Texas A & M University.

The original stocks of reference test oils are now nearly exhausted, although demand from researchers for samples continues. The usefulness of these reference test oils in facilitating comparisons of research results among various laboratories, including API contractors, has been established. API has initiated action to choose and to characterize a new, expanded set of reference test oils. The United States Environmental Protection Agency (EPA) has indicated interest in jointly sponsoring the new reference test oil repository. The EPA would store and distribute the oils, which would be provided and analyzed by API.

It is now contemplated that the original group of four reference test oil types would be expanded to include two additional crudes -- a light Arabian crude and an Alaskan (Prudhoe Bay) crude -- and perhaps a second No. 2 fuel oil of lower aromatic content.

Analytical Method Development

Studies of the biological effects and toxicity of petroleum hydrocarbons on marine animals inevitably involve coordinated chemical analyses of the amounts of those hydrocarbons in target animal tissues and in their habitats -- water column and sediments. The necessary chemical analyses include determinations of whole oil, of classes of hydrocarbon compounds, and even of individual compounds in some cases. With respect to analyses for hydrocarbons in animals, both whole animal and specific organ measurements may be needed.

The most important analytical method development was performed for the API program by Dr. J. S. Warner, Battelle Columbus Laboratories.(Refs. 2,3,4) His approach was to emphasize the following criteria:

      a.   Ease of performance, so as to minimize time and costs.
      b.   Completeness of recovery.
      c.   Reproducibility.
      d.   Avoidance of contamination and interferences.
      e.   Sensitivity.

Parallel to and supportive of the analytical method development was the working out of adequate and careful sample handling and shipping procedures. Twelve specific methods were developed during this program.

Another important task of developing analytical methods was performed for API by Drs. R. A. Brown and R. J. Pancirov (Refs. 5,6,7) of the Exxon Research and Engineering Company. They specified a method for identifying and quantifying individual polynuclear aromatic hydrocarbons in petroleum and in animal tissues, in some cases to a level of 0.1 parts per billion (ppb).

FATE

Mathematical Modelling of the Fate of an Oil Spill

The fate of oil in the marine environment is of great importance and pertinence. Except as a matter of scientific study, one might not think that there is a need for knowledge of the fate of oil; i.e., in what portions of the environment it finally resides and how it is transported to them. After all, of greatest importance are the effects of oil; if there were none, we hardly would be concerned. As a practical matter, however, the effects of oil are consequential, and thus it is essential to know the fate of oil that appears in the marine environment, from whatever sources, in order to be able to evaluate and predict the effects.

The API Fate and Effects of Oil Committee has completed an elaborate oil spill modelling project at the University of Southern California (USC), (Ref. 8). The researchers there developed a highly sophisticated mathematical computer model for oil spills. The model not only considers the movement of the slick from place to place but also calculates a three dimensional fate based on oil composition, environmental data such as temperature and sea state, and all known mechanisms of spill dispersion and degradation, including such more important factors as evaporation, dissolution, emulsification, biodegradation, and photo-oxidation.

While the USC model is too elaborate and complex to be used in the short time frame of a spill emergency, it is of very great importance as the first effort to include all known operative mechanisms that actually work to disperse and to degrade a spill.

Experimental Oil Spills

Two problems arose which required experimental oil spills in the ocean for their resolution. The first one evolved from the above-described project to develop a mathematical model, and which had as one of its purposes an assessment of the quality of the data that are available. These data are required to aid in formulating algorithms which adequately describe the various physical, chemical, and biological processes that disperse and degrade the oil slick as a function of time. The most important deficiency discovered was the lack of real time data describing the post-spill variations with time of the concentrations of various hydrocarbon components from an oil slick both on the water surface and in the water column.

The second problem concerned the use of dispersants in United States waters for control and dissipation of oil spills in appropriate situations. In the aftermath of the Torrey Canyon oil spill of 1967, where some portion of the disaster was attributed to the improper application of toxic dispersant chemicals, the U.S. Environmental Protection Agency (EPA) adopted a very conservative policy governing dispersant usage. With the development of low toxicity dispersants and more sophisticated knowledge of how to use them effectively, both EPA and API realized the need for data that would verify the effectiveness and safety of the new materials and techniques.

Both of the above two problems were addressed in research on three separate series of experimental oil spills.

The problem of data acquisition on the fate of spilled oil was the subject of the first series of experimental spills. API's contractor was the JBF Scientific Corporation of Wilmington, Massachusetts (Ref. 9). In the fall of 1975 four spills of about $10\frac{1}{2}$ barrels each were made from a research vessel off the U.S. east coast. A light gravity crude oil (Murban) and a medium gravity crude (LaRosa) were used. The slicks were tracked by the vessel and an airplane and were sampled both on the surface and in the water column at depths of 5 and 10 feet for up to two days. Associated oceanographic data were obtained at the same time. Chemical analyses of the slick samples resulted in some important observations:

(1) Low molecular weight hydrocarbons were undetectable in the water column after 20 minutes.

(2) All soluble hydrocarbons through C_{10} were lost from the surface slicks in 6 to 8 hours.

(3) Total hydrocarbons in the water column beneath the slicks returned to background levels in about 4 hours.

This latter observation gives substance to the conclusion that -- at least in the open ocean -- toxic concentrations of hydrocarbons do not persist long enough in the water column to constitute a significant threat to marine life that may be present.

The problem related to dispersants was addressed with two additional series of spills. These were again performed off the U.S. east coast by the JBF Scientific Corporation under contract to API, with partial support by the EPA and the U.S. National Aeronautical and Space Administration. These spills took place in the fall of 1978 and 1979, using the same two crudes as in 1975 to facilitate comparisons. Simultaneous biological studies of the effects of the spills were carried out in 1978 by the Virginia Institute of Marine Sciences and by Dr. D. T. Boyles of British Petroleum. In 1979 the biological experiments were performed by the University of Southern California. Detailed supporting chemical analyses were made by the Exxon Research and Engineering Company, Chevron Oil Field Research Company, and West Coast Technical Laboratories.

The 1978 spill series demonstrated essentially 100% dispersion of the light crude, as verified by a mass balance. The medium gravity crude oil dispersed satisfactorily, but not as well as the light crude oil. The 1979 spill study appears to have been performed successfully, although data from chemical analyses are as yet incomplete.

Further information, discussion, and details concerning the 1978 series of controlled - dispersed oil spills will be reported by Dr. C. D. McAuliffe et al, elsewhere in this session of the Eurocean Petromar '80 Conference (Ref. 10).

LABORATORY STUDIES OF EFFECTS

Effects of Oil and Chemically Dispersed Oil -- Battelle Pacific Northwest Laboratories

This was a pioneer research project sponsored by the API Fate and Effects of Oil Committee. Battelle investigated the acute and chronic (i.e., lethal and sublethal) effects of API standard reference test Kuwait and South Louisiana crudes and No. 2 fuel oil. All were tested in both the dispersed and undispersed states. Both flow-through and static bioassays were performed with algae, Pacific oysters, Dungeness crabs, 11 fish species, and some eggs and larvae. The important conclusions (Ref. 11) were:

(1) Actual animal exposures to oil were very much lower than the metered inflow amounts.

(2) The No. 2 fuel oil was more toxic than the crudes.

(3) While the dispersed oils entered the water column in greater amounts than did the undispersed ones, there were no differences in toxicities based on oil concentration in the water column.

(4) Uptake and depuration were demonstrated. In other words, organisms exposed to oil-contaminated water built up tissue burdens of petroleum hydrocarbons, which were rapidly purged when the organisms were placed in clean water.

Effects of Dispersions and Water Soluble Fractions of Oil -- Texas A & M University

In a sense, this project was a continuation and amplification of the above reported work by Battelle Pacific Northwest Laboratories. Oil-water dispersions and water-soluble fractions of the four API standard reference test oils were used in response to a requirement to standardize exposures in bioassay systems. A wide variety of experimental species was used, including fish, crustaceans, and phytoplankton. The results of this research have been reported extensively in the scientific literature; some thirty-seven technical papers have been published.*

The most significant conclusions of the Texas A & M University research were (Ref. 12):

(1) Both oil in water dispersions and water soluble fractions of oils were significantly more toxic than crude oils to phytoplankton, crustaceans, and fish.

(2) In assessing physiological stress due to sublethal concentrations of hydrocarbons, respiration rates gave inconsistent results, but heart beat rates of fish embryos as well as growth rates correlated most consistently with hydrocarbon concentrations.

---

* See Oil and the Sea published in 1979 by the American Petroleum Institute for a listing of these research articles in various technical and scientific journals and compendia.

(3) In agreement with the studies by Battelle Pacific Northwest Laboratories, animals exposed to petroleum hydrocarbons took up these materials in their tissues, but released them when placed in clean water.

(4) Naphthalenes turned out to be especially toxic (in the low parts per million range), were accumulated to the greatest extent in all test species, and were retained the longest after the animals were returned to clean water; but depuration did occur to below detectable levels.

Both the previously described work of Battelle Pacific Northwest Laboratories and Texas A & M University have demonstrated the ability of very many types of marine organisms to purge their body tissues of petroleum hydrocarbons taken up from their surroundings. Once the organisms have been removed from the source of contamination and have been placed in clean seawater, this purging or depuration occurs at different rates which depend on the species.

There is great significance to this phenomenon beyond the mere fact that oil-contaminated sealife can purge its tissues to background levels in clean water. The hypothesis of bioaccumulation of petroleum hydrocarbons in marine organisms, a process which can pose a threat to human health, depends upon permanent retention of hydrocarbons by such organisms. The depuration process interrupts the food chain transfer mechanism and renders this hypothesis invalid.

Oil Effects On Lobster Larvae -- Westinghouse Ocean Research Laboratory

Floating oil slicks resulting from spills generally do not seem to pose the same threat to marine life as does oil entrained in sediments, where it can persist for long periods of time. Yet many planktonic life forms, including larvae and eggs of some species, conceivably could be subject to damage from floating oil should the two coincide. The American lobster, Homarus americanus, is an especially important commercial species in the United States. The eggs and first four larval stages are planktonic species that reside in ocean waters just below the surface.

The South Louisiana crude API reference test oil was used as the hydrocarbon source for exposing the first four lobster larval stages in a flow-through system (Ref. 13). Oil concentrations of 0.1 and 1.0 parts per million (ppm) were used in the form of an emulsion-like mix. Survival, growth, feeding, and activity of the lobster larvae exposed to the 0.1 ppm oil level were normal and comparable to the unexposed controls. However, those exposed to 1.0 ppm exhibited reduced survival, lethargic feeding and behavior, and increased time for development. Even those larvae exposed to 0.1 ppm showed an abnormal reddening, however, and this effect was much more pronounced in the lobster larvae exposed to 1.0 ppm of the crude oil.

Used Drilling Muds -- Texas A & M University and Bowdoin College

Two projects have been undertaken by API to determine the biological effects of drilling muds. Researchers at Texas A & M University have addressed their studies to effects of used drilling muds in warm waters, whereas the Bowdoin College group has dealt with cold water effects. Both groups used five typical used drilling muds: (1) a spud mud, (2) a seawater chrome lignosulfonate mud, (3) a lightweight lignosulfonate mud, (4) a midweight lignosulfonate mud, and (5) a high-weight lignosulfonate mud. Both research teams have reported their results to date at the recently held symposium -- Research on Environmental Fate and Effects of Drilling Fluids and Cuttings, held in January, 1980, at Lake Buena Vista, Florida (Refs. 32,33,34,35,36,37). In addition to experimenting with the effects of whole muds, suspensions of fine particulates and an aqueous fraction containing the water soluble components were used. A wide variety of marine species was subjected to toxicity testing using various of the five used muds.

In the warm water studies (Texas A & M University), it was revealed that the chrome lignosulfonate mud, the mid-weight, and high-weight lignosulfonate all exhibited relatively low acute toxicities. The used spud mud was completely non-toxic. The water-soluble fraction contained the most toxic components of the muds. An exception was the case of the opossum shrimp where exposures to the aqueous fraction of the four lignosulfonate types of muds resulted in reduced production and growth. Bioavailability to bivalve mollusks of heavy metals (chromium, lead, and zinc) from used drilling muds proved to be quite limited.

The results from the cold water (7°-22°C) experiments (Bowdoin College) were roughly comparable to those from warm waters. Ninety-six-hour exposures of a wide spectrum of animals to all five used muds, as well as to the mud aqueous fraction, showed mortalities generally less than 10% of the test populations -- and often no mortality occurred. In addition, the researchers concluded that toxic and sublethal effects on cold water marine organisms would be expected to occur only close to the source of discharge of the used drilling muds. In studies in the ocean, small amounts of whole muds (up to 33% v/v) do not significantly affect recruitment to benthic infaunal communities over an 84 day period.

Effects of Dispersed Oil -- Battelle Pacific Northwest Laboratories

A key question to be answered about oil spills that are dispersed with chemical agents concerns the effect of dispersed oil on marine organisms as compared to what the effect might be if the oil were not dispersed. In order to approach this most important concern, API commissioned Battelle Pacific Northwest Laboratories to investigate the biological effects of dispersed vs. undispersed oil, and to develop appropriate flow-through bioassay methods for evaluating dispersed oil. The experimental animal chosen was the coonstriped shrimp (Pandalus danae). Rather than using one of the API reference test crudes, Alaskan Prudhoe Bay crude oil was used because it is one of the most important crudes shipped along the Pacific coast. Two dispersants were used, and they were reported to be non-toxic to the shrimp at the test concentrations (20 oil: 1 dispersant) (Ref. 14). Several interesting results were reported:

  (1) Addition of dispersants effected important alterations in the types of hydrocarbons in the water column, as compared to those present from crude alone. Most important, this change was shown to be an increase in the proportion of relatively insoluble saturated compounds in the water column.

  (2) Probably as a result of the above, on the basis of parts per million of oil in the water, dispersants reduced the toxicities of the solutions by about one-half, or did not reduce them as compared to undispersed crude, depending on the season.

  (3) Food consumption by the shrimp turned out to be a far more sensitive indicator of the toxicity of oil or of dispersed oil than was toxicity testing.

(4) A great seasonal variability in feeding rates occurred. Also size and age of animals were factors producing important variability of results.

(5) A toxicity index in ppm-days was developed that was the product of exposure concentration and duration. This allows a single number characterization of toxicity that accounts for varying levels of exposure (as occurs in natural spills) and of cumulative and accommodating effects of the toxicant on the test organisms. Seasonal effects are not included in the toxicity index.

FIELD STUDIES OF EFFECTS

Chronic Exposure -- University of Southern California

As discussed above, many factors make it quite difficult, if not impossible, to duplicate in the laboratory the variability and complexity of the natural environment and the action of many natural mechanisms that degrade and disperse spilled oil. At Coal Oil Point, near Santa Barbara, California, there is a large scale natural oil seep from fissures in the ocean bottom. About 50 to 100 barrels per day of a heavy, tarry crude oil pollute the surrounding waters and beaches. This natural laboratory was the locale of two long-term studies by the Allan Hancock Foundation of the University of Southern California (Refs. 15,16). In the first study, the experimental scheme consisted of studying the growth and reproductive rates; hydrocarbon concentrations in the water column, sediments, and animal tissues; and species abundance for the seep area. The data from the oil-impacted area were then compared with data from similarly populated nearby control areas. The researchers found that, for most species studied, the chronic exposure to the oil seeps did not affect the community structure and other possible indicators of chronic toxicity in any significant way. Moreover, there was no evidence of abnormal growths. The second study by the University of Southern California at Coal Oil Point examined in depth the interactions of the seep oil and the community structure of the mussel, Mytilus californianus, which dominates most of the lower rocky intertidal areas of the west coast of North America. Adjacent unoiled areas were studied as controls. The study showed that there were community differences attributable to petroleum exposure. Other natural variables -- amount of sediment trapped in the community, and intertidal height -- proved to be more important than petroleum in affecting mussel community structure and species distributions.

Pelagic Oil -- Bermuda Biological Station for Research

Another natural ocean laboratory is the Bermuda shoreline which is heavily impacted by floating tar balls. They originate primarily from tanker washings and bilge discharges from ships in the mid-Atlantic, and reach Bermuda through the South Atlantic circulation and the so called gyre. The research group reported (Refs. 17,18,19) that the most abundant life in the intertidal zone is concentrated below the splash zone where most of the tar accumulated. Snails living in the splash zone had the highest hydrocarbon levels in their tissues of all animals analyzed, while the animals from immediately adjacent tidal pools and from the intertidal zone did not contain any petroleum hydrocarbons. In the intertidal zone it was reported that, despite the presence of tar on the rocks, neither the abundance, size distribution, or reproductive capability were adversely affected.

Marshland Exposure -- Virginia Institute of Marine Sciences

Of all types of areas that are subject to damage from oil spills, marshlands are perhaps the most sensitive and important, because they serve as the breeding grounds for many marine species whose early life forms have been demonstrated to be the most sensitive of the life cycle. Thus an API investigation of the recovery processes of an oiled marshland was undertaken.

Through the cooperation of the U.S. Navy, a tidal salt water marsh area was made available to the Virginia Institute of Marine Sciences, API's research contractor (Refs. 20,21,22). Five rectangular pens were constructed in the marsh and partially into the adjacent creek. Each pen extended 100 feet along the creek and 200 feet deep into the marshland. The pen wall prevented surface flow back and forth between the creek and the marsh but allowed tidal water flow underneath the pen wall where it extended into the creek. Thus, oil dumped into the pens was retained while tidal water flow was permitted.

The experimental plan involved setting aside one pen as a control while dumping three barrels of South Louisiana crude oil into each of the other four pens. Two of these latter pens were treated with fresh oil while the other two pens were treated with artificially weathered oil. Over the course of the next four years the marshland pens were systematically sampled on a seasonal basis. The ecosystems in each pen were evaluated primarily in terms of changes with time of population densities of benthic organisms, periphyton, phytoplankton, key bacterial species such as petroleum and chitin degraders, and the marsh grasses.

Decreased marsh grass production and declines in population of benthic animals, phytoplankton, and fish were the initial observations of the effects of the experimental spills. Periphyton increased, as did petroleum and chitin degrading bacteria. Some of the initial effects of the weathered oil were more severe than those of the unweathered oil. Also, mortalities of fish held in live boxes occurred only in the weathered oil pens. Gradual recovery of both marsh grasses and of animal population levels has taken place, although a precise statement of how long recovery took is not possible.

West Falmouth (Buzzards Bay) Oil Spill of 1969 -- Marine Biological Laboratory at Woods Hole, Massachusetts

The spill of about 4,000 barrels of No. 2 fuel oil in 1969 in Buzzards Bay near West Falmouth, Massachusetts, was perhaps the worst in U.S. history in terms of severity of biological effects. Although the spill was small, the extraordinarily severe results were due to the combination of three unfavorable circumstances: (1) the spilled oil was a No. 2 distillate fuel (much higher in aromatics content than crudes and hence about the most toxic petroleum product); (2) the spill impacted narrow enclosed waters where the oil volume was significant with respect to the size of the affected area; and (3) storms entrained the oil into the sediments as the water was shallow.

The first extensive studies of the West Falmouth oil spill were made by the Woods Hole Oceanographic Institution (Ref. 23). Realizing the long term nature of the effects of this spill, the API engaged the services of the

Marine Biological Laboratory at Woods Hole, Massachusetts, to perform an in-depth study of the recovery of the heavily oiled area over a period of several years. Samples of sediments and their benthic populations were studied through 1975, the sixth year after the spill (Ref. 24). Progress of recovery of most of the offshore sampling sites was indicated by similarity of population densities to those of unoiled control areas, although numbers of species were slightly but significantly lower at the offshore stations as compared to the controls. A boat basin was still heavily affected. Also, adjacent marshland still had fewer species and lower population densities than a control marsh. Finally, the researchers concluded that recovery is continuing.

LABORATORY AND FIELD STUDIES OF EFFECTS

Effects of Oilfield Brine Effluents on Benthic Life and of Drilling Muds on Corals -- Texas A & M University

Discharges from oil field platforms are of two general types. From producing platforms the characteristic effluent is a brine containing small (parts per million) amounts of oil. The brine is produced along with the oil; they are separated on the platform before the oil is sent ashore via pipeline and the brine is discharged to the ocean. This is a continuous and long-term low level source of oil to the environment. An exploration or well-drilling operation, by contrast, has short duration and occasional overboard discharges of used drilling muds that may contain minor amounts of entrained oil.

In the case of the brine discharges, Texas A & M University conducted a survey of effects on the benthic community adjacent to an oil-producing platform in the Gulf of Mexico (Ref. 25). The oil separator platform from which the discharge occurred was located in shallow turbid water (8 feet in depth) and the entire study area was also quite shallow (6-9 feet in depth). Thus, this study represented a "worst case" from the point of view of oil concentrations that could reach the benthic life. Three transects were run surveying the benthic animal populations starting from the platform and extending radially outward for 2½ to 3½ miles. As might be expected, hydrocarbon concentrations in the sediments within 50 feet of the outfall were four times higher than in the discharge and, correspondingly, the bottom was almost completely devoid of organisms. Levels of benthic animals up to 500 feet away were still severely depressed. Samples taken 1,500 feet away from the platform revealed normal levels of population and species of bottom dwellers. It can be concluded from this worst case study that discharges of oilfield brines can depress benthic populations in the immediate vicinity of producing platforms. Obviously, in deeper waters, due to the dilution and dispersion of the oil entrained in the brine discharge, concentrations of petroleum hydrocarbons will be much lower and biological effects should be diminished.

Drilling muds discharged from platforms from which wells are being drilled could impact sensitive species, although such muds disperse and spread very rapidly in the water column, and are intrinsically low in toxicity. Coral reefs pose a special case because they can grow close to the ocean surface and are sensitive to various forms of pollution. Thus, API asked Texas A & M University to investigate effects of drilling mud discharges on coral reefs (Refs. 26,27). The researchers discovered an interesting phenomenon by means of time lapse photography. When a coral reef was coated with whole, undiluted drilling mud, the coral would periodically "pulse" to produce a cleansing action that shook the mud loose. In the laboratory it was discovered that concentrated muds exerted sufficient toxicity to prevent the corals from cleansing themselves. Also in the lab studies, fish that characteristically fed on a mucus exuded by corals could take up minor amounts of methylnaphthalenes from the mucus of corals that had previously been exposed to oil slicks floating above them.

Hyperplastic Disease of Soft-Shell Clams -- University of Rhode Island

Starting in 1971 in the United States (and subsequently elsewhere in the world) reports appeared of hyperplastic, or possibly neoplastic, disorders in the soft-shell clam, Mya arenaria. A U.S. report in 1971 referred to a disease of the gonadal tissues of the soft-shell clam at Searsport, Maine, where the clam bed had been contaminated by a chronic discharge of jet fuel (JP-5) and No. 2 fuel oil. The immediate inference was that petroleum hydrocarbons had caused the disease. In order to investigate the abnormal growths, API sponsored a study at the University of Rhode Island (Refs. 28,29,30). The study involved an interdisciplinary team of biologists, histopathologists, analytical chemists, and biostatisticians. A virologist joined the team later. The first effort was to survey the incidence of the clam hyperplasia along the east coast of the United States. The disease showed up in very many areas. Surprisingly, some of the locations apparently had never experienced oil pollution, or had been polluted by municipal discharges, pesticides, and heavy metals as well as by oil. The clear implication at this stage was that, at worst, oil could only be a secondary factor in the etiology of the disease. Furthermore, some enzyme studies of the diseased clams indicated that a viral factor was implicated in causing the hyperplasia. Thus, a cross infection biology experiment was devised. Two groups of healthy clams were exposed to waters flowing through a separate head tank for each group. One head tank contained diseased clams; the other head tank contained healthy clams. The unequivocal result of the experiment was that only clams in the waters passing through the diseased clam header tank acquired the disease. Because of the importance of this result, the cross infection experiment is now being repeated. Also underway is an intensive series of virological experiments to identify a specific virus which is now presumed to be the cause of the disease. It is thus now apparent that the blood hyperplastic (or neoplastic) disease of the soft-shell clam, Mya arenaria, is definitely not caused by petroleum hydrocarbons.

Polynuclear Aromatic Hydrocarbons in Marine Animals -- Exxon Research and Engineering Company

As noted above in the discussion of Analytical Method Development (Refs. 5,6), Exxon Research and Engineering Company, acting as API's research contractor, has developed an accurate and extremely sensitive method to determine individual polynuclear aromatic hydrocarbons (PNA's) in the tissues of marine animals. Levels at the parts per billion level of specific PNA's in the tissues of shellfish and finfish were measured. Both contaminated and uncontaminated areas were studied (Refs. 7,31). Concerns have been expressed that PNA's present in the tissues of marine animals might constitute a threat to human health. Thus an understanding of the levels of PNA's present in typical foods containing significant amounts of such compounds is important.

The study showed that most of the PNA's generally found in shellfish and finfish tissues were not derived from petroleum. Instead, the PNA analytical techniques developed during this project proved that these PNA's were produced from combustion or other thermal sources (e.g., forest fires). Comparison of the levels of PNA's in marine animal tissues with PNA levels reported in the literature for common foods such as smoked meats, lettuce, potatoes, and mushrooms showed that the latter common food PNA levels exceeded the levels in the seafood.

SUMMARY

Some of the more important conclusions that have been developed or re-substantiated by API-sponsored research are:

1. Crude oils generally are less toxic in the marine environment than is No. 2 (home heating) fuel oil. Heavy fuel oils such as No. 6 (Bunker C) are much less harmful than No. 2.
2. While marine animals are capable of taking up petroleum hydrocarbons in their tissues, they rapidly purge themselves once placed in clean water.
3. Because of the above mechanism of uptake and depuration of petroleum hydrocarbons, there is no food chain threat to man of toxicity from consuming petroleum hydrocarbons in his food.
4. Observed levels of polynuclear aromatic hydrocarbons (PNA's) in seafood are much less than typical levels in foods such as smoked meats, lettuce, potatoes, and mushrooms. In any event, PNA's in shellfish and finfish do not originate from petroleum sources, but rather from combustion processes.
5. A widely prevalent neoplastic disease of the soft shell clam, although originally observed in areas contaminated by petroleum hydrocarbons, is not caused by petroleum but instead appears to have a viral etiology.
6. Used drilling muds for the most part are low in toxicity to a wide variety of marine test animals. Only in the immediate vicinity of a platform discharge pipe have there been any observable effects of used drilling mud discharges.
7. The planktonic early life stages of American lobster show only very minor effects from floating oil at concentrations up to 0.1 parts per million.
8. Low-level continuous exposure of many marine species to seep oil from the ocean bottom was not harmful.
9. Levels of total hydrocarbons in the water column beneath an experimental oil slick dissipated in less than 4 hours, indicating that the threat to marine life in the water column beneath oil spills should be of short duration.
10. Dispersants effectively break up spills of light gravity crudes and are moderately effective on a medium gravity crude oil.
11. Action of dispersants greatly increases the percentage of relatively insoluble saturated hydrocarbons in the water column, and toxicity of dispersed crude oil to a shrimp species tends to be less than that of the same oil in undispersed form.
12. After even the most severe oil spills, gradual recovery occurs, although it may take many years.

REFERENCES*

1. R. J. Pancirov, Compositional Data on API Reference Oils Used in Biological Studies: A #2 Fuel Oil, A Bunker C, Kuwait Crude Oil and South Louisiana Crude Oil, (API, 1974).

2. J. S. Warner, Chemical Characterization of Marine Samples, (API Publication No. 4307, December, 1978).

3. J. S. Warner, "Determination of Aliphatic and Aromatic Hydrocarbons in Marine Organisms", Analytical Chemistry, 48, (1976), 578-83.

4. J. S. Warner, "Determination of Sulfur - Containing Petroleum Components in Marine Samples", Proceedings, Conference on Prevention and Control of Oil Pollution, API Publication No. 4245, (1975), 97-101.

5. R. J. Pancirov and R. A. Brown, "Analytical Method for the Measurement of Polynuclear Aromatic Hydrocarbons in Marine Tissue", Preprints, Division of Petroleum Chemistry, American Chemical Society, 20(4)(1975), 812-823.

6. R. J. Pancirov and R. A. Brown, "Analytical Methods for Polynuclear Aromatic Hydrocarbons in Crude Oils", Proceedings, Conference on Prevention and Control of Oil Pollution, API Publication No. 4254, (1975), 103-113.

7. R. J. Pancirov and R. A. Brown, "Polynuclear Aromatic Hydrocarbons in Marine Tissues", Environmental Science and Technology, 11(10), (1977), 989-992.

8. R. L. Kolpack, Fate of Oil in a Water Environment -- Phase II: A Dynamic Model of the Mass Balance for Released Oil, (API Publication No. 4313, 1977).

9. JBF Scientific Corp., Physical and Chemical Behavior of Crude Oil Slicks on the Ocean, (API Publication No. 4210, April, 1976).

10. C. D. McAuliffe, G. P. Canevari, T. D. Searl, J. C. Johnson, and S. H. Greene, "The Dispersion and Weathering of Chemically Treated Crude Oils on the Sea Surface", Proceedings, Petroleum and the Marine Environment, Association Europeene Oceanique, Monaco, (1980).

11. B. E. Vaughan, Editor, Effects of Oil and Chemically Dispersed Oil on Selected Marine Biota - Laboratory Study, (API Publication No. 4191, November, 1973).

12. J. W. Anderson, Editor, Laboratory Studies on the Effects of Oil on Marine Organisms: An Overview, (API Publication No. 4249, March, 1975).

13. J. M. Forns, "The Effects of Crude Oil on Larvae of Lobster Homarus americanus", Proceedings, Oil Spill Conference - Prevention, Behavior, Control, Cleanup, API Publication No. 4284, (1977), 569-73.

14. J. W. Anderson, S. L. Kiesser, R. M. Bean, R. G. Riley, and B. L. Thomas, "Acute and Chronic Effects of Oil and Oil-Dispersant Mixtures on Pandalus danae", to be published in Helgolander Wissenschaftliche Meeresuntersuchungen (1980).

15. D. Straughan, Sublethal Effects of Natural Chronic Exposure to Petroleum in the Marine Environment, (API Publication No. 4280, October, 1976).

16. D. Straughan, Analysis of Mussel (Mytilus californianus) Communities in Areas Chronically Exposed to Natural Oil Seepage, (API report to be published, 1980).

17. N. G. Maynard, "The Effects of Pelagic Hydrocarbons on the Rocky Intertidal Flora and Fauna of Bermuda", Proceedings, Oil Spill Conference - Prevention, Behavior, Control, Cleanup, API Publication No. 4284, (1977), 499-503.

18. A. Zsolnay, "Lack of Correlation between Gas-Liquid Chromatograph and UV Absorption Indicators of Petroleum Pollution in Organisms", Water, Air and Soil Pollution 9, (1978) 45-51.

19. A. Zsolnay, "The Weathering of Tar on Bermuda", Deep Sea Research 25, (1978) 1245-52.

20. M. E. Bender, E. A. Shearls, R. P. Ayers, C. H. Hershner, and R. J. Huggett, "Ecological Effects of Experimental Oil Spills on Eastern Coastal Plain Estuarine Ecosystems," Proceedings, Oil Spill Conference - Prevention, Behavior, Control, Cleanup, API Publication No. 4284, (1977), 505-9.

21. R. H. Bieri, V. C. Stamoudis, and M. K. Cueman, "Chemical Investigations of Two Experimental Oil Spills in an Estuarine Ecosystem", Proceedings, Oil Spill Conference - Prevention, Behavior, Control, Cleanup, API Publication No. 4284, (1977), 511-15.

22. H. Kator and R. Herwig, "Microbial Responses after Two Experimental Oil Spills in an Eastern Coastal Plain Estuarine Ecosystems", Proceedings, Oil Spill Conference - Prevention, Behavior, Control, Cleanup, API Publication No. 4284, (1977) 517-22.

23. M. Blumer, J. Sass, H. L. Sanders, J. F. Grassele, and G. R. Hampson, "The West Falmouth Oil Spill", Woods Hole Oceanographic Institution, Reference No. 70-44, (1970).

24. A. D. Michael, C. R. Van Raalte, and L. S. Brown, "Long-Term Effects of an Oil Spill at West Falmouth, Massachusetts", Proceedings, Conference on Prevention and Control of Oil Pollution, API Publication No. 4245, (1975), 573-82.

25. H. W. Armstrong, K. Fucik, J. W. Anderson, and J. M. Neff, Effects of Oilfield Brine Effluent on Benthic Organisms in Trinity Bay, Texas, (API Publication No. 4291, December, 1977).

26. J. H. Thompson and T. J. Bright, "Effects of Drill Mud on Sediment Clear-
    ing Rates of Certain Hermatypic Corals", Proceedings, Oil Spill Confer-
    ence -Prevention, Behavior, Control, Cleanup, API Publication No. 4284,
    (1977), 495-98.

27. J. M. Neff, Sublethal Effects of Petroleum on Reef Corals, (API Report to
    be published, 1980).

28. R. S. Brown, R. E. Wolke, S. B. Saila, C. W. Brown, "Prevalence of Neo-
    plasia in 10 New England Populations of the Soft-Shell Clam (Mya
    arenaria),   Aquatic Pollutants and Biologic Effects with Emphasis on
    Neoplasia, Annals of the New York Academy of Sciences,   298,   (1977),
    522-34.

29. R. S. Brown, "A Disease Survey of New England Softshell Clams Mya aren-
    aria" Proceedings National Shellfish Association 68 (1978) 75.

30. R. S. Brown, R. Appledoorn, C. W. Brown, and S. Saila, "Bioassay System
    to Determine the Cause of a Neoplastic Disease in New England Soft Shell
    Clams", 4th American Society for Testing and Materials Symposium on Aqua-
    tic Toxicology, ASTM to publish (1980).

31. R. A. Brown and P. K. Starnes, "Hydrocarbons in the Water and Sediment of
    Wilderness Lake, II", Marine Pollution Bulletin, 9 (1978), 162-5.

32. J. M. Neff, W. L. McCulloch, R. S. Carr, and K. A. Retzer, "Comparative
    Toxicity of Four Used Offshore Drilling Muds to Several Species of Marine
    Animals from the Gulf of Mexico", Proceedings of Symposium - Research
    on Environmental Fate and Effects of Drilling Fluids and Cuttings, to be
    published by Courtesy Associates, 1629 K Street, N.W., Suite 700,
    Washington, D.C. 20006 (1980).

33. R. S. Carr, L. A. Reitsema, and J. M. Neff, "Influence of a Used Chrome
    Lignosulfonate Drilling Mud on the Survival, Respiration, Growth, and
    Feeding Activity of the Opossum Shrimp, Mysidopsis almyra," ibid.

34. W. L. McCulloch, J. M. Neff, and R. S. Carr, "Bioavailability of Heavy
    Metals from Used Offshore Drilling Muds to the Clam Rangea cuneata and
    the Oyster Crassostrea gigas", ibid.

35. E. S. Gilfillan, R. P. Gerber, S. A. Hanson, D. S. Page, and J. B.
    Hotham, "Effects of Used Drilling Muds on Recruitment to Soft Bottom
    Benthic Communities", ibid.

36. R. P. Gerber, E. S. Gilfillan, B. T. Page, D. S. Page, and J. B. Hotham,
    "Short and Long Term Effects of Used Drilling Fluids on Marine Organ-
    isms", ibid.

37. D. S. Page, E. S. Gilfillan, R. P. Gerber, B. T. Page, and J. B. Hotham, "Bioavailability of Toxic Constituents of Used Drilling Muds", ibid.

---

*   Most publications of the American Petroleum Institute (API) cited here are obtainable from that organization's Publications and Distribution Section, 2101 L Street Northwest, Washington, D.C. 20037. Available publications are listed in the annual catalog issued by this Section.

SELECTIVE OIL SPILL-COMBAT PLANNING FOR OFFSHORE OPERATIONS

J. Ph. Poley and D. Callaghan
Shell Int. Petroleum Mij., The Hague

1. INTRODUCTION

The Ekofisk-Bravo and the IXTOC-1 blow-outs, and the coastal pollution following the wrecking of the Amoco Cadiz have not only stressed the need for timely organisation of combat-measures, but also highlighted the dramatic difference in pollution consequences between a large oilwell blow-out on the high seas, and a large tanker spill on a rocky coast.

An offshore blow-out provides for space and time, for nature to take its toll of the spilled oil and for man to mount combative action. Optimum use can be made of the available space and time through advance diagnosis of specific coastal pollution hazards. Advance consensus on measures to be taken, both near the source and to protect the threatened coast, should be reached between government, industry and other interested parties. Such an advance scenario will minimize time-delays, problems and expenses during an offshore oilwell blow-out (See Ref. 1).

An estimate of the coastal pollution risk from an offshore well blow-out will have to take into account: the location of the specific source, the composition of the oil, the natural fate of that oil at sea, and the expected movement of the slick.

2. THE LOCATION OF THE SPECIFIC OIL SPILL SOURCE

The location of the potential source (its distance from coasts, its seasonal local weather, wind, and current patterns) largely determines the time scale and the regulatory cadre within which combat action has to be fitted. For blow-outs from, for instance, fields in the central North Sea area, a minimum of one to two weeks generally will be available for combat reaction along endangered coast-lines, while an international agreement on near-source measures (such as the use of mechanical clean-up equipment, or use of dispersants) will have to be reached well in advance. On the other hand, blow-outs from near-shore fields such as Beatrice (off Scotland) or Castellon (off Spain) most probably will remain a national problem, but will require scenarios adjusted to quite different time schedules. Advance knowledge of the (minimum) time span available will promote a well-timed action scenario, and will - which is as important - prevent rash actions and unnecessary expenses. Such knowledge would certainly have saved a major portion of the expenses (estimated at more than U.S.$2 million) made for emergency mobilisation and preparation of coast-line protection in Denmark and Norway following the Ekofisk blow-out.

3. OIL COMPOSITION AND THE FATE OF OIL AT SEA

The composition of the oil blowing out of the well and its temperature will set the scene for the physical phenomena determining the natural fate of the oil under the prevailing weather and sea state.

a.   Specific density

In the area near the source where mechanical or chemical combat measures
can be used, the gravity differential between the oil and the sea water
will, in practice, govern the rate at which the slick expands. Higher
gravities will involve slower spreading, and thus most times offer more
possibilities for effective near-source combat. The (flash) spreading of
sheens must be considered unimportant for slick-combat purposes. Ambient
water temperature is another factor governing slick-expansion. At lower
temperatures (North Sea) expansion is slower (than for instance in the
Mediterranean), offering more combat opportunities.

b.   Evaporation, dissolution and weathering

Oil gushing out and landing at the surface is immediately subject to evaporation. The amount and rate of evaporation essentially depends on the
percentage light (volatile) components. Fig. 1 shows the evaporation rates
of several crude-oil constituents, the difference between lighter and heavier
ends is obvious.

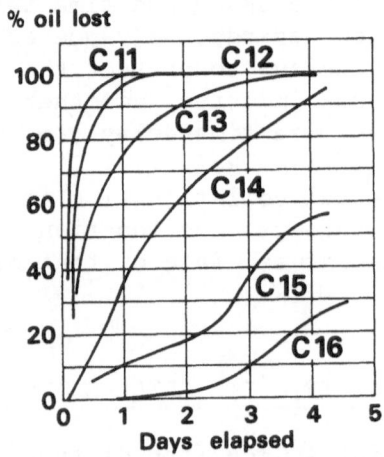

FIG. 1 RATES OF OIL EVAPORATION (REF.3)

     Data from the Ekofisk blow-out (1977) indicated that after a few days
total evaporation amounted to some 65%, for instance (Ref. 2). The degree
of dissolution of a crude also depends on these differences, the lighter
components dissolving better than the heavier ones.
     Sea state, evaporation, etc. make the oil weather rather quickly at
the sea surface. This manifests itself in a rapid disappearance of lighter
components, while a quick increase in density and viscosity results.
Composition and properties of weathered crude, after, say, two days after
landing, are essentially different from those of the original crude: those
components that are difficult to attack, physically or chemically, remain.
How much weathered crude will remain, depends on the original percentage
of heavy components (gas-oil and "residue"). Through evaporation of the
volatile components, weathering forms a natural protection of marine organisms
against exposure to potentially damaging crude-components.

c.   Emulsification, natural dispersion

Under the conflicting influences of wind and current the breaking waves will at larger distances from the source (beyond, say, one mile) break the slick up into wind-rows, slick-fingers and -patches. An emulsion of water-in-oil is formed, which often is very viscous and can contain up to 75% water. There are indications that the pour point/waxiness/naphtene contents of the oil is related to the stability of such w/o emulsions. Weathered crudes with a.o. higher pour point/higher wax content seem to produce stabler emulsions ("chocolate mousse"). The ambient temperature is also very important: emulsions would form only at temperatures lower than the pour point of the (weathered) crude.

d.   Degradation

Real degradation of the crude takes place through bio-oxidation to the water mass, and through photo-oxidation to the atmosphere.

Bio-degradation is the main factor in the long-term disappearance of the spilled crude. Bacteria that take care of such degradation are available anywhere at sea. They use the crude oil as an energy source for their activity. In addition, they need much oxygen (greatest activity therefore in the turbulent upper part of the water mass), and other nutrients like phosphor, nitrogen and potassium.

As such bio-degradation only takes place at the oil-water interface, it can be a slow process. Increase of the oil-water interface (by spreading or by smaller droplets) will therefore increase the efficiency of the process considerably. Thin surface films appear to be completely degraded after 2-3 months. Dependent on the type, bacteria feed on paraffins, on aromatic compounds or on naphtenes. Higher temperature leads to higher activity. Under aerobic conditions, per day 0.5 grams oil per m^2 o/w interface can be oxidised by bacteria. Biological research is being carried out on oxidation products, to assess whether any products with detrimental effect on the marine ecosystem are created. As there are indications that the lighter cyclic compounds (f.i. aromatics) would be the oil components that could be most harmful to marine life, knowledge of their relative abundance in a specific crude is most important.

Surface oil is also broken down by photo-oxidation. This, like bio-degradation, is again a slow process. Its effectiveness depends on the oil composition: vanadium components will act as catalyst, while a high sulphur contents will diminish oxidation. The oxidation products (acids, alcohols, carbonyl-compounds, or metal-salts) are usually soluble in water and dilute away naturally. Sometimes compact waxy polymers form at the sea surface. Because this process is active at the sea surface, it is often rather in-effective. Non-toxic chemicals that promote spreading at the surface will be quite useful in this respect.

*   *

From the above it is obvious that the following characteristics of crude oil will play an important role in oil spill combat scenarios:

a)   specific density: in slick expansion (near-source combat)
b)   ratio of light vs heavy components: in evaporation and weathering (mass balance)
c)   pour point/waxiness/naphtenes content of weathered crude: in stability of w/o emulsions
d)   relative wt % paraffins/naphtenes/aromatics: toxicity.

We have given in Table 1 a review of these characteristics in crude oils from several producing fields. From this review the fate of a specific oil at sea can be estimated.

Table 1   Examples of estimated crude oil characteristics

	AUX (U.K.)	DAN (DENMARK)	EKOFISK (NORWAY)	CASTELLON (SPAIN)	CHAMPION (BRUNEI)	KUWAIT
CRUDE PROPERTIES						
SPEC. GRAVITY 15 /4°C	0.837	0.877	0.847	0.849	0.913	0.870
GRAVITY API, 60°F	37.5	29.8	35.6	35.0	23.4	31.1
VISCOSITY CS	3.65	23.6	4.25		23.4	7.6
POUR POINT °C	-10	-30	-9	-12	-40	-20
WAX CONTENT(% WT)	7	2	6.5	9(est)		5.5
COMPOSITION (% WT)						
TOPS (5-100°C)	9.5	5.7	7	4.2	0.7	8.3
% WT P/N/A (est)	65/32/3	75/20/5	75/19/6	65/33/2	40/51/9	85/12/3
NAPHTA (100-160°C)	10.6	7.7	11.1	9.0	4.0	9.1
% WT P/N/A (est)	50/35/15	25/70/5	47/35/18	48/46/6	25/65/10	63/23/14
KEROSINE (160-250°C)	16.6	17.5	15.1	21.6	20.8	12.4
% WT P/N/A (est)	30/50/20		/ /18	40/50/10	60/20/20	53/26/21
GAS OIL (250-350°C)	18.8	19.7	19	23.1	40.4	16
% WT P/N/A (est)	/ /	/ /	/21	/ /		/ /32
POURPOINT °C (est)	-9	-30	-9	-6	-30	-12
RESIDUE (350°C)	44.5	48.9	47.8	41.2	34.1	54.2
% WT WAX	13	4	11	16		21
POURPOINT °C (est)	21	6	25	33	9	15

4.   ESTIMATING THE SLICK-VOLUME DECAY

For a realistic estimate of the amount of crude landing in the sea, dependable estimates have to be available of both the (maximum) flow rate and the gas/oil ratio of the production from a specific well. Such an oil-flow volume estimate is necessary first of all to quantitatively assess the potential risks of pollution of specific coast-lines in the case of a blow-out of prolonged duration. It is also important for a balanced judgement on the necessity, nature, and size of any near-source combat measures. The relative contribution of, for instance, the mechanism of spreading to the total slick dispersion will depend on the rate at which oil is spilled.

The Ekofisk blow-out volume estimates came down from an initial 22,500 tons of oil and 1.7×10^6 m^3 of gas, to some 10,000 tons of oil and 0.5×10^6 m3 of gas, illustrating the importance of realistic estimates at an early stage.

The actual conditions of wind, sea state, and current will largely decide the movement and fate of the oil after it is in the water. Advance knowledge of the (seasonal) current patterns is most important to be able to estimate their impact on the slick's track. The circulatory current pattern together with the prevailing winds, kept the Ekofisk oil patches within the same broad central North Sea in May/June 1977, whereas available current and wind estimates indicate that in different seasons slick tracks could have been expected to be more directly headed toward specific coast-lines. Although the main physical factors governing the weathering of spilled oil at sea are reasonably well-known, the relative contribution of such factors as spreading, evaporation, dissolution, emulsification, dispersion, and degradation to the breakdown of an oil slick is only roughly known under fairly ideal circumstances. Under the actual circumstances, the sea state probably will be the dominating factor, a higher sea state near the source promoting, for instance, a much more effective emulsification, dispersion, and dissolution.

An useful scenario for describing the various factors determining the fate of a specific crude oil spill has been developed by the E & P Forum (Ref. 3) see Fig. 2.

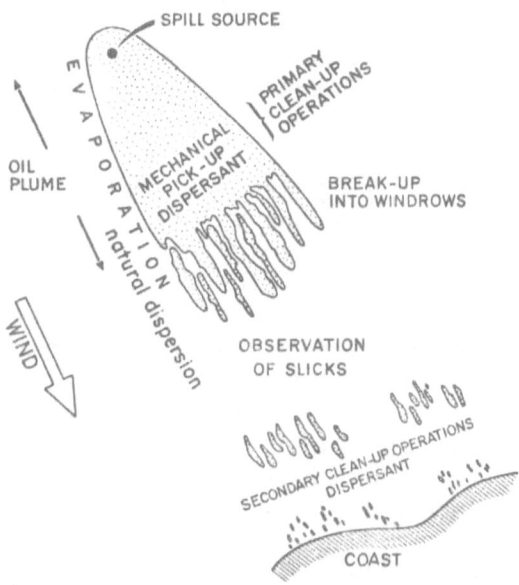

FIG. 2 OILSPILL COMBAT SCENARIO (REF.3)

In the computer program 'SLIKTRAK' these factors are quantified to arrive at estimates of risk/volumes of pollution for specified coastal areas.

The most dependable attempt to arrive at a quantitative description of slick weathering has been made by the IKU, Norway, in their analysis of Ekofisk slick volume observations (Ref. 2). In fitting their breakdown model with oil volume observations they arrive - under the prevailing conditions for Ekofisk oil at the surface - at a characteristic breakdown time (half-life time) of seven to ten days. For the time being, their aggregate accounting for all phenomena that contribute to the weathering of the oil seems to be the best available approach to quantitative estimating of the volume of oil at the sea surface as a function of time elapsed.

In their model, the total breakdown of a slick is described by two main parameters.
The first, λ_e, is the "evaporative" time constant (which could basically be taken to   include also a dissolution contribution) and governs the volume decrease as a function of time due to evaporation (and dissolution). This phenomenon is active from the moment of spillage into the water, and operates

until all volatile and soluble (lighter) compounds have disappeared, and a weathered oil of gas-oil/residue characteristics remains. λ_e will depend on the actual oil composition, the sea state and wind, and the prevailing temperature.

The second parameter, λ_d accounts for the volume decrease due to dispersion, and will basically remain active continually. Its magnitude will depend mainly on the sea state, and somewhat (through the stability of the water-in-oil emulsions) on the composition of the oil. Its magnitude probably will have to be estimated from actual slick case histories.

Adding to their model the efficiencies C_M and C_C of near-source mechanical resp. chemical clean-up, an estimate describing the slick-volume decay would follow from

$$V_t :: (1-C_M)(1-C_C) \left\{ C_1 + (1-C_1)\, e^{-\lambda_e t} \right\} e^{-\lambda_d t}\, V_o$$

where C_1 is a constant, depending on the oil composition, and V_o is the volume of oil, landed at the surface.

In this model, the stability of the emulsions formed and the time allowed at sea will determine the actual chances of survival of an oil patch under the prevailing circumstances. Any means to lengthen the time of stay at sea could pay off quite nicely.

The chances of survival of slick patches at sea depend - as will be clear from the above review - mainly on the volume spilled, the stability of the emulsion (i.e. its half-life-time) under the prevailing (seasonal) sea state, and its residence time at sea. To illustrate this, some results are shown (in Table 2) of pollution risks, estimated arrival times and volumes for several coast-lines during the Ekofisk blow-out, computed with SLIKTRAK. Maximum flowrates at some 200 t/h, together with six years Ekofisk weather history during April and May, and a relatively low efficiency of near-source clean-up were used for the input as the best estimates available at the time of the accident. Fig. 3 is an example of the arrival times and volumes, as computed for the Dutch Waddenzee.

Table 2   Expected shore pollution from Ekofisk-Bravo blow-out

	Chance of any pollution (%)	If pollution, the expected quantities reaching shore (m³)		First oil on shore after (days)
		Total	Possible daily peak	
U.K.	2	55	25	40
Norway	38	310	50	30
Denmark	58	150	35	30
Germany (excl. Wadden Sea)	31	40	5	40
Netherlands excl. Wadden Sea)	18	115	20	55
Wadden Sea				
German Sector	21	150	15	40
Dutch Sector	28	135	10	35

The results turned out to be too pessimistic: while SLIKTRAK indicated a non-negligible pollution risk for several coasts, no reports of beached Ekofisk oil have been received. The data provided in these examples illustrate the usefulness of this approach in putting any shore-protection measures into a realistic perspective. The available input into SLIKTRAK has since then been updated with recent results of investigations, such as those from IKU (Ref. 4). Following international consultations, a.o. with Warren Spring Laboratory, their aggregate breakdown rate (half-life time) of about seven days, at least at North Sea latitudes, appears to be the most reliable value presently available. It is supported, for instance, by observations of the fate of Amoco Cadiz crude at sea, where under prevailing Channel conditions, no oil sightings were reported anymore after 14 days.

5.   COASTAL POLLUTION RISK

For an adequate planning of protective measures a realistic estimate of the pollution threat as provided, for instance, by the above mentioned SLIKTRAK program, is required. To illustrate the potential of such an approach we have carried out some simulation exercises, in the North Sea and in the Mediterranean.

We have assumed a specific blow-out with oil composition and actual flow rates from a field in the central North Sea area, and taken the Dutch Wadden coast area as the specific area threatened. Mechanical clean-up with sea-state dependent efficiencies derived from Ekofisk has been assumed in near-source combat. Evaporation, and the information and stability of water-in-oil emulsion together with its half-life, were estimated based also on Ekofisk evidence (Ref. 2). Using historical weather input for that sea area with the corresponding sea states, the expected risks, arrival times and volumes of Wadden coast area pollution have been computed for several seasons. For this specific simulation the impact risk (of any pollution) turned out to be: for the spring season   abt. 40%, relatively small volumes
        for the summer season   abt. 15%, larger volumes.

Figs. 4 and 5 show the volume vs time results for the spring and summer seasons. The differences in volumes and arrival times demonstrate how important it is to make a detailed quantitative analysis of the threat for a certain area. Although certainly manageable, the quantities involved are not negligible.

The arriving oil-emulsion, if any, will propably be in patches or lumps of a weathered highly-viscous seawater-in-crude emulsion with roughly the following estimated characteristics: water content some 70 percent by volume, crude content some 20 percent by weight. The threatened area chosen (the Dutch Wadden coast) is a well-known area of very high ecological importance. It is not only a protected area for very large numbers of migrating fowl and of fish in many varieties, but also a nursery for a sizeable part of the North Sea standing stock for quite a few species, and a breeding ground for more than 65 percent of the Dutch mussel cultivation. The beaches of the Waddenzee islands are well-known holiday resorts, and therefore carry a fairly high amenity value in the season between April and October.

It must therefore be expected that the risk of arrival of patches of oil emulsion in quantities as indicated will require extensive consultation between the parties responsible for and interested in the protection of their sensitive areas.

Similar exercises are being made for wells located in the Mediterranean coastal waters offshore Spain (Castellon/Amposta) and offshore Tunisia (Gulf of Hammamet). At the time of writing this paper only limited historical weather data was available for these two areas, covering a period from approx. May to August and preliminary results, based on this limited data, are presented in the final three figures. It must be recognised that at this stage, the same credibility cannot be assigned to these findings as to those for the North Sea discussed previously.

The results are not surprising considering the locations and circulatory patterns, and show large pollution risks for the coasts of Spain, Libya and Pantelleria and sizeable risks for the Tunisian shore. The data indicates that an incident off the coast of Spain should remain a national problem, whereas for a similar incident off the Tunisian coast there is a significant chance of other countries being affected. The exercise so far takes no account of secondary combat measures, but it does indicate that for example, in the case of an incident offshore Tunisia there should be considerable time available to implement measures to protect the Libyan coast.
This exercise is being updated as more relevant weather data becomes available and such results should stimulate advance consultation between governments and industry on joint measures to protect the environment.

To provide a general quantitative basis for decisions between offshore industry and governments on adequate measures for near-source combat as well as for later shore-protection, a general programme should provide both a short-term slick prediction (based on real weather/volume input), and a best forecast of the (longer-term) pollution risk for specific coast-lines, with estimates of arrival times and volumes of oil involved.

The Oil Industry E & P FORUM has taken the initiative to provide such a general purpose programme. It is sponsoring a cooperative project at IKU, Trondheim, Norway, which will integrate IKU's "OILSIM", a deterministic slick-prediction programme, and "SLIKTRAK", the above-mentioned probabilistic coastal pollution-risk forecast programme, into a new programme "SLIKFORCAST". For further detail see Ref. 5.

REFERENCES

1. J.Ph. Poley, "Selective Oilspill Combat Planning for offshore E and P operations in the North Sea", Proc. Oilspill Conference (1979), p. 641-647.

2. T. Audunson, "Bravo Blow-out" I.K.U. Trondheim, Norway (1977). Report No. 90.

3. D.R. Blaikley, G.F.L. Dietzel, A.W. Glass and P.J. van Kleef, "SLIKTRAK - a computer simulation of offshore oilspills, effects and associated costs", Proc. Oilspill Conference (1977), p. 45-52.

4. A.W. Glass, B.R. Horton, "SLIKTRAK - a computer simulation of surface pollution problems arising from offshore oilspills" (1979), Proc. 10th World Petr. Congress, Sec. PD 14, p. 1-10.

5. T. Audunson, "SLIKFORCAST - a simulation programme for oilspill emergency tracking and long-term contingency planning" (1980). Paper submitted to PETROMAR 1980.

FIG. 3

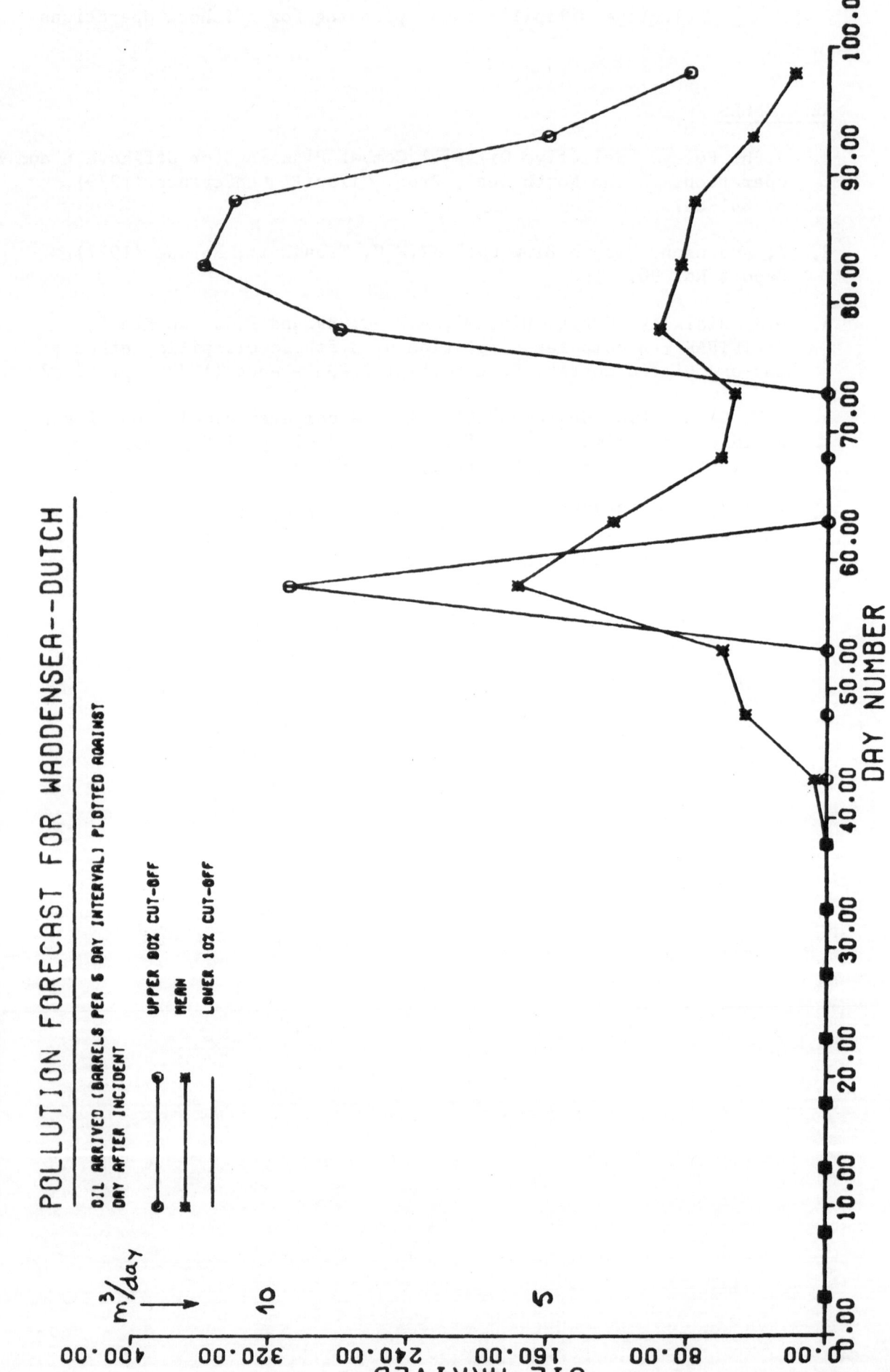

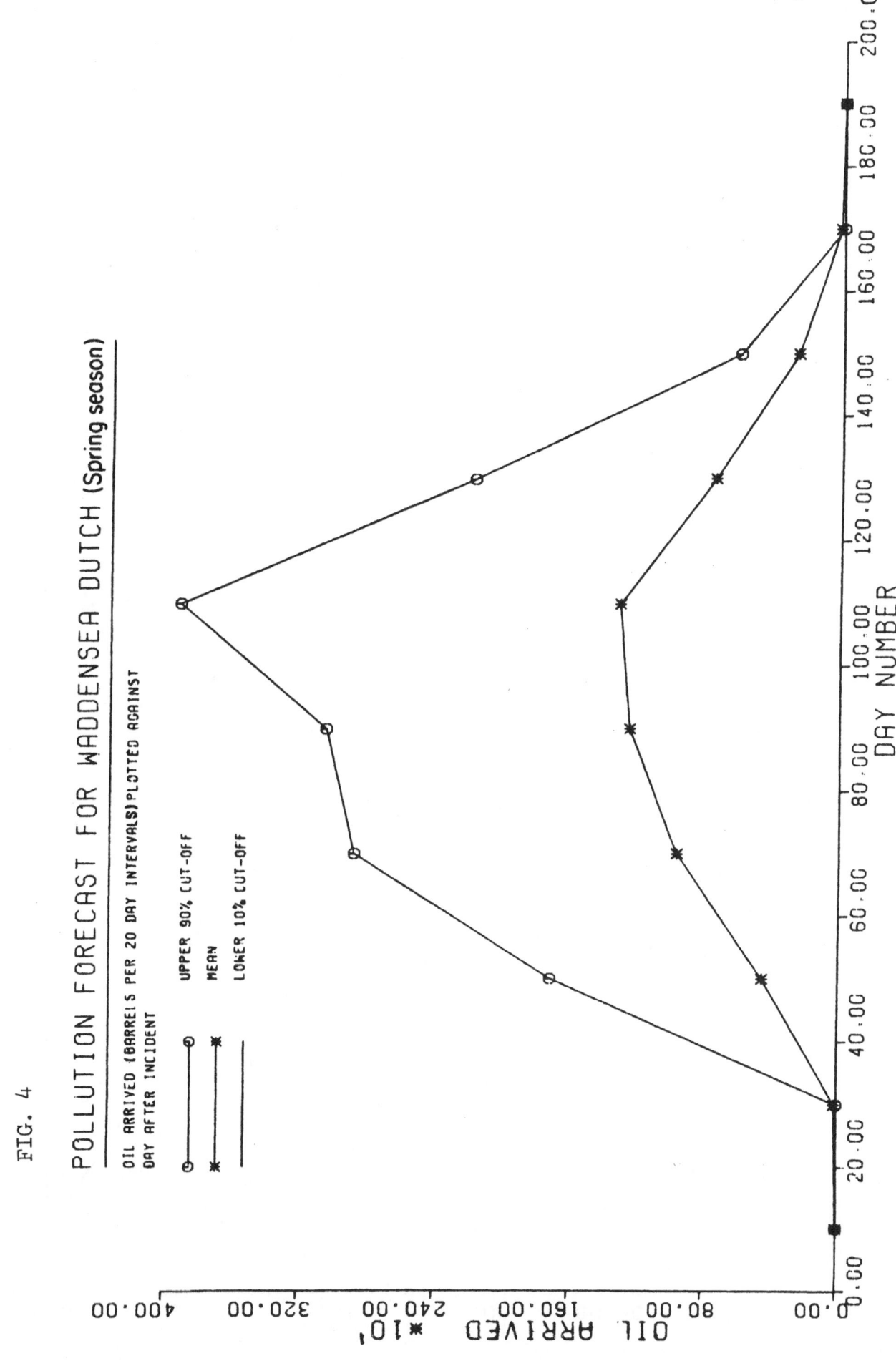

FIG. 4

POLLUTION FORECAST FOR WADDENSEA DUTCH (Spring season)

OIL ARRIVED (BARRELS PER 20 DAY INTERVALS) PLOTTED AGAINST
DAY AFTER INCIDENT

UPPER 90% CUT-OFF

MEAN

LOWER 10% CUT-OFF

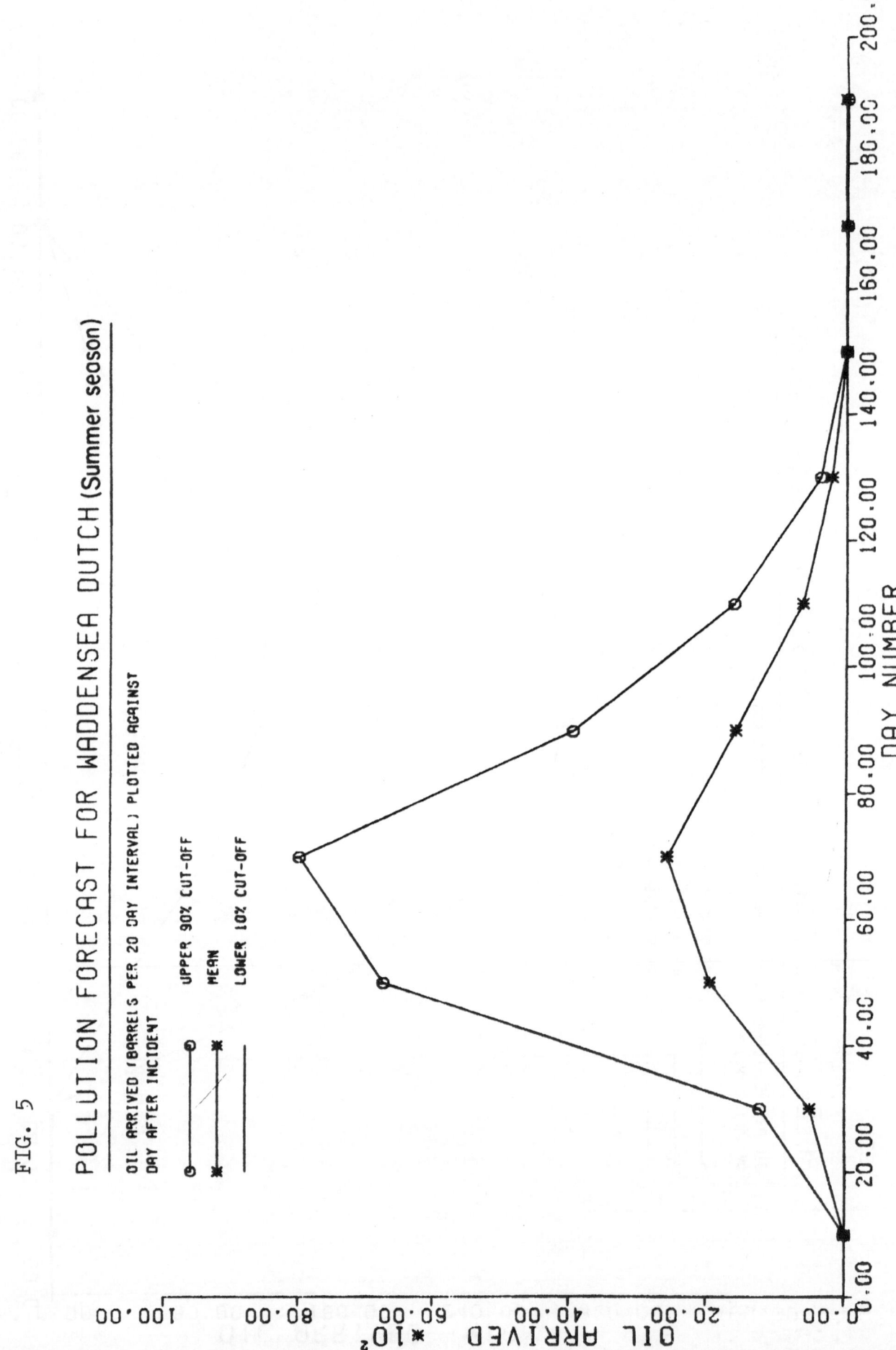

FIG. 5

POLLUTION FORECAST FOR WADDENSEA DUTCH (Summer season)

OIL ARRIVED (BARRELS PER 20 DAY INTERVAL) PLOTTED AGAINST
DAY AFTER INCIDENT

UPPER 90% CUT-OFF
MEAN
LOWER 10% CUT-OFF

FIG. 6

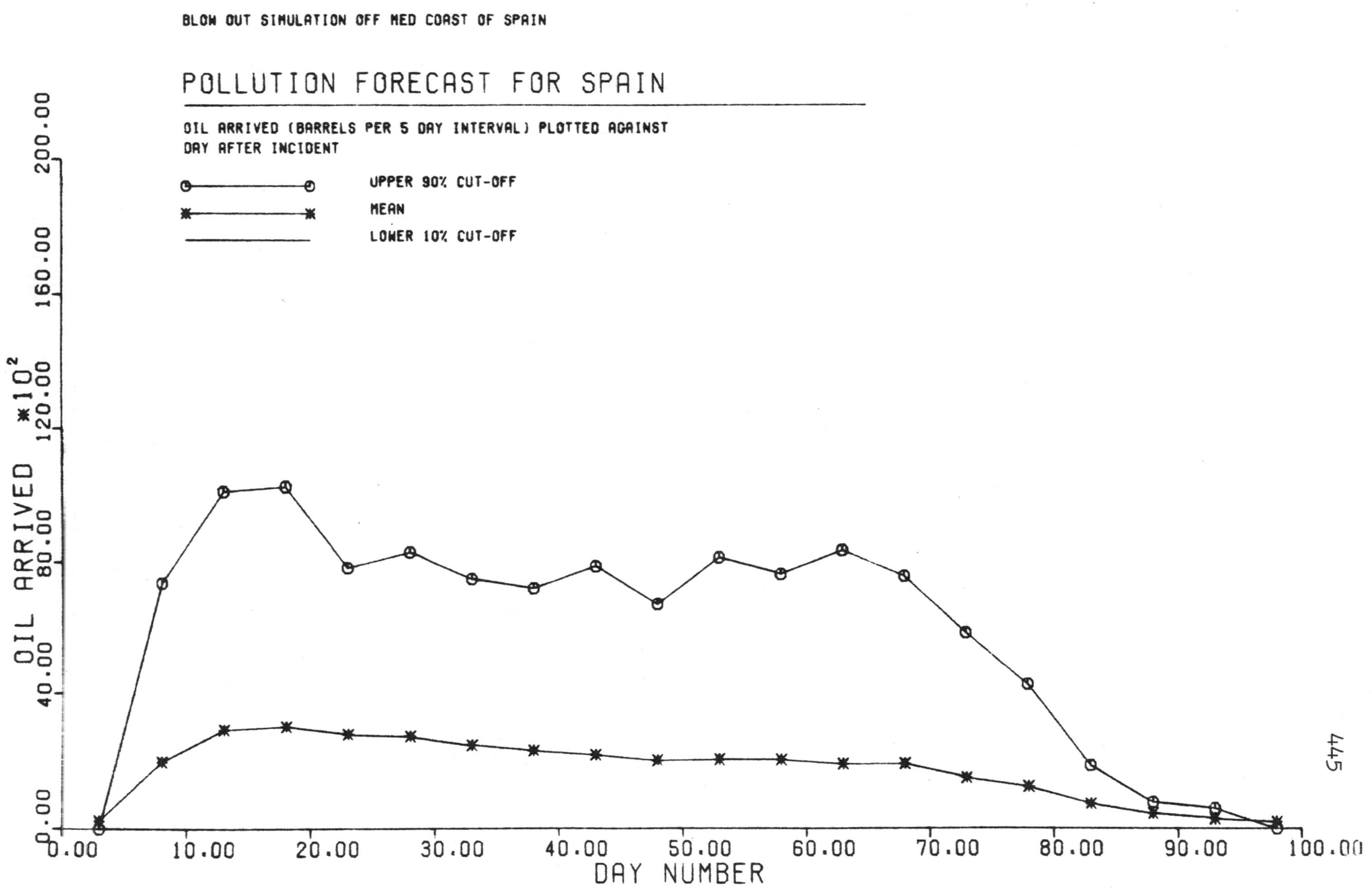
BLOW OUT SIMULATION OFF MED COAST OF SPAIN
POLLUTION FORECAST FOR SPAIN
OIL ARRIVED (BARRELS PER 5 DAY INTERVAL) PLOTTED AGAINST
DAY AFTER INCIDENT
UPPER 90% CUT-OFF
MEAN
LOWER 10% CUT-OFF
OIL ARRIVED *10²
0.00
40.00
80.00
120.00
160.00
200.00
0.00
10.00
20.00
30.00
40.00
50.00
60.00
70.00
80.00
90.00
100.00
DAY NUMBER

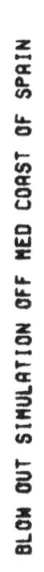

FIG. 7

BLOW OUT SIMULATION OFF TUNISIAN COAST

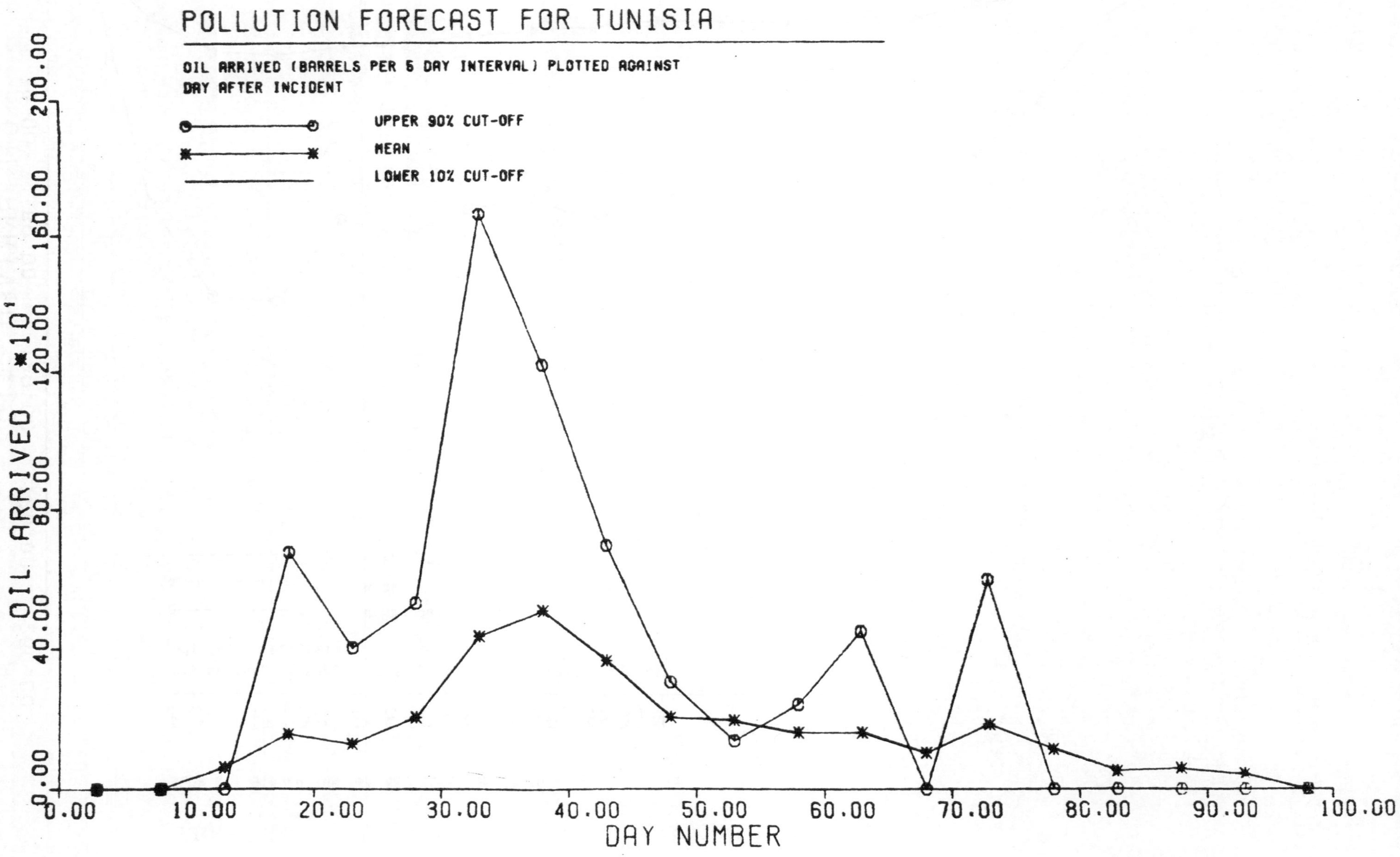
POLLUTION FORECAST FOR TUNISIA
OIL ARRIVED (BARRELS PER 5 DAY INTERVAL) PLOTTED AGAINST
DAY AFTER INCIDENT
UPPER 90% CUT-OFF
MEAN
LOWER 10% CUT-OFF
OIL ARRIVED *10¹
0.00
40.00
80.00
120.00
160.00
200.00
0.00
10.00
20.00
30.00
40.00
50.00
60.00
70.00
80.00
90.00
100.00
DAY NUMBER

FIG. 8

BLOW OUT SIMULATION OFF TUNISIAN COAST

POLLUTION FORECAST FOR LIBYA

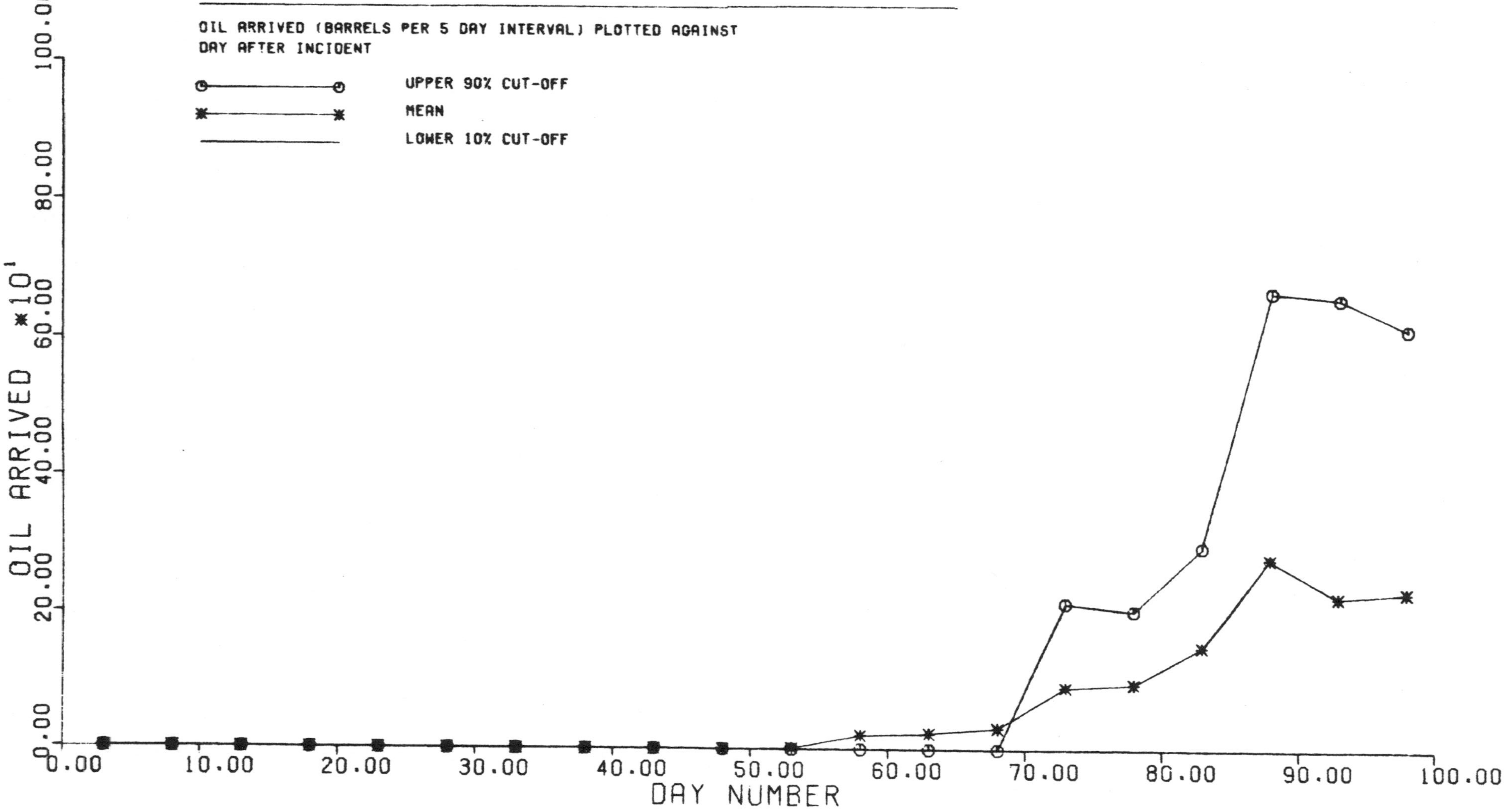
OIL ARRIVED (BARRELS PER 5 DAY INTERVAL) PLOTTED AGAINST
DAY AFTER INCIDENT
UPPER 90% CUT-OFF
MEAN
LOWER 10% CUT-OFF
OIL ARRIVED *10¹
0.00
20.00
40.00
60.00
80.00
100.00
0.00
10.00
20.00
30.00
40.00
50.00
60.00
70.00
80.00
90.00
100.00
DAY NUMBER

UNEP REGIONAL SEAS PROGRAMME

Stjepan Keckes
Director, Regional Seas Programme Activity Centre
United Nations Environment Programme
Geneva, Switzerland

In accordance with resolution 2997 (XXVII) of the United Nations General Assembly, the United Nations Environment Programme (UNEP) was established "as a focal point for environmental action and co-ordination within the United Nations system". The Governing Council of UNEP defined this environmental action as encompassing a comprehensive, transectoral approach to environmental problems which should deal not only with the consequences but also with the causes of environmental degradation.

The UNEP Governing Council has designated "Oceans" as a priority area in which it will focus effort to fulfil its catalytic role. In order to deal with the complexity of the environmental problems of the oceans in an integrated way, it has adopted a regional approach as exemplified by its Regional Seas Programme.

Although the environmental problems of the ocean are global in scope, a regional approach to solving them seemed more realistic. By adopting a regional approach, UNEP felt it could focus on specific problems of high priority to the States of a given region thereby more readily responding to the needs of the Governments and helping to mobilize more fully their own national resources. It was thought that undertaking activities of common interest to coastal States on a regional basis should, in due time, provide the basis for dealing effectively with the environmental problems of the oceans as a whole.

Two elements are fundamental to the Regional Seas Programme:

a) Co-operation with the Governments of the regions. Since any specific regional programme is aimed at benefiting the States of that region, Governments are encouraged to participate from the very beginning in the formulation and acceptance of the programme. After acceptance, the implementation of the adopted programme is carried out by national institutions which have been nominated by their Governments.

b) Co-ordination of the technical work through the United Nations system. Although the regional programmes are implemented predominantly by Government-nominated institutions, a large number of the United Nations specialized organizations are called upon to provide assistance to these national institutions. UNEP acts as an overall co-ordinator although in some cases this role is limited to the initial phase of the activities. Thus the support and experience of the whole United Nations system contributes to the programme.

The components of a regional programme are outlined in an "action plan" which is formally adopted by the Governments before the programme enters an operational phase.

Each action plan consists of three standard components as adopted by the United Nations Conference on Human Environment (Stockholm, 5 - 18 June 1972) and endorsed by subsequent meetings of UNEP's Governing Council. They are:

a) Environmental assessment. The assessment and evaluation of the causes, magnitude and consequences of environmental problems are essential activities providing the basis for assistance to national policy-makers to manage their natural resources in an effective and sustainable manner.

b) Environmental management. A wider range of activities requiring regional co-operation falls under this component: rational exploitation of living resources, utilization of renewable sources of energy, management of freshwater resources, disaster preparedness and co-operation in cases of emergency, etc. Regional conventions, elaborated by specific technical protocols, usually provide the legal framework for the action plan and have in many regions proved to be an excellent tool in the hands of environmental managers.

c) Supporting measures. The national institutions are the institutional basis for the implementation of the action plan. Large-scale technical assistance and training are provided to them where necessary to allow their full participation in the programme. Existing global or regional co-ordinating mechanisms are used when appropriate. However, specific regional mechanisms may be created if Governments feel they are necessary. Public awareness of environmental problems is stimulated as an essential supporting measure for the action plan. Financial support is initially provided by UNEP and other international and regional organizations, but, as the programme develops, it is expected that the Governments of the Region assume increasing financial responsibility.

At present there are eight regional seas areas where action plans are operative or are under development: The Mediterranean (adopted in 1975), the Red Sea (adopted in 1976), the Kuwait Action Plan Region (adopted in 1978), the Wider Caribbean Region (adoption expected in 1980), the West African Region (adoption expected in 1981), the East Asian Seas (adoption expected in 1981), the South-East Pacific (adoption expected in 1981) and the South-West Pacific (adoption expected in 1981) (figure 1).

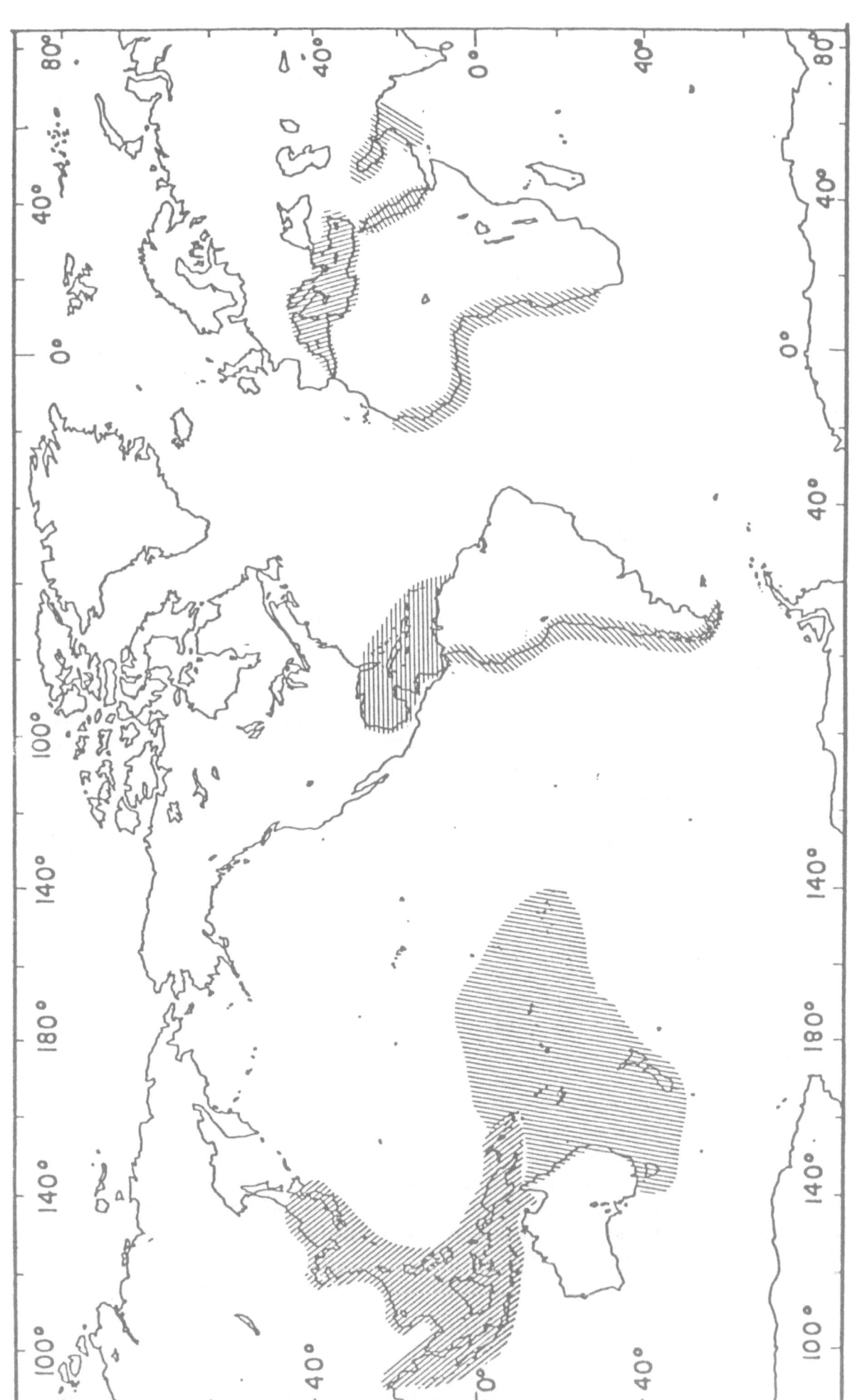

Figure 1 : UNEP Regional Seas Programme Areas

The following is a brief review of the recent progress in each of the eight
Regional Seas.

The Mediterranean

The Mediteranean was an ideal place to begin UNEP's Regional Seas
Programme, since the Mediterranean Sea is subjected to pollution stress of
every variety. It is a relatively well-known sea, scientifically speaking,
and has great importance due to the dependence of millions of its coastal
inhabitants on its resources. In 1977, 830,000 tons of fish worth $1.3
billion were caught in the Mediterranean. More than 100 million tourists
migrate to its coasts every year, to join the 100 million people already
inhabiting the coastal zone.

The main pollution problems of the Mediterranean coastal areas are caused
by a general lack of adequate treatment of domestic sewage and industrial
discharges, the use of pesticides in agriculture, and oil pollution from
accidental and operational discharge from vessels. These factors
contribute to a general pollution load of toxic chemicals in sediments and
biota, an overload of nutrients in certain areas with a resulting increase
in BOD, and occurence of pathogenic organisms in waters and shellfish. The
state of open waters is not yet critical but many coastal zones are
considered badly polluted.

In September 1974, under UNEP sponsorship, the Intergovernmental
Oceanographic Commission (IOC) of UNESCO, the General Fisheries Council for
the Mediterranean (GFCM) of FAO, and the International Commission for
Scientific Exploration of the Mediterranean (ICSEM) convened an
International Workshop on Marine Pollution in the Mediterranean, at which
scientists from Mediterranean marine research centres reviewed the state of
Mediterranean pollution and the facilities already available throughout the
region for its assessment.

At a meeting in Barcelona, January 1975, representatives of the
Mediterranean Governments approved an action plan for the protection and
environmentally-sound development of their eco-region. The content of the
plan is summarised in four inter-dependent chapters: scientific, legal,
integrated planning and institutional/financial.

On the basis of the recommendations emanating from the 1974 workshop, the
1975 Intergovernmental meeting in Barcelona approved a Mediterranean
Pollution Monitoring and Research Programme, known as MED POL, which is the
most comprehensive pollution research effort ever attempted in the area.
Designed to draw upon existing research facilities of the region, the
original programme consists of seven pilot projects. Four of these entail
monitoring of the environmental quality for the level of selected
pollutants, and three deal with research on the behaviour and effects of

pollutants in the marine environment. At present, 84 institutions in 16 Mediterranean countries are participating in MED POL.

At a joint workshop on pollution of the Mediterranean, co-sponsored by ICSEM and UNEP and convened in Antalya in November 1978, papers were presented to the scientific community at large, most of which were based on the initial results of MED POL.

Additional scientific projects on the assessment of riverborne and airborne pollutants, the build-up of modelling capabilities of Mediterranean scientists, the creation of a central data bank for all information and statistics generated by the action plan were begun. The monitoring of open waters of the Mediterranean, the assessment of the role of sedimentation in Mediterranean pollution and development of conceptual and predictive models of biogeochemical cycles and water mass movement were also initiated or are under consideration. A survey of land-based sources of pollution has already been completed, resulting in comprehensive information on the sources, types and amounts of various pollutants entering the Mediterranean from rivers and coastal sources.

In 1976, at a Conference of Plenipotentiaries held in Barcelona, representatives of 16 Mediterranean States adopted three international agreements: the Convention for the Protection of the Mediterranean Sea against Pollution, the Protocol for the Prevention of Pollution of the Mediterranean Sea by Dumping from Ships and Aircraft, and the Protocol on Co-operation in Combating Pollution of the Mediterranean Sea by Oil and Other Harmful Substances in Cases of Emergency.

Following the signature of the Barcelona Treaties, the Regional Oil Combating Centre was established in Malta in December 1976. The primary objectives of the Centre are to provide a mechanism for exchange of data relevant to oil pollution, training in oil pollution control techniques, and assistance in designing contingency plans.

In February 1978, the Convention and Protocols entered into force, and to date 15 countries and the European Economic Community have ratified them.

Since data from MED POL have confirmed the suspicion that most pollution in the Mediterranean comes from land-based sources, Mediterranean countries have pursued the development of a new treaty to control these sources. Following intensive preparatory work at meetings in Athens (February 1977), Venice (October 1977) and Geneva (June 1979), the text of the protocol is ready for the conference of plenipotentiaries planned for Athens next May.

The environmental management activities are centered around the Blue Plan and the Priority Actions Programme (PAP). The Blue Plan, as the study of major development issues requiring regional co-operation, is the conceptual backbone for the PAP, which includes inter-country co-operation on development of mariculture, application and renewable sources of energy, protection of soil from degradation, improvement of human settlements, utilization of freshwater resources and development of alternative patterns of tourism.

Since 1975, when the action plan was adopted, annual intergovernmental meetings have been organized to review the progress and plan future activities. At the meeting held in Geneva, February 1979, which was also the First Meeting of the Contracting Parties to the Barcelona Convention, the programme and budget for the 1979/1980 biennium was agreed. The Governments also pledged $3.2 million for a Trust Fund supporting the programme, thereby assuming major financial responsibility for the action plan.

UNEP, as the organization discharging secretriat functions for the Barcelona Convention, co-ordinates the implementation of the Mediterranean Action Plan with the co-operation of 14 international organizations, most of them specialized agencies of the United Nations.

The Red Sea and Gulf of Aden

The Red Sea is of unique geological and biological interest. It seems to be a new ocean in the first stages of formation and contains the northmost coral reefs in the world. The possible effect of its elevated temperatures and salinities on its marine inhabitants, the presence of endemic species, the relatively low species diversity compared with that of the Indian Ocean, and the possible effect of the re-opening of the Suez Canal on species composition on either side are all questions which render the sea especially intriguing to scientists.

Today this region is relatively unpolluted, but a rapid increase in oil exploitation, production and shipping is taking place, and the environmental impact is expected to be considerable. Pollution by oil from accidental spills and intentional discharge from ships and land-based sources, as well as the physical disturbances resulting from construction and dredging operations, may place undue stress on marine ecosystems.

A preparatory conference to discuss the elements of a regional action plan was convened by the Arab League Educational, Cultural and Scientific Organization (ALECSO) in Jeddah in late 1974. The scientific workshop organized by UNESCO earlier that year provided the necessary scientific input into the meeting.

In January 1976, again in Jeddah and under the aegis of ALECSO, an intergovernmental meeting adopted the action plan for the protection of the Red Sea and the Gulf of Aden. The action plan calls for intensive development of the region's capabilities in the field of basic and applied marine sciences and for the adoption of a regional convention for the protection of the Red Sea and the Gulf of Aden.

Under the co-ordination of ALECSO, and with the active support of UNESCO, UNEP and other organizations, activities have concentrated on the training of marine scientists, development of the marine science laboratories in the region and organizing of study tours. A number of small seminars, workshops and training courses have been held and others are planned for the future.

The Kuwait Action Plan Region

The Region includes the coastal area of Bahrain, Iran, Iraq, Kuwait, Oman, Qatar, Saudi Arabia and the United Arab Emirates as well as the body of water encompassed by this coastline. This area is undergoing the world's most rapid economic development: average investment per kilometre of coastline is between $20 and $40 million. About 60 per cent of all the oil carried by ships throughout the world-around a billion tons per year - is exported from this region, passing through the narrow Strait of Hormuz.

An interagency mission visited the region for three months in 1976 to collect information, define problems and advise on the content of an action plan. The possible elements of the action plan were discussed at three meetings of government-designated experts (Kuwait, December 1976; Bahrain, January 1977; Nairobi, June 1977). The results of their deliberations were reviewed by the Kuwait Regional Conference of Plenipotentiaries on the Protection and Development of the Marine Environment and the Coastal Areas, held in Kuwait, April 1978. At this conference, Governments adopted the action plan, including two legal instruments: the Kuwait Regional Convention for Co-operation on the Protection of the Marine Environment from Pollution, and the Protocol Concerning Regional Co-operation in Combating Pollution by Oil and Other Harmful Substances in Cases of Emergency. In addition the Conference approved:

a) the establishment of a marine emergency mutual aid centre to co-ordinate action against pollution by oil and other harmful substances;

b) the creation of a regional trust fund (close to $6 million) for an initial period of two and a half years; and

c) the establishment in Kuwait, after entry into force of the Convention, of a Regional Organization for the Protection of the Marine Environment to manage the action plan and to co-ordinate all projects executed thereunder.

The action plan calls for a large number of projects to deal with such matters as the origin and magnitude of oil pollution, an evaluation of all industrial wastes and municipal sewage reaching the sea, stock assessment of commercially-important species of fish and shellfish, and the ecological effects of coastal engineering and mining. It also calls for activities related to contingency planning for accidents arising from oil exploration, exploitation and transport, and to environmental engineering, public health problems, aquaculture, marine parks, port pollution and freshwater management. Intensive training programmes and public awareness campaigns are included.

A programme document describing the details of these projects was prepared under the sponsorship of UNEP, which was designated as the interim secretariat for the Convention. The document was reviewed by an expert group meeting in Kuwait, November 1979. Another expert group meeting convened in December 1979 advised on actions to be taken for the establishment of a marine emergency mutual aid centre. As a first step towards programme implementation, the office of the interim secretariat was established by UNEP in Kuwait (January 1980) and an interdisciplinary mission to all countries of the region will be launched in early 1980. An IMCO/UNEP Workshop on the combating of pollution from ships will be held in mid-1980.

The West African Region

The West African region has been defined for the purposes of the Regional Seas Programme as including the marine environment and coastal area of 19 States from Senegal in the north to Namibia in the south. These States, whose environment comprises both tropical and subtropical systems, face many common environmental problems. Among these problems are pollution from petroleum hydrocarbons (mainly from maritime transport), from agricultural and industrial waste, and from untreated or insufficiently treated sewage. All these sources of pollution may affect, directly or indirectly, such interests as the health of the coastal population, fisheries resources marine and coastal ecosystems, and the tourist industry.

Concrete preparatory activities for the development of a comprehensive programme for the West African region began in 1976, when a UNEP exploratory mission visited 16 States of the region to survey enviromental problems and to make a preliminary assessement of the States' interest in developing an action plan.

In 1977 an IMCO/UNEP workshop was held in Douala on the problem of pollution from ships in the region, and in 1978, FAO, IOC, WHO and UNEP jointly sponsored a scientific workshop in Abidjan. Both workshops were attended by technical experts from the region. The latter served to define more clearly the environmental assessment component of the draft action plan, including the need for training of local scientists and technical co-operation in strengthening of national institutions.

Immediately after the workshop, UNEP convened a meeting of representatives from the United Nations system and other organizations concerned with the region to exchange views as to what activities could effectively and practically be included in a draft action plan and what contribution each organization was willing to make to a regional programme. A particularly important input to the meeting was provided by the participation of 13 UNDP Resident Representatives, each of whom was posted in one of the West African States.

After additional consultations with the Governments and interested organizations, including a second UNEP mission to the region in late 1978 - early 1979, the process of developing an action plan seemed to be at a sufficiently advanced stage to call a meeting of Government-nominated experts at which the entire programme (assessment, management and legal components), as outlined in the draft of the plan, might be discussed for the first time on a region-wide basis. This meeting was held in Libreville in November 1979.

The meeting revised the first draft of the action plan which follows the example set by other regions such as the Mediterranean and the Kuwait Action Plan Region. In substance, however, the plan reflects the priorities and needs particular to West Africa.

The chapter of the plan concerned with environmental assessment calls for the assessment of the origin and magnitude of oil pollution, of suspended and dissolved matter in rivers, of chemical residues and of domestic wastes. The need for studies dealing with coastal lagoons, estuaries and mangroves as important features of the West African coast is recognized. The training of local scientists and technicians to carry out the research and monitoring incorporated into the programme is stressed in the plan and, as an initial step, a survey of the capabilities of scientific institutions in the region will be undertaken.

The environmental management chapter of the action plan reflects the widespread concern that development of the coastal and adjacent sea area of the region be accompanied by sound management and control of industrial, agricultural and domestic wastes, rational exploitation of marine living resources (including aquaculture), management of coastal lagoons and mangrove ecosystems, and development of non-polluting alternative sources of energy. A programme is envisaged to assist the Governments to identify development opportunities that are in harmony with their natural environment. In addition, the formulation of contingency plans for dealing with pollution emergencies caused by maritime accidents is recommended as well as the control of oil pollution from deballasting of ships. Training of technical personnel in environmental management practices is prescribed as a priority.

When discussing the legal activities to be included in the action plan, the experts gave their full support to the development of a regional convention for co-operation on the protection of the marine environment from pollution.   It was suggested that such a convention should be supplemented by protocols specifying detailed obligations of contracting parties, and UNEP was requested to prepare the negotiation of a first protocol concerning co-operation in combating pollution in cases of emergency.

In the light of the recommendations of the meeting, UNEP plans to complete preparatory studies and activities in 1980 so that the action plan, including its legal component, may be adopted in 1981. These preparations will   include surveys of oil pollution problems with particular emphasis on pollution from shipping activities, of industrial pollution from land-based sources, and of river inputs to the marine environment, to be carried out in co-operation with IMCO, UNIDO and UNESCO, respectively.   Initial activities related to coastal lagoons, estuaries, and mangroves will be begun with the support of UNESCO, FAO, and IUCN. A regional seminar on coastal area management and development will be organized by UN/DIESA in mid-1980. FAO and IMCO will co-operate with UNEP in preparing guidelines on which the experts may begin their negotiations towards the final text of the regional convention and the first protocol. Two meetings of legal experts from the region will be held in 1980 to discuss these texts.

Once the action plan is "mature" enough it will be submitted to an intergovernmental conference for adoption.   It is hoped that such a conference may be held in early 1981. Thereafter the regional programme should become fully operational.   It is expected that funds for these programmes will be made available partly by the Governments concerned and partly by the resources at the disposal of the UN system.

The Caribbean Region

The Caribbean region comprises an assortment of complex and fragile tropical and subtropical ecosystems which extend throughout the Caribbean Sea and Gulf of Mexico west of the Greater and Lesser Antilles. The region has 19 island nations which share the common problem of the extreme fragility of their ecosystems.   Just as islands dare not risk following patterns of development created for continental land masses, care must be taken when applying development patterns first pursued by countries in temperate climates to the Caribbean's distinctive environment.

Most of the Caribbean States are developing countries whose major problem is to fulfil their peoples' basic needs, and their desire for an action plan reflects a conviction that environmental management is a necessary feature of development, regardless of the stage that development has reached.

The first steps in developing a comprehensive action plan for the wider Caribbean region were to identify the area's environmental activities already underway, and to fix objectives for future action.

Towards this end:

A small group (Caribbean Environment Project team) was established early in 1977 by UNEP and the United Nations Economic Commission for Latin America (ECLA) to provide the focal point for the development of an action plan for the region.

An International Workshop on Marine Pollution, convened by IOC, FAO and UNEP in Trinidad in December 1976, reviewed the marine pollution problems of the wider Caribbean region and recommended a set of projects with a view to obtaining information needed for a proper understanding of the causes and consequences of marine pollution.

- An International Advisory Panel was established early in 1978 to assist and guide preparations for the action plan.

- An Interagency meeting of 16 international organizations working in the region was convened in Mexico City in August 1978, to review the first draft of the Caribbean Action Plan and to discuss their involvement in its preparatory phase.

- In October 1978 experts from the region met in Colombia at an IMCO/UNEP workshop. The workshop focused on oil pollution, considered to be the most prevalent problem associated with maritime transport. It examined problems of clean-up and containment after a spill, appropriate disposal methods and possible preventive measures.

- A three-year project on marine pollution investigation and control was started in March 1979 by the Government of Cuba, UNDP and UNEP. This national project is expected to become a model or "pilot project" for subsequent regional marine pollution monitoring and control activities within the Caribbean Action Plan.

- A Caribbean Disaster Preparedness Seminar was co-sponsored by 11 organizations in St. Lucia in June 1979. The seminar reviewed the natural and man-induced disasters relevant to the region, such as volcanic eruptions, hurricanes and oil spills, and formulated specific recommendations for measures to be taken to mitigate their effects. These recommendations will be included in the Caribbean Action Plan.

- A Conference on Environmental Development and Economic Growth was held in Barbados in September 1979. Co-sponsored by, among others, UNEP and UNESCO, and attended by the representatives of Caribbean islands having an area of less than 10,000 km or less than half a million population, the Conference stressed the need for close co-operation as a tool for development of the natural resources of the small Caribbean islands on an economically sound and sustainable basis.

- Direct assistance was or is being provided to individual countries having environmental problems. Recent examples include a UNEP/FAO/ IMCO/IUCN mission to Mexico to assist in the assessment of the environmental consequences of the IXTOC-I blow-out and to advise on measures which might mitigate these consequences.

- A large number of documents has been prepared with the co-operation of the United Nations system as supporting documentation for the action plan. These include a directory of Caribbean marine research institutions, a proposal for a conservation strategy, and overviews on specific subjects such as human settlements, agriculture, fisheries, human health, marine pollution and energy.

The preparatory phase of the Caribbean Action Plan, which was jointly co-ordinated by the Economic Commission for Latin America (ECLA) and UNEP, is nearing its end. The draft Action Plan and the supporting documentation have been reviewed by a meeting of government-designated experts in Venezuela in January 1980. The revised draft action plan will be submitted for adoption to a meeting of Governments, planned for September 1980 in Jamaica.

The implementation of the activities agreed upon at the meeting in Jamaica will begin almost immediately after that meeting.

The South-West Pacific Region

Discussions on an action plan for this region, a wide area extending from Australia and New Zealand to the islands of the south Pacific as far east as the Tuamotu archipelago, were initiated by the South Pacific Commission (SPC) in 1974. The South Pacific Bureau for Economic Co-operation (SPEC), the Economic and Social Commission for Asia and the Pacific (ESCAP), UNEP and many other specialized organizations of the United Nations system participated in the discussions.

Based on a broad consensus on the part of all interested parties, preparations have been launched for the development of an environment programme for the South-West Pacific. These preparations are to include:

- the compilation and provision of information and expertise not available in the region;

- assistance in the planning and implementation of national environment programmes;

- technical assistance in the evaluation of the environmental implications of development programmes;

- education and training in environmental management, including the environmental implications of development;

- exchange of information on environmental legislation, programmes, standards, and on monitoring of environmental conditions;

- preparation of regional approaches to contingency planning for environmental disasters such as oil spills, tsunamis, etc; and

- development of regional services to assist States and Territories in their own environment management programmes.

A joint SPEC/SPC/ESCAP/UNEP co-ordinating group is supervising the preparation of documents for review and revision by experts of the region (in late 1980) with a view to their final consideration and approval at a Regional Conference on the Human Environment in the South-West Pacific Region, planned for early 1981. These documents will be based primarily on country reports. The four main working documents being prepared are:

a) Report on the State of the Environment in the South-West Pacific Region.

b) Draft Declaration of Principles on the Management and Improvement of the Environment in the South-West Pacific Region.

c) Draft for a Regional Plan on development and environmental protection with proposals for specific co-operative projects to be implemented at the international, regional, national and local levels.

d) Proposals for the administration and financing of the Action Plan.

The South-East Pacific Region

The South-East Pacific region refers to the coast of South America stretching from Panama in the North to the southern tip of Chile, and the adjacent coastal waters. The physical and biological complexity of an area which comprises tropical, subtropical, temperate and sub-antarctic systems, and hosts a variety of dramatic oceanographic phenomena, gives pause to those who favour a uniform, systematized approach to the environmental problems of the area.

Nevertheless, in an attempt to begin development of a regional action plan, scientists, managers and legal experts attended an international workshop held in Santiago, Chile, in November 1978. Organized by the Permanent Commission for the South Pacific (CPPS), with the co-operation of FAO, IOC, WHO and UNEP, experts from Chile, Peru, Ecuador, Colombia and Panama gathered to discuss the area's environmental problems in general, the state of marine pollution in particular, and legislation relevant to the region's environmental protection. The representatives of the five countries adopted the following:

a) a short-term plan of action which includes the assessment and monitoring of the sources, levels and effects of pollutants contributing to the environmental degradation of the region, and suggestions for an institutional framework for the implementation of the action plan;

b) guidelines for a convention on the protection of the marine environment against pollution in the South-East Pacific;

c) a draft agreement concerning regional co-operation on emergency measures against pollution of the South-East Pacific by hydrocarbons and other harmful substances;

d) the establishment of a special fund to assist national institutions participating in the action plan.

Since the Santiago Workshop, consultations have been conducted between CPPS, the Governments of the region, the relevant specialized organizations of the United Nations and UNEP on the ways and means of supporting the full development of the action plan and its formal adoption by the Governments.

Preparatory activities, which should lead to an intergovernmental meeting planned for early 1981, include:

- in-depth review of available institutional infrastructure for the marine pollution monitoring and research programme;

- updating and completing available information on sources, levels and effects of marine pollution in the region;

- formulation of a programme for monitoring of, and research on, marine pollutants found in the region, including the preparation of technical guidelines applicable to the implementation of the programme;

- assistance (training and intercalibration) to national institutions which may participate in the marine pollution reseach and monitoring programme;

- training of regional experts in prevention and abatement of marine pollution by petroleum hydrocarbons, and other harmful substances;

- formulation of a draft convention on the protection of the marine environment against pollution in the South-East Pacific;

- formulation of a draft agreement (protocol) on regional co-operation in cases of pollution emergencies;

- formulation of a draft action plan for the protection of the South East Pacific.

The CCPS is playing a central role in the day-to-day co-ordination of preparatory activities.

The East Asian Seas

In the East Asian Waters, the complexity of marine pollution problems and their impact on the coastal environment were first recognized at the international workshop on marine pollution in East Asian Waters, held in Penang, in April 1976.   This regional scientists' meeting established a comprehensive list of priority issues and, in addition, identified common needs such as:

- more baseline studies of the marine environment with emphasis on marine ecosystems (e.g. coral reefs and mangroves), including the determination of present levels and distribution of pollutants in the various media (water, sediments and organisms);

- greater understanding of the physical processes in the coastal waters, particularly water circulations, as mechanisms for the distribution of pollutants;

- research on problems such as toxic effects, pollutant degradation, transfer of pollutants and ecological effects;

- standardization of analytical methods, including suitable intercalibration exercises;

- intensive training of scientists and technicians in all aspects of marine pollution research and monitoring;

- mechanisms for the exchange of data, information, specimens and samples on a regional basis; and

- efforts to increase public awareness of environmental pollution and its significance.

The Penang workshop proposed several regional projects and, recognizing the large geographical dimensions of the region, prepared individual projects for six subregions, namely the Bay of Bengal, the Strait of Malacca, the Gulf of Thailand, the South China Sea, the Sea of Japan, Yellow Sea and East China Sea, and the seas of the Eastern Archipelago. The subregional approach was also endorsed by UNEP's Governing Council which decided, in

1977, that steps were urgently needed to formulate and establish a scientific programme involving research, prevention and control of marine pollution and monitoring for the waters of Indonesia, Malaysia, the Philippines, Singapore and Thailand.

Substantive proposals contributing to the development of a first subregional action plan for East Asian Seas were received through regional meetings such as:

- Third Session of the Indo-Pacific Fisheries Council (IPFC) Working Party on Aquaculture and Environment, Bangkok, August/ September 1976;

- ESCAP/UNEP Intergovernmental Meeting on Environmental Protection Legislation, Bangkok, July 1978;

- UNESCO Regional Seminar on Human Uses of the Mangrove Environment and Management Implications, Dacca, December 1978;

- ASEAN Expert Meeting on the Environment, Jakarta, December 1978;

- UNESCO/IOC Workshop on the Western Pacific (WESTPAC), Tokyo, February 1979.

- Second Meeting of ASEAN Experts on the Environment, Penang, September 1979.

In anticipation of regional action plans in Asia and the Pacific, the Indian National Institute of Oceanography, Goa, prepared in 1978 a directory of Indian Ocean marine research centres. An intensive training programme was launched by UNEP and United Nations agencies to support the national institutions interested in the regional programme, including:

- national seminars on the protection of the marine environment and related ecosystems;

- workshops on aquatic pollution in relation to protection of living resources;

- an international workshop on the prevention, abatement and combating of pollution from ships;

- a regional workshop on coastal area development and management; and

- a regional seminar on environmental impact assessment.

In addition, several pilot projects are now commencing in collaboration with ASEAN countries, covering subjects such as regional oil spill contingency planning, toxicity of oil dispersants, land-based pollution sources, river pollution transport, impact of pollution on mangrove ecosystems and of oil pollution on living aquatic resources, and environmental problems of off-shore exploration and exploitation. Legal instruments for marine pollution control are also under consi- deration and preparation.

The Draft Action Plan met with agreement at the second meeting of the ASEAN Experts on the Environment in Penang, September 1979. This meeting also established a list of seven priority items for immediate action under the regional programme dealing with the assessment of:

- silt and sediment load and impact on marine biota;

- organic and nutrient load in ASEAN seas and its impact;

- oil pollution;

- levels of heavy metals and chlorinated hydrocarbons;

- impact of pollution on mangroves and coral reefs;

- impact of thermal pollution;

- adverse effects of off-shore seabed exploration and exploitation on marine ecosystems.

The Draft Action Plan will be further reviewed and refined by meetings of scientific and legal experts in the summer and autumn of 1980, leading, it is hoped, to its adoption by an intergovernmental meeting in early 1981. The present five-country programme is intended to serve later as the nucleus of a wider regional programme covering adjacent areas of the East Asian Seas as identified by the Penang workshop in 1976.

THE ROLE OF A LARGE DATABUOY IN SATISFYING ENVIRONMENTAL DATA

REQUIREMENTS IN A NEW UK EXPLORATION AREA

W.B. Woollen, B.Eng.
Marine Technology Support Unit, AERE, Harwell, UK

INTRODUCTION

At the time exploration for hydrocarbons started in the North
Sea, there were distinct limitations to the understanding of the
climate, although visual weather reports had been received from
ships and meteorological data had been collected at a number of
coastal sites for many years. Few wave measurements had been
made before 1970 and the techniques of estimating waves from
winds were only just being developed. The position in the
N. Sea is now completely different since several years of wave
data have been collected at a number of stations using techniques
which give relatively high accuracy.
   The main doubts in the N. Sea arise from the errors in
extrapolation from the short data base of say three or five years
to obtain the 50 or 100 year maximum wave height. In addition,
owing to the absence of suitable instruments, there is virtually
no information on wave directional spectra, which is now
attracting greater interest, and the detailed understanding of
currents is limited.
   Environmental data in the N. Sea is being collected at deep
water sites by the UK Offshore Operators Association (UKOOA)
with 30% support from the UK Department of Energy (D.En), and
by the Natural Environment Research Council's Institute of
Oceanographic Sciences (IOS) at near coastal sites, fully
funded by the D.En. In addition to the locations where data
collection is continuing there are stations where work has
terminated. Both are shown in Fig. 1 & 2. Stations where
measurements are being made by individual oil companies are not
shown.
   As would be expected, measurements have concentrated in the
areas where oil exploration and production is most active. In
the S.W. Approaches, no instrumental wave measurements had been
taken routinely prior to 1978, although measurements using a
weathership had been taken by UKOOA at the Boyle Station for
the three years up to 1977 and by IOS at Seven Stones Light
Vessel since 1961 (Fig. 1). Two of the principal reasons for
the delay in extending data collection to the S.W. Approaches
were that a weathership was considered to be expensive and a
UK databuoy was not yet available. That position has now
changed and one of the most advanced databuoys in the world is
now routinely taking a comprehensive set of measurements in the
area.
   If there is little success in the search for oil in the
S.W. Approaches, it is likely that funding of the present
measurements will not continue beyond about 1982. However,
the data has more widespread uses than determining the local
climate for oil exploration purposes and so this paper attempts

by illustration, to give an overview of the importance of good environmental data in the S.W. Approaches to oil companies and other users.

If oil companies cease to be interested beyond 1982, other sources of funding will be required, possibly by a number of organisations in a joint programme. Such multi-funding is difficult to organise, so that if the present programme were to stop, it is likely to be difficult to re-start.

USES OF ENVIRONMENTAL DATA

Good environmental data is now required for a variety of purposes, the principal ones are listed below:
    Ship design and operations
    Coastal defence design
    Input to emergency warning systems
    Offshore oil structure design and operation
    Research on alternative energy devices
    Prediction of pollution behaviour
    Helicopter operation
    Input for the general meteorological services

METEOROLOGICAL DATA REQUIREMENTS OFFSHORE

During the last few years, the UK Meteorological Office (Met. Office) has provided support to the offshore oil industry which requires environmental data more detailed and accurate than has been available in the past(1).This requirement is only one of several which the Met. Office attempts to satisfy and in order to do this the meteorological organisations throughout the world have established complex data gathering and forecasting organisations.

For example, the Met. Office is connected to the highly sophisticated Global Telecommunication System of the World Meteorological Organisation and is equipped with one of the most powerful computer complexes in Europe. Complex mumerical prediction models are run on a real time operational basis. However sophisticated, forecast systems are useless without an adequate data base and because of the nature of weather systems and the large expanses of water, it is essential that the existing network of on-shore data collecting stations is complemented by high quality and regular measurements offshore.

The basic surface data measurement recommended by the World Meteorological Organisation should be made at points on a network not more than 300 miles apart for forecasts up to 18 to 24 hours ahead. This condition is largely met in the N. Sea but this is not the case to the west of the U.K. For operational, short period forecasting requirements, i.e. up to 6 to 18 hours, a denser network of observations is necessary in order to analyse the fine detail of weather systems more precisely. Frequency of observations should be at least 3 hourly and preferably every hour in accord with most land stations (2). The observations should include the following parameters :
    Wind speed and direction
    Pressure
    Visibility
    Temperature
    Cloud Base
    Present Weather
    Significant waveheight and
    Swell.

In general, the quality of offshore observations is variable.
Some of the observational sources find it difficult to produce
the desirable high quality observations which the forecasters
can trust. The main exceptions are those where the Met. Office
and IOS have been able to influence directly the design and use
of the measuring instruments. The Atlantic Weatherships,
automatic weather stations and stations installed by UKOOA,
amongst others, normally give good, reliable real time data.
        The Met. Office has also developed a numerical model (3)
to forecast wave conditions covering the N. Atlantic Ocean north
of 18°N, which covers the majority of atmospheric systems which
generate waves affecting the UK. The model is run twice a day
to provide a starting point principally to help the London
Weather Centre with forecasts for the offshore oil industry.
The model incorporates a spectral representation of the wave field
and water depth effects within the Continental Shelf. The sea
state output gives values at points on a network 100 Km apart
for periods up to 36 hours ahead, twice daily. The output
consists of both wind, wave and swell height, direction and
period as well as a scalar field of significant wave height.
Although the model is in operational use, its detailed formulation
is under constant review and subject to occasional modification
as a result of feedback from forecasters. The predicted spectral
information consists of the energy distributed in 30 degree
directional sectors for wave periods ranging from 6 to 20 seconds.
The spectral form of output is required for sophisticated support
vessels which have on-board computer facilities to calculate the
vessel heave, pitch and roll.
        Ideally, the wave forecasts should be validated with
observations. There are, however, very few observations of wave
conditions in the N. Atlantic, some of which are unreliable and
with one exception none give the spectral detail needed to
validate the model. The exception is the measurement made on
DBI to be described later.

OCEANOGRAPHIC DATA REQUIREMENTS

In contrast to the requirements for meteorology, the other
principal requirements for data are for oceanographic statistical
purposes and are not normally required in real time. In new
exploration areas such as the SW, where there is a limited amount
of wave data, there is also a requirement for local on-line
information. Data is required in order to predict the largest
wave that a structure will encounter in its lifetime and it is
normal for the 50 or 100 year wave to be used for this purpose.
This enables the structure to be designed with adequate clearance
above the sea and with adequate strength. More detailed
description of the waves is required including single point and
directional spectra, to be used, for example, in fatigue calculations.
For operational planning, knowledge about weather windows and
various weather patterns is required including the duration and
extent of calms and storms.
        Oceanographic measurements are more difficult than meteoro-
logical ones because of the hostile conditions which exist at
the sea surface. The measurements taken are often influenced
greatly by what is possible at the particular site, and also the
cost of collection. Most observations are limited to measurements
of the surface elevation at a point and the sea temperature. At
certain locations, the current speed and direction at a number of

depths between the sea surface and the seabed are also measured.
At present there is no routine system for obtaining wave
directional spectra other than DBI. Similarly the only routine
and accurate system for measuring surface current is that on DBI.
    The doubt about extreme wave height and period in the S.W.
Approaches is well illustrated by a comparison of the 50 year
storm wave heights (Fig. 3) and periods (Fig. 4) in the area,
calculated for the years 1972 and 1977 and published in 1st and
2nd editions respectively of the D.En's Guidance Notes (4).
The 1977 heights and periods in the Celtic Sea are significantly
different to those in 1972, principally because of the additional
information obtained by the weathership at the Boyle Station
operated for UKOOA. Particular notice should be taken of the
25m contour line. At Boyle, the 50 year wave height was reduced
by a few metres and period by a small amount. However, the short
data base of 3 years at the Boyle Station gives wide confidence
limits in these extrapolated values. The wave height at the DBI
location was also reduced by about 2 metres but will be
recalculated when sufficient data has been obtained.

DIFFERENCES BETWEEN THE S.W. APPROACHES AND THE N. SEA

Meteorological forecasting and oceanographic conditions in the
S.W. Approaches and the N. Sea differ considerably.. The
N. Sea is surrounded by land. Sites are rarely more than about
150 miles from a data collection point onshore and weather
systems are monitored several times before arriving over the
N. Sea. The situation in the S.W. Approaches is completely
different in spite of its importance to those UK activities
affected by the weather, listed earlier. The most distant
points of interest are the potential oil producing sites in the
S.W. Approaches which are almost 200 miles SW from Cornwall, a
similar distance from Brittany and Ireland, and almost 400 miles
N of Spain. Most of the weather systems approach from the
Atlantic Ocean to the west where apart from ships of passage
and high flying aircraft, environmental data is only available
from the three or four weatherships over 500 miles to the west.
Ships of passage which provide weather reports are few and far
between and often do not report during the night.
    The limited amount of offshore data round the UK which the
meteorological forecasters have available is illustrated in Fig. 5.
which shows the frequency of observations per reporting period
of three hours. The two observations in the Sole area include
one from DBI and contrasts with the six observations in the Viking
and four in the Forties areas of the N. Sea, all from oil platforms.
The other observation in the Sole area will be from ships
travelling through the area and therefore less reliable.

CHOICE OF MEASURING PLATFORM FOR THE S.W. APPROACHES

During 1977, the UKOOA Oceanographic Committee decided that there
was sufficient interest developing in the SW to start collecting
environmental data there. No oil platforms had been constructed
in the area, sites of interest were too far from land to permit
a Waverider to be operated and weatherships were considered to
be expensive. The alternative was a larger databuoy but at that
time, MAREX had not had sufficient operating experience with their
buoy 30 miles west of Foula, (Fig. 1), to clear all its teething
troubles. This buoy was the first to use a new and advanced

data collection system which incorporated the first micro-processor for on-board data processing. Obtaining this experience had been hampered by accidental and possibly deliberate interference, the repair of which was made difficult by the combination of severe weather and long lines of communication. The longer range HF Piccolo was also not then available on the buoy. The teething troubles should have now been resolved and a number of MAREX databuoys are now operating in the world (Fig. 6).

At the same time, there were suggestions that the role of the National Databuoy DBI (Fig. 7) developed in 1974 for the Department of Industry (DoI) by IOS and constructed by the industrial consortium SEATEK (5), should be changed from instrument and concept development to one where its immense capability was used to collect data for the oil industry and other groups.

UKOOA with some misgivings at the time decided to deploy DBI in the SW, the doubts being that the buoy was not entirely ready for full commercial use, in spite of 18 months of trials in the N. Sea. EMI, one of the original SEATEK group, undertook to refurbish and operate the buoy in the SW to collect data, and the DoI made a substantial contribution to the costs of refurbishment and redeployment.

Firstly, however, it was necessary to ensure that the buoy would not overturn in the most severe seas to be expected in the SW. Model tests were conducted and a satisfactory three point resilient mooring was devised by IOS which would maintain an adequate righting force yet not interfere with its wave following characteristics. UKOOA had also to be satisfied that the 7.6m buoy would respond to waves of a period of at least 4 sec, and at one stage, a satellite Waverider to measure the short period waves was being suggested. This was shown to be an expensive addition which could not be justified. Data recovery from DBI had been designed to rely on the relatively complex Piccolo HF telemetry link but UKOOA, as a result of Foula experience thought that all measurements should also be recorded on-board. Recorder limitations at the time necessitated a compromise solution so that it was agreed to record semi-processed rather than the raw data using a processor developed by IOS.

LOCATION OF BUOY

The meteorologists preferred to have the buoy located as far west as possible whilst the oil companies wished to have it located at the centre of potential oil fields which stretch from the English Channel to the edge of the Continental Shelf. Some measurements were already available at Boyle and Seven Stones (Fig. 1) and so it seemed reasonable to place it close to the edge of the Shelf. The main disadvantages of such a site were the longer sailing times for maintenance visits and the possible reception difficulties over the radio link at that particular range. However, transmission tests were made from such a location in December 1977 which indicated that the strength of the signal to be expected at the Culdrose receiving station in Cornwall using the ground wave, would give a sufficiently low error rate. A site for the buoy was eventually chosen at 48°43'N 8°58'W close to the 100 fathom contour line, 10 miles from the edge of the Continental Shelf. A side-scan sonar survey and camera

studies were made and grab samples taken, which indicated good
holding ground for the anchors in 168 metres of water (Fig. 8).
    Trinity House Light House Service was contracted to lay the
three moorings (Fig. 11) which they did with great precision
and expertise, the final leg being laid within 3° of the specified
bearings. Wind statistics had shown that the direction of the
highest-winds would be from SW and so the two westerly mooring
arms were placed with their bisector pointing due west, the
importance of which has subsequently been demonstrated and will
be mentioned later.

FEATURES OF BUOY

The general layout of the buoy, made to an overall specification
prepared by IOS, is shown in Figs. 9 & 10. Its steel hull,
7.6m in diameter, a depth of 2.1m and a weight of 35 tonnes is
capable of supporting a 2 tonne payload with a free board of 1m.
The aluminium mast carries a platform 3.9m above the deck giving
access to the top of a retractable mast and which can be extended
to a height of 7.5m upon which some of the meteorological sensors
are mounted. Access to the centre compartment which contains
the electronic equipment is through a 1m diameter hatch where
maintenance of the internal equipment can be done in a standing
position. The outer portion of the hull is divided into six
watertight compartments, two of which are floodable ballast
tanks used whilst the buoy is being towed and a further two
house the two 100 gallon propane gas cylinders for the thermo-
mechanical generator. The remaining two house the main and
emergency batteries.
    The design of the moorings has been validated by scale model
tests to withstand the anticipated 50 year wave for the S.W.
Approaches, including breaking waves. Compliance in the mooring
is provided by the 81mm nylon braidline in each of the three
mooring legs, the braidline being terminated in polished thimbles
cast from aluminium - silicon bronze. Each leg is held by a
3 tonne single fluke anchor. The length of the braidline is
such that accurate wave measurements are still made when the most
adverse meteorological and oceanographic conditions occur. The
layout of the mooring between extreme conditions is shown in
Fig. 11. The moorings are attached to the buoy via mooring
tension load cells capable of measuring loads up to 30 tonnes.
    The buoy and its three point moorings are designed so that
the buoy follows the slope of the sea surface thus permitting
directional wave spectra to be computed from measurements made
every 1.2 sec by the installed Datawell heave and pitch/roll
sensors. Surface current measurements are made with a long path
acoustic meter designed by AERE Harwell using transducers
mounted on three spars approximately 3 metres below the surface.
Component velocities are measured along two perpendicular axis
and the amplitude and direction of the current are computed
after allowing for the compass reading. The fact that the current
velocities are being measured on a surface - following buoy is
believed to ensure that the surface current velocities can be
differentiated from the orbital wave particle velocities.
    The main power source on the buoy is a Stirling cycle, thermo
mechanical generator (TMG) which burns propane and float-charges
four banks of nickel cadmium cells to provide the peak power
required during transmissions. The TMG has no rotating or
sliding parts and has sufficient gas to work unattended for over

Measured Variable	Sensor type & Manufacture	Range	Location on buoy	Time, Averaging period or Sampling Rate	Calibration Accuracy	Printed resolution	Field Accuracy
Wind Speed 1	Cup counter Vector Instruments A100 R	0 to 150 kn	6.0m above sea level	mean value during last 10 minutes of hour	±0.5kn	0.1kn	±2kn or ±5%
2	Weather Measure W 102		8.7m above sea level				
Wind Direction 1	Wind vane - self referencing Vector Instruments SRW1G	0 to 360°	6.0m above sea level	mean value during last 10 minutes of hour	±2°	1°	±10°
2	Weather Measure W102 Referencing from digital compass						
Air Pressure (one static pressure head) (2)	Aneroid capsule RDG 8190	925 to 1050mb	Head at 6.0 but measurement made at sea level.	mean value during last 45 seconds of hour	±0.2mb	0.1mb	±1mb
Air Temperature (2)	Platinum resistance Rosemount Ltd. E13418	-10 to 40°C	Stevenson's Screen 5.8m above sea level	mean value during last 45 seconds of hour	±0.1°C	0.1°C	±0.2°C
Relative Humidity (2)	Chemical Hygrometer Phys-Chemical Res. Corp. PCRC11	0 to 100% RH	Stevenson's Screen 5.8m above sea level	mean value during last 45 seconds of hour	±3%	0.01%	±5% 0 to 85% ±3% over 85%

TABLE 1 : Specifications of Meteorological Sensors

Measured Variable	Sensor type & Manufacture	Range	Location on buoy	Time, Averaging period or Sampling Rate	Calibration Accuracy	Printed resolution	Field Accuracy
Sea Surface Temperature (1)	Platinum resistance	-10° to $+40^{\circ}$C	Base of hull	45 secs mean every hour	$\pm 0.1^{\circ}$C	0.1°C	$\pm 0.2^{\circ}$C
Sea Surface Temperature (2)	Platinum resistance	-5 to $+20^{\circ}$C	Base of hull		$\pm 0.001^{\circ}$C	0.06°C	$\pm 0.06^{\circ}$C
Heave Amplitude	Accelerometer Datawell HIPPY	$-20m$ to $+20m$	Centre well	Continuous record for 20min.	3% up to 15s period	0.1m	$\pm 0.2m$
Pitch and Roll	Gravity stabilised platform Datawell PIRO M402	0 to $\pm 60^{\circ}$	Main battery compartment		1% up to 30° 2½% up to 60°	0.1°	$\pm 1^{\circ}$
Surface Current Magnitude EW & NS	Acoustic pulse velocity AERE	0 to 2.55 m/s (no lower threshold)	3 metres below sea level	5 min. mean about the half hour	± 0.01 m/s	0.01 m/s	± 0.02 m/s
Direction	Magnetic Compass Colnbrook Instruments Ltd.	0 to 360°	5.2 metres above sea level		$\pm 2^{\circ}$	0.4°	$\pm 2^{\circ}$

TABLE 2 : Specifications of Oceaonographic Sensors

two years. Primary batteries are carried as an emergency
back-up and will give over three months of normal operation.

BUOY SENSORS

The buoy is equipped with sensors to measure the following
meteorological and oceanographic variables several of which
are duplicated in order to improve overall reliability.

Meteorological variables (all duplicated)
Wind speed and direction
Barometric pressure
Air temperature
Humidity

Oceanographic variables
Heave giving surface elevation
Pitch and roll giving wave direction
Sea surface temperature (duplicated)
Surface current speed and direction

The specification of each sensor measurement is given in Tables
1 and 2. and the locations of the sensors is shown in Fig. 9.
Measurements are made at various times throughout each hour and
transmitted four times on the hour virtually eliminating any
risk of undetected error. On every third hour, this hourly
report is followed by a continuous 20 minute real-time
transmission of the heave, pitch and roll and bearing made every
1.2 sec. The hourly reports and semi-processed heave data in
the form of four parameters characterising the sea surface
during the hour, are recorded by a special high-density cassette
recorder on the buoy. No wave directional data is recorded
on-board. In addition various 'housekeeping' and status parameters are monitored and transmitted as part of the 90 channel
message to shore every hour, which enable faults to be diagnosed
remotely and maintenance planned before embarking on the buoy.
    The transmissions from the buoy are received at RNAS Culdrose
near Falmouth by two independent receivers using separate and
space diversified aerials. The data is then sent to EMI Woking
over two separate private land lines where two separate dedicated
computers check for consistency and convert the data to
engineering units which are then recorded on magnetic tape. The
London Weather Centre can assess this hourly data via a modem
interface to one of the two computers. The two separate magnetic
tape recordings are later 'merged' and if necessary 'merged'
also with the recordings recovered from the buoy during a three-
monthly maintenance visit. The final data tape is then transported by hand to the processing centre where the following data
and analyses are prepared for publication.

DATA ANALYSIS AND PRESENTATION

The data which is confidential to UKOOA Oceanographic Committee
participants is distributed quarterly in a detailed report
giving tables and graphs containing the results of a wide range
of analyses. Annual summaries of the data will also be available.
The following analyses are prepared:

<u>Meteorological data</u>
Frequency of occurrence of 10 minute mean wind speed and direction
Wind rose diagram
Wind speed exceedence
Distribution of gust ratios
Persistence diagram of 10 minute mean wind speeds
Time histories of 10 minute mean wind speed and direction and
  3 second gust speed
Time histories of barometric pressure
Time histories of air temperature, sea temperature and relative
  humidity

<u>Oceanographic data</u>
Ratio of H_s (spectral) and H 1/3 (deterministic)
Scatter diagram of H_s and mean zero up crossing period
  (derived spectrally and determinatically)
Wave height exceedence diagrams
Persistence diagrams of H_s
Scatter diagram of H_s and wave height and wave direction
Scatter diagram of mean zero crossing period and wave direction
Time histories of H_s, maximum wave height (3 hours) and mean
  zero crossing period
Time histories of peak energy wave direction and directional
  spread
Time histories of high frequency wave direction and directional
  spread
Current rose diagram
Current speed exceedence diagram
Frequency of occurrence of current speed and direction
Time histories of N-S and E-W current vectors
Time histories current spread and direction

<u>SERVICING THE BUOY</u>

The importance of getting regular and reliable environmental
data from the S.W. Approaches is illustrated by the difficulty
that can be experienced in forecasting, at times other than the
summer months, an acceptable period of weather for servicing
the buoy. Routine service visits are made to the buoy at
intervals of three months and additional visits can be arranged
to clear unexpected faults. During the summer months, the
68 ft research motor vessel Fathomer is an excellent vessel for
the service visits but at other times, a larger vessel is
required for the 36 hour journey to the buoy. During the winter
months, the periods when the strength of the wind and the height
of the waves are less than the limit for carrying out servicing
work are short and infrequent. Forecasting their occurrence
and length over 36 to 48 hours ahead is difficult, yet if an
emergency service visit is required to repair a serious fault,
it is important that this takes place as soon as possible in
order to minimize the loss of data. EMI are also anxious not
to lose data since there is a cash penalty under their contract
with UKOOA which soon applies. On the other hand abortive
visits can add significantly to the overall cost. The limiting
weather conditions are reached when boarding the buoy is
difficult and heavy spray starts to enter the hatch. Work inside
obviously becomes difficult when the motion becomes excessive.

An emergency visit

The difficulty experienced in deciding to embark upon the first additional visit early in 1979 illustrates the weather forecasters problems. On Friday 23 February, it was decided to carry out an additional visit and planning commenced immediately. Contact with the London Weather Centre late on Sunday 25 February reported :

"Wednesday and Thursday, sea expected to be relatively calm at around 6 ft, with winds generally not more than 15 knots. Will tend to increase through Thursday. Two fronts expected through DBI area, one each day, but these should take no more than 2 hours to pass through. Beyond Thursday winds will strengthen further and sea will become unsuitable".

This forecast seemed promising but unexpected mobilisation problems created delays. After checking with the London Weather Centre at 0300 on Tuesday and still obtaining a forecast which indicated that conditions should be within the limits set, the vessel started on its journey. However within a few hours the forecasts started to deteriorate further and by 0930 the forecast was :

"Winds for Wednesday now expected 25-30 knots occasional 35 knots, WNW backing SW later. Waves 12 ft increasing to 16-20 ft. Similar severe conditions continuing through Thursday with no immediate prospects for improvement thereafter".

Since these expected conditions would be too severe for boarding the buoy, it was decided to abort the service visit and return to port. It was not possible to organise a visit because of unsuitable weather until 16 March, three weeks later, and even then the period of suitable weather was too short to complete all secondary tasks.

A repair to the mooring

Integrity of the moorings is obviously important and the advantage of a three-point over a single-point mooring is illustrated by the findings during the service visit in September 1979. During this visit, divers carrying out a routine check of the equipment beneath the buoy were surprised to discover the braidline in the east mooring had been cut 100 ft below the surface and was dangling free. Unfortunately the depth was too great for the divers to inspect the end at close range. Although the east mooring is not expected to carry a large load, it was surprising that inspection of the records for the previous weeks failed to identify any event in the data which would indicate the time of occurrence.

Whilst the weather continued from the prevailing south-west direction, the two remaining moorings were capable of holding the buoy but its behaviour and the fate of the two remaining moorings if the weather changed to a north-east direction was unknown (Fig. 11). An emergency visit was commenced on 7 October with a larger vessel capable of lifting the 8 ton anchor and chain. In case grappling for those components on the sea bed was unsuccessful, a complete set of mooring components including anchor was taken out on hire. In spite of the problem of wrecks found close to DBI, the anchor, chain and thimble were successfully found and lifted from the seabed. They were accurately re-laid at the end of a total period of 2½ days. Inspection of the two parts of the recovered braidline showed clearly that it had been cut by a trawl line. Abrasion marks were present on the line below the cut and a section of 1" mesh fishing net, illegal in the UK, was attached to the cut end of the line.

Other local problems

On two occasions, the buoy has been found to be covered in thick oil which has made work on it difficult and hazardous. On the visit following the first occasion, further difficulty was caused by the coating of weed feeding on the oil. The underside of the buoy is also a favourite place for goose barnacles. The marker buoys above the end of the moorings disappear very quickly and are thought to be cut free by trawl wires.

In spite of problems such as the ones illustrated in the foregoing paragraphs, the buoy has performed well and regularly collects the data as is illustrated by Fig. 12 which shows in bar-chart form the data returns for September, October and November 1978.

DATA OUTPUT

As with all UKOOA data gathering projects, great care is taken to ensure that good, reliable data is produced which involves not only specification of the requirement but advice from IOS and the Met. Office on quality control and the statistical form in which the oceanographic and meteorological data should be presented.

An example of the consistency and potential usefulness of the data can be seen from Fig. 13 where a number of telemetered parameters from DBI have been plotted by IOS over a period of 10 days in November 1978 (6). The plots include the passage of a depression between 6th to 8th November when the pressure fell to 1,005 mb and the wind increased to almost 40 knots, veering from SE to SW round to NE. As the pressure rose, the cold front dropped the temperature by $3^{o}C$ in approximately 14 hours. (Winds are plotted as hourly "stick" vectors with the vectors drawn in the direction to which it is blowing). Maximum wave crest height rose to 5.5m. The residual current plotted for the period 5th to 10th November, from which the tidal component has been removed by using a sliding 13 hour average, appears to follow the direction of the wind. IOS are examining the surface current data more rigorously since it is a unique, high quality data set.

EVENTS OF SIGNIFICANCE IN S.W. APPROACHES

Two examples will be given which illustrate the importance of improving data measurement and forecasting in the S.W. Approaches. In such cases, DBI has two important roles, both to help to improve the forecasts in the area, one directly with real time measurements and the other indirectly by helping to improve the operational numerical models used by the forecasters.

Chesil Bank overtopped

An unusually high wave activity occurred along the central and western English coasts of the Channel at about the time of high tide during the morning of 13 February 1979. The damage which occurred as a result of the sea over-topping the famous Chesil Bank at the point of connection with Portland, received wide publicity in the national newspapers.

According to the analysis made by IOS (7), the source of the

wave energy appears to have been a depression which at noon GMT
on 11 February was situated about 1500 miles due west of Brest
with a central pressure of 956 mb. (Fig. 14). The westerly winds
on its southern flank were estimated to be of 50-60 knots. It
moved eastward and filled and by noon the following day had
ceased to exist. Such winds are not uncommon in an Atlantic
storm but the waves which are generated usually travel faster
than the storm and so move out of the storm area. What appears
to have been unusual in this case, was that the depression moved
with roughly the same speed (30-40 knots) and in the same direction
as the longer period waves of about 18-20 seconds which it had
generated. It, therefore, continued to feed energy preferentially
into those longer period waves for a considerable time creating
high and unusually long-period waves. The depression filled, but
the packet of wave energy it had produced continued eastwards as
a swell at 30-40 knots, and was detected at about 2400 hours on
12 February by DBI.

    DBI is the only deep water instrument providing evidence to
support this type of analysis. During 12 February, it continuously
recorded waves with a period of around 12 seconds and a significant
wave height of about 4m but by midnight the wave period had
increased rapidly to nearly 17 seconds and by 0100 hours 13 February
to over 18 seconds with a significant wave height of 7 metres.
It continued to record these high values until about noon on
13 February. Other wave recorders operated by IOS at near Cornish
coastal sites also recorded these exceptional conditions.

    This situation has been also interestingly illustrated by
results (3) subsequently obtained using the wave model developed
by the Met. Office and mentioned earlier. Fig. 15 indicates the
energy at midnight on 11, 12 and 13 February. The wave energy,
which is proportional to the square of the wave height, has been
plotted for the particular period of 20 secs and for the direction
required to penetrate the English Channel. The successful
predictions of an event such as this one, requires reliable
environmental data over a period of two days. DBI makes a
valuable contribution to this data.

Fastnet Race Storm
During the August 1979 Fastnet Race between Cowes on the Isle of
Wight and Fastnet Rock off Southern Ireland many yachts were
engulfed by heavy seas with a loss of 15 lives. Draper of IOS (8)
using windfields provided by the Met. Office has analysed the
situation and says that the primary cause of the high waves seems
to have been a lenticular area of strong winds of about 50 knots
which approached from the west along the line of the 50° latitude
on 13 August. The shaded portion indicating winds greater than
50 knots on Fig. 16, continued along 50° latitude, growing in
size and extending slightly north of the latitude (Fig. 17). In
the area of most severe weather, within about 50 miles north of
50°N, on the morning of 14 August, the waves probably achieved
a significant wave height of almost 10m. If one accepts yacht
reports of force 11 and over they might have achieved a
significant height approaching 14 metres. The most likely highest
individual wave would be close to about twice the significant
wave height. Draper says that considering the periods of the two
principal wave system experienced, 10 seconds from the south and
12 to 13 seconds from the west, the extreme waves could have
possessed steep or near-vertical-sided profiles and the individual
wave crests would have been travelling at speeds of about 30-40
knots.

DBI was important both in real time in that it allowed the forecasters to improve the accuracy of the pressure gradient between its position and Ireland and subsequently in the reconstruction of the weather leading up to the storm. DBI was on the edge of the system but its wave measurements increased from a significant wave height of 4 metres at 0200 on 14 August to 6 metres at 0400 which fits in well with the wind field data. They remained at around 6m until about noon on 14 August and then decreased. This case also illustrates the necessity of measurements on a smaller grid than 300 Km if accurate short term forecasts are important.

ACKNOWLEDGEMENTS

A large number of people in several organizations have been involved in the development of the DBI and its subsequent refurbishment and deployment in the S.W. Approaches. The majority of the development funds has been provided by the DoI, owners of the buoy. It was constructed by Hawker Siddeley Dynamics Ltd (now part of British Aerospace), EMI Ltd and R & H Green and Silley Wier members of the Seatek Consortium. Instrumentation was provided by IOS, the Met. Office and AERE, Harwell. EMI was responsible for its refurbishment and deployment in the S.W. Approaches for UKOOA.
Much of the information in this paper has been provided by Mr Martin Morris, Head of LWC, Dr Stuart Rusby, Head of Mechanical Engineering, IOS, who also provided invaluable support to EMI in the preparation of the buoy and its deployment, Dr Brian Golding of the Met. Office, who has developed the Met. Office wave model, Mr Laurie Draper of IOS and Mr Stephen Waites, project manager at EMI. UKOOA has supported the DBI by deploying it in the S.W. Approaches to collect data and permitted a small amount of that data to be published.

REFERENCES

1. M. Morris, London Weather Centre. Presentation to Society of Underwater Technology Seminar 23 October 1979. "Oceanographic and Meteorological implications of oil exploration and production in the S.W. Approaches and the Western English Channel". SUT, 1, Birdcage Walk, London SW11 9JJ.

2. "Guidelines concerning the meteorological and oceanographic observations to be made on offshore installations". Meteorological Office, Beaufort Part, Easthampstead, Wokingham, Berkshire RG11 3DN.

3. B. Golding, "Computer calculations of waves from windfields". Proceedings, The Institute of Mathematics and its Applications Conference. "Power from the sea waves". Edinburgh 26-28 June 1979.

4. "Offshore Installations: Guidance on the design and construction". UK Department of Energy, Thames House South, Millbank, London SW1P 4QJ.

5. Contributions to the UK Data buoy (DBI) Symposium. 23 November 1976. Institute of Oceanographic Sciences, Brook Road, Wormley, Surrey GU8 5UB.

6. J.S.M. Rusby, B.S. McCartney, "The deployment and operational performance of a large discuss buoy at 40°43'N, 8°58'N". Proceedings of COST 43 Seminar, Lisbon 4-6 December 1978.

7. L. Draper, internal report, Institute of Oceanographic Sciences, Brook Road, Wormley, Surrey GU8 5UB.

8. Annex 2A, The Fastnet Race Enquiry Report, The Royal Ocean Racing Club, 20, St. James Place, London SW1A 1NN.

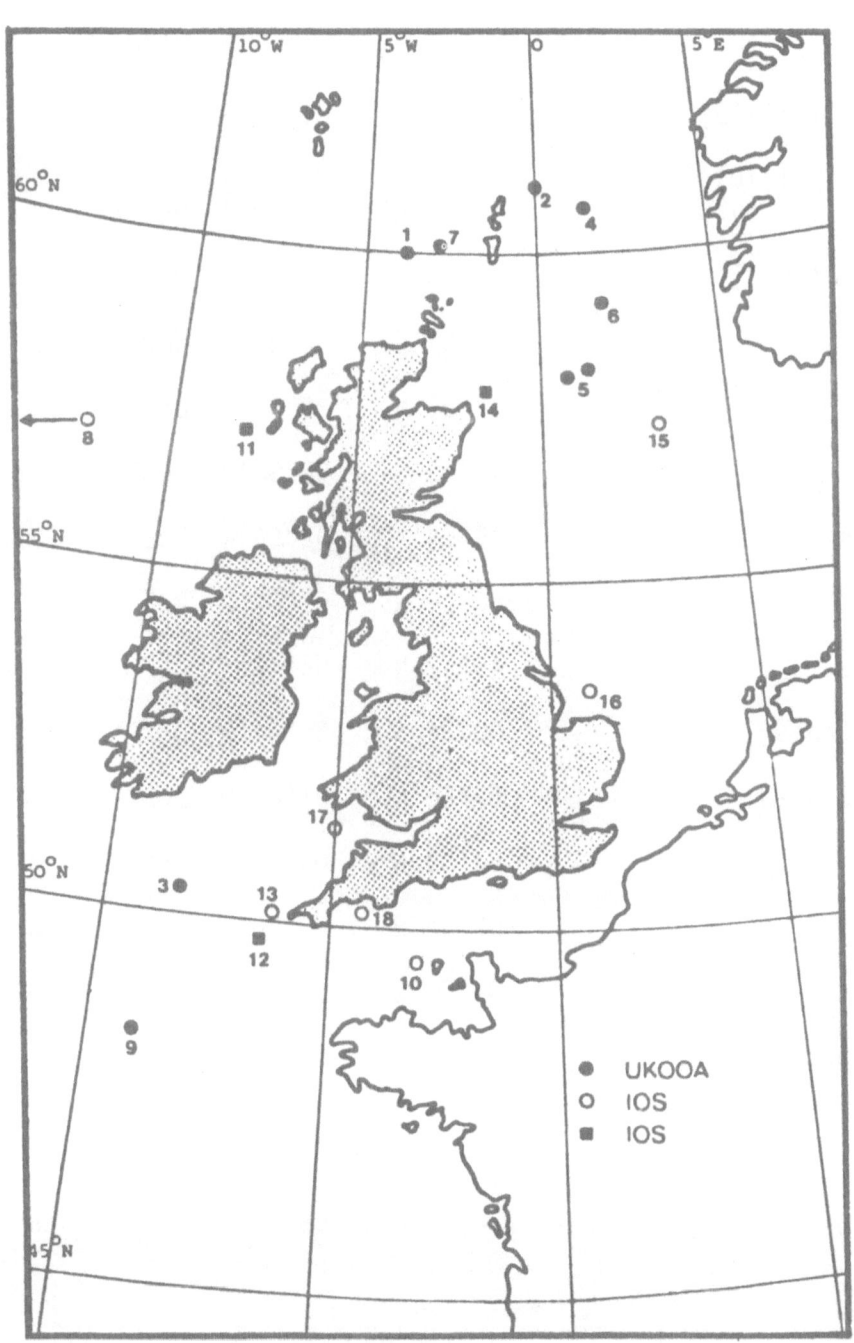

Fig 1. Data collection stations funded by DEn

No	Location	Lat/Long			Contractor
1	Fitzroy (Block 202/5)	60°00'N	4°00'W	UKOOA	MAREX
2	Stevenson (210/21)	61°20'N	00°00'E	UKOOA	MAREX
3	Boyle (92/13)	50°40'N	07°30'W	UKOOA	MAREX
4	Brent C (211/29)	61°04'N	01°43'E	UKOOA	SHELL/MAREX
5	Forties (21/10)	57°40'N	0°50'E)	UKOOA	BP/MAREX
	(22/6)	57°50'N	01°10E)	UKOOA	
6	Frigg (10/1)	59°55'N	02°05E	UKOOA	ELF
7	Foula	60°08'N	02°59'W	UKOOA	IOS/MAREX
8	Lima	57°00'N	20°00'W	DoE	IOS
9	DB1	48°43'N	08°58'W	UKOOA	EMI
10	Casquets	49°55'N	02°56'W	DoE	IOS
11	S.Uist	57°20'N	07°29'W	DoE	IOS
12	Scillies	49°52'N	06°41'W	DoE	IOS
13	Seven Stones	50°04'N	06°04'W	DoE	IOS
14	Kinnards Head	57°56'N	01°54'W	DoE	IOS
15	Famita	57°30'N	03°00'E	DoE	IOS
16	Dowsing	53°34'N	0°50'E	DoE	IOS
17	St Gowan	51°30'N	5°00'W	DoE	IOS
18	Eddystone	50°10'N	4°16'W	DoE	IOS

Fig. 2.   Key to Fig. 1. Wave Data Collection Stations

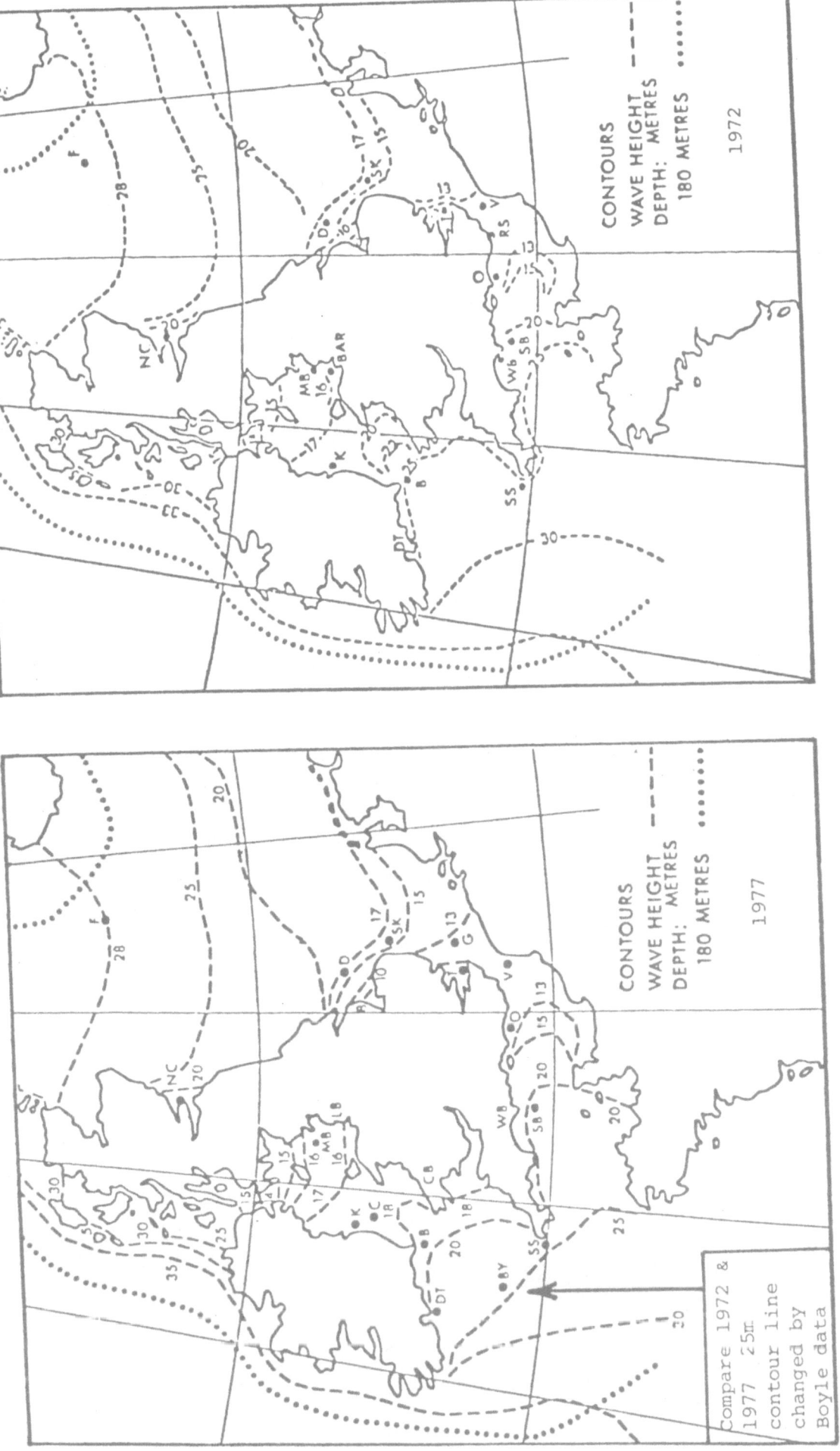

Fig 3. 50 year design wave heights for 1972 and 1977 (D En Guidance Notes)

484

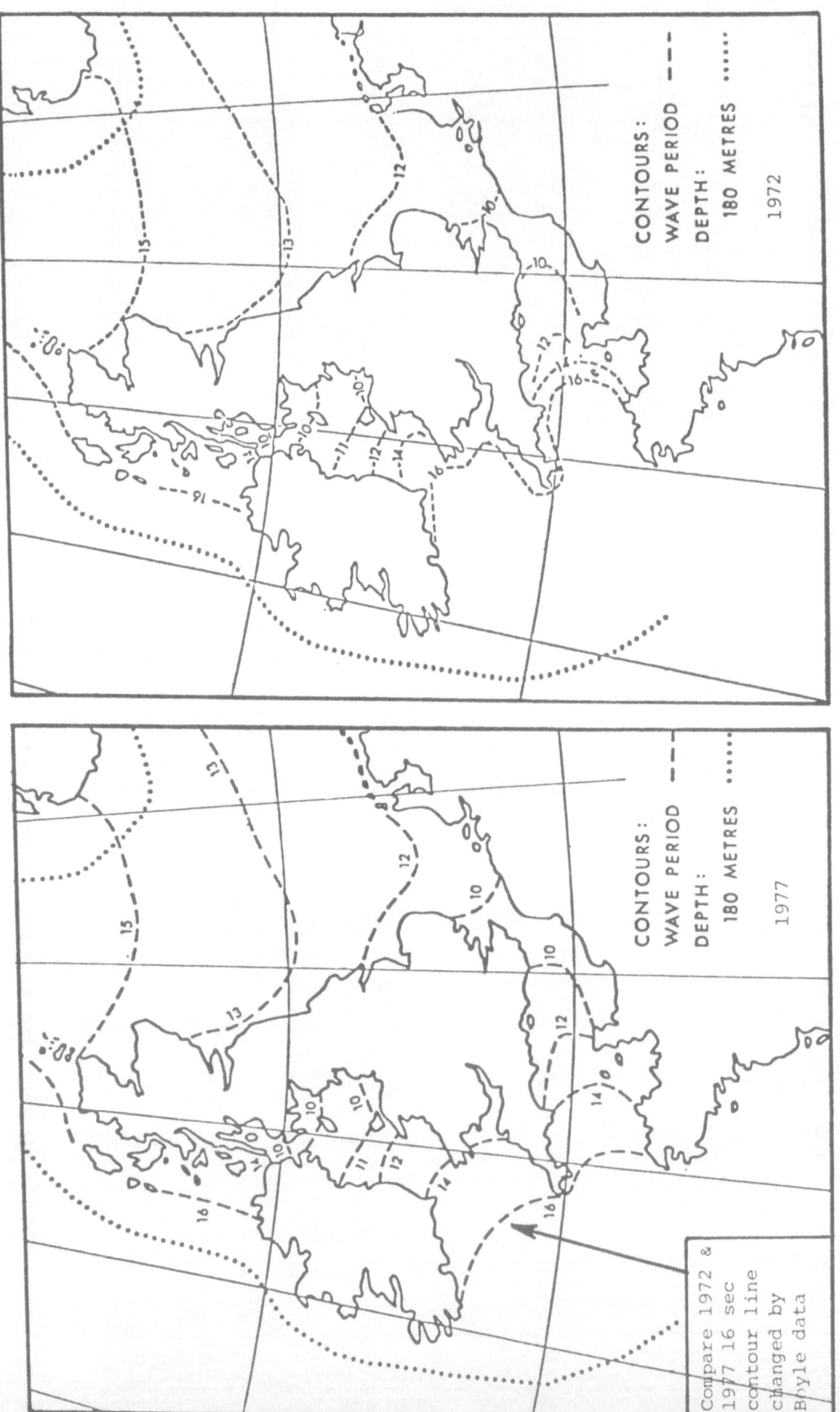

Fig 4. 50 year design wave periods for 1972 and 1977 (D En Guidance Notes)

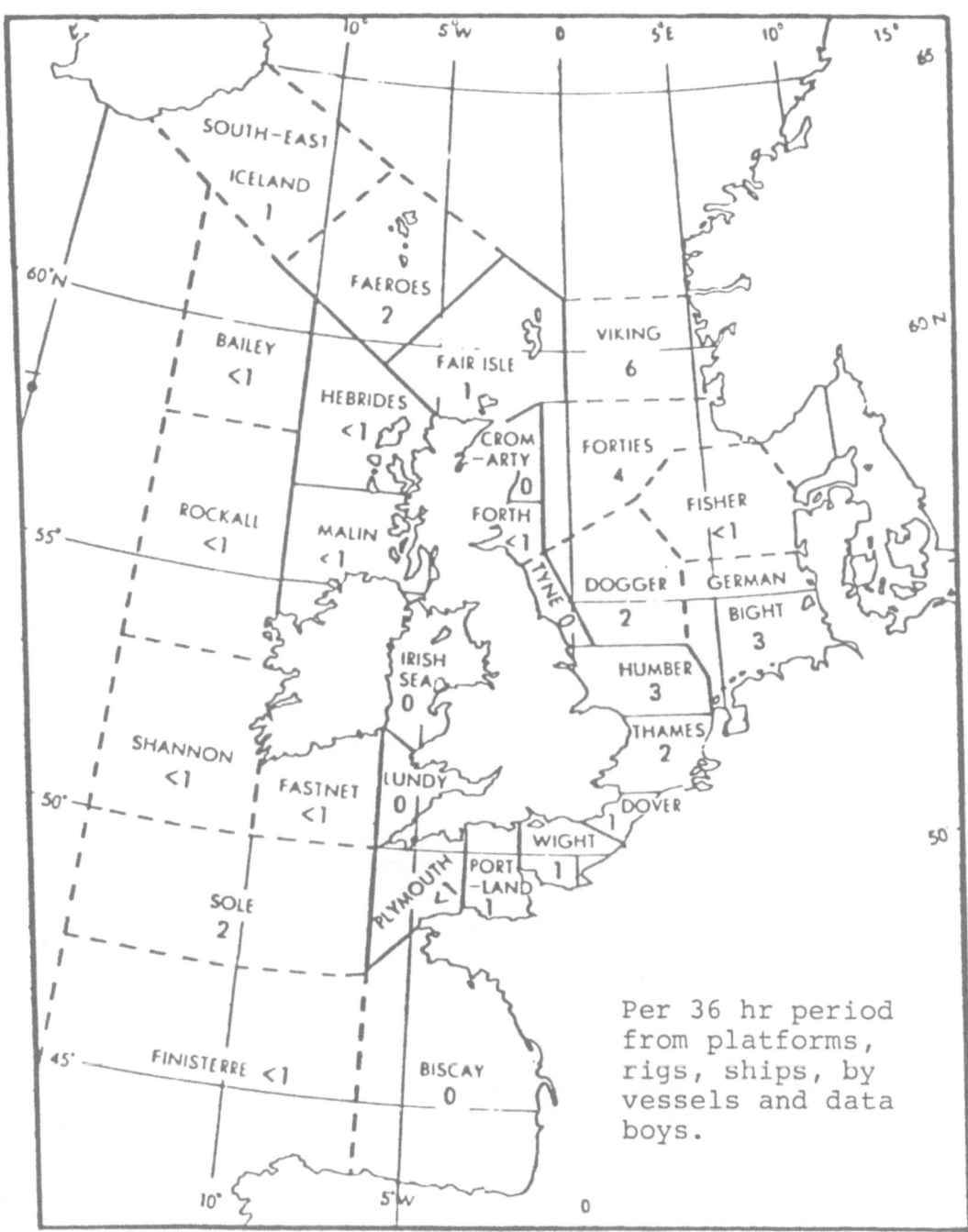

Fig 5. Typical frequency of offshore observations.

UK NATIONAL DATA BUOY

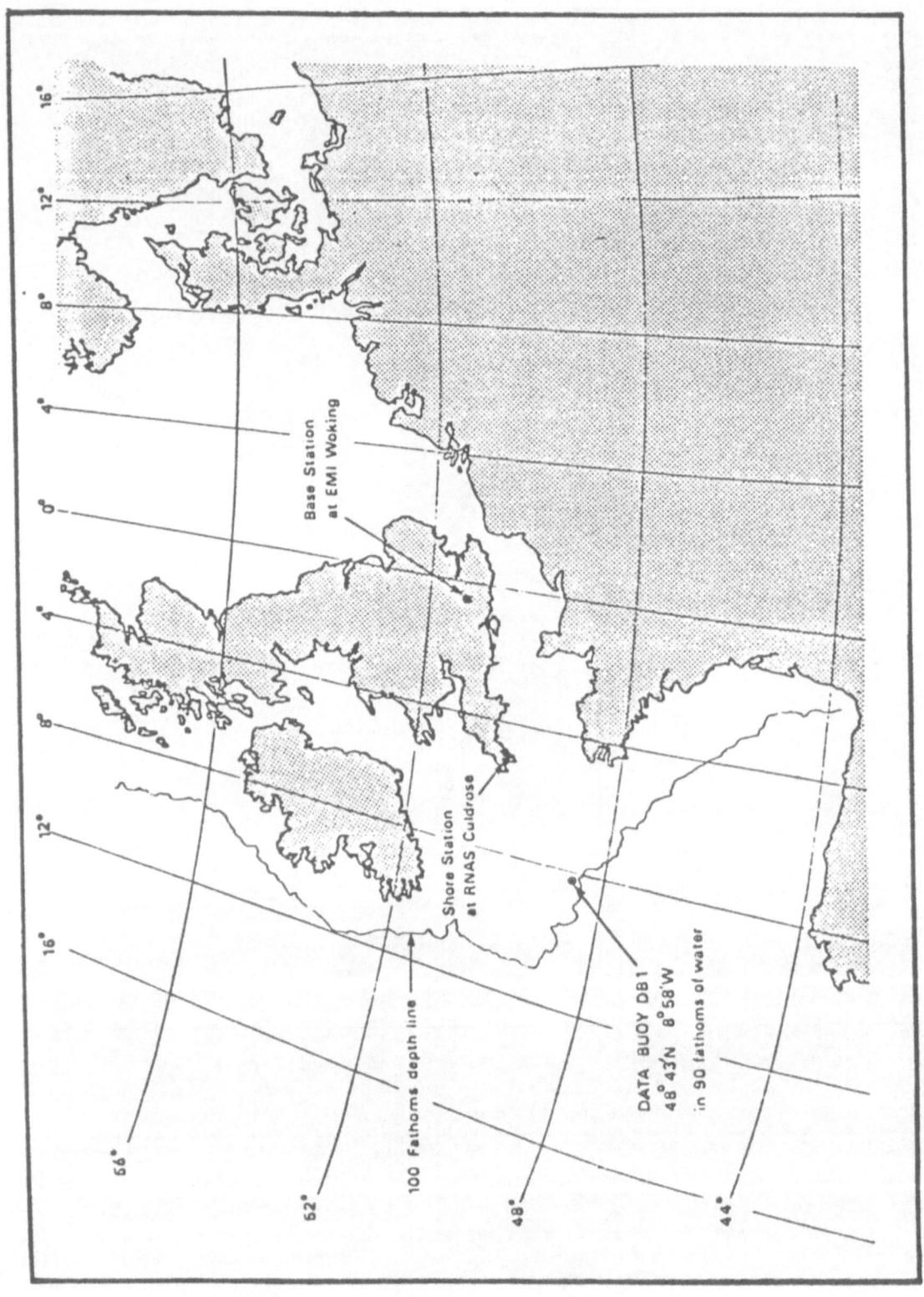

Fig 8. Location of DBI in SW Approaches

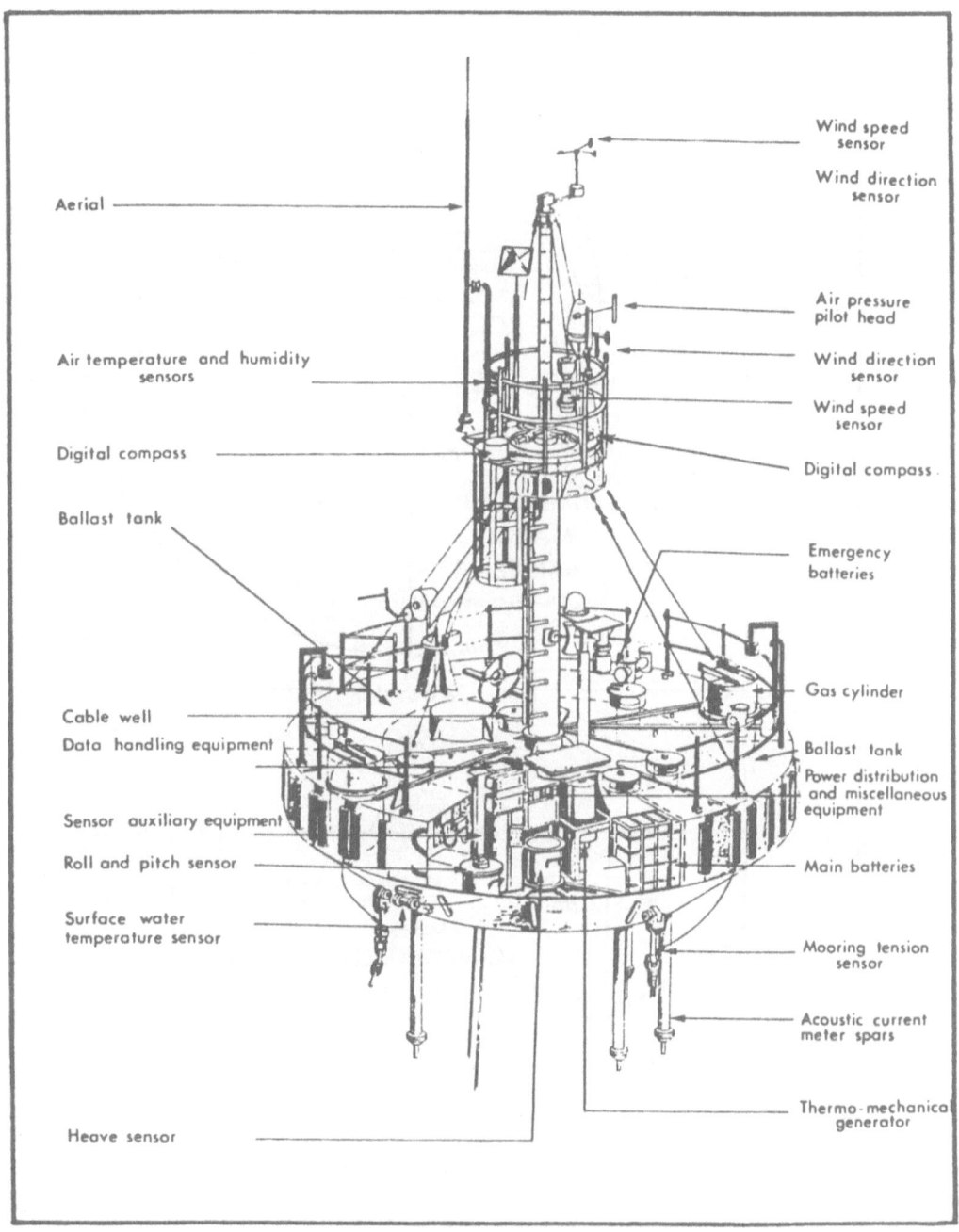

Wind speed sensor

Wind direction sensor

Aerial

Air pressure pilot head

Wind direction sensor

Air temperature and humidity sensors

Wind speed sensor

Digital compass

Digital compass

Ballast tank

Emergency batteries

Cable well

Gas cylinder

Data handling equipment

Ballast tank

Power distribution and miscellaneous equipment

Sensor auxiliary equipment

Roll and pitch sensor

Main batteries

Surface water temperature sensor

Mooring tension sensor

Acoustic current meter spars

Thermo-mechanical generator

Heave sensor

Fig 9. Location of sensors and other key items on DBI

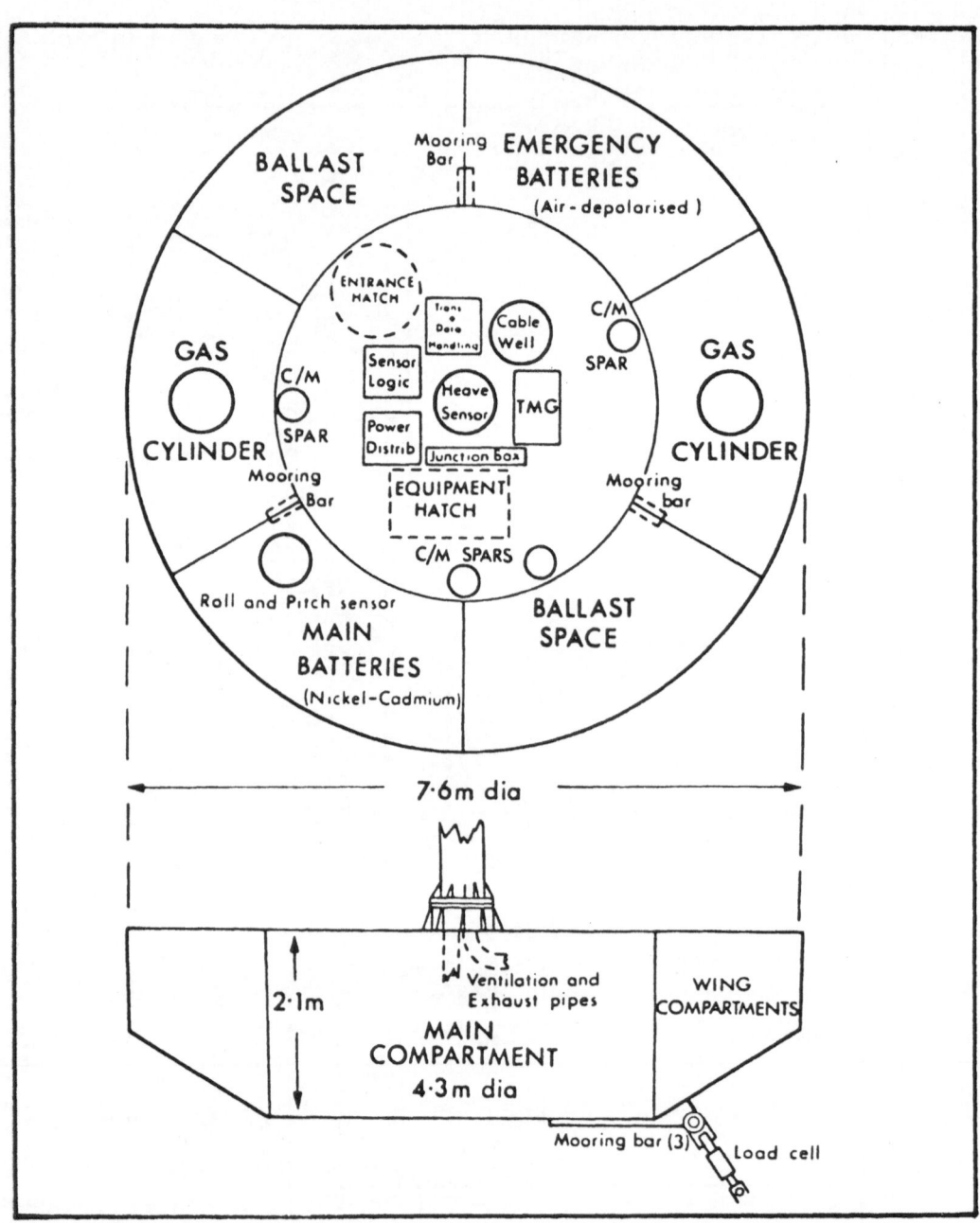

Fig 10. Layout of the components in DBI hull

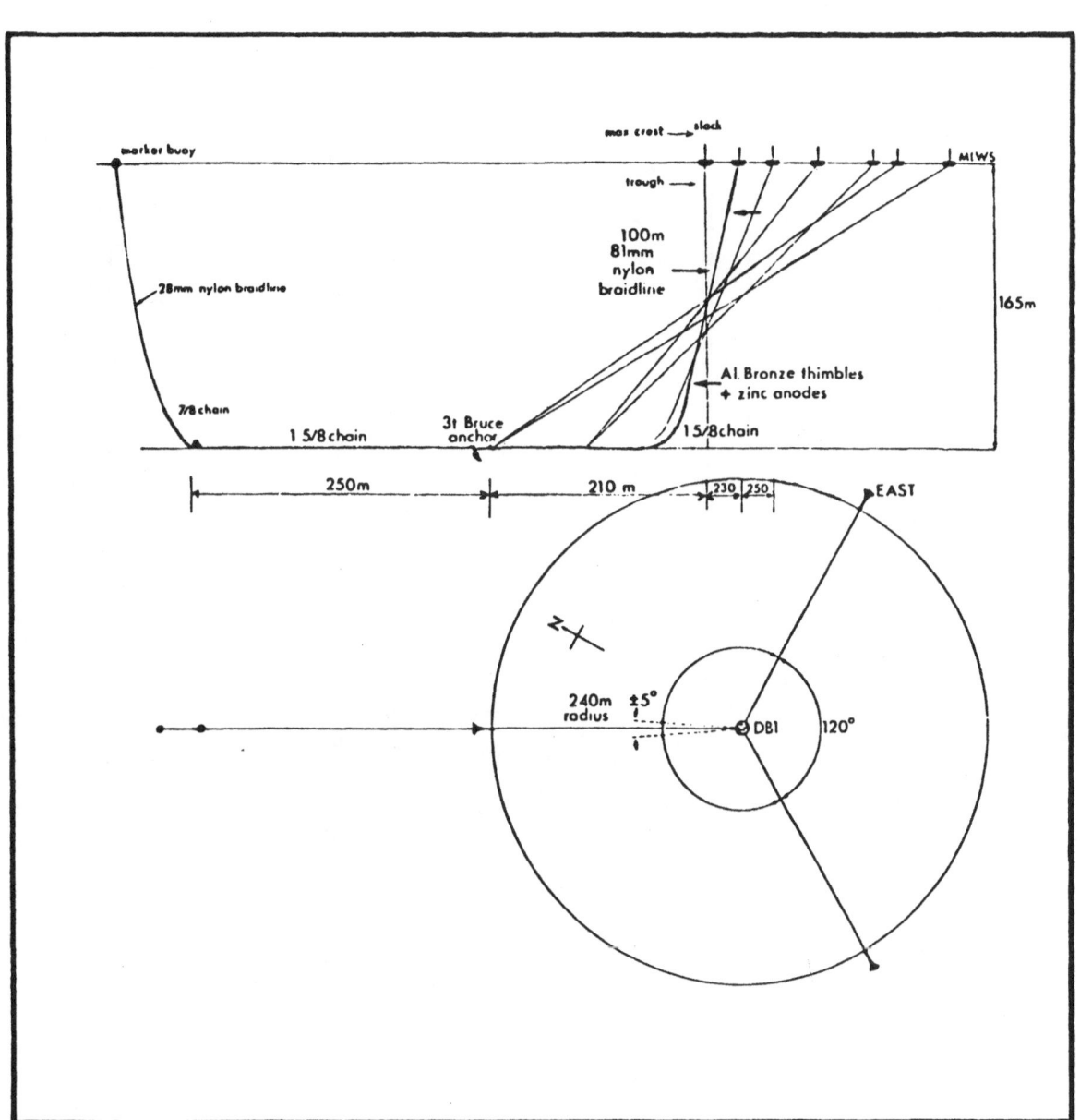

Fig 11. 3 point mooring layout of DBI

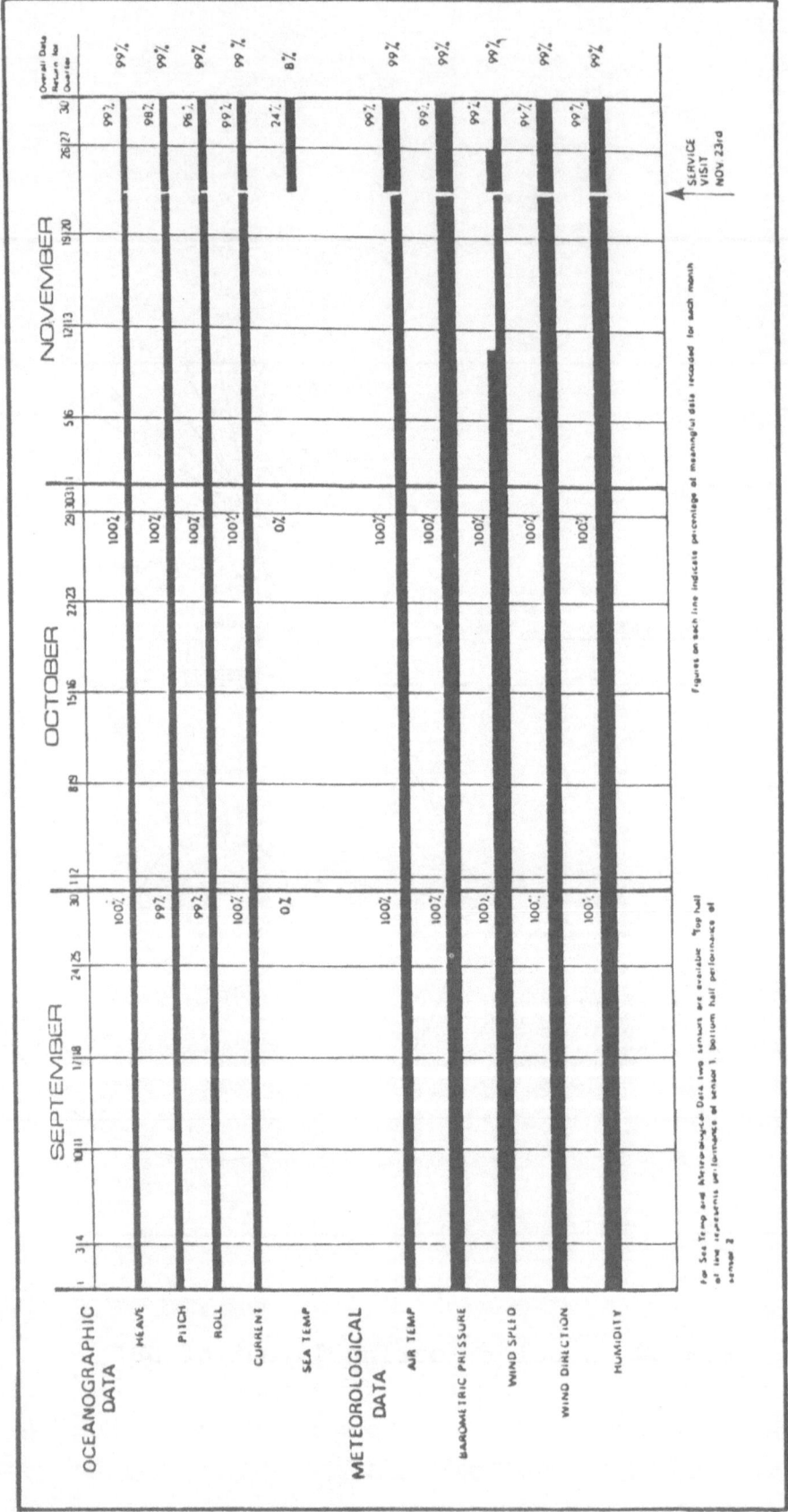

Fig 12. Data return for Autumn 1978 from DBI

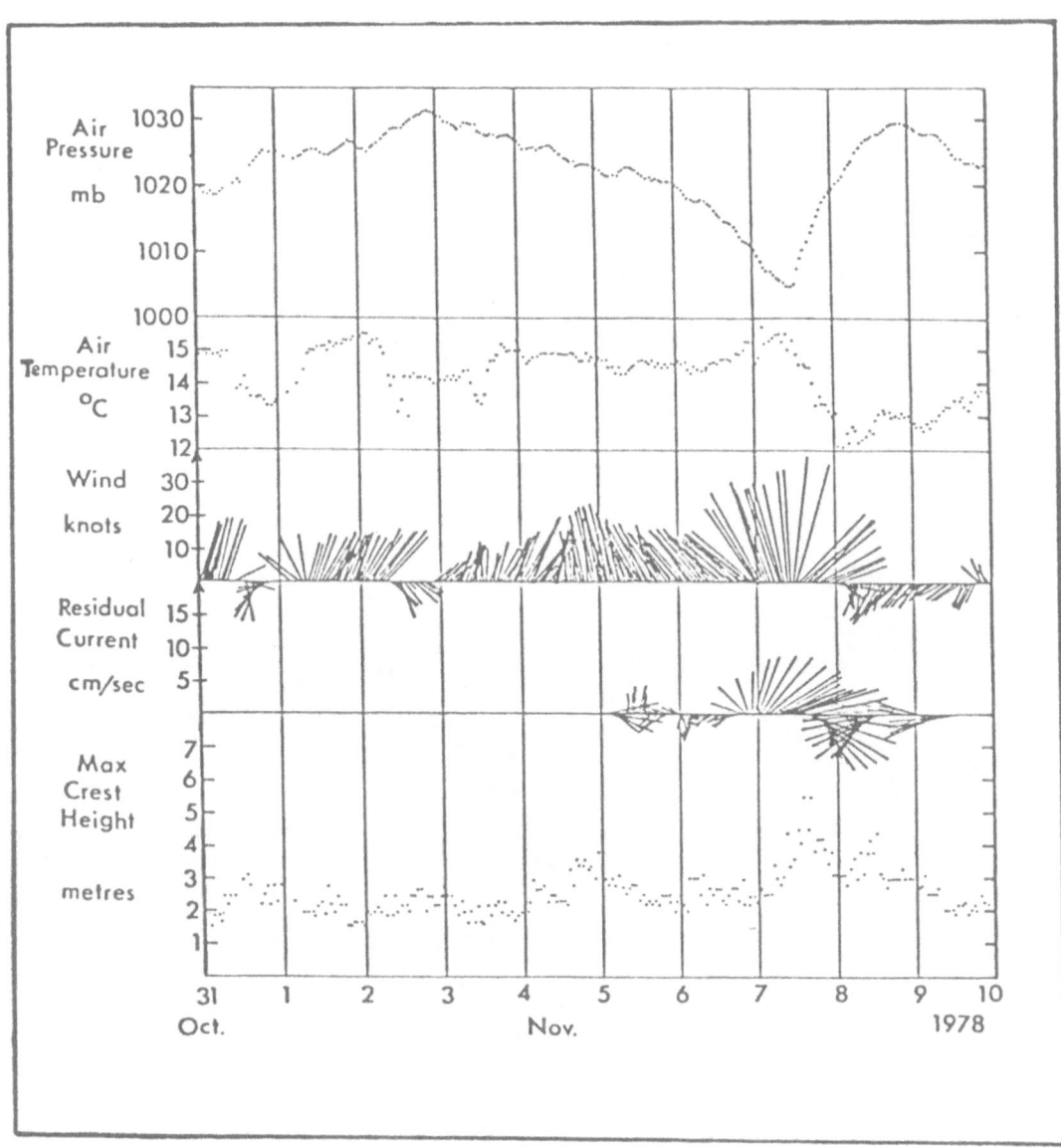

Fig 13. Sample plot of principle measurements on DBI

494

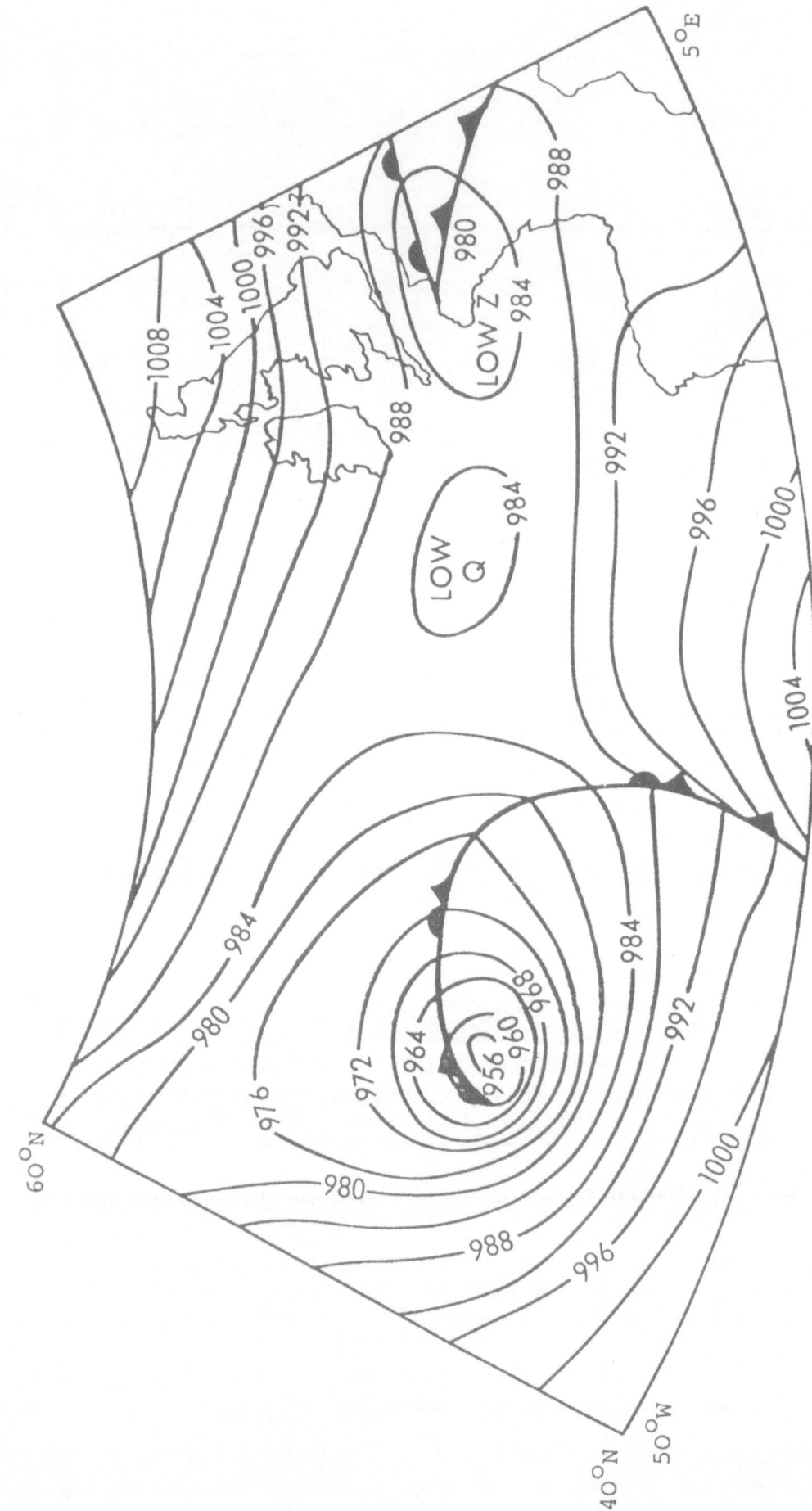

Fig 14. Atmospheric pressure chart 11 February 1979 1200 GMT

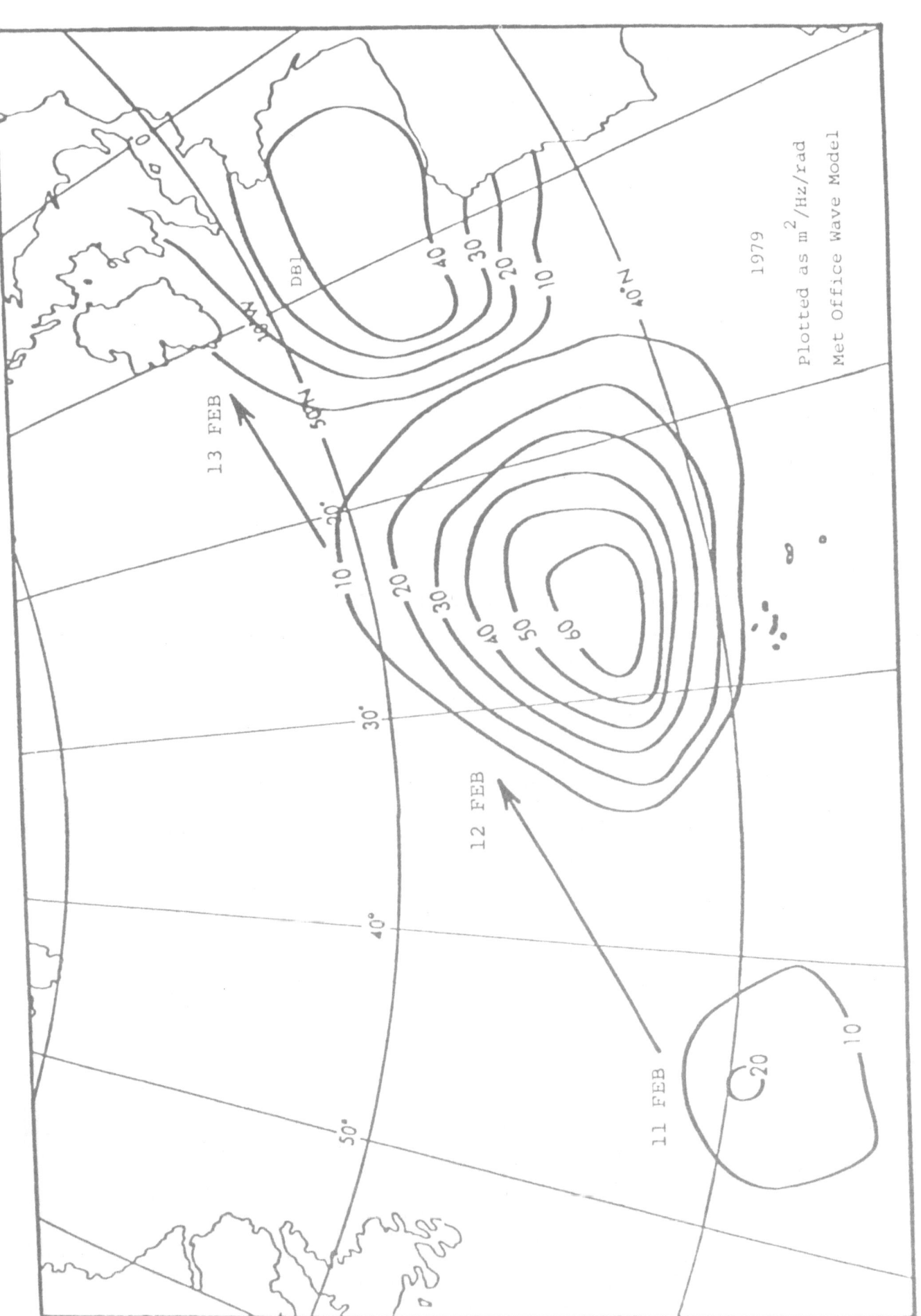

Fig 15. Wave energy at 0001 GMT each day

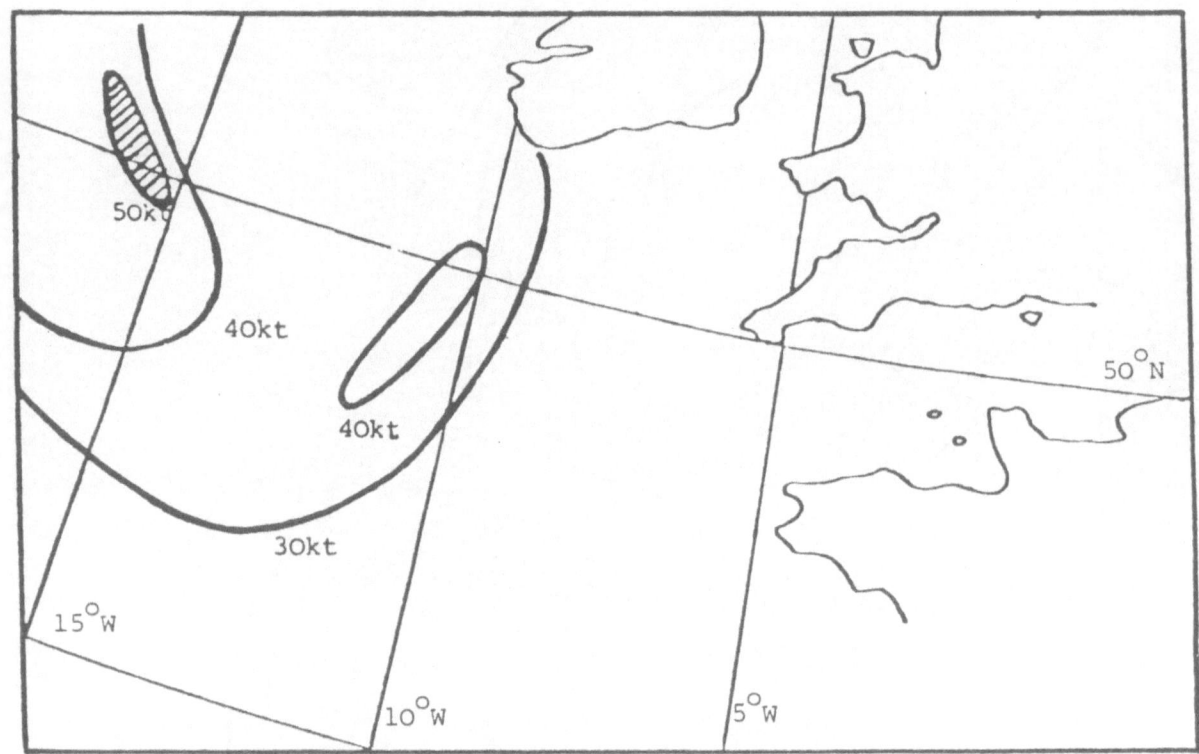

Fig 16. FASTNET RACE. Windfield 1800 GMT 13 Aug 1979

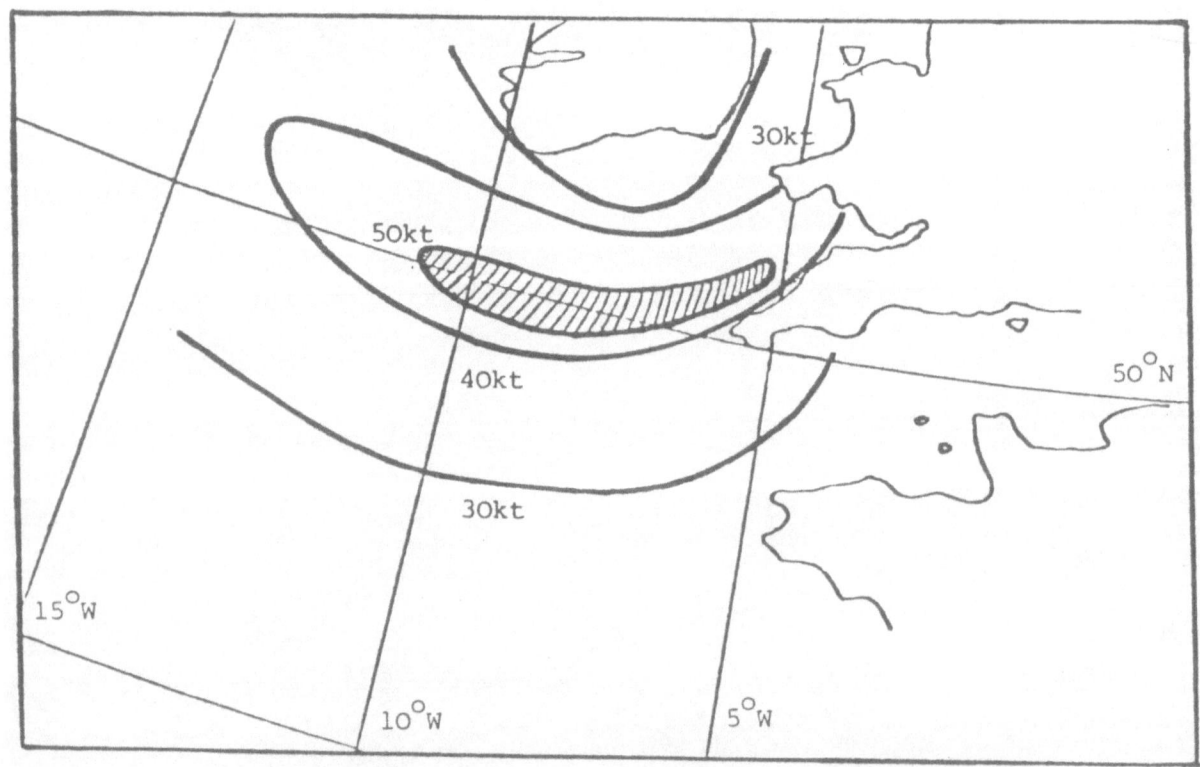

Fig 17. FASTNET RACE. Windfield 0600 GMT 14 Aug 1979

THE IMPORTANCE OF OCEANOGRAPHY TO THE OFFSHORE OIL INDUSTRY

M J Tucker
Institute of Oceanographic Sciences, Taunton, Somerset, UK.

1. INTRODUCTION

Since this conference is concerned with oil pollution, an attempt
will be made to illustrate how recent research in physical
oceanography, particularly by the author's colleagues in the
Institute of Oceanographic Sciences, has relevance to this field.
Two main physical oceanographic processes appear to be important:
sea-surface roughness which can break up an oil slick, and
surface currents which can transport it.

2.0 GENERAL

Depending on circumstances, it seems that a rough sea can produce
an oil-in-water or a water-in-oil emulsion. If the former, then
the surface area available for bacterial attack is increased
and the chances of sinking and dispersal are improved. If the
latter, the resultant 'oil-balls' become almost completely
resistant to bacterial or chemical attack or to further mechanical action by the sea. A further effect of this sort of process
is to carry the oil below the surface, and as explained in the
section on currents, this can have a major effect on its rate of
transport.
 It seems likely that the crucial agent in breaking up the
oil and in particular in emulsifying it is the action of breaking
waves, though boundary layer turbulence and other phenomena may
also contribute. This section will illustrate how recently-
acquired wave data and understanding of wave processes can be
applied to further our understanding of breaking waves in this
context, by demonstrating that the effect of swell is small and
can probably be neglected for practical purposes.
 As explained in Appendix 3, the approach in previous work
on the frequency of breaking waves cannot be applied to this
problem, and a different approach had to be used. This is based
on a non-dimensional "roughness parameter".
 It should, perhaps, be pointed out straight away that the
parameter under discussion here is a measure of the "roughness"
of the sea as, for example, a sailor might perceive it, and not
the "roughness length" which governs the drag of the wind on the
sea-surface.

2.1 The roughness parameter
It is argued in Appendix 1 that a roughness parameter relevent
to the breaking up of oil slicks by waves can be defined by

$$R = (2\pi)^4 g^{-2} \int_{f=0}^{0.9 H_z} f^4 S(f) \, df \qquad (1)$$

where g is the acceleration due to gravity.
 $S(f)$ is the spectral density of the wave system at a
frequency f.

This parameter is non-dimensional and represents the contribution to the mean-square slope of the water surface carried by wave components with frequencies below 0.5 Hz.

The contribution due to all the higher frequencies is not calculable but can be considered to be constant for wind speeds over about 5 m/s.

Fig 1 shows the value of R as a function of wind speed calculated from the Pierson-Moskowitz spectral formulation for a fully-arisen sea, and the curve is physically reasonable.

2.2 The roughness parameter calculated from measured spectra

A year's set of contoured spectra is available from a site approximately 8 miles west of the Outer Hebrides, Scotland, based on measurements from a Waverider in approximately 40 m depth of water (Ref. 1). From this set a number of individual spectra were selected (Fig. 2) representative of the following conditions.

(a) A fully-arisen sea with no swell (Serial Nos. 1 to 4)
(b) Sea plus swell in well-separated bands (Serial Nos. 5 to 8)
(c) Swell only (Serial Nos. 9 and 10)

The process of selection and the characteristics of the records are described in Appendix 2. From these spectra   was calculated using the formula

$$m'_4 = \sum_{f=F_1}^{F_2} f^4 \phi(f) \Delta f \tag{2}$$

where $\phi(f)$ is the measured estimate of $S(f)$ at frequency f

$\quad F_1 = 0.0444$ Hz

$\quad F_2 = 0.4937$ Hz

$\quad \Delta f = 0.00976$ Hz is the separation of the estimates

Then $\quad R = (2\pi)^4 g^{-2} m'_4 \tag{3}$

The results are given in Table 1 and plotted in Fig. 1.

2.3 Conclusions about sea-surface roughness

The most obvious aspects of the results are

(a) The measured values of R for a fully arisen sea are approximately 50% higher than values calculated from the Pierson-Moskowitz spectrum. Inspection of the spectra shows that this is because the energy at higher frequencies drops off more slowly than the P-M formula predicts. Possible reasons for this are complex and will not be discussed here, but strong local tidal currents could be a factor (Ref. 2).
(b) The values of R for the local-sea components in records 5 to 8 are much lower. It happened by chance that these all correspond to offshore or along-shore winds with limited fetches (see Appendix 2).
(c) The swell components in the mixed records contribute negligibly to the roughness parameter.
(d) Even the high swell in record 10 has a roughness factor equivalent to that of a sea generated by a local wind of only about 5 m/s, and some of the measured roughness in this record will in fact be contributed by the local wind of 3.3 m/s.

Perhaps the most useful result of this study is to demonstrate that even a high swell has a comparatively small effect in breaking up an oil slick compared with locally-generated waves. Thus, in calculating the effect of wave activity in a real case,

it is almost certainly adequate to consider only those waves
generated by winds within, say, a 200 mile radius of the point
of interest.

Of course, within the slick itself the oil will damp out the
higher-frequency waves, but the wind-waves from the clear sea
will slowly eat into the edge of the slick, and this is the
phenomenon considered here.

The arguments developed here could be extended. For example,
the mean-square wave slope can be calculated to the correct
frequency limit from the output of pitch-roll buoys. These will
also give a probability distribution for the wave slopes, but
not including the contribution of the very highest frequencies.
However, Cox and Munk (Ref. 11) and Saunders (Ref. 12) have
measured this by optical methods. The mean-square wave slopes
obtained can be compared with the above calculations, but more
importantly the probability distributions will give a numerical
relationship between the mean-square slope and the area of the
sea-surface covered by breakers. Unfortunately the author has
not had time to follow up this line of thought.

It may be also worth remarking that IOS is trying to
develop buoys which will measure the frequency of breaking waves
direction and, of course, the spatial distribution has been
photographed (Ref. 13).

3.   CURRENTS

3.1 General

If there is enough oil present in a slick to cover most of the
sea-surface, then the movement of the oil is likely to be very
different from that of the surface of the water in the absence of
the oil. However, if the oil is broken up or covers only a
small proportion of the surface, then the motion of the surface
water is probably little affected.

The pattern of movement of the surface waters of the sea is
very complex. If we consider only those components of it which
are driven by the local wind, the mechanisms are still not
properly understood even qualitatively. It seems that both wind
drag on the sea surface and breaking waves put momentum into
the near-surface waters, and there is momentum put in by non-
linear effects in non-breaking waves. The resultant shear
produces instabilities such as Helmholz instabilities and Langmuir
cells resulting in local over-turning, as well as the usual sort
of boundary-layer turbulence. Horizontal variations in the
depth of wind-mixing in warmer surface layers, or in the upwards
mixing due to bottom turbulence generated by currents can produce
local fronts with their own circulation patterns. The resultant
currents are subject to the geostrophic force due to the earth's
rotation which can convert changes in current into oscillations
with a period of half a pendulum day (about $14\frac{1}{2}$ hours in the
latitude of the UK). A major attack on this whole problem was
made during the international JASIN experiment which took place
in Autumn 1978 in the North East Atlantic (Ref. 7) and an
improved understanding of the processes at work is beginning to
emerge as the results are analysed. It is likely to be another
2 years or more before the full benefits of this work appear in
published papers.

However, with oil pollution we are usually concerned with
movements over a period of days rather than minutes or hours,
and in this case many of the shorter-term phenomena can be

averaged out. In terms of measurements, one can track the movement of drogues at various depths as was done in the JASIN
exercise, or average the vector velocities as measured by a
suitable current meter. Such measurements are beginning to show
some consistent patterns which conflict in terms of magnitude
with conventional wisdom. The reason for this advance is that
near-surface measurement in the presence of waves is difficult
and techniques for doing so accurately have only recently been
developed.

3.2 Some empirical results

A well-known theory due to Ekman predicts that in a homogeneous
ocean with a constant eddy-viscosity, the current vector at the
surface in the northern hemispere is 45° to the right of the wind
and decreases exponentially with depth while rotating further
in a clockwise direction. There has been little convincing
evidence that such "Ekman spirals" actually exist in practice.
One of the author's colleagues, P Saunders (private communication),
tracked the relative motion of a number of drogues at different
depths during the JASIN exercise, using a precise positioning
system. He averaged the motion over the period of inertial
oscillations (14 hours). He was then able to look at the shear
in the surface layer due to the wind. Relative to the deepest
drogues, which were at between 25 and 30 m, he obtained a convincing Ekman spiral, with the current at a depth of 1 m being
between 30° and 60° to the right of the wind. However, the most
intriguing result was that the current speed at 1 m depth was
between only 1/3% and 2/3% of the wind speed at a height of 10 m
(wind speeds in the region of 10 m/s and, as mentioned above,
relative to the velocity at between 25 and 30 m depth).

An entirely different source gives results in general agreement with the above. The UK Data Buoy, DB1 (Ref. 8) has a
unique vector-averaging acoustic current meter which will give
true average currents at a depth of 3 m independent of wave effects
and routinely over long periods of time. The buoy was initially
deployed approximately 5 km away from the coast near Lowestoft,
Suffolk, England, for testing. The alongshore wind-driven
component of current in this situation has been shown (Ref. 9)
to be 0.87% of the wind speed. This could clearly be influenced
by the presence of the coast and by the shallow water. However,
the buoy has been deployed at a position 48° 42' N, 08° 58' W
since June 1978 in a depth of 170 m of water and approximately
200 km from the nearest land. (Financed by the UK Offshore
Operators Association, (UKOOA) to whom the author is grateful
for permission to publish Fig. 3). One of the author's colleagues,
J S M Rusby (private communication), has analysed a particular
event in November 1978 by taking a running average over 13 hourly
samples to remove the tidal components. The results are shown
in Fig. 3. It is instructive to consider the changes between
00.00 on 7 November and approximately 03.00 on 8 November, when
the wind velocity changed through 180° in direction, and 15 cm/s
between the ends of the vectors. The current was consistently to
the right of the wind and the change was approximately 0.75% of
the change in wind speed. This is an intriguing glimpse of the
information contained in this data set, but scientists are
reluctant to use it because it is confidential to members of
UKOOA.

The message is coming through from results such as the above
that the wind-driven component of near-surface currents in open

water is well under 1% of the wind-speed, compared with the figure of 2.7% often used by engineers. However, work on oil slicks and measurements using cards drifting very near the sea-surface give velocities in the range of 2.6% to 5.5% (Ref. 10). The implication is that there is tremendous shear in the top metre of the water. This would be predicted by normal boundary-layer theory in which the mixing length and therefore the eddy viscosity decreases rapidly as the boundary is approached, and perhaps something of this sort actually happens, though the concepts of normal boundary-layer theory seem to be too simple for the real state of the surface waters. This shear is in the direction of the wind, so that although the current at 1 m depth may be at a considerable angle to the wind, the deviation from the wind direction in the top few centimetres is small.

3.3 Numerical modelling of surface current movement

Dr A M Davies of the Tides and Surges Group at IOS has been trying to compute surface currents using three-dimensional models (private communication). The variation of current in the vertical is modelled by a continuous function and can therefore be computed right up to the surface, though the resolution is in effect limited by the number of parameters of this function which can be carried through the computations.

Perhaps the most important feature of these models in the present context is that they allow the effect of restricted waters to be taken into account whereas the discussion in 3.2 above was restricted to open waters where coasts and other restrictions could be neglected.

The wind drag on the surface is probably modelled reasonably well since the model gives correct values for the heights of storm-surges. However, the values of surface current obtained are critically dependent on the assumptions made about the physical processes in the near-surface layers of the water. His work suggests that the commonly-used assumption that the water velocity is a constant proportion of the wind speed may not be even approximately valid.

The numerical values of surface current which he has obtained so far are not well correlated with local wind speed, but average in the region of 5% of it.

3.4 Conclusions regarding surface current

It is clear that there is a lot remaining to be learned about wind-driven currents in the surface layers of the sea. There is a strong suggestion that for engineering applications where the currents at depths of 1 m or more are the relevant ones, the velocities due to local winds are considerably less than had been thought.

When considering the dispersion of oil the situation is more complicated. There is a large body of evidence which taken together must be regarded as convincing that the top centimetre or two of the water travels at a much higher velocity, and it is in fact not unreasonable to expect a steep velocity gradient in the top metre of the water column. However, if wave action breaks up the oil so that it is carried down to depths of a metre or two, then it will travel at a greatly reduced velocity.

ACKNOWLEDGEMENT

Apart from the acknowledgements made in the text, the author wishes to thank Dr R T Pollard for discussion on the results coming from the JASIN work, R Gleason and I Finn for help with the computations and preparing data, and numerous other colleagues for advice and discussion.

APPENDIX 1 Arguments leading to the definition of the Roughness Factor

A1.1 Progressive waves break when the slope of the water surface exceeds 30° from the horizontal. For full standing waves the slope has to reach 45° before breaking. However, anything approaching a full standing-wave system occurs only rarely, though in passing it may be remarked that it seems likely that such an occurrence was a factor in the recent Fastnet Race disaster. Thus, a reasonable measure of the proportion of the sea-surface which is covered by breaking waves is given by the probability of the slope of the water surface exceeding 30°.

The non-linearity of the wave process is clearly important when calculating this probability, and it will not be attempted here. However, it seems reasonable to assume that for any given mean-square surface slope there is a unique probability distribution for individual surface slopes and thus a single value for the probability of the slope exceeding 30° at any chosen place. If this is so, then the mean square slope is a measure of the proportion of the sea-surface covered by breaking waves, though not, of course, proportional to it.

To be precise, if $\vec{3}$ is the slope vector, that is, the direction and magnitude of the slope at a point on the water surface, then the mean-square slope is defined as $\langle |\vec{3}| \rangle$ where the symbol $\langle \; \rangle$ represents an average for all points on the sea-surface.

Although wave breaking itself is a very non-linear process, it can be shown that in calculating the mean-square wave slope from the wave spectrum there is negligible error for the present purposes in assuming linear theory, that is, that the sea consists of a linear superposition of random wave trains travelling independently of one another. This is because most of the sea-surface is covered by comparatively small slopes. On this assumption it can easily be shown that over deep water:

$$\text{mean-square surface slope} = m_4 (2\pi)^4 / g^2 \tag{4}$$

where
$$m_4 = \int_o^\infty f^4 S(f) \, df$$

Equation (4) is true for any directional distribution of wave energy flux.

When it comes to computing m_4 from either actual spectra or from formulations of wave spectra, a problem arises because of the big lift that has been given to the high frequency end of the spectrum. This can be illustrated by reference to the Pierson-Moskowitz spectral formulation, which is, for a fully-arisen sea,

$$S(f) = A f^{-5} \exp -B(fU)^{-4} \tag{5}$$

where
$$A = 5.20 \times 10^{-6} g^2$$
$$B = 4.75 \times 10^{-4} g^4$$

U is the wind speed at 19.5 m above the sea surface.

Thus
$$m_4 = A \int_o^\infty f^{-1} \exp -B(fU)^{-4} \, df \tag{6}$$

At high frequency the term inside the integral tends to f^{-1}, so that the integral does not converge.

The first point to note is that the Pierson-Moskowitz formula does not apply to indefinitely high frequencies, and in reality the spectrum is terminated by viscous and surface tension effects.

The second point to note is that at frequencies well above that of the spectral peak, $exp -B(fU)^{-4} \simeq 1$ and $S(f)$ is no longer dependent on wind speed: that is, the spectrum is "saturated". Thus taking these two points together, for all wind speeds above about 5 m/s

$$\int_{0.5Hz}^{\infty} f^4 S(f) df = \text{constant} \qquad (7)$$

and for all wind speeds relevant to the present problem one can write

$$m_4 = m_4' + \text{a constant}$$

where

$$m_4' = \int_0^{0.5Hz} f^4 S(f) df \qquad (8)$$

m_4' is finite and calculable (see below).

Thus, going back to equation (4) a roughness parameter R may be defined by

$$R = m_4' (2\pi)^4 / g^2 \qquad (9)$$

When computing m_4 from measured spectra, the enhance importance of the high frequency components again causes difficulty because there are nearly always uncertainties due to instrumental responses and/or noise at high frequencies. For example, the Waverider has a resonance at approximately 0.8 Hz. Thus, again, it is necessary to truncate the integration and the figure of 0.5 Hz used above was, in fact, chosen because it is suitable for use with spectra measured by a Waverider. There is also sometimes a problem with low frequency noise, but this is not critical in the present application and can be solved by starting the integration from a low frequency limit determined by inspection of the spectrum concerned.

A1.2 The roughness parameter as a function of wind speed: calculation from the Pierson-Moskowitz spectrum

Using equation (5) put $x = B(fU)^{-4}$

Then

$$m_4' = \frac{A}{4} \int_X^{\infty} \frac{1}{x} e^{-x} dx \qquad (10)$$

Where

$$X = B(0.5U)^{-4} \qquad (11)$$

The solution of equation (10) is tabulated in standard books of tables such as Abramowitz and Stegun (Ref. 3). The roughness parameter derived from it is plotted in Fig. 1.

Remembering that R is the sum of the mean-square wave slope components carried by spectral components with periods exceeding 2 seconds, the values are physically reasonable.

APPENDIX 2 Criteria for selection of example records

The process and results of selection illustrate some well-known wider principles in a rather elegant way, so they will be described briefly.

The record set used was the first year of recording (5 March 1976 to 28 February 1977) at the Institute of Oceanographic Sciences Waverider installation approximately 14 km west of South Uist, Outer Hebrides, Scotland, in approximately 40 m of water. 2680 valid spectra were available, 92.9% of the possible number. These spectra are listed, together with Hs and Tz, in Ref. 1. This report also contours the spectra as a time series, and this is a great aid in the selection process. Initial selection was made by visual inspection of this report, and further selection

504

by applying a number of criteria based on computations and on
inspection of the meteorological charts.
   It should be said straight away that the North Atlantic
ocean is big enough and its meteorology is sufficiently complicated for the situation to be complex from the wave point of view
for a high proportion of the time. Situations in which effectively
all the wave energy is generated by local winds are rare, and
there are many occasions on which what appear at first sight to
be locally generated waves are in fact generated by nearby wind.
The criteria used for selection of the four locally generated
wave systems were as follows:
(a) Neglibible swell existed before the sea developed.
(b) The spectra show no obvious swell, and show saturated spectral
densities at high frequencies.
(c) Hs/Tz^2 0.085. This is a measure of the steepness of the
sea, and is roughly constant for a locally-generated sea as
measured by a particular instrument system with a reasonable
good high-frequency response (see also Appendix 3).
(d) Uf max $\simeq$ 1.4
   where U is the wind speed at 19.5 metres height over the sea
      f max is the frequency of the peak of the spectrum
(Note that Ewing (Ref. 4) has shown that the winds measured at
the nearby airport of Benbecula have to be multiplied by a
factor of 1.28 to get the correct value for U).
   For a Pierson-Moskowitz spectrum for a fully-arisen sea,
Uf max = 1.37. For a fetch-limited sea it is higher, and for
swell it will be lower.
(e) The local wind direction is between 210° and 360°, in which
range of directions there is no fetch limitation due to land.
(f) The meteorological charts show a fairly simple local wind
system.
   Table 2 shows some of the parameters for the records chosen,
and the spectra themselves are shown in Fig. 2. Looking at
Hs/Tz^2, this varies less than 10% from a value of 0.09 for the
locally-generated seas, but has values in the region of 0.05
for sea plus swell, and for even quite a heavy pure swell
(Hs = 2.6 m) has a value of only 0.03.
   Uf max has values in the region of 1.4 for the locally
generated seas in records 1 to 4, with onshore winds. It so
happens that the selection process has produced records in
category 2 which all occur with long-shore or offshore winds and
for which the wind-sea components are therefore fetch-limited.
For these the peak-frequency is higher than that of a fully
developed sea and values of Uf max are much higher than those
for the onshore winds. For swell, however, the opposite is
strikingly true.
   A detailed discussion of some of these topics as related to
the South Usit spectra is given in Ref. 4.

APPENDIX 3 The Hs/Tz^2 parameter and the frequency of breaking
of indiv   of individual waves
A pure unidirectional single-frequency wave train with a crest-
to-trough height H and with period T has a non-dimensional
"steepness" defined as:

   height/wavelength = $2\pi H/gT^2 \simeq 0.64 H/T^2$

Theory shows that the wave breaks when the steepness $\simeq$ 1/7. Thus,
it is not unreasonable to expect that the value of Hs/Tz^2 in a
real sea is limited to a constant value by breaking, and indeed

the Pierson-Moskowitz formula gives $Hs/Tz^2 = 0.0793$ m s^{-2} independent of wind speed. Measured values depend to some extent on the frequency response of the system used, but do tend to a specific upper limit for a given installation. For this reason contours of Hs/Tz^2 are often shown on the bivariate probability diagrams of Hs vs Tz.

This sort of argument has led some workers to use this parameter when calculating the frequency of individual breaking waves; in effect, they consider each wave between successive zero up-crossings to have the form of a simple wave but to have a height probability following the Rayleigh Distribution (eg Refs. 5 and 6). However, when applied to a sea with swell present, this approach leads to a physically unreasonable answer. This arises as follows:

It can be shown that $Hs/Tz^2 = 4 m_2 m_0^{-1/2}$

where

$$m_n = \int_o^\infty f^n S(f) df$$

$m_2 m_0^{-\frac{1}{2}}$ has the dimensions of acceleration, and it is interesting to note that m_4 used in the present paper has the dimensions of (acceleration)^2. However, if a swell component is added to a local wind sea $m_2 m_0^{-\frac{1}{2}}$ is decreased and the theory predicts a resulting decrease in the number of breaking waves, which is physically unrealistic.

The sensitivity of Hs/Tz^2 to the presence of swell can be illustrated by a simple model. Suppose that there is a "sea" whose spectrum consists of a rather narrow band centred on 0.2 Hz and with unit area. That is, $m_0 = 1$, $m_2 = 0.04$ and therefore $m_2 m_0^{-1/2} = 0.04$. Now add a band of "swell" also of unit area but centred on 0.1 Hz: m_0 becomes 2 and $m_2 \simeq 0.04+0.01 = 0.05$. Thus, $m_2 m_0^{-1/2} = 0.035$. This effect can also be seen in Table 2, where Hs/Tz^2 for local wind-sea only is in the region of 0.09, whereas values for sea plus swell are in the region of 0.06. Hs/Tz^2 is thus a useful indicate of whether one is dealing with a pure locally-generated sea or not.

REFERENCES

1. B C H Fortnum, J D Humphrey and E G Pitt 'Contoured wave data off South Usit' Report No 71 (1979) Institute of Oceanographic Sciences
2. H E Krogstad, L I Erde, K Torsethaugen and S Tryggestad 'Analysis of wave spectra from the Norwegian continental shelf' Proceedings: POAC 79 (Norwegian Institute of Technology, Trondheim) 1 (1979) 547
3. M Abramowitz and I A Stegun 'Handbook of mathematical functions' (Dover Publications, New York)
4. J Ewing 'Observations of wind-waves and swell at an exposed coastal location' Proceedings: Conference on Sea Climatology, Paris, 3-4 October 1979 (Societe des Editions Techniques, Paris) (In press).
5. J A Battjes 'Probabilistic aspects of ocean waves' Report No. 77-2 (1977) Laboratory of Fluid Mechanics, Department of Civil Engineering, Delft University of Technology, Netherlands
6. J H Nath and F L Ramsey 'Probability distributions of breaking wave heights emphasizing the utilisation of the JONSWAP spectrum' Journal of Physical Oceanography 6 (1976) 316
7. R T Pollard 'The Joint Air-Sea Interaction Experiment – JASIN 1978' Bulletin of the American Meteorological Society 59 (1978) 1310

506

8. J S M Rusby, C A Hunter, R F Kelly, J Wall and J Butcher
    'The construction and offshore testing of the UK Data
    Buoy (DBI Project)' <u>Proceedings of Oceanology International</u>
    <u>1978</u> Technical Session J (BPS Exhibitions Ltd, Great Portland
    Street, London W1N 5HH) 64
9. P G Collar and J M Vassie 'Near-surface current measurement
    from a surface following data buoy (DBI) - II' <u>Ocean</u>
    <u>Engineering</u> 5 (1978) 291
10. P Lange and H Huhnerfuss 'Drift response of monomolecular
    slicks to wave and wind action' <u>Journal of Physical Oceano-</u>
    <u>graphy</u> 8 (1978) 142
11. C Cox and W H Munk 'Statistics of the sea surface derived
    from sun glitter' <u>Journal of Marine Research</u> 13 (1954) 198
12. P Saunders 'Radiance of sea and sky in the infra-red window:
    800 to 1200 cm^{-1}' <u>Journal of the Optical Society of America</u>
    58 (1968) 645
13. E C Monahan 'Ocean white caps' <u>Journal of Physical Oceano-</u>
    <u>graphy</u> 1 (1971) 139

TABLE 1   The Roughness Parameter calculated from measured spectra

Type	Identification number	Wind speed m/s	Wind direction degrees T	m'_4	R	Wind-sea component only m'_4	R
				All are × 10⁻³	All are × 10⁻²	All are × 10⁻³	All are × 10⁻²
Category 1	1	9.2	210	0.80	1.30		
Local wind-	2	13.2	270	0.92	1.49		
sea only	3	21.1	230	1.48	2.40		
	4	16.5	300	1.20	1.94		
Category 2	5	8.6	10	.371	.601	.362	.586
Sea plus	6	6.5	90	.328	.531	.319	.517
swell	7	10.5	80	.479	.777	.452	.732
	8	15.8	70	.685	1.109	.673	1.090
Category 3	9	2.0	180	.126	.204		
Swell only	10	3.3	180	.216	.350		

TABLE 2 Data for spectra chosen as illustrations

Type	Identifi- cation number	Date	Time	Hs m	Tz s	Hs/Tz^2	f max Hz	Wind speed U m/s *	Wind direction degrees T	$U f$ max
Category 1	1	17 Jly 76	5.59	2.63	5.55	.0854	.125	9.2	210	1.15
Local wind-	2	2 Aug 76	8.59	3.42	6.20	.0890	.11	13.2	270	1.45
sea only	3	28 Nov 76	8.59	8.29	9.34	.0954	.072	21.1	230	1.52
	4	3 Feb 77	23.59	5.39	7.44	.0974	.090	16.5	300	1.48
Category 2	5	26 Apr 76	16.56	1.28	4.74	.0567	.29†	8.6	10	2.48
Sea plus	6	7 Jly 76	8.59	.87	3.54	.0694	.31†	6.5	90	2.02
swell	7	27 Jan 77	11.59	2.80	7.98	.0440	.28†	10.5	80	2.94
	8	11 Feb 77	2.59	2.25	5.78	.0673	.25†	15.8	70	3.97
Category 3	9	10 Aug 76	17.59	2.56	9.24	.0300	.065	2.0	180	.122
Swell only	10	20 Oct 76	5.59	1.82	12.25	.0121	.088	3.3	180	.289

* Wind speed at 19.6 m over the sea calculated by multiplying Benbecula values by 1.28

† Of local sea component

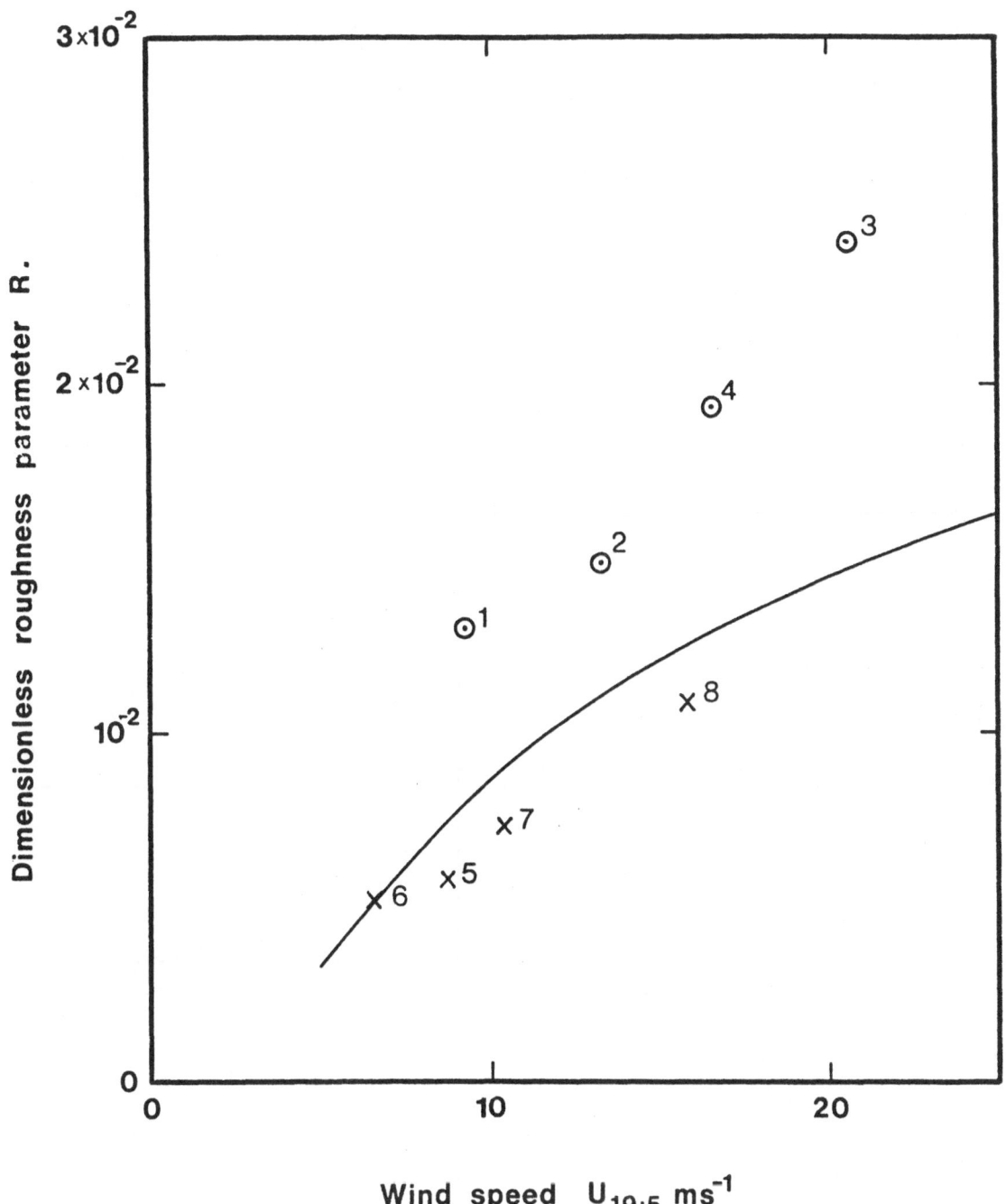

Fig.1 The roughness parameter as a function of wind speed.

——— From Pierson-Moskowitz spectrum for a fully-arisen sea

⊙ From measured spectra : Fully arisen sea.

✗ From measured spectra : Fetch limited sea

Sérial numbers correspond to Fig.2 and tables 1 & 2

510

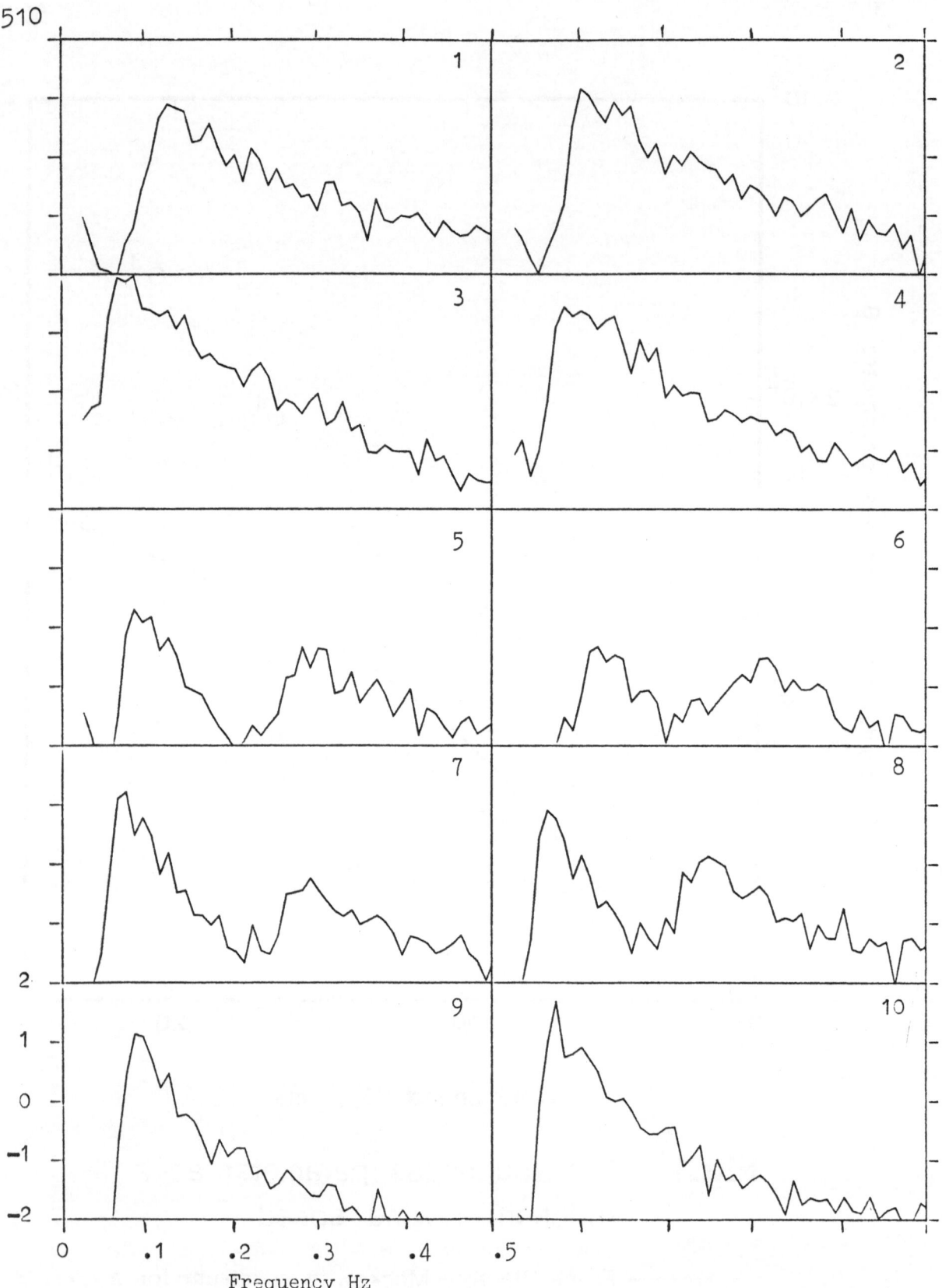

Fig. 2 Spectra of the 12 records used as examples. Serial
numbers as in Tables 1 and 2. Scales all as marked for No. 9.

The ordinate is $\log_{10} \phi(f)$ where $\phi(f)$ is the measured spectral
density in $m^2 Hz^{-1}$

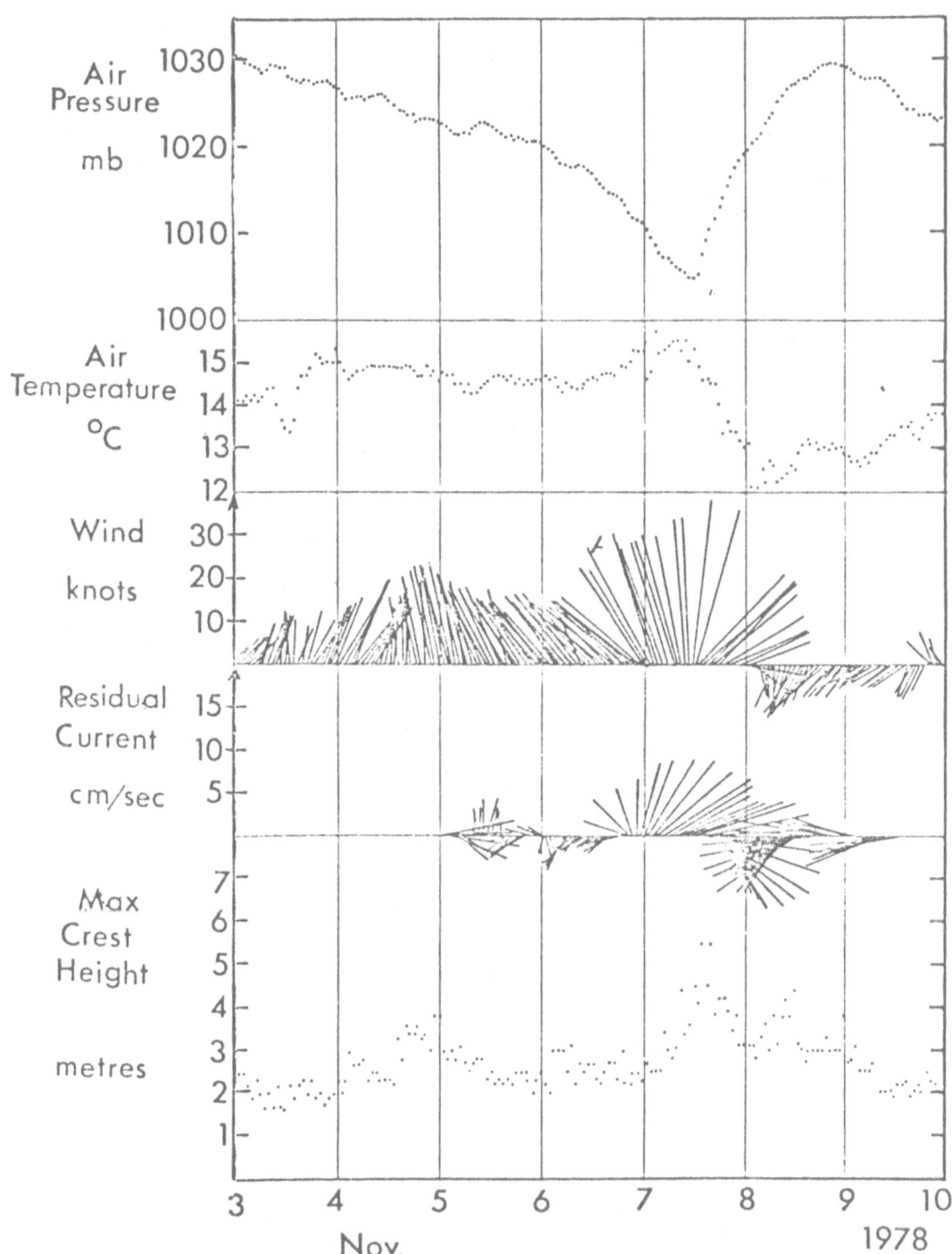

Fig. 3 Data from DB1 in the S.W. Approaches.
Wind & current vectors are running averages
over 13 hours.

SLIKFORCAST - A SIMULATION PROGRAM FOR OIL SPILL EMERGENCY TRACKING AND LONG TERM CONTINGENCY PLANNING

T. Audunson,
V. Dalen,   J.P. Mathisen.
Continental Shelf Institute, Trondheim, Norway

J. Haldorsen,   F. Krogh.
Det norske Veritas, Oslo, Norway

INTRODUCTION

The development of offshore production and transportation facilities has been accompanied by a growing concern for the possibility of oil spills and the associated potential for adverse impacts upon coastlines and the biologically important areas of ocean waters.

Regulatory frameworks are under consideration for most oil production areas for the purpose of balancing the risks of possible oil spill damage against the benefits of proposed developments. Decisions must be based upon predictive impact evaluations of the fate of possible oil spills that are adequate for risk assessment. In addition, when an oil spill occurs, planning and execution of protective measures to minimize impacts likewise require the capability to forecast the short term and long term behaviour of the spilled oil.

Several models for simulating the drift and spread of oil on the sea have been developed. A review of such models may be found in several publications (Ref. 1 and Ref. 2).

In this presentation we shall present a model which we refer to by the name "Slikforcast". The model is composed of two oil spill models, one deterministic model (Oilsim (Ref. 3)) and one statistical model (Sliktrak (Ref. 4)). The detetministic model was originally developed jointly at Det norske Veritas and at the Continental Shelf Institute in Norway, whereas the statistical model was developed at Shell International Maatschappij in the Netherlands. In addition to these two oil spill simulation models, "Slikforcast" also contains a hydrodynamic vertically integrated numerical model for generating tidal transport and a wind data preprocessor suitable for handling wind data from meteorological wind models and/or observations.

The development of the "Slikforcast" composite model is financed by E & P Forum. The Forum also took an active part through its "Subcommittee f, - Working group 3" in the initial planning of the project. The working group has also acted as a steering committee for the work. The model as such is available to all the Forum members.

In the development of the model care has been taken to make it general enough for application to most areas. In what follows, we will give a brief presentation of the different components of the model and also illustrate its use with one example. The general layout of the model is illustrated in Fig. 1.

2. OIL SPILL MODELING

Drift and spreading

Both the deterministic and the statistical model compute the drift and spread of an oil spill by following the drift of individual oil lots ("slick-lets") released at a given source at prescribed time intervals. Each oil lot is treated individually and the sum of all the lots form the spill. It is thus a LaGrangian type simulation.

Oil Drift. The drift of the oil is computed by adding vectorially the wind induced drift, the drift resulting from residual currents and for the deterministic model also the drift induced by tidal currents. (See also Sec. 3.) The drift velocity (U_D) is therefore, expressed as follows:

$$\vec{U}_D = \alpha\vec{U}_w + \vec{U}_{res} + \vec{U}_{tidal} + \vec{\varepsilon} \tag{1}$$

where $\vec{U}$ denotes velocity vector and the subscripts w, res and tidal denotes wind, residual currents and tidal currents respectively. In the above expression the wind factor α is set to 2.7 % of the wind at 10 m (U_w) and the wind induced current is deflected 12° to the right to account for the Coriolis effect. The symbol ε denotes a fluctuating velocity component.

The wind velocities used may be taken from adequate wind observations, from a simulated wind field on a gridded net or from statistical wind information. A more detailed description of this is given in Sec. 3.

The residual currents in Eq. (1) will have to be supplied from available information from the area of interest.

The tidal currents may be represented by their M_2 and S_2 harmonic constituents. These must be supplied to the model either as observational data or, alternatively, the tidal velocities may be calculated for the region of interest using the hydrodynamic simulation model described in Sec. 3.

The effect of waves upon the surface drift is not accounted for in the simulations.

It is further assumed that the drifting oil responds instaneously to a changing wind and that the water currents transfer their full velocities to the oil.

Spreading. The almost "classical" approach assuming a continuous slick and then compute the horizontal spread from a balance of gravitational and inertial forces, of frictional and inertial forces and finally surface tension induced spreading is included in the deterministic Oilsim model (Ref. 3, 5, 6). Field observations of crude oil spills in the open sea have demonstrated, however, that the spreading mechanisms are more complex and not described by such simple expressions (Ref. 7, 8, 9, 10 and others).

The observations referred to above showed that crude oil very quickly forms two distinct phases on the sea; a thick oil phase (1-20 mm thick) consisting of highly viscous emulsified oil and a thin oil or blue shine phase (0.01 - 0.001 mm thick). Thick oil, furthermore, appeared on the surface not as a continuous slick but as several small slicks or strips of oil which gradually broke down into smaller and smaller lumps of oil floating on the surface. The major portion (> 90 %) of the oil on the surface was estimated to be within this "thick" oil part.

In the deterministic computations the total sea surface area affected by the drifting oil is estimated by enveloping the area covered by the trajectories of all the oil lots up to the given time. This area is defined as the thin oil or the blue shine area within which mainly thin oil film may be found.

The instantaneous position of each individual oil lot is combined to form the main slick area. Within this area the major portion of the spilled oil is found. The principle of these simulations is shown in Fig. 2. A further discussion of this is given in Ref. 8.

For the statistical computations the spread of the oil slick is not computed.

Fate of Oil

The amount of oil on the surface at any time is a function of many factors. Waves and surface turbulence are dominant factors governing the natural dispersion of oil into the water column. Evaporation is the dominant factor governing the loss of oil to the atmosphere. Other phenomena governing the amount of oil on the water surface are dissolution, biological degradation, photochemical oxydation, sinking due to adherence of suspended sediments, sea spray etc.

Of the several phenomena causing removal of oil from the surface by "natural" effects only evaporation and natural dispersion have been included in the model. These factors are dominant factors governing the amount of oil on the sea.

Evaporation. In the deterministic model evaporation of oil from the sea surface may be computed by means of a simple model for evaporation of crude oil. In this model the oil is represented by at least two components, one inert and one evaporation part, or up to 16 different components, of which one is inert. The choise of weather or not to use a multi- or a two-component model will depend upon the amount of information available.

The general formulation of the evaporation model is as follows:

$$\frac{dQ_i}{dt} + \lambda_i^{evap} Q_i = x_i \rho_i q \tag{2}$$

$$\Sigma x_i \text{ and } \Sigma z_i = 1 \tag{3}$$

$$\Sigma Q_i = Q_{tot} \tag{4}$$

where Q_i is the weight of oil of component i on the surface, Q_{tot} is the total amount of oil on the surface, t is time, x_i is the volume fraction of each component, z_i is the molecular fraction of each component and q is the volume rate of oil discharge per unit of time. The evaporation parameter λ_i^{evap} is a function of the wind velocity $((W)(m/s))$, the gas pressure of each constituent (p_i), the thickness of the major portion of the oil slick $(\delta$, assumed constant), the oil density (ρ) and the molecular weight of the oil (M_o).

The expression for λ_i^{evap} reads:

$$\lambda_i^{evap} \ (s^{-1}) = (7.4 \times 10^{-3} + 1.87 \times 10^{-3} W) \frac{p_i M_o}{RT \delta \rho} \tag{5}$$

where R is the universal gas constant and T is the temp in ^{o}K of the evaporating gas.

In the above Eq. (5) the expression in paranthesis accounts for the effect of the wind. The expression has been taken from Throne (Ref. 12) assuming similarity in evaporation of oil and water from the sea surface.

It is, furthermore, assumed that the oil evaporates into an infinite reservoir and that the simple Raoult's law applies to the evaporating gas. The value of the gas pressure p_i for each constituent is computed by an empirical relationship as a function of the oil temperature and the boiling temperature of the constituent in question.

Solutions to the above Eq. (2) have been carried out for Ekofisk crude and comparisons between observations and simulations are shown in Fig. 4. If evaporation simulations are not feasible, experimental curves as the ones shown in Fig. 4 may be used directly.

For the statistical program the evaporation curve is represented by two straight line segments, one for the initial phase and one for the asymptotic phase.

Natural dispersion. The importance of natural dispersion upon the amount of oil on the surface at any time has been clearly demonstrated both in experimental spills (Ref. 10, 9) and in the recent larger oil spills (f.ex. the Bravo blowout, the Intoxic blowout and the Amoco Cadiz oil spill).
    In the deterministic model, the natural dispersion is calculated by adding a dispersion term to Eq. (2). The equation then reads

$$\frac{dQ_i}{dt} + \lambda_i^{(evap)} Q_i + \alpha \, Q_i = x_i \rho_i q \qquad (6)$$

where α is the natural dispersion coeffisient. The dispersion coeffisient α is expressed as a function of the wind speed according to the following equation:

$$\alpha = \alpha_o \, (\frac{W}{W_o})^2 \qquad (7)$$

where α_o is the dispersion rate at W_o. For Ekofisk crude the value of α_o is estimated to $0.1 \rightarrow 0.15$ day^{-1} for $W_o = 8.5$ m/s (Ref. 8). In Eq. (7) the windspeed W is the wind speed at time t.
    For the statistical computations the rate of vertical natural dispersion is expressed more simply as the percentage of oil dispersed per day. The value of the daily dispersion rate (α') for different ranges of wind velocities has been estimated from Eq. (7) for Ekofisk crude. The form of the dispersion equation for the oil budget is then

$$Q_{dissipated \ per \ day} = Q_{surface} \ \alpha' \qquad (8)$$

where Q now refers to the total amount of oil on the surface.
    The range of values for α' used in the statistical calculations for different wind velocities are listed in Table 1 below:

Table 1: Examples of dispersion constants used for the statistical simulations (Ekofisk crude).

Wind force m/s		0-8	7-13	13-20	>20
Dissipation constant λ'_n in % per day	First 10 days	1-9	5-23	20-46	40-59
	Remaining drift time	0-7	0-23	0-46	0-60

In the above table the upper and lower limits of the dispersion rates during the first 10 days correspond to values estimated from Eq. (7). For the following days the lower limit has conservatively been set equal to 0 %.

Spill clean up. The ability to undertake spill clean up operations and their effectiveness will of course, be a function of sea state or the wind conditions. In the deterministic model clean up operations is roughly included by

simply assuming that a given weather-dependent percentage of the oil is removed by primary combat actions. As the oil approaches the coast, a similar
technique is used for the secondary or near coast combat operations.

In the statistical model a range of combat efficiencies is given for
each state or wind force range. These values will have to be adjusted according to the level of clean up technology in a given application.

For illustration combat efficiencies used in simulations applicable off
the Norwegian coast is shown in the table below:

Table 2: Examples of combat efficiencies in per cent of oil on the sea at
various wind velocity ranges, used in the statistical computation.

Wind force	Near the platform	Near the coast
0- 8 m/s	10 % - 67 %	10 % - 50 %
8-14 m/s	10 % - 25 %	5 % - 20 %
14-20 m/s	1 % - 10 %	0 % - 5 %
> 20 m/s	0 % - 1 %	0 % - 1 %

Oil-in-water concentration estimates. The natural dispersion of oil into
the underlying water masses will result in hydrocarbon concentrations in
these waters. In the models estimates of these concentrations are made for
the given critical sea areas shown in Fig. 13. The estimates are based upon
a simple box model approach for the water masses in the top layer. Each
critical sea area is divided into a maximum of 25 square boxes.

For each box a mass conservation equation is derived as follows:

$$\frac{d}{dt} (V_{i,j} C_{i,j}) = (C_{i-1,j} - C_{i,j}) F_{i-1,j} + (C_{i,j-1} - C_{i,j}) F_{i,j-1}$$

$$+ D_{i-1,j} + D_{i,j-1} - D_{i+1,j} - D_{i,j+1} - D_{i,j,z}$$

$$+ Q_{dispersed} \tag{9}$$

where $V_{i,j}$ is the volume of box i,j
$C_{i,j}$ is the hydrocarbon concentration of box i,j
$F_{i,j}$ is the net advective flux in the horizontal direction
$D_{i,j}$ is the diffusive flux in the horizontal direction
$D_{i,j,z}$ is the diffusive flux in the vertical direction down into
the underlying water masses

The dimensions of each box is assumed to be 10-15 km in the horizontal
and with a depth down to the characteristic depth of the pychnocline (~20 m).

For simplicity any differences in advective fluxes through the region
in question have been neglected. The advective fluxes for each time step
are computed from the wind induced current corrected for depth by an assumption of Ekman current.

The diffusive losses are computed using ordinary difference techniques
and assuming equal diffusivities in the two horizontal directions (value
according to a time step of one day) and a different and much smaller value
in the vertical direction.

The value of $Q_{dispersed}$ for each block is taken from the solution of
Eq. 6 or Eq. 8.

Numerical solutions to the above Eq. 9 are carried out for given initial conditions, and assuming zero or constant hydrocarbon concentrations in the waters adjacent to the critical sea area. (see Ref. 12.)

Blowout characteristics

In the deterministic model the amount of oil being released per unit of time is given as an input parameter. The rate do not have to be constant with time. A moving source such as a ship may also be specified. The shortest time step (and consequently the smallest individual oil lots which may be handled by the program) is 6 minutes. This implies that we may simulate a spill by a finite amount of oil released at most every 6 minute.

The statistical model accepts a range of blowout rates and blowout durations. The upper and lower limit for a blowout may be specified according to assumed or known characteristics of the well.

If one expects bridging to occur, this will usually occur early in the blowout period, and a minimum duration range may, therefore, be specified (typical one to two weeks). For the case of no bridging a maximum duration range may be given (typical 3 to 6 months for drilling of relief wells). The choise of one or the other blowout duration range depends upon a given probability of bridging. An average value of this factor is estimated to 60 % (Ref. 4). The actual value, however, is dependent upon many factors both of geological and technological nature.

Once the blowout duration is chosen, at random, within the given probability range, then the size of the blowout is chosen, at random, within the given blowout ranges.

3. AMBIENT INPUT DATA

The main input data to the model simulations is wind and current. Since Slikforcast is a composite program built up around the two aforementioned oil spill simulation models, great care has been taken in order to ensure equal requirements on common input data for the two models. In what follows, we will briefly describe the different modes of input data which may be used.

Wind

The program will accept three types of wind data information, i.e.: data from one or more observational stations; data from meteorological simulation models on a gridded net; statistical wind data. An illustration of the handling of the weather data is shown in Fig. 4. The major purpose of the wind preprocessor set up is to generate time series of wind on a format consistent with the requirements of the model. If historical wind data is used, there is no limitations with respect to the permissible length on the time series. The statistical program will accept two years as well as 10 years of wind data. For the statistical simulations it is also possible to pick out a certain time window in the given data set, f.ex. in order to "phase in" the simulations with a given blowout date and/or period.

Data from meteorological simulation models: Both models permit the use of simulated wind data from meteorological models. In fact, it has been assumed that when constructing the grid net for the oil spill simulation model, the grid should be aligned with the computational grid for the most convenient meteorological forcasting model available for the region in question. If there for a given region is more than one major forecasting centre, data from one additional simulation source may be accepted by the model.

This is illustrated in Fig. 5, where the North Sea has been chosen for demonstration only. Here the oil spill model grid is aligned with the meteorological grid from the Norwegian Meteorological Institute (NMI). Wind simulations for this region are, however, also carried out at the London Weather Centre (LWC). The orientation of the LWC-grid is shown in Fig. 5. In the model one may use the LWC-data in addition to or instead of the NMI data. The necessary data transforms from either grid into the oil spill simulation model grid system is incorporated in the preprocessor. No extra programming is required if the grid projections in the two weather grids both are polar stereographic.

For illustration we have in Fig. 6 shown the correlation between wind force simulations from LWC and NMI for December 1975. Although the main purpose of the "double wind source facility" is to make the program as flexible as possible, the results in Fig. 6 also demonstrate that it might be advantageous, whenever possible, to actually use two different data sources in order to quantify variations in the oil spill simulation results due to wind field uncertainties.

Commonly the grid size used in most meteorological models are fairly coarse, typically of the order of 100-300 km. The grid size of the numerical oil spill models are usually smaller and interpolation routines are, therefore, included in order to compute wind force and direction at points within the meteorological grids.

Data from observational stations: Wind data from observational stations may be used instead of simulated wind data if this is more convenient. Any number of stations may be included, and the program user will have to specify the region of representativity for each station used. No interpolation routines is used between each region. Fig. 7 illustrates the division of a region into different observational areas.

In the deterministic model, observational data may also be used to correct the computed velocity field according to the following expression.

$$u_{corr} = u_{sim} + \Delta u \tag{10}$$

where $\Delta u = u_{sim}^{(o)} - u_{obs}^{(o)}$ (11)

where subscript sim refers to simulated value and, superscript (o) refers to the given location of observation, subscribt obs refers to observed value and subscript corr refers to corrected values.

Statistical wind generator. For the case when historical weather data is available in statistical form such data may also be used to generate time series of wind. The generated time series will have the same statistical properties as the time series from which the data were originally formed.

From statistical information on wind force duration (autocorrelation), on wind force distribution and on distribution of wind direction, we may construct first order transition probably matrixes and from these generate the desired time series. The principle is illustrated in Fig. 8.

The transition probability matrixes are generated using the assumption that for each time step from which the statistics is generated, the wind condition may only change from a given state to the neighbouring state. The results is a tri-diognal matrix. From this matrix, however, transition probability matrixes for time steps which are multiples of the initial time step may be constructed by simple matrix multiplication. Thus if the initial matrix $||A||$ is based upon three hour time steps, then transition probability matrixes for 24 hr. time steps is $||A||^8$.

Current

As shown in Sec. 2, the drift of the oil is the vector sum of the wind and the current induced drift.

In the oil spill simulation model, the current induced drift is assumed to be the linear sum of the various current components such as the residual current, the tidal current and a stochastic or fluctuating current component.

The residual current. Both the deterministic and the statistical models employ a residual current. The residual current pattern must be obtained either from general knowledge of the area in question, from measurements, or from computations. The statistical model will accept four different current patterns. These current patterns may represent seasonal variations or variations of shorter duration.

For the deterministic model there is no limitations to the number of residual current patterns which may be employed. An example of the residual current pattern used in the North Sea is shown in Fig. 9.

The tidal current: The tidal current is used only in the deterministic model. The longterm nature of the statistical simulations does not warrant the use of such a detailed and cyclic current pattern.

The tidal current may either be introduced as a simple harmonic expression using two harmonic components, the M_2 and the S_2 component, or using a hydrodynamic model to compute the tidal velocities. Such a model is included as part of the program package.

The formulation of the hydrodynamics equations in spherical coordinates is shown below.

Continuity

$$\frac{\partial \xi}{\partial t} + \frac{\partial Ud}{R\partial\psi} + \frac{1}{R\cos\psi}\frac{\partial Vd}{\partial\kappa} - \frac{dU}{R}\, tg\,\psi = 0 \tag{12}$$

Momentum

$$\frac{\partial U}{\partial t} + \frac{U}{R}\frac{\partial U}{\partial\psi} + \frac{V}{R\cos\psi}\frac{\partial U}{\partial\kappa} + \frac{V^2}{R}tg\psi + \frac{g}{R}\frac{\partial\xi}{\partial\psi} - 2\omega\sin\psi V + g\frac{U(U^2+V^2)^{\frac{1}{2}}}{c^2d} = 0 \tag{13}$$

$$\frac{\partial V}{\partial t} + \frac{U}{R}\frac{\partial V}{\partial\psi} + \frac{V}{R\cos\psi}\frac{\partial V}{\partial\kappa} - \frac{UV}{R}tg\psi + \frac{g}{R}\frac{\partial\xi}{\partial\kappa} + 2\omega\sin\psi U + g\frac{V(U^2+V^2)^{\frac{1}{2}}}{c^2d} = 0 \tag{14}$$

where U, V = ψ and κ components of the velocity vector averaged over the depth d. (U positive southward and V positive eastward)

ξ = water level above an arbitrary horizontal datum.

d = depth of the water column.

c = the Chézy coefficient of friction.

g = acceleration due to gravity.

R = mean radius of the Earth, assumed spherical.

ω = the angular speed of the Earth's rotation.

The equations are solved numerically using a modified version of the original two-dimensional Leendertse scheme (Ref. 13).

The numerical equations have been applied to the North Sea, and results for the M_2 tidal currents are shown in Fig.10. The computed values are found to be in good agreement with other results (Ref. 14).

In order to compute the S_2 tidal component, a rather simple approach is used. Its validity must be checked for each area in question. Assuming a complete linear response of the ocean to tidal forcing, we may use the expressions:

$$\frac{h(S_2)}{h(M_2)} = \frac{V(S_2)}{V(M_2)} \qquad (15)$$

$$\Theta(h(S_2)) - \Theta(h(M_2)) = \Theta(V(S_2)) - \Theta(V(M_2)) \qquad (16)$$

where $h(S_2)$ = diurnal tidal amplitude for the S_2 component.
      $h(M_2)$ = diurnal tidal amplitude for the M_2 component.
      $V(S_2)$ = diurnal tidal velocity for the S_2 component.
      $V(M_2)$ = diurnal tidal velocity for the M_2 component.
      $\Theta(h(S_2))$, $\Theta(h(M_2))$ = Phases of the S_2 and the M_2 diurnal tide amplitudes.
      $\Theta(V(S_2))$, $\Theta(V(M_2))$ = Phases of the S_2 and M_2 diurnal tide velocities.

Eq. (15) and (16) are assumed constant for the region, and the amplitude
ratio and velocity difference may be determined at a tidal reference station.
Using the values of $h(M_2)$, $V(M_2)$, $\Theta(h(M_2))$ and $\Theta(V(M_2))$ simulated from Eq.(13)
(14) and (15), the corresponding S_2 values of h, V and Θ may be computed for
the entire region.

The relationships in Eq. (15) and (16) have been computed for the North
Sea. The values were 0.35 and 42° for Eq. (15) and (16) respectively. The
values were constant to within a factor of approximately 10 %. Note that
other tidal components may be included if so desired.

A current pre-processor is used to transfer the results from the tidal
simulation model to the deterministic oil spill simulation model.

Fluctuating component. In order to simulate horizontal diffusive spreading of
the oil slick, a stochastic velocity component may be added to the velocity
field (Eq. (1)). The amplitude variation of such a stochastic component may
be obtained from measurements or from assumed functional relationships. In
the present version of the deterministic model provisions are made to include
such a stochastic component.

4. MODEL OUTPUT

Both the deterministic and the stochastic model can be used either alone or
in a combined mode, the "Slikforcast" mode. In order to facilitate this last
option it has been necessary to insure that the two models have identical
grid systems and that the input/output options are compatible.

For both models, therefore, one may specify 15 different shore line
areas, five sea straits and five critical sea areas for which oil spill
characteristics are computed. Fig. 13 shows an example of how a region may
be divided into the different areas mentioned above.

Stochastic model.

As previously mentioned, the stochastic model accumulates results for a
number of simulations, maximum 5000, in order to generate the probabilistic
results for given blowout location. For each simulation the value of the
various parameters discussed before is chosen at random within the given
range. The output thus gives the "average" impact forcasts and its uncertainty range resulting from the different combinations of the uncertainties
in the blowout conditions.

Some of the major output parameters are:

- Statistics for tons of oil on shore (min, max, avr.).
- Rate of oil shoring (avr., max).

- Drift times to shore (min, avr.).
- Tons of oil collected at primary and secondary combat.
- Tons of oil dispersed in critical sea areas.
- Drift times to critical sea areas.
- Concentration estimates in critical sea areas.
- Accumulated spill trajectory plot.
- Pollution cost estimates.

The outputs may be presented on a line printer or as drawings.

Deterministic model

The deterministic model aims at oil spill simulations for a specific spill situation.

The model is structured into modules, each simulating the various physical and chemical processes as described in the previous sections.

An efficient and easy operational procedure is of great importance for the use of the model. This is especially true for its implementation in emergency situations, i.e. for an operational tracking and prediction of oil slick movements in case of an actual blowout, tanker accident or pipeline rupture. Special emphasis has, therefore, been given a user-oriented man-machine interface, allowing the operator to update the simulations against field observations.

Some of the major output parameters are:

- Accumulated amounts of oil on shore.
- Accumulated amounts of oil dispersed and evaporated.
- Accumulated amounts of oil collected by primary and secondary combat actions.
- Accumulated number of hits of drifting oil in each gridpoint.
- Drift times to various shore locations.
- Position of drifting oil at every time-step.
- Amount of oil on the surface at every time step.
- Estimates of hydrocarbon concentrations in given critical sea areas.
- Accumulated surface area affected by drifting oil.

The output may be presented on a line printer on TV-screens or on hard copy drawings.

Slikforcast option.

An unique feature of the Slikforcast model is the possibility to transfer simulation results from the deterministic model to the stochastic model. This makes possible statistical forcasts of the probable fate of oil spills during a given emergency situation. The forcast estimates are based upon historical weather and current data, given blowout rate and location, estimated ranges for continued blowout duration and the deterministic results up to and including the latest deterministic forcasts. As time moves on, the stochastic or probabilistic forcasts can be updated starting each time from the latest deterministic results. The transfer of data from one simulation mode to the other is illustrated in Fig. 11.

The appropriate wind data must be taken from the historical weather data file for the time window corresponding to the given oil spill situation. In Fig. 12 is shown a schematic illustration of the selection of wind data. The same figure also illustrates the basic feature of the Slikforcast model, i.e. as more and more information is gained about the specific spill situation, the uncertainty range in the estimates gradually decreases.

SOME EXAMPLE RESULTS

In what follows we shall briefly present some example results from simulations with both the deterministic, the statistical and the composite
model. We have for this illustration chosen as example the conditions
during the Ekofisk blowout.

In Fig. 13 is shown an average yearly pollution forcast for the
Ekofisk area based upon five years of gridded wind data from the Norwegian
Met. Inst. and blowout conditions as listed in the figure. The wind data
covers the periode from 1973 to 1978.

As it appears from the results the largest shore impact measured in
tons of oil may be expected for a winter season spill. The results for the
winter season also exhibit the shortest drift time to the shore. As may
be seen from the diagrams in Fig. 13 the Danish coast may expect the
highest oil impact for both season. Measured in per cent the amounts, however, does not appear very large.

In Table 3 below we have also shown the estimated average maximum
concentrations in the three critical sea areas shown in Fig. 14.

Table 3: Estimates of average max. hydrocarbon concentrations in criticalsea
areas, C_{max}, and average min. drift times, t, to the indicated area.

Sea area	Winter		Summer	
	$\bar{c}_{max}$ (ppb)	$\bar{t}$ (days)	$\bar{c}_{max}$ (ppb)	$\bar{t}$ (days)
Ekofisk	260	1	220	1
Great Bank	31	11	25	15
Coral Bank	28	17	13	18

According to the results in Table 3 we may expect almost an order of
magnitude lower concentration levels at the Great Bank and Coral Bank areas
than in the sea area adjacent to the spill site. The computed concentration
levels near the platform (~250ppb) seems reasonable compared to the measured
values during the Bravo blowout (Ref. 8). It should also be noted that the
computed drift time to the Coral and Great Bank areas is of the order of
two weeks. The oil will then be of an age where it is generally considered
to represent a lesser threat to fish resources.

In Fig. 14 is shown results of the deterministic simulations for four
different days during the blowout period. Both the instantanious position
of the "thick oil" area and the total area of oil impact is shown together
with observed oil drift and spreading. The results demonstrate a typical
short-time forcast during a blowout situation. It is worth noting that the
width of the simulated "thick" oil slick is resulting mainly from the tidal
effect. This because the various oil lots are situated at different phases
of the tidal cycle.

Turning next to the results for the Slickforcast simulation an example
of this is shown in Fig. 15. The Bravo oil spill is again used for illustration purposes. The starting conditions for the slickforcast simulation
are taken from the latest results in the Oilsim simulations shown in Fig. 14
in accordance with the technique illustrated in Fig. 11 and Fig. 12a. In the
Slikforcast simulation the size of the spill is known and set equal to
2000 tons/day. The assumed duration is set equal to 80-180 days if bridging
does not occur and to 5-15 days if bridging occurs. The results in Fig. 13
and Fig. 15 may, therefore, be directly compared. It should be noted that
other conditions indeed also could have been set in order to investigate

their effect on the pollution forcast.

The results in Fig. 13 shows that the Danish shores also under these conditions remains the area of highest impact. The whole drift pattern has shifted more southwards in comparison to the results for the summer season, shown in Fig. 13. Less oil is shoring in Norway more in Sweden, Denmark and Germany. The average minimum drift times are also longer in the Slickforcast simulation than in the probabilistic summer season simulation.

In Table 4 below we have also shown the estimated average maximum concentrations in the critical sea areas shown in Fig. 15.

Table 4:   Estimates of average max. hydrocarbon concentrations in critical sea areas, $\bar{C}_{max}$, and average min. drift times, $\bar{t}$, to the indicated areas.

Sea area	$\bar{C}_{max}$ (ppb)	$\bar{t}$ (days)
Ekofisk	50	1
Great bank	6	15
Coral bank	1	35

The results in Table 4 again gives the highest concentrations in the area surrounding the platform. The results appears very reasonable when compared to actual measurements in the area. As also may be seen from Table 4 the input on the Great bank area is much smaller than the input on the Coral bank area, illustrating again the more southerly driftpattern for these simulation conditions.

SHORT SUMMARY

The oil spill simulation program Slikforcast is an integrated program system where the main components are a deterministic oil spill simulation program, a statistical oil spill simulation program, a hydrodynamical model for tidal currents, a data preprosessor, and various output routines. Wind data and data on residual currents must be supplied from outside sources. The wind data may be from (i) meteorological wind models with winds on a gridded net; (ii) from time series of wind from one or several observational stations; (iii) from statistical wind data from given locations. The data preprocessor is used to arrange the data on a unified format suitable for the simulation models.

Simulation of oil spill drift and fate may be carried out using either the deterministic or statistical model or using the two simulation models in a coupled mode. In the latter case the results from the deterministic simulations are used as starting conditions for the statistical model. This permits oil spill forcasting both on a short term and long term basis. The deterministic model is interactive allowing for continuous updating of the simulation results against field observations.

The program package may be used both for contingency planning and for oil spill forcasting and hindcasting during an emergency situation. Great care has been taken to assure a general program structure suitable for implementation for different geographical regions of interest. The access to sufficient and relevant ambient data and boundary conditions is, however, paramount to the quality of the simulation results.

We would like to acknowledge the E&P Forum both for the permission to use the results from the Slikforcast project in this publication, and for their encouraging support during the entire Slikforcast project work.

REFERENCES

(1)   M.H. Fallah and R.M. Stark, 'Movement of spilled oil at sea' Marine Technology Society Journal, (1976) 1-10.

(2)   K.D. Stoltzenbach, O.S. Madsen, E.E. Adams, A.M. Pollack and C.K. Cooper, 'A review and evaluation of basic techniques for predicting the behaviour of surface oil slicks' MIT report no. MITSG, 8 (1977).

(3)   F. Krogh, J. Haldorsen, T. Audunson, 'OILSIM - a computer program simulating the fate of oilspills' Norwegian Maritime Research, No.4 (1978).

(4)   G.F.L. Diezel, A.W. Glass and P.J. Kleef, 'A computer simulator for the prediction of slick movement, removal by natural means, clean up and potential damages arising from oil spills originating from offshore oil well blow-outs. Its development and application to the North Sea' Shell Report, EP-47436 (1976).

(5)   T.K. Fanneløp and G.D. Waldman 'Dynamics of oil slicks', AIAA Journal, 4 (1972) 10.

(6)   J.A. Fay, 'The spread of oil slicks on a calm sea.' Oil On the Sea, (D.P. Houlst (Ed.)), Plenum Press, New York and London, (1969)

(7)   T. Audunson, 'Dispersion of pollutants in water' In: The Environmental Aspects of the Petrochemical and Light Refinery Industry, Trondheim, Norway, (1975).

(8)   T. Audunson, 'The fate and weathering of surface oil from the Bravo blowout', Marine Environ, Res., 3 (1980) 35-61.

(9)   T. Audunson, J. Haldorsen, J.P. Mathisen, P. Steinbakke, 'Oljesøl langs Norskekysten - Delrapport 4, Utslipp av olje på Tromsøflaket Sept. 1978', IKU Rapport P195/4 (1978).

(10)   P.G. Jeffery, 'Large-scale experiments on the spreading of oil at sea and its dissapperance by natural forces', Proc. Joint Conf. on Prevention and Control of Oil Spills, Washington, DC., (1973).

(11)   R.F. Throne, 'How to predict lake cooling action', Power, September vol. (1952).

(12)   T. Audunson, et al., 'SLIKFORCAST - A multi-purpose oil spill simulation system', SLIKFORCAST - Main report, IKU, Trondheim, Norway, (1980).

(13)   J.P. Mathisen, 'NEPTUN - C 21 - A dept-integrated numerical model for the calculation of tidal currents! SLIKFORCAST - Subreport no. 3, IKU, Trondheim, Norway, (1980).

(14)   R.A. Flather, 'A tidal model of the north-west European continental Shelf', Memorires Sociéte Royale des Sciences de Liége, 6e série, tome X. (1976).

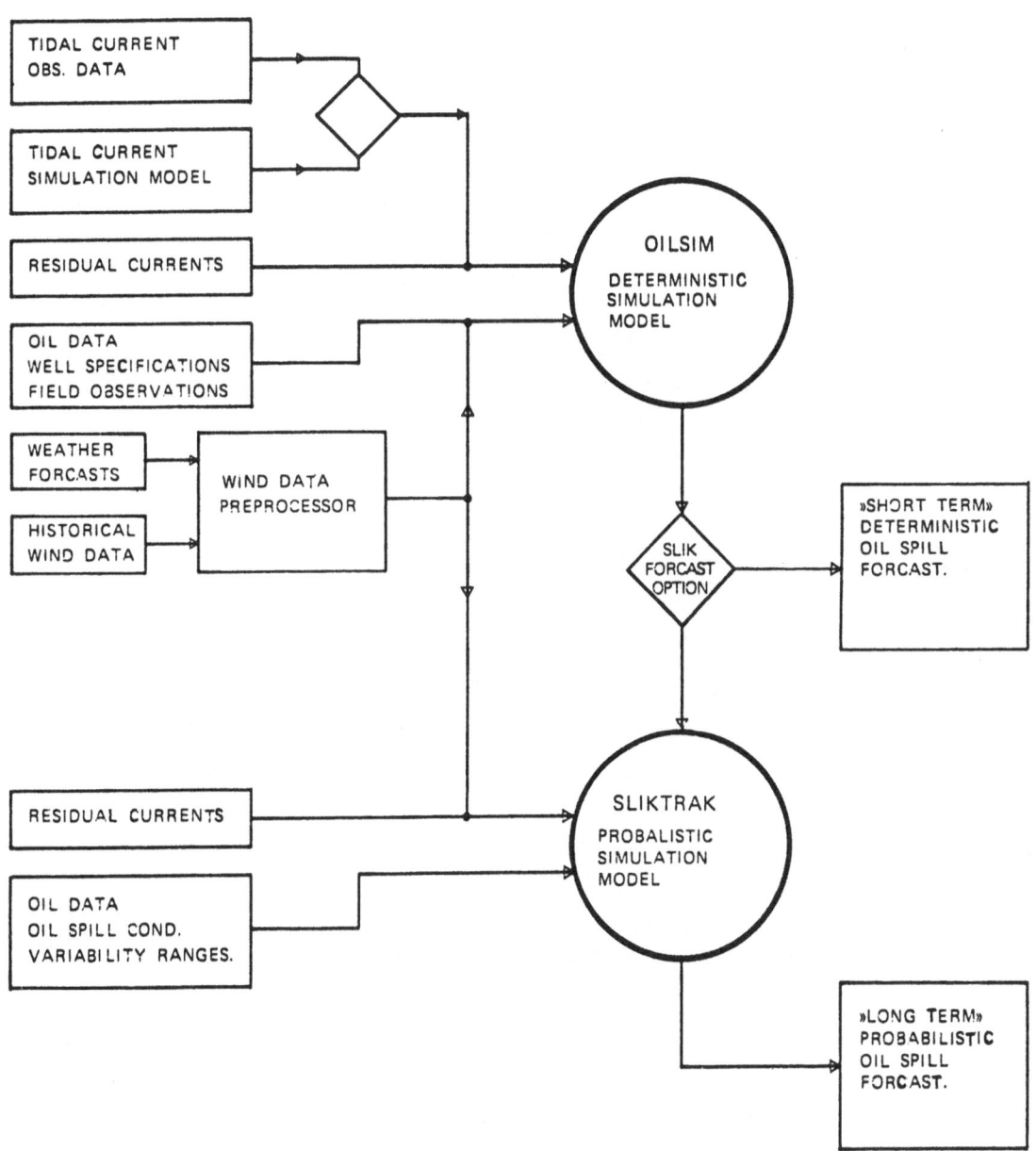

Fig. 1. Block diagram showing the general organisation of the "Slikforcast" program.

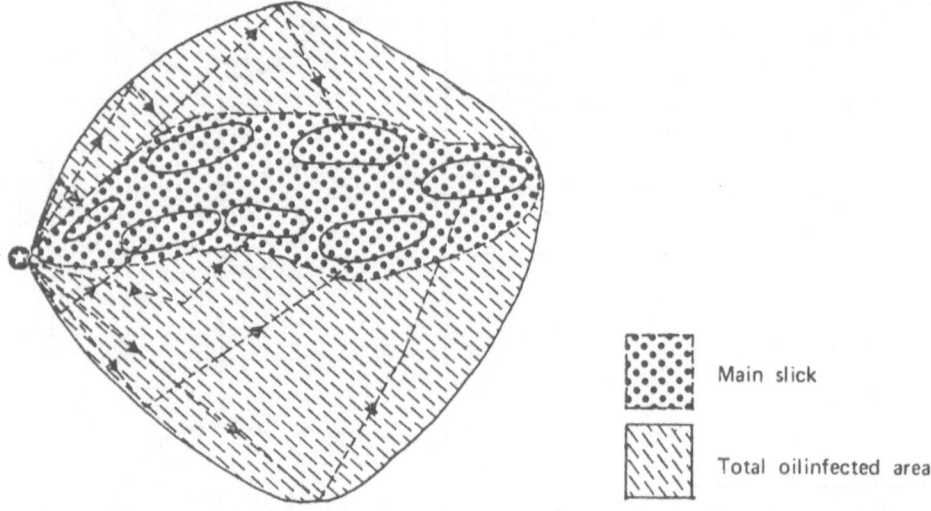

Fig. 2. Sketch illustrating the simulation of the main slick area (thick oil)
and the total oil infected area (thin oil).

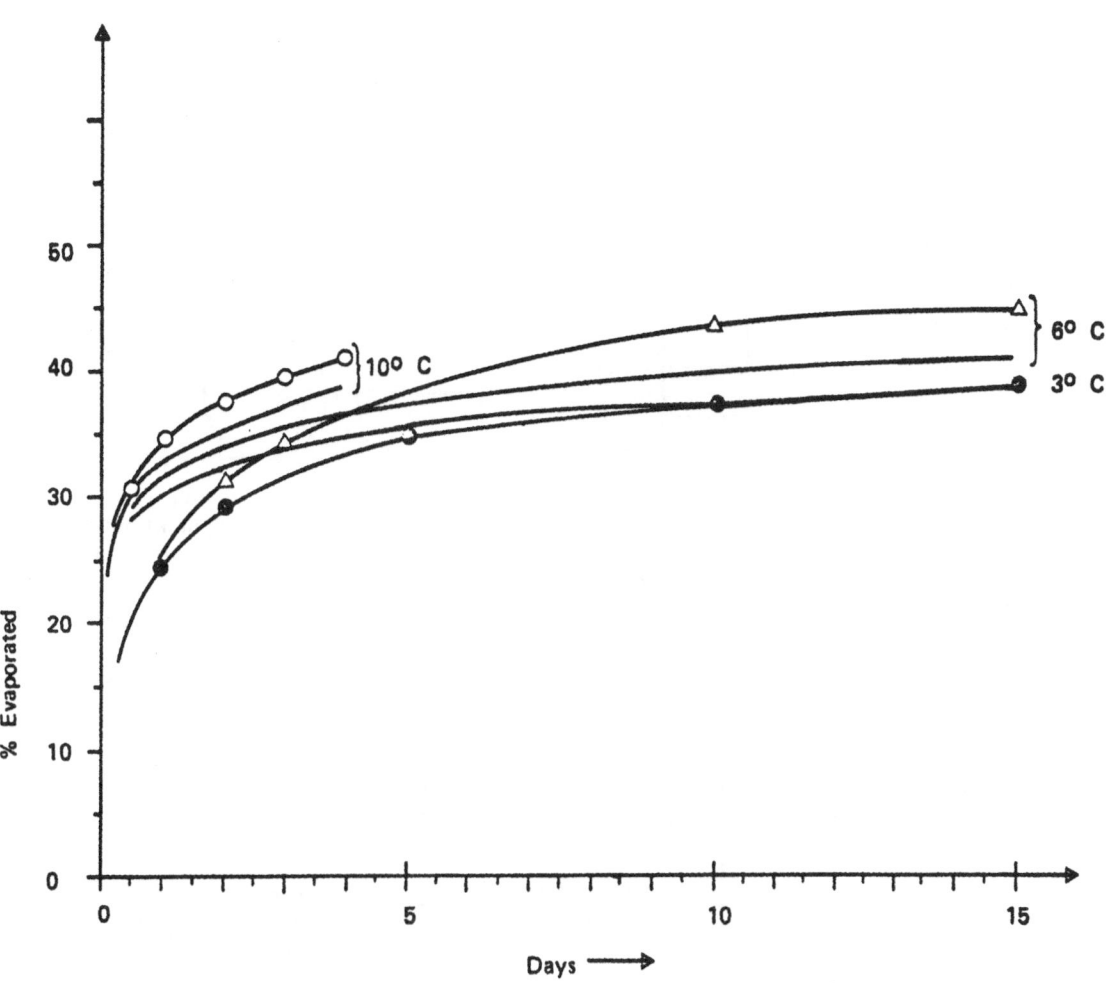

Fig. 3. Evaporation curves for Ekofisk crude: ——●—— Laboratory experiment,
——△—— Observations from the Bravo blow out, ——○—— Observations
from Tromsøflaket, ———— Simulated results.

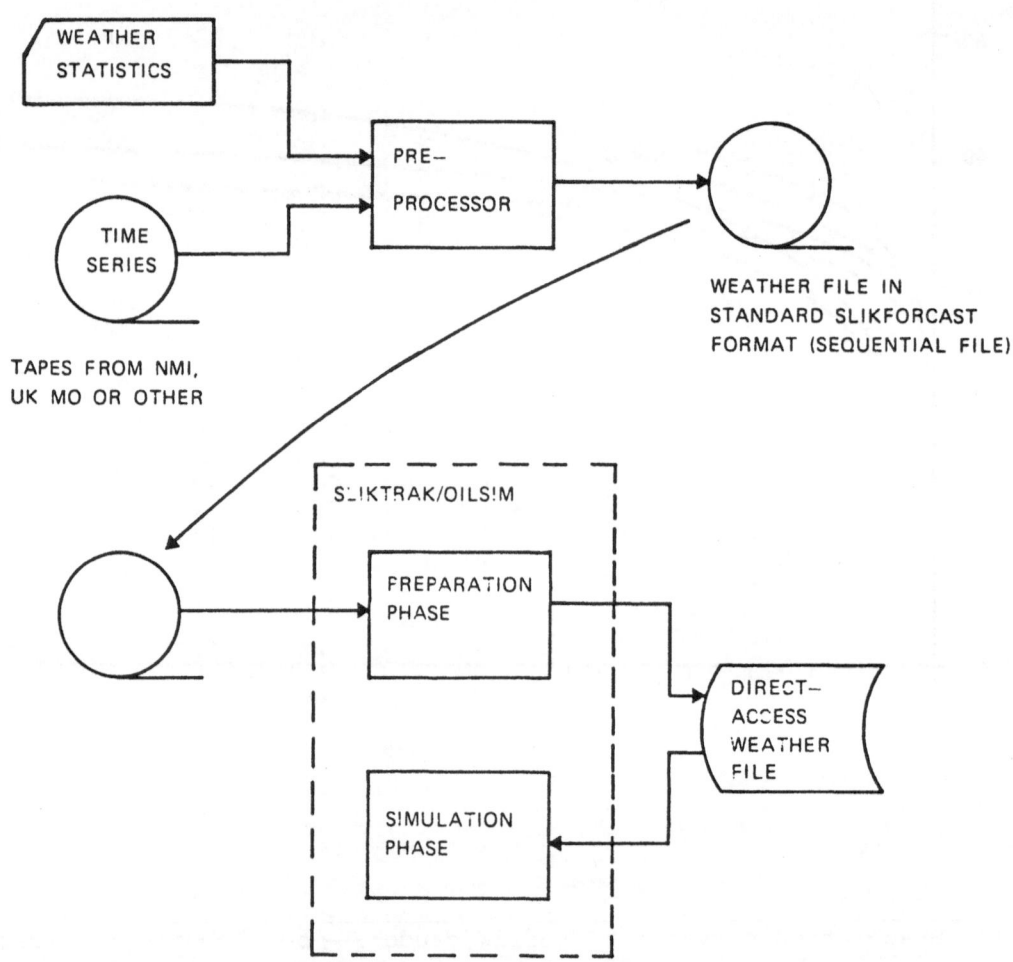

Fig. 4. General layout of the wind data preparation in the model.

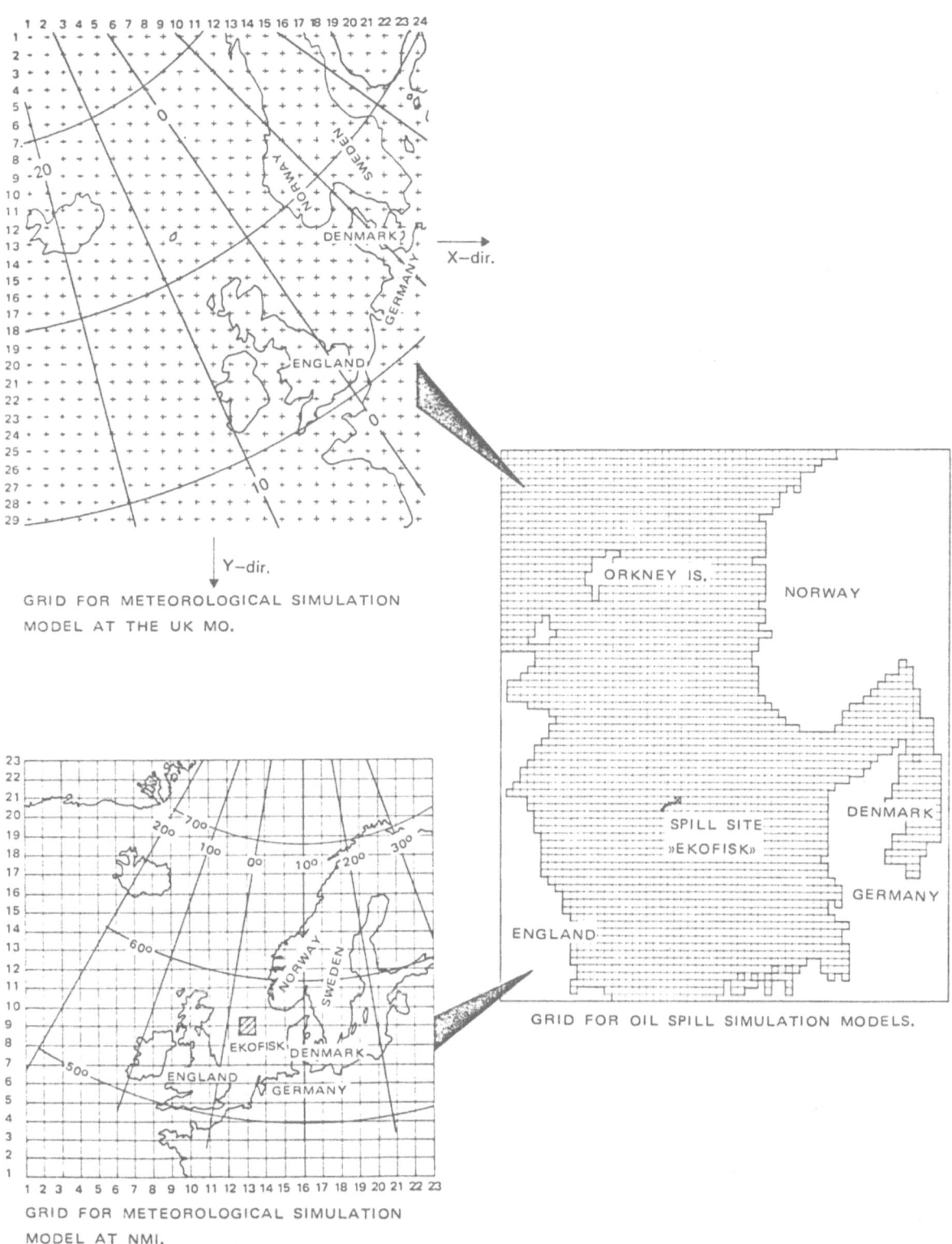

GRID FOR METEOROLOGICAL SIMULATION MODEL AT THE UK MO.

GRID FOR METEOROLOGICAL SIMULATION MODEL AT NMI.

GRID FOR OIL SPILL SIMULATION MODELS.

Fig. 5. Example of gridded wind data input accepted by the Slikforcast model.

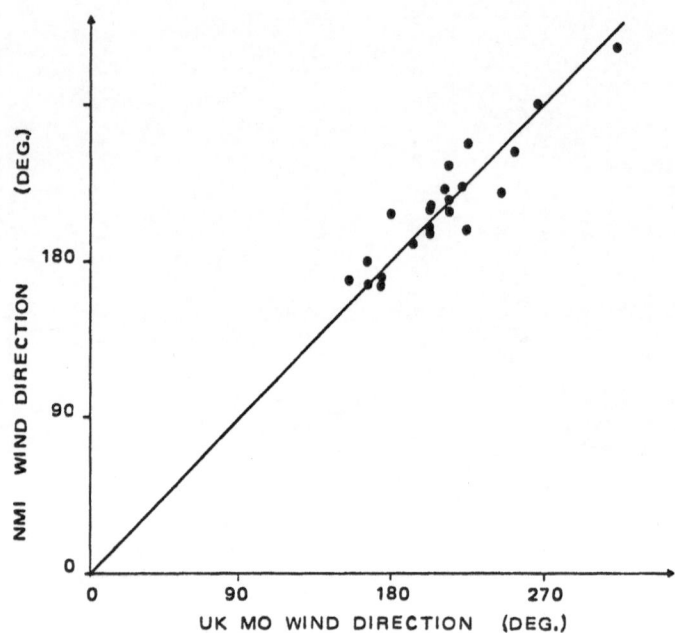

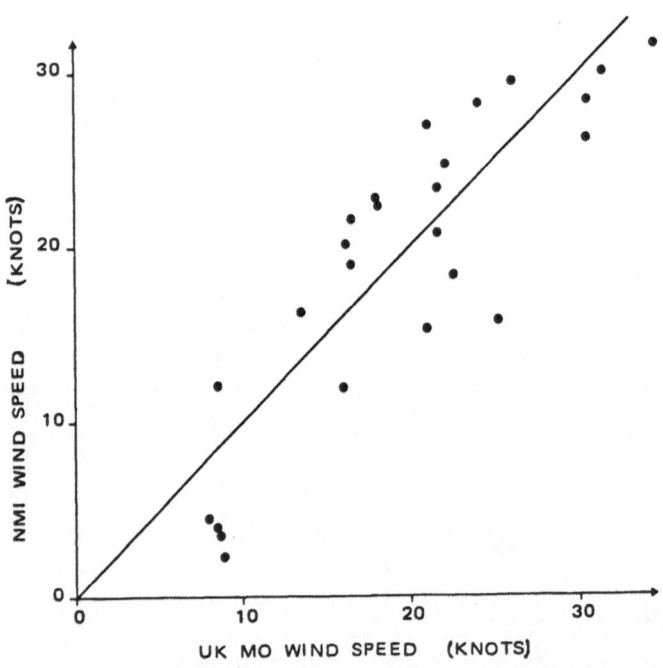

Fig. 6. Correlations between wind direction and wind speed simulated at
NMI and UK MO for Dec. 1977.

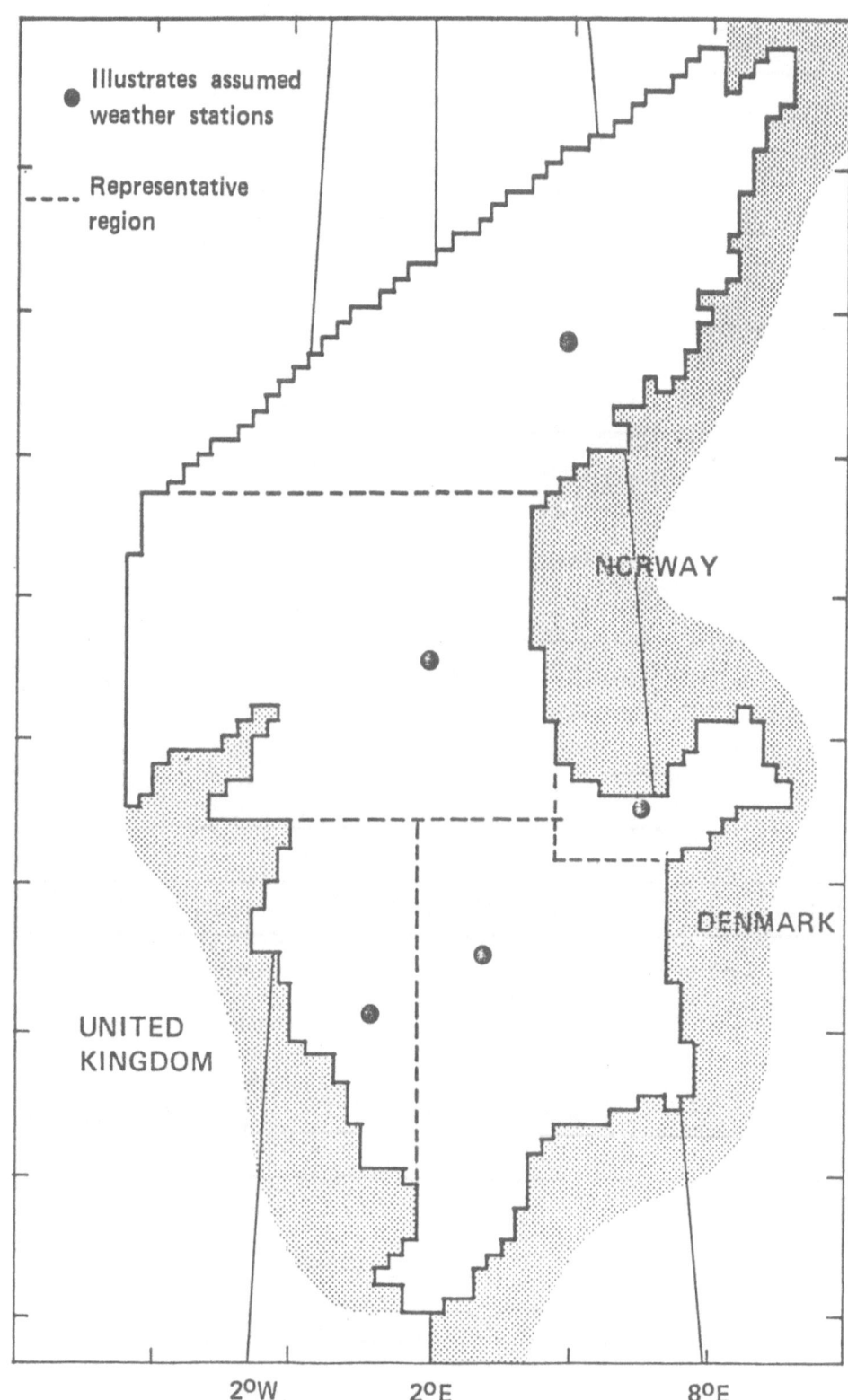

Fig. 7. Division of an area into different observational regions.

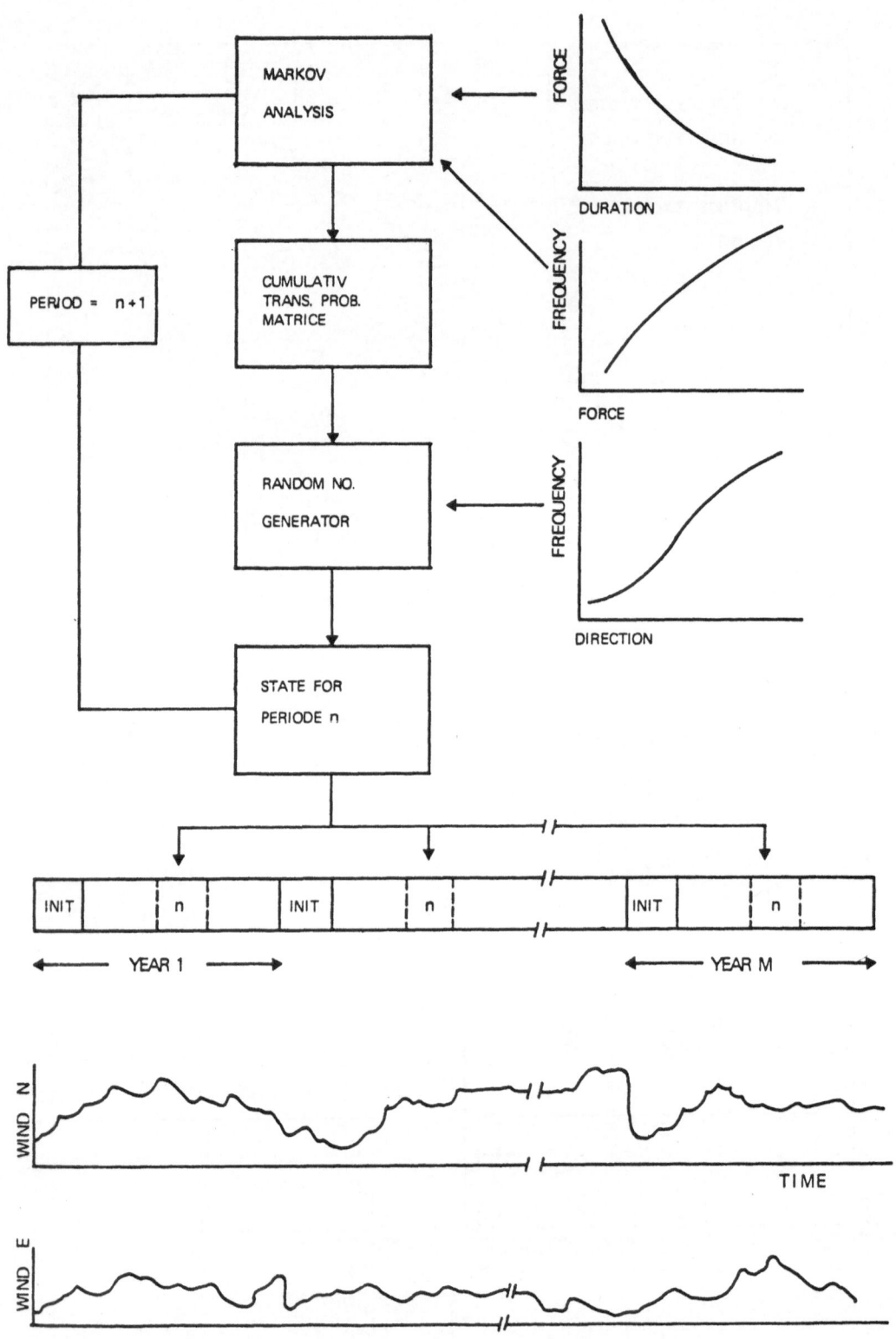

Fig. 8. Generation of time series of wind from statistical wind data.

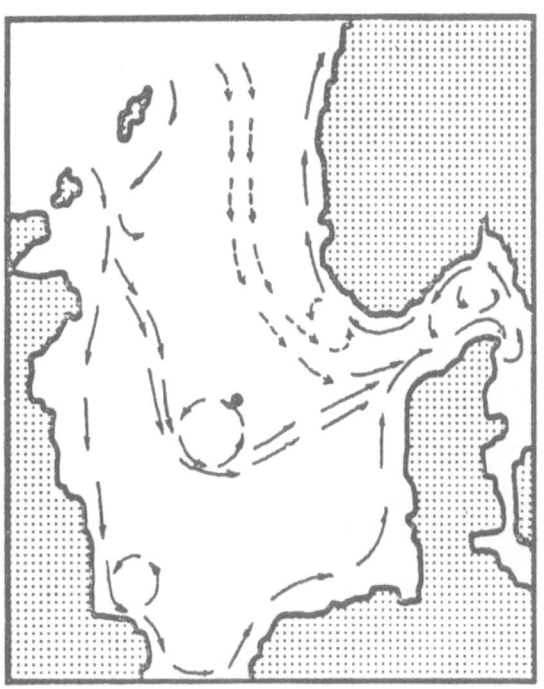

Residual current pattern (june)
(0.1, 0.2, 0.3, 0.6 knots).

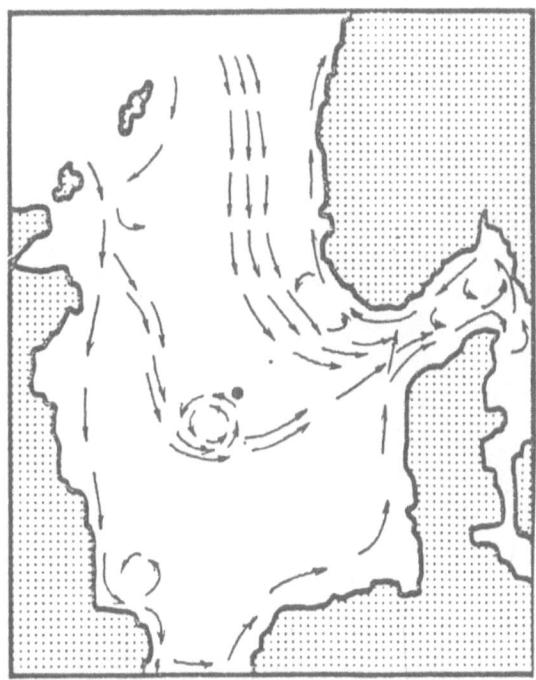

Residual current pattern (dec.) (0.2, 0.3, 0.6 knots).

Fig. 9. Examples of residual current pattern for the North Sea.

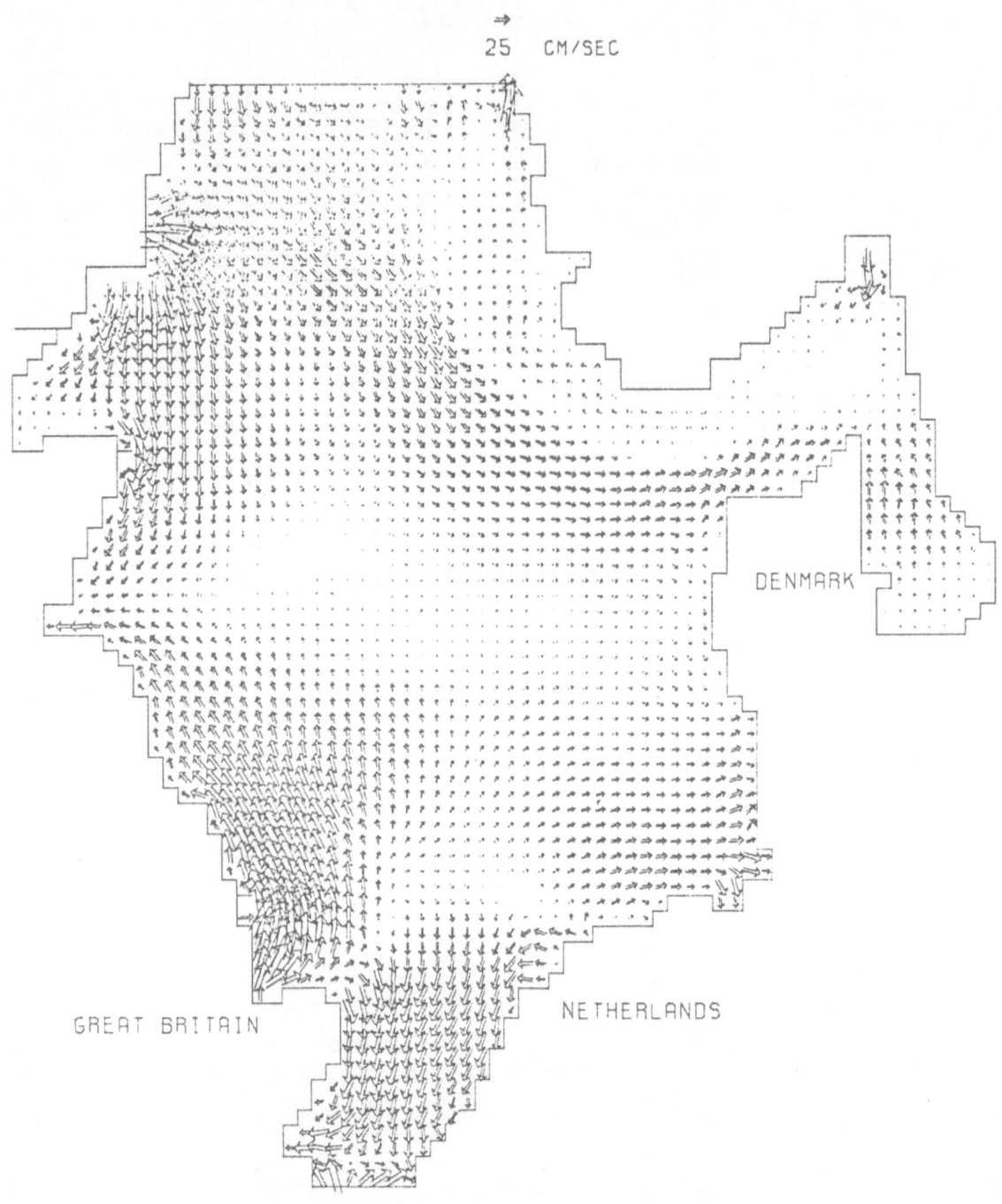

Fig. 10. Example of tidal (M_2) velocities simulated for the North Sea.
(High tide at Dover.)

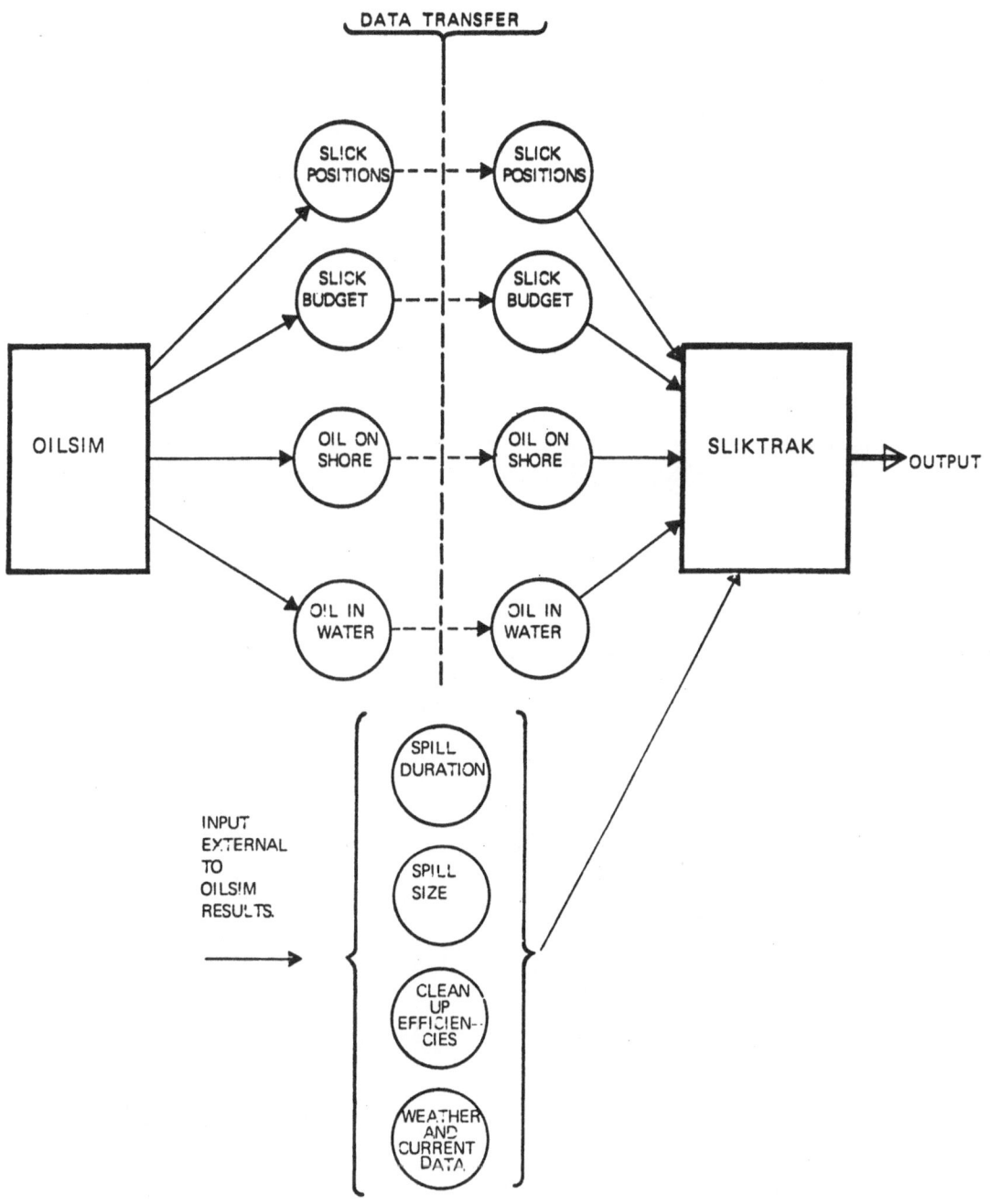

Fig. 11. Illustration of the data transfer from the deterministic to the probabilistic mode.

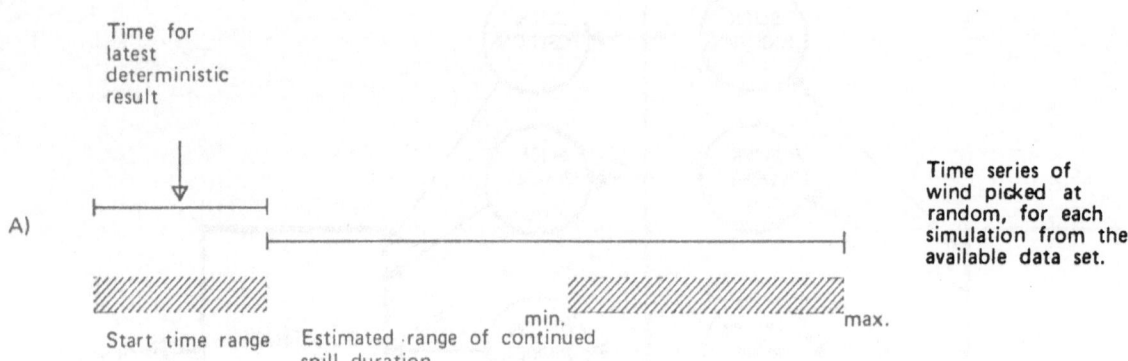

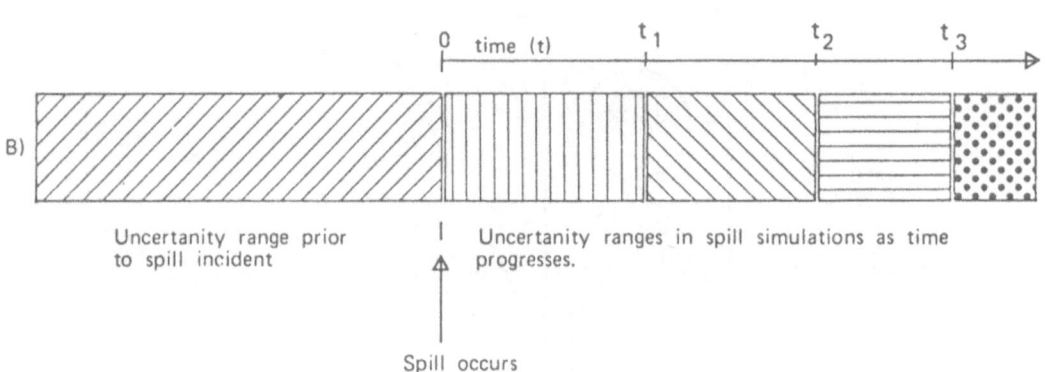

Fig. 12. a) Illustration of the principle used when selecting a time series
            for wind used in the probabilistic "Slikforcast mode".
         b) Illustration of the narrowing down of uncertainty limits for a
            given spill.

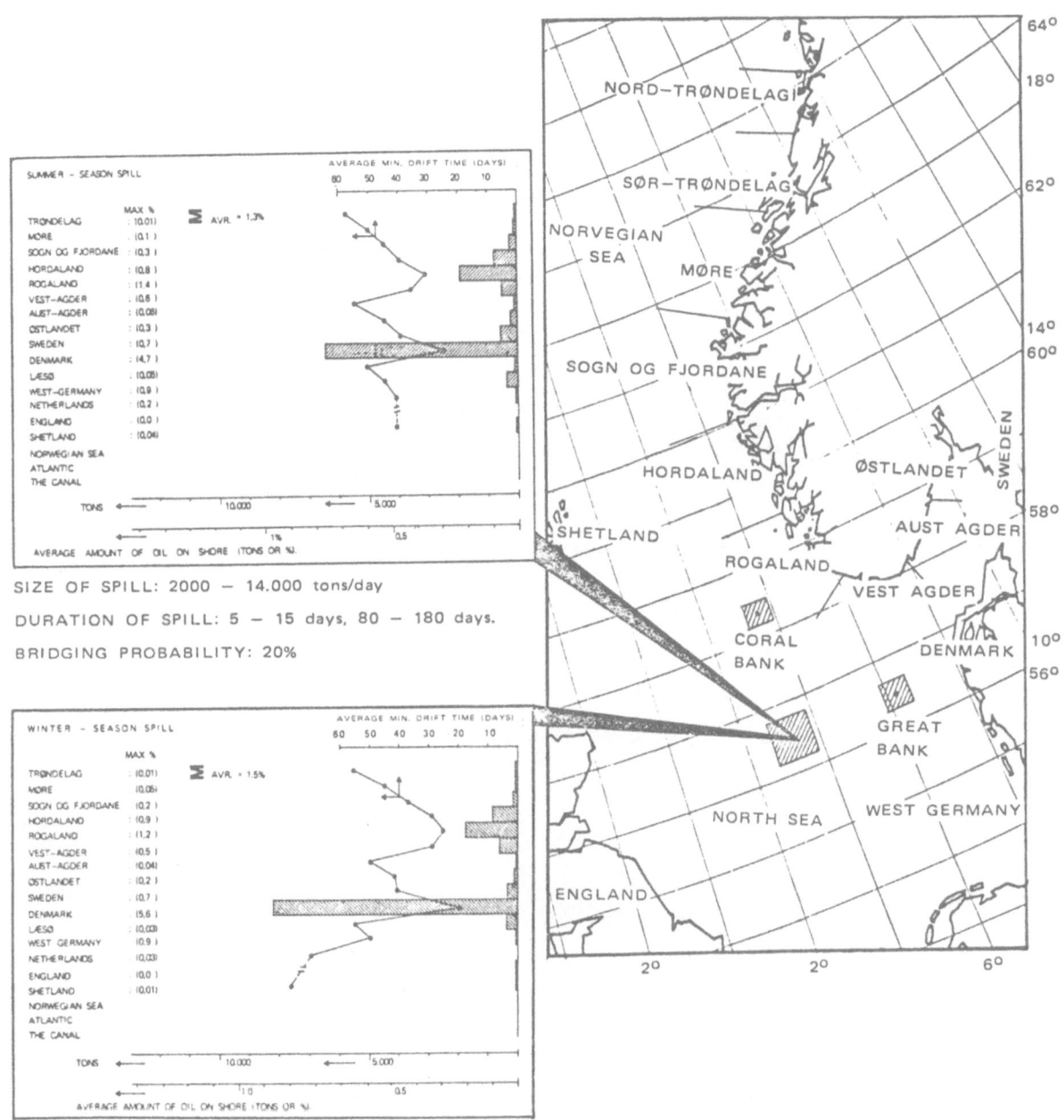

Fig. 13. Probabilistic general pollution forcast for an assumed North Sea
blow out during the winter or summer season.

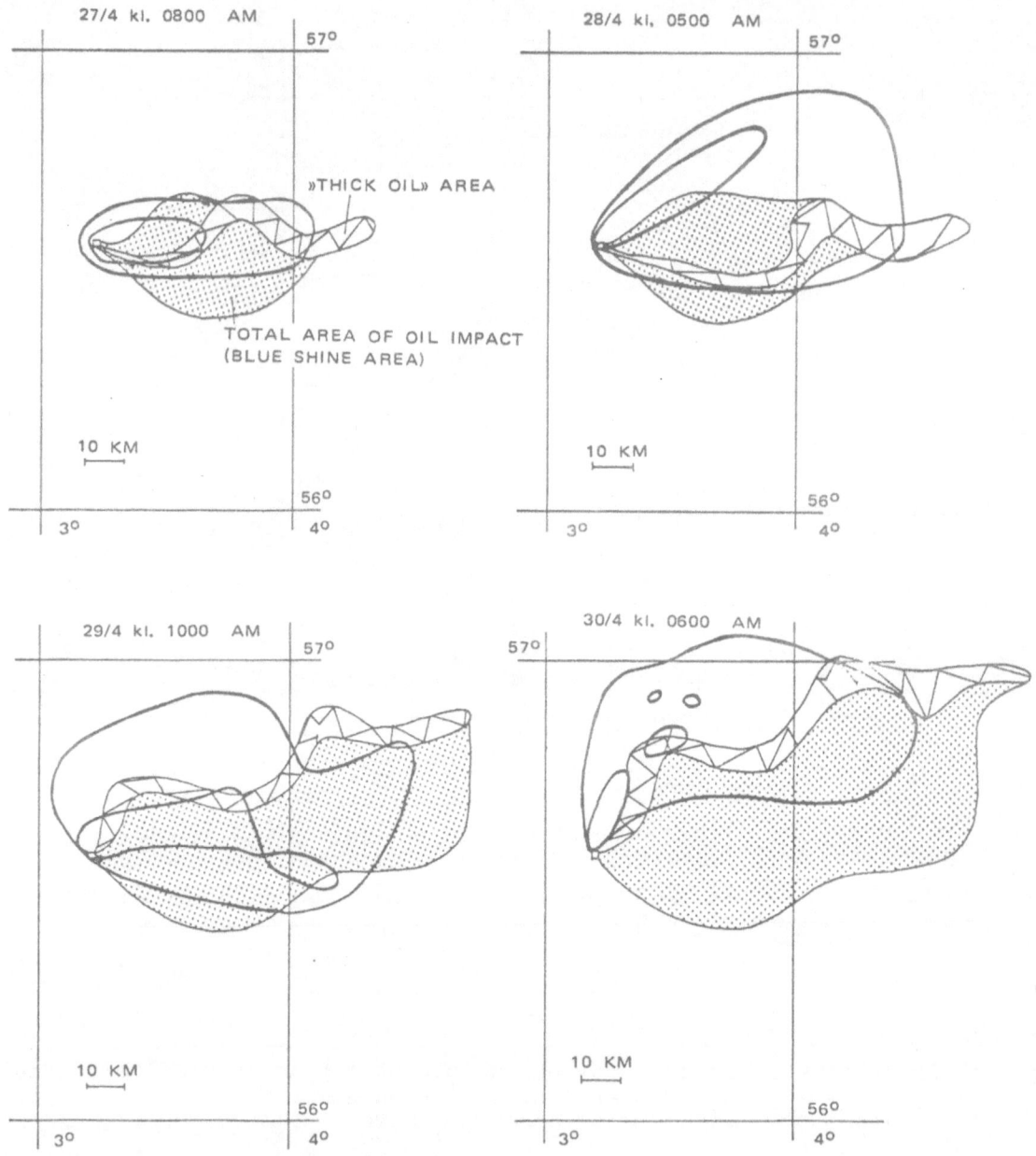

Fig. 14. Deterministic oil spill simulations for the Bravo oil spill in the
North sea.

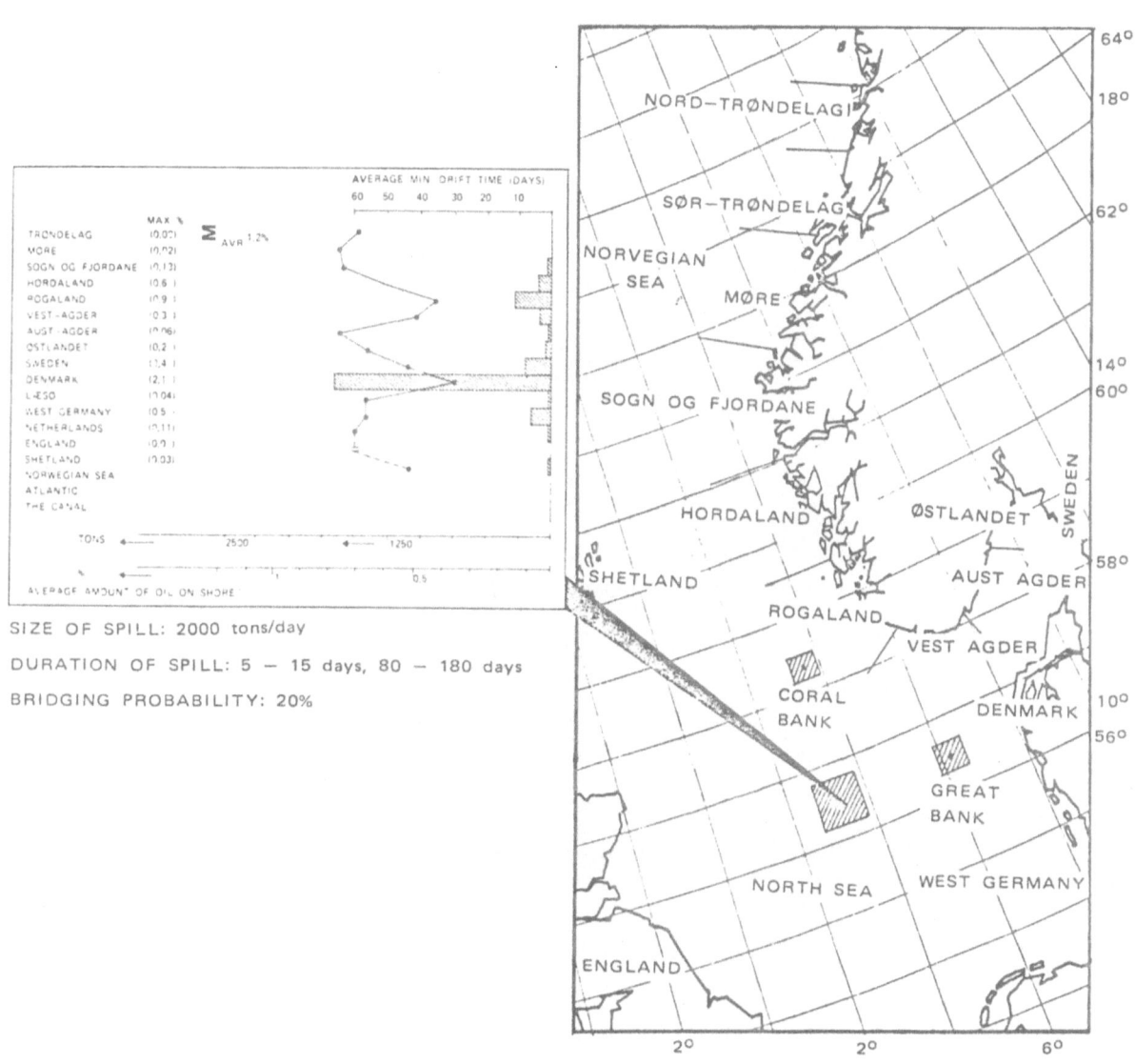

SIZE OF SPILL: 2000 tons/day

DURATION OF SPILL: 5 — 15 days, 80 — 180 days

BRIDGING PROBABILITY: 20%

Fig. 15. Slikforcast pollution forcast for the Bravo oil spill assuming
the oil spill to last for a maximum of six months.

AN APPROACH TO OIL POLLUTION RESEARCH AND MONITORING

R Hardy
Torry Research Station, 135 Abbey Road, Aberdeen AB9 8DG

K J Whittle
Torry Research Station, 135 Abbey Road, Aberdeen AB9 8DG

P R Mackie
Torry Research Station, 135 Abbey Road, Aberdeen AB9 8DG

A D McIntyre
Marine Laboratory, Victoria Road, Aberdeen AB9 8DB

Concern about the presence of oil in the marine environment
is not new. As long ago as 1922 the Congress of the United
States requested the President to convene an international
conference on the prevention of oil pollution. Fossil fuel
utilization has increased dramatically since then and public
concern, especially with regard to visible pollution, effects
on amenities and damage to sea birds, has grown with it.
Subsequently, many years of research, initiated as a result
of that concern, have not only provided some of the answers
and raised further questions but have also vividly illustrated,
especially to the analyst, the difficulties hindering a complete
understanding of the presence, fate and effects of oil in the
seas and the problems of devising adequate and meaningful
methods for the monitoring or surveillance of oil.
Crude oil is a complex mixture comprising tens of thousands
of components (Ref. 1) which cover a wide boiling point range
from volatiles to waxy residues and have not yet been characterised completely in any oil. Despite some 50 years research
on Ponca City crude oil only about 60% by weight of it, representing a mere 260 components, has been characterised (Ref. 2).
The principal elements of crude oils are carbon and hydrogen
with lesser amounts of sulphur, nitrogen, oxygen and various
metals such as nickel and vanadium. These elements are combined to form various chemical classes ranging from the simple
n-paraffins, branched paraffins, cycloalkanes (naphthenes),
aromatics, naphtheno-aromatics, acids, phenols, organo-sulphur
compounds and so on to the highly complex ashphaltenes whose
structures are not properly elucidated but are thought to
consists of 10 to 20 fused aromatic rings, some heterocyclic,
with aliphatic and naphthenic side chains (Ref. 3).
Once this 'complex chemical soup' (crude oil petroleum
products or oily water discharges) has entered the sea its
composition changes progressively due to physico-chemical
factors such as dissolution, evaporation, adsorption and photo
oxidation (Fig. 1). Then there are the effects of the biota,
selective or non-selective uptake and depuration, metabolism
and biodegradation at different rates depending upon the nature
of the substrate. Added to this input and its subsequent
modifications are the hydrocarbons from marine and terrestrial
biosynthesis as well as pyrolysis products. Thus it can be

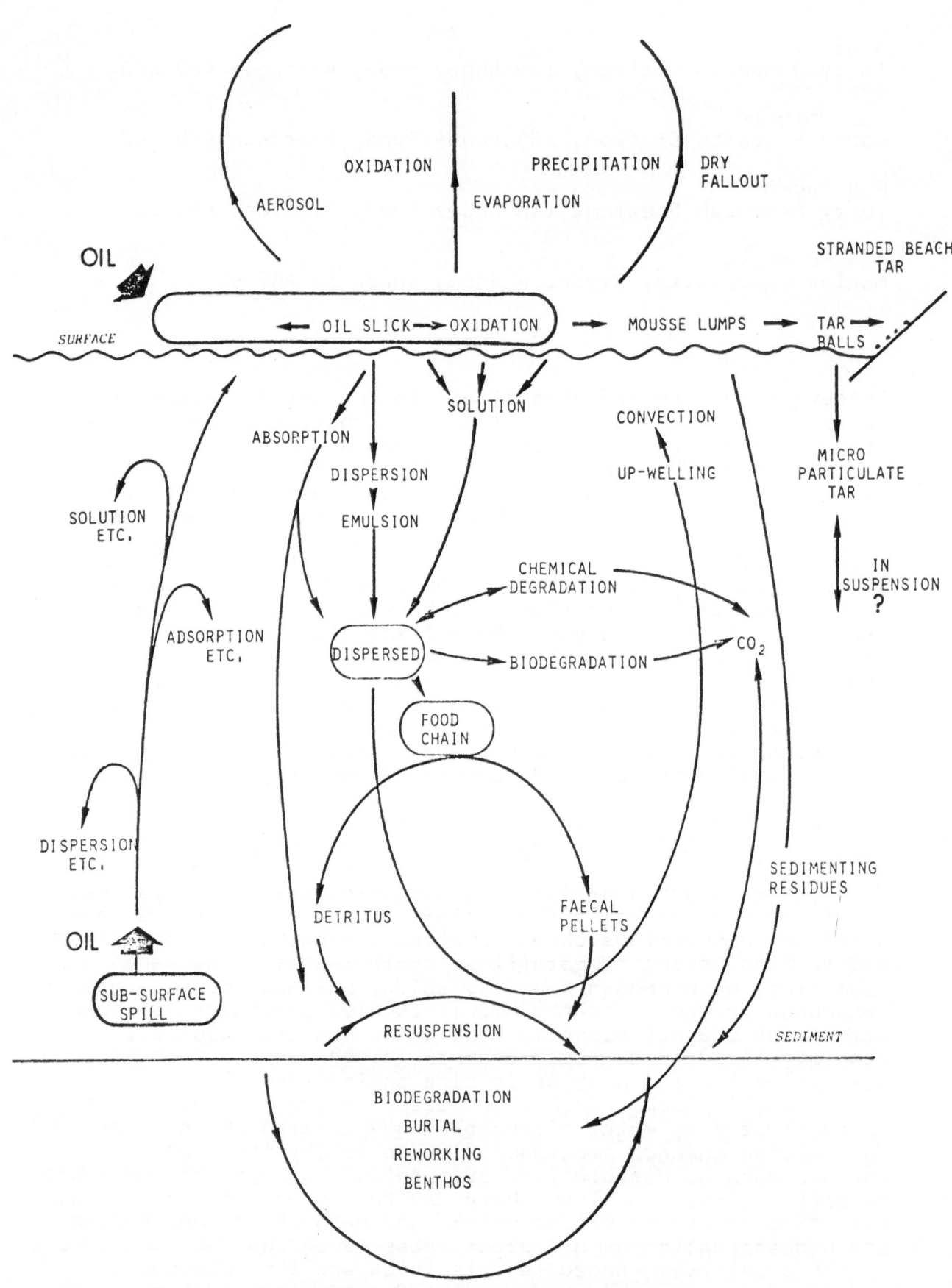

MODIFIED AFTER BURWOOD AND SPEERS (REF. 30)

seen that the analyst concerned with monitoring or surveillance faces formidable tasks in not only finding suitable methods for analysis but also in having to decide what compound(s) to measure and where to measure them. This contrasts sharply with the chemist concerned with other forms of chemical pollutants such as radionuclides and pesticides who deals with relatively well defined substances the sole source of which is man's activities. Discounting the obvious gross effects of large spills and floating oil, a study of the chemical basis of the effects of oil in the marine environment must be preceded by deciding which of the multitude of oil components are of significance before any meaningful experiments can be initiated to examine the consequences of oil pollution.

It is not possible to measure amounts of 'oil' as such in the environment which correctly represents what remains of the actual input of oil since it is such a complex mixture and its composition changes so markedly with time. The methods of determining 'oil' are dependent on the amount of oily material present. When visible as a film, slick, tar balls or dispersion the techniques of sampling, isolating and quantifying are relatively simple, indeed several methods have been proposed for the remote sensing of oil slicks (e.g. Ref. 4). A variety of methods can be selected not only to determine the amount of the residue (e.g. gravimetry) but also to provide information on the likely sources by such techniques as gas chromatography, fluorescence spectroscopy and vanadium/nickel assay. Even so the total picture is distorted if the volatile material is lost during the stages of analytical preparation.

However, at lower levels a number of problems of scale and concentration arise but adventitious contamination must also be prevented throughout the analyses otherwise more 'oil' can be introduced to the sample, at all stages from the sampling device employed to the solvent used for extraction, than is already present in the sample. At these low levels we have to decide which oil derived components are to be analysed since oil per se cannot be determined. Basically, there are two approaches to this problem. Firstly a specific property of the oil such as fluorescence, ultra-violet or infra-red absorption may be determined and extrapolated to provide an estimate of 'oil units' using a standard oil for reference but this is relatively non-specific as far as the components measured are concerned. Alternatively a single component or class of components may be quantified and, either reported as such or again extrapolated using standard oil. The analyses of specific properties are simple, sensitive and relatively cheap in terms of manpower and equipment costs. However, they rely on two basis assumptions: that the property measured is solely due to the pollutant and that the 'oil' isolated is identical to the standard. It is extremely doubtful if either of these conditions is ever met entirely. Both these assumptions also have to be made in the case of analyses of components if a figure for 'oil content' is desired. However, the approach most commonly adopted is to quote the individual compound(s) with no reference to 'oil' content and where necessary to distinguish between contributions of petroleum and biogenic origin. This type of measurement requires more complex analytical procedures and is expensive in labour, equipment and materials. For instance, Goldberg et al (Ref. 5) estimated that the analyses of a single mussel sample would cost ⊄800 and it is obvious that any large scale

monitoring programme involving detailed analysis would be both
inordinately expensive and of dubious value for determining the
extent of contamination with oil.

The approach we have adopted has been to analyse initially
the paraffinic and latterly the polycyclic aromatic (2 to 5
ring) hydrocarbons (PAH) by means of gas (GC) and gas chroma-
tography/mass spectrometry (GC/MS). The paraffins are one of
the major components of oil and their analysis is relatively
simple. The interest in the PAH arose because of their
established toxic, carcinogenic and mutagenic properties. More
recently we have established a trained sensory panel of judges
to evaluate oily taints.

PRESENCE AND ORIGIN

At the start of the work in this field some 8 years ago very
little information existed on the occurrence, levels and origin
of hydrocarbons throughout the marine environment. The first
step was to rectify this deficiency by conducting a series of
surveys, mainly in UK waters, of the aliphatic hydrocarbons
present in the water, sediment and representative biota
including species of commercial value. No suitable methodology
existed at that time to extend the analyses to routinely cover
the PAH. The results obtained have been reported in detail
elsewhere (Ref. 6) but together with experimental studies,
which will be discussed later, they have provided a basis for
more selective sampling regimes as the needs have arisen and
have given a clear indication of the precision and predictive
value of such a survey.

Briefly, normal paraffinic hydrocarbons with from 15-33
carbon atoms plus pristane and phytane were observed at generally
low levels in all samples taken, surface film, sub-surface
water, sediment, fish and plankton, from sites around the UK
coast and the sub-antarctic island of South Georgia, selected
to provide a range of supposed petroleum contamination and
including for comparison areas which might be regarded as pristine.
Excepting samples taken in areas where oil pollution was visible,
the levels of alkanes found were remarkably constant - a few
parts per billion (ug/l) in water samples and a few parts per
million (ug/g) in the sediments and biota (Ref. 6 and 7). The
distribution of the observed alkanes with respect to carbon
number varied from sample to sample. Sediment and fish livers
had an excess of odd over even carbon numbers, the hydrocarbons
in the former being thought to be derived from land plants
(Ref. 8). The remaining samples in general exhibited a smooth
envelope peaking around $n-C26$ with no odd/even preference.
Interestingly, some weathered crude oils show a similar
distribution of alkanes, although the peak is generally some
3 to 4 carbon atoms lower. This, together with the ubiquitous
nature of this type of alkane distribution tended to suggest
that it was not derived from specific oil inputs but did not
rule out the possibility that it represented a widely distributed
low background of petroleum derived hydrocarbons.

Clearly, there were doubts about the origin of the hydrocarbons which had to be resolved, and this was examined in a
series of experimental studies. The biosynthesis of hydrocarbons by phytoplankton was examined by cultering various
unicellular marine algae using $^{14}CO_2$ as their sole carbon source.
The radio-labelled phytoplankton were fed to natural zooplankton

populations in the laboratory. Hydrocarbon production was also followed in natural plankton populations in experimental enclosures spiked with $^{14}CO_2$. The algae produced only a limited array of aliphatic hydrocarbons, predominately polyunsaturated alkanes which do not occur in crude oil. The radioactivity associated with the zooplankton was mainly in the form of pristine produced by the metabolic breakdown of the algal chlorophyll together with lesser amounts of polyunsaturated alkanes transferred unchanged from the algae. In no instance did the labelled hydrocarbons produced resemble in any way those found in the survey although a low background, smooth envelope of unlabelled n-paraffins was present in the experimental cultures. No evidence of biosynthesis of aromatic hydrocarbon structures was found. Similar findings have been observed by other workers (Ref. 9) suggesting that the marine biosynthetic input of hydrocarbons is limited to a small range of branched saturated and olefinic compounds.

Our analyses of samples from a site on the remote sub-antarctic island of South Georgia with a well documented history of local pollution from fuel oil and whaling operation suggest that there is a world-wide dissemination of abiogenic hydro-carbons presumably derived predominantly from an airborne input (Ref. 10). No hydrocarbons ascribable directly to an oil source were found in the upper few centimetres of the sediment, deposited since the cessation of the local whaling industry, below that however residual hydrocarbons from whale processing and fuelling operations were found to a depth of at least 30 cm corresponding to approximately 100 years of deposition assuming present day sedimentation rates (Ref. 11). The oil input to this cold environment is still easily recognisable a minimim of 20 years later.

This observation is of some importance concerning the environmental consequences of exploiting oil in such areas especially as studies carried out in the aftermath of the Ekofisk blowout (Ref. 12) and a number of tanker disasters indicated that sedimentation is a major pathway for the removal of oil from the surface and that the sediments are major sinks. Experimental studies on areas of the sea bottom enclosed with cylinders off the West coast of Scotland indicated that after a single input of an oil coated sediment (6 kg of sediment containing 2.5% weathered N. Sea crude distributed over 1.8 sq. m) the alkane content returned to a normal concentration after six months. These results together with those of other workers indicate that although sediments have a capacity to mobilise or redistribute oil components, their ability to do so may be severely limited and when it is long term, contamination results.

From this work it does appear that most of the alkanes in the sea are of exogenous origin and as their composition does not resemble those of land biota except possibly in sediments, it would seem that they must arise from man's activities, seeps or natural fires. Differentiation between these possible inputs has not been made.

UPTAKE BY BIOTA

The uptake of hydrocarbons from food or water by representative demersal and pelagic fish have been studied (Refs. 13 and 14) using both crude oil and specific radio-labelled compounds. Codling were fed 1 mg Kuwait crude oil/day in their normal diet

for six months and then kept for a similar length of time
without oil present, samples being removed every two months
throughout the experiment. Oil derived hydrocarbons were found
only in the liver and their uptake was apparently selective;
30% of the n-C26 alkane fed was retained by the liver, the
corresponding figure for the n-C16 alkane was only 0.5%. In
short term experiments (4 days) with fish fed radioactive
n-C16, the low retention was confirmed and no evidence for
the metabolism of hexadecane was found. A similar result was
observed in fish kept in water pumped from below an oil slick
i.e. uptake was only observed in the liver and was discriminatory.
^{14}C-Benzo (a) pyrene on the other hand remained predominantly
in the stomach, the small amount that did cross the stomach wall
apparently followed the route, stomach-liver-bile-intestine-
excretion.

Juvenile herring presented a completely different picture.
Approximately 60% of the recovered activity from ^{14}C-hexadecane
was found in the muscle with only 7% being detected in the liver
and the same in the mesenteric fat. Metabolite(s) were present
in the liver, gills, intestine and pyloric caecae. The distribution was confirmed by the results of ingestion of small amounts
of crude oil from which the alkanes were observed mainly in
the muscle. As in codling, the uptake was selective but in this
instance the n-C15-20 alkanes were predominant. Again as in
codling, the bulk of the benzo (a) pyrene radioactivity (80%)
was recovered from the stomach. In fact 98% of the activity
was recovered in the digestive system as a whole, i.e. stomach,
pyloric caecae and intestine. The benzo (a) pyrene absorbed
followed the same excretory pathway, largely as metabolites.

We concluded from these experiments that cod and herring
selectively absorb ingested n-alkanes from crude oil and deposit
them in their lipid store, the liver in cod and muscle in
herring. This discrimination in uptake vividly illustrates the
further modification of oily residues incorporated into tissues
and the pitfalls associated with extrapolating the data from
experiments using single compounds to illustrate the behaviour
of such a complex mixture as oil.

The results we obtained with hexadecane would have indicated
at their face value that codling do not assimilate paraffinic
hydrocarbons to any marked extent whilst herring retain 60% of
them. The turnover of alkanes by the codling appeared to be
slow. Oil derived alkanes were still present in the liver at
the end of the 6 month depuration period. Again this contrasts
sharply with planktonivorous species such as herring which show
from field analyses that turnover is probably more rapid since
the hydrocarbon composition is related closely to dietary intake.

The behaviour of benzo (a) pyrene suggested that in both
species absorption of dietary benzo (a) pyrene was relatively
inefficient and a barrier to uptake in contrast to the rapid
absorption of benzo (a) pyrene across the gills of marine fish
noted by Lee et al (Ref. 15). Nevertheless, it is unwise to
generalise about the behaviour of PAH on the basis of these
few experiments. Indeed, other evidence suggests that the
behaviour of different PAH varies as noted with the alkanes.

MONITORING

Our experience of analyses of field samples of biota and the
wide range of experimental studies on uptake all point to the

fact that organisms probably act more as indicators of recent
exposure to oil contamination rather than integrators of
pollution due to the observed differences in uptake, deposition,
depuration and metabolism of the various oil components. Thus
measurements of water or more expecially sediments may be more
appropriate to indicate the presence and persistence of oil or
oil components since changes with time might be expected to be
more predictable.

Nevertheless, it was suggested as part of the so-called
Mussel Watch Programme to monitor water quality that mussels
may also act as integrators of oil pollution because it was
believed that those bivalves did not metabolise hydrocarbons.
It has been shown since that metabolism of aromatic hydrocarbons does occur in bivalves, albeit at a very slow rate.
Nevertheless, to examine the validity of this approach we
carried out qualitative and quantitative analyses of the
paraffinic and aromatic hydrocarbons of mussels collected from
around the Scottish coastline (Ref. 16). The industrialised
areas contained the higher levels of both types of hydrocarbons
but, some of the stations expected to be relatively free from
petroleum contaminants also contained elevated high molecular
weight PAH (4, 5 and 6 rings) which appeared to be pyrolysis
products. Some industrial sites did show suspected oil derived
hydrocarbons superimposed on the combustion products, however
this species was not found to be a reliable indicator of oil
contamination over a period of more than a few days.

At this point, some attention should be given to the PAH
of the pyrolysis products, which include known carcinogens such
as benzo (a) pyrene, because it should be emphasised that they
are now generally believed to make a much greater contribution
to the environmental load of these compounds than does petroleum
itself of which they represent only a tiny fraction. The PAH
content and composition of the pyrolytic products of the most
commonly combusted materials i.e. coal, oil and its products and
wood have been determined by several workers (Ref. 17). Qualitatively they are very similar although there are marked quantitative differences, depending largely on the combustion temperature
(Ref. 18). Indeed, some of the constituents, e.g. phenanthrene,
are common to pyrolytic and crude oil sources which can, in some
cases, make their differentiation more difficult.

An annual survey of a given station or a series of stations
might reasonably be expected to reflect any sustained increase
or decrease in hydrocarbon contamination in relation to the
general and specific levels of usage and/or spillage of petroleum
or petroleum products in the site areas. The paraffinic hydrocarbons in the water at a number of stations on the River Forth
(upstream of a refinery), through the Firth of Forth, the North
Sea, to the Forties oilfield in the North Sea have failed to
show a significant increase or decrease over the last 5 years
(unpublished results). However it must be emphasised that the
variance in the water sample analysis is large (Ref. 19) compared with the expected changes in input except at the Forties
oilfield. There it is impractical to sample sea water adjacent
to the production platforms at Forties and an immense dilution
would have occurred at our sampling station some 1000 m distant.
A change of about 100% in the oil input to the sampling stations
would have to occur before a statistically valid difference
could be detected in the analyses above the normal sample
variability. The refinery on the River Forth at Grangemouth

emits large quantities of hydrocarbons but much of the non-volatile portion is incorporated in the sediments around the outfalls and concentrations in the water diminish rapidly with distance from the outfall. The same is true of other refinery oily water effluents such as those at Fawley on Southampton Water (P. Williams, pers. comm.).

The sediments act as a sink at least for the non-volatile fraction of oil input and so changes in the hydrocarbon composition of the surface sediments should provide some clues to the influence of a changing input. We examined the extent to which such changes might be detectable in the surface sediments at the Sullom Voe oil terminal development which show the characteristic and easily recognisable odd carbon predominance in their alkane patterns. Calculation indicated that doubling the alkane concentration by addition of weathered North Sea crude oil to the surface sediment would be easily detected simply by comparison of the alkane profile (Ref. 20). On this basis, even the lowest estimated annual levels of contamination of the area due to the terminal operations would be detectable if it was assumed that biodegradation could be ignored.

The sample replicate variation in the field, whether in tissues, sediments or water is an inescapable fact affecting the precision of the measurements but their comparison with those from other laboratories is rendered difficult in the extreme due to differences in sampling techniques, analytical procedures and levels of sophistication as well as units of calculation.

To rationalise the experimental and analytical approaches in different laboratories having varying technological and scientific capabilities we have proposed useful areas of study for chemical and biological research in the context of oil spill incidents (Ref. 21). More importantly, these proposals (see Appendix) on the development of an analytical protocol covering samples, sampling procedures and methods of analyses which would be followed rigorously to allow more useful and comparative analysis of chemical and biological data from different laboratories are equally applicable in the context of oil pollution research and monitoring.

Different laboratories are capable of varying degrees of analytical sophistication and their programmes will be dictated by their capabilities and the questions and problems which are relevant to their national needs. So far there has been little uniformity between programmes in terms of observations, sampling, analyses and assessment but it is important that a degree of standardisation be introduced at all levels into the methods used in order to improve comparability of results between the various techniques as an essential step towards interlaboratory comparison. One approach is to set up intercalibration exercises based on scientific techniques but another is to encourage progress towards standardisation of a range of analytical methods from the simple to the more sophisticated to take account of the widely differing capabilities of research groups. Inevitably, some of the proposed methods will be less satisfactory than those already in use in certain laboratories. These are matters for discussion. Once agreed, the protocol should be followed exactly and vigorously for a period of assessment and subsequently revised if necessary in the light of experience gained.

Complementary towards satisfying the protocol we have drawn up an example of a sample preparation of graded complexity

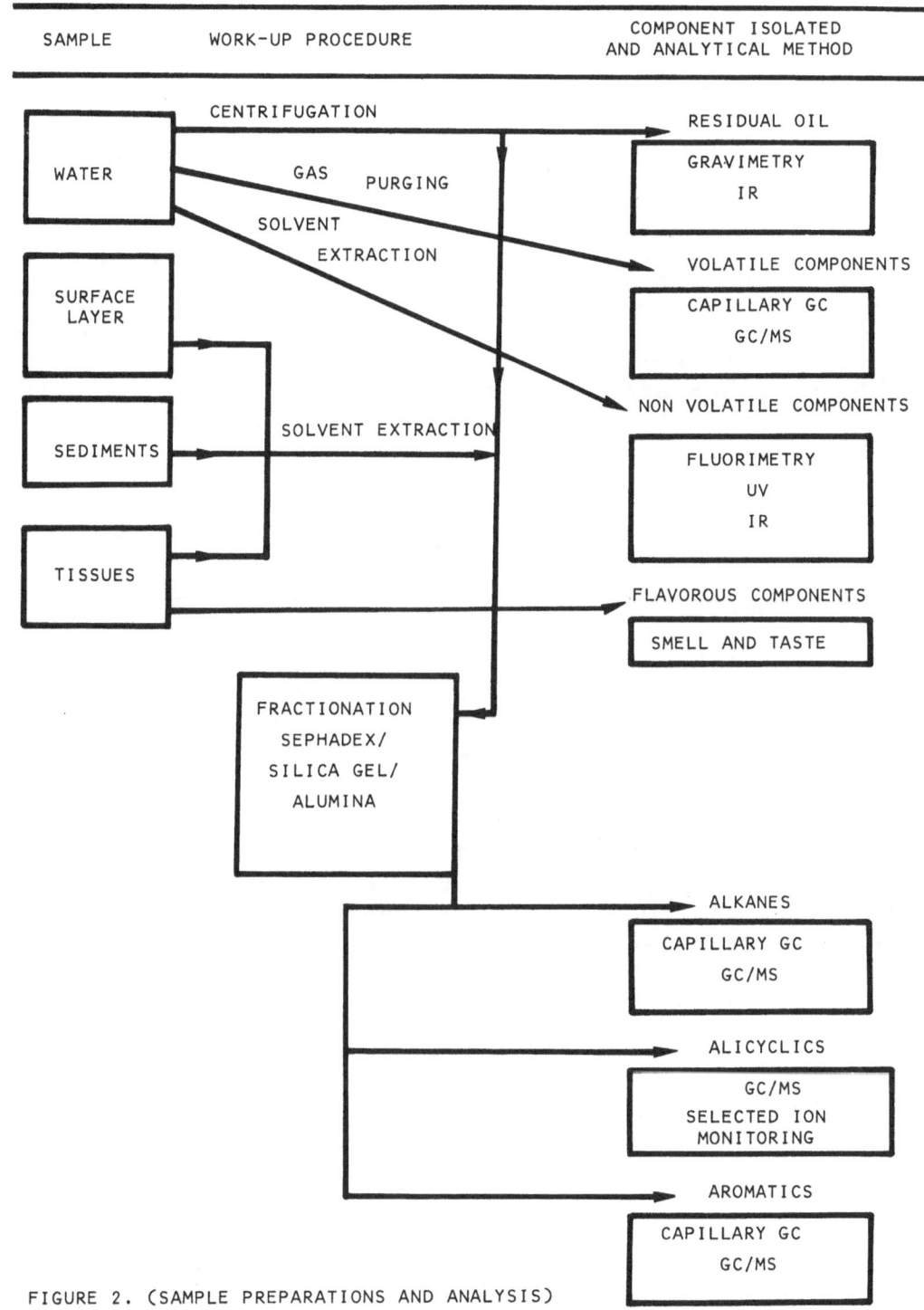

FIGURE 2. (SAMPLE PREPARATIONS AND ANALYSIS)

552

(Fig. 2) which all analysts would be encouraged to follow to
the level necessary for their particular analytical method to
be applied, whether it be simple gravimetry or spectrophotometry
or highly sophisticated computerised gas chromatography/mass
spectrometry.

REFERENCES

1. G C Speers and E V Whitehead, 'Crude Petroleum', In :
   Organic Geochemistry (Springer, Berlin, 1969)
2. B J Mair, Annual Report Am. Petr. Inst. (Carnegie Inst.
   Technol., Pittsburg, 1965).
3. J Posthuma, 'The Composition of Petroleum', In : Petroleum
   Hydrocarbons in the Marine Environment. Rapp.P-v. Reun.
   Cons. Int. Explor. Mer, 171 (1977), 7-16.
4. D E Wright and J A Wright, 'Evaluation of an IR oil film
   monitor'. Interim Report No. CG-D-51-74. U.S. Coastguard
   Office of Research and Development, 1973.
5. E D Goldberg et al, 'The Mussel Watch'. Environ. Conserv.
   5 (1978), 101-125.
6. R Hardy, P R Mackie, K J Whittle, A D McIntyre and R A A
   Blackman, 'Occurrence of hydrocarbons in the surface film,
   sub-surface water and sediment in the waters around the
   United Kingdom'. In Petroleum Hydrocarbons in the Marine
   Environment. Rapp. P.-v Reun.Cons.Int.Explor.Mer, 171
   (1977), 61-65.
7. K J Whittle, P R Mackie, R Hardy, A D McIntyre and R A A
   Blackman, 'The alkanes of marine organisms from the United
   Kingdom and surrounding waters'. Ibid. p. 72-78.
8. R Hardy, P R Mackie and K J Whittle, 'Hydrocarbons and
   petroleum in the marine ecosystem - a review'. Ibid.p.17-26.
9. R F Lee and A R Loeblich, 'Distribution of 21:6 hydrocarbon
   and its relationship to 22:6 fatty acid in algae'. Phytochem.
   10 (1971), 593-602.
10. P R Mackie, H M Platt and R Hardy, 'Hydrocarbons in the
    Marine Environment II. Distribution of n-alkanes in the
    fauna and environment of the sub-antarctic island of
    South Georgia'. Est.Coast. Mar. Sci., 6 (1978), 301-313.
11. H Platt and P R Mackie, 'Analysis of aliphatic and aromatic
    hydrocarbons in Antarctic marine sediment layers'. Nature
    (Lond.) 280 (1979), 576-578.
12. P R Mackie, R Hardy and K J Whittle, 'Preliminary Assessment
    of the Presence of oil in the ecosystem at Ekofisk after
    the blowout, April 22-30, 1977'. J.Fish.Res.Bd.Can., 35
    (1978), 544-551.
13. R Hardy, P R Mackie, K J Whittle and A D McIntyre, 'Discrim-
    ination in the assimilation of n-alkanes in fish'. Nature
    (Lond.) 252 (1974), 577-578.
14. K J Whittle, J Murray, P R Mackie, R Hardy and J Farmer,
    'Fate of hydrocarbons in Fish'. In Petroleum Hydrocarbons
    in the Marine Environment. Rapp.P.-v.Reun.Cons.Int.Explor.Mer,
    171 (1977), 139-142.
15. R F Lee, R Sauerheber and G H Dobbs, 'Uptake metabolism
    and discharge of polycyclic aromatic hydrocarbons by marine
    fish'. Mar.Biol. 17 (1972), 201-208.
16. P R Mackie, R Hardy, K J Whittle, C Bruce and A S McGill,
    'The tissue hydrocarbon burden of mussels from various sites
    around the Scottish coast'. In Proceedings 4th Intl.
    Symposium on PAH, Columbus, Ohio (1979) In Press.

17. A Hase, P H Lin and R A Hites, 'Analysis of complex polycyclic aromatic hydrocarbon mixtures by computerised GC/MS'. In Carcinogenesis Vol. 1 (Raven Press, 1976).

18. R E La Flamme and R A Hites, 'The Global distribution of polycyclic aromatic hydrocarbons in recent sediments'. Geochim et Cosmochim. Acta. 42 (1978), 289-303.

19. P R Mackie, R Hardy and K J Whittle, 'Sampling and extraction methods and their associated problems'. In Petroleum Hydrocarbons in the Marine Environment. Rapp. P.-v.Reun.Cons.Int.Explor.Mer., 171 (1977), 27-32.

20. J M Davies, R Johnston, K J Whittle and P R Mackie, 'The origin and fate of hydrocarbons in Sullom Voe'. In Proc. Symp. 'The Marine Environment of Sullom Voe and the implications of oil development'. (Scottish Marine Biological Association meeting, Oban, 1979. Royal Soc. Ed. In Press).

21. K J Whittle, R Hardy and A D McIntrye, 'Scientific studies at future oil spill incidents in the light of past experience'. I.C.E.S. CM 1978/E:33 Marine Environmental Quality Committee.

30. R Burwood and G C Speers, 'Photo-oxidation as a factor in the environmental dispersal of crude oil'. Est. Coastl. Mar.Sci., 2 (1974), 117-135.

554

APPENDIX

PROPOSED ANALYTICAL PROTOCOL

SAMPLING

Samples	Method	Comments
Surface film	eg a)stainless steel screen (Ref.19) b)Teflon disc (Ref. 22)	Calm seas essential Other devices available
Water from 1m depth and 10m depth intervals	a)manual filling (Ref. 19) b)remote filling (Ref. 12)	Sub-slick samples obtained last (1)
Tar balls	Neuston tow (Ref.23)	Two of 1 naut. mile
Sediment cores	Collected by diver, Van Veen grab or box corer	(2)
Sedimenting particles	Aberdeen trap (Ref. 24)	Other types available
Plankton	Oblique, vertical or horizontal tows (250u mesh net with bucket type cod end for live organisms	(3)
Edible fish	Pelagic and bottom trawl	
Benthos	a)Diver or b) Agassiz trawl	
Mussels	Mussel cages (Ref.25) (4)	

(1) To prevent contamination at surface, devices should be protected or towed submerged to the sampling site.
(2) Sampling by diver preferred because mechanical methods displace surface sediments and cause mixing.
(3) In the presence of a visible slick plankton net and sample should be protected from contamination at surface.
(4) Can be located on the mooring lines of the sediment traps.

EXTRACTION

Oil component	Sample Type	Method	Comments
Total oil	Surface film, water and sediment	Centrifugation	For high oil concentrations only (1) (2)
Volatiles	Liquid	Gas purging (eg Ref. 26)	(2)
Non-volatiles	Liquids Solids	Solvent extraction :dichloromethane :chloroform:methanol (Ref. 27)	(2) (3)

(1) Method can be applied to solids by the addition of water.
(2) No internal standard is completely satisfactory. Squelane
    is suitable for non-volatile aliphatic analyses.
(3) A number of solvents have been used to extract liquids. Dich-
    loromethane is recommended for ease of processing. Freon
    is used particularly for the analysis of the more volatile
    components.

SENSORY ANALYSIS

Sample	Method	Comments
All, including site	Visual description	
Edible portion of commercial species	Smell and taste (Ref.29)	(4)

(4) Many people can detect very low concentration of oil by
    smell and taste. This technique can be used to provide a
    simple qualitative assessment of the presence of freshly
    spilt oil.

FRACTIONATION AND CHEMICAL ANALYSIS

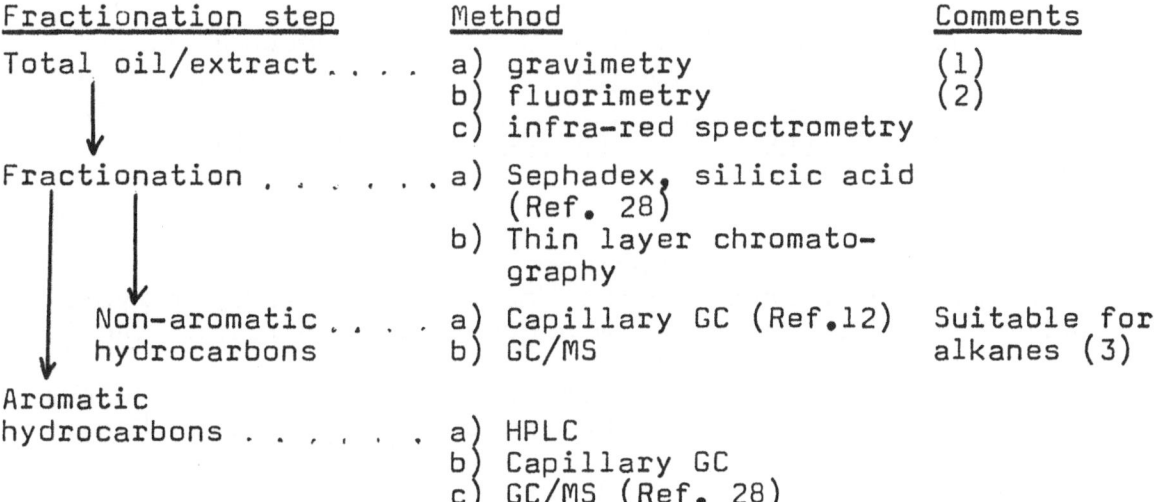

Fractionation step	Method	Comments
Total oil/extract....	a) gravimetry b) fluorimetry c) infra-red spectrometry	(1) (2)
Fractionation	a) Sephadex, silicic acid (Ref. 28) b) Thin layer chromatography	
Non-aromatic . . . . hydrocarbons	a) Capillary GC (Ref.12) b) GC/MS	Suitable for alkanes (3)
Aromatic hydrocarbons	a) HPLC b) Capillary GC c) GC/MS (Ref. 28)	

(1) Gravimetry or fluorimetry is only suitable for large amounts
    or oil. When the concentration is low other extractants
    interfere.
(2) The standard (IGOSS) method relies on single wavelength
    measurement. When oil concentration is high method is
    satisfactory, otherwise scanning fluorimetry should be used.
    The ease and rapidity of the method allows on-site analysis
    and thus rapid decisions on the advisability of further
    sampling.
(3) GC/MS used to confirm identity of chromatograph eluates.
    Can be used as semi-specific for mixtures, e.g. alicyclic
    hydrocarbons and alkanes.

BIOLOGICAL EFFECTS

Sample	Method	Comments
Micro-organisms	1) Counts: saprophytes, hydrocarbon degraders 2) Mineralisation rate using radio labelled hydrocarbon substrates	Will only show how microbial population responds to oil
Copepods	1) Compare survival in clean and oily water 2) Compare feeding rates in clean and oily water enhanced with phytoplankton	Useful and simple to determine sublethal effects
Fish eggs	Compare viability	Seasonally dependent. Laboratory testing using site water may be more appropriate
Fish larvae	1) Compare feeding in clean and oily water 2) Examine for defects	
Fish and shellfish	Standard 96 h LC50's, eg Crangon crangon, Pleuronectes platessa or Agonus cataphractus	Response unlikely except in acute pollution or presence of dispersants

REFERENCES

22. R Miget, H Kator, C Oppenheimer, J L Laseter and E J Ledet, 'New sampling devices for the recovery of petroleum hydrocarbons and fatty acids from aqueous surface films'. Anal. Chem., 46 (1974), 1154-1157

23. E M Levy and A Walton, 'Dispersed and particular petroleum residues in the Gulf of St Lawrence'. J.Fish.Res.Bd.Can., 30 (1973), 261-267.

24. R Payne and J Davies, 'The Aberdeen sediment trap and its moorings'. Scott.Fish.Res.Rep., No. 8 (1977).

25. K J Whittle, P R Mackie, J Farmer and R Hardy, 'The effects of the Ekofisk blowout on hydrocarbon residues in fish and shellfish'. Proc. of the Conf. on Ecological Impact of Oil Spills, Colorado (June 1978). In Press.

26. J N Swinnerton and V J Linnenborn, 'Determination of C_1 to C_4 hydrocarbons in sea water by gas chromatography'. J. Gas.Chrom., 5 (1967), 570-573.

27. P R Mackie, K J Whittle and R Hardy, 'Hydrocarbons in the marine environment I n-alkanes in the Firth of Clyde'. Est.Coast.Mar.Sci., 2 (1974), 359-374.

28. W Giger and C Schaffner, 'Determination of polycyclic aromatic hydrocarbons in the environment by glass capillary gas chromatography', Anal.Chem., 50 (1978), 243-249.

29. P Howgate, A D McIntyre, A Eleftheriou, P R Mackie, K J Whittle and J Farmer, 'Petroleum tainting in fish'. In Petroleum Hydrocarbons in the Marine Environment. Rapp. P.-v.Reun.Cons.Int.Explor.Mer., 171 (1977), 143-146.

COASTAL MORPHOLOGY AND OIL SPILL POLLUTION: THE AMOCO CADIZ EXPERIENCE

Laurent D'Ozouville[1], Serge Berné[1], Erich R. Gundlach[2], and Miles O. Hayes[2]

[1] Centre Oceanologique de Bretagne (CNEXO), Boite Postale 337
29273 Brest Cedex, France.

[2] Research Planning Institute, Inc., 806 Pavillion Avenue, Columbia,
South Carolina, USA 29205.

INTRODUCTION

The breakup of the supertanker AMOCO CADIZ off the coast of Brittany, France, on 17 March 1978 resulted in the discharge of 233,000 tons of light Arabian crude oil. To date, this is the largest tanker spill ever to occur. This report concerns the interaction of spilled AMOCO CADIZ oil within various coastal environments, particularly emphasizing the residence time or persistence of oil within each shoreline type. This information has specific relevance in planning counter strategies to effectively control future oil spills.

Data presentation follows the format of a previously developed oil spill vulnerability index (Ref. 1) which classifies coastal environments on a scale of 1 to 10 in terms of the predicted probability of being damaged during a spill. The higher the index value, the greater the likelihood of long-term, spill-related damage. As field data from the AMOCO CADIZ spill support this shoreline vulnerability index, a discussion of index applicability to spill-prone areas of the world is also included. Summary reports of the AMOCO CADIZ oil spill are contained in references 2-5. Papers particularly concerned with the shoreline impact of AMOCO CADIZ oil are included in references 6-11.

METHODS OF STUDY

In order to determine the short- and long-term persistence of AMOCO CADIZ oil, field studies of the spill site were conducted during March, April, July/August, and November 1978, and March and November 1979. During these investigations, two types of stations were repetitively surveyed (Fig. 1):

Detailed Study Sites - 23 in total. These stations were selected as representative of various shoreline types within the spill site. At each station, stakes were placed to enable repetitive measurement of a beach profile. Along the profile, the concentration of surface oil was determined by considering oil thickness, surface coverage, specific gravity, and water content. Subsurface or buried oil was measured along a trench dug parallel to the beach profile. Oil concentration at these stations was integrated over the entire oiled shoreline to yield an estimate of the total oil quantity.

Rapid Inspection Stations - 147 in total. These stations were set up to monitor the rapidly changing nature and wide distribution of the spill. Notes and photographs were taken at each site to document surface and sub-surface oil distribution and obvious biological effects.

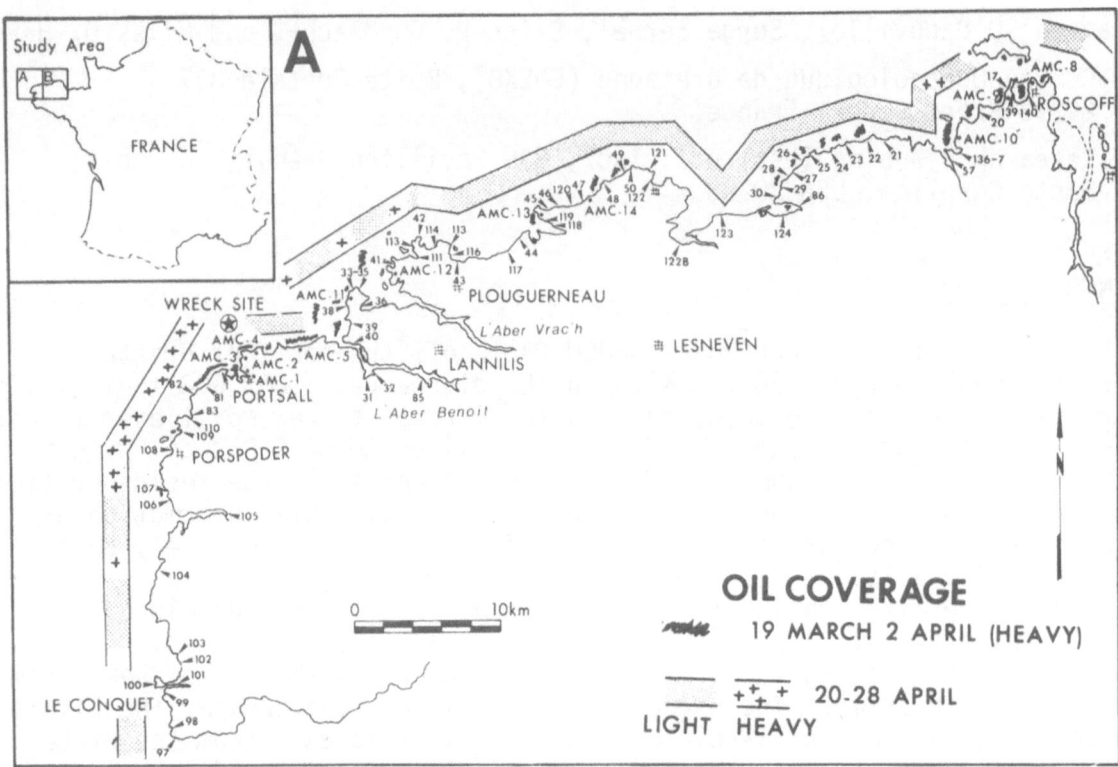

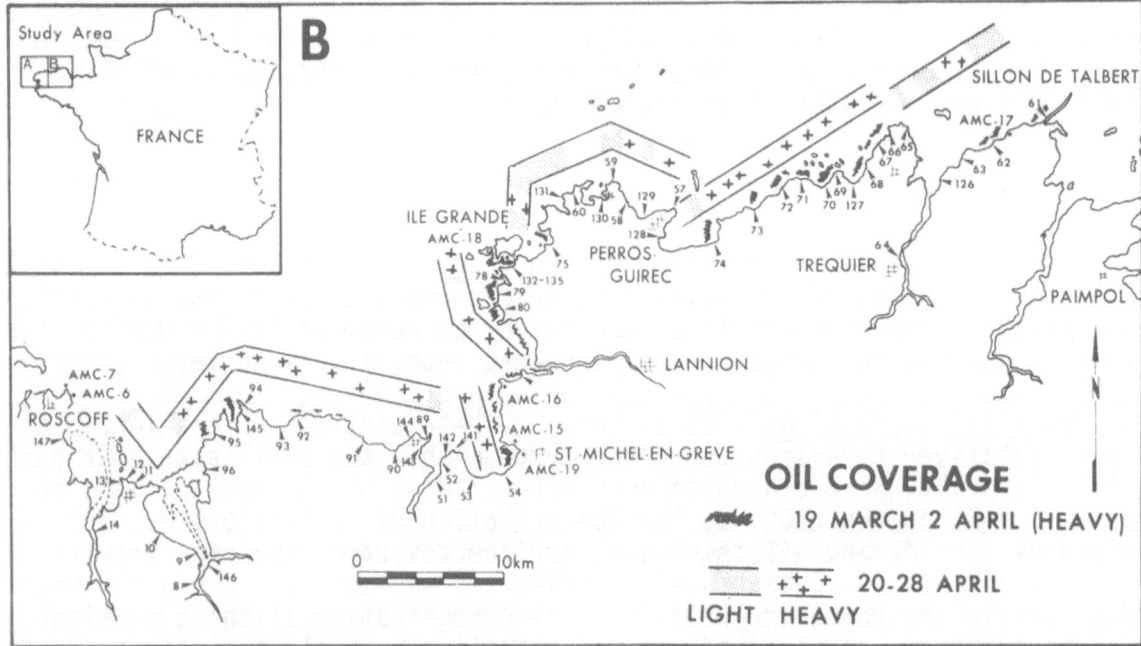

Fig. 1. Station locations and oil coverage within the AMOCO CADIZ spill
site in Brittany, France. Detailed study stations are designated by
AMC- . Rapidly inspected stations are without prefix.

In addition to the ground surveys, numerous low-altitude overflights and high-resolution aerial photographs (taken by the Institut Geographique National) were used to create a time history of the spill.

RESULTS

Extent and Quantity of Oil along the Shoreline

During the first weeks of the spill, approximately 62,000 tons, or slightly more than one quarter of the cargo, were deposited along 72 km of shoreline. Oil was primarily concentrated in large patches. One month after, the onshore quantity of oil was reduced to less than 10,000 tons, primarily due to natural processes (waves and tides) rather than cleanup efforts. However, even though the quantity of oil was much less, changing wind conditions increased the extent of oiled shoreline to 330 km. Table 1 shows the results of follow-up studies extending to one year after the spill. Areas of heavy pollution radically decreased over the year. However, light quantities of oil were still visible along 69 km of shoreline during April 1979.

TABLE 1. Evolution of surface pollution resulting from the AMOCO CADIZ spill from March 1978 to March 1979 (adapted from Ref. 11).

	STATE OF SHORELINE OILING		
	Heavy (km)	Light (km)	Tons Present
End of March 1979	72	0	62,000
End of April 1979	175	155	< 10,000
End of May 1978	109	123	*
November 1978	54	156	*
March 1979	8	69	*

* Difficult to determine because oil was thinly scattered along most of the shoreline.

Interaction and Persistence of Oil within Coastal Environments

Oil interaction with the shoreline is presented in terms of environments listed in the oil spill vulnerability index (Ref. 1). This shoreline index, primarily developed through investigation of the METULA and URQUIOLA oil spills (Refs. 12 and 13), ranks shoreline types on a scale of 1 to 10 in terms of potential damage by spilled oil. The index is based mainly on the persistence or longevity of oil within each environment, but also takes biological considerations into account. The higher the index value, the greater the predicted persistence of oil within that environment.

The following section presents the observed reaction and persistence of AMOCO CADIZ oil within the environments along the Brittany coast. A data summary is presented in Table 2. Environments are presented in order of increasing oil persistence.

1. Exposed Rocky Headlands. -

These areas are very common along the rocky and high-energy coast of Brittany. During the spill, the reflection of waves off the steeply dipping rocks caused most of the oil to remain offshore (Fig. 2). Oil that did im- pact these areas (where the shore was very irregular) was rapidly removed by normal wave activity a short time after.

Fig. 2. Arrow indicates oil held offshore by waves re- flecting off a rocky headland south of the wreck site on 28 April 1978. These areas are ranked lowest on the oil spill vulnerability index. In contrast, the sheltered environment behind the headland became heavily oiled.

2. Eroding Wave-Cut Platforms. -

Wave-cut platforms in Brittany commonly contain large boulders. Oil tended to persist for several months within the pockets and sheltered areas created by the boulders (Fig. 3). In contrast, oil was removed rapidly in areas directly exposed to the waves.

3. Fine-Sand Beaches. -

This beach type is common in Brittany in areas of exposed, relict sed- imentary deposits. During the spill, most of the surface oil was removed relatively rapidly within a few weeks. The compact surface of this beach type inhibited penetration of the oil, thereby keeping it exposed to waves and tidal currents. In some localities, cleanup equipment was effectively

Fig. 3A. Heavily oiled boulder beach south of the wreck
      site on 21 April 1978.

Fig. 3B. The same site on 7 November 1978. Despite the
      exposure of this area to high wave energy, oil per-
      sisted along the upper part of the beach (marked by
      arrow). The dark color lower on the beach is caused
      by algae.

used. After one year, most of these beaches were entirely clean, except for
some scattered, buried oil layers in depositional areas.

4. <u>Coarse-Sand Beaches</u>. -

Coarse-sand beaches within the spill site are located in more sheltered
environments than fine-sand beaches and, consequently, retained the effects
of the spill longer. Oil-stained, surface sediment was still common six to
nine months after the spill. One year after, the beachface was entirely
clean, although some buried oil still remained.

5. <u>Exposed Tidal Flats</u>. -

Oil rapidly passed over the compact surface of exposed tidal flats and
was deposited along the upper portions of the flat. The biological effects
of the oil varied. On most exposed sand flats fronting fine-sand
beaches, impact was minor; however, at the very large and productive St.
Efflam sand flat, several million heart urchins, razor clams, and cockles
were killed by the oil (Fig. 4). Nine months after the spill, oil was
still obvious within trenches dug along this area as part of the cleanup
effort. One year later, oil was not evident as discrete buried layers;
however, the interstitial water of the flat was still contaminated.

6. <u>Mixed Sand and Gravel Beaches</u>. -

Mixed sand and gravel beaches are common within sheltered and semi-
sheltered environments, but not along the exposed coast of Brittany.
Within the sheltered areas, scattered discontinuous asphalt pavement,
formed of desiccated mousse and sediment, persisted throughout the study
period. Along the semi-sheltered but active beaches, oil was removed
within nine months after the spill.

7. <u>Gravel Beaches</u>. -

Gravel beaches are placed fairly high on the index because oil rapidly
and deeply penetrates into the sediment. At the gravel beach at Pointe
de Sehar in France (Fig. 5), oil penetrated 65 cm into the substrate.
Even after cleanup operations, oil remained mixed into the beach and persisted for longer than a year after the spill. However, by November 1979,
most of the remaining oil at this locality was gone.

8. <u>Sheltered Rocky Areas</u>. -

The last three environments of the vulnerability index are areas sheltered from major storm and wave activity. Within sheltered rocky areas,
oil persisted for at least a year as a black coating on the rocks (Fig. 6),
or as asphalt pavement located between them.

9. <u>Sheltered Tidal Flats</u>. -

After one and one-half years, oil was still common on the surface (Fig.
7) and/or mixed into the sediments of most of the sheltered tidal flats
initially oiled during the spill. Oil that was mixed into the sediments
still emitted a fresh-appearing surface sheen. The interstitial water
of most of these areas also remained contaminated. The lack of any ero-

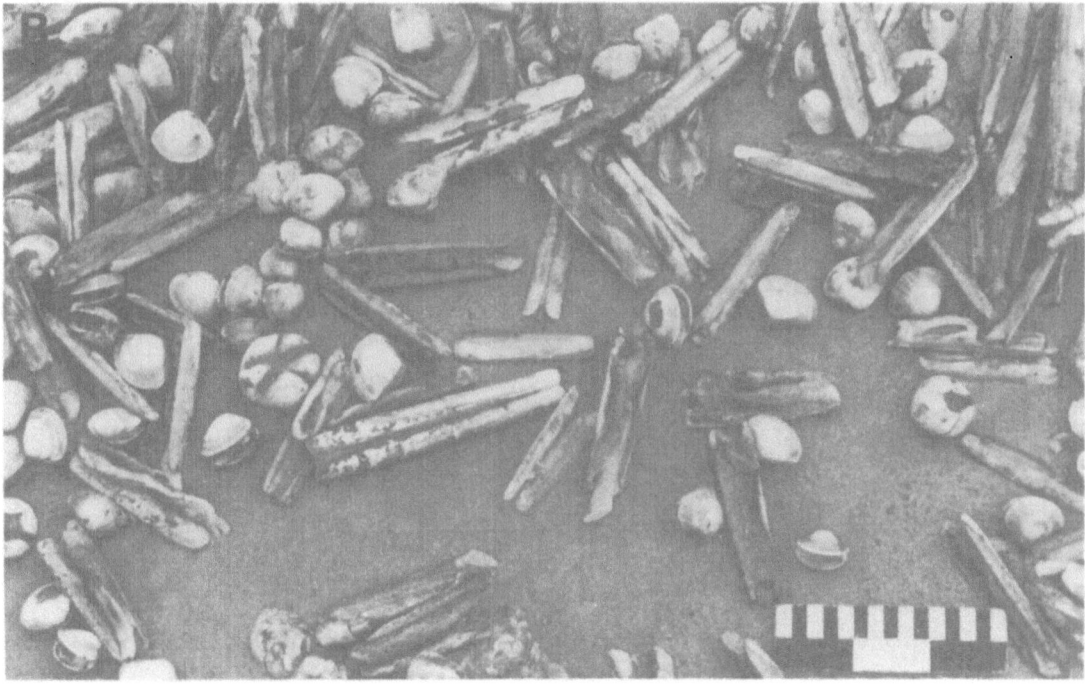

Fig. 4. (A) Overview and (B) close-up of swashlines of
   dead heart urchins, razor clams, and cockles along
   the St. Efflam sand flat on 2 April 1978. Rocks on
   the right of the top photograph are heavily oiled.

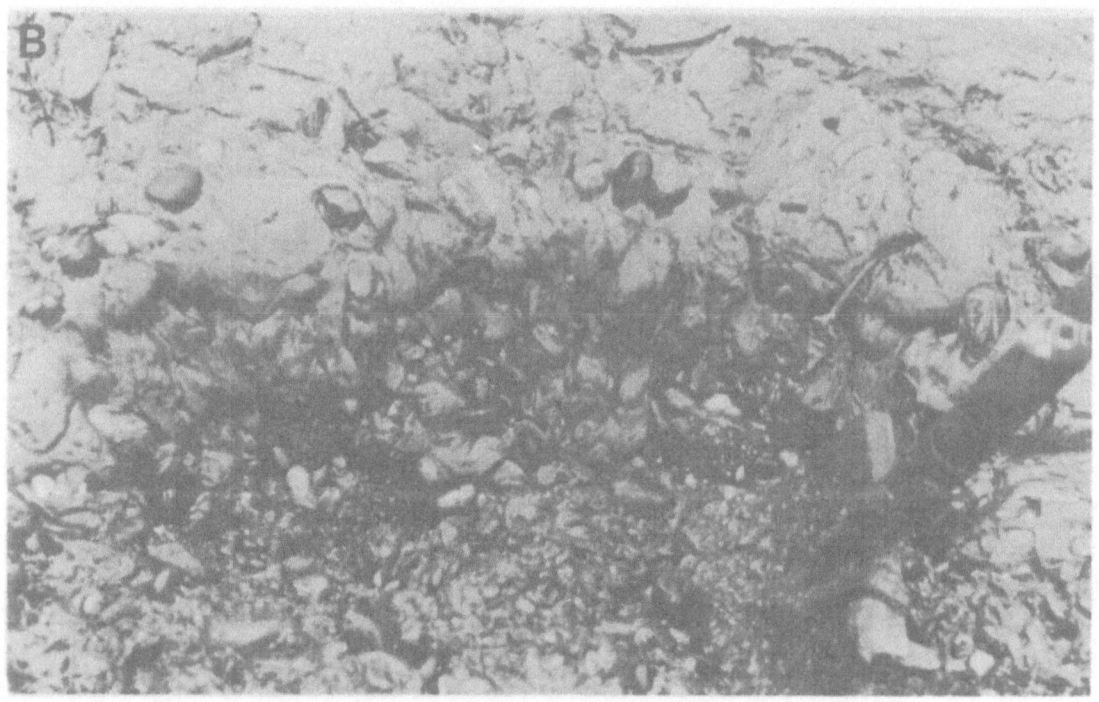

Fig. 5. (A) Overview and (B) close-up of the heavily
    oiled gravel beach at Pte. de Sehar (station AMC-16
    in Fig. 1) on 28 March 1978. During the spill, oil
    penetrated over 65 cm into the loose gravel. De-
    spite cleanup attempts, oil persisted for over a
    year at this location.

Fig. 6.   Sheltered rocky area within Portsall harbor,
        very close to the AMOCO CADIZ wreck site.
        (A)   Heavily oiled algae and rocks on 22 March 1978.
        (B)   The same site on 7 November 1978. Despite
              cleanup activities, which also removed all the
              algae, the rocks remained oiled. However,
              substantial recovery of the algae was noted by
              November 1979.

Fig. 7. A heavily oiled sheltered tidal flat (station
    F-75) on (A) 29 March 1978, and (B) 4 November 1978.
    A trench used during cleanup remains to the left of
    photograph B. During November 1979, oil was still
    common throughout the area.

sive waves or currents enables oil to persist in these areas long after the initial event.

10. Sheltered Marshes. -

Several marshes within the AMOCO CADIZ spill site became oiled. By far the largest within the impact area is the Ile Grande Marsh. During the spill, the entire marsh was covered by 2-5 cm of oil (Fig. 8A). A large cleanup operation succeeded in removing the thick pools of oil, but as the effort included several scores of men and heavy machinery, the marsh was physically damaged by these activities. Vegetation along the upper marsh showed significant recovery since the spill; however, lower marsh grasses have not returned at all (Fig. 8B). The AMOCO CADIZ spill significantly altered the physical (drainage patterns, etc.) and biological characteristics of this site.

DISCUSSION

The AMOCO CADIZ oil spill illustrates the great impact that a single incident can have along a very large section of coastline, and the long-term persistence of oil within certain coastal environments. Proper contingency planning against the shoreline impact of future spills must take these two factors into account. But rather than expending our energies and resources to protect all the shorelines of the world, it is important to focus our attention on those parts of the globe where major spills are most likely to occur. Based on previous spill histories (Fig. 9), high risk areas include the coastlines of South Africa, western India, South Korea and Japan, parts of the Caribbean, and the Atlantic shorelines of Europe and North America.
   Within the high-risk areas in particular, the shoreline should be mapped in detail and classified according to the vulnerability index. Local areas of special biological consideration (e.g., bird rookeries, seal haul-out areas, etc.) should also be placed on these maps. This information becomes critical during a spill to rapidly and effectively place men and matériel to counter the impending spill. Along the U.S. coast, vulnerability index mapping has been completed for many oil-development areas of Alaska (Refs. 14-17). Recently, the entire coast of Texas was completed for the U.S. Coast Guard and National Oceanic and Atmospheric Administration as part of preparatory activities in relation to oil coming north from the IXTOC I blowout in the Gulf of Mexico. Mapping has also been completed for the New Zealand coast (Ref. 18).
   Additional pre-spill contingency measures include trajectory analyses to delineate local areas having high probability of impact, and thorough current studies of major inlets or channels in order to effectively deploy booms or skimming systems. In the case of the AMOCO CADIZ, booms were ineffective due to strong currents and a high tidal range. Previous study would have uncovered this problem and enabled a proper line of defense in accordance with known boom technology and efficiency. Of course, for any of these pre-spill contingency activities to be practical and cost-effective, containment devices and cleanup equipment must be stockpiled and ready for transport, and personnel must be properly trained. As more lawsuits are brought against the spiller and the financial penalties for spills increase, functioning regional contingency plans will become a necessity.

Fig. 8.   Heavily oiled marsh at Ile Grande on (A) 2 April
1978 and (B) 18 November 1979. Upper marsh grasses
recovered in many areas, but the lower marsh was still
completely denuded. Erosion of the channels has
occurred due to the lack of stabilizing plants.

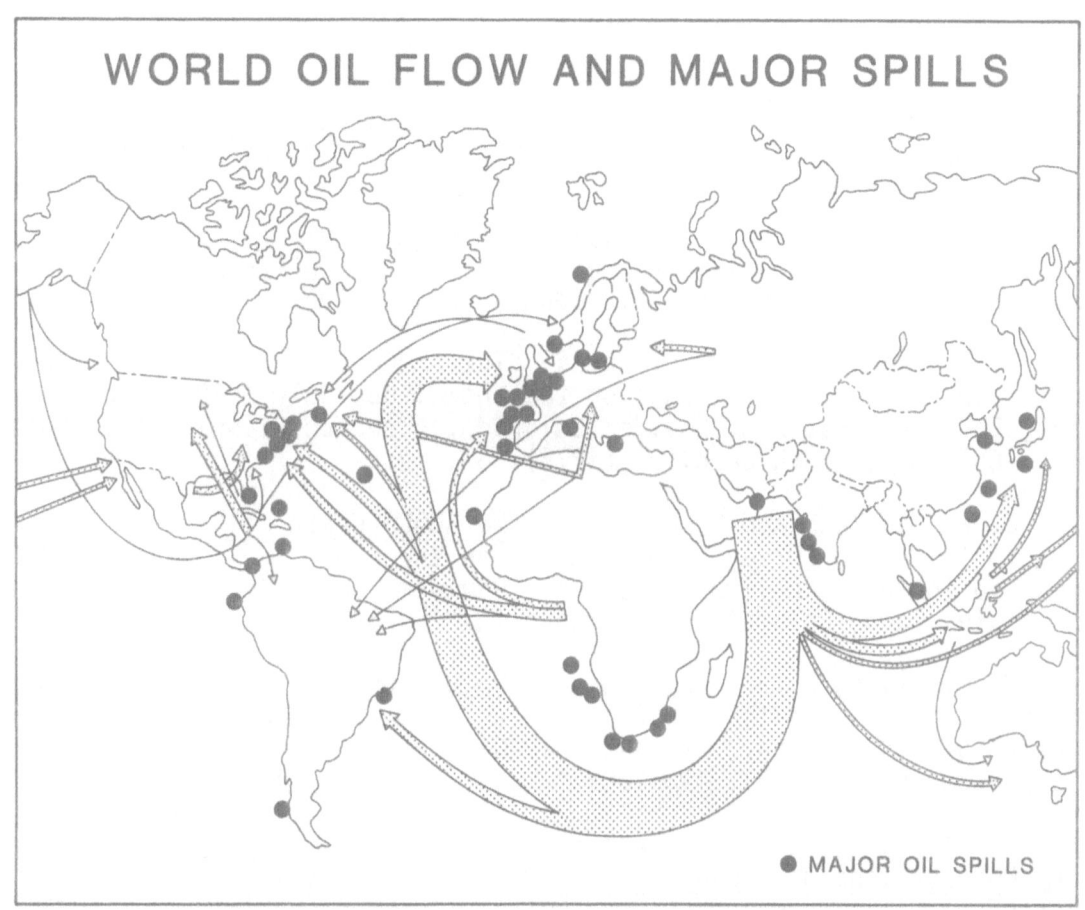

Fig. 9.   World oil flows and major spills (adapted from Ref.19).   High-
risk areas are located in southern Africa, western Europe, the
Atlantic coast of North America, western India, and in the Far East.
It is proposed that these areas receive priority for contingency
planning and coastal vulnerability index mapping.

REFERENCES

E. R. Gundlach and M. O. Hayes, "Classification of coastal environments
in terms of potential vulnerability to oil spill damage," Mar. Technol.
Jour., 12 (1978), 18-27.

W. N. Hess (editor), "The AMOCO CADIZ oil spill, a preliminary scientific report," NOAA/EPA special report.   Environ. Res. Lab, Boulder, Colorado (1978), 347.

G. Conan, L. D'Ozouville, and M. Marchand, "AMOCO CADIZ, preliminary
observations of the oil spill impact on the marine environment," Publ. du
Centre National pour l'Exploitation des Oceans (CNEXO), Actes de Collogues
No. 6, Centre Oceanologique de Bretagne, B. P. 337, 29273 Brest, France
(1978), 239.

REFERENCES (continued)

M. F. Spooner (editor), "AMOCO CADIZ oil spill," Mar. Pollut. Bull., 9(11) (1978a), 281-311.

"AMOCO CADIZ: consequences d'une pollution accidentelle par les hydrocarbures (Fate and effects of the oil spill), resume de communications," CNEXO, Centre Oceanologique de Bretagne, B. P. 337, 29273 Brest, France (1979), 120.

E. R. Gundlach and M. O. Hayes, The AMOCO CADIZ Oil Spill, "Investigation of beach processes," (W. N. Hess, editor, NOAA/EPA Special Rept., Environ. Res. Lab, Boulder, Colorado (1978), 85-196.

E. R. Gundlach and M. O. Hayes, "The AMOCO CADIZ Oil Spill - Third follow-up survey of oil impact on the shoreline, July 1978," prepared for NOAA, Environ. Res. Lab, Boulder Colorado, Research Planning Institute, Inc., Columbia, South Carolina (1978), 67.

E. R. Gundlach and M. O. Hayes, The AMOCO CADIZ Oil Spill, (Part II) "Fourth follow-up survey of oil impact on the shoreline, November 1978," prepared for NOAA, Environ. Res. Lab, Boulder, Colorado, Research Planning Institute, Inc., Columbia, South Carolina (1978), 29.

M. O. Hayes, E. R. Gundlach, and L. D'Ozouville, "Role of dynamic coastal processes in the impact and dispersal of the AMOCO CADIZ oil spill (March 1978)," Brittany, France, 1979 Oil Spill Conf., Amer. Petrol. Inst., Washington, D.C. (1979), 193-198.

L. D'Ozouville, E. R. Gundlach, and M. O. Hayes, AMOCO CADIZ, Preliminary Observations of the Oil Spill Impact on the Marine Environment, "Effect of coastal processes on the distribution and persistence of oil spilled by the AMOCO CADIZ - preliminary conclusions," Publ. du Centre National pour l'Exploitation des Oceans (CNEXO), Actes de Collogues No. 6, Centre Oceanologique de Bretagne, B. P. 337, 29273 Brest, France (1978), 69-96.

S. Berné and L. D'Ozouville, "Cartographie des apports polluants et des zones contaminées," CNEXO, Centre Oceanologique de Bretagne, B. P. 337, 29273 Brest, France (1979), 175.

A. Blount, "Two years after the METULA oil spill, Strait of Magellan, Chile - oil interaction with coastal environments," Tech. Rept. No. 16-CRD, Coastal Research Division, Dept. of Geology, Univ. of South Carolina, Columbia, S.C. (1978), 214.

E. R. Gundlach, C. H. Ruby, M. O. Hayes, and A. E. Blount, "URQUIOLA oil spill, La Coruna, Spain; initial impact and reaction on beaches and rocky coasts, " Environ. Geol., 2(3) (1978), 131-143.

M. O. Hayes, P. J. Brown, and J. Michel, "Coastal morphology and sedimentation, lower Cook Inlet, Alaska: with emphasis on potential oil spill impact," Tech. Rept. No. 12-CRD, Dept. of Geology, Univ. of South Carolina, Columbia, S.C. (1976), 107.

REFERENCES (continued)

J. Michel, M. O. Hayes, and P. J. Brown, "Application of an oil spill vulnerability index to the shoreline of lower Cook Inlet, Alaska," Environ. Geol., 2(2) (1977), 107-117.

C. H. Ruby, "Coastal morphology, sedimentation and oil spill vulnerability, northern Gulf of Alaska," Tech. Rept. No. 15-CRD., Dept. of Geology, Univ. of South Carolina, Columbia, S.C. (1977), 223.

D. Nummedal and C. H. Ruby, "Spilled oil retention potential - Beaufort Sea coast of Alaska," Conf. Proc., Port and Ocean Engineering under Arctic Conditions, Norwegian Inst. of Technol., August 1978, (in press), 9.

M. Gregory, A Minor Oil-Spills Workbook, "The New Zealand shoreline and application of an oil spill vulnerability index," Centre for Continuing Education, Univ. of Waikato, New Zealand (1979), 42-72.

Oil Companies International Marine Forum, "Higher risk sea areas, a risk analysis" (1979), 24.

THE DISPERSION AND WEATHERING OF CHEMICALLY TREATED CRUDE OILS ON THE SEA SURFACE

Clayton D McAuliffe
Chevron Oil Field Research Company, La Habra, California 90631

Gerard P Canevari
Exxon Research and Engineering Company, Florham Park, New Jersey 09732

Thomas D Searl
Exxon Research and Engineering Company, Linden, New Jersey 07036

Jaret C Johnson
JBF Scientific Corporation, Wilmington, Massachusetts 01887

Stephen H Greene
JBF Scientific Corporation, Wilmington, Massachusetts 01887

Four research crude oil spills discharged on the open ocean were chemically treated with a dispersant. The underlying water was then analyzed to determine (1) the dispersion of oil into the water column, and (2) the rate of loss (weathering) or low-molecular-weight hydrocarbons from the dispersed oil. These tests, funded by the American Petroleum Institute and the U.S. Environmental Protection Agency, were conducted in a manner similar to those for untreated spills conducted in 1975 (Ref. 1, 2). The current tests were designed to compare the dispersion and weathering of chemically treated and naturally dispersed oils.

The untreated oils (Ref. 1, 2) showed relatively low concentrations of nonvolatile hydrocarbons in the water column under the slicks, generally less than 1 mg/L. These samples containing naturally dispersed oil showed very rapid weathering of the C_1 to C_{10} hydrocarbons (30 min). The C_1-C_{10} hydrocarbons detected were residual in the oil droplets, and truly dissolved hydrocarbons were apparently not present. Samples of oil collected over time from the surface slicks showed slower weathering (7 hrs for trimethylbenzenes).

Chemical dispersion is thought to accelerate the natural weathering processes. This would result in higher concentrations of oil penetrating to greater depths, and accelerated escape of volatile hydrocarbons to the atmosphere. The mechanism for this behavior was expected to be the mixing of dispersed droplets having high specific surface areas in near-surface water, causing rapid loss of volatile hydrocarbons. An untreated slick, although constantly exposed to the atmosphere, may be less susceptible to evaporation than dispersed oil because its lower surface-to-volume ratio tends to retard transport (by diffusion) of volatile hydrocarbons.

Oil emulsified in water is removed from most of the wind's influence, so that it does not travel as far as a surface slick. This minimizes the possibility of oil stranding or entering

573

biologically sensitive areas. A review and discussion of the
alteration of oil on a water surface is given in Reference 3.

The current study also involved extensive aerial remote
sensing, and a limited biology program. This report, however,
covers mainly the chemical results, plus limited observations
made visually and from aerial photographs.

EXPERIMENTAL METHODS

General Operations

In November 1978, four spills were conducted approximately 40 km
(25 mi) off New Jersey and 96 km (60 mi) south of Long Island,
New York. Each spill was approximately 1.67 m^3 (440 gal) of one
of two crude oils (Murban from Abu Dhabi and La Rosa from
Venezuela). These were the same crudes used for the 1975 untreated tests. Composition of the naphtha fraction is given in
Reference 2.

Each spill was discharged from a 1.9 m^3 (500 gal) tank mounted
on the research vessel through two 7.6 cm (3 in) hoses. The ends
of the hoses were on floats, causing the oil to discharge horizontally on the water surface. This minimized both evaporation
losses due to discharge above the water, and vertical descent of
the oil into the water. The less viscous Murban (0.83 specific
gravity, 40° API) discharged in approximately 3 min; the La Rosa
(0.91 specific gravity, 23.9° API), in 6 min.

The oils were treated by aerially spraying a self-mix dispersant from a pod and spray booms mounted above the skids on a
helicopter. The helicopter flew approximately 10 m above the
water surface. One slick of each oil was dispersed immediately,
and one each after 2 hr.

The immediately dispersed slicks were sprayed with 150 L
(40 gal) of chemical dispersant; the slicks sprayed after 2 hr
with 360 L (95 gal). In all cases, there was over-spraying
(outside the slick) and a percentage loss due to wind draft.
The major experimental conditions are summarized in Table 1.

Table 1   General experimental summary

	Murban 1	La Rosa 1	La Rosa 2	Murban 2
Date of spill	2 Nov 1978	3 Nov 1978	9 Nov 1978	9 Nov 1978
Time of spill	1153	1014	1019	1404
Time of dispersion	1350	1200	1028	1411
Spill location				
latitude	40°09'09"N	40°09'12"N	40°09'18"N	40°09'30"N
longitude	73°30'39"W	39°33'40"W	73°32'00"W	73°34'45"W
Conditions				
wave height m	0.3 to 1.0	0.3 to 1.0	0.3 to 1.0	0.3 to 1.0
wind (m/s)Knots	8 to 11	8 to 11	5 to 12	5 to 12
air temp, °C	15-20	15-20	12-14	14-17
water temp, °C	14	14	13	13

Sample Collection

The sampling program was designed to obtain water samples at
approximately equally spaced stations on transects through the
surface slicks and emulsion plumes. Figure 1 is a schematic
diagram of a typical sample run. Samples of all four tests were
taken at 1 and 3 m depths at all 10 stations, and at 6 and 9 m
at Stations 3 and 8. Surface samples were taken, with a small

bucket, at all stations during sampling runs through dispersed oil. A sampling run took about 45 min.

For the immediately dispersed slicks, the first run was started a few minutes after dispersion, and the second after about 1.3 hr. For the two delayed dispersion tests, one sampling run was made before dispersion (untreated oil), and two after. The two sampling runs after dispersion were immediate and after about 1 hr.

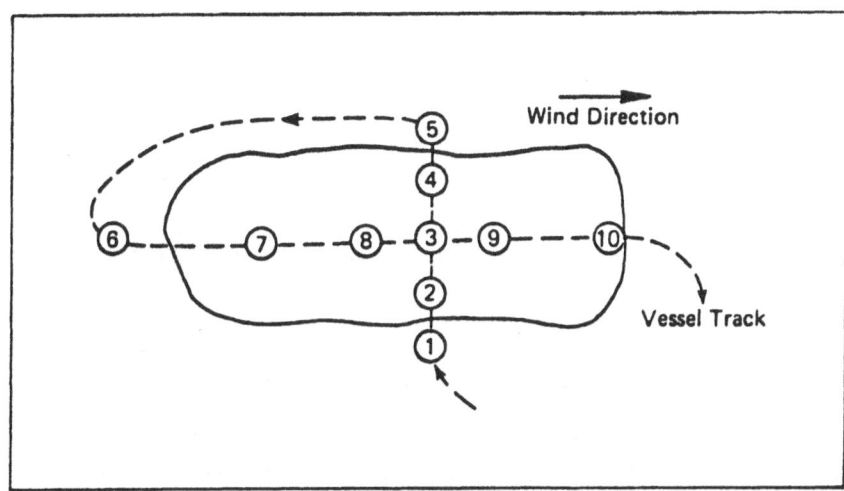

Fig. 1 Schematic of immediately dispersed oil slick and location of sample stations for typical 10—station sample run

For all of the spills a few samples were also taken 2, 3 and 4 hr after dispersion, at Stations 3 and 8.

The subsurface samples were collected with small submersible pumps discharging through polypropylene tubing, at approximately 4 L/min. The pumps were attached approximately 0.5 m below a floating 115 L (30 gal) steel drum towed 3 m lateral to the bow of the research vessel. In this position, the ship's bow wave did not cause water mixing at the sample inlets. These were 1, 3, 6 and 9 m below the water surface, along a line suspending a 23 kg weight from the bottom of the float. The sample gear was lowered and removed from the water outside the observed slicks to avoid surface oil contamination.

The two types of samples were collected at each station and depth: one 1.5 L sample in 1.9 L (0.5 gal) flint glass jug and duplicate completely filled 300 mL (10 oz) "soft drink" bottles with crown caps. The 1.9 L jugs had been cleaned by rinsing three times with distilled-in-glass carbon tetrachloride (CCl_4) that was checked for purity by infrared (IR) spectroscopy. Immediately after collection, 50 mL of this CCl_4 was added to each jug from an all-glass dispensing pipet. The jugs were sealed with teflon-lined metal screw caps, and hand-shaken for about 10 sec to initiate the solvent extraction of organic matter including the dispersed oil. The CCl_4 also prevented bacterial degradation of the hydrocarbons. In the laboratory, the samples were shaken 2 min to complete the extraction.

Prior to sample collection, about 30 mg of mercuric chloride ($HgCl_2$) was added to each 300 mL bottle to prevent biodegradation prior to analysis. Each bottle was then flushed with reactor-grade helium and sealed with a crown cap (polyvinyl chloride seal). At time of sample collection, each bottle was uncapped, filled to within 3 mm of the top, and resealed with a crown cap. The small air space minimized loss of volatile hydrocarbons (as

well as CCl_4 vapors) that may have been in the atmosphere during
sample collection.

Samples of each crude oil were taken form the spill tank in
glass bottles with teflon-lined screw caps. In the laboratory
these oil samples were equilibrated with sea water collected
outside the spill area, to provide equilibrium dissolved hydrocarbon concentrations in sea water.

Aerial control and photography

A small twin-engine high-wing aircraft served as a control
platform from which to direct the dispersant-spraying helicopter,
to direct the research vessel to each sampling station, and to
provide visual and photographic documentation of the oil slicks
and their chemical dispersion. Periodic color photographic runs
were made over each slick using a vertically mounted camera in
the floor of the aircraft. Each exposure recorded the time and
Loran C co-ordinates.

Chemical analysis

Total extractable organic matter was measured on the single
50 mL portion of CCl_4, with an IR instrument, as absorbancy at
2930 cm^{-1}. This method measures other CCl_4 soluble compounds
such as organic acids, esters, and alcohols in addition to the
crude oil. The CCl_4 extracts of a few samples were further
analyzed for total nonvolatile ($C_{14}+$) hydrocarbons, by removing
polar organic compounds with a silica gel column and reanalysis
by IR. Details of these techniques are given in Reference 4.

Volatile hydrocarbons (C_1 to C_{10} fraction) in the water
samples were analyzed by a gas equilibrium method (Ref. 5).
Forty millilitres of Murban and La Rosa oil samples were equilibrated with 140 mL of sea water collected prior to the oil spills.
The oil and water were hand shaken gently and periodically for
24 hr or more. Mercuric chloride added at the time of water
collection prevented possible biodegradation of dissolved hydrocarbons during equilibration and prior to analysis. This water
was filtered (from one 50 mL glass syringe into a second) to
remove any separate-phase oil that may have been dispersed
during oil-water mixing. Twenty-five millilitres of this water
was gas equilibrated five times.

These successive analysis were used to measure the equilibrium concentrations of individual C_1 to C_{10} hydrocarbons for
the two crude oils, and to calculate individual hydrocarbon
distribution coefficients.

The water samples collected at the various stations and
depths were then analyzed with a single equilibration using
the measured distribution coefficients to calculate concentrations.
This gives sufficient accuracy and saves time and cost of multiple
equilibrations. For those samples that contained significant
separate phase oil, the duplicate sample was filtered and
analyzed. Separate-phase oil contributes hydrocarbons to the
gas phase in concentrations higher than if the hydrocarbons
were only in solution. Method details are given in References
1 and 5.

RESULTS AND DISCUSSION

Visual and photographic observations

Application of dispersant after two hours of weathering appeared
to have little effect on Murban crude oil, based on visual and

photographic observations. Dispersal of weathered La Rosa crude oil did not appear effective. However, some oil reappeared within 10 to 15 min after dispersant application.

When dispersant was applied to fresh La Rosa, no sudden change was apparent. However, in time this oil became a thin sheen, as contrasted with the thick, black, asphaltic appearance of untreated La Rosa. Also, the track of the research vessel remained visible for a considerable period of time as contrasted with the quick closing behind the vessel with the untreated oil.

Murban crude oil changed dramatically when dispersant was immediately applied. A distinct whitish-brown subsurface plume appeared quickly. Over several hours, this plume dispersed in the water column, growing in area and diminishing in color and visibility. A thin-transparent surface oil sheen gradually appeared during this time period as some of the emulsion droplets resurfaced and broke. These visual observations give qualitative indication of the dispersion of crude oils by chemical treatment, but chemical analysis is needed for quantitative interpretation.

Oil dispersion as determined by infrared analysis

The large number of chemical analyses prevents a complete tabulation of the results for total extractable organic matter (OM). Some of the analyses will be presented in graphical form to document chemical dispersion of these crude oils. As expected, the highest concentrations in the water column were attained after immediate dispersion as compared with dispersion after two hours. Interpretation will concentrate on immediate dispersion results.

The crossed transects of a sampling run permit a three-dimensional analysis of plumes of dispersed oil (i.e., in crossed vertical planes). However, the limited sampling at 6 and 9 m (only at Stations 3 and 8) may give a distorted (narrow) view of dispersion at these depths. Based upon natural dispersion of oil into the water column (Ref. 2) and the lack of chemically dispersed oil at 6 m (Ref. 6), significant dispersion of oil down to 6 and 9 m was not expected. These depths were added only to verify the prior results, but we were surprised to find measurable oil at 6 and 9 m. In subsequent studies conducted in September and October 1979, all stations were sampled through 9 m.

Fig. 2 shows the extractable OM concentrations with depth along the two transects of the first sampling run following the immediate dispersion of La Rosa crude oil. The vertical scale exaggeration is about 45X. The contour for 0.25 ppm was at approximately 9 m at its deepest point; for the 1.0 ppm contour, 4 to 5 m.

The shape of the 2.0 ppm contour on the transect of Stations 6 through 10 is interesting in its asymmetry. Relying as it does on one data point for its asymmetric shape, this might be suspected as an experimental artifact. However, the second set of transects, approximately 1 hr later (Fig. 3) produced the same type of contour. Fig. 3 also shows the lower concentrations brought about by dilution of the plume in a larger volume of water.

Petroleum hydrocarbons ($C_{14}+$) were determined on three of the extracts. Extractable OM was 2.24, 1.25, and 2.54 mg/L; $C_{14}+$ hydrocarbons were respectively 1.43, 0.72 and 1.97 mg/L.

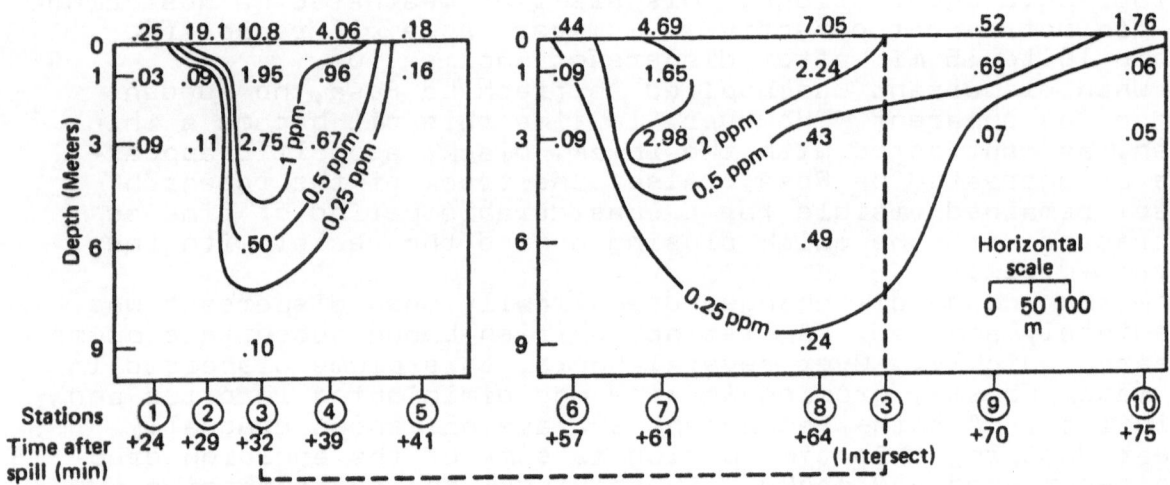

Fig. 2 Total extractable organic matter (ppm) in water samples collected during first sample run through La Rosa crude oil spill immediately dispersed (oil spilled 1019, dispersed 1028–1035)

Petroleum hydrocarbons averaged 76% of the total extractable OM. This is in the range previously observed for a much larger number of analyses (Ref. 8). However, the actual crude oil content of the original CCl_4 extracts is higher because hydrocarbons C_{14} are lost when the CCl_4 is evaporated to 1 mL prior to adding to the top of the silica gel column. Thus, the C_{12} and C_{13} with lesser amounts of C_9 to C_{11} hydrocarbons are present in the original CCl_4 extract.

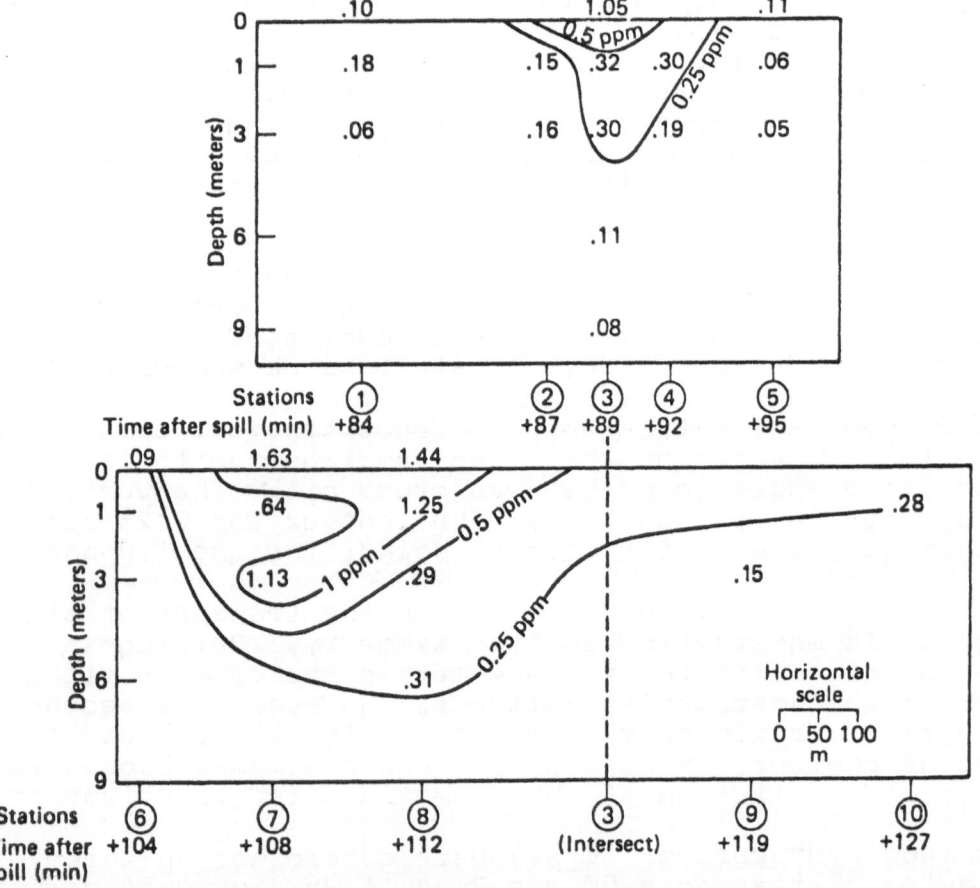

Fig. 3 Total extractable organic matter in water samples collected during second sample run through La Rosa crude oil spill immediately dispersed

This may amount to 10 to 15%, thereby raising the oil content to 85 to 90%. The percent oil may be even higher for those samples with the highest oil content (5 ppm).

The polar organic compounds removed by silica gel appear to exceed the extractable OM from background water samples outside the oil spill areas (particularly noticeable when the extractable OM ranges from 0.2 to 1 ppm). A possible explanation is that crude oil acts as an organic solvent, extracting and concentrating natural organic compounds in sea water.

Another way to view the dilution is shown in Fig. 4 for the immediately dispersed La Rosa spill. Concentrations at Station 3 in the center of the plume, are plotted with depth over time. Because each depth concentration is a single analysis, great reliance should not be placed on an individual point. As expected, a steady decrease in concentrations toward background values occurred over time.

Concentration lines for the immediately dispersed Murban spill (Fig. 5) show higher values than with La Rosa (Fig. 2). Dispersed oil was also found in higher concentrations at greater depths (almost 1 ppm at 9 m).

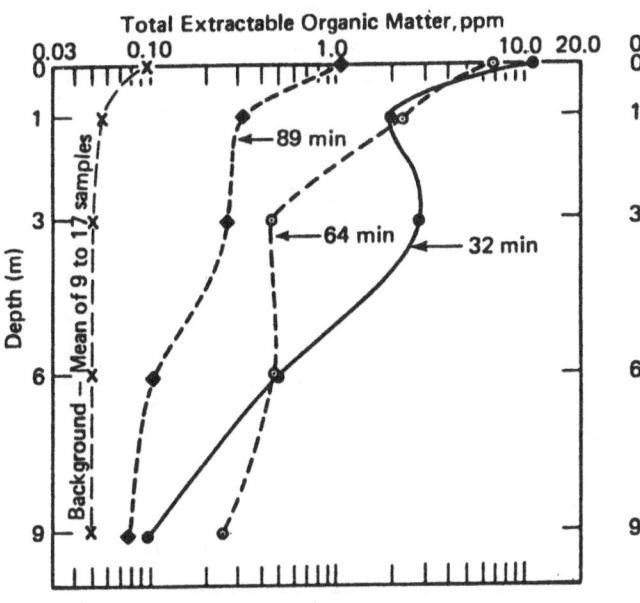

Fig. 4 Comparison of concentration — depth profiles at one station for various times under the immediately dispersed La Rosa crude oil spill

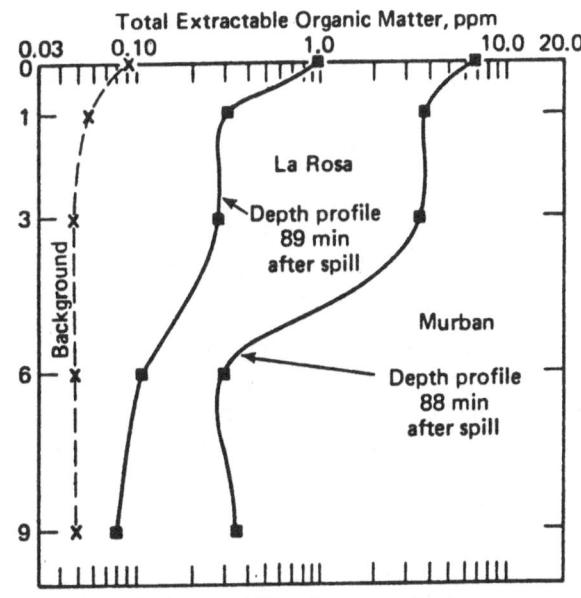

Fig. 6 Comparison of concentration — depth profiles for La Rosa and Murban crude oils at about the same time following discharge and dispersion

Fig. 6 compares concentration-depth profiles of the two crude oils, for samples from the center of the plume at similar times after oil discharge and dispersion. Again, each concentration is a single data point. A rough material balance calculation indicates that the Murban crude oil was almost completely dispersed, whereas the La Rosa was about half dispersed. These evaluations concur with visual impressions of effectiveness.

The data for the two spills that were allowed to weather for two hours before dispersion do not allow such clear graphical display. Most values for total extractable OM were much lower than those from the immediately dispersed spills. One explanation is the larger area to be treated after two hours, with a consequently larger water volume available to dilute an equivalent amount of oil.

580

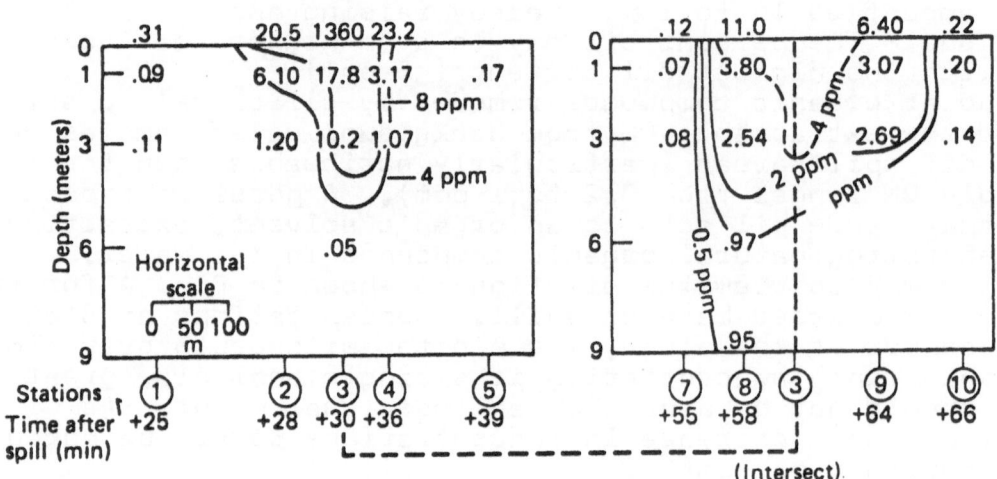

Fig. 5 Total extractable organic matter (ppm) in water samples collected during first sample run through immediately dispersed Murban crude oil spill (oil spilled 1404, dispersed 1411–1416). The dashed contour for 4 ppm is based on the station 1 to 5 transect

Most of the oil was concentrated in the leading (downwind) part of the slick, in perhaps only 10% of the total slick area, as observed by Hollinger and Mennella (Ref. 8). The dispersant was applied uniformily over the whole slick rather than concentrated on the area of heavy oil. Therefore the dispersant to oil application rate was not as high as for the immediately dispersed spills, which were treated before appreciable spreading had occurred. Weathering also would have increased oil viscosities and thereby would have decreased dispersant effectiveness.

A summary of the total extractable OM in water under the four research oil spills is shown in Table 2. It includes only values exceeding 0.10 ppm (approximately two times background). Untreated oil dispersed naturally in the water to a lesser extent than chemically treated oil. Immediate dispersion was more effective than after two hours, but most of the difference may be attributed to differences in application rate of dispersant to the oil.

Table 2 Summary of carbon tetrachloride extractable organic matter in water from under 4 research oil spills (concentrations in mg/L, ppm) *

| | | La Rosa | | | Murban | |
	n**	Maximum	Mean	n	Maximum	Mean
Not dispersed						
1 m	4	0.22	0.13	1	0.95	–
3 m	3	.51	.26	2	.16	0.14
Dispersed at 2 hr						
1 m	7	.23	.15	8	.18	.13
3 m	7	1.05	.27	4	.11	.10
6 m	2	.65	.38	1	.14	–
9 m	1	.29	–	1	.12	–
Dispersed within 10 minutes						
1 m	16	2.24	.69	13	17.80	3.10
3 m	14	2.96	.67	9	10.20	2.45
6 m	5	.50	.31	4	1.00	.45
9 m	1	.25	–	4	.95	.40

*Background concentrations(ppm); 1m, 0.061; 3m 0.050; 6m 0.048; 9m 0.051; ** number of samples

The greatest difference between oils was evident when they were dispersed immediately. Murban oil concentrations were higher at all water depths than for La Rosa. The slightly higher concentrations for La Rosa compared with Murban following delayed dispersion may reflect differences in chemical application and/ or sampling locations.

Oil Weathering as Measured by C_1 to C_{10} Analysis

Infrared analysis of CCl_4 extracts provides a measure of total oil in water samples, but is relatively insensitive. It is also complicated by the presence of background hydrocarbons and CCl_4 extractable organic compounds such as acids, alcohols, and esters in sea water. As used in this study, the method had a limit of detection of about 0.02 mg/L. The method also does not give information on individual hydrocarbons, classes of hydro- carbons or degree of weathering (loss of low-molecular-weight hydrocarbons).

A gas equilibrium method (Ref. 1, 5) using gas chromatography permits the measurement of most individual hydrocarbons in the C_1 to C_{10} fraction with a limit of detection of 2 ng/L (ppt) for alkanes and cycloalkanes and 10 ppt for aromatic hydrocarbons (weathering) to be followed with time (Ref. 1-3).

If adverse biological effects (immediate toxicity) result from oil spills, they are thought to be produced principally by the more soluble low-molecular-weight hydrocarbons (princi- pally aromatics such as benzene and toluene). Of importance, therefore, are the concentrations of the dissolved hydrocarbons and the duration of organism exposure to them. When water is equilibrated with crude oils, the C_1 to C_{10} soluble fraction comprises over 98% of the total soluble hydrocarbons (Ref. 9). For typical crude oils, benzene plus toluene constitute 70 to 80% of the aromatic hydrocarbons, and 62 to 78% of the total C_{6+} hydrocarbons (saturates plus aromatics).

Gas Chromatograms. Gas chromatograms of (1) dissolved hydrocarbons in sea water equilibrated with an excess of Murban crude oil from the spill tank; and (2) C_1 to C_{10} hydrocarbons residual in dispersed oil droplets in a water sample collected under the chemically treated Murban oil spill are shown in Fig. 7.

The GC column was 6 m of 3.2 mm stainless steel tubing packed with 10% UCW-98 silicone fluid on chromosorb W-HP. The column was temperature programmed from 60° to 145°C at 6°C/min. A 30 cm precut (backflush) column was in a sample valve oven at 100°C. The column was backflushed at 4 min which prevented C_{10} hydrocarbons from entering the 6 m column. A 2.0 mL (1.6 mm, diameter) sample loop in the sample valve oven intro- duced 1.5 mL (at 100°C) of the 20 to 23 mL of gas flowed from the 50 mL equilibration syringe through the sample loop.

The numbers over or near the individual hydrocarbon peaks are the relative retention times in hundredths of minutes. Each principal hydrocarbon peak has been named, and the GC amplifier attenuation is given. Fig. 7A is the gas chromato- gram (GC) of dissolved hydrocarbons in sea water equilibrated with Murban crude oil at attenuations of 1×10^{-9} (methane through pentanes) and 500×10^{-10} for the remaining hydro- carbons. The partial GC (Fig. 7A) is from another analysis with less attenuation, to better show the characteristic di- and trimethylbenzene peaks.

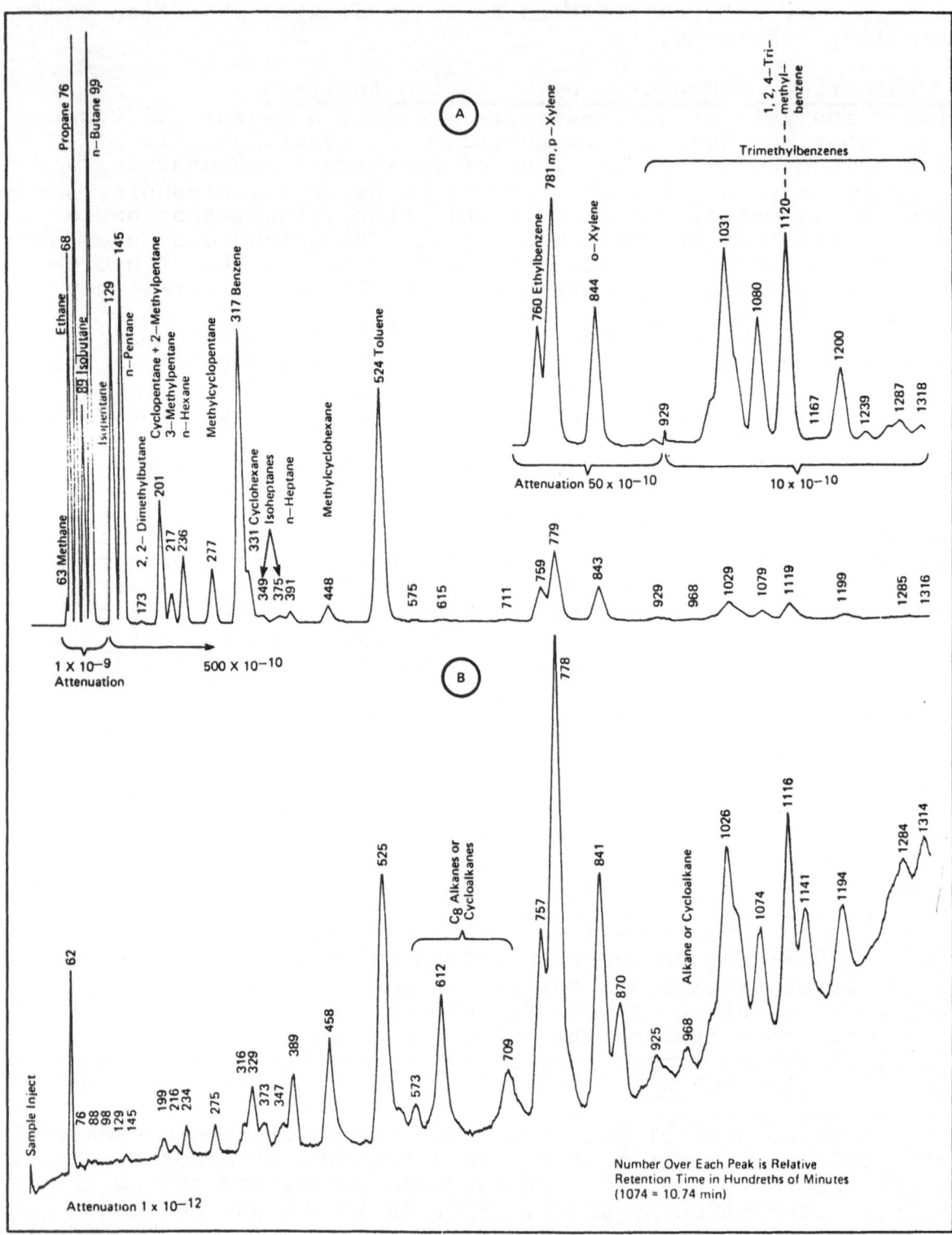

Fig. 7 Gas Chromatograms: (A) Equilibrium concentrations of dissolved hydrocarbons in sea water mixed with an excess of Murban crude oil from the spill tank. Inset is from second chromatogram with less attenuation to show more clearly the di— and trimethylbenzenes. (B) C_1 to C_{10} hydrocarbons found in 1 m water sample collected 49 min after immediate dispersion of Murban crude oil spill (total extractable organic matter was 3.8 ppm). See text for details of analytical procedures.

Fig. 7A shows the marked decrease in concentration of hydro-carbons with increase in molecular weight (carbon number), and the much greater solubility of aromatic hydrocarbons relative to the saturated hydrocarbons of the same carbon number (cyclo-alkanes are more soluble than alkane hydrocarbons). In particular note the large benzene and toluene peaks. The decrease is due to the lower concentrations of individual hydrocarbons in crude oils (higher carbon numbers than toluene for aromatics) as carbon number increases. An increase in number of isomers occurs with an increase in carbon number.

For pure hydrocarbons, normal alkane solubility decreases six to seven orders of magnitude between carbon numbers 1 and 12. For aromatics, the solubility decreases similarly between carbon numbers 6 and 24 (Ref. 9, 10). For example, hexane, cyclohexane, and benzene, each with six carbon atoms in the molecule, have respective solubilities of 9.5, 60, and 1,750 mg/L (Ref. 10, 11). Thus benzene and cyclohexane are respectively 185 and 6 times more soluble than hexane. The aromatic to n-alkane solubility ratio increases (Ref. 9), so that dimethylnaphthalenes are over 600 times more soluble than $n-C_{12}$.

Most of the gas was separated from the crude oil. Thus, the peaks in Fig. 7A for methane through pentanes (particularly methane, ethane, and propane) are lower than if the crude oil was "live" (gas not removed). Figure 7B represents the C_1 to C_{10} hydrocarbons in a water sample collected at 1 m near the center of the immediately treated Murban spill 46 min after dispersion. The attenuation is 500 times less (1000 times for C_1 to C_5) than in Fig. 7A, and the peak areas (concentrations) are entirely reversed (methane through the trimethylbenzenes). This qualitatively shows not only the very low concentrations of these low-molecular-weight hydrocarbons (Fig. 7B), but progressively greater loss with decrease in carbon number. Weathering of these low-molecular-weight hydrocarbons was very rapid. Quantitative data are presented in tables that follow.

Weathering of Murban crude oil. Table 3 shows C_1 to C_{10} hydrocarbons in water samples from 0 to 9 m depths at the center of the Murban slick 46 min after spraying with a self-mix dis-persant. The first numerical column "Oil max" gives the equili-brium concentrations of dissolved hydrocarbons in sea water that was thoroughly mixed with an excess of Murban crude oil from the spill tank. Note, as discussed above, the decrease in con-centration with increase in carbon number, and the high concen-tration of benzene and toluene. The alkane and cycloalkane hydrocarbons (7 carbon atoms) have become so low that they are difficult to identify and separate from aromatic hydrocarbons (Fig. 7). Thus n-heptane and methylcyclohexane are the highest carbon number saturate hydrocarbons shown in Table 3. Methane-through-pentane hydrocarbons (Oil max.) are lower than if the gas had not been separated. In essence, only aromatic hydro-carbons were measured in solution from toluene through trimethyl-benzenes. In addition to those peaks designated as alkane or cycloalkane (Fig. 7B), peaks 870 and 1141 are also non-aromatic (compare Fig. 7A and 7B). These peaks arise from presence of nondispersed oil and presumably droplets smaller than the filter used to remove most of the separate phase oil.

The concentrations of the individual hydrocarbons found in the dispersed (emulsion) plume of the Murban crude oil confirm the distribution and values indicated in Fig. 7B. They are very

Table 3 Low-molecular-weight hydrocarbons in water samples from various depths collected 46 min after immediate dispersion of Murban crude oil

Depth, m		0		1		3	6	9
Extractable OM, mg/L		11.0		3.8		2.54	0.97	0.95
Hydrocarbons, µg/L (Oil max.)*								
Methane	102	.077		.070		.072	.070	.073
Ethane	1560	.004	.0003**	.002	.0001	–	–	–
Propane	2360	.009	.0004	.004	.0002	–	–	–
Isobutane	940	.006	.0006	.003	.0003	–	–	–
n-Butane	2720	.004	.0002	.002	.0001	–	–	–
Isopentane	870	.002	.0002	.002	.0002	–	–	–
n-Pentane	1080	.007	.0006	.005	.0005	–	–	–
Cyclopentane	510	.026	.005	.012	.002	–	–	–
3-Methylpentane	125	.023	.018	.006	.005	–	–	–
n-Hexane	290	.072	.025	.022	.008	–	–	–
Methylcyclopentane	280	.092	.033	.027	.010	–	–	–
Benzene	6080	.260	.004	.095	.002	.041	.041	–
Cyclohexane	270	.205	.076	.066	.024	.011	.018	–
n-Heptane	65	.34	.52	.067	.103	.008	.020	–
Methylcyclohexane	140	.59	.42	.135	.096	.040	.022	–
Toluene	5630	3.75	.067	.140	.025	.395	.48	.26
Ethylbenzene	610	2.25	.37	.80	.13	.285	.23	.075
m, p-Xylene	1550	7.70	.50	2.55	.16	1.05	.95	.50
o-Xylene	900	5.45	.61	1.85	.21	.75	.57	.24
926*** Trimethylbenzene	68	.42	.62	.12	.18	.050	.035	.023
1027 Trimethylbenzene	800	6.50	.81	1.55	.19	.74	.50	.27
1077 Trimethylbenzene	370	2.75	.74	.65	.18	.30	.16	.035
1,2,4-Trimethylbenzene	920	6.20	.67	1.40	.15	.75	.31	.065
1197 Trimethylbenzene	300	3.20	1.07	.66	.22	.38	.13	.040
Total saturates	11,300	1.46		.42		.13	.13	.07
Total aromatics	17,200	38.5		11.1		4.75	3.41	1.51
Total hydrocarbons	28,500	40		11.5		4.90	3.54	1.58

*Equilibrium concentrations of dissolved hydrocarbons when an excess of Murban crude oil from the spill tank was mixed with sea water

**Underscored value is percent hydrocarbon found in water sample compared with equilibrium concentration of dissolved hydrocarbon (Oil max.)

***Number is relative retention time (see Fig. 7)

low, and the lowest carbon numbers are present in the lowest concentrations. This is the opposite of that expected if solution were an important process.

Consider the hypothetical situation of oil on a water surface with (1) evaporation prevented, and (2) a limited volume of water maintained in contact with the oil (i.e., the laboratory conditions for mixing a sample of crude oil from the spill tank with sea water in a sealed glass bottle). Under equilibrium conditions, one would expect to find the concentrations and relative concentrations as shown in Oil max., Table 3. Removing the restriction on water movement, but preventing evaporative loss would result in non-equilibrium solution of hydrocarbons into the water, and the rate of solution of individual hydrocarbons would become important (just as for evaporation).

The rate of solution increases with decrease in carbon number, and with class of hydrocarbon (i.e., aromatic vs alkane

for the same carbon number). Under nonequilibrium conditions,
methane would go into solution faster than ethane, ethane
faster than propane, etc. Similarly, benzene would go into
solution faster than toluene, toluene faster than xylenes, etc.
    The concentration of each hydrocarbon becomes progressively
lower as the degree of departure from equilibrium increases.
Thus, the shorter the contact time of oil and water, the lower
the concentration of each hydrocarbon in water, and the higher
the relative concentrations for those hydrocarbons having the
lowest carbon numbers for each class of hydrocarbons (alkane,
cycloalkane and aromatic). Because this was not observed in
the water samples under the slick, leads to the conclusion
that solution is apparently not a very important process, even
when crude oil is chemically dispersed and emulsion droplets
penetrate the water column.
    It appears that evaporation is the dominant process. The
low-molecular-weight hydrocarbons that do dissolve apparently
quickly evaporate to the atmosphere or dilute to very low
concentrations. The hydrocarbons in solution measured in the
water samples (Table 3) apparently were not in true solution
at the time of collection, but residual in separate-phase oil
droplets. After collection, they equilibrated between the oil
droplets and water. The equilibrium solubility of $C_{10}+$ hydrocarbons in crude oils is very low, probably less than 10 ppb.
Thus, the separate oil phase in the samples ranged from about
11,000 ppb in the surface sample to 940 ppb in the 9 m sample,
300 to 600 times the total dissolved hydrocarbon concentrations.
    The data in Table 3 show that the residual hydrocarbons are
low in concentration, even for the surface-collected sample.
Thus, the biologically toxic low-molecular-weight hydrocarbons
have been quickly lost. The concentration of the least volatile,
trimethylbenzene (1197), is only 1.07% of the equilibrium
solubility for unweathered oil (Oil max.). The remaining percentages (Column 3) show a generally progressive decrease with
carbon number (0.0002 to 0.0006% for ethane through pentane
hydrocarbons).
    The percentages found at 1 m (Column 5) are even lower,
showing that oil emulsion at this depth is more weathered than
oil droplets at the surface. Accelerated weathering is noted
with increasing depths as shown by decreasing concentrations
and by percentages if calculated for 3, 6 and 9 m. For example,
1,2,4-trimethylbenzene concentrations in the dispersed oil droplets for 0, 1, 3, 6 and 9 m are respectively 0.67, 0.15, 0.08,
0.03 and 0.007%.
    Because the samples were collected simultaneously, the
accelerated weathering with increasing depth apparently relates
to smaller droplet sizes. Evaporation and solution are diffusion
processes; and the shorter the diffusion pathway, the higher the
rate. As droplet size decreases, the surface-to-volume ratio
increases, with resulting faster loss of volatile and soluble
hydrocarbons.
    It seems reasonable to expect smaller droplets at greater
depth. Oil-in-water emulsions have a size distribution that
ranges from 0.1 to 100 m (Ref. 12). The larger droplets
(some may be even larger than 100 m, Ref. 13) will be buoyant
(Murban crude oil has a specific gravity of 0.83), and will rise
toward the water surface after mixing downward by wave action.
However, below diameters of about 2-3 m, gravitational effects
are balanced by Brownian forces. These small droplets move by

Brownian motion and will disperse in all directions, just as
clay-sized (2 m) mineral particles stay indefinitely suspended in water.

The percents for the aromatics benzene and toluene in
Column 3, Table 3 are 0.004 and 0.067 whereas methylcyclopentane,
cyclohexane, n-heptane, and methylcyclohexane are 0.033, 0.076,
0.52 and 0.42 respectively. Thus benzene and toluene are very
much lower. This is also shown to a lesser extent for the
percents of these hydrocarbons in Column 5. This also suggests
that the hydrocarbons found in waters associated with the dispersed slick were residual in the droplets and not in true
solution at the time of collection. These data indicate that
benzene and toluene were lost more rapidly by solution (although
evaporation greatly predominates) than the corresponding carbon-
number-saturates; and that once lost, were subsequently evaporated or quickly diluted. Had these aromatics been in true
solution at time of collection, their concentrations should
have been higher than the corresponding-carbon-number alkane
and cycloalkane hydrocarbons.

The concentration of methane (Table 3) is constant at
about 70 ppt. This reflects the expected equilibrium concentration of methane in sea water with that in the atmosphere for
this region of the Atlantic Ocean (Ref. 14).

Table 4 presents the concentrations of C_1 to C_{10} hydrocarbons in water samples collected at 1 m over the time (18 to
110 min) that measurable oil could be detected. These data
show the rapid loss of volatile hydrocarbons with time. Even
at 18 min, the trimethylbenzenes average a little over 1%
remaining in the dispersed oil droplets. This 18 min sample
had the highest observed oil content (17.8 ppm) of all the sub-
surface samples collected.

The decreasing percentage with decrease in carbon number
(Column 3) confirms data in Table 3 showing that these measured
hydrocarbons are residual in the emulsion droplets. Note again
that solution preferentially removed benzene and toluene (and
probably the higher aromatic hydrocarbons, but to a lesser
extent) from these droplets compared with corresponding carbon
number saturates.

Weathering of La Rosa crude oil. Table 5, for immediately
dispersed La Rosa crude oil, shows the concentrations of low-
molecular-weight hydrocarbons in water samples collected at
0 and 3 m, and 47 and 94 min after dispersion. Also shown are
the equilibrium concentrations of dissolved hydrocarbons
attainable when an excess of La Rosa crude oil was thoroughly
mixed with sea water (Column 1, Oil max.), and the percent of
hydrocarbons remaining in the dispersed droplets (Columns 4
and 7).

The equilibrium concentrations (Oil max.) for La Rosa are
somewhat different from Murban, reflecting the differences in
specific gravities and viscosities. La Rosa has less C_5+
hydrocarbons than Murban, but comparable C_1 to C_4. The lower
C_5+ concentrations reflect the lower naphtha fraction (La
Rosa, 11 volume %; Murban, 19%, Ref. 2). The comparable C_1
to C_4 concentrations are probably related to less complete
separation of gas from the more viscous La Rosa.

The hydrocarbons in water samples also reflect the apparently slower diffusion (evaporation and solution) from the more
viscous La Rosa crude oil, particularly for the C_1 to C_4 fraction.

Table 4 <u>Low-molecular-weight hydrocarbons in water samples collected over increasing time at 1 m under the immediately dispersed Murban crude oil spill</u>

Time after dispersion, min	18	46	72	110
Extractable OM, mg/L	17.8	3.8	1.55	0.31

Hydrocarbons, µg/L

	18	46	72	110
Ethane	.004	.0003*	.002	–	–
Propane	.016	.0007	.004	–	–
Isobutane	.008	.0008	.003	–	–
n-Butane	.006	.0002	.002	–	–
Isopentane	.003	.0003	.002	–	.008
n-Pentane	.007	.0006	.005	–	.006
Cyclopentane	.004	.0008	.012	–	.004
3-Methylpentane	.010	.008	.006	.008	.008
n-Hexane	.024	.008	.022	.006	.008
Methylcyclopentane	.036	.013	.027	.018	.006
Benzene	.117	.002	.095	.020	.024
Cyclohexane	.139	.051	.066	.041	.040
n-Heptane	.128	.196	.067	–	–
Methylcyclohexane	.315	.225	.135	.014	–
Toluene	3.50	.062	.140	.55	.150
Ethylbenzene	2.20	.360	.80	.42	.090
m, p-Xylene	8.85	.57	2.55	1.80	.429
o-Xylene	5.65	.63	1.85	1.13	.300
926 Trimethylbenzene	.50	.73	.12	.07	.029
1027 Trimethylbenzene	8.95	1.12	1.55	1.06	.051
1077 Trimethylbenzene	3.90	1.05	.65	.51	.078
1,2,4-Trimethylbenzene	7.52	.82	1.40	1.13	.175
1197 Trimethylbenzene	4.60	1.53	.66	.68	.120
Total saturates	.70		.35	.09	.08
Total aromatics	45.8		11.1	7.4	1.45
Total hydrocarbons	46.5		11.1	7.5	1.53

*Underscored value is percent hydrocarbon found in water sample compared with equilibrium concentration of dissolved hydrocarbon (Table 3, Oil max.)

There was also a slower change in concentration with time and depth, compared with Murban.

This slower weathering may be due not only to higher viscosity, but also to larger droplet sizes. Observations and extractable oil reported previously show La Rosa to have been less effectively dispersed, compared with the almost complete dispersion of Murban. However, the generally lower concentrations were more uniformly dispersed to 6 m (Table 2). The larger La Rosa droplets being less buoyant may well have mixed downward by wave action more easily than Murban.

Although the weathering of La Rosa was slower, it should also be noted that the concentrations of low-molecular-weight hydrocarbons were very low in the water samples. The highest concentrations of an individual hydrocarbon was 1.5 ppb toluene at 3 m, 47 min after dispersion; and 0.6 ppb at 3 m, 94 min after dispersion. Total low-molecular-weight hydrocarbons were

Table 5 <u>Low-molecular-weight hydrocarbons in water samples collected at two depths and at two times following immediate dispersion of La Rosa crude oil</u>

Depth, m		0	3		0	3	
Time after dispersion, min		47	47		94	94	
Extractable OM, mg/L		4.70	2.96		1.63	1.13	
Hydrocarbons, µg/L	(Oil Max.)*						
Methane	170	.073	.053		.077	.076	
Ethane	1740	.070	.083	.005**	.064	.045	.003
Propane	2360	.38	.36	.015	.30	.19	.008
Isobutane	620	.25	.25	.040	.20	.12	.019
n-Butane	1510	.57	.59	.039	.48	.26	.017
Isopentane	470	.47	.53	.113	.38	.21	.045
n-Pentane	480	.35	.45	.094	.33	.17	.035
Cyclopentane	330	.43	.52	.16	.34	.17	.051
3-Methylpentane	72	.13	.18	.25	.12	.05	.069
n-Hexane	125	.13	.23	.18	.17	.07	.056
Methylcyclopentane	230	.46	.57	.25	.34	.18	.078
Benzene	2870	.60	.50	.017	.37	.20	.007
Cyclohexane	270	.63	.76	.28	.37	.19	.070
n-Heptane	23	.09	.09	.39	.07	.03	.13
Methylcyclohexane	120	.49	.51	.43	.36	.17	.14
Toluene	2370	1.80	1.50	.063	1.10	.57	.024
Ethylbenzene	300	.66	.59	.20	.41	.20	.067
m, p-Xylene	680	1.75	1.40	.21	1.05	.51	.075
o-Xylene	360	1.30	1.05	.29	.77	.40	.110
926 Trimethylbenzene	24	.15	.14	.58	.08	.04	.17
1027 Trimethylbenzene	170	.94	.66	.39	.61	.43	.25
1077 Trimethylbenzene	55	.35	.20	.36	.18	.10	.18
1,2,4-Trimethylbenzene	125	1.15	.65	.52	.67	.34	.27
1197 Trimethylbenzene	75	.90	.47	.63	.51	.27	.36
Total saturates	8520	4.52	5.18		3.60	1.93	
Total aromatics	7030	9.60	7.16		5.75	3.06	
Total Hydrocarbons	15,500	14.1	12.3		9.3	5.0	

*Equilibrium concentrations of dissolved hydrocarbons when an excess of La Rosa crude oil from the spill tank was mixed with sea water
**Underscored value is percent hydrocarbon found in water sample compared with equilibrium concentration of dissolved hydrocarbon (Oil max.)

less than 15 ppb for all samples.

The percent benzene and toluene (Columns 4 and 7) show as for Murban crude oil, that these aromatics were preferentially removed by solution from the oil droplets. However, once removed, they apparently very quickly diluted or evaporated to the atmosphere, as previously discussed.

SUMMARY AND CONCLUSIONS

Four research oil spills (1.7 m³ each) of two crude oils (Murban, 0.83 specific gravity; and La Rosa, 0.91) were made 40 km off New Jersey. Two spills were immediately sprayed by helicopter with a self-mix dispersant; two, after 2 hr.

Water samples were collected over time at 1, 3, 6 and 9 m under the nontreated slicks and following dispersion. The dispersant application and the sampling were directed from another aircraft. This plane also provided a platform for observation of dispersant effectiveness and taking of color photographs.

Water samples were analyzed by IR for total oil content of a carbon tetrachloride extract; and for weathering of the C_1 to C_{10} hydrocarbon fraction, by gas chromatography.

Total oil under the immediately dispersed slicks at 1, 3, 6, and 9 m were respectively: La Rosa - 0.7, 0.7, 0.3 and 0.2 mg/L; Murban - 3.1, 2.4, 0.5, and 0.4 mg/L. The highest concentrations (30 to 90 min after dispersion) were La Rosa, 3 mg/L; Murban, 18 mg/L.

Oil concentrations for dispersion delayed 2 hr were lower (1.1 mg/L), and only slightly higher than found under non-dispersed oil (highest concentration for La Rosa was 0.5 mg/L; for Murban, 0.9). The less effective dispersion after delayed treatment reflects lower and less efficient dispersant application for these small spills, as well as increased oil viscosities due to weathering.

Samples collected 2 to 4 hr after dispersion contained no more than 2 to 3 times background concentrations of about 0.06 mg/L.

Rough material balance calculations, supported by visual and photographic evidence, indicate that Murban crude oil treated immediately was almost completely dispersed; for La Rosa, about half was dispersed. It follows that oil removed from the influence of wind will not travel as far, and thereby reduce the likelihood of oil stranding or entering biologically sensitive areas.

The dispersed oil in the water column weathered very rapidly. Evaporation of C_1 to C_{10} hydrocarbons greatly exceeded solution. Relative concentrations of the individual C_1 to C_{10} hydrocarbons show that dissolved hydrocarbons (including benzene and toluene) were not present at 0.01 g/L detection limit. Apparently the more soluble hydrocarbons quickly evaporate or dilute to even lower concentrations.

The measured C_1 to C_{10} hydrocarbons were residual in dispersed oil droplets, and did not exceed 50 g/L, even for samples collected at 1 m and 18 min after dispersion. After 2 hr this had decreased to 2 g/L.

Weathering increased from the surface to 9 m depth for samples collected at the same time, indicating decreasing droplet sizes with increasing depth. Weathering also increased with time for samples collected at the same depth.

Murban crude oil dispersions weathered more rapidly than La Rosa, reflecting Murban's lower viscosity (and possibly smaller droplet sizes).

The rapid weathering of low-molecular-weight hydrocarbons from dispersed crude oil droplets should quickly reduce the biological toxicity from hydrocarbons such as benzene and toluene.

The observed changes in concentrations and weathering of chemically-dispersed crude oils provide real-world data that can assist in the design of initial concentrations and dilutions for realistic laboratory bioassays.

590

ACKNOWLEDGEMENT

The work reported here was conducted under contracts from the
American Petroleum Institute. Financial assistance from the U.S.
Environmental Protection Agency under grant number R806056 is
gratefully acknowledged.

REFERENCES

1. C D McAuliffe, 'Evaporation and solution of C_2 to C_{10}
    hydrocarbons from crude oils on the sea surface', in Fate
    and Effects of Petroleum Hydrocarbons in Marine Ecosystems
    and Organisms, D A Wolfe, Ed., (Pergamon Press, New York,
    1977), 363-372.
2. J C Johnson, C D McAuliffe and R A Brown, 'Physical and
    Chemical behavior of small crude oil slicks on the ocean',
    in Chemical Dispersants for the Control of Oil Spills,
    ASTM STP 659, L T McCarthy, Jr., et al, Eds., (American
    Society for Testing and Materials, 1978), 141-158.
3. C D McAuliffe, 'Dispersal and alteration of oil discharged
    on a water surface', in Fate and Effects of Petroleum Hydro-
    Carbons in Marine Ecosystems and Organisms, D A Wolfe, Ed.,
    (Pergamon Press, New York, 1977), 19-35.
4. R A Brown, J J Elliott, J M Kelliher, and T D Searl, 'Sampling
    and analysis of nonvolatile hydrocarbons in ocean water',
    in Advances in Chemistry Series, 147, T R P Gibb, Jr., Ed.,
    (American Chemical Society, Washington, 1975), 172-187.
5. C D McAuliffe, 'GC determination of solutes by multiple
    phase equilibration', Chem. Technol., 1 (1971), 46-51.
6. C D McAuliffe, A E Smalley, R D Groover, W M Welsh, W S
    Pickle and G E Jones, 'The Chevron Main Pass Block 41 oil
    spill: Chemical and biological investigations', Proceedings
    1975 Conference on Prevention and Control of Oil Pollution,
    (American Petroleum Institute, Washington, 1975), 555-566.
7. API Publication 4290, Physical and Chemical Behavior of Crude
    Oil Slicks on the Ocean, (American Petroleum Institute,
    Washington, 1976), 1-98.
8. J P Hollinger and R A Menella, 'Oil spills: Measurements of
    the distributions and volumes by multifrequency microwave
    radiometry', Science, 181 (1973), 54-56.
9. C D McAuliffe, 'Oil and gas migration - Chemical and physical
    constraints', Amer. Assoc. Petrol. Geol. Bull., 63 (1979),
    761-781.
10. C D McAuliffe, 'Solubility in water of paraffin, cyclo-
    paraffin, olefin, acetylene, cyclo-olefin, and aromatic
    hydrocarbons', Jour. Phys. Chem., 70 (1966), 1267-1275.
11. L C Price, 'Aqueous solubility of petroleum as applied to
    its origin and primary migration', Amer. Assoc. Petrol.
    Geol. Bull., 60 (1976), 213-244.
12. P Becker, Emulsions: Theory and Practice, (Reinhold Publishing
    Corp., New York, 1976), 2
13. C D McAuliffe, 'Oil-in-water emulsions and their flow prop-
    erties in porous media', Jour. Petrol. Technol. 23 (1973),
    727-733.
14. J W Swinnerton and R A Lamontagne, 'Oceanic distribution of
    low-molecular-weight hydrocarbons - Baseline measurements',
    Environ. Sci. Technol., 8 (1974), 657-663.

TREATMENT OF PRODUCED WATER DISCHARGED FROM U.S. COASTAL AND OFFSHORE PLATFORMS: TECHNOLOGY/REGULATORY REQUIREMENTS

W L Berry
Senior Staff Environmental Specialist, Shell Oil Company, New Orleans, Louisiana, U.S.A.

INTRODUCTION

The overboard discharge of saline water produced in association with crude oil and natural gas (hereafter called "produced water") is generally allowed in both saline coastal and offshore areas of the United States. However, the water must first be treated to minimize the entrained oil and grease content. Such discharges are regulated by the individual states or appropriate agencies.

The purpose of this paper is twofold :

1. Review the development of U.S. federal regulatory requirements controlling produced water discharges.

2. Describe the "state-of-the-art" of treatment technology to removed oil and grease from produced water.

Deck drainage discharges are also briefly addressed. Ballast water treatment is not covered. In the United States ballast water is not handled with produced water. Furthermore, the characteristics of the two types of water are completely different. Therefore, while the same types of treating schemes generally are adaptable to both produced and ballast water streams, the treatment experiences are not necessarily correlatable.

The paper includes no new data or conclusions. It is a synopsis of available information. The references given are major works pertaining to the specific subjects addressed, and should be of value to those interested in these areas. However, they comprise only a partial listing of available material on processing produced water to remove entrained oil and grease.

DISCUSSION

Background
Prior to passage of the United States Federal Water Pollution Control Act Amendments of 1972*, the individual states exercised control of discharges to state surface waters, including produced water, within their jurisdiction - inland waters and seaward out to three miles from the coastline (three leagues in Texas). The United States Geological Survey (USGS) regulated discharges from oil and gas activities seaward of state waters - federal waters of the Outer Continental Shelf (OCS). For example, the USGS required that produced water discharged into the Gulf of Mexico

* Additional amendments were enacted in both 1977 and 1978. The act as it now stands is Public Law 95-217. It is generally referred to as the Clean Water Act and will be so called throughout the remainder of the paper.

EPA developed effluent limitations for the following discharges
associated with drilling and production activities in coastal
and offshore areas: produced water, deck drainage, drilling
fluids, drill cuttings, well treatment fluids, produced sand,
sanitary wastes and domestic wastes. These effluent limitations
have been promulgated as enforceable federal regulations. The
Agency also made a similar evaluation of discharges from true
onshore operations, and developed effluent limitations, such
as no discharge of produced water except in those few situations
where the water is fresh and is being put to a beneficial use
(livestock watering, etc.). The scope of this paper, however,
is limited to a discussion of produced water and deck drainage
in coastal and offshore regions.

Industry took exception to certain provisions included in
the regulations implementing the various effluent limitations
for drilling and production operations. As a result, the
American Petroleum Institute (API) with a number of oil companies
joining as co-petitioners filed suit in federal court to obtain
relief from the overly restrictive provisions. Except, where
pertinent to the subjects covered here, the law suits are not
discussed further. One other matter not covered is the National
Pollutant Discharge Elimination System (NPDES) permitting system
EPA has developed to enforce the previously mentioned effluent
limitations. It too is beyond the scope of this paper.

Development of Produced Water BPCTCA

EPA selected oil and grease as the only produced water pollutant
parameter for which BPCTCA effluent limitations are necessary.
The rationale for this determination is given in Section VI,
pp. 55-63, of the Development Document (Ref. 3). The Development Document also contains a detailed discussion of the Agency's
derivation of the specific produced water BPCTCA oil and grease
limitations (Section VII, pp. 65-94 and Section IX, pp. 125-133).

The basic data bank of oil and grease effluent concentrations
considered by EPA in developing BPCTCA was that contained in the
previously mentioned Offshore Operators Committee report of
December 1972 (Ref. 2). However, EPA did obtain from the U.S.
Geological Survey supplemental data on facilities in OCS waters.
Also, the OOC provided additional information on other systems
(64 in Texas and 16 in California), as did several states and
EPA regional offices.

Since most information was obtained from industry, EPA
conducted field studies to verify the data for statistical analysis
purposes prior to the determination of BPCTCA. These verification studies were conducted for those discharges off California,
Louisiana and Texas. A major area of concern was whether or not
the various effluent oil and grease concentrations included in
the data bank could be statistically correlated since they were
measured by different analytical methods including gravimetric,
colormetric, infra-red, ultra-violet fluorescence, and variations
of these techniques. The analytical technique employed in the
verification studies was the method specified by EPA in Chapter
40 Code of Federal Regulations (CFR) Part 136 "Test Procedures
for the Analysis of Pollutants": liquid-liquid extraction with
trichloro-trifluoroethane (Freon 113)/gravimetric.

The results of the EPA study on coastal and offshore Louisiana
facilities supported use of the data collected over the years
by industry (Refs. 4,5). Such correlation was not obtained in
the California and Texas verification studies (Refs. 3,6). Thus,

OCS have an average oil and grease content of 50 ppm or less with a maximum of 100 ppm.

Surveys by the Offshore Operators Committee (OOC)* showed that as of December 1972 there were 93 produced water secondary treatment facilities (flotation systems, plate coalescers, fibrous media coalescers and loose media coalescers) in coastal and offshore Louisiana fields operated by the seven companies** then represented on the OOC Sheen Technical Subcommittee (Refs. 1,2). Those same companies also had tested or were testing 32 various research and development devices. Obviously, at that point in time these installations did not represent all such facilities in operation in the Gulf of Mexico or in the U.S. Further, they did not include the many locations where secondary equipment is not needed; i.e., produced water can be treated to a low oil and grease content in a skim tank, retention pit, etc. However, these data do indicate that a substantial number of secondary systems were operational when the federal legislation was enacted late in 1972. Therefore, since the OOC surveys included performance data, adequate information was available which could be used to develop produced water effluent limitations required by the CWA.

Effluent Limitations for Existing Sources

Under the terms of the Clean Water Act, all point source discharges of pollutants, other than publically owned treatment works, were required by July 1, 1977 to achieve effluent limitations based on the application of "Best Practicable Control Technology Currently Available" (BPCTCA). The Environmental Protection Agency was charged with the responsibility to develop effluent limitations representative of BPCTCA considering these factors :

1. The total cost of application of technology in relation to the effluent reduction benefits to be achieved from such application.

2. The age of equipment and facilities involved.

3. The process employed.

4. The engineering aspects of the application of various types of control techniques.

5. Process changes.

6. Non-water quality environmental impact (including energy requirements).

* The OOC is an organization of companies engaged in crude oil and natural gas exploration and production activities in the Gulf of Mexico and the Atlantic OCS of the US. The Committee was organized over 30 years ago to work with the problems associated with oil and gas operations in the marine environment. Virtually, all of the companies who operate in these two areas are OOC members.

** These companies were: Chevron-USA; Conoco, Inc.; Exxon Company, U.S.A.; Gulf Oil Exploration and Production Co.; Mobil Oil Exploration and Producing Southeast, Inc.; Shell Oil Co.; and Texaco, Inc.

only Louisiana data were considered in the produced water
BPCTCA determination.

EPA utilized 2262 effluent oil and grease concentrations
from 27 flotation cells reported by the OOC (Ref. 2) plus
supplemental data provided by the USGS in the BPCTCA statistical
analysis. Some of the reported data were averages of up to four
grab samples taken in a 24-hour period, analyzed separately and
the results averaged. The rest were the results of individual
grab samples. EPA determined that the grab samples had a higher
variance than the composites and that the compositing technique
gave more representative results. Thus, statistical techniques
were used on the grab sample analyses to simulate composite
sampling.

Based on the EPA statistical analysis of the flotation
system data, BPCTCA for the oil and grease concentration of
existing produced water discharges into the coastal and offshore waters of the United States was established as :

> 72 mg/l daily maximum (The daily maximum is determined by
> the obtaining four grab samples in any 24-hour period,
> analyzing them separately, and averaging the results).

> 48 mg/l maximum monthly average (The monthly average is
> determined by obtaining a set of four grab samples as des-
> cribed under daily maximum on a weekly basis, four sets in
> a month, and averaging the results).

Two important points concerning the EPA effluent limitations
should be noted.

Any Treatment Process Can Be Used - Even though: (1) the performance of exemplary flotation systems is representative of BPCTCA
and (2) they provide a reasonable basis for establishing numerical
effluent limitations, EPA also specified (Ref. 3) that operators
can utilize any available process they choose to meet the
numerical limitations. Normally, this would be either gravity
separation (tank or pit), other physical separation (flotation
system, plate coalescer, fibrous media coalescer, or loose media
coalescer) or a combination of gravity separation and other
physical separation.

Proper Sampling Schedule Required - Since the 48/72 mg/l oil and
grease limitations were established by statistical analysis of
data directly related to sampling frequency, it is essential that
monitoring requirements for compliance with NPDES discharge
permits be consistent with the same sampling scheme. For example,
if sampling is required by a regulatory agency on a monthly basis,
the monthly average limitations which cannot be exceeded would be
the same as the daily maximum of 72 mg/l. Further, the operator
must obtain four grab samples in any 24-hour period and have them
analyzed separately. The average of these four readings is the
value which cannot exceed 72 mg/l. One sample a day, two samples
a day, three samples daily or any number other than four, cannot
properly be equilibrated to the 72 mg/l requirement. Another
illustration is the situation where weekly sampling is required
by a regulatory agency. In this case the monthly average limitation which cannot be exceeded would be 48 mg/l. Also, any scheme
other than four sets of four samples taken weekly in a month's
time cannot be taken as representative of a 48 mg/l monthly
average requirement.

Brown and Root Study

Concurrent with EPA's BPCTCA efforts, the Offshore Operators Committee commissioned Brown & Root, Inc. to conduct a similar analysis (Ref. 7). To establish BPCTCA, Brown and Root made a comprehensive analysis of the performance of the various treatment processes employed by the member companies of the OOC Sheen Technical Subcommittee in coastal and offshore Louisiana operations.

As previously mentioned, these processes are: pits, tanks, flotation systems, plate coalescers, fibrous media coalescers, and loose media coalescers. Effluent oil and grease data from 166 installations (over 6000 analytical measurements) were evaluated. All data used were obtained from Reference 2 for all processes expect pits and tanks. Information on the latter two types of systems was obtained from a subsequent survey by the OOC (unpublished). Results of the analysis are summarized on Tables 1 and 2.

Based on an in-depth evaluation of the data, Brown & Root recommended that effluent limitations for oil and grease in produced water commensurate with BPCTCA are a long-term average of 44 ppm with a daily maximum of 100 ppm. These values are reasonably comparable to EPA's determinations. The basis for this definition of BPCTCA is as follows.

Of the processes studied, loose media coalescers and flotation systems produced the lowest mean effluent oil concentrations. The flotation process was selected to represent exemplary produced water treatment performance. Flotation is a widely used mechanism and is almost universally applicable. Loose media coalescers were not selected to represent exemplary performance because of operating problems at many locations associated mainly with suspended solids in the produced water. These factors make loose media coalescers impracticable for discharges containing high suspended solids concentrations. Subsequent to publication of Brown & Root's report, the Offshore Operators Committee conducted an in-depth study of the field performance of all coalescers operating in the coastal and offshore waters of Louisiana. The results of this study confirmed Brown & Root's conclusions concerning loose media coalescers(Ref. 8).

After the exemplary treatment system was established, effluent limitations corresponding to exemplary operation of that system were defined. The highest average effluent concentration observed for an individual flotation unit cannot be used since some of the existing units are not exemplary. However, effluent limitations based on the unit with the lowest average effluent oil concentration are also unacceptable. Variations in the characteristics of the crude oil, limitations in removing dissolved and emulsified oil, and other factors greatly influence effluent oil concentrations. For these reasons, even exemplary flotation units will exhibit a wide range of mean effluent oil concentrations. Therefore, the average effluent oil content maintained by 75 percent of the flotation systems (total of 19 systems and 1218 oil and grease effluent concentrations) was selected as being representative of a properly designed, well operated BPCTCA system. This value, 44 ppm, also approximates the mean value plus one standard deviation. It should be noted that the statistical methods and the criteria for editing the data employed by Brown and Root were not the same as those used by EPA. These differences contribute to the variation in the number of units and data points between the Brown & Root and EPA studies.

In deriving effluent limitations for the petroleum refining

596

category of point sources, EPA used the single sample concentration which is exceeded only two percent of the time as the
basis for the daily maximum. The value for the concentration
which was achieved 98 percent of the time by the flotation
systems evaluated in the Brown & Root study is 100 ppm. Thus,
this value was recommended as a daily maximum.

It is interesting to note that a 1974 literature survey
conducted by the University of Tulsa for EPA confirmed the
Brown & Root study numbers (Ref. 9): "The overall conclusion
based on the average data is that the oil content of produced
brine cannot be reduced to levels much below 50 ppm with existing
equipment." Except for minor improvements, there has been no
major change in equipment performance or concepts since these
conclusions were made.

Similar conclusions were reached in a recently published
world-wide study by the Oil Industry International Exploration
and Production Forum (E & P Forum) (Ref. 10): "Under normal
operating conditions treatment units, based on best practicable
technology, are able to process water to a monthly average oil
content of 50 ppm with extreme values of up to 100 ppm."

Upset/Bypass Considerations
As the effluent limitations promulgated by EPA basically are
comparable to those determined by Brown & Root, industry has
not objected to the 48/72 mg/l oil and grease limitations.
However, industry believes the method EPA used to establish the
limitations does not adequately take upsets into consideration.
Industry's position is that since no mechanical equipment can be
relied upon to function properly 100 percent of the time, operators
should not be penalyzed for those occasions when effluent limitations are exceeded for causes beyond their reasonable control.
Examples of the types of episodes which might result in excusable
upsets include: equipment start-up, changes in the number of
producing wells or in reservoir or strata produced, breakthrough
of water as a result of enhanced recovery injection or natural
water drive, preventive or corrective maintenance, downhole
equipment or process failures, etc.

To insure that effluent limitations can be consistently
achieved, water treating equipment must be bypassed occasionally
for brief periods to perform corrective or preventive maintenance.
Otherwise costly redundant standby treating equipment must be
installed or production shut-in during periods of maintenance.
Bypassing for maintenance is presently not allowed by EPA. The
only conditions under which bypassing is acceptable is to prevent
loss of life or severe property damage (including permanent loss
of crude oil and natural gas reserves).

The upset and bypass issues are two of the points being
contested in the previously mentioned American Petroleum Institute/
Industry litigations against EPA. Extensive negotiations have
failed to resolve either issue satisfactorily and the cases are
now before the court. A court decision is anticipated sometime in
1980.

Monitoring-Compliance/Optimizing Equipment Performance
The only oil and grease analytical technique approved by EPA for
compliance monitoring purposes (to insure that effluent limitations
are being met under the terms and conditions of an NPDES discharge
permit) is the gravimetric method (extraction with Freon 113)
previously discussed. The major drawback of this method is that

the analysis must be conducted in a laboratory. Therefore, by
the time a sample is obtained, transported to a lab, analyzed,
and the results transmitted back to the field, at least several
days have elapsed. Usually the time lag is more like two weeks.
Thus, while the results of the Freon 113/gravimetric analysis are
adequate for compliance monitoring purposes, they are of little
value for day-to-day control purposes. To insure optimum equipment performance, most operators routinely use other techniques
which provide oil and grease content data directly in the field.
A commonly used tool is an infra-red analyzer.

Continuous monitoring is not done by operators since it is
not required to properly access equipment performance and
optimize waste treatment facility operation. This can readily
be achieved with systems such as an infra-red device mentioned
above. Further, continuous oil-in-water monitoring equipment is
impractical in oil and gas production operations for the following
reasons:

1. Sophisticated electronics cannot tolerate exposure to the
   marine environment.

2. Platform vibrations are detrimental to the fragile complicated devices.

3. Entrained solids, scales and corrosion products can foul
   the optics.

4. Oil droplet size, association of the oil droplets with
   solids, solids contents, entrained gas bubbles, color,
   etc., all interfere with obtaining proper readings.

5. Specialized technicians, not normally employed on a
   routine basis in production operations, are necessary to
   properly operate and maintain calibration in the field.

6. Solvent can easily become contaminated which leads to
   false readings.

EPA is cognizant of the problems associated with continuous monitoring. The Agency also recognizes that such monitoring is not
necessary to insure compliance with applicable effluent limitations.
For a number of reasons, the effluent oil and grease content from
any produced water treatment facility will fluctuate, even from
exemplary units operated in an optimum manner. However, these
fluctuations are not of sufficient magnitude or duration to
warrant EPA's requiring continuous monitoring of oil-water
separators handling produced water. Performance data contained
in Reference 2 bears this out.

Environmental Aspects of Produced Water Dischages
A panel of industry experts reviewed all available scientific
information on produced waters for the Offshore Operators
Committee. The consensus conclusions reached by the panel
were (Ref. 11):

   "This report reviews the constitutents of produced waters and
   their effects on the marine environment both in offshore and
   coastal areas. The considerable data which are at hand and
   which are discussed in this report show clearly that the
   toxicity of produced waters is low. Consequently, produced
   waters do not have a detrimental effect on the marine environment or on marine biota. References are listed which provide
   an extensive background and contain much further, detailed
   information.

"A review of the large amount of data discussed in the report leads to the following conclusions:

1. "Natural forces, including dilution, evaporation, and chemical and biological reactions, rapidly act to reduce the concentration of hydrocarbons and inorganic components in produced waters.

2. "Because of these natural forces, produced waters have not been found to cause measurable effects on the composition, appearance, or the biota of the marine environment.

3. "Field studies in petroleum-producing areas show that the low-level discharge of hydrocarbons and inorganic components which arise from petroleum operations do not have a detrimental effect on the marine environment or the marine biota."

Based on these conclusions, industry believes there is no environmental justification for produced water effluent limitations more stringent than those published by EPA as BPCTCA; i.e. 48/72 mg/l oil and grease concentration in the discharged effluent. However, even these limitations should include adequate provisions for upset and bypass conditions.

Produced Water Treatment - "State-of-the-Art"

As indicated by the above discussions, it is generally recognized the "state-of-the-art" for removing entrained oil and grease from produced water is such that, under normal operating conditions, the oil and grease content can be reduced to a monthly average of about 50 mg/l with a maximum on the order of 100 mg/l. Other pertinent conclusions were also reached by Brown & Root, Inc., in the BPCTCA study conducted for the Offshore Operators Committee concerning produced water treatment. These are enumerated below.

Brown & Root's analysis of over 6000 effluent oil and grease concentrations for 166 facilities representing six basic process schemes (Tables 1 and 2) led to the following conclusions (Ref. 7):

1. While the performance of exemplary flotation systems is representative of BPCTCA, in actual practice the EPA specified effluent limitations can be achieved by either gravity separation (tank or pit), other physical separation (flotation system, plate coalescer, loose or fibrous media coalescer), or a combination of gravity separation and other physical separation.

2. Not all treatment processes are applicable at all locations. For example, because of operating and maintenance difficulties, the use of fibrous and loose media coalescers is rather limited, particularly at remote locations, such as coastal and offshore platforms. These and other specific process limitations are summarized on Table 2.

3. Performance of water treatment systems is not only a function of retention time, design parameters, use of chemical coagulants and influent oil concentration but is also influenced by a number of factors most of which the operator has little or no control over, including: water characteristics, crude type and solubility, degree of entrained oil emulsification, sizes of the entrained oil droplets, and suspended solids concentrations and characteristics. All of these should be considered in designing a unit for a specific application.

4. Because of the factors given in Item 3, the average
    effluent oil concentration achieved by exemplary control
    (with any process system) may vary widely from discharge
    to discharge. In other words, if identical treatment
    systems are installed at two different locations, the
    average produced water effluent oil content may be quite
    different.

5. Also, the performance (efficiency) of a particular unit
    may vary from day to day as a result of fluctuations in
    flow rates and influent oil concentrations, short term
    water characteristic variations and equipment upsets.
    For example, at one installation the effluent oil concen-
    tration was observed to be 23 ppm one day while on the next
    day it was 230 ppm.

6. An overall average performance for a given type of treat-
    ment system can be misleading when evaluating its applica-
    tion to a specific produced water discharge because of
    variations in treatability covered in Items 3 and 5.

It is interesting to note that both EPA (Ref. 3) and the E & P
Forum (Ref. 10) have confirmed most of these conclusions.
    One other important point emphasized in the Offshore Operators
Committee initial survey which is still a valid consideration in
selecting produced water treating equipment for a specific
application is (Ref. 1):

   "One of the main points brought out by our experience with
   water clarifiers is that most of the equipment manufacturers
   tend to overstate what their equipment is capable of doing
   and that none can provide an acceptable performance guarantee.
   We also know from experience that a flat across-the-board
   claim by any manufacturer that his equipment will produce an
   effluent with a certain oil discharge content at any and all
   locations cannot be accepted as being correct. Experience
   indicates that a unit which produces results within design
   limits at one location probably will not perform in the same
   manner elsewhere. It is well recognized within the oil producing industry that a reputable manufacturer of this type of
   equipment will not make broad claims for his product but
   recognizes that the conditions vary and likewise the results."

Deck Drainage

Deck drainage is either rainwater or washdown water used to clean
equipment that is captured in areas where overboard oil spillage
from process equipment, tankage, etc., is prevented by the use of
curbs and gutters and drip pans.
    Initially, EPA established deck drainage effluent limitations
the same as for produced water; i.e. 48/72 mg/1 oil and grease.
This arbitrary decision was based on an erroneous assumption by
EPA that deck drainage and produced water are commingled at all
facilities and treated together. Therefore, EPA concluded they
should be subject to the same discharge limitations. In reality,
this is not possible at many coastal and offshore facilities,
since, while practically all of them have deck drainage, only some
also process produced water. Further, commingling is not usually
practiced, even at facilities where such mixing could be done,
for the following technical reasons:

1. Addition of deck drainage would result in fluctuating

flow through produced water treating equipment which could significantly increase the effluent oil content; i.e. by reducing the efficiency of the equipment.

2. Oxygen-type corrosion of produced water treating equipment could result by adding deck drainage to the system since deck drainage is generally saturated with oxygen (produced water is maintained in an oxygen-free environment).

3. Mixing of produced water and deck drainage is also avoided because of incompatabilities of the waters. Mineral deposits, such as calcium carbonate and corrosion products, such as ferrous sulfide, ferrous carbonate and ferric oxide, can foul water treating equipment resulting in reduced operating efficiencies.

4. It is essential that decks be kept clean. Considering fire hazards and personnel safety, the use of detergents in washdown water offers the only practical and safe procedure for cleaning decks. Consequently, deck drainage may contain small amounts of detergents that would present severe problems with produced water treating system efficiencies.

Not only are the two types of waters not normally treated together for the reasons enumerated, but there are three basic reasons why produced water effluent limitations are not applicable to deck drainage.

Different Treating Technology - Deck drainage treatment technology is significantly different from that used for produced water. Deck drainage is in almost all cases treated by simple gravity separation. EPA's effluent limitations for produced water are based on an analysis of 27 flotation systems. The technology differs because the nature of the wastes differs. Deck drainage is intermittent, occurring only during rainfall or washdown. In an American Petroleum Institute survey, it was noted that many deck drainage incidents lasted less than one hour and some only 15 minutes. In addition, deck drainage is discharged on numerous unmanned platforms. As a result, sophisticated treatment technology is simply not applicable to deck drainage. For example, if flotation cells were used to treat deck drainage, operators would have to be dispatched to the many unmanned facilities during each rainfall. By the time the operator arrived at the facility and adjusted the flotation system, chances are the rain would have ended. For this reason, simple and trouble-free gravity systems are usually employed.

De minimis Source - Deck drainage is a de minimis pollution source. There is no discharge except during infrequent periods of rainfall or washdown. During these events, only water collected in drip pans or curbed areas enters the system. An API survey of 23 facilities showed that the amount of oil and grease in deck drainage discharges generally averages less then 0.1 gallons per calendar day and in most cases is several orders of magnitude less than this value.

Sampling and Monitoring Impractical - Routine sampling of deck drainage discharges is impractical because of their intermittent nature. Sampling would have to be timed to co-incide with the infrequent rainfall or washdown events. This would present a

severe logistical problem due to the remote location of many of
the unmanned platforms. Further, it would be virtually impossible
to obtain a proper set of samples for NPDES permit compliance
monitoring; i.e., four samples in a 24-hour period. In addition,
most deck drainage discharges are not equipped for sampling.
Many are even underwater and substantial modification would be
necessary to allow sampling.

For these reasons, industry has long maintained that effluent
limitations are not necessary for deck drainage discharges.
However, as a negotiated settlement in the before-mentioned
litigation, a "no discharge of free oil to surface waters" criteria
has been accepted as a reasonable compromise. This standard has
been promulgated by EPA as the applicable effluent limitation
for existing discharges.

Standards of Performance for New Sources

EPA is also responsible under the Clean Water Act to develop
effluent limitations for new pollutant sources achievable through
the application of the "Best Available Demonstrated Control
Technology." In response to this charge, EPA initially proposed
the following limitations on the oil and grease content of new
produced water discharges:

1. In 1975 for federal waters (waters seaward of state waters)
    52 mg/l daily maximum, 30 mg/l maximum monthly average.

2. In 1976 for state waters (inland waters and seaward out
    to three miles from the coastline - three leagues in
    Texas) - no discharge; i.e., reinjection required.

As neither proposal was finalized within 120 days after initial
publication as required by CWA, they are not now in effect.
However, industry believes neither is necessary to protect the
marine environment (see "Environmental Aspects of Produced Water
Discharge") and is greatly concerned that similar excessively
stringent standards could ultimately be reproposed. Industry
supports new source standards for produced water discharges
comparable to BPCTCA for all waters; i.e., 48/72 mg/l oil and
grease.

Concerning the "no discharge" criteria for state waters,
industry does not argue that the technology for reinjection of
produced water does not exist. Industry long ago developed
methods to inject produced water underground. The technology
is widely used in areas where surface discharges are not environmentally suitable; i.e., freshwater regimes which encompass most
of the United States except saline coastal and offshore areas.
However, the application of reinjection technology requires a
significant cost and also consumes substantial amounts of energy
and other resources such as steel (Ref. 12). These impacts are
particularly severe in water operations where facilities must be
installed on elevated platforms and where costly drilling and
construction equipment is required. Industry believes such
measures are completely unwarranted for state waters in light
of the fact that no environmental gains would be achieved.

EPA derived the 30/52 mg/l oil and grease new source standards
for OCS activities by conducting a statistical analysis of the
effluent oil content data for the ten "best" of the 27 flotation
units included in the BPCTCA analysis. Statistical techniques
used were the same as in the BPCTCA study. The term "best" is
defined here as those flotation systems exhibiting the lowest
effluent oil concentrations regardless of any other factors

contributing to the treatability of the waste streams analyzed; i.e., hydraulic loading, influent oil concentration, and other waste stream characteristics. EPA's basic rationale was that new sources can employ the most recent technology. Therefore, they can perform as the "best". Industry disagreed for three reasons: (1) a reduction from 48/72 mg/l is not necessary to protect the marine environment, (2) EPA's technical arguments are basically unsound, and (3) optimum waste treatment technology is already being applied.

Mainly because of industry's objections to the 30/52 mg/l oil and grease standards, EPA and the Offshore Operators Committee are presently conducting a joint offshore field study to obtain data for establishing rational produced water new source performance standards. In exchange for OOC's co-operation, EPA will not promulgate new source standards until the program is completed, and will use the study results as a basis for developing such standards. Costs estimated to be about $750,000 are being borne by EPA. OOC has no direct financial commitment. However, the various operators whose platform facilities are included in the field surveys are indirectly providing financial support by supplying meals, quarters and transportation for EPA contract personnel. The project was initiated in June 1979. Estimated completion date is the end of 1980. Rockwell International, Los Angeles, California, is the primary contractor. Crest Engineering Incorporated, Tulsa, Oklahoma, is the subcontractor for the field work.

REFERENCES

1. Sheen Technical Subcommittee, Offshore Operators Committee, "Status Report Water Treating Facilities Louisiana Coastal Producing Operations", (February 1972)
2. _____,"Addendum Status Report Water Treating Facilities Louisiana Coastal Producing Operations", (December 1972)
3. Environmental Protection Agency, "Development Document for Interim Final Effluent Limitations Guidelines and Proposed New Source Performance Standards for the Oil and Gas Point Source Category", (September 1976)
4. L H Myers, B L DePrater, T E Short, Jr., and B B Shunatona, Jr. "Offshore Crude Oil Waste Water Characterization Study", Robert S Kerr Environmental Research Laboratory, Ada, Oklahoma, Final Draft Under Review, (August 1975)
5. Sheen Technical Subcommittee, Offshore Operators Committee, "OOC Analytical Results EPA Field Sampling Program Water Treating Facilities Louisiana Coastal Producing Operations", (April-May 1974)
6. _____,"Analytical Results EPA - Texas Produced Water Oil in Water Verification Program", (August 28,29,30, 1974)
7. Brown & Root, Inc., "Determination of Best Practicable Control Technology Currently Available to Remove Oil from Water Produced With Oil and Gas", Prepared for Sheen Technical Subcommittee, Offshore Operators Committee, (March 1974)
8. Sheen Technical Subcommittee, Offshore Operators Committee, "Supplemental Performance Data on Fibrous and Loose Media Coalescers, and Magnetic Oil-Water Separation Apparatus", (May 1974)
9. F S Manning, N D Sylvester and G W Reid, "Effluent Limitations for Onshore and Offshore Oil and Gas Facilities", University of Tulsa, Tulsa, Oklahoma, (May 1974)

10. The Oil Industry International Exploration and Production Forum, "Treatment of Production Water - The State of the Art", (London, May 1979)

11. Sheen Technical Subcommittee, Offshore Operators Committee, "Environmental Aspects of Produced Waters from Oil and Gas Extraction Operations in Offshore and Coastal Waters", (September 30, 1975)

12. Brown & Root, Inc., "Potential Impact of EPA Guidelines for Produced Water Discharges from the Offshore and Coastal Oil and Gas Extraction Industry", Prepared for the Sheen Technical Subcommittee, Offshore Operators Committee, (October 1975)

Table 1

Summary - produced water treatment system performance

Process Type	Number of (1) Units Analyzed to establish Means	Number of Effluent samples in Analysis	Average of Mean Effluent Oil Concentrations (PPM)	Mean Effluent Oil Concentration achieved by 75% of units Analyzed	Effluent Oil Concentration achieved 98% of the time (PPM)
Pits	55	1800	35.4	43	140
Tanks	52	1399	41.9	52	155
Flotation systems	19	1218	33.5	44	100
Plate coalescers	21	995	56.8	78	200
Fibrous media coalescers	6	191	42.0	55	180
Loose media coalescers	13	467	21.8	24	60

(1) Units were selected based on the availability of at least ten effluent observations per unit.

Source : Reference 7

Table 2

Comments - produced water treatment system performance

Process Type	Remarks
Pits	Large land area required. Usually used for total treatment where high retention times can be provided. Average effluent oil concentrations do not always show a relationship to retention time.
Tanks	Versatile and widely used. Effluent oil concentrations often show little relation to retention time.
Flotation systems	Widely applicable. Three general types employed: full stream pressurization, recycle stream pressurization and induced gas units. Influent concentrations have only a minor effect on effluent oil concentrations unless influent oil contents are less than 150 ppm. Also, conservative design flow rates improve performance. However, as designs become more conservative, the gain in performance becomes smaller.
Plate coalescers	Subject to upsets due to rapid changes in flow rates. Buildup of oil/water emulsions, fine sand and other materials on plates necessitates periodic cleanout. Performance highly variable.
Fibrous media coalescers	Not widely applicable. Frequent filter replacement usually necessary. Filter disposal presents an operational problem. Units require frequent backwashing to maintain hydraulic conductivity.
Loose media coalescers	Not widely applicable. Frequent backwashing or flushing and subsequent disposal of suspended solids presents a significant operational problem. Installations require tanks for handling backwash or flush water and influent flow equalization. Sophisticated instrumentation requires frequent attention creating a problem at remote locations. Large variations in performance observed. Influent and effluent oil concentrations correlate poorly.

Source : Reference 7

COSMOS - THE NEW OIL COMBAT DREDGER

Kees d'Angremond, M.Sc.
Director
Cosmos Dredging v.o.f.,
Hook of Holland, Netherlands

Eric Zijlstra, B.Sc.
Member working Committee
Cosmos Dredging v.o.f.,
Hook of Holland, Netherlands

INTRODUCTION

After the "Torrey Canyon" disaster of 1967, the Netherlands Ministry of Transport and Public Works assigned Rijkswaterstaat (State Waterways Board) with the task of organizing oil combat procedures. Since 1971, the North Sea Directorate of Rijkswaterstaat is charged with detection and combat of oil spills. A detailed description of the operational organization is given in Ref. 1. At present, this organization can supply a number of vessels, applying various oil combat techniques, such as:
- dispersion (by spraying chemicals, such as finasol)
- recovery
- sedimentation (by spraying a chemically treated sand-water mixture)

The capacity and applicability of the available methods is restricted. It was therefore considered to increase the capacity of the fleet. An additional one or more oil recovery vessels were envisaged. The cost of construction and operation of such a vessel is, however, almost prohibitive in view of the very limited use that is made of it. These considerations form the background to the decision to build the slicktrail "COSMOS".

PRESENT METHODS AND EQUIPMENT

Dispersion method

Five vessels of the North Sea Directorate are equipped to spray detergents for dispersion of an oil spill. Although the oil remains in the marine environment after dispersion, the negative effects of its presence are greatly reduced and the possibilities for further biological degradation are enhanced.

The method is less attractive in shallow waters, because of the high remaining concentration of oil in the water column and because of the great mobility of oil and chemicals resulting from the tidal currents. An increase in the capacity of spraying detergents is anticipated from the present dispersion rate of 170 m3 of oil per hour to around 450 m3 oil per hour.

Sand sink method

The sand sink method was developed in 1970 by Volker Stevin Dredging in close co-operation with Royal Dutch Shell. The trailing-suction hopper dredger "GEOPOTES VII" has been modified to sink the oil spill to the seabed by means of a chemically treated sand-water mixture. The "GEOPOTES VII", fitted with two spraying booms, is able to dredge pure sand from the seabed, storing the wet sand in the hopper compart- ments and sailing to the oil slick site. One load of sand is sufficient to sink 2,500 tons of oil. This load can be spray- ed over the floating oil within one hour.

After the collision of the tanker "PACIFIC GLORY" off the Isle of Wight, the "GEOPOTES VII" was mobilized and re- mained standby during the salvage operation. (Fig. 1)

Fig. 1 "GEOPOTES VII" spraying chemically treated sand

The main drawback of this method is the fact that water surface pollution is merely converted into seabed pollution. The oil sand mixture covering the seabed after application of this method may destroy biological life and even prohibit a recovery unless the oil sand mixture is removed lateron.

Despite this drawback, the method was kept in reserve, because of its high capacity and because of the possibility to immobilize the spilled oil in order to prevent greater environmental damage elsewhere.

Recovery method
From an environmental point of view, the recovery of oil from

the marine environment is the optimum solution. In this way, no further harm is done to flora and fauna. A comprehensive review of existing and future recovery methods is given in Ref. 2.

   A problem associated with the oil recovery method is the restricted storage capacity of the available equipment and the limitations imposed by the weather conditions. At present, the vessel "SMAL AGT" of the North Sea Directorate is equipped with sweeper skimmers of the Hydrovac system. (Fig.2)

Fig. 2   m.v. "SMAL AGT"

   The performance of the skimmer is satisfactory up to sailing speeds of 2 to 3 knots, depending on the pump capacity. When the ship sails at higher speeds, the oil passes underneath the skimmers, and cannot be recovered. The "SMAL AGT" is owned and operated by Rijkswaterstaat and supplied and manned by Bos Kalis Westminster under contract with the North Sea Directorate. The ship is on standby on a round the clock basis. The sweeping capacity is 370 m3 of oil water mixture per hour. This capacity is considered inadequate by the North Sea Directorate in view of the intensity of the tanker traffic to and from the harbours of Rotterdam and Amsterdam. A considerable increase in recovery capacity was therefore required.

FUTURE METHODS AND EQUIPMENT

Capacity
As mentioned previously, a considerable increase of oil recovery capacity is required. The aim of the North Sea

Directorate of Rijkswaterstaat is to be able to cope with a
spill of 15.000 m3 of oil within a three day period.

This size of spill requires a recovery vessel, which is
considerably larger than the "SMAL AGT". The recovery capacity is determined partly by the sweeping capacity of the
skimmers, but to a far greater extent by the storage capacity
of the recovery vessel. The latter aspect becomes more significant as the sailing distance to and from the spill area
increases. To cope with the demands of Rijkswaterstaat, a
storage capacity of 5000 m3 is considered to be a minimum.

Cost analysis
The annual costs for the operation of an oil recovery vessel
of this size are extremely high.
These costs consist of:
- depreciation and interest
- maintenance and repairs
- personnel (crew)
- insurance
- fuel and lubricants
- harbour dues
- overheads and supervision
and are of the order of magnitude of Dfl. 6 to 8 million per
year for an estimated 500 effective working hours per year.

Based on the experience with the operation of the m.v.
"SMAL AGT", it is clear that a costly vessel and the complete
crew would be idle for most of the time. Since the greatest
part of the total cost are fixed costs, an enormous saving
could be achieved if the idle time could be made productive.

A study with this in mind was performed by Baggermaat-
schappij Holland, joined by Bos Kalis Westminster and Volker
Stevin Dredging, as COSMOS Dredging v.o.f.. This study proposed to combine maintenance dredging in the Rotterdam harbour
entrance with oil recovery operations. In this way, the
annual cost could be reduced to roughly 25% of the operational cost of a single purpose recovery vessel. Maintenance
dredging will not be adversely affected by occasional inter-
ruptions and hence is ideally suited to be combined with
emergency oil recovery operations.

MAINTENANCE DREDGING WITH TRAILERS

To fully appreciate COSMOS Dredging's proposal to combine
maintenance dredging and oil recovery, the maintenance dredging activities in the Rotterdam harbour entrance are briefly
explained.

Siltation in the access channel, the harbour entrance
and the harbour basins consists mainly of fine sand and silt.
The annual amount of siltation is in the order of 20 million
m3 per year. Due to the busy traffic and the meteorologic
conditions maintenance dredging is generally not carried out
by stationary equipment, but with the aid of trailing suction
hopper dredgers.

This type of dredger is a seaworthy vessel, which performs dredging by pumping sediment into its hopper through a

draghead and articulated suction pipe. (Fig. 3)

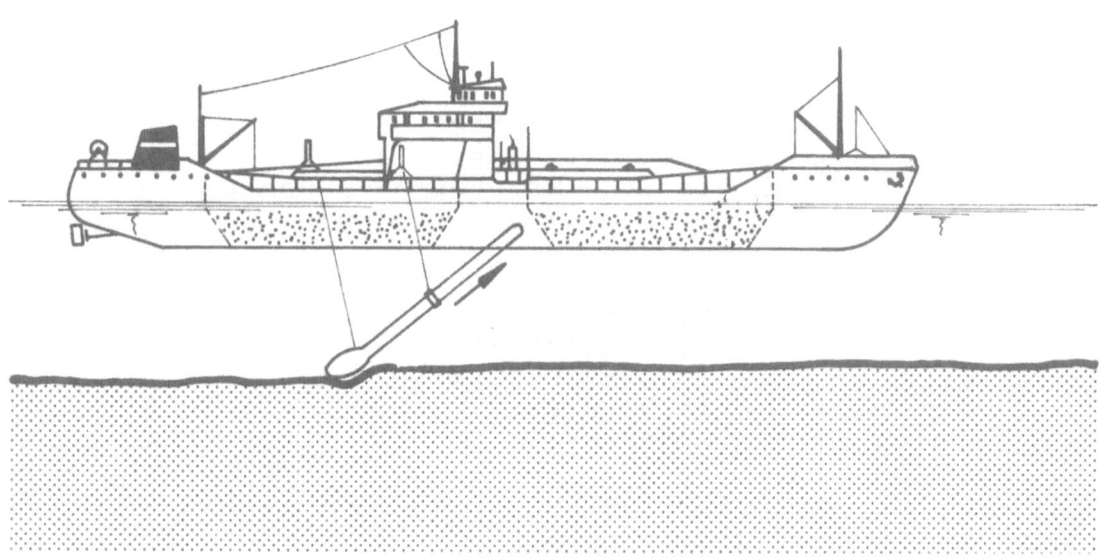

Fig. 3 Trailing Suction Hopper Dredger - loading

After loading, the vessel sails to a dumpsite and unloads by
opening bottom valves or bottom doors. (Fig. 4)

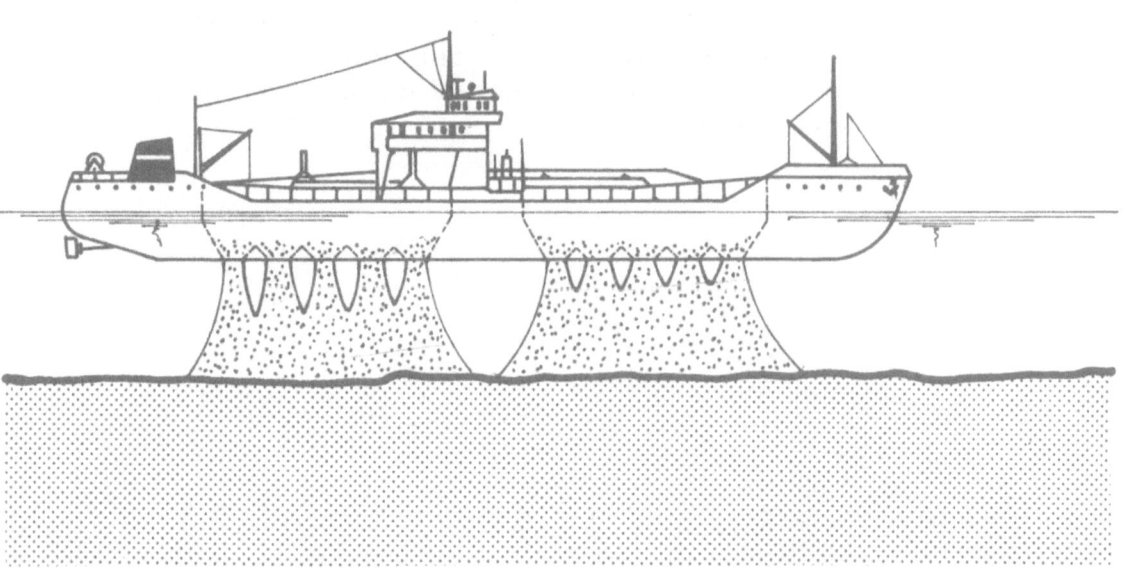

Fig. 4 Trailing Suction Hopper Dredger - unloading

In the Rotterdam harbour entrance, maintenance dredging is
carried out by private contractors under contract with
Rijkswaterstaat. The dredging operations are performed on a
round the clock basis during five days per week.

SLICKTRAIL "COSMOS"

Dredging Aspects

When the slicktrail "COSMOS" performs maintenance dredging, she operates as an ordinary trailing suction hopper dredger. For this purpose the vessel is equipped with two suction pipes, one on either side of the vessel. Contrary to the traditional trailing suction hopper dredgers, the dredge pumps are not situated in the hull but on the suction pipes. This position improves the suction characteristics of the pumps. On the other hand it requires a special design of pump and pump drive since both are in a very exposed position under water. The main characteristics of the vessel and its dredging capabilities are:

- length   : 113.60 m.
- beam   : 20.00 m.
- draught   :   7.28 m. (8.35 m.)
- hopper capacity   : 5375 m3.
- speed   : 13.7 knots
- accomodation   :   38 pers.
- dredging depth   : 32.00 m.
- propulsion   : 2 x 5200 hp.
- pump drive   : 2 x 1450 hp.
- bow propeller   :   750 hp.
- shore discharge pump : 2300 hp.

Oil Recovery Aspects

When the ship operates as an oil recovery vessel, two sweeping arms are used which are normally stowed on deck. The sweeping arms have been designed by Hydrovac in accordance with the experience obtained by Rijkwaterstaat on board of the "SMAL AGT". The design was again modified during discussions between Hydrovac, Rijkswaterstaat, IHC and Cosmos Dredging. The sweeping arms float independently from the vessel and are kept in position by a flexible ladder attached to the afterdeck and a wire to the foredeck. (Fig. 5)

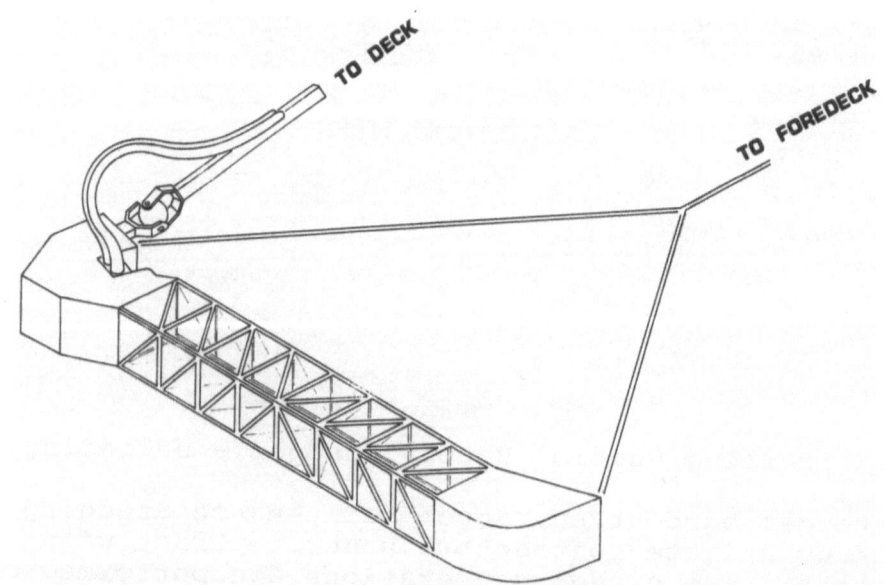

Fig. 5 Lay-out of sweeping arms

During oil skimming, a forward velocity up to 2 knots is
maintained. When operating, a single path width of 50 m. is
cleaned. Combined with the forward velocity of 2 knots
approximately 180,000 m2 of sea surface can be cleaned per
hour. The surface oil caught by the booms is guided by collec-
tor blades towards hydraulically operated pumps. These pumps
have a capacity of 1000 m3 of oil water mixture per hour. They
pump the mixture on board the ship through flexible hoses. For
storage purposes, the 5375 m3 hopper is available together
with the additional storage tanks with a total capacity of
960 m3. The hopper works as a giant oil separator. Water is
drawn from the lower part of the hopper and discharged
through an additional separator. For safety purposes, the oil
surface can be covered with a foam blanket to reduce the risks
of explosion. The complete design of the vessel, including
it's electrical installation, meets tanker requirements as
set forth by Bureau Veritas and the Netherland's Shipping
Inspectorate.

Pumping of the oil water mixture on board the vessel and
discharging the mixture afterwards is seriously hampered by
the formation of an emulsion, "chocolate mousse". The dynamic
viscosity of the emulsion may be as high as 100,000 cP, which
reduces the pump discharge considerably due to the high
resistance. Although this problem could be solved partly by
switching the oil pumps in line instead of parallel, a major
breakthrough was achieved after the development of a
demulsifying agent by Shell. Application of this agent reduces
the viscosity to approximately 4000 cP and further enhances
segregation of the water and oil phases. Following succesful
tests by Shell and Rijkswaterstaat, it was therefore decided
to modify the additional storage tanks of the "COSMOS". An
injection system will be installed to distribute the
demulsifier from these tanks either to the sweeping arms or
to the hopper.

The workability and manoeuverability of the ship have
been tested by the N.S.M.B at Wageningen. The design of the
booms allow continued operation up to sea state 5 (wave
heights up to 1.5 m.) The bow propeller provides excellent
manoeuverability even at the lowest speeds.

Operational Aspects

During working days, the time for mobilization of the oil
recovery vessel is negligible. Since the crew is already on
board and the oil recovery equipment is fixed to the ship,
the ship can sail immediately after an emergency has been
reported. En route, the load, if any, can be dumped and the
required safety measures can be taken. Specialists and/or
additional equipment can be flown to the ship as a drop
area is available. During weekends, a special alarm system
is in operation which guarantees availability of the ship
within a period of 8 hours. The slicktrail "COSMOS" will be
commissioned in November 1980. She is contracted for a 5-year
period by Rijkswaterstaat for combined use as a maintenance
dredger and oil recovery vessel. However, the contract allows
for other clients to hire the vessel for emergency oil combat

614 Cosmos - the new oil combat dredger

operations as well.

ACKNOWLEDGEMENT

The design of "COSMOS" has been realized by the close
co-operation of many authorities and organizations. The efforts
of Rijkswaterstaat, IHC, and Hydrovac are especially acknowledged. An artist impression of the "COSMOS" is given in
Fig. 6.

Fig. 6 Artist impression slicktrail "COSMOS"

REFERENCES

1.   H.M. Menagie, "Het bestrijden van olieverontreigingen
      op de Noordzee", OTAR, 6 (1979), 367 and OTAR, 7 (1979),
      439. (In Dutch).

2.   J. Cranfield, "What's new in oil spill cleanup equipment",
      Ocean Industry, 13 (1978), 115.

STATEMENT

Mr. L. Andrén
Assistand Secretary, IOC Marine Pollution
Research and Monitoring Unit, on behalf
of the Intergovernmental Oceanographic Commission

Mr. Chairman, with your permission I shall briefly explain the
main features of an on-going programme for monitoring background levels
of petroleum pollution in the oceans, and, as it has a certain relation
to the UNEP Regional Seas Programme, I shall pose a related question
to the UNEP representative;

The Intergovernmental Oceanographic Commission is an autonomous body
of Unesco charged with the co-ordination of international oceanographic
research programme, in a wide sense, and active in the development of ocean
services. Prompted by a recommendation by the UN Conference on the Human
Environment in 1972, the IOC, jointly with the World Meteorological
Organization, and supported by UNEP, instituted in 1975 a pilot project
for monitoring of petroleum pollutants at sea and on beaches. Approaching
the end of the project's pilot phase (by 1 July 1980 it will be converted
into a regular marine pollution monitoring programme MARPOLMON, covering
initially oil only), more than 100,000 reports related to visual
observations of oil have been gathered in special data centres. Further,
about 6000 samples of floating particulate petroleum residues, and more
than 4000 samples of near-surface water for analysis of dissolved/
dispersed petroleum hydrocarbons, were analyzed. The latter were
analyzed by UV fluorescence spectrophotometry, and a special study was
made by gas chromatography/mass spectrometry which confirmed that the
fluorescence method was not affected by recent biogenic hydrocarbons in
teh water. The method gives a measure of "environmental quality" so that
related data can be used when establishing an "oil pollution baseline",
and for future monitoring of background levels (useful when studying for
example relationship between oil pollution from tankers and that of other
sources).

Several countries also contributed data from systematic studies of tar on
beaches.

Although intended mainly to test all operational aspects of such a
programme, the project has yielded tangible results, and patterns of oil
pollution, as evidenced by the data gathered, could be related to maritime
traffic patterns (especially that of tankers), to certain dynamic processes
in the ocean surface water, and to wind regimes.

The oil industry may wish to demonstrate its interest not only in oil
pollution combatting but also in long-term monitoring of background levels
of oil pollution in the marine environment.

The programme has an important element of support to developing countries, funds for which have to come from extra-budgetary sources. Funds are therefore solicited from the oil industry and interested member countries.

From among the operational and planned scientific projects under the UNEP Regional Seas Programme, a project on petroleum pollution monitoring is already being operated in collaboration with IOC in the Mediterranean, and similar activities are planned in other Regional Seas. It is of great value that data generated under such projects are intercomparable with those of the "global" programme (MARPOLMON), and that data are submitted to the "Responsible Oceanographic Data Centres" for MARPOLMON in Washington and Tokyo.

In this connection it would be interesting, also perhaps for other participants, to hear an estimate by the UNEP Representative as to the future time scale and phasing of the active implementation of the research components in the various regional programmes of UNEP.

SESSION 5

Tropical Zone

GULF AREA OIL COMPANIES MUTUAL AID ORGANISATION

"INCEPTION, FORMATION AND OPERATION"

Captain T P Hebden
Port Captain, Bahrain Petroleum Company Limited, Bahrain,
Arabian Gulf - Former Chairman of GAOCMAO

INTRODUCTION

The Gulf Area Oil Companies Mutual Aid Organisation (GOACMAO)
was formed to pool the resources of all member Companies for
joint capability to clean up oil spills, which are beyond the
capability of a single party.

Any member affected by an oil spill may call on other members
of the organisation for assistance. The member requesting
assistance shall reimburse other members for any costs they have
borne providing this assistance.

At the present time, the members of the Organisation are
those Companies operating on the Arabian side of the Gulf.
Map (2) outlines the areas and the location of the Companies.
In many cases, through Government participation, Companies
have changed or amended their names since the formation of
GAOCMAO in 1972.

The following list of Companies, moving from North to South,
are the members of the Organisation :

Present Name	Former Name	Joined	
Iraq National Oil Co.	(Basrah Petroleum Co)	Founder	Member
Kuwait Oil Co.		"	"
Kuwait National Petroleum Co. (Mina Abdulla Refinery)	(American Independent Oil Co.)	"	"
Getty Oil Co.		"	"
Arabian Oil Co (Japan)		"	"
Aramco	(Arabian American Oil Co.)	"	"
Bahrain Petroleum Co.		"	"
Qatar Petroleum Producing Authority on-shore	(Qatar Petroleum Co)	"	"
Qatar Petroleum Producing Authority off-shore	(Shell Co. of Qatar)	"	"
Abu Dhabi Marine Operating Co.	(Abu Dhabi Marine Areas)	"	"
Abu Dhabi Oil Co (Japan)		"	"
Abu Dhabi Company for Onshore operations	(Abu Dhabi Petroleum Co)	"	"
Amerad Hess Oil Corp. of Abu Dhabi		1979	
Total ABK		1974	
Zakum Development Co		1979	
Dubai Petroleum Co		Founder	Member
Petroleum Development (Oman) Ltd		1977	

THE NEED FOR GAOCMAO

Each of these Companies has at least one tanker loading terminal
with associated under sea pipelines. A number of the Companies
operate offshore, producing and drilling platforms with many
miles of submerged pipelines. The Dubai Petroleum Company has
a complete offshore operation which includes three, one half
million barrel submerged crude oil storage tanks.
    In addition to the producing and terminal operations hand-
ling some 19 million barrels a day there is, of course, the
threat of a major oil release from tanker traffic in the area,
resulting from collision or grounding damage.
    As far as tanker traffic is concerned, there are some 20
tankers per day entering and leaving the Gulf carrying some 20
million barrels per day produced by the Companies operating in
the whole Gulf.
    Additionally, there are about 12 freighters entering and
leaving the Gulf each day. All this traffic is funnelled through
the Straits of Hormuz at an average of a vessel every 22 minutes.
While this is not as dense a traffic pattern as the Straits of
Dover, the size of the tankers tends to be larger. The overall
picture presented, therefore, is of an area where a huge oil
spill could occur which would require an immense co-ordinated
effort by all Companies to effectively minimize the resulting
damage.

THE INCEPTION OF GAOCMAO

In December 1970 at the request of the Iranian Oil Exploration
and Producing Consortium, a meeting was held at their Kharg
terminal with the ARAMCO Oil Spill Task Force Co-ordinator to
discuss offshore oil spill clean up procedures. As a result of
this meeting, it was proposed that urgent steps should be taken
to develop a co-operative agreement among the producing
Companies of the Gulf Area to prepare to deal with the type of
catastrophic oil spill which could occur at any time, and the
probability of which steadily increased with production growth.
Map (1) shows the current circulation, prevailing wind and the
most vulnerable areas.
    Initially ARAMCO and IOE & PC planned to work together on
this development, drawing in other Companies when expedient to
do so. Shortly after this, however, in January 1971, a meeting
was held in London of representatives of the major Parent
Companies of the Gulf Area producing Companies to discuss the
same problem. At this meeting, it was proposed that O.C.I.M.F.
(The Oil Companies International Marine Forum) should set up a
specialist group to recommend lines of action to be taken.
O.C.I.M.F. did consider this proposal, but felt strongly that
the detailed discussion involved in any co-operative agreement
should be handled by the local operating Companies, with policy
guidance from their Parent Companies. Meanwhile the ARAMCO/IOE
& PC initiative had gained momentum and in March 1971 ARAMCO
officially approached BAPCO requesting their co-operation in
hosting an initial meeting of Companies at interest in the near
future. BAPCO readily agreed and appointed their Environmental
Conservation Engineer to co-operate in handling the arrangements
for the meeting which was finally scheduled for June of that year.
    Work began immediately on a draft agenda. The aim of this
agenda was twofold :

(1) to brief participants on the extent of the problem they all faced, and the equipment and materials available to combat a major spill

(2) to propose a form of agreement to draw the Companies together into a co-operative agreement.

The draft agenda was completed by mid-April and distributed to all the Gulf Companies, as a follow-up to a cable invitation to attend the proposed meeting sent out two weeks before.

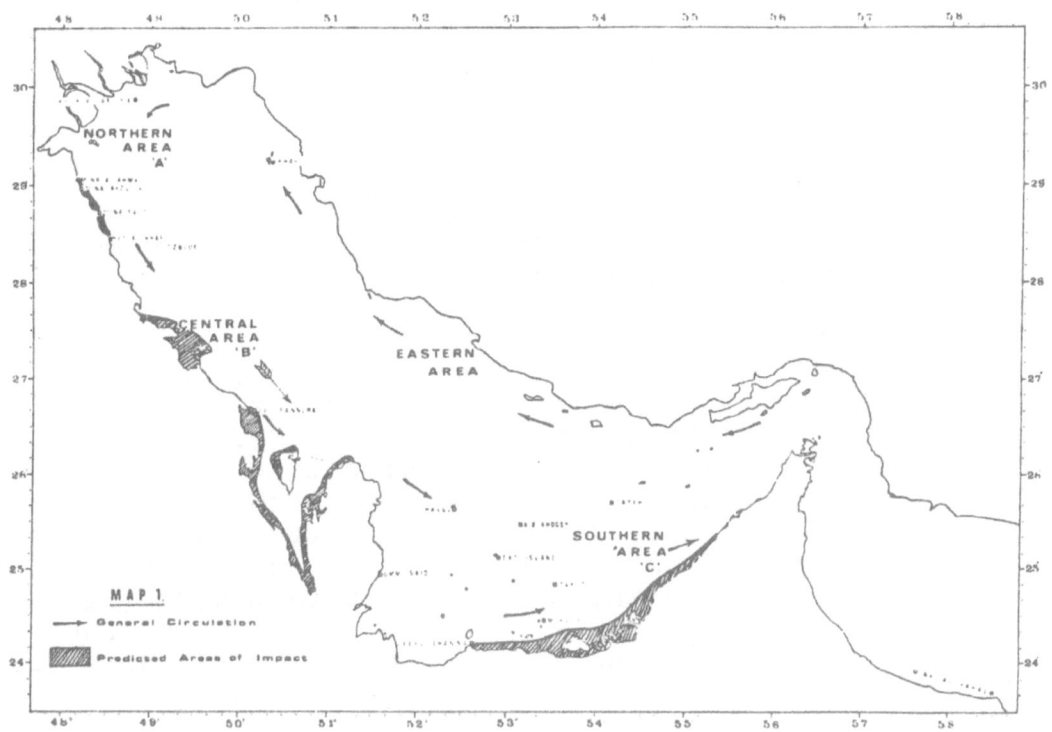

ORGANISING THE MEETING

The agenda was based on a meeting spanning a day and a half, commencing with a review of the present situation, continuing with technical briefings interspersed on the following day by a session on the form of co-operative agreement that Companies would be willing to enter into. Two speakers from outside the Gulf were invited: one from SOCAL's Richmond Refinery to review oil spill clean up technology in the U.S.A. and to describe the practicalities of a major spill clean up operation in the sophisticated environment of San Francisco Bay, and the other from U.K. Government's Warren Spring Laboratory to describe the virtually 'grass roots' development of clean up techniques following the 'Torrey Canyon' disaster. To broaden the scope of technical briefings, locally based Communications and Navigation experts were also invited.

It was considered of great importance in the initial planning of the meeting to encourage maximum participation by the 40 or so Senior Company representatives expected to attend.

To assist this aim, Moderators for the majority of the meeting sessions were appointed from among the representatives who had accepted the invitation to attend. These Moderators set about the tasks of researching background information relevant to their topics and analysing results of questionnaires on equipment, materials and communication facilities which has been sent out with the draft agenda. The venue chosen was not so

large as to overwhelm those representatives present, or so small
as to be uncomfortable. Conference reporters used successfully
at Caltex Maintenance Conferences were hired from the U.K. and
a meeting administrative office was set up.
     Arrangements proceeded smoothly, but unfortunately the
Companies operating in Iran finally did not join the meeting
for reasons of their own, so the meeting finally went ahead
with 13 Companies represented, all from the Arabian side of the
Gulf. Apart from this one unfortunate set-back, arrangements
progressed very well, with good co-operation from all the
Companies who had agreed to attend.

THE FIRST MEETING (JUNE 1971)

The meeting opened on June 20th 1971 addressed by the President
of BAPCO and the Senior Vice President of ARAMCO welcoming
representatives and setting the theme of the meeting; then it
immediately got down to the business in hand.
     The Map of the Gulf (2) which shows potential participants,
clearly demonstrates the large area involved. Total oil movements (not shown) are also very large (average for Arabian
side alone over 9,000,000 bbls/day in 1971), 19,000,000 bbls/
day in 1978. The discussions therefore set out to brief the
representatives on the extent of the problem, with some telling
practical local experiences by ARAMCO to illustrate what could
well happen, and a summary and analysis of the equipment and
materials available in the area, which showed that there was
not very much, and that not very efficient! There were, and
still are, no Contractors in the area set up to assist effectively.

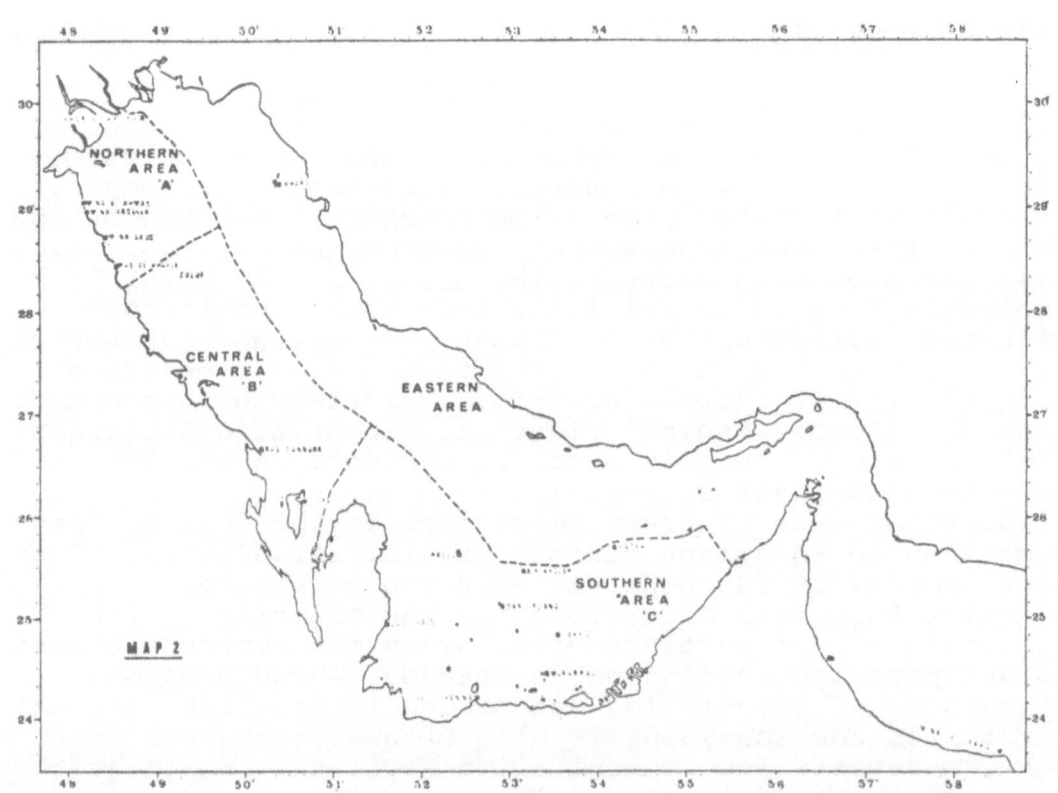

MAP 2

The presentations that followed outlined :
    (1) Communications.
The network in the Gulf, (and this at least seemed encouraging) with a good basis for fast communication between
strategic points, and pointing fairly conclusively to Bahrain
(and hence BAPCO) being a Communications centre in the event
of a really catastrophic spill in the area. This has been
borne out by the Kuwait Regional Agreement choosing Bahrain
as their reporting centre.

  (2) Oil Spill Clean Up Equipment & Technology.
A SOCAL representative reviewed the current 'state-of-the-
art' in the United States, and set this as the background
to a very lucid account of the mammoth clean up operation
by skimmers of all sorts - booms, absorbents and sheer
manual hard labour, following the collision of two tankers
beneath the Golden Gate Bridge in San Francisco Bay.
The Warren Spring Laboratory representative (Dept. of Trade
& Industry, British Government) described the development
of the chemical oil spill dispersant spraying and agitating
equipment undertaken by his Department, following the
chaotic attempts to clean up the massive 400,000 bbls
'Torrey Canyon' spill, where indiscriminate use of highly
toxic dispersants and the attendant prodigious damage to
the marine ecology of the adjacent coastal areas led to
stringent legislation, and public protests against their
use in many areas of the world, including, of course, the
United States.
Warren Spring Laboratory probably represented the best single
authoritative source of common sense and experience about
oil spill clean up anywhere in the world at that time, and
their presence at the meeting coupled with a readiness and
capability to answer the many questions following his
presentation, was of great value to the meeting.

There followed two ARAMCO presentations detailing their improvised,
but effective response to several major oil spills which had
recently occurred in their crude pipeline systems, and the
exhaustive researches carried out on their behalf by a Marine
Biologist in an attempt to assess the after effects of these
spills on the ecology of the area.
    The morning of the second day of the meeting brought a
different situation. ARAMCO had drafted a proposed form of co-
operative oil spill clean up agreement loosely based on an API
model which had also been distributed with the agenda for consideration prior to the meeting, but in the discussions which
now began, it was soon apparent that any agreement was a long
way off, and the Companies present first wanted to digest the
vast amount of technical information already presented to them
the previous day, before committing themselves to any sort of
co-operative agreement. Accordingly a Steering Committee was
chosen, comprising six members with the task of consulting
with other members and outside experts to come up with a
workable and acceptable draft form of agreement, as soon as
possible, for submission to Members' Parent Companies for
review and, hopefully, approval!

FORMULATION OF THE AGREEMENT

The Steering Committee held its first meeting in BAHRAIN three
weeks after the General Meeting and initially attempted to (1)
define the parameters of the spill the co-operative would set
up to handle, and (2) draft an acceptable Form of Co-operative
Agreement to cover (1).

The smaller, more manageable number in the Steering Committee and the channels of communication to all major Parent
Companies involved enabled good progress to be made, and a
second meeting held in DHAHRAN early in August ended with
approval of a revised draft, largely based on the original
ARAMCO proposal discussed with the June General Meeting. At
this stage, the ARAMCO General Attorney was co-opted to the
Steering Committee to assist with the legal aspects of the
Agreement (having prepared the original Agreement draft).

A basic premise of the proposed Agreement was that the Co-
operative was not in the unlimited business of cleaning up
all the oil spills in the Gulf. The decision as to whether a
spill should be cleaned up would rest solely with the member
Company whose operations the spill threatened (subject of
course to the situation that the Company's host Government
might insist on a clean up operation which Company would not
otherwise have undertaken). This Company could call upon
assistance from the other members of the Co-operative and
would reimburse these members promptly for any costs incurred
thereby. The recovery of all such costs of any clean up
operation would be the responsibility of the Company (or
Companies) initiating the clean up, the Co-operative as such
would not be involved in any such claims.

Points which still provoked much discussion at this time,
however, were :
   (1) The minimum requirements for oil spill clean up equip-
       ment and materials to be held by each Co-operative
       member, and additional quantities to be held jointly
       by the Co-operative, all of which came back to the
       difficult question of what size/type of spill should
       be planned for, this still being unresolved.
       It was unanimously recommended, however, by the Steering
       Committee that major oil spills in the Agreement areas
       would be cleaned up by chemical dispersants using the
       Warren Spring Laboratory designed spray sets. The big
       question that remained was, how much dispersant should
       be held by the Co-operative?
       After prolonged discussions on this expensive item
       (figures of up to 30,000 drums were proposed, versus
       total U.K. Government stocks of approx. 100,000 drums
       at that time, by comparison), it was proposed that
       Warren Spring Laboratory should be approached to under-
       take a wide ranging study of the proposed Co-operative
       area to determine, among many other things, recommended
       dispersant quantities, and storage locations, etc.
       BAPCO, who had first made contact with Warren Spring
       as part of their own Environmental Conservation prog-
       ramme, undertook to investigate this proposal and
       report back.
   (2) The second point at issue was the 'third party liability
       question'. BAPCO was, because of its geographical
       position (see Map 1), particularly sensitive to this

situation, where a significant oil spill occurs in a member Company 'A' operations and enters member Company 'B' area of operation before 'A' can effectively clean it up, thus 'B' is possibly forced to undertake the clean up of 'A's oil. We wanted it clearly expressed in the Agreement that 'A' would reimburse 'B' for this clean up (subject to reasonable scrutiny of the costs involved), without obtaining prior permission to commence clean up operations from 'A'. This was considered unduly restrictive by some members of the Steering Committee, who favoured a 'looser' type of Agreement altogether. However, the concensus favoured the more formal Agreement and this was issued for consideration by all the potential member Companies on August 10th. Comments were requested by mid-October and a further General Meeting hopefully to sign the Agreement, planned for November, again in BAHRAIN. This meeting would also consider the proposed Warren Spring Study, and a site Communications Study by Cable and Wireless which the Steering Committee also thought important.

It was also proposed to invite a Senior Representative of TOVALOP to brief the November meeting; ARAMCO undertook to arrange this, and the Managing Director of TOVALOP came to the meeting. TOVALOP (The Tanker Owners Voluntary Agreement Concerning Liability for Oil Pollution) embracing over 97% of the world tanker fleet, is concerned with reimbursement to responsible authorities of costs involved in cleaning up an oil spill involving any tanker owned or chartered by a TOVALOP member. Consequently it was of considerable interest to the proposed Co-operative to determine TOVALOP's attitude to repayment of spill clean up operations undertaken by member Companies.

The next Steering Committee meeting was held in KUWAIT, hosted by the Kuwait Oil Co., to finalise arrangements for the November General Meeting.

Until this time, it was generally expected that the Organisation would be called Gulf Area Environmental Conservation Co-operative or something on those lines. However, KOC wished the word 'Co-operative' eliminated from the title as it had a special connotation in Kuwait Law which would be embarassing to Kuwait based member Companies. Thus the phrase 'mutual aid organisation' was eventually arrived at.

THE SECOND MEETING

The second general meeting took place on 29th/30th November and resulted in some progress, but not as much as had been hoped, unfortunately. The 3rd Party clause remained in contention. However, the Steering Committee was now replaced by an Operators Committee and Executive Committee, nominated from the members attending the meeting, in accordance with the terms of the proposed Agreement. These 'de facto' Committees were instructed to work for an early adoption of the Agreement, whereupon they would be confirmed in office.

THE THIRD MEETING

A further general meeting of the Operators Committee was held in BAHRAIN on 7th/8th February of 1972, but contrary to expectation, signature of the Agreement seemed now to be further from realisation as three more Companies refused to sign with the

contentious 'third party liability' clause incorporated in the
Agreement. This impasse, however, as it later turned out, was
the turning point in the long negotiations to achieve an
acceptable form of Agreement.

During the meeting, however, two important steps were agreed
upon :

    (1) Prospective GAOCMAO members were to be asked to approve
        and finance the Warren Spring Laboratory oil spill clean
        up equipment and materials study which had 'hung fire'
        for several months, as soon as possible, and in advance
        of signature of any Agreement.

    (2) The 'third party liability' clause was slightly modified
        to overcome some of the objections of the 5 dissenting
        Companies, especially their concern about giving a third
        party 'carte blanche' to carry out an expensive clean
        up operation for their account with no requirement for
        prior consultation. This was now modified to ensure
        that the member initially involved in the oil spill was
        kept fully informed from the outset in any clean up
        initiated by a fellow member as a result of the spill.

Meanwhile, the Warren Spring Laboratory Study was unanimously
approved within 3 weeks of the meeting, and commenced in March.

In April, objections to the 3rd party clause were dropped,
and at an Executive Committee Meeting in Dubai late in April,
the <u>final</u> draft form of Agreement was agreed upon and sent to
all 13 potential participating Companies, for signature by
<u>June 1st.</u>

Companies were sent 3 copies, and asked to sign 2 copies
and return to the Chairman of the Operators Committee for
ratification, whereupon each Company would receive a counter-
signed copy back.

The Agreements were <u>all</u> signed and returned by July 11th,
at which date GAOCMAO came formally into being, 13 months after
that initial meeting in Bahrain and with all the 13 potential
participating Companies that had attended the meeting, as
founder members. PD Oman had also attended the first two
meetings, but from the outset of detailed discussions on the
scope of the Agreement, it was decided with PD Oman's full
concurrence, that membership of the Agreement was not a practical
proposition in their case, owing to the isolated position of
the PD Oman terminal beyond the Straits of Hormuz. This view
was reversed in 1977 when PD Oman considered the above was no
barrier to membership due to the vast improvement in transportation and communication within the area, also, with the
increased world wide awareness to environmental conservation,
it became essential for PD Oman to belong to GAOCMAO for the
actual advantages such an organisation could provide.

TECHNICAL ASPECTS

During the long and sometimes frustrating period that an
acceptable form of Agreement took to develop much fruitful work
was carried out on the technical aspects of oil spill clean up,
spearheaded by ARAMCO, but with active co-operation from other
potential GAOCMAO members, to the limit of their capabilities.

Main interest centred on the choice of an acceptably
efficient and economic dispersant, which would also be readily
available in very large quantities in the event of a catastrophic
spill.

Following the guide lines given by Warren Spring Laboratory at the June 1971 meeting, trials were arranged in August by ARAMCO, using Warren Spring Laboratory dispersant spraying equipment, to which all Steering Committee members were invited. These trials were not conclusive owing mainly to lack of experience in testing procedures. However, later on in the year (November) ARAMCO carried out a further series of trials and invited a representative of W.S.L. to supervise the exercise, bringing any specialised test equipment required with him. These trials were entirely successful in qualitatively evaluating the three dispersants then under consideration, and although detailed results remained confidential between ARAMCO and Warren Spring Laboratory, informal briefings were given to other prospective member Companies at the November 1971 General Meeting. Other members, including KOC and BAPCO, also carried out small scale trials, mainly to familiarise themselves with the dispersant spraying equipment.

So, technical co-operation among prospective members on evaluation of dispersant materials and equipment proceeded very smoothly despite the long period of argument and counter argument concerning the terms of the Agreement.

Nevertheless, the need for a definitive study by Warren Spring Laboratory became quite apparent as more was learnt about equipment and materials, especially, as previously remarked, to give guidance on the optimum size and location of dispersant stockpiles in the Agreement areas.

In a further attempt to eliminate uncertainty and potential problems GAQCMAO also commissioned the study on local Communications by Cable and Wireless, who operate extensively in the Gulf, that was mentioned previously. This study determined the compatability between communications systems installed on member Companies aircraft, helicopters and Marine craft in the event of their joint use in a major spill clean up operation, and recommended what changes/improvements/procedures would be needed to ensure that an efficient system of communication be provided to control a clean up operation at any location covered by the Agreement.

The need for such a study was initially brought home to members at the first General Meeting in June 1971, when the representative of SOCAL vividly described the chaotic communications situation which arose in San Francisco Bay during their clean up operation, due to lack of a clearly understood emergency communication network, and general incompatability between portable shore based transceivers and the multitude of Marine craft involved. The study concluded that the type of systems available was adequate for oil spill clean up work in general. Since that time there has been a continual improvement in communications in the area.

CONDITIONS OF MEMBERSHIP

1. Within 30 days of signing the Agreement each PARTY shall :
   (a) Place an order for, or have immediate access to, at all times 500 drums of dispersant or 100 drums of dispersant concentrate exclusive of the supply held by any other PARTY.
   (b) Order and thereafter keep on hand, in operable condition, a reasonable amount of clean up equipment which initially shall include two Warren Spring Kits or other similar

equipment approved by W.S.L. and capable of spraying dispersant concentrate.

(c) Pay a membership fee of ø3,000. The Bahrain Petroleum Company shall act as a temporary custodian of all funds until the Operators Committee makes other arrangements.

2. Furnish to each PARTY at such times as the Operators Committee of the Executive Committee shall specify :
    (a) An inventory of equipment and materials in a form to be specified.
    (b) Information on its communications systems in a form to be specified.
    (c) The names of individuals authorised to release equipment and materials and the means by which they may be contacted in an emergency.

3. Furnish to the Chairman, in each January, offshore production and marine terminal export information for the previous calendar year to calculate each members contribution to the annual budget. Each barrel of persistant oil exported and imported is registered. The required budget is then divided by the total number of barrels to give a monetary figure per barrel. The number of barrels registered by each Company is multiplied by that figure and the resultant is that Company's contribution.

PARTIES operating in each Area shall make arrangements among themselves to stock or have access to, at all times, within their Area a minimum of 3,000 drums of dispersant or 600 drums of dispersant concentrate. The amounts that each PARTY must stock or have access to pursuant to Article IV shall not be included within this figure. These 3,000 drums of dispersant or 600 drums of dispersant concentrate shall be released immediately upon request to any PARTY experiencing a major spill. The stock of dispersant shall then be replenished as soon as practical, at cost of the PARTY using them. PARTIES operating in each area may make separate costs as to the purchase or lease of any materials or equipment.

GAOCMAO TODAY

The Organisation has grown from 13 to 17 members and the possibility of more Companies joining soon.
   The Organisation is divided into 3 areas, Northern, Central and Southern and operates in 7 different countries. As can be seen in comparing Map No. 2 with Map No. 3, there has been a change in the area covered by the Central and Southern Areas. This was due to increased membership in the Southern Areas and to make the areas more balanced.
   The Organisation has an Operators Committee which comprises delegates from all member Companies. At the Annual Operators Committee Meeting a Chairman and 2 Vice Chairmen are elected to form the Executive Committee for the following year. The Chairmanship and Vice Chairmanship rotates around the three areas. One area under the Chairman and one Vice Chairman in charge of each of the other two areas. These functions are carried out as additional duties by personnel holding full time jobs within their Companies. The Organisation has no exclusive staff.

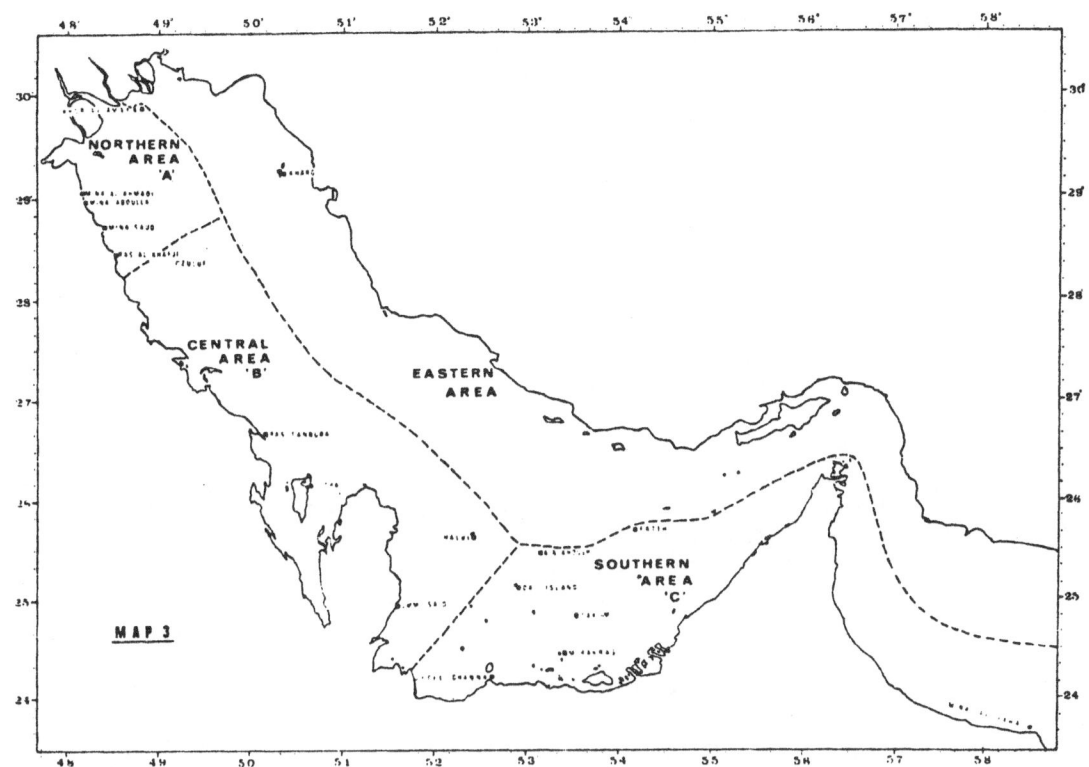

MAP 3

GAOCMAO commenced its life playing a very low profile and for
a very good reason. It was felt that the Organisation should
obtain its inventory of equipment, dispersants, etc. and gain
some knowledge and experience otherwise the Organisation could
be made to look very foolish if called upon before being sufficiently organised, but in the last few years the Organisation
has emerged more into the public eye. The Chairman for 1977/78
went to Nairobi to the Conference for the formation of the
Kuwait Regional Agreement and stated that; whatever support
GAOCMAO could give the Regional Agreement, it would. Our aim
is to work with and alongside the Governments. At the inception
of GAOCMAO all the Companies were commercial companies, but in
the intervening years the majority have become partly or totally
Government owned.

Each year a practice alert is carried out to test communications
and evaluate what would be available in terms of assistance.

Training courses have been held with the assistance of Warren
Spring Laboratory and a Senior Management Seminar was held in
1979 to enable the level of Management, who make policy, to be
brought up-to-date on Environmental matters and organisations.

The Equipment Inventory shows that the available equipment
has grown to reach a significant quantity and the Organisation
probably has the largest stockpile of dispersant outside Western
Europe and the United States. Dispersing has become the standard
approach to clean up techniques by all participating Companies.

EQUIPMENT INVENTORY

W.S.L. spray units (Deep sea) 40 units.
   "   "   "   (Inshore) 32 units.
Boom. 28 units - 3,500 meters.
Skimmers. 17 units.

Pneumatic fenders. 26 units.
Pillow tanks. 11 units - 198 bbls capacity.
Skid Mounted tanks. 6 units - 109 bbls.
Oil Storage barges. 8 units - 15,600 bbls.
Dracone. 1 unit - 286 bbls.
Sea-going marine craft normally available. 39 units.
Inshore marine craft normally available. 40 units.
Many of the above marine craft have dispersant tanks and spray
equipment fitted. More marine craft can be made available by
hiring or by reducing operations. The W.S.L. spray units
would be fitted to these craft.
Beach cleaning units - 9.
Helicopter spraying equipment. 4 units.

During 1979, there was strong feeling within GAOCMAO that the
Organisation should make more effort to improve its efficiency
and progress along with the changes taking place in the Region.
    To this end an Extraordinary General Meeting was held in
Bahrain on the 27th/28th October, 1979. At this meeting it was
decided that activity could not be increased while the Organisation did not have its own full time staff. Accordingly it
was agreed in principle that a full time Executive Secretary
should be appointed to assist the Chairman and the Executive
Committee.
    It was also agreed in principle that the purposes of the
Organisation should be amended to :-
(a) Arrange for joint capability to prevent and clean up oil
    pollution, if requested, which requires effort beyond the
    resources of a single party.
(b) Acquire, compile and disseminate (to members) information
    on Gulf area oil industry-related marine pollution and
    maintain links with concerned international organisations.
(c) Promote technical discussion between members and provide
    co-ordinated advice and assistance if requested by individual
    members.
The Executive Secretary will devote himself to the fulfillment
of these three purposes. As each member will have equal right
to call on his services, it was decided that the system of budget
by barrelage of persistant oils handled, was now obsolete and
the Organisation's operating budget will be equally shared by
all parties.

CONCLUSION

Experience during the last 8 years has shown that the need for a
mutual aid organisation in this part of the world was well-
founded.
    The quantity and variety of equipment and materials currently
available to the Organisation are a reflection of the progress
made since that first exploratory meeting in Bahrain in 1971.
    In retrospect, the Organisation has worked well and, on
several occasions, has been put to the test when emergency
situations have arisen.
    Contacts and Communications are a vital component of the
scheme. The first occasion when the Organisation was called
upon was not for oil spill clean up, but when a serious fire
occurred in the BAPCO Refinery in Bahrain. Direct contact
between GAOCMAO members was put to the test and enabled large
quantities of fire fighting equipment and materials to be

mobilised and shipped to Bahrain in a short space of time. A
significant part was played in this by the Kuwait Government who
helped GAOCMAO by providing the service of two Airforce Transport
aircraft which demonstrates that Government and GAOCMAO together
can be a very effective force. It may sound irrelevant that this
success was in fire fighting rather than in oil spill clean up
but it is not so. It tested the Contacts and Communications
of the Organisation which so soon after its formation were not
found wanting. This incident indicated that the vagueness of
the title "Mutual Aid Organisation" may well have advantages.

In some areas of the world oil companies have formed full
time clean up organisations to take care of spills, but these
organisations have to rely on contractors, otherwise to be large
enough to deal with major spills they would be extremely costly.
For areas not having contractor availability, GAOCMAO provides
an organisation on the lines of a volunteer fire brigade, together with an Information Forum and advisory service at
moderate cost for investment of materials and running costs.
It is really a system of co-operation, in fact International
Co-operation (GAOCMAO comprises 17 multinational members in
7 different countries) for the purpose of protecting the
environment and maintaining trust between the Industry, Government and the people of the region.

THE GULF WEATHER FORECAST SCHEME - 21 YEARS ON

David Hibbert, Director, Imcos Marine Ltd., London, U.K.

INTRODUCTION

More than any other modern industry, the offshore oil industry has its investments continuously at risk from the weather and in attempting to minimise this risk the industry has contributed more to knowledge of weather
over the sea, wave forecasting and wave forces than any other. Modern meteorology is founded upon the work of mariners such as Fitzroy and the first
weather services were directed to meet seamens' needs. Their subservience
to the weather diminished and aviation's requirements became most important
and have generally remained so, insofar as state meteorological services
are concerned. More recently these organisations have become interested
in the needs of the offshore industry, endeavouring to emulate the private
practitioners in meteorology who have been long established offshore: in
the Arabian Gulf since 1958. State meteorological offices are not, however, organised to offer such specialised weather services where the need
is for wind and wave and general forecasts designed around and altogether
"tailored" to meet the industry's requirements and given by oil industry
orientated meteorologists, and it is here that the value of an independent,
supplementary weather service becomes apparent. An example of such a service is that provided for oil company operations in the Arabian Gulf during
the past 22 years. This forecast service is unique for the way in which the
oil companies themselves participate in its organisation and provide weather
observing and reporting stations in over half a dozen countries and the
states around the Gulf.

Within recent years some state services, notably the U.K. Meteorological
Office and, latterly, some Gulf states within the Regional Marine Meteorological Plan have attempted to follow this example, one of several services
built up by Imcos Marine around the world in the past 25 years. However,
adoption of the terminology and attempts to copy techniques of application
may not completely compensate for some lack of credibility which State
service weather forecasting has earned during the many decades that government organisations have been custodians of the art. The place of the independent oil-orientated specialist service is secure, unless, of course,
special powers are invoked to protect the State meteorological organisation.

THE OIL COMPANIES' WEATHER CO-ORDINATION SCHEME

The oil companies' offshore weather forecasting service in the Arabian Gulf
was initiated by British Petroleum and Shell in 1957. For the early movements of the Shell exploration platform reliance had to be placed on the
very sketchy weather information which could be gathered from an R.A.F.
meteorological office 100 miles away, and from temporary observation points.
The loss of this platform in 1956 brought home to all concerned the necessity
of some form of specialist meteorological service. The planners of Abu
Dhabi Marine Areas (ADMA) saw this need in connection with their impending
offshore operations and BP with Shell enlisted the services of meteorological

consultants who would carry out the specialist marine forecasting work and,
as a necessary preliminary, advise on the setting up of a weather reporting
network without which an adequate forecast service was impossible. Shell
and the (then) Qatar Petroleum Company agreed to help in the first positioning of the ADMA "Enterprise" by having weather observations sent from weather
stations established at their bases, to Das Island where the first oil company offshore forecast office of its kind was set up. This joint reporting
venture worked well, the weather observation coverage in the southern Gulf
being more than doubled. It was still not sufficient to cater for the requirements of the forecaster and the only method of achieving anything like
the desired network of observing stations was to take advantage of all oil
company activities in the Gulf. The operating companies had oilfields,
exploration camps, airstrips or marine installations which were manned and
suitably equipped with communications facilities to permit them to take on
weather observing and reporting duties, and these companies were invited to
participate in a major scheme which would provide Gulf-wide weather observation and forecast coverage.

Weather Scheme Members
In 1959, eight companies were participating in this scheme, the Oil Companies' Weather Co-ordination Scheme (OCWCS). Now, in 1980, the membership
totals 18 companies as below.
     Abu Dhabi Marine Operating Company
     Abu Dhabi National Oil Company
     Abu Dhabi Oil Company Ltd.
     Abu Dhabi Company for Onshore Oil Operations
     Arabian American Oil Company
     Arabian Oil Company Ltd.
     Bahrain Petroleum Company
     Dubai Petroleum Company
     Iran Pan American Oil Company
     Iranian Marine International Oil Company
     Kuwait Oil Company (K.S.C.)
     National Iranian Oil Company
     Qatar Petroleum Producing Authority
     Saudi Arabian Meteorological Department
     Total Abu al Bukhoosh
     Umm Al-Dahlk Development Company
     Wintershall Aktiengesellschaft
     Zakum Development Company

Weather Scheme Co-ordination
The work of the oil company weather stations throughout the Gulf is co-
ordinated, and specialist wind and wave forecast services are provided, by
Imcos Marine, International Meteorological Consultant Services.
     There are at present some 50 weather reporting oil company stations
(WROCS - pronounced ROX) between Muscat on the Indian Ocean and Masjid
Soleyman in Iran, 17 on land, 34 on offshore installations or rigs. Not
all WROCS are permanent. Several, as on offshore rigs or inland at desert
exploration sites, are transient. These may, during a relatively brief
sojourn in one locality, contribute to knowledge of meteorology in the area,
as well as providing synoptic observations for forecasters' use (Fig. 1).
From Qoin Island, a station maintained by MENAS, weather reports are received
daily and continuous wind recording is carried out.

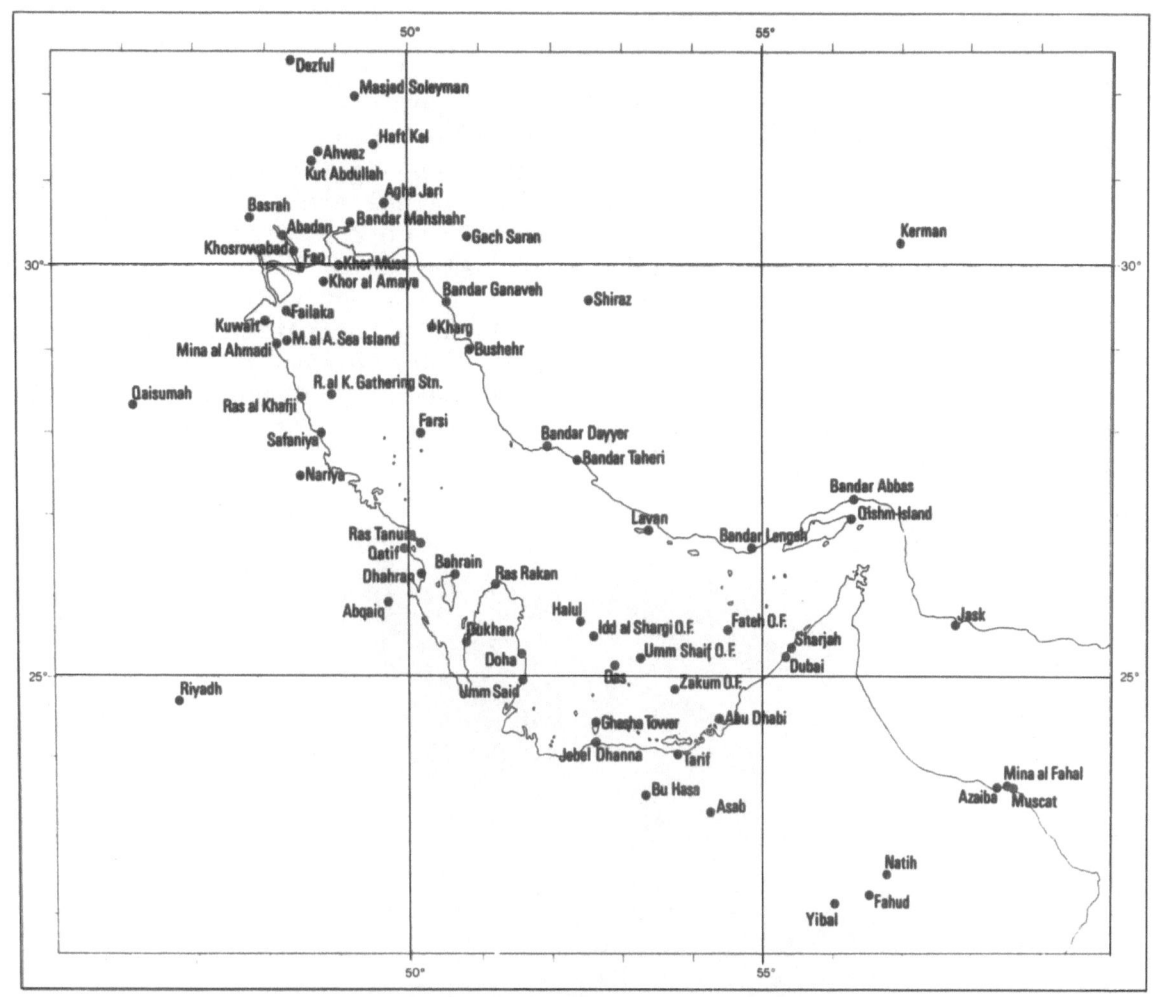

THE GULF AREA

Excluding offshore rigs, more than 40 of some 60 Weather Stations
shown have past or present OCWCS affiliation

Fig. 1

Forecasting offices are established at several places. These offices
provide forecasts for critical operations such as rig moving, well jacket
placings and the weather-sensitive stages in submarine pipe laying, and all
offshore exploration and production operations are covered by the Weather
Scheme forecast umbrella. The network of stations is kept under scrutiny
and their work co-ordinated by Imcos Marine meteorologists who are stationed
in the Gulf or who may be sent from the U.K. on Gulf field supervisory tours.
     The representatives of the oil company members of the Weather Scheme
meet annually in the Gulf to review, discuss and decide upon recommendations
about the Scheme made by Imcos Marine. Few oil companies are not members
of the Scheme; it is perhaps significant that in over 21 years of the oil
companies' joint effort the major accidents which have occurred in the Gulf
have been to rigs and marine operations not covered by forecast services
given under the Weather Scheme. All members contribute to the costs of operating the Scheme, sharing the expenses of field co-ordination work. Members
needing forecast services pay for these direct under separate and individual
contract with the Co-ordinators, Imcos Marine.

Long Term Data

The Weather Scheme depends upon a foundation of goodwill by all the participants and even those who may have had few offshore interests and have put in more by way of observational work than they might seem to have got out of it, are now drawing on the fund of meteorological and allied oceanographic data acquired and held within the Weather Scheme during the past 22 years. This valuable bank of wind, wave, weather and allied data is possibly the most comprehensive for any marine area in the Middle East, and has facilitated the solution of many problems, not only theoretical problems of wind/wave relationships in the Arabian Gulf and in similar waters elsewhere, but also those concerned with planning and design. Authoritative design data can now be provided for every part of the Gulf for all the marine development work now going on (Fig. 2).

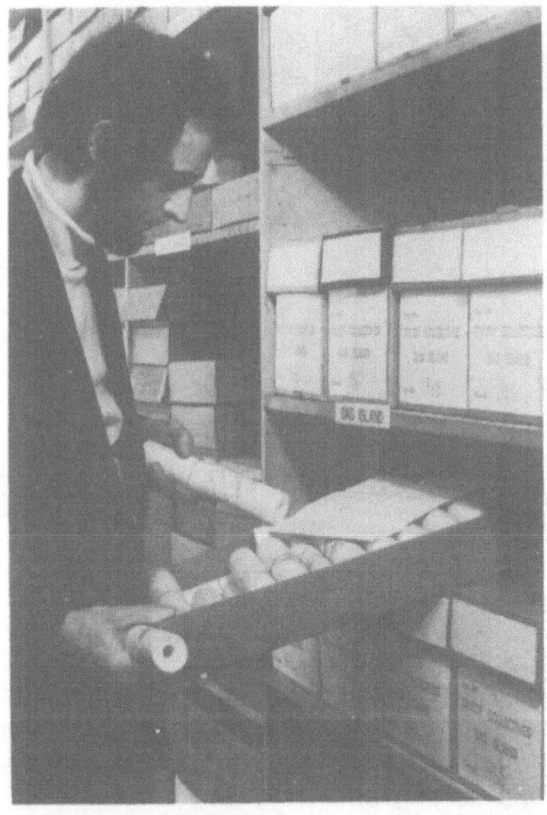

The data bank with over
twenty years of records

Fig. 2

Forecast Services

In all forecasting for any critical operation the principle is observed that the forecaster be stationed at the operational shore base or even offshore if forecast office communications facilities permit. This is fundamental to the success of forecasting for offshore operations. There must be constant liaison between the forecaster and the Drilling and Marine Superintendents as only in this way can the proper application of weather and wave forecasting be developed. The operator comes to appreciate the forecaster's need for more and frequent observations, while the forecaster

develops a sense of the operator's precise requirements and is able to express
his forecasts in terms of those requirements. The aim is for a forecast
service which is so integrated with the offshore operations that the meteorologists will regard themselves, and hope to be regarded as technicians in
the oil industry and not only as members of a somewhat exotic outside pro-
fession.

Weather Observation

The forecaster is entirely dependent upon the surface weather and upper air
observations which he receives and which are the back-bone of all weather
maps. No forecasts of future weather are possible without them.

The number of State weather stations around the Gulf has increased in
recent years but is still inadequate for offshore operations forecasting;
sea state reports are virtually non-existent and Ships weather reports within
the WMO procedures are few and far between in the Gulf. Satellite imagery
cannot fill many gaps; when it is available it is inappropriate to surface
wind and wave forecasting in all but very few situations. These information
deficiencies the oil companies endeavour to make good through their own
efforts.

Before the advent of the Weather Scheme, weather observation in the
Gulf by the oil companies was mainly for climatological record purposes,
limited observations being made once or twice daily. The observations had
general distribution only in monthly meteorological returns; there was no
exchange of observations between companies and there were of a wide disparity
of content, time and method. Improvement of observing procedures to the required standard was perforce a gradual process as the observations were still
to be made by oil company staff not specifically trained for this work. A
simple, standard report for synoptic (forecasting) purposes requiring as
little subjective observation as possible, while yet being of value to the
forecaster, was devised. This simplification was possible in the Arabian
Gulf because the weather features which require particular skills for their
observation, and which are somewhat complicated to codify in a synoptic
report, e.g. cloud height and type, are relatively infrequent and of lesser
significance in wind field and wave height prediction. Observations had, initially, to be confined to daylight hours.

The observation is made up into a message of 5 groups, including the
station name, as : Name DDDFF PPPP $T_dT_dT_wT_w$X and, (for example), cloudy.
The final group can be expanded upon if necessary. This WROCS message is
virtually a plain language version of the observation and requires no coding.
Time and date of observation are passed as part of the preamble in the message which also states that it is a WROCS (Weather Reporting Oil Company
Station) observation being sent, as distinct from a standard SYNOP observation. Offshore drilling rigs and installations include wave height and
period in their observations, differentiating when possible between sea and
swell.

The procedures for observing, reporting and recording are to World
Meteorological Organisation standards, with internationally approved instrumentation to match. Autographic wind recording is standard for all but the
most transient of stations and record analysis services are provided by the
Coordinators.

As observers have become used to working to these standards, many have
been given further instruction and at some stations the full international
synoptic weather observation is reported. In Iran, where the National Oil
Company's observers at airfields and other bases are trained by Imcos Marine
meteorologists, several of the oil company stations have World Meteorological

Organisation status and station numbers have been allocated to them by the
Iranian Meteorological Organisation. WROCS reports are sent to the State
meteorological offices of the area in which they are located. The 10.00 LT
observation at all WROCS is designated the climatological observation and
more elements are observed for this than at other times. All observations
are recorded and retained for processing by Imcos Marine on behalf of the
Weather Scheme.

Sea State Observations

Observations of state of sea from offshore installations and drilling rigs,
usually a statement of maximum wave height around the time of observation
based on a visual estimate are, with wind speed and direction, of paramount
importance. Wave length and period observations are generally less reliable
than the height estimate, the two often being irreconcilable. Estimates of
maximum wave (H max) are usually somewhat greater than the true significant
wave height (Hs) and nearer to H10, the average height of the highest ten
percent of waves. This is doubtless due to the observer endeavouring to
report Hmax but seldom achieving this due to brevity of observation time and
coarseness of measurement. The "cheap and simple" wave recorder which operators frequently call for and which would permit instrumental observations
from all offshore platforms seems as far away as ever. The Datawell Waverider system is the most acceptable to OCWCS members but as yet few are being
operated. Record interpretation services are offered by the Co-ordinators.
Wave height and wind observations may be all that is asked of those offshore
rigs where there is limited enthusiasm and personnel for weather work, and
where it is felt that to ask for more detailed reports could be counter productive. Barometric pressure readings which need to be reduced to m.s.l.
can be an observer problem where mobile jack-up rigs are concerned. Wind
speed readings are not reduced to standard height, 10m above m.s.l., by the
observers. This is left to the forecaster who in addition to knowing from
experience the idiosyncracies of the rigs in his "parish" will be kept advised by the Co-ordinators of instrument exposure variations on offshore rigs
in other areas.

Tanker Weather Reports

When in the area of the Arabian Gulf, west of 58°E, Shell, BP, NITC and
tankers of the fleets of other Weather Scheme members send weather and state
of sea reports to the Co-ordinators, Imcos Marine, at IMCORTT Bahrain.
Tankers which are "selected" ships report the full ships SYNOP message,
others may report in a modified "Supplementary Ships" code. Observations
are received on most days and the observing standard is high. In one unbroken period of 11 years 6559 of these special tanker observations were
received. This private arrangement between OCWCS and its members' tanker
fleets does not conflict with the ship Reporting System of the WMO but rather
supplements it through this high level of activity in one small critical sea
area. All OCWCS affiliated ships are invited to send in observations to
IMCORTT Bahrain including those operating coastally in the Region.

IMCORTT - Imcos Radio Teletype Transmission

WROCS reports are sent within the hour by radio, often by oil company
channels, to one of two major collecting centres, IMCORTT Bahrain in the
south and Abadan in the north. The Abadan collective is broadcast on a
special radio channel for interception at Bahrain. Reports from Kuwait
(South Pier) are, when feasible, transmitted as part of the tanker advisory
service and may thus also be picked up at Bahrain. It is at Bahrain that

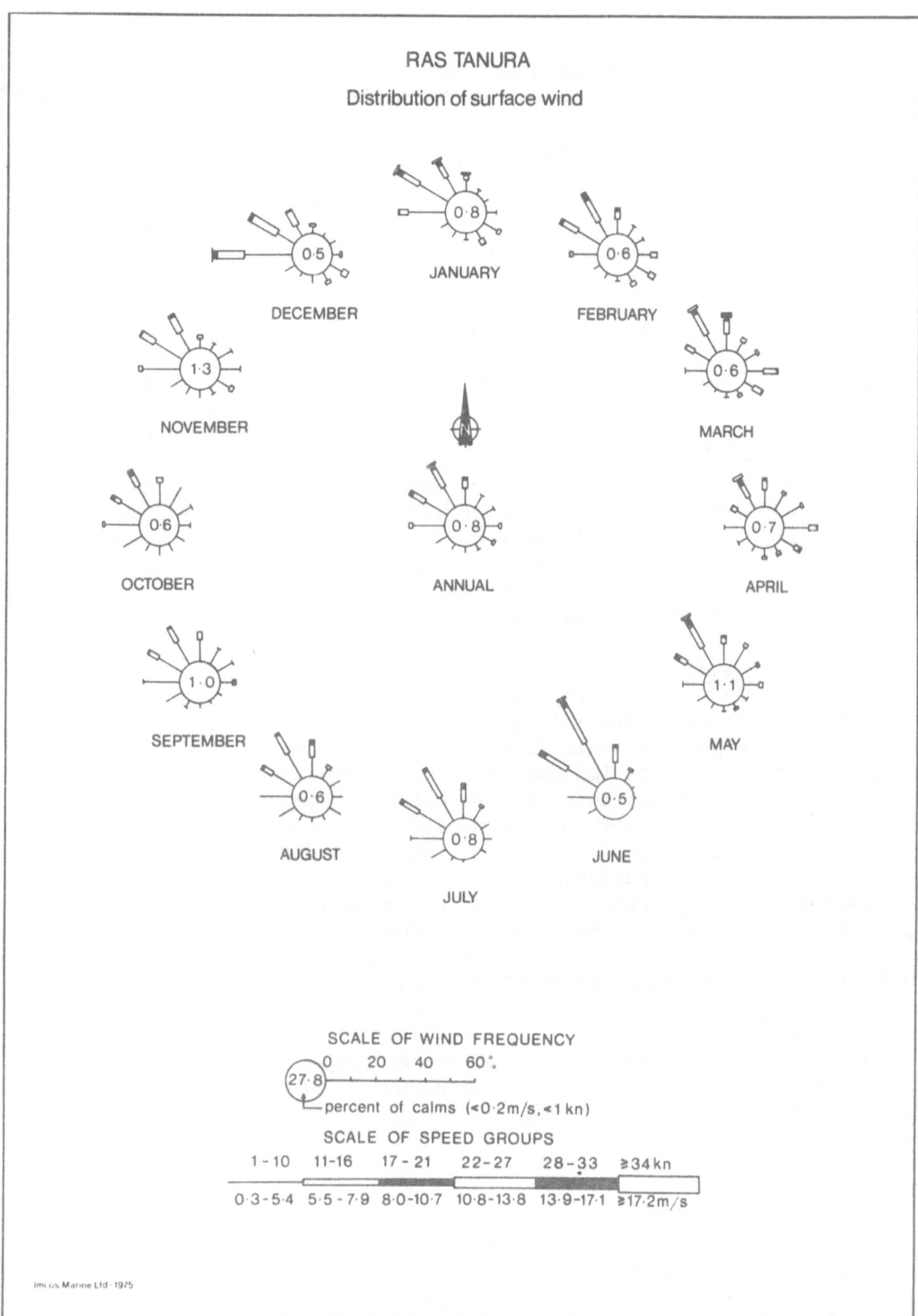

Training School Display Poster

Fig. 3

all WROCS are eventually collected and enter another area of operation,
unique in that it is the only privately run meteorological broadcast of any
significance in the world, Imcos Radio Teletype Transmission, or IMCORTT.
        At the forecast offices established at oil company bases around the
Gulf the meteorologist must receive not only the weather observations, upper
air data and ship reports from the Gulf, but those also from the wider
region; from all Iran and Iraq and Saudi Arabia, west and north to the
Eastern Mediterranean, Turkey and Russia and east to Pakistan and India.
To obtain all this information demands staff and equipment to receive a wide
range of international weather data broadcasts from surrounding states;
Athens, Karachi, Moscow, Ankara and Jeddah to name but a few. To eliminate
this costly requirement at each forecast office, IMCORTT was established to
receive, edit and then broadcast in radio teletype all the information that
any Middle East forecast office needs.   Very considerable staff and equipment cost savings are effected. Established in 1965, IMCORTT operates out
of Bahrain and the broadcast contains information sufficient for any offshore operations office in the Middle or Near East. It is also the channel
by which all OCWCS observations are speedily disseminated to all users who
need the information at their forecast offices. IMCORTT is a non-profit
organisation backed by many oil companies in the region and subscribers pay
at-cost fees. The broadcast is intercepted by State users who to date have
been slow to acknowledge its value to them.

Training of Local Staff
It was envisaged that some of the burden of weather observing work being done
by members at their oil installations would eventually be borne by state
organisations. Progress in this direction is slow and it is now questionable that the policy can be implemented at offshore observing stations, which
are probably the more important. In the meantime personnel are drawn from
the local staff of member companies and given instruction in weather observing, recording and reporting, on instrument maintenance and general
meteorological assistant duties at varying levels. The training of staff
for weather forecasting duties is an aim of the Weather Co-ordination Scheme
but prospective forecasters,having been trained in the somewhat demanding
discipline of meteorology,are soon attracted to the higher posts and greater
rewards which business and government in their home countries can offer.
That all such trainees have prospered may be significant and a tribute to
their training.
        Educational posters for display at member companies' training esta-
blishments are part of an exercise to inculcate an interest in weather and
the environment.   (Fig. 3, page 7).

METEOROLOGICAL SATELLITES

The satellite is one of the two most significant tools ever to be placed at
the hand of prognostic meteorology. In the mid-1800s the works of Chappe and
Morse had made it possible by telegraphy to collect weather information at
a speed which permitted the plotting of reliable weather charts from which
forecasting techniques could be developed, and now the meteorological satellite repeats the promise to put before the forecaster a wealth of data previously unimaginable (Ref. 1).
        Two types of meteorological satellite are in use, the geostationary
satellites of which METEOSAT, of the European Space Agency (ESA), is an

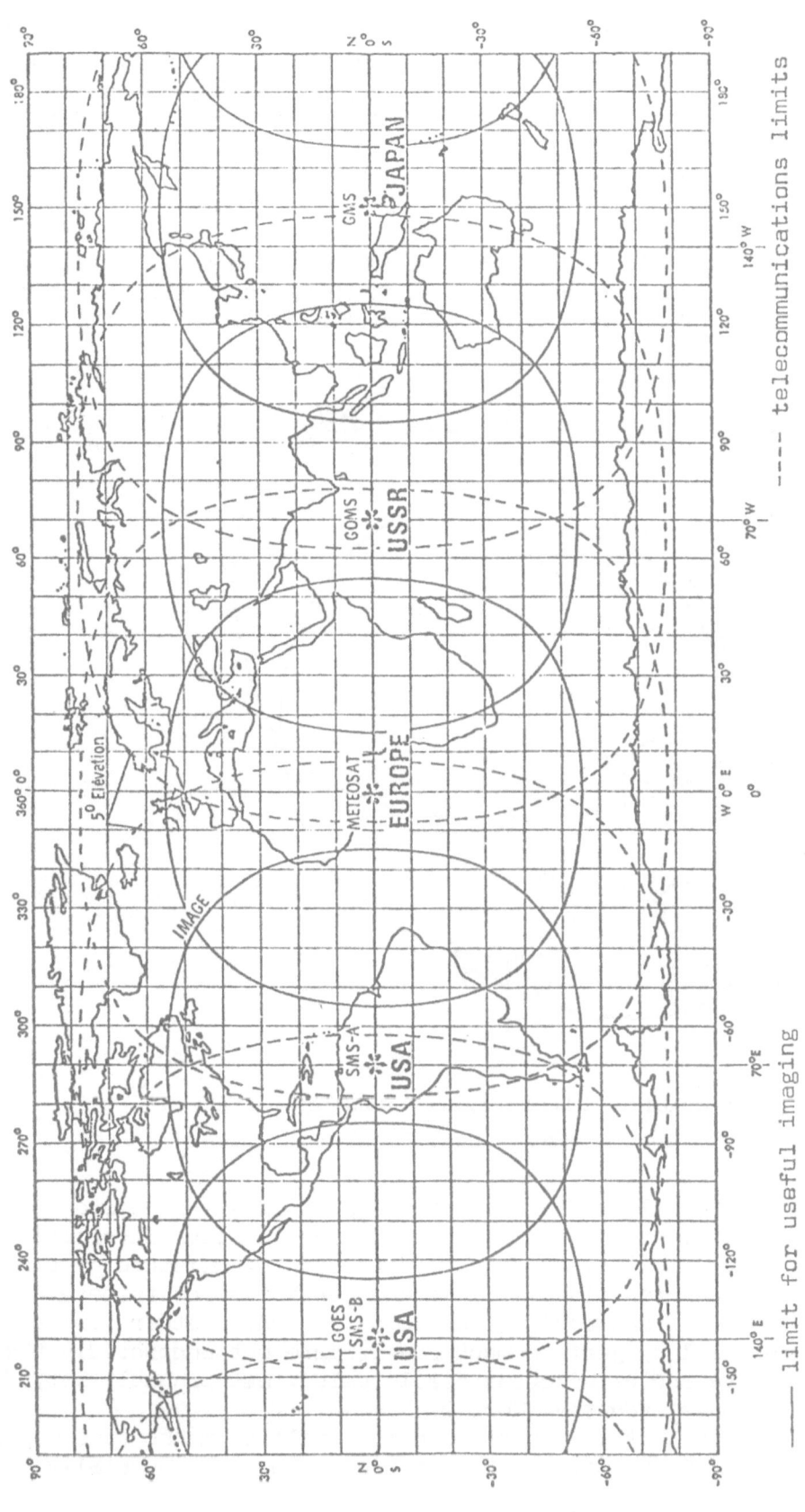

Area of Coverage by Meteorological Geostationary Satellites

(see page 10)

Fig. 4

example, and the polar orbiter type. The two types are complementary. The
geostationary satellites, five in number, remain in their allotted places,
spaced around the equator at a height of 36000 km (Fig. 4). With their un-
blinking stare at that region of the earth for which each is responsible,
they can transmit every 30 minutes a picture, in the visible (except at
night) light band and in the infra red light band, of the full earth disc;
the useful field of view is somewhat less than this.

The polar orbiters, by comparison, hurry around the globe via the poles
at a height of 850 km. Each circumnavigation takes two hours. The earth
spins beneath their flight path and they pass over the same earth location
twice every 24 hours (Fig. 5). They offer unique opportunities for closely
monitoring the earth environment especially when there is co-ordination of
effort by the designers of the observing systems (Ref. 2).

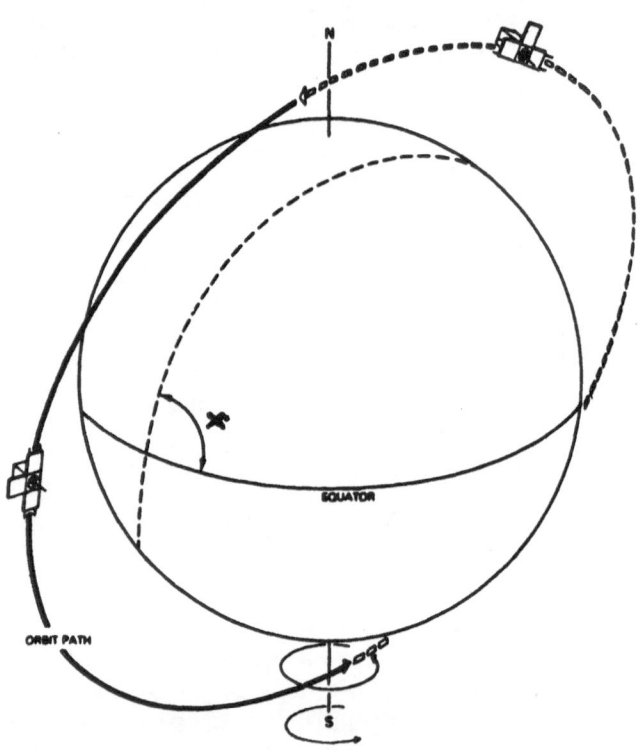

Polar orbiter "flight path"

Fig. 5

The geostationary satellites (GOES) are the platforms for remote sensing
and picture imagery equipment intended to provide quantitative information
based on remote vertical soundings of the temperature and moisture of the
atmosphere, winds derived from cloud motions and earth surface temperature.
This material has little direct application to day-to-day weather forecasting
as yet; the data are still subject to much guesstimate but undoubtedly the
information will improve and be of quantifiable value, but probably not, for
some time, to the offshore industry (Ref. 3).

A major contribution that the geostationary satellite makes to synoptic
analysis and forecasting is by way of its pictures. This is especially so
with METEOSAT, which covers Europe.

Limited Application in the Arabian Gulf

Claims made for the economic importance of American GOES include reference
to the use of data in crop-kill frost warnings to farmers in Florida. More
credible is the value attached to the location and avoidance of icebergs and
the provision of clear information about major weather features such as the
nature and direction of travel of hurricanes. Almost all of the surface
weather features in which the offshore industry weather forecaster is keenly
concerned are outside the scope of the satellite. Sea states are not revealed, nor surface winds, both items of the highest importance. Cloud
information, of some importance in European latitudes has much lesser importance in most places elsewhere outside cyclone or hurricane regions.
Within the Arabian Gulf satellite pictures have been received from the Indian
Ocean geostationary satellite, GOES (IO), an American spacecraft at temporarily at about 57°E in place of the Russian GOMS eventually to be stationed
at 70°E, above the equator. These satellites make little contribution to
synoptic analysis within the Gulf and surrounding region insofar as forecasting for offshore oil operations is concerned. Outside the Gulf the
pictures are of greater value; the presence of an Indian Ocean cyclone may
be revealed and its movement westwards and perhaps to the southeast coast
of the Arabian peninsula is of concern, both in respect of its immediate
effects locally and its influence upon water levels in the Arabian Gulf.
The value of satellite imagery in associated disciplines such as hydrology,
where snow cover may be of concern, and to oceanography in the Region is
undisputed even if the application of received data is not yet obvious. The
geostationary system may be used in remote surface-located automatic weather
station networks for transmitting data through the satellite to central
ground stations.

Wave Height Observations

It is disappointing that shipping and the offshore oil industry are less than
well served in respect of satellite data. The ill-fated recent SEASAT polar
orbiter experiment in which wave height measuring equipment was to be evaluated is unlikely to be repeated for some time. Synthetic Aperture Radar
(SAR) which it was hoped might be developed with the experiment for wave
measurement may be incorporated in COMSS (Coastal and Ocean Monitoring Satellite System) programme, the foreseen launching dates of which are mid- or
late 1980's. NASA plans preclude the flight of any SAR system except for
experimental missions on shuttle/space laboratory craft. The National Oceanographic Satellite System, NOSS, now being planned, will have microwave
sensors as on SEASAT except for the SAR. No alternative remote sensing
system to SAR seems to be available for development and ultimate routine use
for wave height measurements. A greater contribution to marine meteorology
than the facility to plot daily synoptic charts of wave heights over the
world's seas and ocean is difficult to imagine. The wave height on-site is
a vital but often ill-defined and suspect reference point for the offshore
forecaster, and for the ship-routeing meteorologist it is also of major
concern. For marine engineers and architects, planners and designers the
amassing of valid wave data for the world's oceans for statistical evaluation
would be invaluable.

The meteorologist working in applied forecasting may have an uneasy
feeling that an unwary public are being over exposed to satellite pictures.
Their use on TV sometimes seems more of a gimmick in the courting of public
support for the weather forecast product, rather than as an aid to understanding. The least reliable of all the meteorologist's output, the public

display and broadcast forecast, may well be to blame for the low level of
esteem in which weather forecasting is held in parts of Europe and to put
further products of limited value in the shop window seems a policy of doubtful wisdom. A Director with a multi-million pound budget to justify may feel
that the risk of the display being counter-productive is worth taking. It
may, of course, be held that the viewer will blame the satellite's imagery
rather than the meteorologist whenever the forecast again seems irrelevant
to him.

RESEARCH

The development of techniques for forecasting wave heights and other parameters from predicted wind fields has been the major aim of Imcos Marine
meteorologists in the Gulf. Before 1958 the weather of the region was understood in general and publications of the UK Meteorological Office covered
the ground fairly well except in two major particulars; the causes of some
of the major phenomena were perforce not dealt with and local weather variability, of particular importance for offshore operations being conducted
at the upper limits of benignity, were little appreciated. Experience within the Weather Scheme has resolved many of the problems of the meteorology
of the region, and weather hazards earlier considered inexplicable and, by
mariners, unforecastable have submitted to the disciplines of conventional
meteorology made possible by the observation network. The possibility of
longer range forecasting has been investigated. Gulf water levels, tides
and currents have been found to be influenced by meteorological conditions
to a degree previously unsuspected but now considered critical in the light
of the coastal marine development envisaged in the Region.

Local Winds

Winds are, of course, the major preoccupation of offshore meteorologists in
the Gulf. The network of continuously recording anemometers has permitted
the documentation and examination of the structure of local winds, such as
the Shamal, and their prediction. The Nashi, for example, is a NE wind often
strong and sudden in onset and relatively common near the coast of Iran in
winter, especially near the entrance of the Gulf along the Makran Coast. It
is now thought to be a katabatic wind, associated with an outflow from the
central Asiatic anticyclone extension over Iran, and similar in character to
the Bora of the Adriatic Sea and similarly a hazard to offshore operations.
Speeds of 40 knots and wave heights of 10-12 ft have been reported.
    Until recent years it was not expected that the Nashi extended very far
from the coast of Iran, and it certainly does not do so often enough to
become manifest in the available offshore wind statistics, but evidence has
accumulated which shows that occasionally the Nashi may influence much of
the southern part of the Gulf.

Waves

Prediction of wave height from the forecast wind field is continuously developing and the forecasting of sea, that is the waves generated from the
winds within the subject area, does not now in the Gulf present difficulties
provided that the wind field is adequately predicted. Problems still exist
which preserve the forecaster from over confidence. In the northeast of the
Gulf the apparent "jetting" of the wind in the lower levels during stronger
shamals gives rise to narrow belts of much stronger winds than are to be
expected from the area wind field. The prediction of the maximum wave height

that may be encountered in these circumstances is one such problem; swell, waves from outside the wind generating area, occasionally experienced in situations not fully understood because of the paucity of other surface observations, is another. Over twenty years of observation is leading to a greater knowledge of the wave climate of the Gulf. An example of how this is being pursued with the Weather Scheme is shown in the map of Relative Wave Climate Variation, Southern Gulf (Fig. 6) on page 14.

Longer Range Forecasting

In response to requests from Weather Scheme members, the possibility of longer range forecasts for the Gulf being available has been investigated. Longer range in this instance defines a forecast for any period ahead longer than might be covered by conventional synoptic meteorology. To be acceptable, any longer range forecasting techniques must express the prediction in terms more precise than those derived from climatological reference; it is not sufficient to predict that the "40-day" shamal will occur in summer or that "the winter will be cooler with rain"!

     No technique has so far been developed for the Gulf, or elsewhere in the world for that matter, which will, in terms of offshore operations, consistently give forecasts better than those obtained by synoptic meteorology methods, which are primarily extrapolative. For the Gulf region evidence of cyclical variability, analogous to known sun spot variability, was sought and none found and no correlation of the sun spot cycle with weather records was apparent for the score or so years for which reliable and comprehensive Gulf weather data are available. Analogue/anomaly techniques based on surface pressure and/or temperature were ruled out due to lack of historical data. No trend, periodicity or changes in variability or any other function of the data have so far been revealed but 20 years is an inadequate sampling period. A search for singularities, in this instance the occurrences of surface winds in given speed ranges at Kharg Island, highlighted some event-dates in the ten years examined but these may do no more than confirm the experienced forecaster's own views as to seasonal weather changes he may expect. Ten years is an inadequate sampling period and the singularity technique may be worth further examination. Statements of singularity coupled with probabilities based on statistical evaluation of data may have some value for general longer term planning.

     What is obvious is that it is the dubious practice, followed by some meteorological organisations, of publicly issuing longer range forecasts that has led some operators to assume that they should be generally available. It is perhaps significant that in the U.K. monthly forecasts became available from the U.K. Meteorological Office at a time when attention had been drawn in the press to such forecasts for NW Europe being issued by American meteorologists for the American forces, and that these were becoming available to the general public. That the UK Meteorological Office was pushed down this road without any credible method of travelling along it is arguable; the state of the art today is to be seen in the published forecasts. For practical, operational purposes within the very demanding limits of offshore operations it is doubtful if long range forecasts will ever be applicable.

Oceanographic Phenomena of Meteorological Origin

The diversity of oceanographic phenomena of primarily meteorological origin may be indicated by reference to three events more recently examined and summarised below. These attracted attention because each was considered to be of considerable importance to offshore oil operations and coastal engineering

648

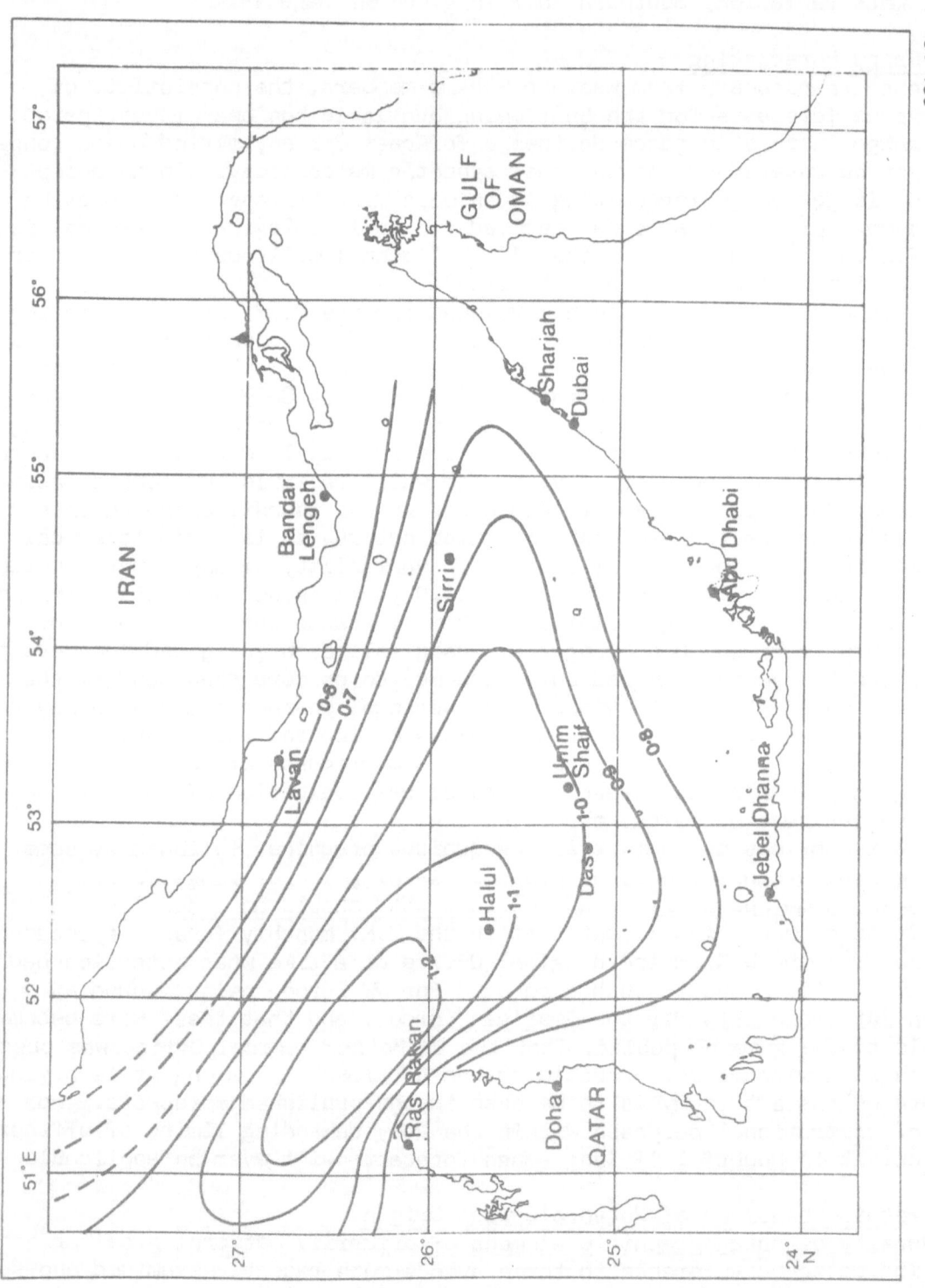

RELATIVE WAVE CLIMATE VARIATION
SOUTHERN GULF

Fig. 6

activities of the Gulf. Each phenomenon is a manifestation of meteorological influences the onset and duration of which can be predicted. It should be possible to issue warnings before and during the relatively short periods, from a few hours to a few days, within which the phenomena are observable. The occurrences outlined below are:

    Tidal Level Anomalies in the Gulf

    Small Boat Harbour Rapid Water Level Oscillations

    Cyclone Water Build-up

References to some special investigations follow.

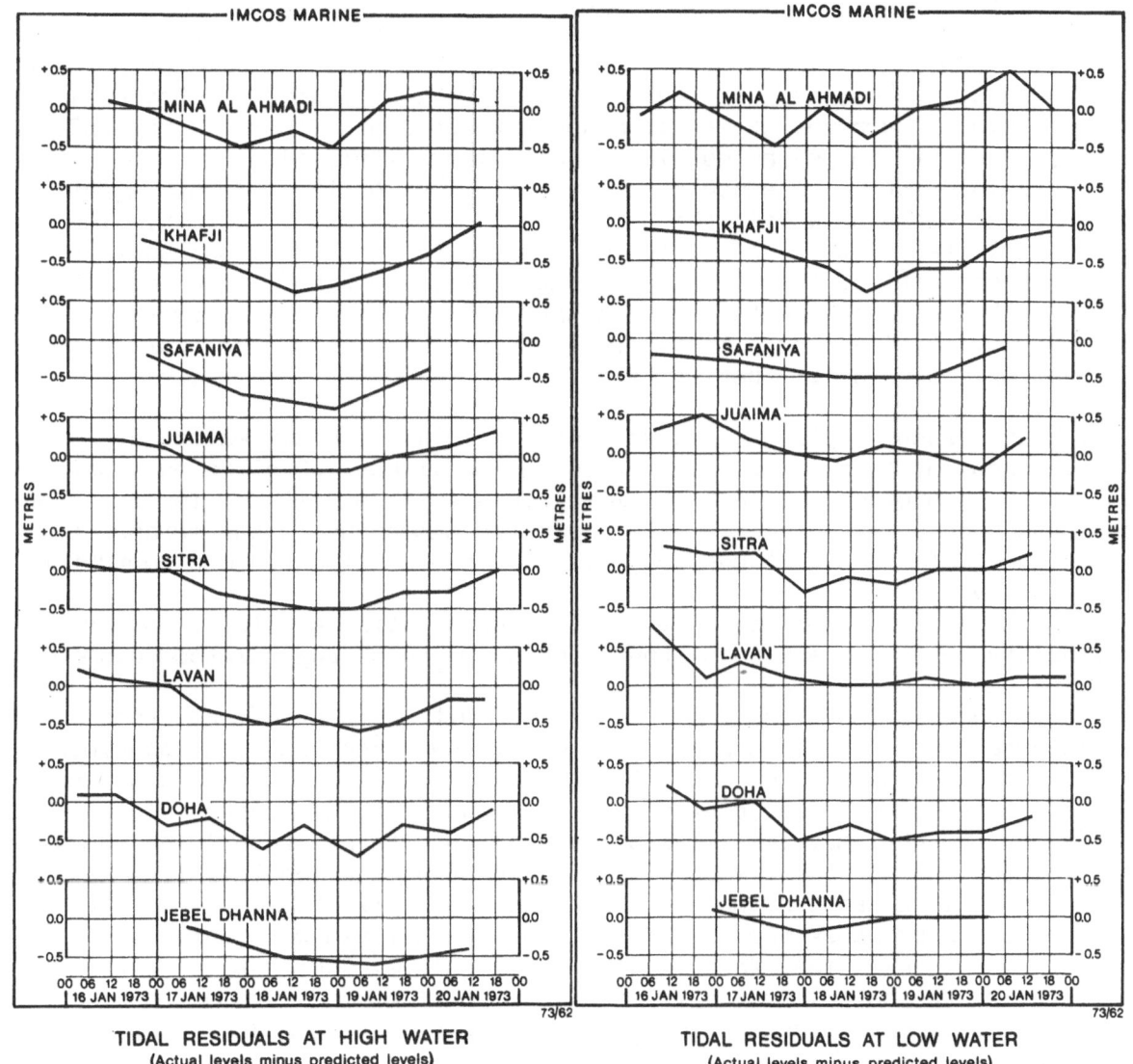

TIDAL RESIDUALS AT HIGH WATER
(Actual levels minus predicted levels)

TIDAL RESIDUALS AT LOW WATER
(Actual levels minus predicted levels)

Tidal Level Anomalies

Fig. 7

<u>Tidal Level Anomalies in the Gulf.</u> The anomalies referred to here are not those frequently observed in the Gulf which are caused by winds persisting for several days from one direction, either the NW(Shamal) or SE (Kaus or Sharki), and piling water up at one end in a manner malefic to the level at the other. Engineering works on the UAE coast may be hazarded by the Shamal

water build-up, usually not more than 1 metre, and low-lying Kuzistan coastal areas around Abadan may be affected by a persistent Kaus, but these anomalies are documented and understood. The anomalies here considered were recorded over six days and were of the same negative sign throughout the Gulf, as would not be the case with storm water build-up above-referred.

Over 17 through 19 January 1973, negative tidal anomalies in the 0.5 to 1.0 metre range were widespread in the Gulf from Bandar Mashahr in the North to Jebel Dhanna in the south, and deficiencies below predicted levels persisted for several days (Fig. 7). It is speculated that this was caused by an episode of persistent abnormally high atmospheric pressure in the Gulf. There is some evidence of a similar phenomenon in the Mediterranean.

It is known that over June through September the southwest monsoon of the Indian Ocean may maintain water levels in the Gulf up to a half a metre higher than during the rest of the year, but persistent and widespread negative anomalies do not appear previously to have been observed. For design and operational purposes it may be prudent to assume offshore in the Gulf extreme "negative surge" of not less than 1 metre (Ref. 4).

It is interesting to speculate what the effect on the currents might be. The water can only have left the Gulf through the Strait of Hormuz. The superficial area of the Gulf is around 240,000 km^2 and the cross-section of the narrowest part of the Strait of Hormuz is around 5 km^2. The loss through Hormuz of 0.5 metres of water in 24 hours implies therefore an average net outflow current of around $\frac{1}{2}$ knot. This is feasible, though no current data are available to confirm that it occurred.

Small Boat Harbour Rapid Water Level Oscillations. Unusual water level oscillations, of 4 to 10 minutes period and amplitude 4.0 ft in one instance and 1.5 ft in another and lasting, with diminishing amplitude, for about an hour have been observed on two well separated occasions in the harbour at Das Island (Ref. 5). That considerable damage did not ensue was fortuitous (Figs. 8 & 9).

The oscillations may have been of tsunami or seismic origin, and this possibility cannot be excluded, though no evidence of any related seismic activity has been obtained.

The meteorological conditions associated with both events have been investigated and on both occasions the water level oscillations coincided with the passage of a front and this strongly suggests that the phenomena of the two events were associated. It is known that the passage of these fronts (or troughs) may sometimes be associated with pressure jumps and wind oscillations and the Das anemograms for these two occasions clearly show wind oscillations associated with the passage of a front.

It appears probable that the fronts were associated with unrecorded pressure jumps sufficient to trigger off water level oscillations within Das Harbour, the calculated oscillation period of about 6 seconds corresponding broadly with pressure jump periods of around 5 seconds observed elsewhere.

The passage of fronts and troughs in the Gulf is quite frequent and wind oscillations accompanying them are relatively frequent and probably are often accompanied by pressure jumps which may trigger off water level oscillations in enclosed waters which have similar critical resonance values. Increasing coastal development, in the Gulf, suggests that more detailed study of this previously overlooked phenomenon is desirable if harbours and marinas are not to be constructed of potentially critical resonance characteristics.

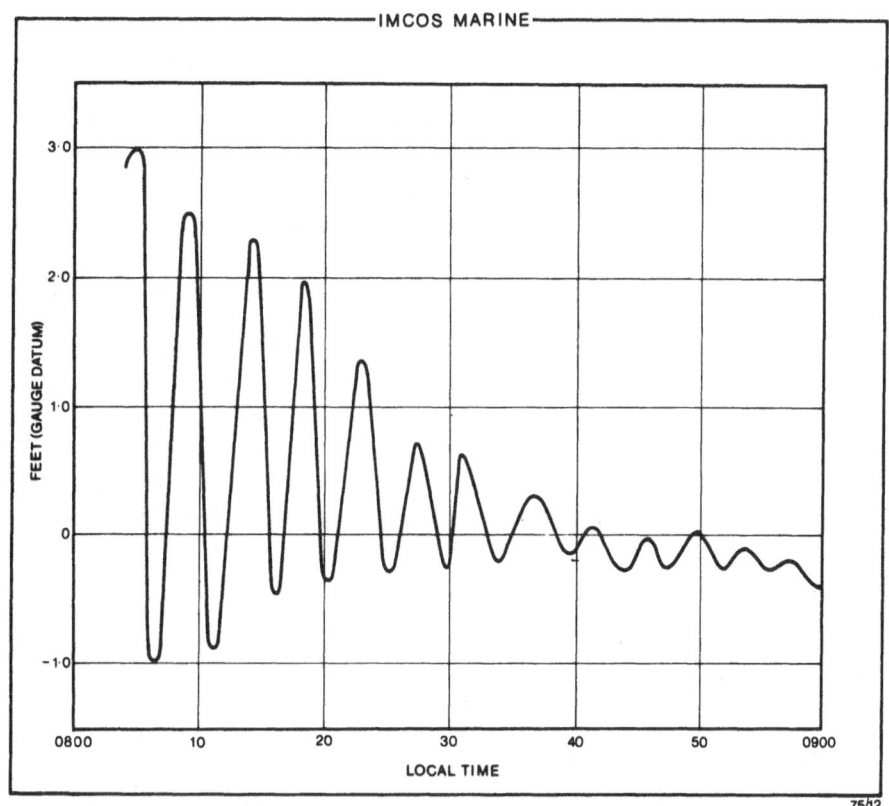

WATER LEVELS: DAS HARBOUR

18 APRIL 1959

Fig. 8

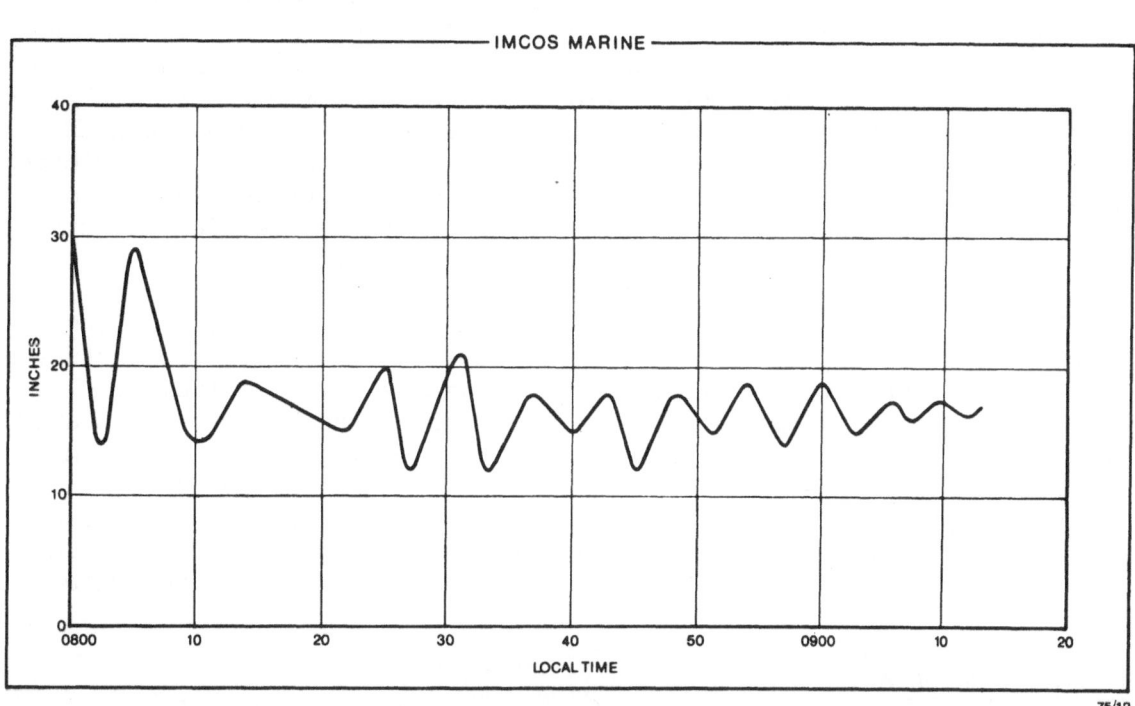

WATER LEVELS: DAS HARBOUR

13 MAY 1961

Fig. 9

Cyclone Water Build-up.   Towards mid-June 1977 a tropical cyclone in the Indian Ocean hit Masirah Island with winds probably well in excess of 100 knots.   The course of this cyclone had been followed for several days by Weather Scheme meteorologists in the Gulf; some concern was felt about the possibility of swell waves in the Strait of Hormuz and possibly further into the Gulf.   No untoward sea states were observed or recorded.

In October 1977 at a meeting of hydrographers in the U.K., mention was made casually of a sudden rise in water level observed in the Qatar offshore area at about the time of the Masirah cyclone.   This water level rise had been recorded by a hydrographic survey team who thought little more about it at the time.   There being no local meteorological phenomena that could explain the event Imcos Marine circularised Weather Scheme members asking for tide records.   That the cyclone was responsible for a significant increase in water level in the Gulf, as far up as Qatar, was evident.   This water level increase was becoming noticeable on 11 June and peaked at over half a metre on 15 June.

Special Investigations
A number of special investigations into matters not necessarily meteorological have been conducted or are still being carried out.

Earth Tremors.   The possibility of significant earth tremors in the region which could seriously affect offshore installations was examined after the report of such a tremor offshore Qatar in 1977.   They are frequently experienced around the eastern coast of the Gulf, for example at Kharg.

Tidal Stream Prediction.   In 1976 an investigation into the possibility of providing tidal stream predictions, based upon tide level predictions, and having application to terminal working and offshore exploration and production operations, was undertaken.   This involved the acquisition, over a period of several weeks, of real-time tidal stream data concurrently with the tide level data for the construction of a computer model.   In 1977, and in collaboration with the U.K. Institute of Oceanographic Sciences at Bidston, the Co-ordinators, Imcos Marine were able to produce tidal stream predictions adequately accurate for operational purposes.   Hourly current speed and direction values can be provided for as far ahead as desired for any location, and the procedure is applicable to any part of the Gulf.

Design Extreme Wind Speeds.   From an increasing number of Gulf stations the years of wind recording are becoming sufficient for the statistical study of extreme speeds with a view to establishing reliable design extremes.   The latter are now becoming available from an increasing number of places and the preparation of Gulf design extreme wind maps will be possible in the not too distant future.   Also under study are quantified directional design extremes, a type of information increasingly required for design purposes in offshore areas.   Information about wind extremes will be published as supplements to the "Handbook of the Weather in the Gulf, Surface Wind Data".

PUBLICATION OF DATA

"Handbook of the Weather in the Gulf"
At the first meeting of the Weather Scheme in 1958 the members determined that the meteorological and allied oceanographic data acquired over the years

ahead would eventually be published. In 1969 preliminary plans were made
to publish wind and, later, climatological data as these became available
for a minimum 15-year period. A series of volumes within the title "Handbook of the Weather in the Gulf" was envisaged. In 1974 the first of these,
"Surface Wind Data" (Ref. 6) and "Bibliography" (Ref. 7) were published.
These were followed in 1976 by "General Climate, Data" (Ref. 8). A volume,
"Port Meteorological and Hydrographic Data", is planned and a volume on the
Gulf general oceanography, drawing upon the 30 or more years of material in
the archives of oil companies and service companies, several now no longer
operating in the Gulf, will be started when economic constraints can be
overcome. Amendment pages to the Handbook volumes will be published as data
qualifying for inclusion, usually by virtue of length of unbroken recording
period, become available.

    The publication of the data has as its major objective the provision of
definitive information to planners, design engineers and operators as well
as other meteorologists, and the presentation of data has departed in some
ways from that conventionally observed in State meteorological office publications. The form of presentation used in the Handbook series was arrived
at through detailed discussion with the marine management, operators and
engineers within the OCWCS membership and many of the features are novel.
The presentation of wind information by duration (Figs. 10 & 11) has not
previously been used in publications of this nature. The percentage exceedance of wind speed by months (Fig. 12) is similarly unprecedented.

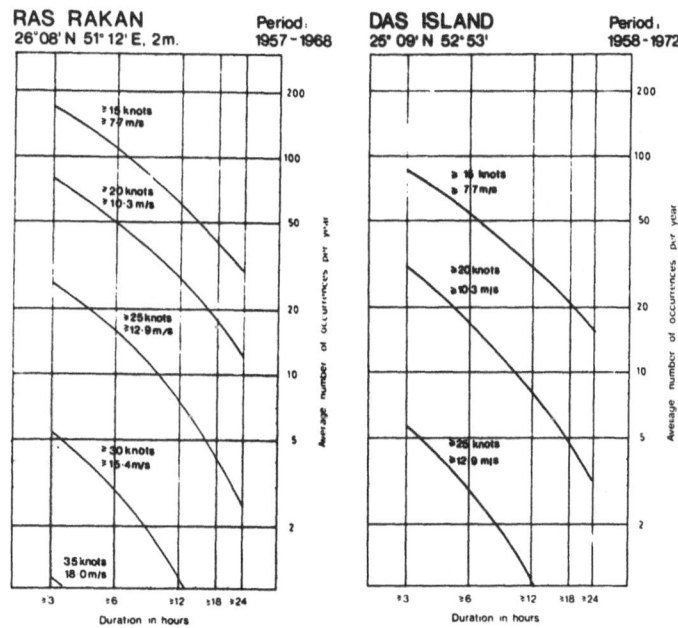

Surface Wind Speed Durations

Fig. 10

HANDBOOK OF THE WEATHER IN THE GULF

Durations of Surface Wind Speed

Annual Average Number of Occasions by Months

RAS TANURA, Refinery Laboratory: 26°42'N 50°05'E, 2 m

Speed knots m/s	JAN	FEB	MAR	APR	MAY	JUN	JUL	AUG	SEP	OCT	NOV	DEC	YEAR
					Duration > 3 hours								
>15 > 7.7	8.9	9.3	9.7	10.1	9.3	10.1	12.4	8.1	5.6	4.1	4.3	9.7	101.8
>20 >10.3	2.1	1.2	2.1	2.3	2.1	0.6	0.8	0.8	0.6	1.0	0.2	0.6	14.5
>25 >12.9			0.4	0.6				0.2					1.2
>30 >15.4				0.2									0.2
>35 >18.0													
>40 >20.6													
					Duration > 6 hours								
>15 > 7.7	6.0	5.4	4.1	4.3	4.1	4.8	7.7	4.8	2.5	2.7	2.5	5.2	54.0
>20 >10.3	0.6		0.6	1.0	0.6	0.4	0.2	0.6		0.4		0.4	5.0
>25 >12.9			0.2										0.2
>30 >15.4													
>35 >18.0													
>40 >20.6													
					Duration > 9 hours								
>15 > 7.7	3.7	3.1	2.7	2.5	2.7	1.9	2.1	2.3	1.0	1.7	1.0	2.9	27.5
>20 >10.3	0.4		0.4	0.2	0.2	0.2	0.2	0.4				0.2	2.3
>25 >12.9			0.2										0.2
>30 >15.4													
>35 >18.0													
>40 >20.6													
					Duration > 12 hours								
>15 > 7.7	1.9	1.0	1.2	1.4	1.9	1.2	0.2	1.0	0.8	1.0	0.4	1.2	13.4
>20 >10.3	0.2		0.4	0.2				0.2				0.2	1.2
>25 >12.9			0.2										0.2
>30 >15.4													
>35 >18.0													
					Duration > 18 hours								
>15 > 7.7	0.8	0.6	0.6	0.8	1.0	0.6	0.2	0.8	0.2	0.6		0.4	6.8
>20 >10.3	0.2		0.4	0.2									0.8
>25 >12.9			0.2										0.2
>30 >15.4													
>35 >18.0													
					Duration > 24 hours								
>15 > 7.7	0.4	0.2	0.6	0.2	0.6	0.4			0.2	0.2			2.9
>20 >10.3			0.4										0.4
>25 >12.9			0.2										0.2
>30 >15.4													
>35 >18.0													

Time of Obs.: 24-hour recording

Period: 1967 - 1972

Authority: Arabian American Oil Company; autographic records, analysed by IMCOS MARINE LTD., London; manuscript data

Fig. 11

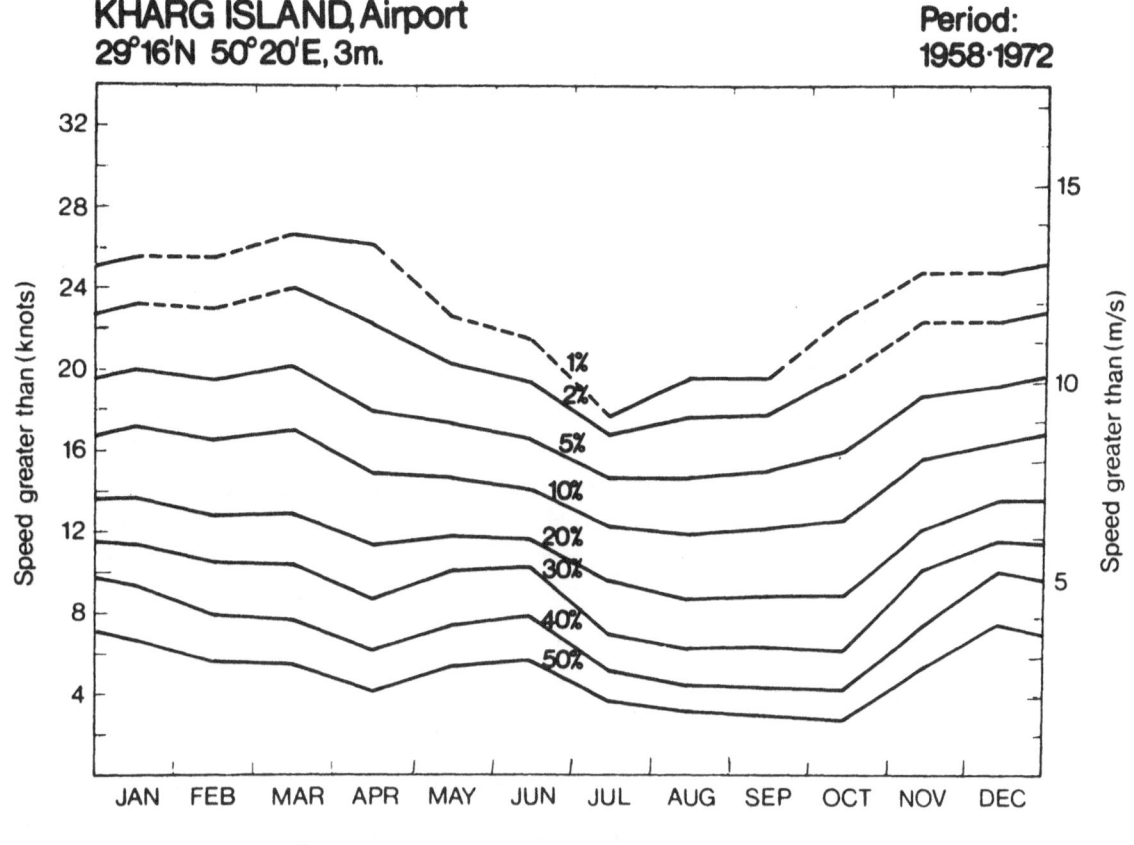

Surface wind speed average exceedance

Fig. 12

        Conventional presentation was rejected for the volume of the Handbook
giving climatological data in favour of presenting, for all stations on one
page, the means, minima and maxima for selected elements (Fig. 13). For
each station individually the full data are also tabulated over two pages
(Fig. 14). The calculation and presentation of extremes for given return
periods (Fig. 15 & 16) is a valued innovation, as is the graphic presentation
of diurnal and seasonal variation of mean air temperature and relative humidity for selected stations (Fig. 17).

Technical Notes
From time to time there occur meteorological and allied oceanographic events
which, because of their novelty or magnitude have operational significance
and are deemed to require further examination. Details of these events are
given in Technical Notes published and available in the public domain as are
volumes of the Handbook.

Information Bulletins
Information Bulletins are issued to Weather Scheme members whenever there is
some occurrence of operational interest but of less general appeal than
merits the publication of a Technical Note, and these Bulletins can be made
available to non-members on application (Ref. 9).

HIGHEST MAXIMUM AIR TEMPERATURE RECORDED

STATION	JAN	FEB	MAR	APR	MAY	JUN	JUL	AUG	SEP	OCT	NOV	DEC	YEAR	NO. OF YRS.
ABADAN (CO.A)	26.8	34.4	35.8	43.2	48.1	49.6	50.0	49.1	47.5	44.0	36.5	27.8	50.0	12
ABADAN (REF.)	27.2	31.1	39.4	42.5	48.9	50.6	50.0	50.8	48.3	44.4	36.9	32.2	50.8	31
ABQAIQ	37.2	35.6	41.1	46.1	46.7	50.0	51.7	51.1	48.9	45.6	38.9	31.7	51.7	22
ABU DHABI	31.4	34.4	40.0	44.5	45.0	47.2	47.4	46.1	43.8	42.0	35.0	29.6	47.4	9
AGHA JARI	30.1	31.9	35.8	41.7	48.4	50.7	53.3	50.6	48.3	42.9	36.2	28.0	53.3	14
AHWAZ	28.3	28.3	35.0	40.6	47.3	49.6	53.5	49.2	48.4	42.8	36.2	27.3	53.5	11
AZAIBA	30.6	33.0	41.0	43.9	46.0	48.0	47.2	45.0	42.5	42.0	41.1	34.0	48.0	6
BAHRAIN MUHARRAQ	32.0	34.7	38.0	42.0	46.7	45.0	44.0	45.0	42.0	41.0	35.0	29.0	46.7	29
BANDAR ABBAS	32.0	29.6	35.0	39.2	44.0	46.5	45.0	45.0	43.0	42.0	36.0	32.0	46.5	15
BANDAR DAYYER	34.5	31.0	34.5	41.0	48.5	50.0	49.0	48.0	47.5	45.0	39.0	36.0	50.0	10
BANDAR LENGEH	27.0	30.4	35.0	40.0	49.0	48.6	46.0	41.8	40.6	41.6	34.0	32.0	49.0	6
BANDAR MAHSHAHR	27.0	29.5	34.7	39.5	46.7	48.8	52.0	49.5	47.5	42.8	36.1	29.4	52.0	14
BU HASA	30.5	36.7	40.2	45.4	46.9	49.0	48.1	48.0	45.6	42.0	35.5	30.0	49.0	3
BUSHEHR	30.0	32.0	35.0	41.0	46.0	46.0	50.0	47.0	46.0	41.0	34.0	30.0	50.0	25
DAS ISLAND	28.7	27.9	31.0	36.7	39.4	39.8	41.7	42.8	43.3	40.0	38.9	30.9	43.3	19
DHAHRAN	29.4	35.6	37.8	44.4	45.6	47.8	47.2	48.3	45.6	41.7	36.7	29.4	48.3	14
DOHA RAS BABUT	29.0	36.0	42.8	45.5	47.0	47.2	48.3	48.3	45.6	43.0	37.2	31.0	48.3	23
DUKHAN	31.7	34.2	40.8	45.3	47.2	48.1	48.9	49.2	48.1	44.2	38.3	30.8	49.2	13
FAHUD	32.0	35.0	41.0	45.0	51.0	52.0	51.0	50.0	48.0	46.0	39.0	34.0	52.0	8
GACH SARAN	26.1	26.6	32.6	37.0	43.2	45.3	48.1	45.4	43.4	38.9	31.6	26.0	48.1	14
HALUL ISLAND	29.0	31.0	32.0	37.0	38.0	40.0	41.0	41.0	39.0	37.0	34.0	29.0	41.0	7
JASK	28.9	32.2	33.3	36.9	43.0	42.0	41.7	42.0	40.0	40.0	33.1	30.9	43.0	21
KHARG ISLAND	27.9	26.8	32.4	37.8	42.5	45.2	47.3	44.1	44.0	38.2	33.0	27.4	47.3	16
KUWAIT	29.8	35.8	41.2	44.2	49.0	49.8	49.2	49.0	46.7	43.2	36.0	30.5	49.8	17
LAVAN ISLAND	28.2	26.8	35.1	40.0	45.7	43.3	44.1	42.7	38.8	39.6	33.3	29.5	44.1	5
MASJED SOLEYMAN	27.8	28.1	33.6	40.3	45.8	48.5	51.6	48.9	46.1	41.7	34.5	27.2	51.6	17
MINA AL AHMADI	27.0	32.8	33.5	40.0	48.3	48.3	50.0	48.9	46.7	42.8	36.7	26.7	50.0	25
QISHM ISLAND	29.0	28.0	33.0	37.0	40.8	45.6	44.0	43.0	41.6	37.4	34.5	29.0	45.6	5
RAS AL KHAFJI	27.0	36.9	36.7	42.8	43.1	46.6	46.4	46.3	47.0	41.5	35.9	28.8	47.0	13
RAS RAKAN	29.4	33.9	41.7	41.7	44.4	46.7	46.1	41.7	42.0	39.5	36.7	30.6	46.7	7
RAS TANURA	26.7	35.0	37.8	41.1	45.0	45.6	45.0	43.9	44.4	40.0	35.0	28.9	45.6	14
SHARJAH	32.8	36.6	40.4	41.4	44.4	45.0	48.3	47.2	44.6	39.9	36.1	32.2	48.3	21
TARIF	34.3	37.5	40.1	44.4	45.5	47.0	48.0	47.2	44.0	43.0	35.5	32.2	48.0	14

Fig. 13

HANDBOOK OF THE WEATHER IN THE GULF

TARIF

24°02'N 53°45'E, 16m

	JAN	FEB	MAR	APR	MAY	JUN	JUL	AUG	SEP	OCT	NOV	DEC	YEAR
AIR TEMPERATURE													
Mean Daily													
Maximum (°C)	22.4	23.9	28.0	32.0	35.6	36.7	38.8	39.1	37.0	33.6	28.5	23.8	31.6
Minimum (°C)	12.7	13.5	16.5	19.2	22.6	24.3	24.9	27.9	25.5	21.8	17.4	13.5	20.0
Mean Monthly													
Maximum (°C)	29.1	30.0	38.2	40.9	42.7	42.6	44.0	44.0	41.2	37.8	33.2	29.2	
Minimum (°C)	8.3	8.7	11.0	13.8	17.3	20.9	23.3	24.8	21.6	18.3	14.1	9.4	
Highest Max. (°C)	34.3	37.5	40.1	44.4	45.5	47.0	48.0	47.2	44.0	43.0	35.5	32.2	48.0
Lowest Min. (°C)	6.5	6.1	8.3	10.0	15.0	18.1	21.0	22.0	19.2	14.4	12.5	7.0	6.1
HUMIDITY													
03 GMT [1]													
Rel. Humid. (%)	75.0	73.7	67.2	65.0	60.1	59.3	52.5	62.0	63.1	63.9	65.7	70.3	64.8
Wet Bulb (°C)	15.9	18.0	19.3	22.2	23.4	26.4	27.1	28.5	27.0	25.1	21.0	17.3	22.6
12 GMT													
Rel. Humid. (%)					No Data								
Wet Bulb (°C)					No Data								
PRECIPITATION													
Mean Monthly (mm)	13.6	5.6	4.4	4.4	4.1	0	0	0.7	0	0.9	3.2	2.8	39.7
Mean No. of Days													
≥1 mm	1.4	0.9	0.7	1.1	0.5	0	0	0.1	0	0.1	0.2	0.5	5.5
≥10 mm	0.4	0.1	0.1	0.1	0.1	0	0	0	0	0.1	0.1	0.1	1.1
Max. Fall in 24 hours (mm)	24.4	10.9	11.0	15.4	20.3	Tr	0.6	8.0	Tr	12.2	26.3	15.2	26.3
GROUND MINIMUM TEMPERATURE													
Mean Daily (°C)						No Data							
EARTH TEMPERATURE													
Mean at 120 cm (°C)						No Data							
ATMOSPHERIC PRESSURE AT M.S.L.													
06 GMT [2]													
Mean (mb)	1020.0	1018.1	1015.2	1012.4	1008.3	1001.8	998.0	1000.3	1006.3	1013.3	1017.9	1019.8	1010.9
Highest (mb)	1028.8	1025.5	1021.2	1019.6	1015.7	1011.7	1002.9	1007.2	1013.2	1019.7	1022.8	1026.2	1028.8
12 GMT													
Mean (mb)						No Data							
Lowest (mb) [3]	1007.8	1009.3	1006.5	1003.7	1000.0	994.8	992.5	994.5	1000.1	1005.5	1013.6	1014.1	992.5
	JAN	FEB	MAR	APR	MAY	JUN	JUL	AUG	SEP	OCT	NOV	DEC	YEAR

Fig. 14

HANDBOOK OF THE WEATHER IN THE GULF

ESTIMATED EXTREME MAXIMUM TEMPERATURES

VALUES FOR 10, 25, 50, 100 YEAR RETURN PERIODS

STATION	OBSERVED EXTREMES	ESTIMATED EXTREMES FOR RETURN PERIODS				n
		10 YEARS	25 YEARS	50 YEARS	100 YEARS	
ABADAN Company Airport	50.0	50.4	51.4	52.1	52.9	12
ABADAN, Refinery	50.8	50.4	51.3	51.9	52.6	30
ABQAIQ	51.7	50.8	52.1	53.0	53.9	23
ABU DHABI	47.4	47.4	48.1	48.7	49.2	7
AGHA JARI	53.3	52.3	53.3	54.0	54.7	14
AHWAZ	53.5	52.6	54.0	55.1	56.2	11
BANDAR ABBAS	46.5	45.6	46.7	47.4	48.2	16
BANDAR MAHSHAHR	52.0	50.8	52.2	53.1	54.1	13
BUSHEHR, Airport	50.0	48.4	50.8	52.5	54.2	17
DAS ISLAND	43.3	41.2	42.0	42.6	43.2	19
DHAHRAN	48.3	48.6	50.0	51.0	52.1	15
DOHA, Ras Babut	48.3	47.8	49.2	50.2	51.2	23
DUKHAN	49.2	49.1	50.0	50.7	51.4	12
GACH SARAN	48.1	47.7	48.6	49.2	49.9	14
KHARG ISLAND	47.3	47.0	49.0	50.4	51.9	16
KUWAIT International Airport	49.8	49.8	50.5	51.1	51.7	15
MASJED SOLEYMAN	51.6	50.6	51.5	52.2	52.9	17
MINA AL AHMADI South Pier	50.0	49.7	51.5	52.9	54.2	22
RAS AL KHAFJI	47.0	47.1	48.0	48.6	49.2	13
RAS RAKAN	46.7	47.4	49.6	51.2	52.8	7
RAS TANURA	45.6	45.6	46.8	47.7	48.6	14
SHARJAH	48.3	47.6	48.9	49.7	50.6	17
TARIF	48.0	48.0	49.1	49.9	50.7	12

n = number of individual yearly values used in computation

Fig. 15

HANDBOOK OF THE WEATHER IN THE GULF

ESTIMATED MAXIMUM PRECIPITATION IN 24 HOURS IN MILLIMETRES

VALUES FOR 10, 25, 50, 100 YEAR RETURN PERIODS

STATION	OBSERVED EXTREMES	ESTIMATED EXTREMES FOR RETURN PERIODS				n
		10 YEARS	25 YEARS	50 YEARS	100 YEARS	
ABADAN Company Airport	57.9	48.7	60.1	68.6	77.1	12
ABADAN, Refinery	68.8	58.0	70.6	80.0	89.2	31
ABQAIQ		No Data				
ABU DHABI		Insufficient Data				
AGHA JARI	78.4	63.5	77.5	87.9	98.1	13
AHWAZ	78.1	82.2	103.2	118.8	134.2	10
BANDAR ABBAS	103.8	80.1	103.2	120.4	137.4	14
BANDAR MAHSHAHR	89.7	65.5	83.2	96.4	109.4	13
BUSHEHR, Airport	155.0	111.0	144.4	169.2	193.8	14
DAS ISLAND		Insufficient Data				
DHAHRAN		No Data				
DOHA, Ras Babut		Insufficient Data				
DUKHAN		No Data				
GACH SARAN	76.9	64.2	75.6	84.1	92.6	14
KHARG ISLAND	196.5	138.2	180.9	212.7	244.1	14
KUWAIT International Airport	39.0	34.1	41.7	47.4	53.0	17
MASJED SOLEYMAN	79.4	79.7	96.7	109.3	121.8	14
MINA AL AHMADI South Pier		Insufficient Data				
RAS AL KHAFJI	42.4	42.2	52.4	60.0	67.6	13
RAS RAKAN		No Data				
RAS TANURA	61.2	49.3	64.1	75.2	86.1	11
SHARJAH		Insufficient Data				
TARIF	26.3	25.8	33.0	38.3	43.6	13

n = number of individual yearly values used in computation

Fig. 16

SHARJAH
25°21'N 55°23'E, 2m

Period: 1958-1970

Diurnal and Seasonal Variation of Mean Air Temperature °C

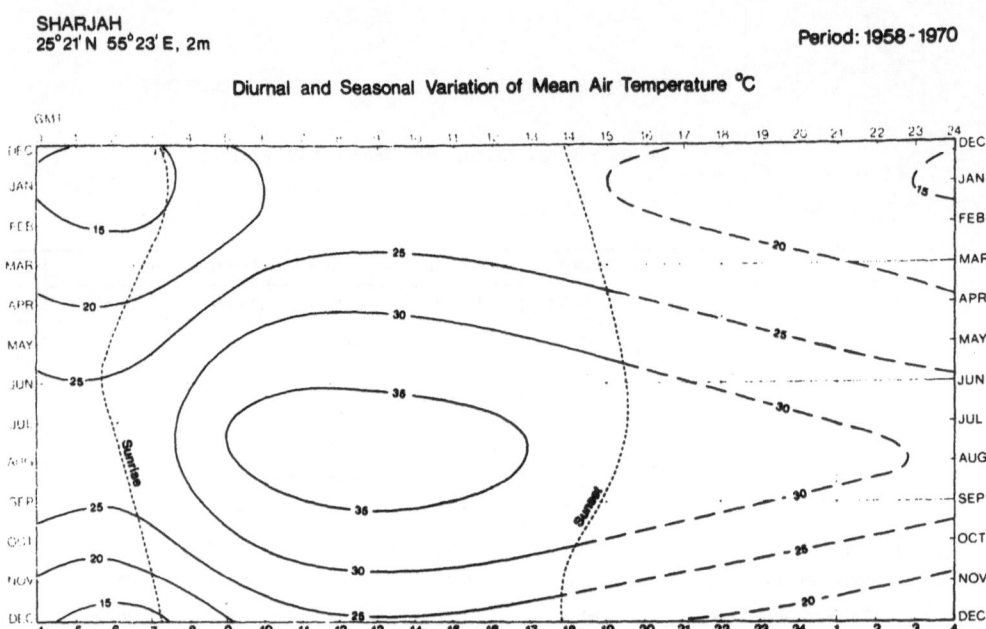

Diurnal and Seasonal Variation of Mean Relative Humidity %

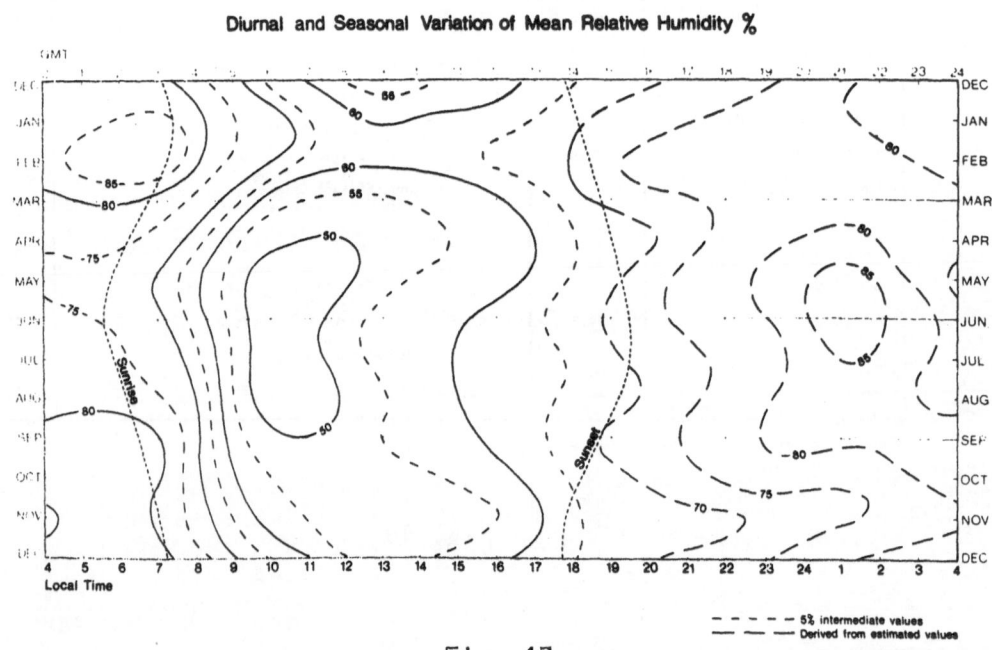

– – – – – 5% intermediate values
– – – Derived from estimated values

Fig. 17

REFERENCES

1. R.K. Anderson et al., "The use of satellite pictures in weather analysis
    and forecasting", WMO Technical Note 124, WMO No. 333, Geneva, Swit-
    zerland, (1973), 273 pp.

2. E.C. Barrett & C.K. Grant, "An appraisal of Landsat 2 cloud imagery and
    its implications for the design of future meteorological observing
    systems", Journal of the British Interplanetary Society, 31, (1978),
    3-10.

3. C.M. Haydem et al., "Quantitative meteorological data from satellites",
    WMO Technical Note 166, WMO No. 531, Geneva, Switzerland, (1979), 102 pp.

4. Imcos Marine Ltd., "Gulf tidal anomalies, 17-19 January 1973", Oil
    Companies' Weather Co-ordination Scheme Technical Note No. 1, London,
    (1977), 8 pp.

5. Imcos Marine Ltd., "Water Level Oscillations, Das Harbour", Oil
    Companies' Weather Co-ordination Scheme Technical Note No. 2, London,
    (1977), 8 pp.

6. Imcos Marine Ltd., Handbook of the Weather in the Gulf, Surface Wind
    Data, (Oil Companies' Weather Co-ordination Scheme, London, 1974),
    103 pp.

7. Imcos Marine Ltd., Handbook of the Weather in the Gulf, Bibliography,
    (Oil Companies' Weather Co-ordination Scheme, London, 1974), 19 pp.

8. Imcos Marine Ltd., Handbook of the Weather in the Gulf, General Climate,
    Data, (Oil Companies' Weather Co-ordination Scheme, London, 1976),
    101 pp.

9. Imcos Marine Ltd., "Storm of January 19th, 1964", Oil Companies' Weather
    Co-ordination Scheme, Information Bulletin No. 11, London, (1964), 7 pp.

OCWCS in session

TROPICAL OIL SPILL CONTINGENCY PLANNING : REQUIREMENTS AND APPLICATIONS

Ronald R Stoner
Marine Environmental *Consultant* , NUS Corporation, Rockville, Maryland, U.S.A.

Special acknowledgement is expressed to Rear Admiral Winford W Barrow (U.S. Coast Guard, Retired) for his technical assistance in preparation of this paper and to many others too numerous to mention.

INTRODUCTION

The potential for oil spill incidents along coastal waters in the Tropics is high because of present and projected large-scale petroleum operations and associated marine traffic. The objective of this paper is to present generic recommendations for development of national oil spill contingency plans that will assure adequate protection of the marine environment and that are commensurate with national requirements and resources. Applicability of these principles to the Tropics is examined.

The urgency for preparing an oil spill contingency plan to respond to such incidents is also associated with the following:

There have been several catastrophic oil spills throughout the world in the past decade, one of which resulted in total costs (including cleanup, property damage, and social costs) estimated at around US $2 billion. In general, cleanup costs alone are estimated to be US $400 to $800 a barrel.

An equally urgent reason for developing and implementing oil spill contingency plans is the need to counter the multitude of lesser accidental and intentional spills which, while not as dramatic as the catastrophic spills, provide a much greater overall cumulative threat to the environment.

The potential for irreversible damage of the marine environment including important coastal fisheries and sensitive ecological areas such as coral reef systems, mangroves, etc.

The dependence of major coastal cities on electrical energy and fresh water from power plants and desalinization plants, respectively, that use coastal waters for their intake.

International commitments, such as anti-pollution measures and regional co-operatives sponsored by the Inter-Governmental-Maritime Consultative Organization (IMCO), and potential spill liabilities, of necessity, require development of contingency plans.

The importance of proper oil spill contingency planning is also illustrated by response efforts for numerous major spills over the last decade (e.g. the Torrey Canyon, Metula, Argo Merchant and Amoco Cadiz disasters). Most of these containment and cleanup efforts were unsuccessful and illustrate that most

authorities are poorly prepared to cope with major oil spills.
The deficiencies are not only in terms of lack of efficient
specialized containment-cleanup equipment but also in the lack
of fundamental response plans, organizations, and policy
decisions on how to deal with spills. Such criticisms of prior
contingency planning/response actions, mainly for spills in
the Termperate Zone, can be used advantageously for developing
effective plans and procedures for application to the Tropics.
It is important to note that the major spills cited above
are all from ships which were grounded/stranded in climates far
more harsh than the tropical regions under discussion. Fifty-
five percent of historical spills accounting for sixty-five
percent of total spill volume have occurred during winter
(November through March) in the Temperate Zone where the average
duration of weather suitable for recovery of oil is less than
three days. Therefore, weather conditions in the Tropics are
more conducive to successful containment and cleanup operations.
The IXTOC 1 Blowout and spill occurred within the tropical
zone, is still flowing, and is now the largest oil spill on
record. The same critical comments made regarding response
activities for ship grounding spills apply to the IXTOC 1 spill.

MARINE ENVIRONMENT OF THE TROPICS

The Tropical Zone has been defined, for the purpose of this
paper, as the land and ocean areas of the world between
approximately 30 degrees North and 30 degrees South latitudes
and includes some regions which are actually subtropical (Ref. 1).
This latitude band roughly corresponds to areas with mean
monthly temperatures greater than 20 degrees Centigrade (68
degrees Fahrenheit) throughout the year (Ref. 2). Winds in the
Tropics are generally steady (especially when compared to the
variable wind direction conditions in the Temperate and Polar
Zones) with moderate velocities in the Northeast and Southeast
Trades. Exceptions are the light and variable winds in the
Doldrums, approximately 10 degrees North to 10 degrees South,
and in the Horst Latitudes, around the 30 to 35 degree parallels
(Ref. 3). Offshore visibility is generally excellent compared
to the Temperate and Polar Zones, with visibilities of less than
two nautical miles occurring less than five percent of the time
annually (Ref. 4).
Ocean surface currents in the Tropics offshore the continents
are considered warm and have sets parallel to the coastline
(Ref. 5). Sea heights are generally less than four feet and
the frequency of higher seas is considerably less compared to
the Temperate and Polar Zones (Ref. 6).
The Tropics, not surprisingly, are seldom affected by extratropical storms, generally referred to as migratory frontal
cyclones (Ref. 7 and 8). However, tropical storms, defined
internationally as cyclones that are over the tropical oceans
and which have wind velocities in excess of 17.5 metres per
second (34 knots), are an important consideration for planning
marine operations in the Tropics (Ref. 9 and 10). Even though
the potential for tropical storms exists, present forecasting
methods and communication systems generally enable the maritime
industry to modify operations so as to minimize the risks from
these relative localized (on a world-wide scale) tropical storms.
This is illustrated by the low number of occurrences of weather
related maritime casualties compared to the Temperate Zone
(Ref 11.) But it is important to note that weather conditions

only accounted for approximately seven percent of all (world-
wide) casualties.

The offshore regions of the Tropics are not generally
considered as areas of high primary productivity, defined as
areas with high rates of creation of organic matter by plants
from inorganic materials and sunlight, compared to the higher
latitudes (Ref. 12). However, there are regions of high
productivity in the Arabian Sea as well as along the western
coasts of Central America, South America and portions of Africa
generally associated with upwelling. The coastal regions of
the Tropics are ecologically sensitive, with rich and varied
marine resources. Important fisheries are located along
portions of the coastal areas in the Tropics, mainly in areas
of high primary productivity as previously discussed (Ref. 13).

Many of the coastal areas of the Tropics have beautiful
beaches which have actual or potential recreational and tourism
values. In addition, there are numerous wetlands and mangrove
coastal areas which are particularly environmentally sensitive.
Coral reef systems, which support fragile marine ecosystems,
are unique to the Tropics (Ref. 14). Also, coastal lagoons are
commonplace and have poor flushing characteristics compared to
estuaries of the Temperate Zone. Therefore, these areas are
extremely vulnerable to the effects of oil spills.

The socioeconomic conditions of the various countries in the
Tropics can be characterized by the Gross National Product (GNP)
which is generally lower than values for the Temperate Zone
(Ref. 15). Relatively low GNP values are typically associated
with a lower degree of industrialization, lower rate of literacy,
and higher population growth rate compared to nations with
higher GNP values (Ref. 16). Thus, these conditions can place
certain limitations on the internal capability for certain
nations to develop and implement an adequate spill response
program. Many of the oil producing nations in the Tropics,
however, are very rapidly developing and have substantial
national technological capabilities and financial resources.

The petroleum resources of the Tropics, especially the
Middle East, represent a significant portion of the world's
reserves (Ref. 17). Therefore, a majority of the worldwide
tanker traffic originates from and/or passes through the Tropic
Zone (Ref. 18). Thus, the potential for oil spill incidents is
high. But a survey of oil spill statistics indicates relatively
few incidents compared to the Temperate Zone (Ref. 19). This
situation, unfortunately, has the potential for fostering a
false sense of security even though the risk for major spills
is high. Also, oil spill statistics need to be viewed with great
care and some scepticism. Accumulation of valid oil spill
statistics is a function of the regulatory and enforcement
approach in effect in the region surveyed. A review of the areas
from which there is a dearth of statistics indicates that many
countries located within the Tropics have minimal or no
organizations established to deal effectively with pollution.

The current status of national oil spill contingency
planning in the Tropics, as indicated above, islless formalized
than in the Temperate Zone. Countries such as the United States,
France, United Kingdom, Sweden, Canada, Norway, Australia, etc.
have fairly comprehensive national and regional plans supplemented
by facility plans developed by the petroleum industry. In the
Tropics, available national plans are generally less formalized
and less institutionalized. However, review of various IMCO

reports, and personal knowledge of the area involved reveals few
national oil spill contingency plans in existence. The same
review reveals a general lack of organizations concerned with
pollution in general, although this situation is rapidly
changing in numerous developing countries. The need exists,
therefore, for comprehensive national planning and response
co-ordination programs. The petroleum industry, though, has
generally taken responsibility for oil spill contingency planning
and response for their own operations. These response resources
are enhanced by such co-operative organizations such as Clean
Gulf (of Mexico), Clean Caribbean, and Gulf (Arabian) Area Oil
Companies Mutual Aid Organization.

PERSPECTIVE ON OIL SPILL RESPONSE

Oil spill contingency planning and response are obviously not
a unique problem of the Tropics. Thus, it is prudent to
acknowledge the fundamental problems (e.g. alternative contain-
ment/cleanup approaches, as well as the fate and effects of oil
spills) involved before developing recommendations for the Tropics.
     The primary objective in responding to an oil spill is to
minimize the environmental damage by limiting the spread of oil
and by removing the oil from the water. In order to accomplish
this, comprehensive contingency plans must be prepared. To
date, there have been few, if any, successful large-scale
operations of this nature. Most often the oil is "lost" or
redistributed through natural evaporation or dispersion and
bacterial action, or it reaches the shore and is removed by
labor-intensive land-based operations. This was the case in the
Torrey Canyon disaster occurring in March 1967, the Amoco Cadiz
grounding in 1978, and the Argo Merchant grounding in 1976.
     Two basic methods exist for the cleanup of oil at sea. These
are mechanical removal and dispersant use.
     In open waters, the problem of conducting response operations
is directly related to the sea state. It is probable that
spilled oil will not exist in a recoverable slick in wave heights
of more than 10 feet (a condition only expected during tropical
storms for the lower latitudes). Currently available equipment
is able to remove oil with varying degrees of effectiveness for
typical wave heights in the Tropics (i.e. wave heights up to
about 4 feet). The effectiveness of a skimmer is dependent to
a large extent on its capability to follow the motion of the sea
and hence is greater on a swell than a chop. There is various
equipment on the market today that has been designed for ocean
or exposed-water use, but most high-capacity units collect large
volumes of water with the oil in the higher sea states. For
large spills, therefore, separation of oil and water and disposal
of the oil become a problem.
     The principles upon which currently available skimmers
operate are based on the differences among properties of oil and
water - such as gravity, viscosity and adhesion. Most appear
relatively ineffective under anything other than calm sea
conditions. Those that can operate in the open seas on a large
oil spill must operate with a substantial fleet of support boats,
barges, booms, and other ancillary equipment in a well co-ordinated
procedure.
     The second way of removing oil from the surface of the water
is by chemically dispersing the oil into the water column.
Proper application is the key to a successful operation; if done

properly, the tension of the oil/water interface is reduced and
the oil slick is broken into small droplets under the mixing
action of the waves (or mechanical agitation). The dispersed
oil presents a much greater surface area for increased rates of
solution, evaporation and biodegradation.

In general, dispersants work better on thinner layers of
oil, while mechanical skimmers work better on thicker layers.
The effectiveness of dispersants is dependent upon the type of
oil treated, the method and efficiency of application and the
desired result. Quite often dispersants have been used with
little advance planning or preparation. Effective use requires
careful designing of the delivery systems, with initial and
follow-up training to provide optimum results. Large stockpiles
of dispersants without readily available equipment, vessels and
aircraft to apply them will result in unsatisfactory performance.
Also, the potential for adverse effects on the marine environment
must be evaluated, especially for ecologically sensitive areas
in the Tropics such as coral reef systems. Use of dispersants
is currently the only known response that can be used in all sea
states in which oil slicks might exist. Therefore, their use
must be considered and planned for in any oil spill contingency
plan.

The movement, transformation and impact of an oil spill is
very complex and dependent on the physical/chemical properties
of the oil as well as environmental conditions. Movement of
an oil spill on water is dependent primarily on wind and current
conditions. The oil may become suspended in the water by wind
and wave actions and/or transported to the coastline, resulting
in environmental damage. Oil in the marine environment will
eventually, but slowly, disappear due to evaporation, biodegradation, petrochemical oxidation and dispersion. These
transformation processes may, under certain conditions, be more
rapid in the Tropics due to high air/water/oil temperatures.

Dissolved and dispersed components of crude oil may have
adverse environmental effects due to their ingestion, absorption
and absorption by marine organisms. The impact of petroleum
in the marine environment may, under certain conditions such as
on beaches, be greater in the Tropics compared to the Temperate
Zone due to synergistic effects of high temperatures and other
stresses. Polluted sediments can be particularly detrimental
to the benthos. At present, there is substantial controversy
over the long-term environmental effects. In general, however,
marine organisms are much less harmed by the heavier ends which
remain after the light ends of the oil have evaporated (Ref. 20).

Early stage (eggs, larvae and juveniles) marine fish,
invertebrates and mammals and certain plants are particularly
vulnerable to environmental damage by oil spills. Adult fish
and mammals have the advantage of mobility and can leave the
spill area unless trapped in beaches or estuaries. Estuaries,
lagoons, marshes, grasslands, shellfish beds, mangroves, reefs,
and beaches, however, are sensitive environmental areas subject
to both severe short-term and long-term ecological damage
(Ref. 20). Since spawning is less seasonally dependent at the
lower latitudes, ecological impact (short to intermediate time
scales) vulnerability from oil spills is somewhat reduced
compared to the Temperate Zone.

Most containment/cleanup efforts, if not all, for the large
oil spills to date have been unsuccessful. However, each of the
major ship casualty spills have been a result of grounding/

stranding where extreme extratropical weather became a factor.
It should be pointed out that given a similar grounding under
typical tropical weather, it is very unlikely that the spill
would have progressed to a catastrophic situation if appropriate
containment/cleanup actions are taken.

NATIONAL OIL SPILL CONTINGENCY PLAN CONCEPTS

Development of a national oil spill contingency plan for use in
the Tropics must include consideration of the specific geography,
environment and resources of the country. In order for
contingency planning to be effective, though, nations must also
institute measures to minimize oil discharges to its waters.
In addition, participation and co-ordination with neighbouring
countries is the most efficient means for assuring adequate oil
spill response/cleanup resources. The national plans implemented
by the United States, France, Canada, Sweden, United Kingdom,
Norway and Australia serve as useful models of plans that have
addressed similar types of considerations. The following
discussions are generic in nature (i.e. applicable to contingency
planning in general) but also are particularly significant for
the developing nations of the Tropics.

Regionalization

Development of a national oil spill contingency plan must reflect
regional differences within a nation between coastal waters.
The concept of regionalization is quite significant in the Tropics
because of the diversity of the marine environment (beaches,
coral reefs, mangrove systems, etc) and coastal development.
For example, Saudi Arabia borders on two large but widely
separated bodies of water - the Gulf and the Red Sea. The Gulf
and Red Sea differ quite significantly in ecology, natural
resources, coastal development and physical oceanographic
conditions. Therefore, it is imperative that a national plan
not only provide a cohesive organization for oil spill response,
but also provide the vehicle for the development of regional
and local plans and organizations. It is the regional and local
plans which should contain the detailed information on response
procedures, etc.

Response Resources

One of the initial steps in developing a national contingency
plan is to conduct a survey of potential response resources.
This is extremely important in the Tropics where a comprehensive
assessment of such resources is frequently not available. Based
on this survey, the national response capability can be determined
and a national plan as well as response organizations can be
developed accordingly. The national plan should also allow for
the effective use of the extensive oil spill response resources
of the petroleum industry. This should be accomplished by not
only placing financial responsibility for oil spill containment/
cleanup on the discharger, but also by affording industry the
first opportunity to implement containment/cleanup actions
commensurate with its capabilities to do so.
    National governmental involvement in contingency planning
and response operations must, of necessity, vary based on
specific national capabilities. For example, the Kingdom of
Saudi Arabia has a wide range of national technical, financial
and manpower resources to apply in response planning/operations

and to supplement response resources of the petroleum industry. However, many countries have more limited technical and response resources. The government role in these cases should be one of co-ordination of available resources and establishment of international co-operation relationships.

International Co-operation

Preparation of a national oil spill contingency plan should include provisions for co-ordination and co-operation among the countries which border common waters. This approach is the most cost-effective means for developing nations in the Tropics to prepare and implement adequate spill contingency plans. Co-ordination efforts should be preplanned in detail since problems such as transport of people and equipment across international boundaries may otherwise be subject to lengthy delays. The importance for international co-operation cannot be over-emphasized since it would not be cost-effective for any one nation to maintain response capabilities for a catastrophic spill (e.g. a major blowout or an Ultra Large Crude Carrier Collision). The United Nations Environmental Progam and IMCO continue to play a leading role in sponsoring international co-operation. An excellent example of international co-operation in the Tropics is the participation of nations which border the Arabian Gulf in the Kuwait Action Plan (Ref. 21). The Kuwait Action Plan, when fully implemented, will provide a comprehensive model plan for international co-operation and co-ordination for response to oil spills on the Gulf. However, implementation of the Kuwait Action Plan is dependent upon the development of effective contingency plans by each of the participating nations. International co-operative activities are also being undertaken in other regions such as Southeast Asia, Gulf of Guinea, Red Sea, South Pacific and the Caribbean Sea.

Plan Scope

A national oil spill contingency plan should provide for co-ordinated national action to limit any environmental damage when discharges occur. The plan should promote the development of co-operation among national, industrial and private organizations to build capabilities for containing and cleaning up discharges. In this respect, requirements for contingency planning in the Tropics are not unique; the differences are basically associated with an application of planning principles to specific environmental and national conditions. Following is a brief summary of the basic components of an effective national plan.

The national plan should specify requirements for associated regional and local plans which provide for :

Identification, procurement, maintenance and storage of equipment and supplies;

Identification and utilization of available ecological, meteorological, oceanographic and other environmental expertise as well as petroleum engineering and maritime expertise;

Conducting of environmental surveys of local areas to identify sensitive coastal sites;

Establishment or designation of response organizations :
(i)   professional and well trained and motivated strike forces to carry out the plan, and

(ii) trained and adequately equipped emergency task forces
at major ports and coastal areas;

A system of surveillance and reporting to facilitate the
earliest possible notice of discharges of oil and hazardous
substances or imminent threats of such discharges;

Procedures for identifying, containing, dispersing, removing
and disposing of oil and hazardous substances. (The need
for pre-planned disposal sites is extremely important but
has been frequently neglected).

The national plan should be applicable to the navigable waters
of the nation, including offshore areas which are subject to
national jurisdiction. Implementation of such a plan is
intended to be complementary to international co-operation and
co-ordination efforts relevant to protection of the marine
environments as previously discussed.

The national plan should assign to appropriate national
agencies the responsibility for minimizing the possibility of
discharges; for developing the capability to respond promptly
to discharges from facilities they operate or supervise; and
for making resources available for national pollution response
operations.

The national plan also should identify requirements for
development of regional and area (coastal/offshore and port)
plans. These plans should be similar in format to the national
plan but should provide more detailed information on the
response resources available (equipment inventory, communications
systems, etc.), identification of emergency force personnel, and
training requirements, as well as oil spill investigation,
containment and cleanup procedures.

Industry and government marine/coastal facilities should be
required (by the national plan) to prepare facility-specific
contingency plans if the facilities have (1) oil transfer
operations to vessels with greater than approximately 250 barrel
capacity, (2) coastal/offshore petroleum exploration wells and
production facilities and (3) sites with a potential for spills
of hazardous substances. The format for preparation of the
facility plans should be specified in the appropriate regional
and area plans as well as review/approval procedures to assure
that such plans for each facility are adequate and are periodically updated, as necessary.

The inter-relationship of national, international, area and
facility plans is illustrated in Figure 1. It should be
emphasized that the national plan is the document which defines
the overall national approach to oil spill response and establishes
the requirement to develop detailed regional, area and facility
plans. A fundamental problem worldwide has been the heavy
reliance on national plans, without developing adequately
detailed regional, area and facility plans. Another extreme is
the availability of only local contingency plans without having
adequate national, regional and area plans to co-ordinate
resources during major spills. Frequently a national plan is
a necessary incentive for the development of detailed local plans.
However, the national plan should be revised, as necessary,
based on a re-assessment of the national response program after
detailed local plans have been developed.

The use of chemicals, such as dispersants, must be rigidly
controlled; therefore, regional and area facility plans should

specify ecologically sensitive areas where the use of chemical dispersants and herding agents are not authorized.

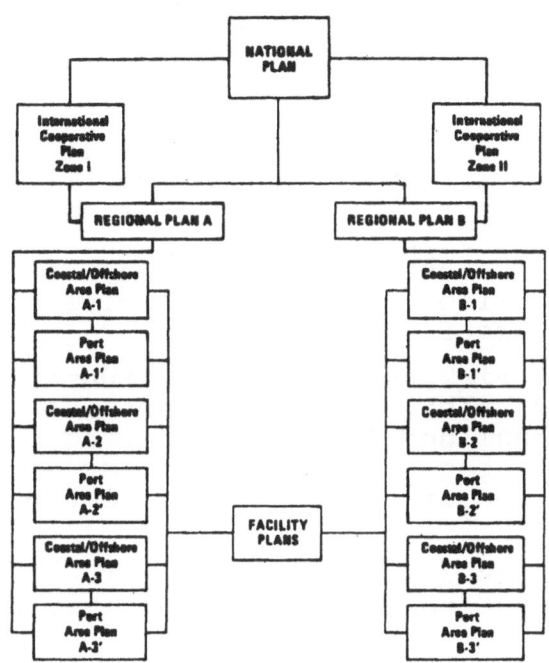

Figure 1.   **Inter-relationship of oil spill contingency plans**

Oil Discharge Policy

The coastal waters of the Tropics along major tanker routes and in proximity to production operations have been subject to oil pollution due to routine oil discharges from ships and petroleum activities. While this has been a worldwide problem, and IMCO (Ref. 22 and 23) has fostered considerable international participation in oil spill prevention agreements, environmental management has in the past been of a lesser priority for many of the developing countries in the Tropics. Therefore, it is important that all coastal nations adopt international pollution prevention measures and rigorously enforce oil pollution regulations to reduce routine oil discharges and to minimize the potential for accidental spills. This should include providing deballasting facilities for ships, as warranted.

In summary, present international commitments to control oil pollution appear to be generally adequate, but more rigorous enforcement, supported by the development and implementation of national regulations, is necessary in order for oil spill response to become an emergency action and not a continuous activity.

PLAN DEVELOPMENT/IMPLEMENTATION REQUIREMENTS

The development and implementation of national and associated regional and local oil spill contingency plans necessitates the conduct of several support projects as well as consideration of requirements for staffing, equipment, and response procedures. These tasks are extremely important in the Tropics since adequate environmental information, as well as time availability of technical and managerial staff may be limited. It is recommended that along with development of a national contingency plan, a comprehensive implementation plan listing equipment, staff size, schedules, etc., be prepared to effectively and efficiently assure adequate national oil spill response capabilities.

Associated Projects

In order for the contingency plans to be initiated with a minimum of delay and to assure efficiency, several specific projects should be undertaken during the initial stages of the overall program. These projects will provide the necessary primary background information for contingency planning and emergency response purposes. Following are brief summaries of the recommended projects in order of priority :

Develop an oil spill risk and marine environmental resource atlas for the nation. This atlas would be useful for the development of regional and facility plans as well as for determining initial response strategy during an actual spill situation. Preparation of an atlas would include the following :

- Compile and summarize existing marine meteorological, marine ecological and physical oceanographic data for the nation.

- Survey and summarize oil pollution sources and quantify oil spill risks. This will indicate assessment of the pollution potential for marine traffic, offshore exploration and development, etc. This would also include a review of oil spills reports for the last several years.

- Identify areas where oil spill impact may be critical (desalinization plants, power plants, sensitive ecological area, etc.).

- Evaluate oil spill risks from various sources and locations as well as the probability of impact on sensitive areas based on use of a climatological oil spill transport model. An example of the use of oil spill risk isopleths is illustrated in Figure 2 (Ref. 24). Such an approach is especially adaptable to the Tropics because of the steadiness of the winds.

Conduct a comprehensive inventory of response equipment in the nation and develop specific equipment recommendations. This would include a thorough evaluation of the state of the art of oil spill containment/cleanup equipment and techniques.

Develop further recommendations for the oil spill prevention regulatory process. This would also include a thorough evaluation of shipping patterns, traffic situations, shipping accidents, industry practices, etc., to develop recommendations to further minimize the potential for oil pollution in the coastal and offshore waters of the nation.

Develop comprehensive training program for national, regional and area response personnel. This would include developing courses and adapting available training resources to the specific needs of the nation.

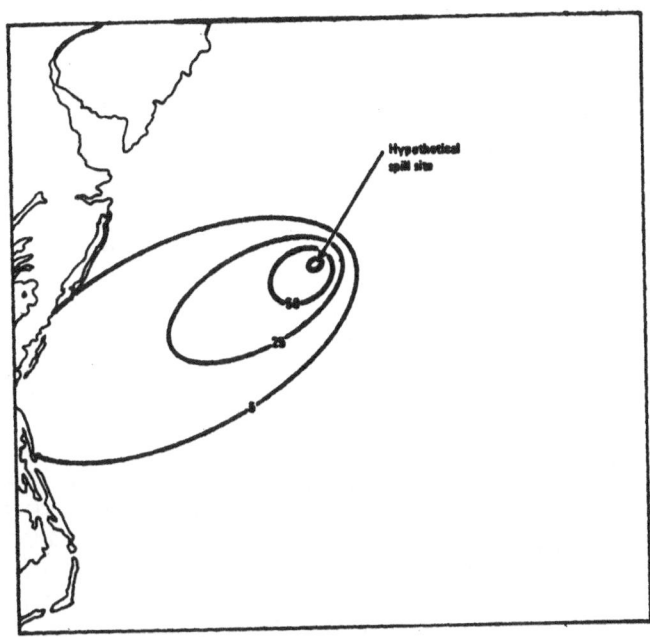

**Figure 2.   Example oil spill risk (%)
presentation (Ref. 24)**

Establish a technical library which will contain the necessary
books and reports, as well as films and videotapes to advance
the knowledge of personnel responsible for response
activities in the nation. The library should also contain
legal and policy documents to enable response personnel to
be fully cognizant of the responsibilities of the various
arms of the oil spill organization, as well as the oil
producing and shipping industries.

Develop requirements for an oil spill reporting system. This
will include development of a data management information
system which receives, reviews and stores oil pollution case
data, such as inspection and violation information, etc. for
fast retrieval.

The projects can be conducted by national institutions, if
capabilities exist, or with the assistance of specialized consultants
as necessary. Many of these items may be best accomplished, or
in fact already underway or completed, by the petroleum industry.
The intent should be to efficiently utilize existing resources
and information based on combined government petroleum industry
capabilities. The use of experienced consultants in an
advisory or co-ordinating role, or in development of special
components of this program, such as the application of models of
oil spill behaviour, will usually be advantageous.

Response Procedures/Equipment
Many countries with sizable oil spill problems and which maintain
conservative views on dispersants have substantial research and
development programs devoted to improving the state of the art
in the containment and mechanical removal of oil. This has
resulted in improvement in containment booms and oil skimming

devices, but to date, the success rate has not been good in
attacking major spills under adverse conditions. However, there
are indications, based on research sponsored in the United States,
Norway and other countries that there have been major advances
in the mechanical cleanup of spills which will be available for
purchase in about two to three years. Use of mechanical cleanup devices, such as especially combined skimming-barriers,
received an excellent testing in IXTOC 1 oil spill and have
proven quite effective in four to five foot seas and held
together in seas much higher.

The recommended approach, especially in the Tropics, to
follow in providing for control, containment and cleanup resources
and for their siting is to adopt a balanced approach for selection
of equipment, materials and their use along the following
principles :

Contain and mechanically remove the oil where possible, both
in the offshore and inshore environment. Weather and sea
conditions in the Tropics are normally conducive to successful
use of presently available mechanical equipment. A rapid
response capability, however, is required in the Tropics
because of environmentally sensitive areas. Once a spill
reaches a coral reef system or mangrove area containment
any cleanup operation becomes extremely difficult.

Use dispersants where safety is a problem and elsewhere
offshore (in deep waters with good mixing conditions) under
a closely held authority, and in accordance with a well
developed environment plan. The expertise of a marine
ecologist is, therefore, essential in the Tropics both for
contingency planning and participation in response operations.
The use of dispersants should be routinely re-evaluated as
additional evidence becomes available concerning the
effectiveness and environmental risks of this approach.

The stockpiling and siting of emergency equipment, supplies and
materials to respond to a major oil spill is not an exact science.
One cannot predict with any degree of accuracy where an oil spill
will occur. High risk areas can and should be identified but a
spill may present no risk to the adjacent coastline and may
threaten the environment much farther away. Vessel collisions
may occur at any point along the coast. The sizes of spills
which might occur are essentially unpredictable. An estimate
may be made of an approximate range of tanker collision scenarios
and the size of the cleanup capacity geared for that eventuality.
However, catastrophic loss of an entire vessel's cargo or an oil
drilling rig's blowout could be an overwhelming containment and
cleanup operation which exceeds current technological capabilities.

Contingency equipment and supplies to meet the needs of a
major spill need to be provided for, but not for a worst case
situation. Stockpiling to meet worst case situations presents
not only major initial costs but unwarranted storage and
maintenance problems. It also involves major replacement costs
as improved materials become available. The suggested approach
is to develop national capability for the particular spill
potential in the region. It is possible to postulate a relationship between frequency and magnitude and the volume of oil
transported or handled. The relationship varies depending upon
whether the throughput is via port operations, transient tankers
and barges, offshore exploration or production oil wells, deep

water ports or lightering. The oil spill potential must be
evaluated prior to determining requirements for national
response resources. Additional resources should be available
via international agreements from neighbouring countries to
supplement the response resources for the nation.

Distribution of the response resources along the coastline should be such as to have major concentrations of supplies
in reasonably close proximity to high risk areas with lesser
quantities at in-between sites. For all oil spill response
sites, there should be provided specific vessels of appropriate
size and conformation to accomodate operation of oil spill skimming
units and associated booms. Shallow draft vessels will be
required in the Tropics for operations in coastal lagoons and
around coral reef areas. Necessary barges, storage containers
and disposal sites should be predetermined. Sufficient vessels
should be preselected and pre-equipped for dispersant spray gear
may be placed on board and ready for use expeditiously. Vessels
selected should have on board carrying capability of 1000 gallons
(or more) of concentrated dispersants. A reliable communication
system must be available to link centralised command centres to
field operations. If it is not possible to provide response
stations near all potential high oil spill risk areas, consideration
should be given to use of high speed delivery systems from the
surface or air.

Staff
The staff responsible for contingency planning and response
operations should involve numerous disciplines. Planning and
advisory (during response operations) functions require technical
specialists in areas such as marine ecology, meteorology, oceanography, petroleum engineering, marine operations, communications,
public relations, law, procurement, etc. National institutions,
universities, consultants, industry and the military are potential
sources for such talent. However, field operations involving
spill containment and cleanup will require very specialized
personnel. The United States, for example, has established
"Strike Teams" composed of personnel with special training and
experience in such operations. The recruitment, training and
retention of personnel with the required expertise and
qualifications can be quite difficult in the Tropics.

Chronic problems have plagued the effectiveness of "Strike
Teams" as utilized by governments, military organizations and
the petroleum industry. The high turnover rate of staff members
has historically limited the performance of such emergency work
forces. Routine transfers and attritions are obvious problems.
In addition, staff members who obtain advance training and
demonstrate responsibility are usually promoted and/or transferred to other jobs not involving oil spill response duties.
Also, many response assignments, especially in industry, are
not full-time jobs and priority is given to "productive" work
which facilitates professional advancement versus emergency
preparedness. This situation is quite common world-wide as well
as in the Tropics.

A possible solution is to establish incentives (professional
advancement within the response organization, pay, liberal leave,
etc. as presently used in the maritime industry) to retain
qualified personnel and foster a high degree of professionalism.
This can be accomplished for organizations within the government,
military and industry. But, especially in the Tropics, it is

also a support service that can be effectively and efficiently
provided by contractors who specialize in oil spill response.

Summary
The potential for oil spill incidents along the coastal waters
of the Tropics is high because of the large scale petroleum and
associated marine activities in the region. The potential short-
and long-term effects of oil spills is a significant concern
since the marine environment, with proper care, can be a long-
term renewable resource (while petroleum must be considered a
finite resource). Tropical coastal areas have great tourism
and environmental value. Coral reefs, mangrove areas, lagoons
and fishery regions are particularly vulnerable to oil spill
damage. However, wind and sea conditions are generally favourable
in the Tropics and, therefore, result in lower risk of weather-
related maritime accidents and facilitate the use of current-
technology containment-cleanup equipment.
     There are numerous developing countries within the Tropics.
Currently, national contingency plans are less formalized and
less institutionalized compared to the Temperate Zone. The
petroleum industry has generally taken responsibility for oil
spill contingency planning of their own operations and has
participated in inter-company co-operatives. Many nations, such
as the Gulf States in the Middle East, are participating in
international co-ordination/co-operation programs with the
assistance of IMCO and the United Nations Environmental Program.
All of these efforts require development of national contingency
plans in order to formalize (and thus facilitate) co-ordination
of national and international response resources during emergency
situations.
     Development of a national contingency plan for use in the
Tropics must include consideration of the specific geography,
environment and resources of the country. In order for contingency
planning to be effective, oil discharge regulations must be
promulgated and enforced. The response resources and expertise
of the petroleum industry should be utilized and international
co-operatives are a necessity since it is cost prohibitive for any
one country to have the capability to handle catastrophic spills.
     The national plan should be the means for specifying the over-
all national policy for oil spill response and requirements to
develop regional, area and facility plans. Such supplemental
plans should include the detailed information necessary to respond
to spills. Development and implementation of a national plan will
require several support projects, many of which usually require
assistance by specialized environmental consultants. The use of
experienced consultants in an advisory or co-ordinating role for
plan preparation and implementation of specialized programs is
usually advantageous. The preparation of a marine resource and
oil spill risk atlas is especially suitable for application to
the Tropics.
     In summary, the problems associated with previous major spills
world-wide have been associated with deficient contingency plans.
Many of these difficulties can be attributed to an inadequate level
of detailed information and procedures in existing plans and to
staffing problems associated with high personnel turnover rates
and conflicting assignments. Progress is being made in the Tropics,
but continued emphasis must be placed on developing and upgrading
national, regional, area and facility plans.

REFERENCES

1. M G Gross, Oceanography, (Charles E Merill Publishing Co.,
    Columbus, Ohio, 1967), 76.
2. G L Cantzlaar, Your Guide to the Weather, (Barnes & Noble
    Inc., New York, New York, 1964), 174
3. S B Cohen, Oxford World Atlas, (Oxford University Press, Inc.,
    New York, New York, 1973), 93
4. H L Crutcher and O M Davis, U.S. Navy Marine Climate Atlas
    of the World - Volume VIII - The World, (Naval Weather
    Service Command, U.S.A. 1969), 4 and 66.
5. Cohen, Op. Cit., 93, 96.
6. Crutcher and Davis, Op. Cit., 22, 53, 84, 115.
7. Ibid., 2 and 64.
8. R E Huschke, Glossary of Meteorology, (American Meteorological
    Society, Boston, Massachusetts, 1959), 216
9. M Bramwell, The Atlas of the Oceans, (Rand McNally and Co.,
    New York, New York, 1977), 30
10. Huschke, Op. Cit., 86
11. R G Quayle, "Weather and Maritime Casualty Statistics",
    Mariners Weather Log, 20 (March 1976), 74-80
12. Bramwell, Op. Cit., 86
13. E P Espenshade, Jr., and J L Morrison, Goode's World Atlas
    (Rand McNally and Co., Chicago, Illinois, 1976), 38
14. Gross, Op. Cit., 23
15. Espenshade and Morrison, Op. Cit., 28
16. Ibid., 24-50
17. J C McCaslin, International Petroleum Encyclopedia, (The
    Petroleum Publishing Co., Tulsa, Oklahoma, 1979), 6.
18. Ibid., 13
19. S Van Gelder-Ottway and M Knight, "A Review of World Oil
    Spillages, 1960-1975", Marine Ecology and Oil Pollution,
    John Wiley & Sons, New York, New York, 1976, 483-520.
20. Texaco, Inc., Oil Spill Contingency Planning and Response-
    Students Handbook, (U.S.A. 1978).
21. United Nations Environmental Program, Final Act of the Kuwait
    Regional Conference of Plenipotentiaries on the Protection
    and Development of the Marine Environment and the Coastal
    Areas, (Geneva, Switzerland, April 1978).
22. G Moore, "Legal Aspects of Marine Pollution Control",
    Marine Pollution, (Academic Press, New York, New York, 1976)
    589-697
23. E D Brown, "The Role of Law in the Prevention of Oil
    Pollution", The Prevention of Oil Pollution,(John Wiley & Sons,
    New York, New York, 1978), 283-294.
24. J M Bishop, A Climatological Oil Spill Planning Guide,
    (National Oceanic and Atmospheric Administration, U.S.A.,
    October 1979 - Draft).

TROPICAL MARINE ECOSYSTEMS AND THE OIL INDUSTRY; WITH A DESCRIPTION OF A POST-OIL SPILL SURVEY IN INDONESIAN MANGROVES.

Dr. Jenifer M Baker
Field Studies Council, Orielton Field Centre, Pembroke, Dyfed,
Wales, U.K.

Prof. Ir Moeso Suryowinoto
Faculty of Biology, Gadjah Mada University, Yogyakarta, Indonesia.

Dr. Paul Brooks
Masspec Analytical (Specialty Services) Ltd., Woodchester, Stroud,
U.K.

Mr Steve Rowland
Masspec Analytical (Specialty Services) Ltd., Woodchester, Stroud,
U.K.

INTRODUCTION

Large areas of shallow tropical sea, coral reefs, intertidal
sand and mud flats, seagrass beds and mangrove swamps are potentially at risk from oil industry influences, but predicting
the effects of oil spills, oil spill dispersants and other
clean-up methods, and effluents is particularly difficult because
relatively little is known about the basic biology of tropical
marine ecosystems. It is apparent, however, that communities
of both plants and animals are characteristically diverse, with
large numbers of species and complex inter-relationships. The
economic importance of shallow tropical seas and mangrove swamps
in the production of fish, shellfish, seaweed products and mangrove products such as timber and tanning bark, is beyond doubt,
though reliable statistics are difficult to obtain and subsistence fishing is often unreported.
    This paper has two main parts. The first part briefly
describes major marine ecosystems and potential oil industry
influences, and is necessarily general and speculative. The
second part describes mangroves in more detail, with reference
to studies in Indonesia to give examples of the fate and effects
of oils. Finally, suggestions are made for assessing and
monitoring possible oil industry effects in tropical marine
ecosystems.

TROPICAL MARINE ECOSYSTEMS

An ecosystem consists of interacting living and non-living
components. For example, a mangrove ecosystem contains several
tree and algal species, birds, insects, fish, crabs, many root
and mud-dwelling invertebrates (e.g. oysters, snails, barnacles,
polychaete worms), bacteria and fungi, organic detritus, sediments,
water of varying salinities, and elements such as nitrogen,
phosphorus and sulphur cycling in organic and inorganic forms.
Ecosystems are rarely closed, for example, mangroves export

dead leaves and organic detritus on the falling tide, and the rising tide brings into the system dissolved nutrients and shoals of fish.

The plant and animal part of the ecosystem form a biotic community, defined by Odum (1971) as "any assemblage of populations living in a prescribed area or physical habitat". Communities may be described in terms of their species, composition and abundance, or in terms of trophic levels (producers, herbivores, predators), food chain relationships and energy flow.

In tropical seas it is convenient to differentiate between the following habitats and their associated communities, though the boundaries between them are rarely sharply defined :

Intertidal Systems

Rocky shores. Unlike many temperate shores that have been described (Lewis 1964), the upper part of the intertidal zone of tropical rocky shores is often dominated by oysters. Barnacles, winkles, limpets, coralline algae and the brown alga Sargassum also occur, and distinct zones can be recognized (see for example Malley et al 1978; Stephenson and Stephenson 1972).

Sandy shores and mudflats. Depending upon degree of exposure to wave action, the sediments may be coarse, mobile sand with relatively few animals or fine sand and mud with relatively large numbers of molluscs, crabs and other invertebrates. Malley et al (1978) briefly describe sandy beach faunas of West Sabah; some (e.g. cockles and other bivalves) are of local economic importance.

Mangrove Swamps. Mangroves are upper shore trees or bushes which occur on sheltered shores and in estuaries throughout the tropics, reaching their greatest luxuriance in parts of south east Asia. Mangrove trunks and prop roots usually support a varied fauna of oysters, snails, barnacles, crabs and other invertebrates. Comprehensive accounts of mangroves are given by Macnae (1968) and Chapman (1977) and a case history description of a mangrove area in Indonesia is given in a later section of this paper.

Shallow Sub-Tidal Systems.

Seagrass beds. Seagrasses grow rooted in sand, silt or mud and play an important role in stabilizing bottom sediments. They affect the local physical environment by absorbing the energy of waves and tidal streams and by removing the sediments from the water column. Grassbeds support large numbers and a large variety of animals, and are the "nursery" for young stages of many important animals such as penaeid prawns (Basson et al 1977). Basson et al give a general account of Western Arabian Gulf grassbeds; and a number of tropical seagrass systems are described by McRoy and Helfferich (1977).

Coral reefs: reef flats (reef platforms). Reef flats consist of sand, rock, coral fragments and small colonies of hardy coral species. Upward growth of these colonies is limited by the level of low spring tides. Reef flats may extend over large areas between the intertidal zone and the reef slope (see Fig. 3) and may be colonised by seagrasses, and a range of algal and invertebrate species.

Coral reefs: reef slopes. Reef slopes are the region of maximum coral development, with the heads, fans or branches of the different coral species and a wealth of invertebrates and fish.
Corals and coral reef communities feature in a following paper by J P Ray so are not described further here.

Bottom Sediments. The distribution of subtidal sediment comm- unities is controlled mainly by sediment particle size (which is related to sediment mobility), organic matter content, salinity and light penetration. Relatively stable bottoms are likely to have many species of burrowing polychaete worms, crustaceans, molluscs and other invertebrates. Sanders (1968) compared a number of benthic samples and found the highest diversities among the tropical shallow marine samples.

Open Water Systems.
Plankton and nekton. Microscopic plants (phytoplankton) and animals (zooplankton) drift in the water more or less passively. Larger, actively swimming animals (nekton) include fish, whales, turtles, squids and crustaceans such as prawns. Data on plankton, fish and crustacean productivity and potential for tropical and other seas are given by Gulland (1971) and FAO (1972).
Some examples of feeding relationships within and between these various communities are shown in Fig. 1.

Productivity and Diversity.
In the open water, the dominant producers (photosynthesisers) are the phytoplankton - unicellular algae such as diatoms and dinoflagellates. In shallow water near the shore, the main producers may be larger attached algae or seagrasses. In the intertidal zone, larger algae may again be important, though mangroves dominate large areas of sheltered shore. Plant material may be consumed directly, though in many cases it enters marine food chains in the form of organic detritus (rotting leaves with associated bacteria, fungi and protozoa). Primary consumers of plant material and detritus include zooplankton, a range of polychaete worms, molluscs, crabs and other inverte- brates, some species of fish, and green turtles (one of the few direct consumers of seagrasses). These animals in turn are fed upon by secondary and further orders of consumers as illustrated in Fig. 1.
Comparative productivities are discussed by Odum (1971). Gross primary productivity is the total rate of photosynthesis during the period of measurement for any specified community. Net primary productivity is the rate of storage of organic matter not used by consumers. Rates of energy storage at consumer levels are referred to as secondary productivities.
Mangrove swamps, seagrass beds and coral reefs have higher gross primary productivities (estimated at 20,000 K cal/m^2/year or more) than most other marine or terrestrial communities (e.g. grasslands and pastures estimated at 2,500 K cal/m^2/year, boreal coniferous forests 3,000 K cal/m^2/year, fuel subsidized (mech- anized) agriculture 12,000 K cal/m^2/year, and the open ocean 1,000 K cal/m^2/year, Odum (1971). Further productivity data are available in Odum (1971) - comparisons of a variety of communities; Golley et al (1962) - mangroves; and McRoy and McMillan (1977) - seagrasses. Contributing factors to the high productivity of these tropical marine communities are sustained high temperatures and the "energy subsidy" contributed by tidal

movements which distribute dissolved nutrients, carbon dioxide and oxygen.

In many tropical systems such as coral reefs or mature mangrove swamps, there may, however, be very little net community productivity because nearly all the gross primary productivity is dissipated through the respiration of the producing plants themselves plus that of complex webs of consumers. Such systems may be described as steady state rather than growth ecosystems, Odum (1971); they tend to have a high diversity of species with complex inter-relationships. In contrast, managed systems such as brackish water fish ponds (described in a later section on "Marine resources vulnerable to oil pollution"), achieve high net community production or secondary production by using small numbers of selected species, by reducing or eliminating other species (which are viewed as pests) and by adding nutrient "subsidies".

Sanders (1968) summarizes and discusses a number of hypotheses which have been advanced to explain differences of diversity with latitude. With particular reference to marine benthic invertebrates, he develops the idea of physically controlled and biologically accommodated communities. In physically controlled environments, physical conditions fluctuate widely, organisms are exposed to severe physiological stress and the number of species is therefore low. Biologically accommodated communities are present where physical conditions are more or less constant for long periods of time. Under these conditions, biological stress (such as competition or lack of equilibrium in prey – predator relationships) is gradually mediated through biological interactions leading to a stable, complex, "buffered" community of a large number of specialized species with narrow physiological tolerances. Tropical shallow water marine communities (together with tropical rain forest and deep sea regions) best represent such conditions. Addy (1979) discussed the important question of stress tolerance (e.g. pollution tolerance) in physically controlled and biologically accommodated communities. As the high specialisation and narrow physiological tolerances of species in biologically accommodated communities does not favour adaptation, it may be speculated that tropical shallow water communities will be less tolerant of pollution than more physically controlled communities such as those of temperate estuaries.

TROPICAL MARINE POLLUTION

The most useful introduction to this subject is the book "Tropical Marine Pollution" (Ferguson Wood and Johannes 1975). They list and discuss a large number of ways (physical, chemical, community structure and biological functions) in which shallow tropical marine ecosystems differ from their temperate counterparts. Regarding the effects and fate of pollutants, several experimental studies (e.g. Cairns et al 1975) have shown that a range of pollutants (e.g. ammonia, cyanide and some heavy metals) are more toxic to test animals (mainly fish) at higher temperatures. Dilution and dispersion of pollutants may be slower on average than in temperature waters because of the comparative shallowness of many tropical seas and their lower tidal amplitudes. On the other hand, it may be expected that biodegradable or chemically unstable pollutants will degrade faster in warmer waters. Generalising, it may be predicted

that acute, local effects of pollutants will be more severe in
the tropics, but that accumulation of many pollutants (including
petroleum hydrocarbons and many constituents of refinery
effluents) is less likely to be a problem. If, however, the
acute local effect of a particular pollutant happens to be the
destruction or partial disruption of a complex biologically
accommodated community,recovery may take a long time for
inherent biological reasons, regardless of how long the
pollutant persists.

The Oil Industry and Tropical Marine Pollution.
Marine communities may be subjected to a variety of stresses
resulting from oil industry activities. These may include spills
from all types of shipping, and discharges from offshore oilfield installations.
     It is difficult to decide how best to deal with oil slicks
in shallow tropical water. There is world-wide interest at
present in the aerial spraying of floating slicks with
dispersant concentrate - this could minimise the fouling of
fishing gear (e.g. nets, fish traps and fishing weirs) and the
possibility of oil being stranded on tourist beaches or in
mangrove swamps, but dispersed oil would enter the water column
where its effects are unknown. Also unknown are the effects
of mis-directed or windblown dispersant sprays on coastal
vegetation.
     Oil may also enter coastal waters from land-based discharges
such as refinery effluents. Treated refinery effluents may
contain oil, phenols, sulphides, mercaptans, cyanides, ammonia,
some heavy metals, and suspended solids. Variation in concentration of these components occurs both between refineries and
within any particular refinery over a period of time. This
reflects differences in age, size or organization of refineries
and the processed being carried out (CONCAWE 1979).

Marine Resources Vulnerable to Oil Pollution.
The following examples are drawn from the seas and shores of
Indonesia. They illustrate the importance of considering
resources on a regional basis, and the occurrence of activities
unfamiliar in the west, which must be taken into account for
contingency planning.
     Marine fish are a staple food in many parts of Indonesia.
Most of the fishing activity is carried out in water no deeper
than 20 m. and traditional shallow water fishing methods such
as those seen by Earl (1837) are still used:".....we anchored
in Semarang Roads.....In the shallow water near the mouth of
the river are fixed a number of fishing weirs, made of bamboo,
each of which has a small watch-box attached to it, erected on
piles".
     Brackish water fish ponds (tambaks) are unique to south
east Asia and particularly common in Java, where they have been
used for the culture of fish for centuries. The elongated ponds
$\frac{1}{2}$ - 2 hectares in area, are cleared out of mangrove swamps,
with seaward fringes of mangroves left as protection against
rough seas. Each pond can be drained or flooded by means of a
sluice gate opening into a tidal canal system. The ponds are
used for the culture of milkfish Chanos chanos which are introduced as fry. The fish feed on algae, and in some areas the
yield is improved by green manuring (often using mangrove leaves).
     Ponds are also used for catching prawns or other crustaceans,

which are admitted in a controlled inflow of seawater. Adult prawns may be caught by trapping them in nets at the sluices on the outgoing tide at night (many species of prawn bury themselves in the mud by day and swim around at night).

Fish ponds have been described by Schuster (1952), Hall (1962) and Macnae (1968). LEMIGAS (1974) give an annual production figure of 52,000 tons of fish from 177,000 hectares of fish ponds in Java and point out that there is potential for increasing this figure. Odum (1971) gives comparative figures for fish productivity (a form of secondary productivity, see section on "Productivity and diversity"). Milkfish cultures in ponds in the Philippines have a productivity of 112 - 202 K cal/m^2/year, which makes an interesting comparison with the North Sea fisheries (5.0 K cal/m^2/year) and the anchovy fishery of the Peru current upwelling area (335 K cal/m^2/year, the world's most productive natural fishery).

The LEMIGAS report (1974) draws attention to the vulnerability of both fish ponds and coastal fisheries to oil pollution. The occurrence of oil slicks at times when ponds are being flooded or used to catch crustaceans, could seriously diminish their use; and other coastal fishing could be affected by fouling of nets, traps, and weirs and subsequent tainting or avoidance behaviour of fish.

Shellfish and seaweeds collected in the intertidal zone are important subsistence foods, and Rao (1965) lists more than 20 species of seaweed which are eaten in Indonesia. Indonesia is also a major producer of the red seaweeds used in the production of agar and some of these, notably species of Eucheuma, are now being cultured with a view to increasing exports. Oiling of shores or seaweed culture rafts could therefore affect subsistence activities or seaweed exports.

Mangroves (which in Indonesia occupy an estimated 6,000,000 hectares - Bardach, quoted in LEMIGAS 1974) are of direct commercial importance in the production of timber, firewood, charcoal and bark for tanning. They are the feeding and breeding grounds for a variety of fish, crustaceans and molluscs. Mangrove leaves are an important source of the detritus upon which many marine food chains are based (see Fig. 1.). Mangroves, by trapping and stabilising sediment, increase land area over the centuries and also protect nearshore corals from being overwhelmed with sediment. Some idea of accretion on mangrove coasts is given by the fact that when Palembang in south eastern Sumatra was visited by Marco Polo in 1292 it was still a coastal or rivermouth port; it is now 50 km. inland (Macnae 1968). The indirect economic value of mangroves (as sediment traps, fish feeding grounds, detritus sources, etc.) is difficult to calculate in the absence of comprehensive food chain, productivity, energy flow and accretion data.

Mangroves (like their temperate equivalents - salt marshes) appear to be "oil traps" in some cases and a range of effects following oiling have been noted. These are discussed in a later section.

Tourism is being increasingly developed in many tropical countries. Two Indonesian areas worthy of note are :

1. The Kepulauan Seribu - a group of coral islands in the Java Sea north of Jakarta, being developed for tourism and recreation under the Jakarta development plan. There are offshore oil wells approximately 20 km. west of the northern segment of the Kepulauan Seribu, and the islands would be

vulnerable to spills occurring during the west monsoon
(LEMIGAS 1974).
2. The island of Bali is a well known tourist attraction;
its beaches would be particularly vulnerable to spills
resulting from tanker accidents in the straits of Lombok.

MANGROVE ECOSYSTEMS AND THE OIL INDUSTRY

Some effects of petroleum hydrocarbons on mangrove systems
reported in the literature are summarised in Table 1. An overall impression from the literature is that following severe
oil spills, the acute short term effects are likely to be
trapping of oil, high mortalities of invertebrates, defoliation
of mangroves, and death of seedlings. In the longer term,
oil is likely to weather comparatively quickly and both mangrove and invertebrate re-colonisation have been observed.
The following summary of biological and chemical work
carried out following an oil spill in Indonesia provides further
information on possible oil effects on mangroves.

A Post-Oil Spill Survey in Indonesia.
The "Showa Maru", en route for Japan with a cargo of Arabian
light, Berri and Murban crude oils, grounded near Buffalo Rock
(Kareng Banteng) on 6th January 1975 (see Fig. 2). It is
estimated that approximately 54,000 barrels of oil were spilled.
The main drift of the slicks (some of which were treated with
dispersants) was to the west and south west, and the following
reports of oil near or on the shore following the accident
were received from local people, officials and a team from the
Indonesian Department of Mines (see Figs. 2 and 3 for locations):
1. Oil at Takong, Pemping, Bulan, Kasu, Ampar and Lumba
Besar.
2. Fish traps oiled, north Kepal Jernih and Pemping.
3. Thick oil on shore, north Kepala Jernih.
4. Tar lumps stranded on shores of Karimun Besar.
A survey was carried out during July 1977 with the following
overall objectives:
1. To find if ecological changes attributable to "Showa Maru"
oil could be seen two and a half years after the grounding.
2. To find if "Showa Maru" oil remained in the area, and if
so where and in what state.
3. To collect information on physical, chemical and biological factors relevant to interpretation and prediction
of ecological changes.
Offshore bottom sediment samples generally showed low level or
negligible petroleum hydrocarbon contamination (see Fig. 2).
A small number of reef flat samples likewise showed only low
levels of petroleum hydrocarbons, and no gross damage to the
reef flat flora and fauna was observed. An extensive reconn-
aisance (by boat, landing at intervals for spot checks and
sediment samples) was made of the intertidal zone. Dead mangroves were observed on the islands of Kepala Jernih and Pemping,
and these areas were therefore studied in more detail.

Mangrove Study Methods.
Mangrove study methods were based upon the previous experience
of the mangrove survey team from the Faculty of Biology, Gadjah
Mada University, Yogykarta. Transects were laid out across the
main areas of interest (see Figs. 4 and 7). Each transect

consisted of a series of 10 m. x 10 m. quadrats, the numbers
of which varied with the width of the mangrove zone. Numbers
of individuals of the various species and their basal areas were
recorded for each quadrat and the data summed and processed to
give for each transect the relative density, relative frequency
and relative dominance of each species. The sum of these three
values is the importance value of each species (Curtis 1959,
Shimwell 1971, Krebs 1978). Importance values of animal species
were found by using 1 m. x 1 m. quadrats within the larger
quadrats, and summing relative frequencies and relative densities.

Transect data and other measurements of distance along the
shore were used to produce maps showing the approximate extent
of damaged mangroves (see Figs. 4 and 7). Sediment samples
(unless otherwise stated these were surface samples from 0 - 10
cm.) were taken using a corer at the sites shown on Figs. 3, 4
and 7. They were placed in pre-cleaned 600 ml. metal cans with
lever lids, and shaken with 2 ml. of quaternary ammonium compound
preservative. The sediment samples were subsequently examined
using ultr-violet spectroscopy, capillary gas-liquid chromatography and computerised gas chromatography - mass spectrometry.
Hydrocarbon analytical methods are described in detail in the
Appendix.

Results.
The main findings are summarised in Figs. 2 - 8. Fig. 2 shows
the general survey area with positions of offshore sampling
stations and crude oil content of bottom sediments (generally
very low) as determined by U.V. Fig. 3 gives the positions of
mangrove survey and sediment sample sites with crude oil content
of Rhizophora zone sediments as determined by U.V. Figs. 4A
and 7 provide further detail for two areas of particular interest
(Kepala Jernih and Pemping). In general the shoreline sediments
contained higher levels of hydrocarbons as determined by U.V.
than the benthic sediments, and particularly high levels were
found in some samples from Kepala Jernih, Pemping and Takong
Besar. Areas of dead and damaged mangrove were found at Kepala
Jernih, Pemping and Ayer Puluh. In the first two cases, these
areas showed correspondence with high sediment hydrocarbon
levels. Figs. 4A, 4B and 7 illustrate how the dead and damaged
mangroves commonly formed pockets behind seaward fringes of
living Sonneratia or Rhizophora. Fig. 5 summarises data from
the Kepala Jernih transects and indicates that crabs (Uca sp.)
and snails (family Littorinidae) were reduced in numbers in areas
with high levels of damaged mangrove/high sediment hydrocarbon
levels. There were some scattered seedlings, generally at a
density of 1 per 100 m^2 or less, rising to 1 per 10 m^2 on transect
11 and 2 per m^2 on transect 8.

The purpose of the glc and c-gc-ms analyses was to investigate
selected sediment samples in more detail. Some of the glc traces
for Kepala Jernih are shown in Fig. 6, for Pemping in Fig. 7, and
for Takong Besar (together with the reference blend trace) in
Fig. 8. A component with a retention time of ca n-C$_{14}$ alkane
was found to be attributable to the preservative.

A summary of the glc data shown in Figs 6 - 8 is given in
Table 2. Analysis of the sediment sample from Ayer Pulah (a
site where patches of dead mangroves were seen) indicated
negligible inputs of petroleum hydrocarbons.

It was not possible from U.V. and glc analyses to attribute
the pollution specifically to oil from the "Showa Maru".

Examination of samples by the more specific technique of c-gc-ms indicated that the closest similarities to the "Showa Maru" blend were shown by the samples from Takong. The triterpane distribution of the Kepala Jernih and Pemping samples were similar to each other but showed dissimilarities to those of the "Showa Maru" blend. The sterane distributions of onshore samples generally showed some features in common with the blend but could not be attributed specifically to the blend on the basis of the analytical methods employed.

Conclusions and Discussion.

On the islands of Kepala Jernih and Pemping, the occurrence of dead and damaged mangroves is associated with comparatively low numbers of crabs and snails and with comparatively high petroleum hydrocarbon residues in the sediments. There is some evidence from the Kepala Jernih samples that relatively undegraded hydrocarbons occur below the surface (20 - 30 cm.) in the anaerobic zone. Petroleum hydrocarbon contamination was negligible at another site (Ayer Pulah) where patches of dead mangroves were found. Conversely, contamination by both relatively undegraded and by degraded hydrocarbons was found on Takong where the mangroves appeared healthy.

This is a complex picture. Eye witness accounts indicate that "Showa Maru" oil entered the area, and while this is not contradicted by the sediment hydrocarbon data, these suggest other petroleum hydrocarbon inputs as well. The areas of greatest damage were the sheltered bays of Kepala Jernih and Pemping, which would tend to act as "oil traps" to a greater extent than the more exposed shores of Takong. The fact that dead mangroves occur in pockets suggests that they were killed by the stranding of slicks at that particular tidal level. Little is known about the mechanisms of damage but it may be suggested that oil on the prop roots and pneumatophores interferes with oxygen supplies to the root system by blocking the air holes. It is worth noting that the existence of living mangroves between dead pockets and the sea means that damage assessment carried out solely from nearshore boats is misleading.

The low levels of seedlings indicated that re-colonisation was at an early stage in July 1977.

ASSESSING POSSIBLE OIL INDUSTRY EFFECTS IN TROPICAL MARINE ECOSYSTEMS.

The post-oil spill survey in Indonesia provides an example of a "one-off" survey. These are useful for locating and mapping areas of obvious damage and the distribution of pollutants or pollutant residues. Single surveys cannot usually be used, however, for assessing trends of damage, recovery rates or natural fluctuations. Such measurements require a surveillance or monitoring programme. The terms monitoring and surveillance are widely used but precise definitions are not unanimously agreed. Monitoring commonly implies measurements taken at intervals and compared with a standard and the term surveillance is used in cases where measurements are repeated at intervals but a standard has not been defined.

Repeated quantitative biological and chemical surveys over thousands of miles of tropical seas and shores would certainly help in the assessment of oil spill damage but are hardly a practicable proposition. Another approach is to carry out "one-off" post-spill surveys to give a broad perspective; and quantitative field experiments involving the application of oils and dispersants to small field plots, to give detail. Monitoring effort may be usefully concentrated round installations and discharges (for example, offshore platforms or refinery effluents) where chronic pollution and progressive effects are most likely to occur.

It is essential to clearly define the objectives of any monitoring or surveillance programme and these of course may vary from place to place. Examples of different objectives and related monitoring programmes are given in the CONCAWE report on the environmental impact of refinery effluents (CONCAWE 1979). Discharge area schemes should take into account a number of sites at different distances from the discharge point of the effluent, so allowing the detection of spatial as well as temporal change. This is the approach which has been adopted for the Cilacap monitoring project in Indonesia. The Cilacap area has the only large mangrove system along the south Java coast, and is an important area for fishing, particularly for prawns. The Cilacap estuary is increasingly industrialised and an oil refinery started operating there in 1976. Refinery effluent is discharged into the estuary and a mangrove monitoring scheme was therefore initiated in 1975. The scheme comprises detailed vegetation mapping based on aerial surveys coupled with ground checks. In addition, quantitative plant and animal data are obtained at permanent check points at different distances from the refinery, and processed as described in the "post-oil spill" section of this paper.

Monitoring schemes like this can provide data on size of effect (if any), the species affected, and whether the area of damage is increasing from year to year, decreasing or remaining stable. Together with field and laboratory experimental data and post-oil spill survey results, they provide the information that is essential as feedback for judging the adequacy of pollution control practice.

ACKNOWLEDGEMENTS.

The authors would like to thank their colleagues in the 1977 survey team, namely Prof. Drs. Kardono Darmojuwono, Prof. Dr. Wahjudi Wisaksono, Prof. Dr. Sumardi Sastrakusumah, Dr. Aprilani Soegiarto, Ir. Jaspar Bilal, Ir. Dedi Sudharma, Ir. Widodo, Ir. Henk Uktolseja, Ir. Bambang Prasetyo, Drs. Soenarto Hardjosoewarno, Drs. Anthon Soekahar, Drs. Agus Pudjoarinto, and Gunawan. Assistance was kindly provided by Bogor Agricultural University, Gadjah Mada University, the Indonesian Embassy in Singapore, the Indonesian Naval Hydro-Oceanographic Service, the Marine Fisheries Research Institute (Jakarta), Pertamina (Pulau Sambu), the Port Authority of Pulau Sambu, Captain Peter Bird (Transworld Marine Ltd, Singapore), Mr J M Bruijn (Caltex Pacific Indonesia, Jakarta), Captain John Pounder (Ritchie and Bisset, Singapore), the master and crew of the "Asiatic Valour", and many other people and organizations. Our thanks to Mr A Aldridge for assistance with hydrocarbon analysis and to Mrs Jo Brazier for typing the manuscript.

The survey would not have been possible without the support of the Department of Justice, Jakarta, and the Britannia Steamship Insurance Association Ltd., London. We are particularly grateful to Mr Robert Seward of the Britannia Steamship Insurance Association Ltd. for his help on many occasions.

Table 1   Some effects of petroleum hydrocarbons on
                           mangrove communities. (Short reviews of
                           this subject are also given in LEMIGAS/
                           Smithsonian Institution 1974, Ferguson Wood
                           and Johannes 1975, and Thia-Eng and Mathias,
                           1978).

Incident/Area	Effects	References
Natural asphalt deposits, Lake Guanoco, Venezuela	*Pterocarpetum rhizophorosus* reported growing adjacent to and on top of deposits.	Lasser and Vareschi 1959.
Spill of c.70,000 barrels of crude oil from the "Argea Prima", Puerto Rico 1962.	Oil accumulated round mangrove roots. Large mortalities of many species of invertebrate, turtles and some fishes.	Diaz-Piferrer 1962.
Spill of 15,000 barrels of diesel fueld and "Bunker C" fueld from the "Witwater", Panama 1968.	Intertidal mud soaked with oil *Avicennia nitida* pneumatophores and *Rhizophora mangle* prop roots oiled (long term effects not detailed). Many invertebrate species reduced or eliminated.	Rutzler and Sterrer 1970.
Spill of c.100,000 barrels of light crude oil from pipeline break, Tarut Bay, Saudi Arabia 1970.	Mangrove leaves oiled on lower half of shrubs, but soil level usually free from oil. Some defoliation observed three months later, but many survived. No evidence of damage to fauna in creeks.	Spooner 1970.
Experimental diesel oil spills (oil mixed with sea water in ratios of 100 ppm, 1,000 ppm, 10,000 ppm, and 100,000 ppm.	Lowest concentrations caused partial defoliation followed by recovery, but the growth form of the plants was frequently altered due to the death of the meristem. At the two highest concentrations, all seedlings died within two weeks.	Mathias 1976.
Spill of crude oil from the "St Peter" Colombia/Ecuador 1976.	Short term effects – thick oil covering on roots and trunks of mangroves. Partial defoliation. Large mortalities of sessile invertebrates (barnacles, mussels, oysters). Reduction of fiddler crab numbers. Longer term effects – wave action removed much of the oil; previously defoliated parts recovered; invertebrate populations recovered well, possibly by migration from unaffected areas.	Jernelov et al 1976.

Incident/Area	Effects	References
Discharge of 1,500-3,000 barrels of crude oil "clingage" Florida Keys 1975.	Rhizophora mangle seedlings sustaining greater than 50% oiling of their leaves were killed. Dwarf Avicennia nitida with greater than 50% oiling of pneumatophores were also killed. Lesser degrees of oil coating resulted in continued growth despite leaf loss and chemical burn scars. Several crab species eliminated from mangrove fringe for several months.	Chan 1976, Chan 1977.
Spill of 37,000 barrels of Venezuelan crude oil from the "Zoe Colocotroni", Puerto Rico 1973.	In one area (1.0 hectare) Rhizophora mangle and Avicennia nitata defoliated and died during the three years following the spill. Mangrove prop root invertebrates were reduced. By 1977 the oil was highly weathered even in the most heavily oiled areas, and some recolonisation was observed.	Nadeau and Bergquist 1977. Page et al 1979.

Table 2   Summary of glc data shown in Figs. 6 - 8.

Site	Low Molecular Weight Range	High Molecular Weight Range	Others	Inferences
Kepala Jernih (0-10cm)	Minor contribution	Definite contribution	Hump	Contamination by degraded pollutants
Kepala Jernih (10-20cm)	Minor contribution	Some contribution	Major hump	Contamination by degraded pollutants
Kepala Jernih (20-30cm)	Major contribution Max. $n-C_{16}$	Minor contribution Max. $n-C_{25}$	No hump	Possible contamination by undegraded pollutants
Pemping	Negligible	Minor contribution	Major hump	Contamination by degraded pollutants
Takong (central site)	Major contribution	Minor contribution	Hump	Contamination by relatively unweathered and by degraded oils.
Takong (northern site)	Minor contribution	Negligible	Hump	Some contamination by degraded pollutants.

691

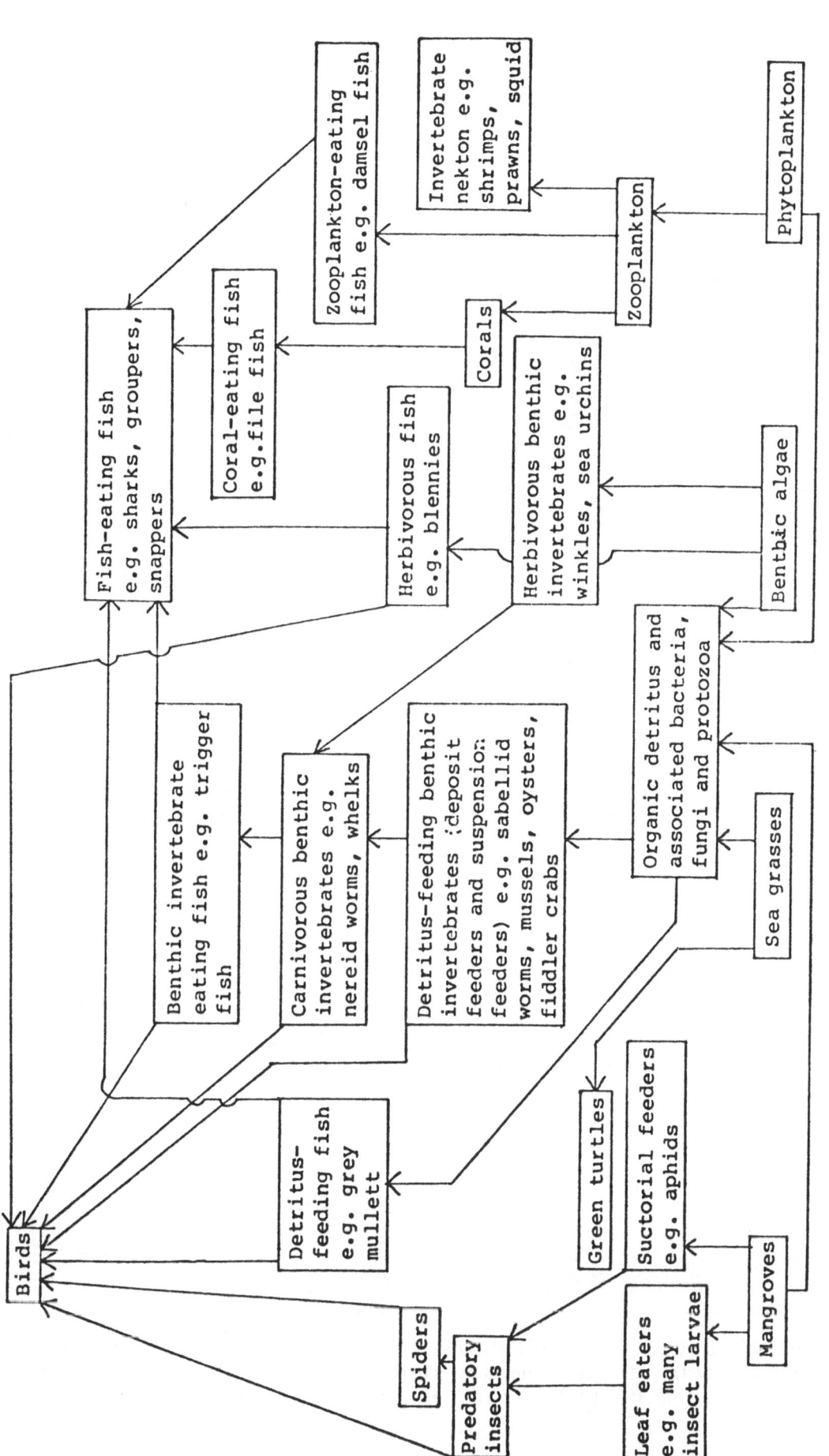

Fig. 1 Some examples of feeding relationships between groups of tropical marine organisms. This diagram is greatly simplified; not all groups of organisms or possible relationships are represented and changes in feeding behaviour (e.g. from herbivore to detritus-feeder) with life cycle stage or food availability are not shown.

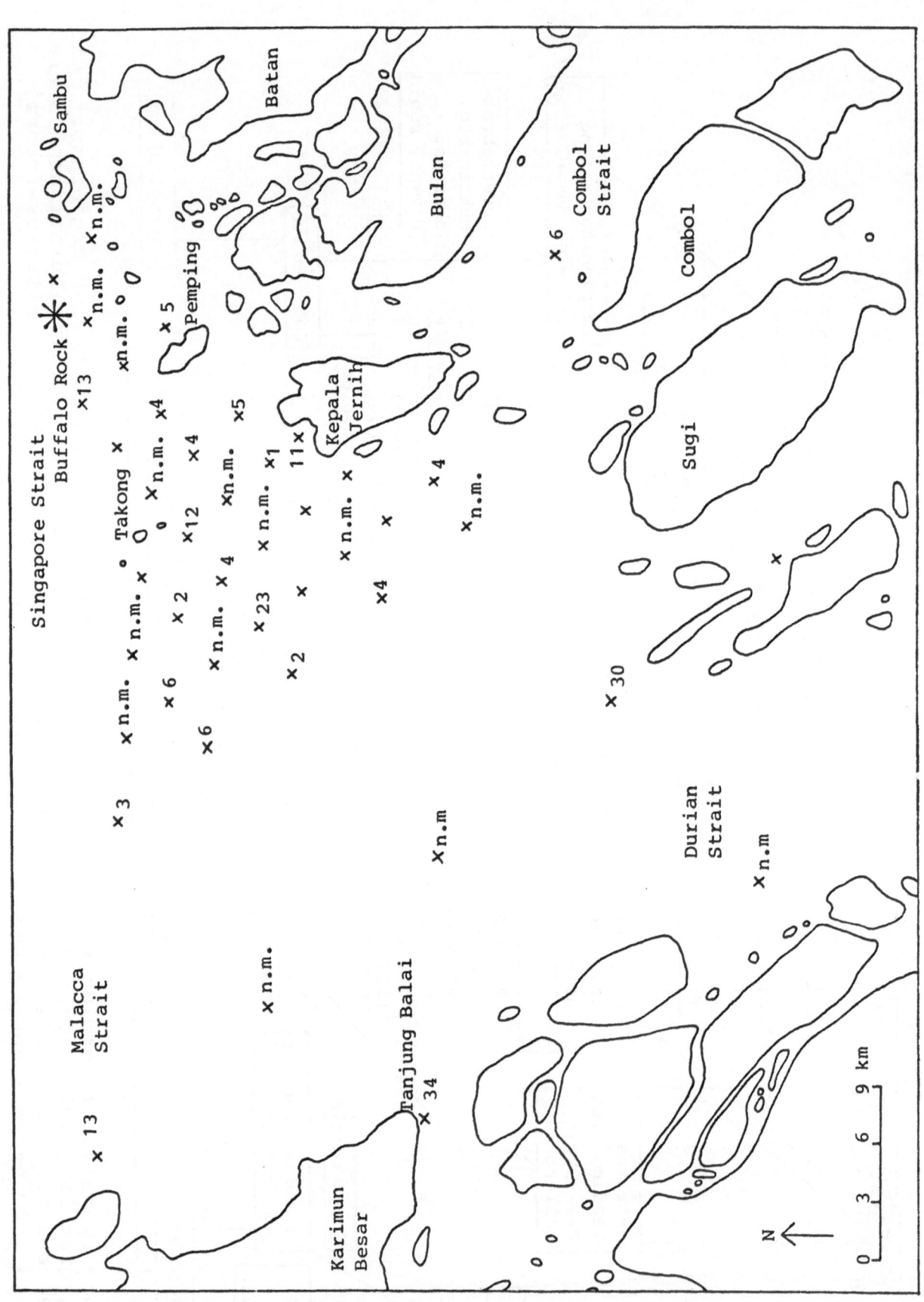

Fig. 2 Survey area, with offshore sampling stations and crude oil content of bottom sediments (ug/g dry wt. sediment as measured by UV). n.m. = not measurable.

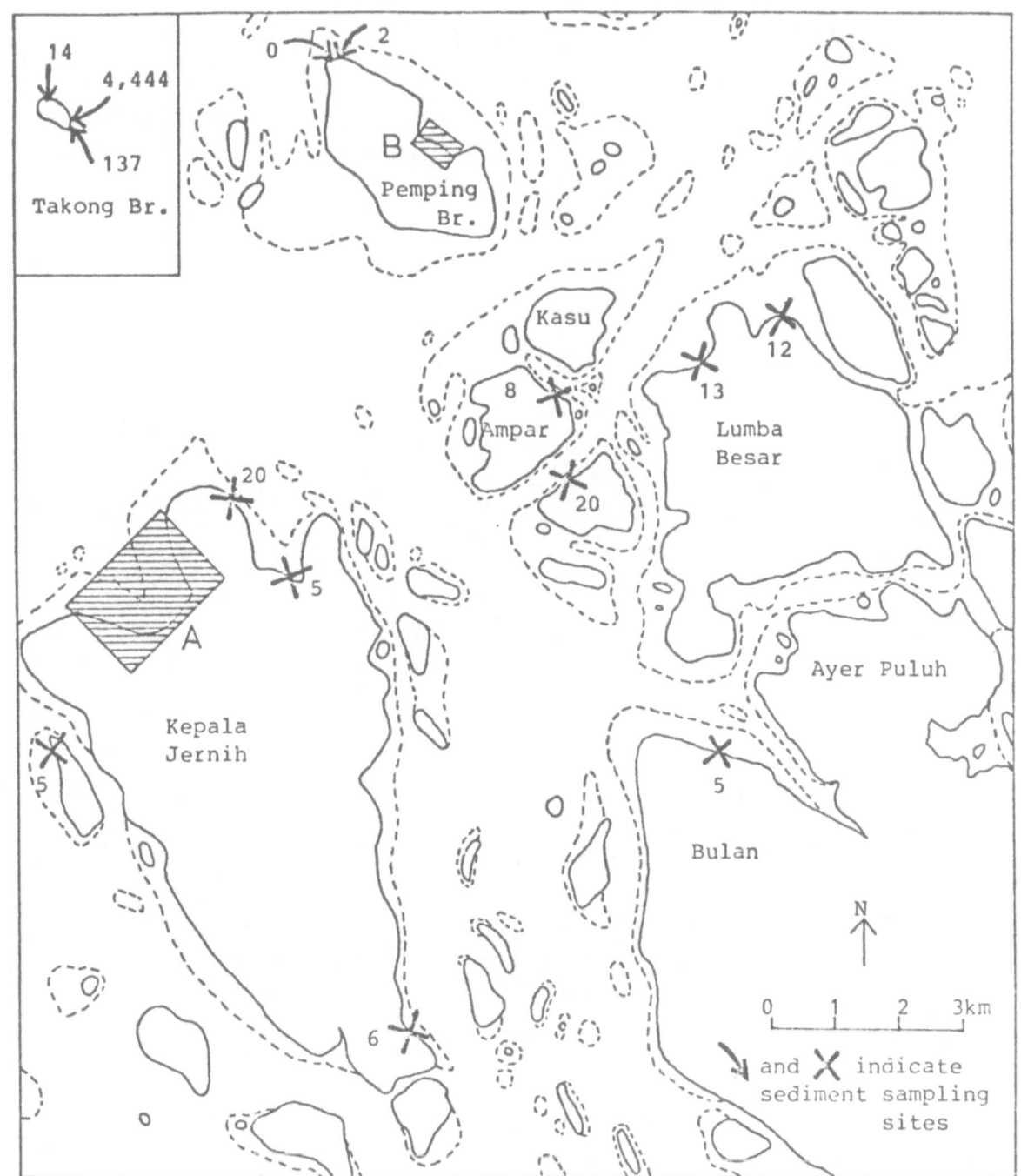

Fig. 3   Shore survey sites, with crude oil content of _Rhizophora_
zone sediments (ug/g dry wt. sediment as measured by
UV).   Areas A and B are shown in more details in Figs.
4 and 7 respectively.   Dashed lines indicate extent of
reef flats.

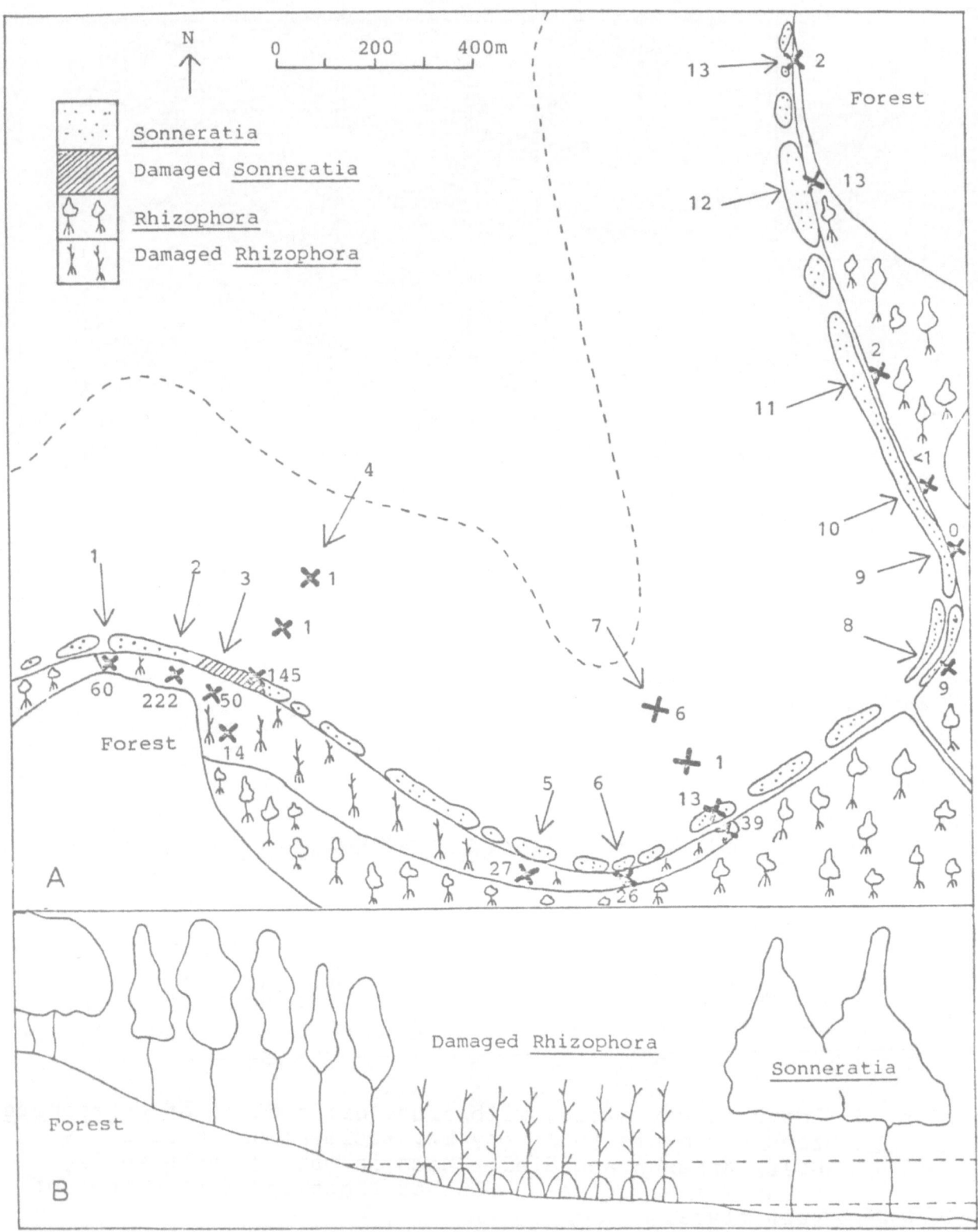

Fig. 4A. Kepala Jernih mangrove study area. Numbered arrows 1 -
13 indicate mangrove and sediment sampling transects.
Crosses and adjacent numbers indicate sediment sample
sites and the crude oil content (ug/g dry wt. sediment
as measured by UV). The dashed line shows the edge of
the reef flat.

Fig. 4B. Sketch profile of transect 2. Dashed lines show high and
low water.

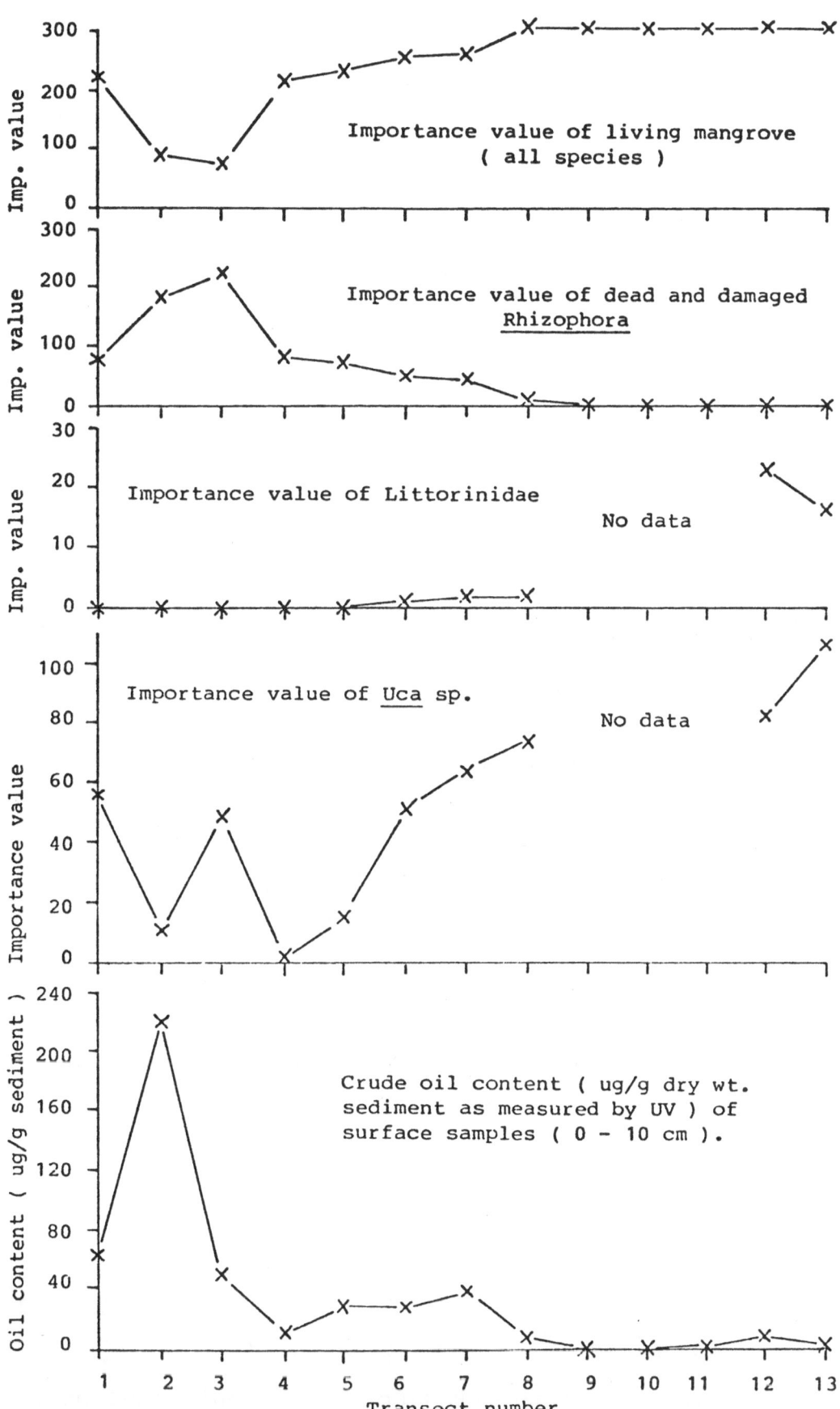

Fig. 5   Kepala Jernih transect data: Importance values and surface sediment oil contents.

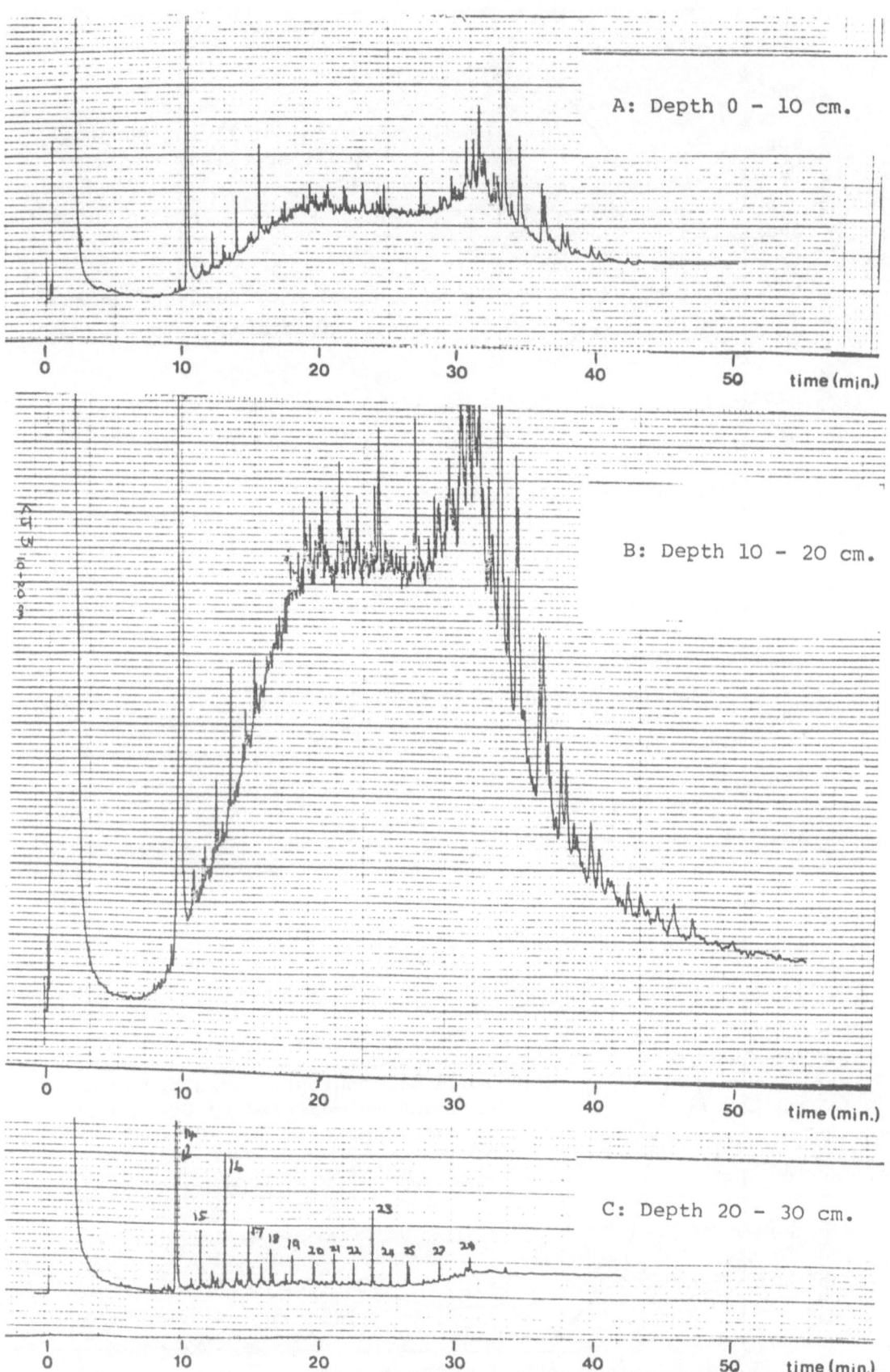

A: Depth 0 - 10 cm.

B: Depth 10 - 20 cm.

C: Depth 20 - 30 cm.

Fig. 6   GLC traces for the transect 3 sediment sampling site, Kepala Jernih.   Sample depths A: 0 - 10 cm; B: 10 - 20 cm; C: 20 - 30 cm.

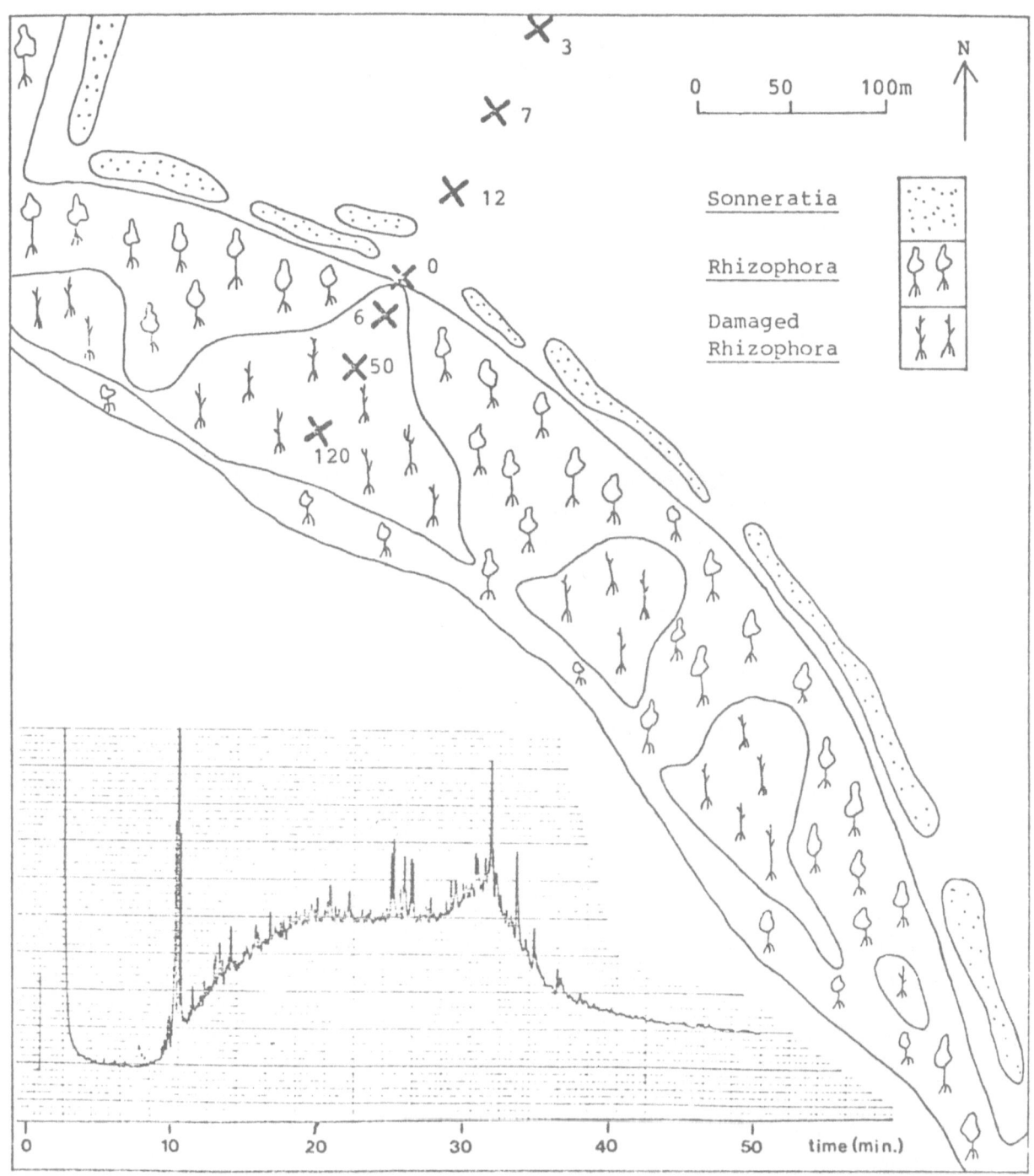

Fig. 7   Pemping mangrove study area. Crosses and adjacent
         numbers indicate sediment sampling sites and the crude
         oil content (ug/g dry wt. sediment as measured by UV).
         Inset : GLC trace for the southernmost sampling site.

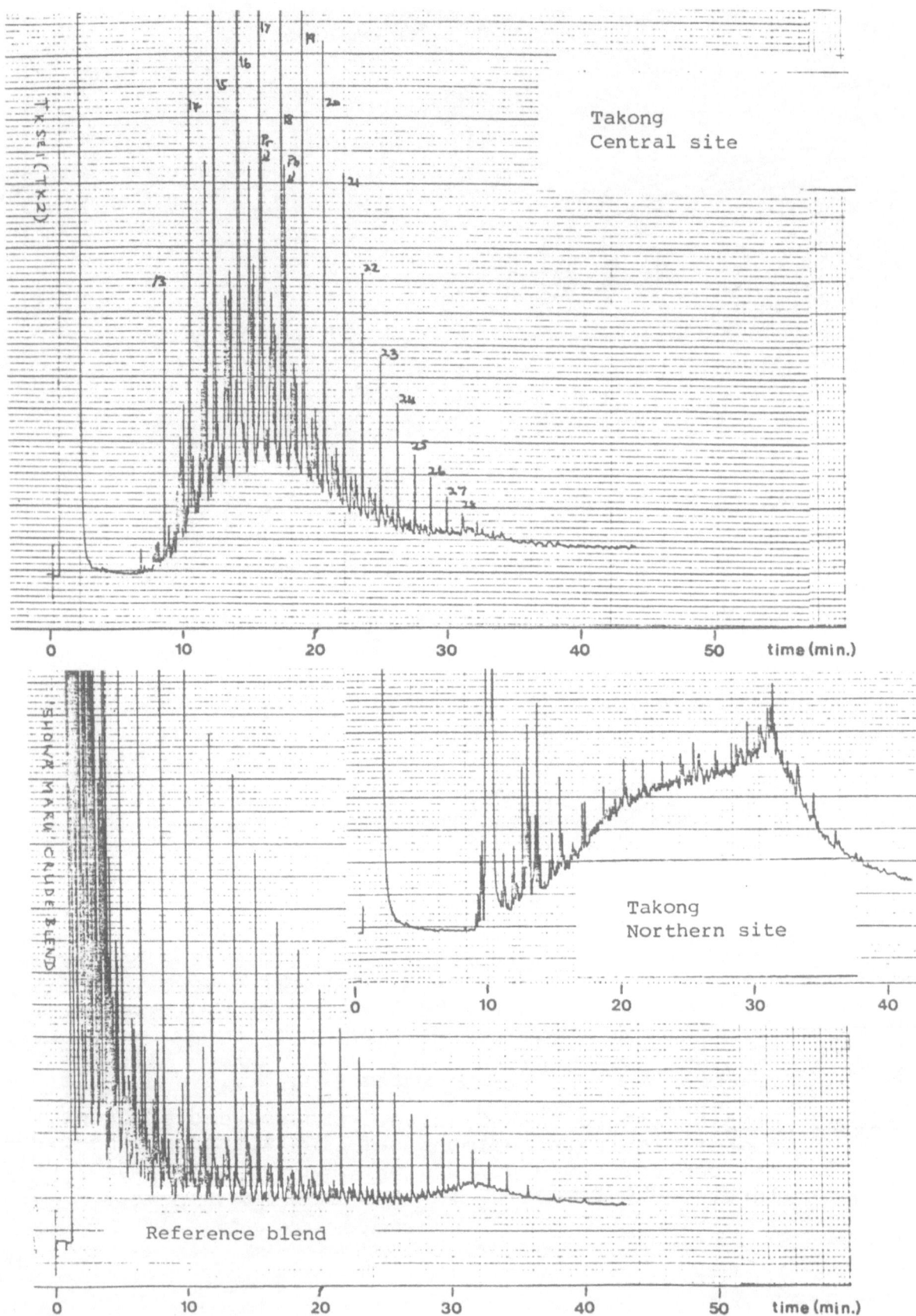

Fig. 8   GLC traces for intertidal sediment (0 - 10 cm.) from
Takong (see Figs. 2 and 3 for locations); and for the
reference blend of crude oils.

REFERENCES.

J M Addy, "Some studies of benthic communities in areas of oil industry activity", Ph. D. Thesis, University of Wales (1979).

P W Basson, J E Burchard Jr., J T Hardy and A R G Price, Biotopes of the Western Arabian Gulf (Aramco, Dhahran, Saudi Arabia, 1977).

J Cairns Jr., A G Heath and B C Parker, "Temperature influence on chemical toxicity to aquatic organisms", Water Poll. Control Fed., 47 (2), (1975), 267-280.

E I Chan, "Oil Pollution and tropical littoral communities; biological effects of the 1975 Florida Keys oil spill", Master's thesis, University of Miami, Rosenstiel School of Marine and Atmospheric Science, Miami, Florida (1976).

E I Chan, "Oil Pollution and tropical littoral communities; biological effects of the 1975 Florida Keys oil spill", Proceedings of the 1977 Oil Spill Conference, API/EPA/USCG (1977), 539-542.

V J Chapman (ed.), Ecosystems of the World, 1, Wet Coastal Eco-Systems, (Elsevier, Amsterdam - London - New York, 1977), 428.

CONCAWE, The Environmental Impact of Refinery Effluents, Report No. 5/79, (CONCAWE, Den Haag, 1979).

J T Curtis, The Vegetation of Wisconsin: An Ordination of Plant Communities, (Madison, Wisconsin, 1959).

M Diaz-Piferrer, "Effects of an oil spill on the shore of Guanica, Puerto Rico", Deep Sea Research, II (5), (1964), 855-856.

G W Earl, The Eastern Seas, (W H Allen, London 1837).

FAO, Atlas of the Living Resources of the Seas, (Prepared by the FAO Department of Fisheries, FAO, Rome, 1972).

E J Ferguson Wood and R E Johannes (eds.), Tropical Marine Pollution (Elsevier, Amsterdam - Oxford - New York, 1975), 192.

F Golley, H T Odum and R F Wilson, "The structure and metabolism of a Puerto Rican red mangrove forest in May", Ecology, 42 (1962), 9 - 19.

J A Gulland, The Fish Resources of the Ocean, FAO, Rome (Fishing News (Books) Ltd., West Byfleet, England, 1971), 225.

D N F Hall, "Observations on the taxonomy and biology of some Indo-West Pacific Penaeidae", Fishery Publs. Colon.Off.No. 17 (1962), 1-229.

C J Krebs, Ecology: The Experimental Analysis of Distribution and Abundance, (Harper and Row, New York - London, 1978), 678.

A Jernelov, O Linden and J Rosenblum, "The St Peter oil spill - an ecological and socio-economic study of effects, Colombia, Ecuador, May - June 1976", Institutet for Vatten-Och Luftvards-forskning, Publication B 334, Stockholm (October 1976), 34.

700

T Lasser and V Vareschi, "La vegetacion del Lago de asfalto de Guanoco", _Acta Biol., Venez._, 2 (1959), 407-452.

LEMIGAS (Lembaga Minyak dan Gas Bumi) and Smithsonian Institution (Office of International and Environmental Program), _Coastal Zone Pollution in Indonesia with Emphasis on Oil_, (LEMIGAS, Jakarta, 1974), 229.

J R Lewis, _The Ecology of Rocky Shores_, (English Universities Press, London 1964), 323.

W Macnae, "A general account of the fauna and flora of mangrove swamps and forests in the Indo-West Pacific region", _Adv. Mar. Biol._ 6 (1968), 73-270.

D F Malley, Ho Sinn Chye and J K Charles, "Beaches, rocky shores and coastal swamps", in C Thia-Eng and J A Mathias (eds.) _Coastal Resources of West Sabah: An Investigation into the Impact of Oil Spill_, (Penerbit Universiti Sains Malaysia, Pulau, Pinang, 1978), 152-170.

J A Mathias, "The effect of oil on seedlings of the pioneer mangrove, _Avicennia_ intermedia in Malaysia" (abstract), Symposium on Ecology and Management of Some Tropical Shallow Water Communities, Jakarta, 27th June - 10th July 1976. Results summarised in C Thia-Eng and J A Mathias (eds.) _Coastal Resources of West Sabah: An investigation into the impact of Oil Spill_, (Penerbit Universiti Sains Malaysia, Pulau, Pinang, 1978).

C P McRoy and C Helfferich (eds.), _Seagrass Ecosystems : A Scientific Perspective_, (Marcel Dekker Inc., New York - Basel, 1977), 314.

C P McRoy and C McMillan, "Production ecology and physiology of sea grasses" in C P McRoy and C Helfferich (eds.) _Seagrass Ecosystems: A Scientific Perspective_, (Marcel Dekker Inc., New York - Basel, 1977), 53-87, 314.

R J Nadeau and E T Bergquist, "Effects of the 18 March 1973 oil spill near Cabo Rojo, Puerto Rico on tropical marine communities", _Proceedings of the 1977 Oil Spill Conference_, API/EPA/USCG (1977), 535-538.

E P Odum, _Fundamentals of Ecology_, (W B Saunders Company, Philadelphia - London - Toronto, 1971), 574.

H T Odum, P Burkholder and J Rivero, "Measurements of productivity of turtle grass flats, reefs and the Bahia Fosforescente of southern Puerto Rico", _Publ. Inst. Marine Sci._, Texas, 6 (1959), 159-170.

W E Odum and R E Johannes, The response of mangroves to man-induced environmental stress" in E J Ferguson Wood and R E Johannes (eds.) _Tropical Marine Pollution_, (Elsevier, Amsterdam - Oxford - New York, 1975), 52-62, 192.

Dr S Page, D W Mayo, J F Colley, E Sorenson, E S Gillfillan and S A Hanson, "Hydrocarbon distribution and weathering characteristics at a tropical oil spill site", _Proceedings of the 1979 Oil_

Conference API/EPA/USCG (1979), 700-712.

G N S Rao, "Use of seaweeds directly as food", Indo-Pacific Fisheries Council, Bangkok (Reg. Stud. Indo-Pacif. Fish Coun. No. 2), (1965).

K Rutzler and W Sterrer, "Oil Pollution damage observed in tropical communities along the Atlantic seaboard of Panama", Bioscience, 20 (4), (1970), 222-224.

H L Sanders, "Marine benthic diversity - a comparative study", The American Naturalist, 102 (925), (1968), 243-282.

W H Schuster, "Fish Culture in brackish water ponds of Java", Diocesan Press, Madras (Indo;Pacif. Fish Coun. Spec. Publ. No. 1), (1952).

D W Shimwell, Description and Classification of Vegetation, (Sidgwick and Jackson, London, 1971), 322.

M Spooner, "Oil spill in Tarut Bay, Saudi Arabia", Mar. Poll. Bull., 1 (11) (1970), 166-167.

T A Stephenson and A Stephenson, Life between Tidemarks on Rocky Shores, (W H Freeman and Co., San Francisco, 1972), 425.

C Thia-Eng and J A Mathias (eds), Coastal Resources of West Sabah: An Investigation into the Impact of Oil Spill, (Penerbit Universiti Sains Malaysia, Pulau, Pinang, 1978).

APPENDIX

Hydrocarbon Analytical Methods.

Samples were extracted and fractionated into hydrocarbon type fractions by the following analytical methods:

To the samples in the cans was added petroleum ether: propan-2-ol (1:4) (100 ml.). Each mixture was then stirred, the can lids replaced and the contents thoroughly shaken. The resulting suspension was allowed to stand for 24 hours. After this period the supernatant liquid was decanted into solvent-rinsed glass jars. The whole extraction procedure was then repeated and the two solvent extracts for each sample combined.

To the extract was added petroleum ether (72 ml.) and water (48 ml.), the mixture was stirred and the layers allowed to separate. The petroleum ether lay was then transferred to a round-bottomed flask (500 ml.), evaporated on a rotary evaporator and transferred to a pre-weighed clean glass vial. The resulting total organic extract (TOE) was weighed.

An aliquot of the TOE was dissolved in toluene and eluted with toleune through a short (5 cm.) alumina column. Approximately 10 ml. of toluene eluent was collected from each sample. The hydrocarbon type fractions thus obtained were evaporated down and transferred to glass vials. An aliquot of each of the hydrocarbon fractions was then examined by the relatively un-specific technique of U.V. spectroscopy. This technique provided an overall estimate of the extent of oil pollution in the area by measurement of the proportion of aromatic constituents in the sediment hydrocarbon fractions.

Measurements were made under the following conditions:

Instrument	Pye Unicam SP800A
Slit width	0.002 mm.
Back-off	0.5
Scan range	300 - 200 nm.
Expansion factor	x 1
Path length	10 mm.

Quantitation was achieved by calibration of the instrument with standard solutions of "Showa Maru" crude oil blend in petroleum ether. Absorbance measurements were made at 248 nm. This wavelength was chosen after consideration of the results of the calibration experiment. Absorbance measurements were corrrected for the solvent blank.

Following U.V. analysis selected hydrocarbon fractions were further fractionated by silver nitrate/SiO_2 (5% w/w) thin layer chromatography (tlc) yielding total saturated hydrocarbon fractions which were analysed by capillary glc under the following conditions :

Instrument	Finnigan 9500
Column	20 m. x 0.2 mm OV-101, programmed from 60-260° C at 6°/min., 2 ml./min. He flow rate.

Subsequent to glc analysis selected saturated hydrocarbon samples were examined by c-gc-ms under the following conditions :

Instrument	Finnigan 3200GC-MS/6100 Data System

GC

Column   10 m. x 0.3 mm. glass coated with Ov-1, programmed from 80-270°C at 4°/min., He carrier flow rate ca. 2 ml./min.

MS   300 uA emission current, 70 eV ionisation energy.

Computer   Continuous scanning from mass 50-560 a.m.u. every 1.7 secs.

THE EFFECTS OF PETROLEUM HYDROCARBONS ON CORALS

James P. Ray, Ph.D.[1]
Senior Staff Specialist - Marine Biology
Environmental Affairs
Shell Oil Company
Houston, Texas
United States

INTRODUCTION

The coral reef community is the most beautiful and complex of the marine ecosystems, only to be rivaled in the terrestrial world by the tropical rain forests. At the center of this unique ecosystem is the small, fragile coral - sometimes represented by small solitary individuals (polyps) and more commonly forming large colonial structures made of numerous individuals.

As the world's need for petroleum hydrocarbons increases, the probability of tropical reef ecosystems (Fig. 1) being affected by fugitive discharges will increase. The objectives of this paper are to establish our present state of knowledge concerning the input, fate, and effects of oil on reef building (hermatypic) corals.

SOURCES OF PETROLEUM HYDROCARBONS IN THE MARINE ENVIRONMENT

Prior to a general discussion on coral/oil interactions, it is important to gain an understanding of the various sources of petroleum hydrocarbons (PHC) and their relative origins as they relate to coral reef areas. Input can range from diffuse chronic inputs (terrestrial runoff and natural seeps), to large point source releases (e.g. tanker spills).

Natural seepage has been occurring from terrestrial and marine sources over geological time. Although the knowledge of seep locations and discharge rates is limited, Wilson et al (Ref. 1) estimated that from 0.2 to 1.0 million metric tons per annum (mta) finds its way into the marine environment (average estimated at 0.6 (mta). This may be a conservative estimate. Recent studies (Ref. 2) conducted in the Caribbean, northwest of the Antilles along a series of stations that stretch for 800 miles, have detected a layer of oil-rich water (3-12 mg/l) at a depth of 200 meters (m).

(1) Environmental Affairs, Shell Oil Co., P.O. Box 4320, Houston, Texas, 77210, United States.

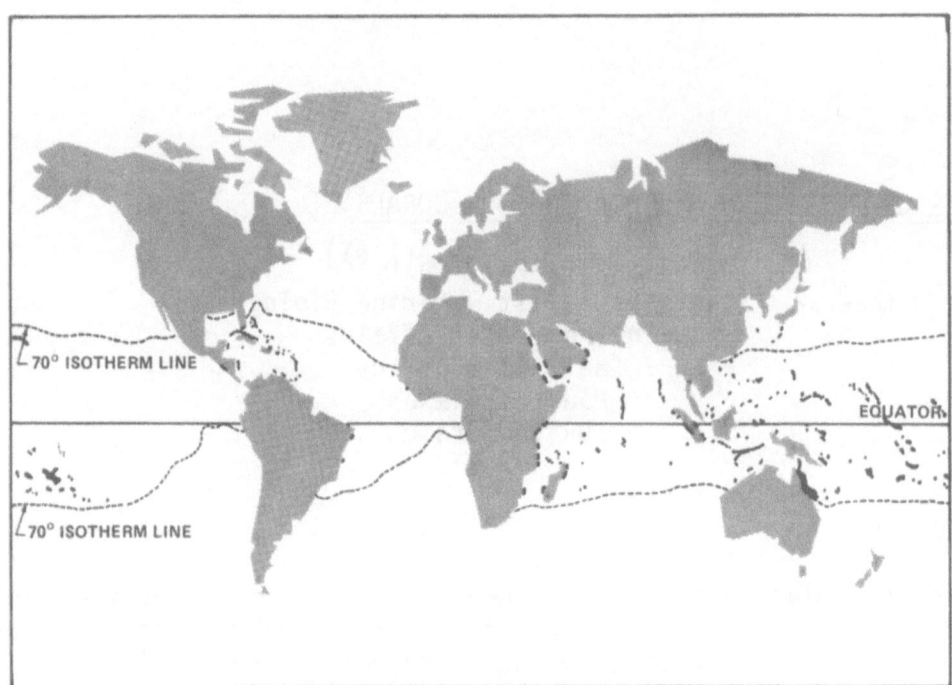

Fig. 1   Worldwide Distribution of Major Coral
Reef Areas in Relationship to 70° Isotherm
(Goreau et al, 1979; Emery, Tracey, and Ladd, 1954)

Geochemistry and ^{14}C activity suggests that the oil is a weathered crude, probably from a submarine seep. Conservative estimates of the layer (exact dimensions were not determined) indicate that more than one million tons of oil could be present, and possibly may represent more than the total estimated input of oil to the oceans from all seeps (Ref. 3)).

Significant PHC inputs are associated with urban centers of the world. Coastal municipal wastes and non-petroleum industries, (0.45 mta), urban runoff (0.3 mta), and river runoff (1.6 mta), represent an estimated major input of 2.35 mta (Ref. 3). Coastal refineries are strategically located along estuaries and coastlines throughout the world. Current best estimates of oil released are 0.02 mta.

The two major inputs directly related to the petroleum industry are discharges related to offshore drilling (exploration and production) and the marine transportation of petroleum. Offshore drilling contributes the least, with an estimated 0.08 mta. The majority of this input is due to minor spills of less than 50 barrels (barrel=42 U.S. gallons= 159 liters) and the low hydrocarbon levels associated with produced water discharge.

The largest industry input, estimated to be approximately 2.13 mta, is due to marine transportation. Spillage can be due to a variety of reasons, including: loading, lightering, terminal operations, bilge pumping, bunkering, and tanker accidents.

During the past decade, many changes have occurred in the marine transportation scene due to the increasing demand for oil by the industrialized nations. The trend has been toward larger and larger tankers which is graphically shown in Figure 2.

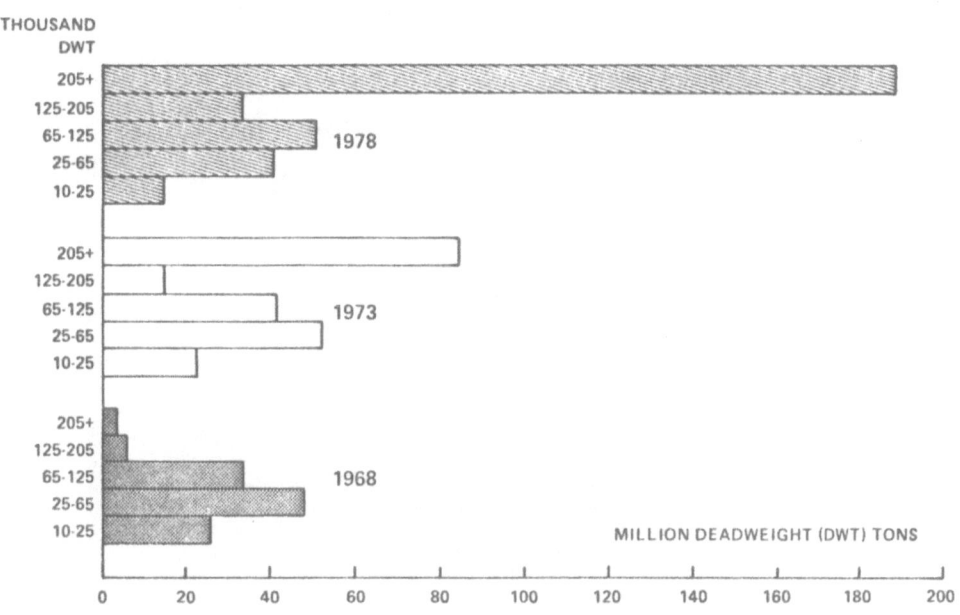

Fig. 2 World Tanker Fleet, 1968, 1973, and 1978 (Source: John I. Jacobs & Co. Ltd.).

In 1968, the majority of the worlds fleet was in the 25,000-65,000 deadweight ton (dwt) size range, which today would be primarily considered a coastal tanker. The day of the super tanker (Very Large Crude Carrier-VLCC; and, Ultra Large Crude Carrier-ULCC) has arrived, with some of the largest behemoths exceeding 500,000 dwt. As can be seen in Fig. 2, the 205,000 dwt (VLCC) and larger category (Ref. 4) represents the majority of tanker tonnage currently transporting oil.

As depicted in Fig. 3, most petroleum transportation (Ref. 4) passes through the tropical zone delimited by the 70° isotherm (dotted lines) within which most coral reef areas are located (Ref.5). This becomes significant to coral reefs in light of earlier statistics from 1956 through 1969 that show most of the 36 major tanker oil spills had occurred within a half mile of shore (Ref. 6).

To summarize, the estimated total input of petroleum hydrocarbons (bio- and anthropogenic) is approximately 4.57 million metric tons per annum, with much of it directly entering tropical waters.

FATES OF PETROLEUM HYDROCARBONS IN THE MARINE ENVIRONMENT

Several important symposia have been held in the past four years concerning the fate and effects of oil. Technical papers included in the proceedings of these meetings cover a broad range of fate and effects information from tropical to arctic environments. This large addition of recent information is evidence of our increasing knowledge on fate and effects of oil in the oceans. For an update on the current literature, References 7-13 should be consulted.

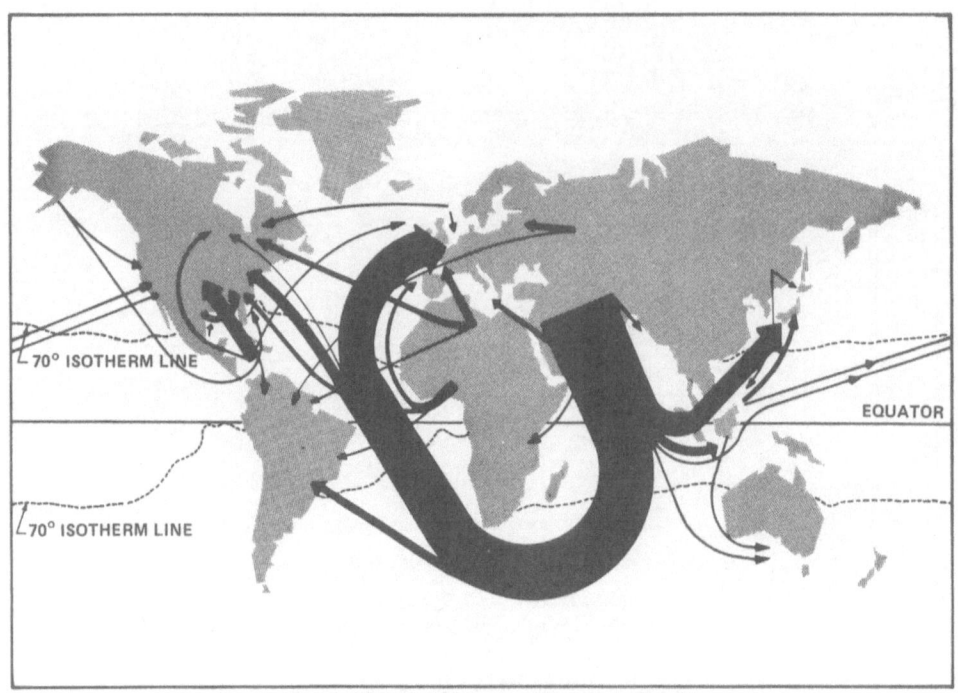

Fig. 3   Main Oil Movements by Sea, 1978
(British Petroleum, 1978)

Various physical, chemical, and biological processes are constantly working to transform and degrade hydrocarbons. Although enormous volumes of oil have entered the worlds oceans in recent geological time, the actual concentrations present in the sediments, water column, and biota are relatively low.

Once oil enters the marine environment, numerous processes act on the oil, including solution, photo-oxidation, sedimentation, and biodegradation. Surface oil is additionally susceptible to evaporation, spreading, air-sea interchange and emulsification.

Evaporation is one of the most important weathering processes due to the selective loss of lower molecular weight hydrocarbons, primarily aromatics, (Ref. 14) which are major contributors to the toxicity of oil (Ref. 15, 16). Several studies have shown that most of the hydrocarbons below C_{12} are lost due to evaporation within the first 24 hours (Ref. 17, 18).

McAuliffe (Ref. 19) showed that the true solubility of hydrocarbons in water dropped exponentially as a function of molecular volume. For example, benzene has a solubility of 1780 parts per million (ppm) whereas the heavier anthracene is soluble to less than 0.1 ppm.

Characteristics of tar ball formation have been reviewed by Butler et al, (Ref. 20) and Mommessin et al, (Ref. 21). The origin of these materials is primarily attributed to tanker bilge discharges. Butler (Ref. 22) predicted the half-life of most tar balls at less than one year.

Oceanic Levels of Petroleum Hydrocarbons

Hydrocarbons are spread differentially throughout the marine environment. Low molecular weight (LMW) hydrocarbons, $C_2 - C_4$, are primarily biogenic in origin and found in the upper 200 meters (Ref. 19). General concentrations typical of marine waters for these LMW hydrocarbons, expressed in nanograms per liter, are: methane, 35; ethane, 0.7; propane, 0.7; butanes, 0.13; ethene, 6; and propene, 2.6. Higher molecular weight (HMW) hydrocarbons appear to be primarily in the upper 10 meters, principally as particles. Approximate concentrations are 0-3 mm(surface layer), 20 ug/l; 1 m, 1 ug/l; and 5 m, 0.4 ug/l. HMW hydrocarbons are generally 3 to 4 orders of magnitude higher in sediments than in the water column (e.g. 40 mg/kg on the continental shelf; 10-20 mg/kg on the continental slope; and 1-5 mg/kg on the deep ocean bottom (Ref. 19).

Brown et al (Ref. 23) reported surface averages of 9 ug/l and 10 meter concentrations of 4 ug/l for non-volatile hydrocarbons along tanker routes from the Caribbean and Gulf Coast to New York. Monaghan et al (Ref. 24) found 4 ug/l in the Mediterranean Sea. Brown and Huffman (Ref. 25) reported 4 ug/l in near surface samples; with depths of 100 meters and more showing 1 part per billion (ppb) or less. In the vicinity of the natural oil seep area off Coal Oil Point (Santa Barbara Channel, California), Koons and Brandon (Ref. 26) reported that concentrations vary with depth from 0.4 to 16 ug/l.

Water samples taken beneath oil slicks in accidental spills such as Chevron's Gulf of Mexico blow out (Ref. 27), Ekofisk blowout, and the Argo Merchant tanker spill, showed that water column concentrations directly below the slick and near the source did not exceed 200 ppb.

EFFECTS OF PETROLEUM HYDROCARBONS IN THE MARINE ENVIRONMENT

A large volume of literature exists (Ref. 7-13) on the effects of oil and its fractions on a multitude of marine species. A general summarization of the literature would show that concentrations of soluble hydrocarbons ranging from 1 to 20 ppm might be hazardous to pelagic species (swimming in the water column). These toxicity ranges are based on laboratory exposures of 24 to 96 hours. To achieve oil/water concentrations of this magnitude, considerable mixing energy is required.

Chronic laboratory exposures have caused various lethal and sublethal effects on the more sensitive species and life stages at concentrations as low as 300 ppb. However, most species of marine animals have biological systems for removing hydrocarbons (Ref. 28). Crustaceans, polychaetes, and fish have a detoxification system in the livers or intestinal tract which degrade aromatics. Bivalve molluscs do not possess this system, and tend to accumulate hydrocarbons more than most animals, although depuration does occur. Zooplankton can both metabolize and excrete (via feces) hydrocarbons (Ref. 29). It is due to these mechanisms for handling hydrocarbons that organisms not only show varying tolerances to hydrocarbons, but often can successfully tolerate short term exposures. The potential for tissue accumulation and magnification was reviewed by Varanasi and Malins (Ref. 30) and Anderson (Ref. 31).

As a result of many years of research, investigators have gained insight into the toxicity of various oil components. Neff et al (Ref. 15) and Rice et al (Ref. 16) have found the monoaromatics to be the least toxic aromatics, with acute toxicity increasing with increasing molecular size, up to 4- and 5- ring aromatic compounds, which have very low solubility. With increasing alkylation of the aromatic rings, toxicity increases. The most toxic compounds in refined and crude oils are the di- and tri- aromatic components (naphthalene being most important).

CORAL REEF BIOLOGY

To the layman, a single coral individual would be a small, inconspicuous organism of no special import. Yet the same small, plant like individual, in concert with millions of other coral polyps, is the greatest marine architect of all. Reef building corals (hermatypic) are responsible for transforming entire coastlines, and building oceanic islands.

The coral animal, known as a polyp, is a member of the phylum Coelenterata. Systematically, he is a close second cousin to the more visible and familar sea anemones, with their stalk like bodies, and waving tentacles. The overall biology of corals is quite complex.

Corals have been extremely successful in typically nutrient poor tropical waters (Ref. 5, 32, 33) and have formed reef structures scattered over 190,000,000 square kilometers (km) (Ref. 34).   The amazing ability of corals to obtain sufficient nutrition in tropical waters was reviewed by Muscatine (Ref. 35), and Taylor (Ref. 36), who attribute their success to polytrophism. This varied approach to obtaining nutrition (Ref. 37) by coral polyps includes: (1) primary production due to an intimate endosymbiotic relationship with the dinoflagellate algae known as zooxanthellae. Taylor (Ref. 38) identified this algae as Gymnodinium microadriaticum. These algae are capable of fixing carbon at levels comparable to free-living algae (Ref. 39); (2) they are primary consumers which utilize photosynthetic products translocated from the zooxanthellae (Ref. 40);   (3) heterotrophic secondary and tertiary consumers capable of suspension feeding on zooplankton, phytoplankton (Ref. 37), and bacteria (Ref. 41, 42); (4) direct deposit feeding by scavenging the surrounding substrate with extruded mesenterial filaments (Ref. 43); and (5) saprotrophic ingestion by direct uptake of dissolved organic substances (Ref. 44).

The key to the hermatypic coral's ability to grow and secrete solid limestone (calcium carbonate) skeletons is the energy efficient symbiotic relationship with the zooxanthellae. Without these important partners, large reefs would not be built. Algae, due to its dependence on light for photosynthesis, make hermatypic corals very restricted in distribution, with depth and/or turbidity determining the vertical range. Also acting as limiting factors are temperature, with maximum success being in the 23°-25°C range (18°C is the lower limit), and low salinity.

The distribution of coral reefs on a worldwide basis is primarily limited within the 70° isotherms, as can be seen in Figure 1. Morphologically, reefs fall into three major categories (Ref. 5). Typical of shallow waters, and common in the Caribbean are fringing reefs, which border shorelines, and are separated from them by a narrow stretch of water. The larger and more continuous barrier reefs are found further from shore. The Great Barrier Reef in Australia, which stretches 2000 km in length and up to 145 km wide is the largest barrier reef. Atolls are coral rings (islands) enclosing a central lagoon and typical of the South Pacific region (Ref. 34).

The reef building corals show a narrow vertical range for successful growth. Indo-Pacific shallow forms are capable of withstanding several hours exposure to air during low tide. Deeper reefs, such as the Flower Garden Banks (Gulf of Mexico) have been successful at depths approaching 50 m. Reproduction can be either sexual (gametes produced) with planktonic distribution of the larvae, or asexual (budding).

EFFECTS OF OIL ON CORALS

There have been very few detailed, quantitative, scientific studies. Much of what is known from the literature is based on qualitative observations by coral reef specialists. However, conditions of both acute dosage and more subtle chronic exposure have been studied to a limited extent.

Oil Spills
In Tarut Bay, Saudia Arabia, healthy growing Acropora sp. in 2.5-4.6 m of water has been observed in an area subjected to chronic oil pollution from a nearby oil terminal (Ref. 45).

Coral proliferation in other oil polluted areas of the Red Sea was documented by Shinn (Ref. 46), especially near the entrance to the Suez Canal which has been repeatedly impacted by crude and fuel oil. He reported apparently unaffected staghorn coral (Acropora) growth around tanker loading terminals at Karg Island, Das Island, and Halul Island (30 km from a natural seep area). Shinn suggests that oil spills may only pose a minor threat to reefs.

Several spills have affected associated coral reef communities, although no coral effects were noted. Gooding (Ref. 47) reported the loss of 6 million gallons of assorted oils off the harbor entrance to Wake Island (Pacific Ocean). No observations were reported on coral response, but numerous invertebrates and more than 2,500 kg of fish were killed. A large spill resulting in mortality to marine flora and fauna in the sublittoral zone (no coral effects noted) resulted from a grounded tanker off Guayanilla, Puerto Rico. (Ref. 54).

Extensive studies have been conducted on several fringing reef tracts on the Gulf of Eilat (Red Sea) south of Eilat, Israel (Ref. 48-52). Following a 90% mortality to the shallow corals due to an unusually low tide episode in 1970, recolonization of this area was studied. The test reef was located 3 km south of an oil terminal which had been a source of frequent oil spillage. Only counting those spills large enough to "blacken" the reef nature reserve, ninety five spills occurred from 1971-1973, with frequency ranging from one to seven per month. In addition to the chronically acute oil input, there was also phosphate eutrophication of the shallow lagoon at the nature reserve due to nearby fertilizer plants. During the three year period, several changes were observed. The community structure of hermatypic corals changed considerably. Dominant species present prior to the low tide killoff were almost entirely absent. Recruitment had been minimal, and algae predominated. Although lacking direct evidence, Loya (Ref. 48) suggested that the failure of opportunistic species such as Stylophora pistillata to recolonize the area was the result of interference with normal recolonization and development. He inferred that oil may damage the reproductive system of corals; interfere with larvae production; or reduce larvae viability and inhibit their normal settling. Rinkevich and Loya (Ref. 53) have confirmed oil effects on reproduction in laboratory experiments, although the exact exposure quantities are not known. On the control reef, 5 km south, which was not exposed to large inputs of oil, recolonization was normal.

Several studies on the effects of an oil spill on tropical communities have been conducted in Panama. Rutzler and Sterrer (Ref. 55) assessed effects on the coral communities of approximately 20,000 barrels of Bunker C and marine diesel spilled from a tanker breakup at Galeta Point, Panama (Atlantic). The least affected community appeared to be the corals. The authors assumed the lack of damage was because the submerged corals did not come into direct contact with the oil.

On July 18, 1975, more than 30 miles of the lower Florida Keys were oiled with an estimated 1,500 to 3,000 barrels of crude oil emulsion released from a tanker passing through the Florida Straits (Ref. 56). Although rocky shores, mangrove swamps and marsh communities were affected, diver surveys conducted at several intervals during the subsequent six months indicated no detectable damage to the coral reefs.

In June 1979, the world's largest oil spill resulted from the blowout of an offshore well on the Campeche Bank off the Yucatan Pennisula, Gulf of Mexico. Since that time, oil has reached the reef areas off Veracruz and Tampico, but no information is currently available concerning effects on these areas.

Shinn (Ref. 46) showed that Montastrea annularis could withstand 2 hours total immersion in Louisiana crude. The staghorn coral Acropora cervicornis, when exposed for 2 hours to a crude-seawater ratio of 1:6 and 1:12, showed immediate polyp retraction followed by complete recovery in 24 hours.

Grant (Ref. 57) floated a Moonie crude oil over specimens of Favia speciosa with no signs of visual effect during a 6 day observation period. Johannes et al (Ref. 58) floated five different types of oil over several species of Hawaiian corals: Porites compressa, Montipora verrucosa and Fungia scutaria for 2.5 hours. No visible damage was observed during the following 25 day observation period. In another experiment, 22 species were exposed to Santa Maria Crude for 1.5 hours. Coral specimens were halfway exposed to the air during the 1.5 hours, at which time the 0.6 mm thick layer of crude was allowed to splash on the corals due to wave action. Corals were then placed back on reef test plots (2 m deep) for four weeks of observation. During air exposure, oil on the coral surfaces caused a 3°C rise in surface temperature (black body effect). Oil had the greatest affinity for branching species of the genus Acropora (most abundant genus in the Indo-Pacific) and Pocillopora (Lewis, Ref. 59, noted that the branching species Porites and Madracis from the Barbados area were more sensitive than the encrusting species Agaricia and Favia). Some oil remained on the corals after four weeks. Other species showed an intermediate affinity for oil: Turbinaria, Favia, Plesiastrea, Favites, Psammocora, Astreopora, Symphyllia, Montipora, and Porites. Areas of coral tissue, where oil patches of more than a few millimeters persisted, underwent tissue disintegration followed by opportunistic colonization of the skeletal surface by macroscopic algae. Although small areas of coral surface were killed, the rest of the colony appeared to remain healthy. Bak and Elgershuizen (Ref. 60) noted that tissue death occurred if oiled sand particles remained on the colony surface for more than two days.

Coral Responses
A broad variety of responses to oil have been observed for hermatypic corals. Unfortunately, due to the numerous exposure techniques and lack of quantification of oil in water, it is often quite difficult to separate the long

term sublethal responses from the short term acute responses. Therefore, no major separation of the literature will be made in this review.

The initial perceptions of oil in water (or associated with particulates) by corals are often observed as feeding and/or rejection responses. Bak and Elgershuizen (Ref. 60) tested 19 species of coral from Curacao, including the common species of Acropora, Porites, Madracis, Agaricia and Montastrea. Carbonate sand grains (mean diameter 1200u) were oiled with Nigerian crude, Tia Juana Pesado crude, Forcados crude, Forcados long residue, and Lagomar short residue. The contaminated sediments were then quantitatively poured on the test specimens. Additional tests included the injection of small drops of Nigerian crude directly into the gastrovascular cavity.

No active ingestion of the oil was observed. Drops introduced into the gastrovascular cavity were extruded through the stomodaeum. Polythoa variabilis (zoanthid) polyps ingested and then rejected filter paper soaked with Bunker C and marine diesel (Ref. 61, 62). Bak & Elgershuizen (Ref. 60) noted that oil was transferred to the peristome (mouth region) and then removed by ciliary currents, and tentacular and polypal movement. Caribbean corals showed tentacular inhibition and depression of feeding activity in test containers containing 350 ml of seawater and concentrations of added crude ranging from 10 ppm for Madracis asperula, 500 ppm for Favia fragum, to in excess of 1000 ppm for Porites porites and Agaricia agaricites. These experiments represent "worst case" exposure, as the containers were sealed so that volatile fractions of the oil could not escape (Ref. 59). This type of exposure would not occur in nature.

Initial polyp retraction is a first line response to strong external stimuli. Shinn (Ref. 46) observed immediate retraction of Acropora cervicornis polyps when exposed to oil-water ratio of 1:6 and 1:12 for 2 hours. Neff (Ref. 63) showed a similar reaction in Madracis decactis and Montastrea annularis when severely stressed with No. 2 fuel oil.

Feeding responses also are affected by oil. Polythoa variabilis exposed to marine diesel shows alterations of polyp response time to food stimuli. Polyps tend to elicit false feeding reactions to untreated filter paper for up to 5 days after exposure to oil (Ref. 61). Results of this research also indicates that oil can activate the feeding behavior. The behavior is abnormal, requiring longer to start and complete, and most importantly, it is followed shortly by egestion. This retention and slow release of oil by zoanthids such as Polythoa and Zoanthus could result in a temporary oil reservoir on an impacted reef.

Alcyonarian (soft) corals, Heteroxenia fuscescens, from the Gulf of Eilat showed a marked decrease in the normal pulsation rate of the polyps when exposed to Agha Jari (Iranian) crude oil. In static tests, with oil added at a ratio of 30 ml/1, a 50% decrease in rate was noted (Ref. 64).

Irregularities in the tentacular pulsation rate are a good indicator of stress. At increasing oil concentrations (10-30 ml/l), polyps showed a decreased response and loss of synchronous folding of tentacles when tactily stimulated.

Typical of hermatypic corals under stress is the expulsion of zooxanthellae (bleaching). Sedimentation, salinity, temperature, and toxic substances can elicit this response. Psammocora, Pavona, and Porites (Pacific coast of Panama) exhibited "bleaching" when exposed directly for 30 seconds to marine diesel and Bunker C. Most bleaching occurred within 5 to 13 days, usually followed by recovery (Ref. 65).

A primary defense system of reef building corals is the ability to produce mucus which is rich in wax esters (cetyl palmitate and triglycerides) as a result of coral metabolism (Ref. 85). The importance of this material in ciliary-mucoid feeding and cleansing was reviewed by Goreau et al (Ref. 43). Marshall (Ref. 66, 67) and Johannes (Ref. 68) showed that particulate organic matter increases downstream from reefs and suggest the importance of mucus to the overall reef food web. Several species of reef inhabitants utilize the food value of mucus, including zooplankton, and fish (Ref. 63, 68-70) and crabs (Ref. 71, 72).

It has been suggested that corals are fairly resistant to surface oil due to the outer mucus layer (Ref. 73). Zoanthids such as Polythoa tend to repel oil from their surface. Abundant mucus producers such as Fungia and Symphyllia show good survival when directly contaminated with oil (Ref. 58). In their studies of oil/sand effects on 19 species of Caribbean corals, Bak and Elgershuizen (Ref. 60) determined that oil was not absorbed by coral tissues. Oil would become trapped in the surface mucus. Due to the potential for food chain transfer of petroleum hydrocarbons Neff (Ref. 63) studied contamination and uptake from Montastrea annularis to the longnose butterfly fish (Chaetodon sp.) from the Florida Keys. Montastrea annularis from the Florida Keys and Bermuda showed a significant incorporation of naphthalene in the mucus layer (was not quantified due to inability to quantify volume of mucus) after exposure to the water soluble fraction (WSF) of South Louisiana crude. ^{14}C-labeled naphthalene also was used in these experiments. The results suggest that coral mucus (mostly mucopolysaccharides) can bind or adsorb aromatic hydrocarbons and may be the primary mechanism of protection from, and rejection of, petroleum hydrocarbons. Butterfly fish (Chaetodon sp) were allowed to feed on mucus that had been exposed for 24 hours. Results indicate a limited potential for contamination of the mucus consuming fish.

Oculina diffusa (Texas) accumulated small amounts of alkanes and aromatics (naphthalenes) during exposure to WSF of ^{14}C-naphthalene. These were readily released when returned to clean water. Hydrocarbons also showed higher uptake during daylight hours which appear to show a link between hydrocarbon accumulation and photosynthesis.

Bak et al (Ref. 60) found no difference in the specific reaction of corals to oil-sand mixtures versus plain sand. Additions of 100 ppm (oil added) to small aquaria containing the Red Sea coral Platigyra elicited copious mucus production (Ref. 74), an increase from 25 to 500 ug per 10 ml of washdown water. Dextrose caused similar mucus production, and copper sulfate stimulated 1200 ug/10 ml in 24 hours. Correspondingly, the bacterial population feeding on the mucus increased dramatically. Excessive production of mucus due to pollution stress resulted in coral death between days 4 and 8. Additional tests with antibiotics to reduce the bacterial population showed increased survival by the corals, indicating the bacterial role in stressed coral mortality. By suppressing bacterial growth, coral survived at oil concentrations of 1000 ppm. It also has been shown that bacteria can consume the coral tissue (Ref. 74). The authors suggest a three step process in microorganism/stress mortality: (1) mucus attracted bacteria attack coral tissue (2) Desulfovibrio grows on mucus at reduced redox potentials produced by other microorganisms, with the resulting production of H_2S. (3) Beggiatoa, a filamentous bacteria, invades and destroys the coral tissue. Mucus plays an important role in the overall rejection of oil (Ref. 60) and sediments (Ref. 75, 76).

Early field evidence (Ref. 49, 50) and subsequent laboratory studies (Ref. 49, 53) indicate possible effects on the reproductive and recolonization success of corals. Studies conducted in the Gulf of Eilat (Red Sea) have provided circumstantial evidence that under conditions of severe chronic exposure (see prior discussion, page 8) recolonization is impaired. Following the unusually low tides near Eilat that killed more than 90% of the exposed corals in 1970, no recolonization by the opportunistic species Stylophora pistillata was observed during a 3 year period. Comparison of the oiled reef (3 km from Eilat) to the control (5 km distant) showed an entirely different community structure after 3 years of recolonization (Ref. 48) with most of the test reef recovery being due to regeneration and not recolonization.

In the control area, colonies of Stylophora had a higher percentage of gonads than were found in the test area. The percentage of actual breeding colonies also was higher (Ref. 49). The number of planulae larvae per coral head was found to be lower in the polluted reef. Using Connell's (Ref. 77) definition, the reduction of eggs and larvae showed that the coral fecundity of the test area is four times lower than the control (Ref. 49).

A water soluble fraction (WSF) prepared from a 1:10 oil-water mixture was used to test viability and settling success of Stylophora planulae larvae. At 10 and 50% of the WSF, more than 50% of the planualae died in 144 hours compared to no mortality in the control (Ref. 40). Settlement also was significantly lower at the same concentration.

Cohen et al (Ref. 64) noted that the alcyonarian coral Heteroxenia released larvae when exposed to Iranian crude. Gohar and Roushdy, (Ref. 78), showed that general stress can trigger this same response in Heteroxenia. The premature release of larvae in Stylophora was also noted in the presence of the WSF of Iranian crude (Ref. 49).

Deposition of calcium carbonate is one of the most important functions of hermatypic corals, and is one of the more sensitive responses to sublethal stress (Ref. 65). Although inter- and intra species rate differences in the uptake of radiolabeled ^{45}Ca have been noted by several investigators (Ref. 79-81), it is still a very useful tool in quantifying changes in deposition rates.

The rate of ^{45}Ca deposition increased with increasing WSF in some tests with Madracis decactis and Montastrea annularis. The same pattern was observed with the WSF of South Louisiana crude. Oculina diffusa and Millepora sp. showed decreases in deposition with increasing exposures of No. 2 fuel oil and phenanthrene.

Porites furcata (Atlantic) from Panama were exposed to Bunker C which was floated immediately above the coral for 1 to 2.5 hours, and then placed back on the reef and observed for 61 days. Differences in growth between exposed and control corals were slight, but statistically significant (Ref. 65). Considerable growth rate differences also were noticed within branches of the same colony, implying that some caution should be used in interpreting growth data (Ref. 63, 65).

As previously stated, the symbiotic relationship between the photosynthesizing zooxanthellae algae and coral polyp makes it possible for corals to grow at rates capable of building reef structures. Influences that can have an inhibitory effect on zooxanthellae can have deterimental effects on the overall success of the corals.

Millepora, when exposed to WSF of South Louisiana crude and phenanthrene showed a dose dependent decrease in ^{14}C fixation (Ref. 63) after 72 hours of exposure. The corals recovered completely when returned to clean seawater. Madracis decactis showed an opposite trend, where ^{14}C fixation paralleled the controls, but showed depression when returned to clean seawater. In contrast, ^{14}C fixation by the zooxanthellae of Favia fragum and Montastrea annularis was unaffected by the WSF of South Louisiana crude.

Free living algae have been shown to be quite sensitive to oil WSF and aromatics such as naphthalene (Ref. 82). The authors noted that the concentration of naphthalene tested (1.0 ppm) was in excess of naphthalene concentrations found in kerosene (0.153 ppm in seawater) and Kuwait crude oil (0.0323 ppm in seawater) (Ref. 83). It was noted that even when inhibited during a 12 hour exposure to naphthalene, carbon fixation returned to normal after the exposure. Goreau et al (Ref. 5) noted that Ca deposition can decrease by 50% on a cloudy day, which indicates the strong relationship between deposition and photosynthesis by zooxanthellae.

Cohen et al (Ref. 64) studied the soft coral Heteroxenia, and noted that after exposure to Agha Jari (Iranian) crude, the coral chromatograms showed the addition of hydrocarbons similar to the oil (e.g., C_{11}-C_{15} and above C_{23}). Although lower molecular weights showed losses, the C_{23}^{15} persisted. The total coral uptake was quite small.

The use of chemical dispersants to accelerate the breakdown and dispersal of surface oil slicks has been quite controversial in the past few years. Unfortunately, perceptions of dispersant toxicity have been based on first generation materials such as the detergents and aromatic solvent based chemicals of the Torrey Canyon era. The new generation of dispersants use less toxic carrier materials and generally show much lower overall toxicity (less toxicity).

Although some dispersant and oil plus dispersant testing has been conducted (Ref. 59, 84) on corals such as Porites porites, Favia fragum, Agaricia agarcites, Madracis asperula, and Madracis mirabilis, enough information was not provided on the dispersants to evaluate the results. Lewis (Ref. 59) did not specify which type of Corexit dispersant was used in his tests (a number of different products are available, for example, 7664 and 9527). Different materials have different toxicities.

DISCUSSION

Coral reefs represent complex tropical ecosystems which are sensitive to environmental change, whether natural or manmade. The exploration, production, transportation, utilization, and disposal of petroleum hydrocarbons presents one potential class of pollutants than can affect these communities. Coinciding with the increasing quantities of petroleum, are technical advances to mitigate its accidental input back into the environment. Advances in technology are resulting in reduced land originated wastes, and new engineering achievements are increasing the safety of offshore drilling and transportation systems.

Large accidents such as well blowouts and tanker accidents, although small in total quantity of oil added to the oceans annually, garner all the press headlines. As noted earlier, a large source of oil is that resulting from river and municipal runoff. Waste crank case oil from automobiles, which is poured down the drain, adds considerably to the total figure.

We are rapidly expanding our understanding of the environmental fate and effects of petroleum hydrocarbons in the environment. Our increasing

knowledge allows us to make a balanced approach between todays industrialized needs and adequate environmental protection. This understanding also
is extremely important in mitigating the effects of oil once spills, both in
containment and cleanup.

Coral and coral/community interaction have not been adequately considered in petroleum hydrocarbon fate and effect research. The literature reviewed in this paper covers most of the published literature regarding the
effects of oil on corals. However, many studies are difficult to extrapolate
to the real world because artificial laboratory testing techniques did not
consider PHC fate.

The type of oil, its aromatic content and composition, state of weathering, and its physical state (e.g., emulsified, dispersed, etc.) all have a
significant bearing on potential impacts. Conditions prevailing in the
environment also are important, including surface conditions (waves), currents,
suspended solids load, and temperature.

Coral reefs present a broad variety of physical conditions that need to
be considered when considering potential oil effects (oil spills). Typical
Atlantic reefs are primarily of the fringing type, often coming right to the
surface. The branching coral species, which show a greater affinity for
oil, are in this shallow, high energy zone. Quite often, the back reef or
lagoon is a shallow, low energy environment with higher suspended solids
loads, and low flushing rates. Deeper water reefs are found in some areas
(e.g., Yucatan, Mexico; Flower Gardens, Texas) where the shallowest corals
are at 15 meters or deeper. The shallow zones of many Indo-Pacific reefs
undergo daily exposure during low tides. This adds an entirely different
dimension when considering surface oil.

The tropical reef area that probably has been subjected to the largest
number of spills and general chronic pollution has been the Red Sea. The
relationship between this and petroleum production and transportation can be
seen in Fig. 3. As noted by several authors (Refs. 45, 46, 55, 56) oil
spills appear to have minimal impact on corals. As pointed out by other
authors, a broad variety of responses appear to be manifest by both physical
contact with oil and by the water soluble components. Although visually the
reefs appear healthy in spill areas, the potential for detrimental sublethal
effects still exists. Although apparently severe effects were studied by
numerous investigators, (Ref. 48-52), it should be noted that this was a
severely impacted area (Gulf of Eilat) which had some 95 spills that blanket-
ed the reef during a three year period. Additionally the area is subjected
to adverse effects from phosphate discharges from a nearby fertilizer plant.
The effects seen on this reef are probably due to the high level chronic oil
pollution, but are very difficult to extrapolate to other locations for any
single spill incident.

Much of the laboratory data presented here are very difficult to analyze
due to the bioassay techniques used and the lack of analytical measurements.

Oil preparation and exposures ran the entire gamut, including: use of water soluble fractions prepared in several manners; bubbling air through oil to disperse it (which drives off the toxic aromatics); injecting pure oil in the coral polyp gut; floating oil on water and inserting oil soaked filter paper to further raise the water oil concentration, and then sealing the container so no evaporation can occur; allowing oil to splash on partly exposed coral; and completely submersing coral pieces in 100% oil. Some of these techniques may simulate actual field exposures, others are unrealistic.

The problem of quantifying the actual oil concentration in the test water makes it additionally difficult to interpret results. With few exceptions (Ref. 63), all of the exposures are reported as "oil added." Due to the tremendous number of variables in oils tested and diluent water conditions, there is no way to determine the concentration or composition of the final test solutions. In almost all tests, the concentrations used probably exceed those that would normally exist in the open environment (there are exceptions).

Predicting actual levels of hydrocarbons that would exist in the water column due to a spill is difficult. In reef areas where shallow reef fronts predominate, considerable oil-water dispersion could occur due to the high mixing energies (wave and current action). For example, Gilmore (Ref. 6) stated that most of the major tanker spills from 1956-1969 occurred within 0.5 miles offshore, during high wind (40% above gale force) and heavy seas (40% of waves over 3 m).

Corals appeared to have numerous built in behaviorial mechanisms to protect themselves from oil. Polyp movements, including tentacular, ciliary, egestion, and polypal swelling all aid the individual in removal of unwanted material. The single most important defense mechanism is the mucus secretions, which tend to either repel, trap or reject oil contacting the surface.

Mitigation
Most importantly, the knowledge that is already available should be used to reduce unacceptable current impacts and to mitigate future problems where possible and reasonable. The single largest threat to coral reef areas is the one time large oil spill. The first line priority should be containment or diversion of oil from the reef proper, the primary purpose being to keep oil away from the shallow, often turbid upper reef area. Once in this zone, the maximum probability for impact exists. In addition, once oil becomes entrained in the sediments of the more quiescent lagoon waters, longer term chronic problems could arise. Spills probably would have very minimal impact to deep reefs.

Because of the recent advent of new, lower toxicity dispersants, serious consideration should be given to using these materials while oil is still distant from the reef in deep water. Dispersal of the oil downward from the influence of surface winds into fine particles in the water column considerably reduces the potential impact on the productive coral reef and accelerates the degradation of the oil. Dispersants should not be used in close proximity to a reef system until more fate and effects data are available.

During a spill, the most damage to the reef could be caused by cleanup personnel and work equipment. Boat traffic and general activity should be held to a minimum near the reef. Resuspension of sediments by propellers would just add to the problem, and mechanical damage to corals by boats, anchors, and people could take years to recover.

SUMMARY

Corals are not only well adapted to living in the nutrient poor tropical waters, but to resisting occasional oil exposure. With the exception of a few isolated areas, oil pollution does not yet appear to pose a major threat to coral reefs. Little research has been done on the effects of petroleum hydrocarbons on corals and much is yet to be learned. We do know that the potential for damage exists, and should actively try to prevent pollution before it occurs.

REFERENCES

1.   R. D. Wilson, P. H. Monaghan, A. Osanik, L. C. Price, and M. A. Rogers, 'Estimate of annual input of petroleum to the marine environment from natural marine seepage', Background Papers, Workshop on Petroleum in Marine Environment, (Ocean Affairs Board, National Academy of Sciences, Washington, D. C., 1973)

2.   G. R. Harvey, A. G. Requejo, P. A. McGillivary, and J. M. Tokar, 'Observation of a subsurface oil-rich layer in the open ocean', Science, Vol. 205, (1979), 999-1001

3.   Petroleum in the Marine Environment, (National Academy of Sciences, Washington, D. C., 1975), 107pp

4.   BP Statistical Review of the World Oil Industry 1978, (British Petroleum Company Limited, London, 1978), 33pp

5.   T. F. Goreau, N. I. Goreau, and T. J. Goreau, 'Corals and coral reefs', Scientific American, Vol. 241, No. 2, (1979), 124-136

6.   G. A. Gilmore, D. D. Smith, A. H. Rice, E. H. Shenton, and W. H. Moser, 'Systems study of oil spill clean up procedures', Vol. I, Analysis of Oil Spills and Control Materials, (American Petroleum Institute, Washington, D. C., 1970)

7.   Sources, Effects, and Sinks of Hydrocarbons in the Aquatic Environment, (American Institute of Biological Sciences, Arlington, Virginia, 1976)

8.   Proceedings of the 1977 Oil Spill Conference (Prevention, Behavior, Control, Cleanup), (American Petroleum Institute, Washington, D. C., 1977)

9.   D. A. Wolfe, (ed.), Fate and Effects of Petroleum Hydrocarbons in Marine Ecosystems and Organisms, (Pergamon Press, Oxford, 1977)

10.   'Proceedings of Symposium on Recovery Potential of Oiled Marine Northern Environments', Jour. Fisheries Res. Bd. Canada, Vol. 35, (1978)

11.   Proceedings of the Conference on Assessment of Ecological Impacts of Oil Spills, (American Institute of Biological Sciences, Arlington, Virginia, 1979)

12.   Proceedings of the 1979 Oil Spill Conference, (American Petroleum Institute, Washington, D.C., 1979)

13.   D. A. Malins, (ed.), Effects of Petroleum on Arctic and Subarctic Marine Environments and Organisms, Vol. I and II, (Academic Press, New York, 1977)

14. Z. R. Regnier and B. F. Scott, 'Evaporation rates of oil components', Env. Sci. and Tech., Vol. 9, (1975), 469-472

15.   J. M. Neff, J. W. Anderson, B. A. Cox, R. B. Laughlin, Jr., S. S. Rossi, and H. E. Tatem, 'Effects of petroleum on survival, respiration, and growth of marine animals', pp.515-539, In: Sources, Effects and Sinks of Hydrocarbons in the Aquatic Environment, (American Institute of Biological Sciences, Arlington, Virginia, 1976)

16.   S. P. Rice, J. W. Short, and J. F. Karinen, 'Comparative oil toxicity and comparative animal sensitivity', pp.78-94, In: Fate and Effects of Petroleum Hydrocarbons in Marine Ecosystems and Organisms, D. A. Wolfe, ed., (Pergamon Press, Oxford, 1977)

17.   R. E. Kreider, 'Identification of oil leaks and spills', pp.119-124, In: Proceedings of the Joint Conference on Prevention and Control of Oil Spills, (American Petroleum Institute, Washington, D. C., 1971)

18.   P. J. Kinney, D. K. Button, and D. M. Schell, 'Kinetics of dissipation and biodegradation of crude oil in Alaska's Cook Inlet', pp.333-340, In: Proceedings of Joint Conference on Prevention and Control of Oil Spills, (American Petroleum Institute, Washington, D. C., 1969)

19.   C. D. McAuliffe, 'Solubility in water of paraffin, cycloparaffin, olefin, acetylene, cyclo-olefin, and aromatic hydrocarbons', J. Phys. Chem., Vol. 70, (1966), 1267

20. J. N. Butler, B. F. Morris, and J. Sass, 'Pelagic tar from Bermuda and the Sargasso Sea', Special Publication No. 10, Bermuda Biological Station for Research, (Harvard University Printing Office, Cambridge, 1973), 346pp

21.   P. R. Mommessin and J. C. Raia, 'Chemical and physical characterization of tar samples from the marine environment', pp.115-167, In: Proceedings of the Joint Conference on Prevention and Control of Oil Pollution, (American Petroleum Institute, Washington, D. C., 1975)

22.   J. N. Butler, 'Evaporation weathering of petroleum residues: the age of pelagic tar', Mar. Chem., Vol. 3, (1975), 9-21

23.   R. A. Brown, T. D. Searl, J. J. Elliott, B. G. Phillips, D. E. Brandon, P. H. Monaghan, 'Distribution of heavy hydrocarbons in some Atlantic Ocean waters', pp.505-519, In: Proceedings of Joint Conference on Prevention and Control of Oil Spills, (American Petroleum Institute, Washington, D. C., 1973)

24.   P. H. Monaghan, J. H. Seelinger, and R. A. Brown, 'The persistent hydrocarbon content of the sea along certain tanker routes - a preliminary report', American Petroleum Institute Report - 18th Annual Tanker Conference, Hilton Head, South Carolina, (American Petroleum Institute, Washington, D. C., 1973), 298pp

25.   R. A. Brown and H. L. Huffman, Jr., 'Hydrocarbons in open ocean waters', Science, Vol. 191, (1976), 847-849

26.   C. B. Koons and D. E. Brandon, 'Hydrocarbons in water and sediment samples from Coal Oil Point area, offshore California', Proceedings 1975 Offshore Technology Conference, Vol. III, (1975), 513-521

27.   C. D. McAuliffe, A. E. Smalley, R. D. Grover, W. M. Welsh, W. S. Pickle, and G. E. Jones, 'Chevron Main Pass block 41 oil spill: chemical and biological investigations', pp.555-565, In: Proceedings, 1975 Conference on Prevention and Control of Oil Pollution, (American Petroleum Institute, Washington, D. C., 1975)

28.   R. F. Lee, 'Fate of petroleum hydrocarbons in marine animals', Mar. Tech. Soc., Oceans '77, (1977), 40C-1 to 40C-4

29.   R. J. Conover, 'Some relations between zooplankton and Bunker-C oil in Chedabucto Bay following the wreck of the tanker Arrow', Fish, Res. Bd. Canada, Vol. 28, (1971), 1327-1330

30.   U. Varanasi and D. C. Malins, 'Metabolism of petroleum hydrocarbons: accumulation and biotransformation in marine organisms', pp.175-262, In: Effects of Petroleum on Arctic and Subarctic Marine Environments and Organisms, D. C. Malins, ed., (Academic Press, New York, 1977)

31.   J. W. Anderson, 'Responses to sublethal levels of petroleum hydrocarbons: are they sensitive indicators and do they correlate with tissue contamination?', pp.95-114, In: Fate and Effects of Petroleum in Marine Ecosystems and Organisms, D. A. Wolfe, ed., (Pergamon Press, Oxford, 1977)

32.   L. Muscatine and J. W. Porter, 'Reef corals: mutualistic symbioses adapted to nutrient-poor environments', Bio Science, Vol. 27, No. 7, (1977), 454-460

33.   D. R. Stoddart, 'Ecology and Morphology of recent coral reefs', Biol. Rev., Vol. 44, (1969), 433-498

34.   J. W. Wells, 'Coral reefs', pp.609-631, In: Treatise on Marine Ecology and Paleocology, (Geological Society of America, 1957), Vol. I, Memoir 67, Ecology, J. W. Hedgepeth, (ed.)

35.   L. Muscatine, 'Nutrition of corals', pp.77-115, In: Biology and Geology of Coral Reefs, Vol. II, Biology I, O. A. Jones and R. Endean, eds., (Academic Press, New York, 1973)

36.   D. L. Taylor, 'The cellular interactions of algal-invertebrate symbiosis', Adv. Mar. Biol., Vol. II, (1973), 1-56

37.   J. W. Porter, 'Zooplankton feeding by the Caribbean reef-building coral Montastrea cavernosa', Proc. Second Int. Coral Reef Symp., Vol. I, (1974), 111-125

38.   D. L. Taylor, 'Ultrastructure of the zooxanthellae Endodinium chattoni in situ', J. Mar. Biol. Assoc. U.K., Vol. 51, (1971), 227-234

39.   L. Franzisket, 'The ratio of photosynthesis to respiration of reef building corals during a 24 hour period', Forma Functio, Vol. 1, (1969), 153-158

40.   D. H. Lewis and D. C. Smith, 'The autotrophic nutrition of symbiotic marine coelenterates with special reference to hermatypic corals, I. Movement of photosynthetic products between the symbionts', Proc. R. Soc. Lond. B. Biol. Sci., Vol. 178, (1971), 111-129

41.   L. Di Salvo, 'Ingestion and assimilation of bacteria by two scleractinian coral species', pp.129-136, In: Experimental Coelenterate Biology, H. M. Lenhoff, L. Muscatine, and L. V. Davis, eds., (University of Hawaii Press, Honolulu, 1971)

42.   Y. I. Sorokin, 'On the feeding of some scleractinian corals with bacteria and dissolved organic matter', Limnol. Oceanogr., Vol. 18, (1973), 380-385

43.   T. F. Goreau, N. I. Goreau, and C. M. Yonge, 'Reef corals: autotrophs or heterotrophs?', Biol. Bull., Vol. 141, (1971), 247-260

44.   G. C. Stephens, 'Uptake of organic material by aquatic invertebrates. I. Uptake of glucose by the solitary coral Fungia scutaria', Biol. Bull., Vol. 123, (1962), 648-657

45.   M. F. Spooner, 'Oil spill in Tarut Bay, Saudi Arabia', Mar. Poll. Bull., Vol. 1, (1970), 166-167

46.   E. A. Shinn, 'Coral reef recovery in Florida and the Persian Gulf', Environmental Conversation Dept., Shell Oil Company, U.S., (1972), 9pp

47.   R. M. Gooding, 'Oil pollution on Wake Island from the tanker R. C. Stoner', Spec., Sci. Rept., U.S. Fish Wildl. Serv. (Fish.), Vol. 636, (1971), 1-10

48.   Y. Loya, 'Possible effects of water pollution on the community structure of Red Sea corals', Mar. Biol., Vol. 29, (1975), 177-185

49.   B. Rinkevich and Y. Loya, 'Harmful effects of chronic oil pollution on a Red Sea scleractinian coral popultaion', pp.585-591, In: Proceedings Third International Coral Reef Symposium, Rosenstiel School of Marine and Atmospheric Science, University of Miami, Miami, Fla., 33149, U.S.A.

50.   Y. Loya, 'Recolonization of Red Sea corals affected by natural catastrophes and man-made perturbations', Ecology, Vol. 57, No. 2, (1976), 278-289

51.   Y. Loya, 'Skeletal regeneration in a Red Sea scleractinian coral population', Nature, Vol. 261, No. 5560, (1976), 490-491

52.   L. Fishelson, 'Ecology of coral reefs in the Gulf of Aqaba (Red Sea) influenced by pollution', Oecologia, Vol. 12, (1973), 55-67

53.   B. Rinkevich and Y. Loya, 'Laboratory experiments on the effects of crude oil on the Red Sea coral Stylophora pistillata', Mar. Poll. Bull., Vol. 10, (1979), 328-330

54.   M. Diaz-Piferrer, 'The effects of an oil spill on the shore of Guanica Puerto Rico', In: Proc. Fourth Meeting, Associated Island Marine Labs, Curacao, (University of Puerto Rico, Mayaguez, 1962), 12-13

55.   K. Rutzler and W. Sterrer, 'Oil pollution: damage observed in tropical communities along the Atlantic seaboard of Panama', Bio Science, Vol. 20, (1970), 222-224

56.   E. I. Chan, 'Oil pollution and tropical littoral communities: biological effects of the 1975 Florida Keys oil spill', In: Proceedings of the 1977 Oil Spill Conference (Prevention, Behavior, Control, Cleanup), (American Petroleum Institute, Washington, D. C., 1977)

57.   E. M. Grant, 'Notes on an experiment upon the effect of crude oil on live corals', Fish. Notes Dept. prim. Ind., Brisbane, Vol. I, (1970), 1-13

58.   R. E. Johannes, J. Maragos, and S. L. Coles, 'Oil damages corals exposed to air', Mar. Poll. Bull. Vol. 3, No. 2, (1972), 29-30

59.   J. B. Lewis, 'Effects of crude oil-spill dispersant on reef corals', Mar. Poll. Bull., Vol. 2, (1971), 59-62

60.   R. P. M. Bak and J. H. B. W. Elgershuizen, 'Patterns of oil-sediment rejection in corals', Mar. Biol., Vol. 37, (1976), 105-113

61.   A. A. Reimer, 'Effects of crude oil on corals', Mar. Poll. Bull., Vol. 6, (1975), 39-43

62.   A. A. Reimer, 'Effects of crude oil on the feeding behaviour of the zoanthid Polythoa variabilis', Environ. Physiol. Biochem., Vol. 5, (1975), 258-266

63.   J. M. Neff (ed.), 'Sublethal effects of petroleum on reef corals', publication in press, (American Petroleum Institute, Washington, D. C., 1980)

64.   Y. Cohen, A. Nissenbaum, and R. Eisher, 'Effects of Iranian crude oil on the Red Sea octocoral Heteroxenia fuscescens', Environ. Pollut., Vol. 12, (1977), 173-185

65.   C. Birkeland, A. A. Reimer, and J. R. Young, 'Survey of marine communities in Panama and experiments with oil', EPA - 600/3-76-028, (Office of Research and Development, Environmental Research Laboratory, Narragansett, Rhode Island, 1976), 117pp

66.   N. Marshall, 'Detritus over the reef and its potential contribution to adjacent waters of Eniwetok Atoll', Ecology, Vol. 46, (1965), 343-344

67.   N. Marshall, 'Observations on organic aggregates in the vicinity of coral reefs', Mar. Biol. (Berl.), Vol. 2, (1968), 50-53

68.   R. E. Johannes, 'Ecology of organic aggragates in the vicinity of a coral reef', Limnol. Oceanogr., Vol. 12, (1967), 189-195

69.   R. W. Hiatt and D. W. Strasberg, 'Ecological relationships of the fish fauna on coral reefs of the Marshall Islands', Ecol. Monogr., Vol. 30, (1960), 65-127

70.   S. L. Coles and R. Strathmann, 'Observations on coral mucus 'flocs' and their potential trophic significance', Limnol. Oceanogr., Vol. 18, (1973), 673-678

71. J. W. Knudsen, 'Trapezia and Tetralia (Decapods, Brachyura, Xanthidae) as obligate ectoparasites of pocilliporid and acroporid corals', Pac. Sci., Vol. 21, (1967), 51-57

72. E. Preston, 'Niche overlap and competition among five sympatric species of xanthid crabs', Ph.D. thesis, University of Hawaii, (1977), 125pp

73. D. Straughn, 'Redressing the balance on the reef', Mar. Poll. Bull., Vol. 1, (1970), 86-87

74. R. Mitchell and I. Chet, 'Bacterial attack of corals in polluted seawater', Microbial Ecol., Vol. 2, (1975), 227-233

75. C. M. Yonge, 'Studies on the physiology of corals. I. Feeding Mechanisms and food', Great Barrier Reef Exped. 1928-1929, Sci. Rep., Vol. 1, (1930), 13-57

76. J. B. Lewis, 'The formation of mucus envelopes by hermatypic corals of the genus Porites', Caribb. J. Sci., Vol. 13, (1973), 207-209

77. J. H. Connell, Biology and Ecology of Coral Reefs, V, part - II, Biology, O. A. Jones and R. Endean, (eds.), (Academic Press, New York and London, 1973), 205-245

78. H. A. F. Gohar and H. M. Roushdy, 'On the embryology of the Xeniidae (Alcyonaria) (with notes on the extrusion of larvae)', Publs. Mar. Biol. Stn. Ghardaga, No. 11, (1961), 247-260

79. T. F. Goreau and N. I. Goreau 'The physiology of skeleton formation in corals. I. A Method for measuring the rate of calcium deposition by corals under different conditions', Biol. Bull, Vol. 116, (1959), 59-75

80. E. A. Drew, 'The biology and physiology of alga-invertebrate symbiosis. III. In situ measurements of photosynthesis and calcification in some hermatypic coral', J. Exp. Mar. Biol. Ecol., Vol. 13, (1973), 165-179

81. T. Stromgren, 'Skeleton growth of the hydrocoral Millepora complanata Lamarck in relation to light', Limnol. Oceanogr., Vol. 21, (1976), 156-160

82. J. H. Vandermeulen and T. P. Ahern, 'Effect of petroleum hydrocarbons on algal physiology: review and progress report', In: Effects of Pollutants on Aquatic Organisms, A. P. M. Lockwood, (ed.), Society for Experimental Biology Seminar Series, Vol. 2, (Cambridge University Press, Great Britian, 1976)

83. D. B. Boylan and B. W. Tripp, 'Determination of hydrocarbons in seawater extracts of crude oil and crude oil fractions', Nature, London, Vol. 230, (1971), 44-47

84. J. H. B. W. Elgershuizen and H. A. M. de Kruijf, 'Toxicity of crude oils and a dispersant to the stony coral Madracis mirabilis', Mar. Poll. Bull., Vol. 7, No. 2, (1976), 22-25

85. A. A. Benson and L. Muscatine, 'Wax in coral mucus: energy transfer from corals to reef fishes', Limnol. Oceanogr., Vol. 19, (1974), 810-814

THE DEVELOPMENT OF A MODEL FOR BIOLOGICAL CONTROL OF SHALLOW

TROPICAL WATERS SUBJECTED TO PETROLEUM THREATS

Dr. J M Jacques
R & D Manager, NAECON N.V., Drongen, Belgium

Cdt. M Renson
Private Secretary to the Minister of the Environment, Belgium

Msc. A Delmotte
Assistant at the Catholic University of Louvain-la-Neuve, Belgium

1.   INTRODUCTION AND PROJECT GOALS

Contemporary workers are endeavouring to create and define
behavioural models of a marine ecosystem suffering physical
pollution, due to surface films of hydrocarbons.
    The goal of this particular model is to determine and quantify
dynamic management for such a polluted ecosystem.
    A marine ecosystem can be briefly represented as follows:

FIGURE 1

A conventional flow-path of a marine ecosystem
according to the ODUM notation (Ref. 1)

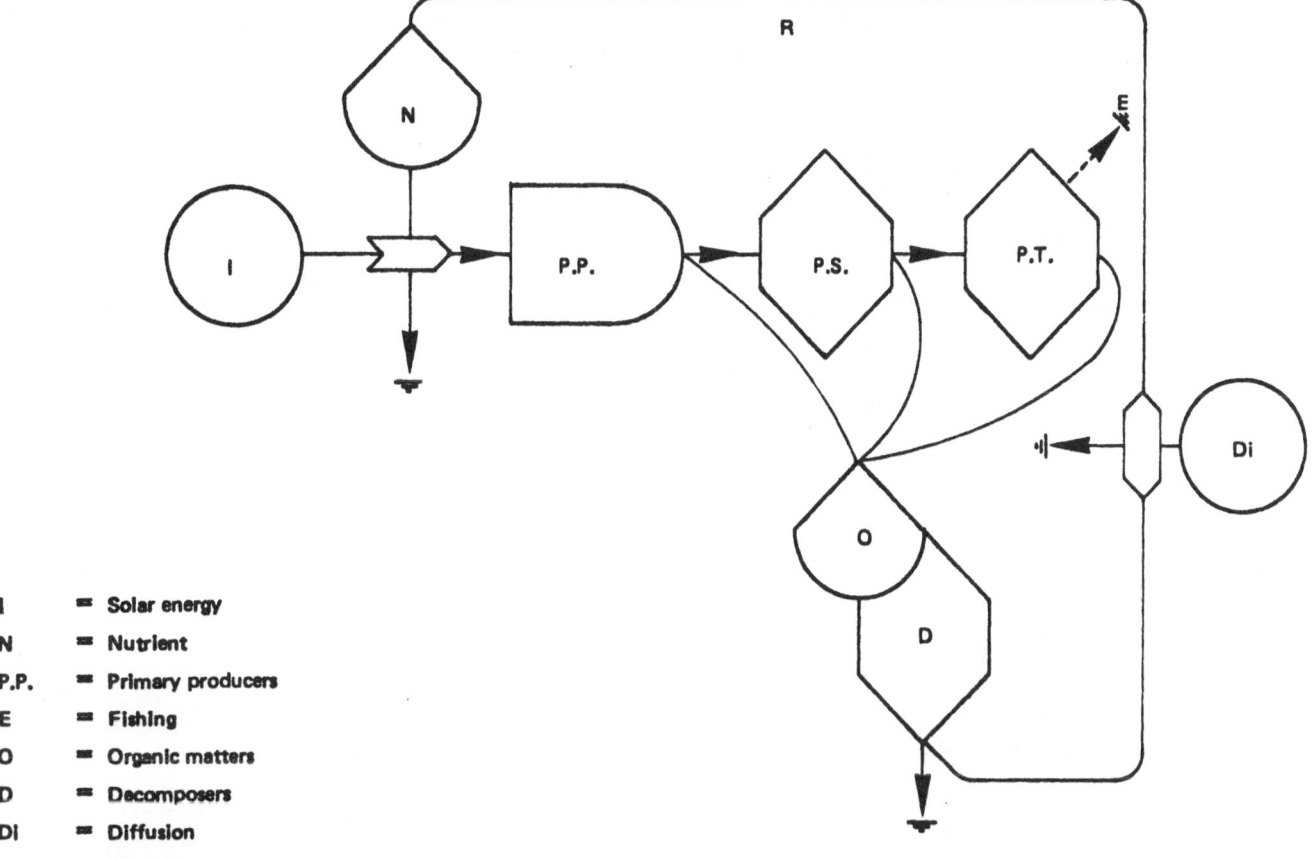

I   = Solar energy

N   = Nutrient

P.P. = Primary producers

E   = Fishing

O   = Organic matters

D   = Decomposers

DI   = Diffusion

R   = Recycling

The motor for this system is the sun which transfers its energy
in photonic form to the stock of the primary producers. This
stock is represented in the pelagic zone by the phytoplankton.
Through photosynthesis, the phytoplankton extract the dissolved
nutrients out of solution. They, in turn, are consumed by the
superseding trophic level species. The final link in the food-
chain often includes organisms which are important to the economy
of man, e.g. fishes. Each evolution into a secceeding trophic
level involves a corresponding loss of matter due to mortalities
and excretions, with energy also lost through respiration. This
lost matter becomes the food supply for a stock of decomposers.
Most of these decomposers are bacteria with the ability to decay
the waste into primary nutrients, so restocking the food supply
to the phytoplankton.
    The goal of this study is to simulate the effect of an oil
layer upon the first two levels of a pelagic food chain. The
direct consequence of the layer upon the intensity of light
penetration is studied. The simulation also determines the time
margin available for human intervention, if the detrimental
effects of the layer on the marine organisms are to be averted.

2. METHODOLOGY AND THEORY

2.1 Brief description of the first two stock groups - theory
Fig. 2 represents the exchanges, transfers and storages of
energy or matter within the primary producer stock (phytoplankton).

FIGURE 2

Exchanges, transfers and storages of energy or matter
within the primary producer stock

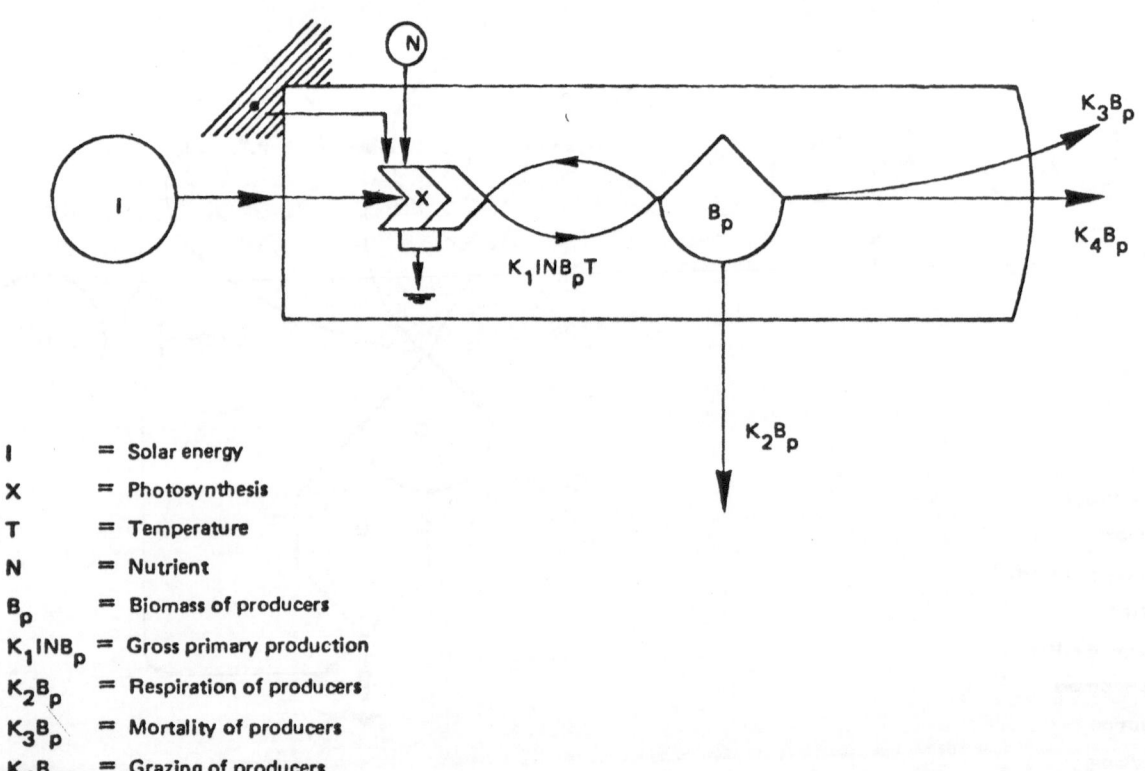

I	=	Solar energy
X	=	Photosynthesis
T	=	Temperature
N	=	Nutrient
B_p	=	Biomass of producers
$K_1 I N B_p$	=	Gross primary production
$K_2 B_p$	=	Respiration of producers
$K_3 B_p$	=	Mortality of producers
$K_4 B_p$	=	Grazing of producers

Fig. 3 represents the exchanges, transfers and storages of energy or matter within the pelagic primary consumer stock (zooplankton).

<u>FIGURE 3</u>

Exchanges, transfer and storages of energy or matter within the pelagic primary consumer stock

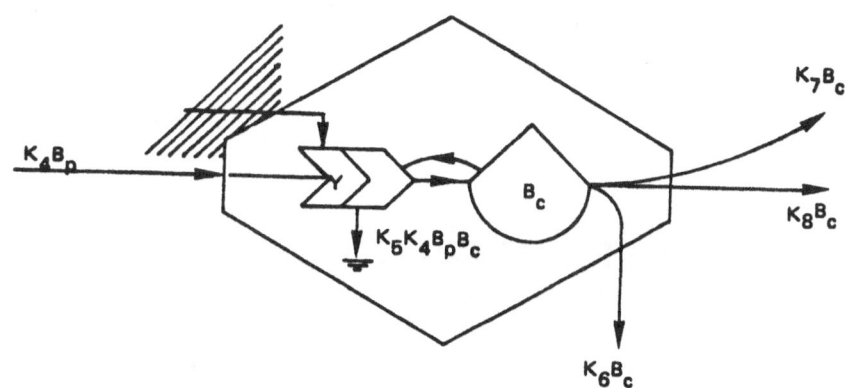

Y	= Biochemical mechanisms
B_c	= Biomass of consumers
$K_5K_4B_pB_c$	= Gross secondary production
K_6B_c	= Respiration of consumers
K_7B_c	= Mortality of consumers
K_8B_c	= Predation of consumers

2

2.2 <u>Theoretical equations of the two first stock groups</u>

The differential equations describing the model are as follows:

For the primary producer stock:

$$\frac{dB_p}{dt} = K_1 1NB_p - (K_2B_p + K_3B_p + K_4B_p) \qquad (1)$$

For the primary consumer stock:

$$\frac{dB_c}{dt} = K_5(K_4B_p)B_c - (K_6B_c + K_7B_c + K_8B_c) \qquad (2)$$

The solution of these differential equations quantifies the change in biomass, or productivity, as a function of the values attributed to their dependent parameters. In addition, it allows the definition of the steady state condition.

2.3 <u>Principle of usage of the system equations within the limits of this study</u>

For the topic which occupies this model, it is feasible to keep constant all the variable affecting the primary producers, excepting one, that of light. The variation of this parameter is described by an elementary analytical function (unitary rectangular increments u(t)) and the designation of the extra-ction factor Y_n is to be reflected as an index X_aY_n of Y_n with respect to X_a.

In this study, the luminous energy will vary as a function of
the thickness of a permanent oil layer overlying the pelagic
milieu within which a simple ecosystem would be free to function.
The envisaged ecosystem is assumed to operate at steady state,
that is to say:

$$\frac{dBp}{dt} = o$$

The response of the ecosystem displayed the change in biomass
and productivity with respect to the variable (-) of the para-
mater "luminous energy", as well as exhibiting variation to the
inverse (+) of the same parameter. These findings initiated
the development of a system to manage the milieux subject to
the hydrocarbon pollution.

2.4 Response potential of the ecosystem
Variations in inlet size, namely the luminous energy, gave a
linear repercussion to the successive stocks foreseen in this
case. Knowing on the one hand that the change (decrease) of the
biomass leaves steady state when the luminous energy disappears,
and on the other hand, that the production conditions and
equations for an infinite population are:

$$N_j = N_d e^{rt} \quad (3) \text{ where } N_j = \text{initial biomass}$$
$$N_d = \text{decreased biomass}$$

one can evaluate the time required to reconstruct the initial
stock after complete disappearance of the hydrocarbon layer,
from the moment that the conditions of the initial luminous energy
reappear. The re-establishment of the stock corresponds to a
recovery to the initial steady state. As a result (3) gives:

$$t = \ln \frac{Ni}{Nd} \quad \frac{1}{r} \quad (4)$$

This shows that the re-establishment to initial stock is a direct
function of N_d, which represented the stock level after the
decrease.

   Fig. 4 indicates schematically the theoretical path of reduction in biomass after steady state is lost, as a function of
light decrease. Consequently one knows that productivity P is
proportional to I, and by approximation one estimates that biomass
will likewise be proportioned to I.

   By assuming in the model a linear reduction relationship,
the theoretical projection indicated a biomass reduction, subsequent to the surface film of petrol being applied. This
response is explained by the reduction of light penetrating into
the ecosystem.

   Fig. 5 indicates schematically the theoretical path of re-
establishment to initial stock as a function of time after the
petrol disappearance and the consequent reappearance of former
light conditions.

FIGURE 4
Biomass reduction as a function of light absorption

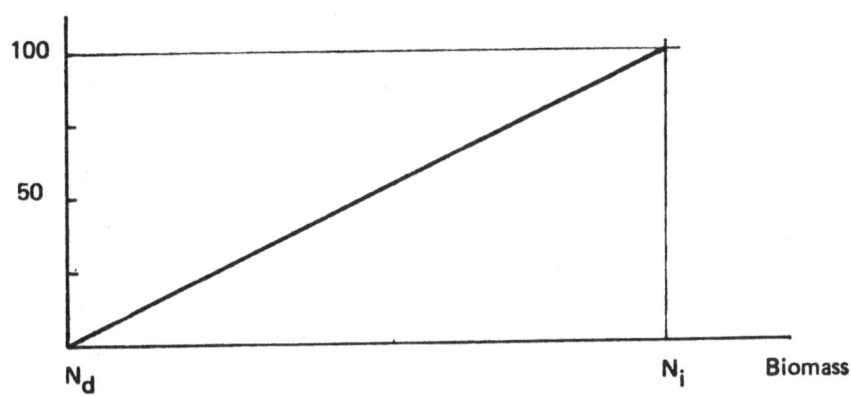

FIGURE 5
Return time to steady state as a function of N_d

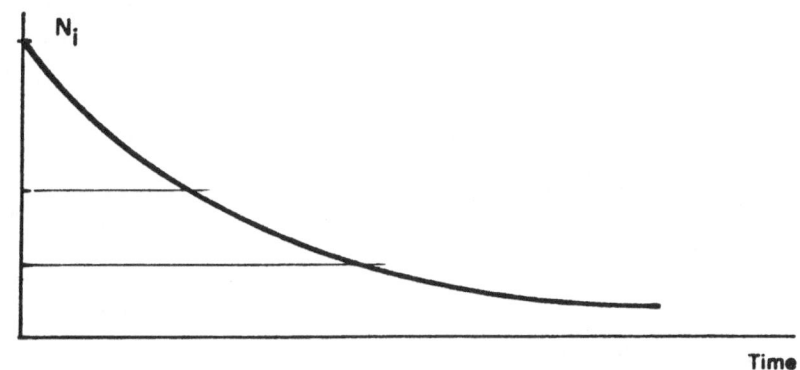

Remarks

It was considered in this instance to formulate population
production under these infinite conditions. It is necessary
however to note that these conditions do not exist, and generally
that the development of a population describes a logistic curve
rather than an exponential. This remark is important because
the stock re-establishment time and the return to steady state
is therefore the minimum time envisaged.

On the other hand, in (3) the variable term 'r' was fixed
throughout the run of the process. We have not therefore incorporated the direct toxic effects of a hydrocarbon on the
individuals themselves. For the more sophisticated management
model it would be necessary to reckon with this, but little
information exists on this aspect at present.

3. APPLICATION TO A WEST AFRICAN UPWELLING

3.1 Introduction

The model described above was applied to a West African upwelling.
This upwelling was chosen for three principal reasons:
1. The upwelling of Western Africa extends from Senegal as far
as Morocco, and is an ecosystem very rich in vegetal and animal
species (Ref. 2). This natural wealth is reflected by the
productivity of its milieux - for which these milieux in turn

deserve ecological recognition.

2. This ecological importance has been highlighted by the wide range of literature that it inspired.

3. This highly-productive milieu holds an extremely important economic advantage because practical fishing of this milieu is essential both to the economy of the coastal countries and to other lands.

An accident in this area would have very damaging consequences.

3.2 Appropriate calculations

Table 1 reviews the local values of the different parameters involved in the calculation.

TABLE 1 - Values of the parameters

Luminous energy I $= 5,700$ Kcal.m^{-2}.day^{-1} (Ref. 3)

Nutrient $=$ Unlimited (o.032g.m^{-3})

Chlorophyll A $= 3$mg.m^{-3} (Ref. 4) (Ref. 7)

Primary biomass (B_p) $=$ Chlorophyll Ax50=150mg.C.m^{-3} (Ref. 5)

Net primary production.
Day^{-1}(PP) $=$ Obtainable from P/B=40.5mg.C.$m^{-3}J^{-1}$

PP/B_p Daily $= 0.27$

Time of doubling primary population $=$ Obtainable from P/B=3.65 days (Ref. 6)

r $=$ Obtainable from (3) and from the time of doubling

Primary consumer biomass (B_c) $= 36$mg.C.m^{-3} (Ref. 7) (Ref. 8)

Net secondary production.
Day^{-1} $=$ Obtainable from PIB=1.47mg.C.$m^{-3}J^{-1}$

Pc/Bc daily $= 0.041$

Time of doubling secondary production $=$ Obtainable from P/B=24.5 days

r $=$ Obtainable from (3) and from the time of doubling secondary population = 0.028

Calculation of the 'K' factor

Note : Since the productivities were expressed in net form, the terms in K_2, K_3, K_6 and K_7 are included respectively in the terms K_1 and K_5.

Combining (1) and (2), one can calculate :

$K_1 = 0.48 \ 10^{-4}$

$K_4 = 0.271$

$K_5 = 0.25$

The gross production of the primary consumers(J^{-1})=10.17mg.C.J^{-1}

Mortality of the primary consumers J^{-1}=3.95mg.C.J^{-1}

Respiration of the primary consumers J^{-1}=4.75mg.C.J^{-1}

By substituting these calculated parameters of this system into (3), one can calculate the variation in time to re-establishment of steady state, after the complete disappearance of the hydrocarbon layer, as a function of the prevailing reduced level of biomass.

Table 2 and Fig. 6 specify the time needed to re-establish initial stock for different values of N_d in the case of the primary producers.

TABLE 2

$N_d$mg.C.m^{-3}	T_p (Days)
125	0.94
100	2.10
75	3.59
50	5.69
25	9.28
1.5	23.86
0.1	25.09

FIGURE 6

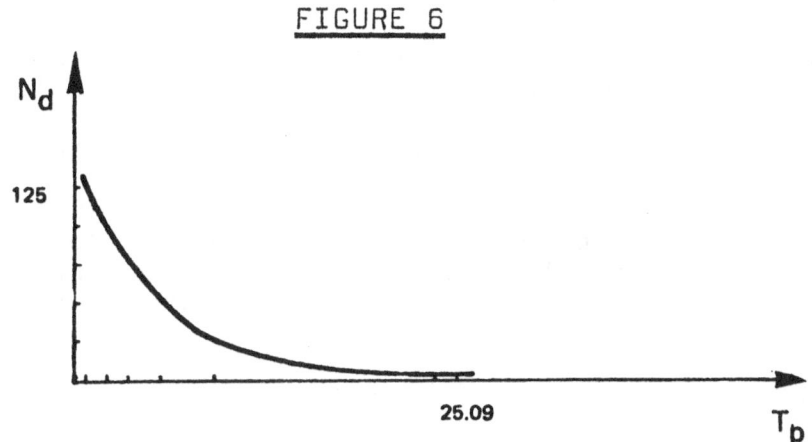

Table 3 and Fig. 7 specify the time needed to re-establish the stock of primary consumers (T_c) for different values of N_d.

TABLE 3

$N_d$mg.C.m^{-3}	T_c (Days)
30	6.53
20	21.01
15	31.28
10	45.77
5	70.52
2	103.25
0.1	128.00

734

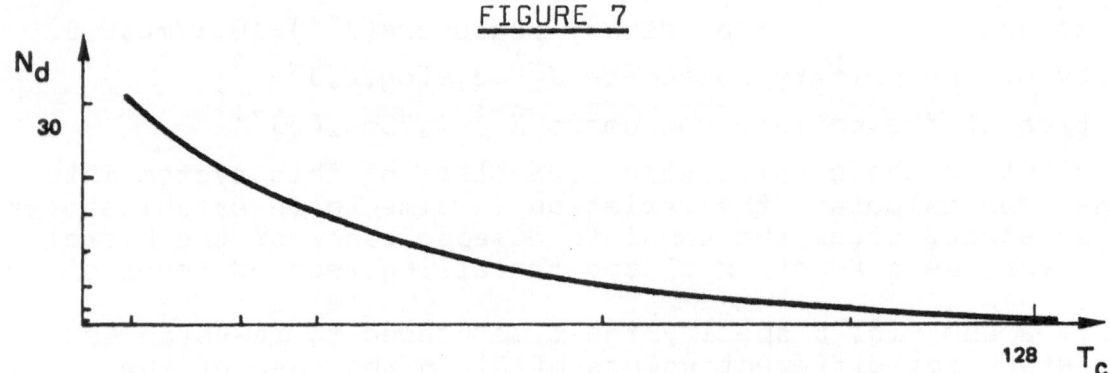

FIGURE 7

4. INTERPRETATIONS, CONCLUSIONS, RECOMMENDATIONS

The results indicate that, even under the most favourable
conditions, the period of time needed before the level of
primary producers is re-evolved, subsequent to the initial stock
being replenished and steady state recovered, is relatively
great. This time becomes considerable if one further allows
for the primary consumers.

The impact of light reduction, due to a hydrocarbon film
above the fish, has not been incorporated into this study. One
is able, however, to contemplate the very damaging effect of
this reduction upon the fish populations, firstly as a relationship between this hitherto excluded factor and the new trophic
levels, and secondly as regards the biomass productivity under
reduced light.

This study confronted an ecosystem, reduced it to a simplified expression, recognised firstly prey from predator, revealed
the light-absorbing relevance of physical pollution for the
complete ecosystem, and then progressed, deepening the interest
into how ecosystems survive under pollution threat, as to how
control systems might be conceived to monitor milieu re-action,
and finally poses the notion of comprehensive management of
selected zones, in which micro-sensors extrapolate readings into
visual form using the differential equations developed by this
paper.

5. BIBLIOGRAPHY

(Ref. 1)   H T Odum, "Marine ecosystems with energy circuit dia-
           grams in 'modelling of marine systems' ", Elsevier
           Oceanography Series 10, (1975), 127-151.
(Ref. 2)   J Cadenat, "Poisson de mer du Senegal", Institut
           Francais d'Afrique Noire, (1950), 344.
(Ref. 3)   A Morel et L Caloumenas, "Variabilite de la repartition
           spectrale de l'energie photosynthetique", Tethys,
           6(1-2), (1974), 93-104.
(Ref. 4)   M Fiala et G Jacques, "Relations entre ATP, chlorophyle
           et production dans la couche euphotique d'une zone
           d'upwelling (Campagne Cinea – Charcot II, 14 mars-
           30 avril 1971)", Tethys, 6(1-2), (1974), 251-258.
(Ref. 5)   J M Ryther, D W Menzel, E M Hulburt, C J Lorenzen et
           N Corwin, "The production and utilisation of organic
           matter in the Peru coastal current", Inv. Pesq., 35,
           (1971), 43-60.
(Ref. 6)   J M Jacques, "Etude comparative de l'ecolgie et de la
           productivite des crustaces copepodes dans deux etangs
           de salmonicultures fertilises et non", projet-IBP-

Unesco, <u>These de Doctorat et Science, Universite Louvain,</u>
(1976),124.

(Ref. 7)   Groupe Mediprod, "Generalite sur la campagne, Cinea -
Charcot II (15 mars - 30 avril 1971), <u>Tethys,</u> 6 (1-2),
(1974), 33-42.

(Ref. 8)   M M Mullin, "Production of zooplankton in the ocean :
the present status and problems", <u>Oceanogr. Mar. Biol.</u>
<u>A. Rev., 7,</u> (1969), 293-314.

OIL SLICK MOVEMENTS IN THE ARABIAN GULF

W.J. Lehr
Department of Mathematical Sciences, University of Petroleum
and Minerals, Dhahran, Saudi Arabia.

H.M. Cekirge
Research Institute, University of Petroleum and Minerals,
Dhahran, Saudi Arabia

1. INTRODUCTION

As the world centre of the oil industry, the Gulf region faces
a high potential danger for oil pollution. One factor in the
location assignment of oil pollution control task forces is the
assessment of the danger of contamination for various shorelines
on the Gulf. Therefore, the authors have constructed a model
for estimating the trajectory of oil spills for various locations
in the Gulf based upon seasonal average climatic data. While
chiefly of use for statistical conclusions, our model could be
used by oil spill detection agencies to provide an expected
spill path on the basis of minimal information about such a
spill. Such a "first guess" trajectory would, for example,
help assess governmental responsibility for clean-up in the
region with its multinational coastline. It could also determine whether such a spill is likely to come ashore in any area
where it could do severe damage and hence must be further
monitored and controlled or whether only minimal risk is
involved and less urgent response required.

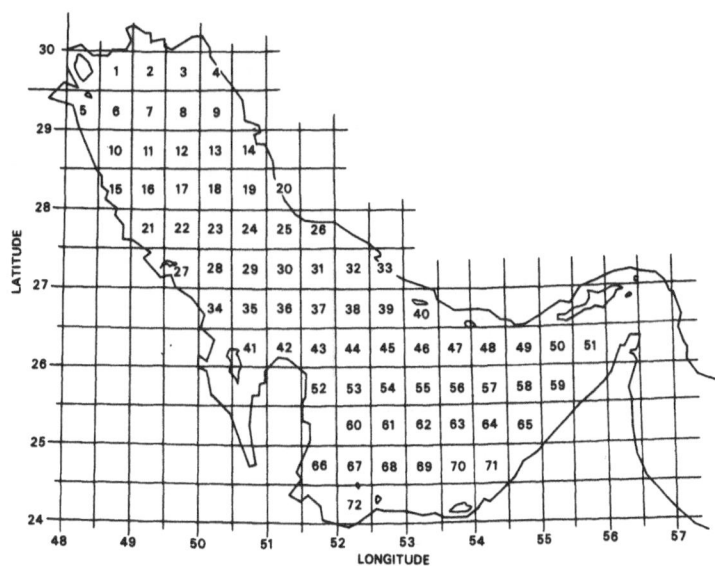

Figure 1. The grid map.

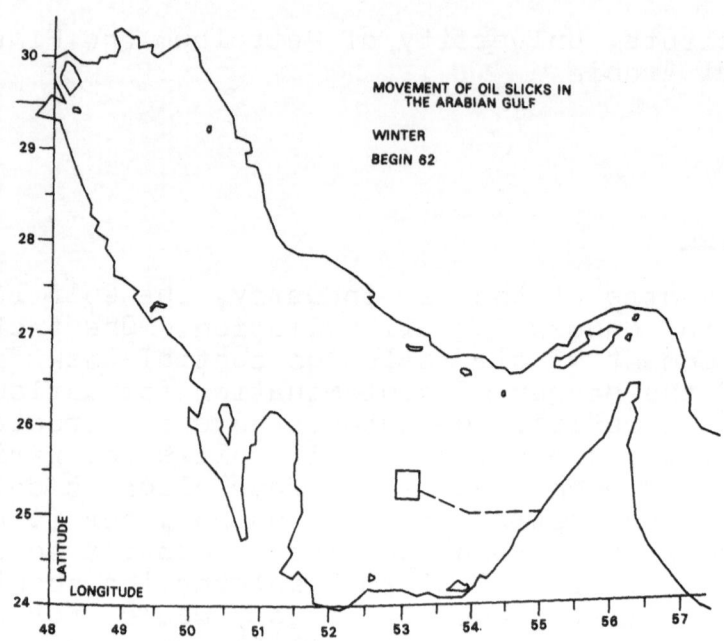

Figure 2. The trajectory of an oil spill
starting from square 62 in Winter

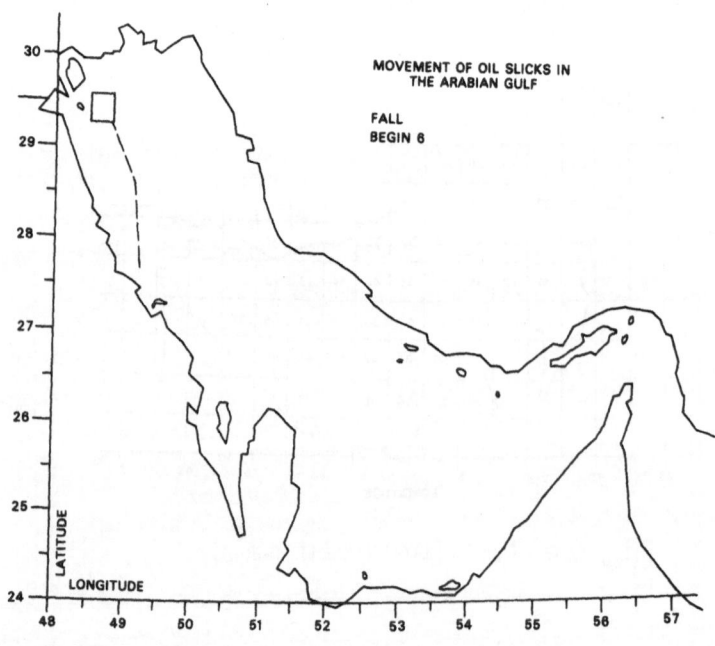

Figure 3. The trajectory of an oil spill
starting from square 6 in Fall

2. DESCRIPTION OF THE MODEL

Due to certain limitations on the available data, only offshore
oil spills were considered. By offshore, we mean spills which
are far enough from shore that their movements will be relatively
unaffected by coastal currents. I.M.C.O. (1) has used as a
basis for such spills the distance of 25 miles but this may be
overly pessimistic. The movement of the oil slicks are presumed
to be determined by the overall drift current, local tidal
currents and wind-induced currents. Current and wind data were
based upon Tetra Tech Co. reports to the Royal Commission for
Jubail and Yanbu (2). Tetra Tech acquired its data from local
observation and sources, and external sources including the
National Climatic Centre in Ashville, North Carolina and U.S.
Naval Hydrographic Office. Using the above data, the Gulf area
was divided into 72 grid blocks each of a half degree longitude
in width and half degree in latitude in height and an overall
seasonal current and wind vector for each grid block was assigned
(see Figure 1). From these vectors a seasonal oil slick drift
vector was computed for each block.
    Considerable dispute exists over the proper determination of
this drift vector. For example, Stokes (3) developed a theory
of wind induced current which predicts a coriolis caused drift
angle between the wind and current directions of up to 45°(3).
Experiments, however, have yielded values deviating significantly
from Stokes' predictions (4). Actual spill observations have
also shown widespread variation in drift angle. Rath and Francis
(5) suggest that field tests for specific sites should be conducted
to determine a proper value. In the absence of such field tests,
we have adopted the formula of Lissauer and Bacon (6) and set
the drift angle equal to zero although this would easily be
modified if observational data yields a different result.
    Similarly, there is no unanimous opinion of the strength
of the surface current induced by the wind, with estimates anywhere from 1 to 10% (5) of the wind velocity. Again following
Lissauer and Bacon, we have adopted a wind factor of 3.5% of
the wind speed at 10 metres above the sea surface. While most
models simply add this wind caused current to the normal drift
current, Schwartzberg (7) has determined emperically that there
is a coupling effect between the two vectors and that a more
appropriate equation for the total oil slick drift vector would
be the wind current vector plus 56% of the normal current vector.
This was the formula used for this model.
    Using these seasonal drift vectors the initial location of
the spill, its expected co-ordinates were tracked every 12 hours,
except during summer season where smaller values for the drift
vectors made a 24 hour period more practical. Thus it was
possible not only to estimate the predicted path of the spill
but also its estimated time of impact with the shoreline as well.
    Figures 2 and 3 present examples of some simulated spill
trajectories. The small square represents the hypothetical
initial spill location and the dash line the predicted path with
each dash representing a 12 hour time span (in the summer season
a 24 hour time span). Since the space between each dash also
represents a 12 hour period, the number of days until estimated
impact with the shoreline can be calculated by simply counting
the number of dashes.

3. CONCLUSIONS

Assuming no prior preference for any site in the Gulf as an oil spill origin, it was possible to determine those coastal areas which have a high risk for oil pollution. Dividing the Gulf coast into six areas (see Figure 4) the following table was constructed using the computer trajectory predictions starting from all the squares except square 20 whose centre lies inland. The numbers represent percent of oil spills which impacted on the specified coastal area.

	Winter	Spring	Summer	Fall
Area A	7	16	4	10
Area B	1	3	39	19
Area C	21	22	14	22
Area D	18	20	21	12
Area E	43	23	17	27
Area F	10	16	4	9

(Numbers may not add to 100 due to rounding)

Coastlines not included in these six areas had no predicted oil spill impacts.

From the table we can conclude that the Southern Iranian coastline (Area E) has the highest danger for oil pollution although the Emirate coasts (Area C and D) also show considerable risk potential. The other coasts are relatively safe except for the Southern Arabia and Qatar region (Area B) which shows high risk for oil drifts in the Summer and Fall. Northern Arabia (Area A), while fairly safe from spills, has its greatest pollution risk season in the Spring. This may suggest that any oil spill task force for Saudi Arabia be stationed in the North during the first part of the year and moved south for the Summer and Fall.

Further refinements of this model which will allow improvement in the estimation of the drift current, expand predicative capabilities to the Gulf of Bahrain and Straits of Hormuz, and predict spread and degradation of the spill by dispersion, evaporation and other natural factors, are currently under development by the authors.

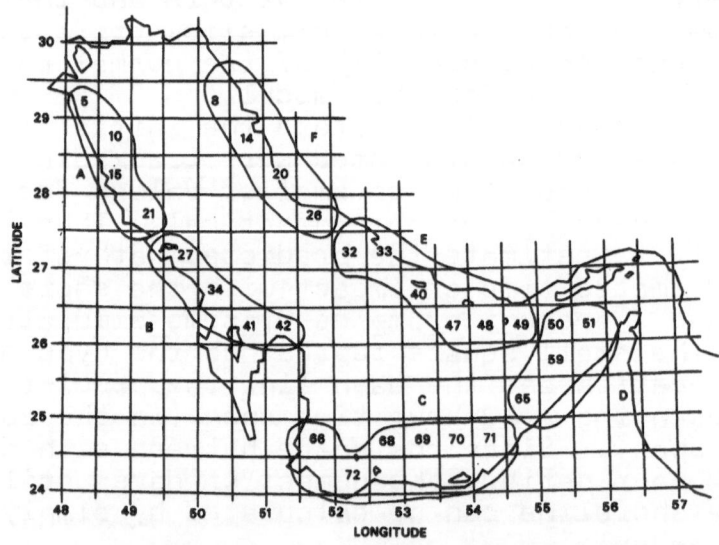

Figure 4. The coastal areas.

4. REFERENCES

1. Inter-Governmental Maritime Consultive Organization, Coastal Oil Pollution Evaluation Study for the Kingdom of Saudi Arabia, 1977.

2. Tetra Tech, Marine and Atmospheric Surveys, Royal Commission for Jubail and Yanbu Contract No. 001-004.

3. G. Stokes, 6, Trans. Cambridge Phil. Soc., 8 and Supplement, Sci. Papers, 1, 1847

4. D. Alofs and R. Reisberg, Journal of Physical Oceanography, 2, p.439, 1972.

5. R. Rath and B. Francis, "Modeling Methods for Predicting Oil Spills Movement", Report to the Oceanographic Commission of Washington, 1977.

6. I. Lissauer and J. Bacon, Predicted Oil Slick Movement from Various Locations off the New Jersey-Delaware Coastline, U.S. Coast Guard Research and Development Centre, 1975.

7. H. Schwartzberg, Proceedings Joint Conference on Prevention and Control of Oil Spills, p. 489, 1971.

IMPLICATIONS OF THE IXTOC I BLOW-OUT AND OIL SPILL

Richard S. Golob
Executive Editor
Oil Spill Intelligence Report
Director
Center for Short-Lived Phenomena
World Information Systems
Cambridge, Massachusetts 02138

Daniel W. McShea
Associate Editor
Oil Spill Intelligence Report

INTRODUCTION

On 23 March, engineers capped the Ixtoc I, a Petroleos Mexicanos (PEMEX) exploratory well in the Bahia de Campeche, about 80 kilometers northwest of Ciudad del Carmen, Mexico (Mex.), almost 295 days after the well blew out and caught fire on 3 June 1979. During the first 7 days of the blow-out, the well lost more oil than the largest spill in U.S. history, the Argo Merchant spill of 7.6 million gallons off the Massachusetts coast in December 1976. By late July, the Ixtoc I had become the largest oil spill in world history, surpassing the 68 million gallons that the tanker Amoco Cadiz spilled off Brittany, France, in March 1978. In mid-January, the well was still spilling oil and had lost more than twice the record amount involved in the Amoco Cadiz incident. By 23 March, the Ixtoc I had lost a total of more than 140 million gallons of oil. PEMEX announced that, during the 295 days of the blow-out, it had recovered 4.5% of the total volume of oil spilled and that 28.5% had formed a slick, 17% had evaporated, and 50% had burned at the well site.

The Ixtoc I may become the most expensive oil spill in history as well as the largest. At current world prices, the 140 million gallons of Ixtoc I oil would cost at least $100 million. Damage suits now pending in U.S. courts total more than $365 million. The loss of the drilling platform amounts to $21 million, and the cleanup in U.S. waters has already cost the U.S. government more than $9 million. PEMEX recently reported that it had spent about $134 million on the entire Ixtoc I spill response and that 25% of its costs involved environmental protection measures, including dispersant spraying, oil recovery, and beach cleanup. Based on these estimates, the Ixtoc I spill may cost a total of more than $629 million.

The Ixtoc I is important not as a statistic, but as a case study for researchers, policy makers, and response teams. The incident has major implications for open-ocean spill response, blow-out prevention technology, international pollution damage compensation, scientific spill research, and outer continental shelf (OCS) development. This paper will explore the problems encountered in capping the well, recovering the spilled oil, and assessing the damages. Since researchers have not yet collected and analyzed all the information on the Ixtoc I, this paper will also discuss some of the issues and questions that deserve further research and investigation.

BLOW-OUT

PEMEX contracted the Mexican drilling firm Perforaciones Marinas del Golfo S.A. (PERMARGO) of Mexico City, Mex., in the spring of 1977 to begin exploring the Chac area in the Bahia de Campeche on the Gulf of Mexico OCS. PERMARGO contracted SEDCO, Inc. of Dallas, Texas, in August 1977 to supply a semi-submersible rig under a bareboat charter. Under a separate agreement with PERMARGO, SEDCO also provided personnel to advise PEMEX and PERMARGO in the operation of the rig.

On 1 December 1978, PEMEX positioned the SEDCO 135 in 48 meters of water about 80 kilometers northwest of Ciudad del Carmen and began drilling the Ixtoc I shortly thereafter. The drilling proceeded without problem or interruption through the end of May, when the well had reached a depth of about 3600 meters. Circulation was lost when the drill bit entered either a cavern or a highly porous section of the formation, and Mexican crews responded by injecting additional mud into the well. The well continued to lose mud, and the drillers quickly injected low-density mud to try to regain circulation, while reducing the downward hydrostatic pressure of the mud column so that the well would lose less mud into the formation.

PEMEX said that, after the well had lost circulation and the drillers had bypassed the shale shaker, geologists could not collect the drill cuttings to analyze the composition of the formation and estimate the proximity of an oil and gas reservoir. The PEMEX drillers were unable to regain circulation and, at 1500 LT, opted to wait and observe the well. After determining that the well was stable, PEMEX elected to pull the drill pipe from the well in order to remove the bit and the drill collars. SEDCO has criticized PEMEX for trying to remove the drill pipe without first regaining circulation. SEDCO said that lifting the drill pipe made it impossible to circulate mud or sea water and sacrificed some degree of control over the well.

As soon as the bit was removed, PEMEX planned to insert the open-ended drill pipe into the well and pump in a gelatinous plug to seal the formation and prevent further loss of mud. At 2200 LT on 2 June, PEMEX engineers began to remove the drill pipe from the well, while monitoring for pressure changes and other signs of instability. As they removed the drill pipe, the engineers pumped mud into the well to fill the space formerly occupied by the pipe, and took measurements of the pressure in the mud column every 300 meters. They also tested the blow-out preventers (BOPs) on the sea floor and a drill-pipe safety valve on the platform.

At 0230 LT on 3 June, mud began flowing up the annulus and drill pipe and spilling onto the floor of the SEDCO 135 platform. When personnel activated the BOPs on the sea floor, only 200 meters of drill pipe remained in the Ixtoc I, and the pipe section with the drill collars had been brought up to the level of the BOP stack. The pipe rams were unable to close completely around the 4.75-inch (1 inch = 2.54 centimeters) drill collars, since the rams were shaped to close around the smaller 3.5-inch drill pipe, and the shear rams were unable to sever the thick steel collars. When the well kicked, the sudden surge of pressure lifted the drill pipe a short distance out of the well and caused the pipe to bend. As a result, the crews could not control the well by lowering or raising the deformed pipe. PEMEX activated the annular BOP which successfully sealed the annulus, but the mud continued to flow up the drill pipe and onto the floor of the platform.

Just prior to the blow-out, workers had been disengaging an adaptor, which joins the drill pipe to the drill collars, in preparation for removing the remaining drill collar sections, the pipe, and the bit. When the flow of mud appeared, workers tried to reconnect the adaptor so that they could install above the adaptor a safety valve that would shut off the flow through the drill pipe. The threads of the connection between the adaptor and the pipe jammed, however, and the workers had to abandon the effort.

Oil and gas began to gush from the drill pipe to a height of 30 meters above the platform floor, and within moments, the oil and gas fumes exploded on contact with the operating pump motors. A fire broke out on the rig, and all PEMEX, SEDCO, and PERMARGO personnel abandoned the rig in lifeboats. No significant injuries resulted. The fire melted the drilling tower and destroyed most of the equipment and machinery on the SEDCO 135. The drilling tower collapsed eventually, and when it sank to the sea floor, it buckled the marine riser and caused some damage to the BOP stack and the well casing beneath the stack.

By early 4 June, the damaged rig had been towed away from the well area, where a large fire was burning. SEDCO said that, at the direction of local Mexican authorities, the rig was towed to an anchorage 50 kilometers northwest of the well. In the following weeks, SEDCO said that investigators from the insurers of the rig boarded the structure and examined it, and then ordered it sunk on 12 July.

Seven months after the Ixtoc I blew out, the Funiwa 5 development well blew out on 17 January about 8 kilometers off the Nigerian coast in the Niger River delta area, and spilled about 8.4 million gallons of crude oil before it bridged in early February. The rig operators lost control of the Funiwa 5 while preparing to convert it into a production well. Unlike the Ixtoc I which was an exploratory well, the Funiwa 5 was a development well in a known oil and gas formation. Also unlike the Ixtoc I, the Funiwa 5 bridged in less than 3 weeks.

The 2 incidents, however, had at least one characteristic in common. Red Adair, the well blow-out expert, said that, when the Funiwa 5 kicked, the drill collars had been raised to the level of the BOP stack and that the shear rams were unable to cut through the drill collars and close off the annular space and the drill pipe. The inability of the BOP to cut through the drill collars at both the Ixtoc I and the Funiwa 5 resulted in massive oil spills, and this causal similarity raises questions about the adequacy of BOP technology to prevent blow-outs.

The statistics suggest, however, that the offshore industry has an excellent pollution record and that a blow-out is a low probability event. The U.S. Geological Survey (USGS) recently compiled U.S. blow-out statistics for the 8 years between 1 January 1971 and 31 December 1978. According to the USGS report, 7553 new wells were drilled on the U.S. OCS during the 8-year period, and 30 of those wells blew out. Among the wells that blew out, 17 were exploratory wells and 13 were development wells. All the blow-outs were gas blow-outs, and no oil spillage resulted. Even though the offshore industry has an excellent record, the Ixtoc I incident shows that a low probability event such as an Ixtoc I can take place.

Comprehensive worldwide information is not publicly available on the causes of historical blow-outs, and the information on the similarity of circumstances surrounding specific incidents, such as the Ixtoc I and Funiwa 5, would likely prove valuable in preventing other blow-outs and spills.

CAPPING OPERATION

In mid-June, PEMEX began drilling the Ixtoc IA, a directional relief well, and the new well had reached a depth of 600 meters when the Ixtoc IB, a second relief well, was begun. PEMEX planned to drill to within about 8 meters of the Ixtoc I at a point below the well casing and above the oil and gas reservoir. After setting the casing in the relief well, the drillers planned to pump sea water or mud into the relief well. The mud would migrate into the formation at the base of the relief well, and some small fraction of the mud would enter the Ixtoc I and begin to fill the well. As the Ixtoc I filled with mud, the weight of the mud would begin to balance pressures within the formation, and the flow of oil and gas would slow and stop. PEMEX could then seal the well with cement.

The following problems hampered the relief well effort:

1) Shale began sloughing from the sides of the Ixtoc IB well bore, and the drilling on the Ixtoc IB was stopped in the third week of September at a depth of almost 3200 meters so that engineers could set casing to prevent the shale from entering the well bore.

2) The precise location of the Ixtoc I bottom was not known, and as a result, the drilling on the Ixtoc IA was stopped at a depth of over 2800 meters so that engineers could plug the last section of the well and redrill that section in a slightly different direction. PEMEX reportedly made inclination surveys but did not make directional surveys while drilling the Ixtoc I, and since the relief well drillers did not know the azimuthal coordinates of the well bottom, they had difficulty finding the well casing at points over 3600 meters below the sea floor. Experts said that the Ixtoc I was a vertical well and that drillers commonly do not conduct directional surveys while drilling vertical wells. Using data from the inclination surveys only, PEMEX was able to determine the location of the bottom of the Ixtoc I well to within 50 meters. PEMEX used a Magrange magnetometer system to detect the casing of the Ixtoc I from the bottom of the relief wells, but the Magrange could only detect the well casing through the formation at distances of up to approximately 15 meters.

3) Relief well drillers were initially unable to recheck the surface surveys to pinpoint the precise coordinates of the well on the sea floor since the SEDCO 135 was no longer positioned over the well and since a fire burned over the spot where the platform had been located. In September, surveyors estimated that the initial determination of the location of the Ixtoc I well on the sea floor was inaccurate, and as a result, the initial determination of the location of the Ixtoc I well at points over 3600 meters below the surface was also inaccurate.

4) In late December, the steel shoe of the Ixtoc IA well casing broke from the bottom of the deepest section of the well and fell into the open uncased hole below. The steel piece lodged in the hole at a point 15 meters below the deepest section of the casing and 135 meters above the bottom of the uncased hole. Engineers were able to use drill pipe attachments to fragment the steel obstruction, and succeeded in early January in extracting the metal from the hole.

When the Ixtoc IB was completed in November, it was located more than 12 meters away from the Ixtoc I, and PEMEX could not deliver mud quickly enough across the 12-meter gap to stop the flow of oil. The Ixtoc IA relief well was not completed until March, and the capping was finally achieved by pumping mud through the Ixtoc IA across a distance of 3 meters into the Ixtoc I and by injecting cement directly into the Ixtoc I well bore.

Other attempts to cap the well or slow the oil flow failed during the blow-out. In June, Red Adair succeeded in closing the BOP that was still intact near the sea floor, and the flow of oil was halted for 4 hours, until the pressure in the well ruptured the well casing below the BOP. Adair then reopened the BOP to prevent further damage to the casing, and the oil began flowing from the Ixtoc I once again.

In August, PEMEX tried to reduce the flow rate by pumping into the well about 100,000 steel and lead balls weighing approximately 1.6 kilograms each. PEMEX claimed that this procedure reduced the daily flow rate from 30,000 barrels to 10,000 barrels, but Red Adair disagreed. He said that the balls were quickly ejected from the well and scattered about on the sea floor and that the balls had not reduced the flow rate.

Throughout the Ixtoc I incident, the control of the spill depended almost entirely on the 2 relief wells that were begun in June and completed the following winter. The problems involved in preventing the blow-out, drilling the relief wells, and then capping the ruptured well helped to emphasize the need for further study into the development of strong drilling safeguards to prevent the blow-out in the first instance and to insure a rapid relief operation thereafter. To reduce the time needed to drill relief wells and control a blow-out, Red Adair suggested that, prior to drilling all "wildcat" wells, drillers should select relief well sites, prepare the site for drilling, and maintain a relief well rig in a standby position. In addition, Adair said that well operators should know at least the bottom location for every well on a platform.

OPEN-OCEAN SPILL RESPONSE: EQUIPMENT

As mentioned in the introduction, PEMEX announced that, during the blow-out, it had recovered only 4.5% of the total volume of oil spilled and that 28.5% had formed a slick, 17% had evaporated, and 50% had burned at the well site. The PEMEX on-scene coordinator said that he did not believe the cleanup technology existed anywhere to deal effectively with an open-ocean spill similar to the Ixtoc I. Although many private contractors and government officials point to poor spill response management as a contributing cause for the low recovery rate, they almost all agree that, given similar weather conditions, they would have had difficulty making a dramatic improvement in the recovery rate of the PEMEX cleanup crews. In fact, based on the Ixtoc I incident, the Canadian Environmental Protection Service (EPS) expressed serious concern about its ability to combat a massive open-ocean blow-out and spill in Canadian waters.

At the well site, PEMEX used a variety of cleanup techniques, including skimmers, dispersants, and the Sombrero oil collection device, to recover the spilled oil. The reasons for the overall failure of these techniques include logistical problems, mechanical failures, weather conditions, and poor response management at the well site. And yet, if the Mexicans, with equipment from the best manufacturers and expertise from the finest spill experts worldwide, were not able to deal with the Ixtoc I spill, then the cleanup problems encountered at the Ixtoc I raise questions about the adequacy of existing equipment and technology to meet the demands of any massive open-ocean spill, whether from a well blow-out or a tanker stranding.

Recovery rates for 6 Oil Mop Inc. (OMI) Mop Machines on-scene ranged from a maximum of 4000 barrels of emulsion per day to no recovery at all on other days. On some days, high waves washed oil onto the deck of the OMI barge, making it too slippery for OMI workers to operate the Mop Machines mounted there. Recovery was also reportedly hampered on some days because

PEMEX would not allow the barge to maneuver during diving operations related to the capping effort. In addition, Mexican workers operated OMI machines without tail pulleys occasionally, and as a result, recovery was less than optimal.

PEMEX obtained from Statoil, the Norwegian state-owned oil company, 1000 meters of boom, 2 Frank Mohn A/S FRAMO ACW-400 skimmers, and assorted support equipment. Statoil said that the booms were not deployed effectively enough during the first few weeks after the blow-out to concentrate the oil into a substantial pool for rapid recovery and recommended that PEMEX use the skimmers in a mobile system, rather than the initial anchored system. PEMEX decided not to use the mobile system, however, in order to protect the divers working near the well, with the result that the frequent current changes carried the oil away from the skimmers much of the time. The barge did not have adequate equipment to permit safe operation of the skimmers at night, and damage repairs following storms also reduced operating time. As a result, according to Statoil, the skimmers were able to pump oil for about 15% of the time on-scene. PEMEX estimated that the FRAMO skimmers were able to recover only up to 1100 barrels of emulsion per day.

PEMEX initially planned to use the OMI Mop Machines and the experimental Shell Oil Co. SOCK skimmer to collect oil escaping from the FRAMO system, but the SOCK skimmer encountered mechanical difficulties. The 3-tier mechanical response planned by PEMEX, involving the FRAMO skimmers, the SOCK skimmer and the OMI Mop Machines, was completely operational for only a short time in June and recovered less than 20,000 barrels of emulsion while it was in operation. On the most effective recovery day, the 3-tier system recovered 6000 to 8000 barrels of emulsion.

The Alsthom Atlantic Inc. (AAI) Cyclonet 150 open-sea skimmer on loan to PEMEX from the Southern California-Petroleum Contingency Organization experienced logistical problems at the well site. The Cyclonet 150 had just begun to collect oil at the well site when the support vessel's main motor clutch broke, forcing a return trip to port for repairs. The skimmer was returned to action 2 weeks later and was collecting oil by early August, but PEMEX workers poured all the recovered oil back into the Bahia de Campeche, reportedly because a storage barge was not available. Other reports indicated that PEMEX considered the recovery efficiency of the Cyclonet 150 too low to warrant transferring the material to a storage barge. PEMEX returned the skimmer to Louisiana in late August. According to an official PEMEX document, the Cyclonet 150 recovered between 1130 barrels and 2260 barrels per hour on one day of operation in 1.5 meter seas from a slick 1 to 10 millimeters in thickness and up to 30 meters wide.

PEMEX also contracted British Petroleum (BP) in late June to supply and operate 2 Vikoma Seaskimmers with 2 Oceanpack inflatable boom systems. BP deployed the skimmers at the well site, but a workboat propeller hit and severely damaged one of the Oceanpack booms, and hydraulic problems prevented the use of the other boom. BP later operated the skimmers along with PEMEX-owned boom, and PEMEX said that the skimmers recovered an average of 500 barrels of emulsion per day. The skimmers recovered more than 10,000 tons of emulsion in total, according to BP.

The U.S. began cleaning up oil at the well site in mid-August, with 2 Offshore Devices Inc. open-water skimming barrier systems (OWSBSs). The U.S. Coast Guard (USCG) began operating one OWSBS, after permanently mooring the system in a U-shape with its 110-meter open end facing the well. The USCG attempted to operate the system for 24 hours per day at one point, but lightning storms and inadequate lighting made nighttime operation dangerous. Storms bringing 40-knot winds later tore loose suction hoses and other couplings on the OWSBS, and 4-meter seas at one point flooded the barge on which the USCG was operating and threatened to sink the vessel.

A few days later, a workboat ran over the OWSBS and broke it into 3 pieces. Crews were able to join damaged sections of the OWSBS and to continue operating the system, but the oil flow was changing direction every 6 to 8 hours, and as a result, most of the oil was escaping the OWSBS. Recovery rates on good days were about 4000 barrels of emulsion per day. A hurricane in September destroyed the anchoring system for the OWSBS, and the Mexican on-scene coordinator instructed the USCG to stop its cleanup operation at the well site and return to the U.S. reportedly because he believed that the continuing rough seas and bad weather would make the USCG efforts ineffective.

At different times throughout the cleanup response, the following factors hindered oil recovery: 1) the recovery equipment broke down or was improperly deployed; 2) periodic changes in current direction shifted the plume emerging from the well, allowing oil to escape past stationary recovery equipment. On one day in June, for example, booms at the well site were floating parallel to the direction of oil movement and were not containing any oil; 3) due to the depth of the spilled oil, barges and workboats were initially sucking oil into their water intakes, impairing the cooling systems on board; 4) diving operations related to the capping effort sometimes forced a suspension of the oil recovery effort. At various times, workboats involved in the capping effort or Sombrero operation damaged cleanup equipment or interfered with the cleanup crews; 5) bad weather, including tropical storms, rough seas, and Hurricanes Henri and Claudette, hampered recovery operations; 6) the oil/water recovery ratios were sometimes so low that PEMEX said it opted to dump the recovered material back into the sea, rather than fill scarce tanker space with a large amount of water and a small amount of oil; and 7) crews did not have lighting and other support equipment necessary to operate equipment 24 hours per day. PEMEX said in October that it had recovered 6 million gallons of oil out of the 103 million gallons spilled and had decided to halt its on-scene recovery operation.

None of the cleanup systems deployed at the well site recovered more than 14% of the emulsion intercepted by the booms deployed around each system, according to an estimate by Research Consultants, Inc. The estimate used the following parameters for each recovery system: 1) 10.5 hours of oil recovery per day, 2) a current speed of 0.75 knots, and 3) a slick thickness of 2 millimeters at the open end of the V-shaped boom arrangement. The estimate did not take into account frequent changes in the direction of the oil plume, which would have resulted in skimmers and booms intercepting no oil at all. The estimate was based on the following maximum recovery rates for emulsion: 1) the OMI Mop Machines - 4000 barrels per day; 2) the FRAMO skimmers - 1100 barrels per day; 3) the OWSBS - 5000 barrels per day; and 4) the Vikomas - 500 barrels per day.

Cleanup contractors said that the total recovery rate of more than 10,000 barrels of emulsion per day was actually obtained on very few days due to equipment damage, mechanical failure, and oil plume shifts. The recovered emulsion contained about 70% water, leaving a maximum oil recovery rate of 3000 barrels per day. For at least 2 months, the well was spilling at about 10 times the maximum daily recovery rate. Contractors involved in the well-site cleanup operation said that PEMEX did not supply sufficient support equipment, did not purchase or borrow enough cleanup equipment, and did not permit the contractors to deploy the equipment in ways that would have taken full advantage of the recovery capability of the equipment. To the extent, however, that on-scene cleanup equipment and booms were not able

to contain and collect oil under adverse weather conditions, and to the
extent that equipment failure prevented successful recovery, the marginal
utility of deploying additional skimmers and booms would have been reduced.
Furthermore, PEMEX restricted the mobility and overall deployment of the
cleanup equipment in order to protect divers working on the capping
operation and to minimize interference with logistical aspects of that
operation. Cleanup contractors said that, while these restrictions reduced
the effectiveness of the cleanup operation, similar restrictions would be
necessary on any open-ocean cleanup operation near a massive well blow-out
anywhere in the world. The contractors also said that the diving conditions
near the Ixtoc I were extremely hazardous, and they did not question the
high priority given to the capping operation or to the PEMEX measures to
protect personnel and equipment.

An incident as catastrophic as the Ixtoc I often forces the development
of new equipment and technology. In early September, PEMEX announced that
it would attempt a new method for collecting the oil that was escaping from
the Ixtoc I. Brown and Root, Inc. had constructed for PEMEX a steel
structure in the shape of an inverted cone, measuring 12 meters wide and 6
meters high. The cone, called the Sombrero, would be lowered over the well
by a steel boom mounted on a drilling platform that PEMEX would bring
on-scene. A 30-inch pipe connected to the top of the Sombrero would carry
oil, gas, and water to the surface of the sea and then up to a platform.
PEMEX said that the Sombrero was designed to handle a flow rate of up to
30,000 barrels per day.

On the platform, a diverter would separate the oil and water from the
gas, and then other equipment would separate the oil from the water. When
the Sombrero arrived on-scene in mid-September, rough seas damaged the steel
support boom. When the device was finally operational in mid-October, the
following problems developed: 1) the oil/water separation equipment was
unable to separate effectively the large volume of water collected from the
relatively small volume of oil. As a result, PEMEX was forced to pump most
of the collected material back into the Bahia de Campeche, although later
PEMEX was successful in burning much of the collected oil; and 2) the
Sombrero failed to contain much of the oil escaping from the Ixtoc I because
the turbulence above the well head carried the oil around the Sombrero and
because some of the oil was flowing from points on the sea floor not covered
by the Sombrero. Rough seas early in December again damaged the Sombrero,
and at that time, PEMEX decided to send the device back to Texas.

During the Ixtoc I incident, cleanup contractors and equipment
manufacturers used the spill to test and modify their equipment and
techniques. For example, Vikoma International Ltd. modified its Oceanpack
hydraulic system to eliminate the problems that the system initially
encountered at the well site. The tropical environment, with air
temperatures of around 41°C in the sun, had caused the hydraulic oil in the
Oceanpack to overheat. Vikoma was able to use its modified Oceanpack at the
well site and said that the Oceanpack operated successfully between -10°C
and 45°C.

BP also took advantage of the Ixtoc I spill to test a prototype for a
weir boom system. Following the Ekofisk well blow-out in the North Sea in
April 1977, the Norwegian government said that it would not issue permits to
companies for drilling above 62°N in the North Sea unless the companies
developed cleanup equipment capable of recovering 15,000 tons of oil per
24-hour day in sea state 5, with wave heights of up to 3 meters and wave
periods of 3 to 12 seconds. BP said that, at Ixtoc I, the weir booms
recovered mousse plus water at rates close to those target specifications.

OPEN-OCEAN SPILL RESPONSE: DISPERSANTS

The Ixtoc I incident involved the largest aerial dispersant spraying operation in history. It proved that airplanes can apply large volumes of dispersants to oil slicks in remote locations in an effort to help prevent the oil from impacting sensitive coastal environments. PEMEX began large-scale dispersant spraying in early June and initially concentrated the spraying on patches of oil that had broken away from the main slick and that threatened sensitive environments. PEMEX dispatched regular observation flights over the spill site to monitor the slick movement and to determine the targets for dispersant spraying. Conair Aviation Ltd. DC-6 airplanes, each fitted with about 34 meters of spray boom, flew 15 meters above the water at a speed of 150 knots and applied the dispersant at a rate of 350 gallons per minute. The plane sprayed from 1.8 to 3.0 gallons of dispersant per acre, and the swath width ranged from 80 to 220 meters, depending on wind speed and other environmental factors.

When the aerial dispersant-spraying operation was stopped in December, Exxon Chemical Co. had supplied over 1 million gallons of dispersant at a cost to PEMEX of about $9.1 million. In addition to the Exxon dispersants, PEMEX began in October to apply more than 7500 gallons per day of a Mexican dispersant, Key Emulsificante, from vessels at the well site. For the Ixtoc I oil in U.S. waters, the USCG and the U.S. Environmental Protection Agency decided not to authorize the use of dispersants because they determined that the oil was too weathered and too widely scattered for an effective dispersant application.

Initially, several logistical problems hampered the dispersant-spraying operation: 1) the pump that was first used to transfer the dispersant from supply drums to the plane worked too slowly and was later replaced with a more powerful pump; 2) the Ciudad del Carmen airport did not have tank trucks available for holding the dispersant; 3) the DC-6 required a special aviation fuel, and the Ciudad del Carmen airport did not have an adequate supply; 4) a shortage of available dispersants in Ciudad del Carmen caused a brief halt in the dispersant spraying in late June. A PEMEX decision to begin shipping the Exxon dispersants from Houston, Texas, to Mexico by barge had created the shortage, but renewed air deliveries eventually allowed Conair to resume dispersant spraying; and 5) Conair continued applying COREXIT 9527 and COREXIT 9517 until the raw material shortages in early July forced Exxon to substitute COREXIT 7664, a less active dispersant. By mid-July, however, Exxon was able to begin producing about 22,000 gallons per day of COREXIT 9527 for shipment to Ciudad del Carmen by plane and barge.

Several research groups used the Ixtoc I incident as a laboratory for testing dispersants in an actual spill. During October, Exxon Chemical Co. collected water and oil samples before and after the aerial application of COREXIT 9527 on Ixtoc I slicks and planned to use the samples to study the effectiveness of the dispersant. The Mexican Department of Fisheries (MDF) laboratory in Ciudad del Carmen conducted studies on the biological effects of dispersants on shrimp eggs, shrimp larvae, oysters, and clams. MDF scientists said that COREXIT 7664, COREXIT 9517, and the 2 Mexican dispersants used during spraying operations were toxic to larval shrimp and fish, and researchers from the Gulf Universities Research Consortium have begun planning a test program to investigate the MDF results. The completion of these studies will help policy makers evaluate the role of dispersants as an integral part of the spill response program.

ENVIRONMENTAL IMPACTS

The Ixtoc I incident resulted in a massive and prolonged exposure of the marine environment to oil. The spill accounted for more than 40% of the oil lost to fire and spillage in major spill incidents worldwide during 1979, and at one point, the spilled oil from the Ixtoc I covered an estimated 10% of the entire surface area of the Gulf of Mexico. Due to a lack of funding, researchers have not yet completed assessments of the impacts of that exposure.

Two years ago, shrimp boats from Ciudad del Carmen caught 6.5 million kilograms of shrimp, worth an estimated $20 million to the local fishermen. The shrimp fisheries extend along the Mexican coast from the tip of the Yucatan Peninsula around the Gulf of Mexico and north to Louisiana. The Campeche Bank, just northeast of the spill site, is one of the richest shellfish fisheries in the Gulf of Mexico, and the Laguna de Terminos, south of the Isla de Carmen, is a major nursery area for the shrimp region to the northeast of the blow-out site.

According to reports from Mexican fishermen, the spilled oil may not have significantly affected fish and shrimp catches overall this year, but shrimpers and fishermen expect reduced catches next year if sunken oil damages the larval shrimp and fish during their migration into the bays and estuaries. Biologists said that shrimping was poor in the Gulf of Mexico this year, but that the heavy rains would have caused a decrease in the shrimp catches this year anyway.

Fishing was also poor in some areas. By September, Mexican authorities had banned fishing along a coastal area 90 kilometers north of Tampico, Mex., and had restricted it along some highly oiled coastal areas south of Tampico, near Tamiahua, Mex. According to reports, fish catches had decreased by 50% from 1978 in coastal waters off Port Mansfield and Port Isabel, Tex., and off the Mexican coast from the U.S. border south to La Pesca. The octopus catch off Ciudad del Carmen had decreased by 70% in 1979 compared to that in 1978. The decline in the catch may have been partially due to the stormy weather which prevented Mexican fishermen from going out to sea on many days and partially due to the spilled oil which forced the adult fish and octopus to move to different feeding areas.

In September, winds and currents in the Gulf of Mexico underwent their seasonal reversal, shifting to northerly and northwesterly and carrying the oil eastward toward the Yucatan Peninsula. Following the reversal, trout, redfish, croaker, and other fish returned to the waters off Tampico, Soto la Marina, the Laguna Madre, Matamoros, Brownsville, and other parts of the northwestern Gulf of Mexico.

Fish catches remained low in the fall, however, in the Laguna de Tamiahua, a bay south of Tampico which PEMEX had reportedly boomed earlier in the spill. Catches of mackerel and pike were minimal off Tuxpan de Rodriguez Cano, Mex., and no snapper or shark were caught off Tecolutla, Mex., while catches of mackerel, grouper, and red snapper off Veracruz, Mex., dropped about 80% from last year. Increased fresh-water runoff from the unusually heavy rainfall this year has improved fresh-water fishing in the rivers near Alvarado, Mex., but the oil almost eliminated finfish fishing in offshore waters along the southern shores of the Bahia de Campeche from Veracruz to Progresso, Mex. East of Progresso, however, fishermen reported catches of octopus and other species as many as 4 times the normal catch size, as the organisms appear to have moved northward and eastward to escape the oil.

In July, U.S. researchers began collecting samples of weathered

Ixtoc I oil and studying the toxic effects of the Ixtoc I oil on adult brown shrimp, polychaete worms, algae, zooplankton, spotted sea trout, and redfish eggs and larvae. The lethal concentration was found to be 18 parts per million (ppm) in 24 hours, while dissolved hydrocarbon concentrations in water under the mousse were found to be approximately 30 ppm. A Texas Parks and Wildlife Department official said that the peak of the redfish spawning coincided with the movement of the slick into Texas waters and that fatal contact of many redfish eggs with the oil was likely since the fertilized eggs float to the ocean surface about 24 hours before they hatch.

The Ixtoc I oil did not cause visible contamination to finfish or shrimp catches, but it may have had some sublethal effects on fish, shrimp, oysters, and other marine organisms. Researchers at Texas A & M University said that oil increases the amounts of metabolic enzyme in various marine species in the Gulf of Mexico and that, by monitoring these amounts and by using baseline data from previous studies, they hope to measure the sublethal uptake of dissolved hydrocarbons by the species.

A USCG spill trajectory model predicted in June that the spilled oil would likely impact the Mexican coast between Tampico and Lower Laguna Madre in July, and would threaten the endangered Atlantic Ridley sea turles that breed along a 25-kilometer stretch of beach near Rancho Nuevo, Mex. The Atlantic Ridley eggs ordinarily begin hatching in mid-June, and young turtles continue to emerge until mid-August. The hatchlings swim west and north in the Gulf of Mexico, primarily on the water surface, during the next 2 months, and experts feared that the young turtles would ingest the Ixtoc I oil if they encountered it. To prevent the potential loss of a year's hatchlings and to reduce the threat of extinction to the Atlantic Ridley sea turtle, Mexican officials airlifted about 9000 turtle hatchlings to a patch of sargassum less than 25 kilometers offshore.

Scientists plan to monitor the Ridley turtle populations off Rancho Nuevo for several years to come. The U.S. Fish and Wildlife Service (USFWS) said that a population of adults next spring of less than 1500, the lowest recorded population in recent years, would indicate that the oil had affected the adult turtles. The impact of the oil on this year's hatchlings will not be known for about 8 years, however, when the year-class has matured and returns to Rancho Nuevo.

Only about 20 birds died as a direct result of oiling from the Ixtoc I spill last year, according to USFWS information. The spilled oil may have caused extensive damage to the intertidal organisms along the Texas shores, however, and that damage may affect several birds, including 3 endangered species, that feed on the worms, crabs, and other organisms in the intertidal zone. The long-range impact of the oil on Gulf of Mexico migratory birds will be difficult to determine, however, because the birds visit so many different and varied habitats outside the Gulf of Mexico each year.

Some reports on the Ixtoc I environmental impacts are already available. For example, the National Oceanic and Atmospheric Administration (NOAA) reported that the combination of the spilled oil and the storms in early September killed as much as 50% of the intertidal populations along the Texas coast. Neither the U.S. nor Mexico has funded, however, a comprehensive assessment of the Ixtoc I spill. As a result, the impact assessment will depend on the results of smaller studies that are currently underway or near completion. Such studies have been conducted by the Mexican Petroleum Institute, the MDF, the University of Mexico, the United Nations Environment Programme (UNEP), NOAA, the University of Texas, the U.S. National Park Service, USGS, and others.

In the U.S., a lack of cooperation among federal agencies has hampered the development of a comprehensive damage study. NOAA collected Ixtoc I water and sediment samples throughout the spill, but many of these samples have remained frozen in storage as NOAA has been unable to obtain adequate funding. The Bureau of Land Management (BLM) has $1 million to start a study but has no samples. NOAA believes that it, rather than the BLM, should conduct an Ixtoc I study because it has already begun the research and has greater scientific expertise for a spill impact study. In addition, NOAA said that it would be a more objective research organization in this case because NOAA, unlike BLM, has no part in the development of U.S. OCS oil and gas reserves.

To understand fully the consequences of the Ixtoc I, researchers will need to assess Ixtoc I damages along the Mexican beaches in the Gulf of Mexico, as well as along the Texas coast more than 950 kilometers northwest of the Ixtoc I well site. In a future offshore blow-out, the environmental impacts in the near-shore area would be most similar to the Ixtoc I impacts in Mexican territorial waters, whereas the environmental impacts in a neighboring state or country would most likely resemble the impacts after the oil has weathered during its movement from the Ixtoc I well site to the U.S.-Mexico border.

DAMAGE CLAIMS

Damage claims resulting from the spill now total at least $365 million, but the U.S. and Mexico do not have a formal agreement for dealing with compensation for damages from a transboundary incident such as the Ixtoc I. The following suits have already been filed in the U.S. against SEDCO, PERMARGO, and PEMEX: 1) up to 3000 U.S. commercial fishermen, shrimpers, and crabbers have claimed at least $155 million for future decreases in catch size resulting from the spill; 2) hotels, motels, restaurants, and other businesses have claimed at least $100 million in revenue losses; and 3) several Texas counties and towns and workers who lost employment as a result of the spill claimed at least $100 million.

In addition, the following suits have been filed against SEDCO and PERMARGO only: 1) the State of Texas filed for at least $10 million for damage to Texas property and natural resources, for tax revenue losses, for expenses incurred in attempting to counter the adverse publicity to Texas resulting from the spill, and for cleanup costs and civil penalties; and 2) the U.S. filed for at least $6 million in cleanup costs and spill response costs and for an unspecified amount in natural resources and property damages.

In response to these claims, PEMEX lawyers have contested the jurisdiction of U.S. courts over a foreign national oil company operating outside U.S. waters, and PERMARGO has also contested the jurisdiction of the U.S. courts, claiming that it does not do business in Texas. Under an 1851 U.S. maritime statute, SEDCO has filed to limit its liability to the present value of the SEDCO 135, claiming that the SEDCO 135 qualifies as a vessel under the statute. The court will begin hearings on these issues in July 1980, more than a year after the blow-out occurred, and the legal battle will likely continue for several years.

Meanwhile, spill victims with valid damage claims have no way to recover their losses. The OCS Lands Act Amendments of 1978 established a $200 million fund to pay for cleanup costs and damage claims from OCS spills in U.S. waters, but not from OCS activities in foreign or international waters. The Superfund legislation that would cover such incidents has been pending in the U.S. Congress for several years, and proposed Ixtoc I damage compensation bills have apparently died in committee. At the international level, the existing oil pollution damage compensation funds apply only to spills from tankers, not from OCS wells. Perhaps the combined effect of the Ixtoc I and the Funiwa 5 will focus attention on the need for new international legislation dealing with OCS well blow-outs and oil spills.

SPILL BEHAVIOR

According to Canadian EPS scientists, oil surfacing at the well head in June had formed a viscous water-in-oil emulsion containing 67% to 70% water. Contractors operating cleanup equipment near the well site estimated that the recovered material was an emulsion consisting of about 60% water and 40% oil. Most of the oil droplets in the emulsion measured several millimeters across. The EPS scientists said that the burn point of the emulsion was 100°C and that 30% of the oil by volume was evaporating immediately after escaping from the well. The EPS concluded that the oil would be difficult to burn or to disperse chemically. Based on this finding, the EPS said it was concerned that oil-burning techniques and dispersants would not be effective in dealing with oil spills from well blow-outs in an arctic environment.

Following a research cruise to the well site in September, U.S. scientists said that water samples from the well area contained small oil droplets and that the mousse formed only after the lighter fractions of the oil had evaporated and the oil/water mixture had been exposed to wind, sunlight, and microorganisms. The scientists from the research cruise and from the EPS said that the composition of the slick may have changed since June for many reasons, including the following: 1) if the flow rate of the oil had decreased in August, as PEMEX reported, the accompanying change in mixing energy could have affected the composition of the slick; 2) as the upper levels of the Ixtoc I oil reservoir were depleted, the oil reaching the Ixtoc I well would likely have migrated through more sediments within the formation and, as a result, might have changed chemically, possibly affecting the character of the slick at the surface.

Other researchers expressed doubt that the chemistry of the oil had changed in ways that would affect the emulsification process. Paul Boehm of Energy Resources Co., Inc. of Cambridge, Massachusetts, suggested that the formation of the emulsion might have been affected by both a change in the flow rate of the oil and a reduction in the saline concentration in the Gulf of Mexico near the well head following heavy rains in August and September. Boehm said that the mousse formed shortly after the oil had travelled about 5 kilometers from the well site, and he speculated that a seeding process might be involved in the mousse formation. According to Boehm, small quantities of mousse might have formed as a result of weathering and other processes, and these small quantities might have stimulated extensive mousse formation. Other researchers have been studying the chemical characteristics of the Ixtoc I oil, mousse, and sheen, and the fate of the Ixtoc I hydrocarbons in the water column.

According to overflights, the oil farthest from the well had formed a thick mousse, similar to that observed following the Amoco Cadiz oil spill off Brittany, France, in March 1978. In U.S. waters, the spilled oil had dispersed into droplets as much as 12 meters down into the water column, and as a result, booms may not have been effective in preventing the flow of these oil droplets into the protected passes along the Texas coast. In addition, the skimmers were not able to recover the weathered or submerged oil.

The spilled oil has formed at least 37 tar mats in the subtidal region off the coast of Texas. These mats measure about 60 meters long and 6 meters wide, and consist of a mixture of sand, shell, organic matter, and weathered oil. Scientists believe that the tar mats may present a continuing pollution problem for the Texas coast, and that tar balls will continue to wash ashore along the Texas and Mexico coasts during the summer of 1980.

Scientists still have not been able to account for the bulk of the 140 million gallons of oil that spilled from the Ixtoc I and that, according to NOAA, had covered up to 10% of the surface area of the Gulf of Mexico in the fall of 1979. An understanding of the fate and behavior of the entire volume of spilled oil will likely provide useful data for assessing environmental impacts and developing new equipment and response strategies.

CONTINGENCY PLANNING

In a crisis situation, the hours immediately following the event's occurrence often provide the greatest opportunity for bringing the crisis under control. In the Ixtoc I incident, adequate contingency planning might have resulted in an early capping and an effective spill response. On 3 June, PEMEX did not have a well-established national contingency plan to respond to a massive open-ocean oil spill. In addition, PEMEX did not have adequate stockpiles of spill response equipment. When the blow-out took place, PEMEX requested equipment and expertise from sources outside Mexico and as far away as Norway.

In July, supply vessels for the cleanup operation were still in short supply. In at least one instance, emulsion that had been recovered was dumped back into the Bahia de Campeche reportedly because a storage tank vessel was not available. As late as August, cleanup equipment operators had difficulty obtaining replacement parts for damaged machines, and the equipment manufacturers were still training PEMEX workers to operate the cleanup equipment.

Norway has developed plans to use mobile skimming systems in the event of a major spill. Statoil has 3000 meters of ocean boom and 14 oil skimmers, operating in 6 to 8 independent mobile oil recovery systems for use in the North Sea below 62°N. According to the Norwegian contingency plan, the skimmers would be able to operate for 24 hours a day, and the mobile systems would allow cleanup crews to adapt the equipment to changing winds and weather.

Current U.S. contingency plans call for the USCG to develop the capability to respond to 80% of all U.S. spill incidents within 6 hours of notification. To achieve this goal, the USCG has developed plans to locate response equipment in at least 11 strategic sites around the U.S. and to increase the availability of booms, skimmers, and support vessels. Currently, the U.S. has 4 OWSBSs, which provide a recovery capability of 1000 tons of oil per day under moderate open-ocean conditions. By 1982, the U.S. expects to have 26 OWSBSs, which will provide a recovery capability of 5500 tons of oil per day.

While the Ixtoc I spill was becoming the largest oil spill in history, 2 VLCC tankers, the Aegean Captain and the Atlantic Empress, collided in the Caribbean Sea off Trinidad, resulting in the largest marine casualty in history as well as the largest amount of oil cargo ever lost. Both the blow-out and the collision took place in the same region, namely the Caribbean Sea and the Gulf of Mexico, and both events emphasized the need for effective spill response plans and equipment stockpiles. According to the Organization of American States (OAS), many countries in the Caribbean do not have national spill contingency programs and likely do not have adequate stockpiles of spill containment equipment. The OAS expects that the blow-out and the collision will provide an impetus for those countries to prepare their national contingency plans and to join together and sign into effect a regional spill response program.

The OAS has already drafted an oil spill control plan for the Caribbean region, including the Gulf of Mexico. Under the plan, a regional coordinator will keep the Caribbean nations up-to-date on spill control resources and techniques, represent the region at conferences on oil transport and pollution prevention, and maintain contact with international sources of assistance. Sub-regional coordinators for the eastern, western and northeastern Caribbean will monitor oil spills and inventory oil spill equipment in their respective areas. In addition, the Caribbean Environment Programme, part of the UNEP Regional Seas Programme, has developed a draft action plan to provide for and coordinate economic and environmental cooperation in the Caribbean region, including the U.S. and Mexico. The plan provides for international response to marine pollution incidents, in addition to natural disasters, energy problems, human health problems, and other concerns.

Planning for marine pollution problems often involves bilateral as well as regional agreements and plans. Canada and the U.S. have a bilateral plan for spill response, as do Canada and Denmark. The 1974 Joint Canada-U.S. Marine Pollution Contingency Plan describes a joint response mechanism for pollution from ships, pipelines, onshore and offshore activities, and other sources. The plan contains provisions for one country to assist the other, if assistance is requested, even when no threat of transboundary pollution is present.

The U.S. and Mexico began discussions on a bilateral plan in 1977, and completed a draft of the plan in 1978. The draft contingency plan has the following objectives: 1) to develop a system for reporting spills that could have an impact on both Mexico and the U.S.; 2) to develop a joint program for containing such spills; and 3) to provide the equipment, expertise, and other resources for cleaning up spills and minimizing their impact. According to the draft plan, the entire cost of the spill response would be covered by the country in which the spill occurs or, for OCS spills, by the country with jurisdiction over the OCS activity responsible for the spill.

Mexico and the U.S. had not finalized the contingency plan at the time of the Ixtoc I spill, and as a result, the legal mechanism did not exist to enable the U.S. to intervene to mount a cleanup effort at the well site. The U.S. Department of State said that the U.S. had no power or desire to force Mexico to accept U.S. assistance and that the U.S. did not want to risk offending the Mexican government with repeated offers of assistance that might appear to the Mexicans as attempts to force the issue. U.S. Senator Lowell Weicker (Republican - Connecticut) said, however, that U.S. reticence in this matter was the result of fears of jeopardizing negotiations between the 2 countries concerning the sale of oil and gas by Mexico to the U.S. Mexico said that extensive U.S. involvement was unnecessary as PEMEX had already contracted for the best available technology and expertise at the well site.

Following the blow-out, the most effective response for the U.S. would have been to collect the freshly spilled oil at the well site, rather than to attempt to collect the weathered oil as it entered U.S. territorial waters. In fact, the efforts to recover the weathered oil were largely ineffective. The U.S. and Mexico still do not have a treaty or an executive agreement to facilitate a future response to another pollution incident similar to the Ixtoc I. At an international level, the experience of the U.S. with respect to the Ixtoc I spill focuses attention on the need for bilateral and multilateral treaties to enable a nation to respond quickly to a major pollution incident, whether an oil spill or a chemical contamination, that takes place outside its borders but, due to environmental conditions such as winds and water currents, eventually impacts its own territory.

The U.S. was not able to send an official team to the spill site until June 22, almost 3 weeks after the blow-out took place. From the first day, the spill had the potential for polluting U.S. waters. Ocean charts indicated that the currents would carry the oil westward and then northward toward U.S. waters, and although the spilled oil originated more than 800 kilometers south of Brownsville, it represented an imminent threat to the U.S.

The U.S. and Mexico reopened bilateral contingency plan discussions during the spill, and PEMEX requested U.S. assistance at the well site in early August. On August 12, more than 2 months after the spillage began, the U.S. sent 2 OWSBSs and USCG operating personnel to the well site and began collecting oil at a rate of about 4000 barrels per day. Although storms and rough seas severely damaged the 2 systems by late September, the U.S. declined to jeopardize its limited involvement by attempting to pressure Mexico into accepting additional equipment and personnel. In early October, the U.S. personnel and equipment returned to the U.S. on the request of PEMEX.

The draft bilateral contingency plan did not provide a mechanism for resolving liability questions or damage compensation claims. In late August, the U.S. sent a note to Mexico proposing compensation talks but made the proposal public the next day. According to former U.S. Ambassador to Mexico Patrick Lucey, Mexico objected to the U.S. making the proposal public and in response to the proposal, said that existing international law did not cover Ixtoc I spill damage liability. In late September, Mexican President Jose Lopez Portillo met with U.S. President Jimmy Carter in Washington, D.C., and both agreed on the need for Mexico and the U.S. to "prevent events or actions on one side of land or maritime boundary from degrading the environment on the other side." The agreement concerned the mitigation of the effect of future incidents only. Lopez Portillo said on his return to Mexico that Mexico did not feel obligated to pay for spill damages from Ixtoc I since the U.S. had not paid for the damage to cropland in the Mexicali Valley, beginning in 1961, caused by increases in salinity in Colorado River water on the U.S. side of the border. With a bilateral contingency plan in place, the U.S. and Mexico might have avoided diplomatic delays and begun sharing cleanup equipment and spill response techniques immediately after the initial blow-out.

CONCLUSION

For all these reasons, the Ixtoc I deserves further attention from government, industry, and the research community. The U.S. needed the Argo Merchant tanker grounding and spill in December 1976 and France needed the Amoco Cadiz tanker grounding and spill in March 1978 to push forward programs to develop spill response plans and cleanup technology to deal with coastal spills. Now the countries around the world have an opportunity with the Ixtoc I spill to learn from the experiences of Mexico and the U.S. and begin improving response programs for open-ocean spills without waiting for a catalytic catastrophe in their own waters. If they do not take advantage of the opportunity, the Ixtoc I will just become "the largest oil spill in history."

REFERENCES

The authors have relied extensively on information that appeared in the Oil Spill Intelligence Report, an international weekly newsletter written by the Center for Short-Lived Phenomena, 138 Mt. Auburn Street, Cambridge, Massachusetts 02138, and published by Cahners Publishing Company, 221 Columbus Avenue, Boston, Massachusetts 02116. The Oil Spill Intelligence Report began reporting on the Ixtoc I incident in its Vol. II, No. 23 issue on 8 June 1979 and, since then, has provided weekly coverage of the relief well operation, spill response, environmental impact, and other issues related to the event. The Vol. III, No. 1 issue on 4 January 1980 provides a special in-depth report on the Ixtoc I, and this paper has excerpted relevant passages for the purposes of its discussion.

SESSION 6

Rapporteurs' Summaries – General Discussion and Conclusion

RAPPORTEUR'S SUMMARY OF SESSION 2 - COST/BENEFIT ANALYSIS OF
ENVIRONMENTAL MANAGEMENT

Mr. Christer Dahlberg
Senior Executive, Marine Technology Programme
STU, Stockholm, Sweden

Mr. Chairman, Ladies and Gentlemen,

Industry

On the one hand industry performs an important role for society i.e.
it produces goods and services for all of us and in this context industry
is the producer of oil which we need so badly.

On the other hand, as industry provides it also pollutes but there is
no reason to believe that industry is a worse environmentalist than the rest
of us. The product - oil - with a market price of some $ 30 a barrel and
the enormous investments are too valuable to be risked. Also, as Mr. George
said in his paper "an unsafe operation is unlikely to be an economic one".
The other incentive for industry to perform in a safe way is that accidents
always call for new regulations, government involvement and in some cases
production stoppage or delays.

Now industry, like us all, has to play according to certain rules. The
framework for this is drawn up by governments but, as was stated by Mr.
Guillaume in his paper, "industry can only operate properly within a system
of well-defined rules which make longterm planning possible...".

As we all know, the basis of industry is profit and industry is therefore
watching costs. Mr. Guillaume put it this way "in view of its
responsibilities, industry must seek to contain the effects of restraints
on productive activity by minimizing the costs of being a good environmental
citizen". At least in the short term, industry has to pay and environmental
protection follows the law of diminishing returns, i.e. the cost of changing
a performance from 50-60 per cent is not nearly as expensive as changing
it from 80-90 per cent.

Governments

The role of governments is to maximize the benefits for today's society
without exploiting or destroying the opportunities of tomorrow. Therefore,
as Mr. Hughes said, "the responsibility of governments is to control the
use of the environment in society's interest and it has to consider society's
needs against the cost".

The normal way for governments to control activities in society is
to make laws and to set standards. However, the question of environmental
standards is a very different one from, for instance, building standards
because we have to decide what standards to set. Mr. Hughes says in his
paper "zero pollution is an unattainable goal... unrealistic goals rather
illustrate the need to use a workable definition of pollution".

The basic question for governments therefore is, to quote Mr. Hughes again, "how clean and at what cost? Or, how clean can we afford to be".

Another major difficulty for governments is that there are no clear frontiers for pollution, but well-defined borders for governmental power. This calls for intergovernmental consultations and cooperation. And, as we have heard from the papers in Session 1 and from Mr. Read in Session 2, there are today a number of bilateral and multilateral agreements in force. But, as we all know, national cooperation is difficult. International cooperation is no easier and has a number of added problems such as differences in governmental policies, evaluations, priorities etc. These problems must be solved if we are aiming at cleaner and safer surroundings.

Then there are the legal questions to be solved. As was stressed by Mme. Rémond-Gouilloud there is an urgent "call to set up more efficient indemnification mechanisms" (compensation pay). And as she said, the size of the disaster and the enormity of the damages have to be dealt with on different levels such as industry, insurance company and national government. Inter-state solidarity is also necessary.

As was mentioned in connection with Mr. Westby's paper, there is a need to initiate and suggest transfer of experience and knowledge and also to push for coordinated R & D actions.

The public

As I stated earlier, in the short-term industry has to pay for a cleaner environment but, in the end, the costs will be borne by the public in one way or another.

The public (i.e. citizens not directly concerned with the oil industry or government policy-making) has become more and more involved in environmental issues mainly due to the increased interest shown by the mass media. We hear nearly every day about new groups formed to protect a region, a country or the world from different types of environmental hazards. Nevertheless, criticism and debate are important for all parties concerned but for a fruitful dialogue a common language must be found and for that the basic information platform must be similar. Here, there is a lack of information transfer to the public. Industry and governments are often criticised and, as Mr. Hughes said, "this is not always unfounded".

The major issue concerning the public I therefore believe is how to make basic information available, and in a balanced way, and, in a clear language, describe existing alternatives.

I have now mentioned the 3 parties involved in the issue of a better marine environment. However, there is one more group I would like to refer to and that is the scientific community.

Dr. Lindstedt-Siva's paper dealt with the question of the ecological effects of oil spills and oil spill response. She said "except when life and limb are threatened ... the primary goal of oil spill response should be to minimize the ecological impacts of oil spills".

To make this possible it is necessary for scientists from different disciplines to be included in the team at an early planning stage as well as when the accident has occurred. I think it is worthwhile to quote Dr. Lindstedt-Siva's answer to the question "what can we do when no biologist or ecologist can be present at an oil spill site"? She replied "I believe that if other specialists are able to go out to the site so could we".

Cost-benefit

Up to now I have given a brief description of the different parties concerned in oil spill prevention planning and recovery. I said earlier that industry has to perform its activities within a framework defined by governments and, with regard to the public, I stated that there is a great need for a common language so that all concerned have a basic knowledge of the issue in question.

How do governments come to their decisions? What is a common language?

Governments have to define their priorities, such as "How clean can we afford to be"? One technique or methodology is cost-benefit analysis.

Mr. Yusuf J. Ahmad dealt with this question in his paper presented here by Mr. Nay Htun.

So what is cost-benefit analysis and to what extent will it solve our problem?

In Mr. Ahmad's paper you will find the following definition: "The marginal cost of environmental quality control measures must be weighed and evaluated against the environmental physical damage that would otherwise take place if the pollution control investments were reduced, postponed or abandoned during the period in question. This is cost-benefit analysis of a sort and involves weighing a series of considerations against each other". In fact, this is a calculation most of us perform every day, e.g. should I take the bus to get to the office 10 minutes earlier, or should I walk instead and thereby get some physical exercise so that I can work more efficiently when I arrive, or will I get a headache from walking in the polluted air and therefore be less efficient and what about the chances of being run over by a car?

Costs and benefits are not necessarily given in monetary terms; it is not even necessary that they appear at the same time. These are the two problems of cost-benefit analysis!

The major constraints are the identification of the "costs" and "benefits" involved and the evaluation of the different items. As this process becomes rather subjective from a personal or national point of view and as UNEP has found in their studies on environmental management "cost-benefit conclusions are very sensitive to assumptions for key parameters".

Another important statement in Mr. Ahmad's and Mr. Guillaume's papers is that cost-benefit analysis <u>assists</u> in the decision-making process. Also, as Mr. Guillaume put it, "... it must be recognized that cost-benefit analysis is a technique for assessing the effect of specific environmental control measures not for assessing the need to apply environmental control per se".

So even if cost-benefit is only one tool it can, if applied in the right way, be a useful tool. It can assist in the choice of different alternatives and if properly used bring to light parameters of importance. And, as Mr. Ahmad stated in his paper, "the economic costs of early environmental planning are substantially lower than the larger costs of subsequent corrective measures". Mr. Ahmad also said that "an important advantage of cost benefit analysis is that it reduces a problem with many aspects and dimensions to one with fewer dimensions".

The big advantage of UNEP studies is that they may bring or assist in bringing into use an internationally acceptable methodology for cost-benefit analysis.

<u>To sum up</u>

As the chairman of our session (2) said, "accidents cost a great deal but they do not benefit anybody". However, one advantage is that we learn from them. This knowledge is the only safeguard we can rely on and from which we should strive to increase our skill and creativity. To make this possible it is necessary to transfer knowledge, experience and ideas between the different partners concerned. And as society's impact on the environment does not respect frontiers, nor should we.

How do we go about increasing the necessary international team work?

Our first step should obviously be to try to prevent accidents and catastrophes. As the chairman of session 2 mentioned, a good start could be to obtain more information from the Australian Counter-Disaster College at Macedon.

Secondly, it has been stressed that we need a better dialogue and cooperation. There do not seem to be any great difficulties here.

<u>Industry</u>, as stated by Mr. Guillaume, "is prepared to contribute to all discussions on a national, regional or global level".

<u>Governments</u> should be quite willing to cooperate as they are not aiming at enforcing regulations which cannot be met by industry.

<u>Scientists</u> are interested and willing to contribute. However, it is my personal feeling that both industry and governments are somewhat hesitant. I do think that this must change. Scientists should perhaps be given better specifications of the problems envisaged and the governmental and industrial priorities. I also think, as has been mentioned, that baseline studies should be carried out regularly as they are the only way of measuring the impact of oil spills and thereby increasing our knowledge for the future.

The public has become more concerned and seems to be quite willing to participate in a dialogue. However, I think we must try to use the media as a means of imparting facts. It is of interest for industry, governments and scientists that the public is kept well informed, but only if accurate information is given. As Mr. Hughes put it, "Industry and governments alike will have to adjust to the reality that the public has become a third dimension in environmental management, with quite a political impact".

We must inform the public of the different alternatives and present them in an understandable way.

Thirdly, what is the use of cost-benefit analysis? As I understand from Mr. Nay Htun, a lot of experience in this field is available at UNEP. And Mr. Ahmad's paper states that "it has been agreed that UNEP should play a stronger catalystic role and help member countries broach environmental problems in an effective manner".

Even if cost-benefit analysis is only one tool I do think that we should make use of experience gained, and industry as well as governments should continue to take an active part in the work of UNEP.

Fourth: The importance of personnel training has also been stressed. Even in this case we should not neglect the increased effects obtainable through cooperation and experience transfer. Just to give one example there are considerable efforts being made today to open an international training school for divers.

Fifth: A further need I think is to acquire a better knowledge of the national technical R & D efforts and results. There is a risk that a lot of overlapping technical development is performed whereas with a better knowledge of national priorities in this field and also through international cooperation, our skilful technicians and innovators could be more efficiently used to the benefit of us all.

Here I think it is of importance to try to give manufacturers specifications or at least better guidelines to enable them to perform their development from a sound economic base. Otherwise there is a risk that they will not put any efforts into solving the problems.

Another important action I believe is to participate in conferences like this where, I hope, useful information is given. Even more important is perhaps the personal contacts and small group discussions that follow a conference such as this. In the future such conferences should perhaps also include workshops.

RAPPORTEUR'S SUMMARY OF SESSION 3 - POLAR ACTIVITIES

Mr. Adrian Cottrill
International Editor, Offshore Engineer Magazine
London, U. K.

Mr. Chairman, Ladies and Gentlemen,

Not surprisingly, in the polar session of the conference, Canadian papers dominated, since the country is the only one to have ventured offshore for oil exploration.

6 of the 10 papers were Canadian.

However, I will start with the paper least specifically tied to the Arctic.

The implications made in the paper from Dr. Donald Malins are of possibly very serious significance for the offshore industry everywhere, not just in the polar regions.

Under the title environmental degradation of petroleum hydrocarbon he said in essence, that current methods of estimating the scale of oil pollution in organisms may severely understate the case because of the problem of identifying the pollutant.

He contended that changes in higher organisms produce suites of new, and often far nastier, pollutants when they react with hydrocarbons, and that huge percentages of error are introduced in estimating pollutant load simply because current investigations are directed towards detecting the wrong chemicals.

Possibly in the next few years we will be able to produce means for gaining a better picture of these pollutants he says.

Judging by some responses in the discussion, many in the oil industry by no means agree with Dr. Malins. They question whether concentrations reach toxic levels and ask if we really do have a problem.

Dr. Malins says most emphatically yes. This debate is obviously only just starting.

A second paper was also not specifically concerned with polar conditions although it contains important conclusions affecting Arctic offshore activity.

Dr. Arthur Grantz went below the ice of the Alaskan Beaufort Sea to describe some of the geological surprises which might be in store on the sea floor.

He spoke of a number of features with the potential to damage the structures which will be built in this potentially very rich petroleum province in the coming years.

His list included earthquakes, seabed slumping and sliding, high pressure shallow gas pockets, and solid hydrates which can decompose to release large volumes of gas during drilling.

All this is offshore from an area where already there are known to be 2 giant oilfields, another 6 oil discoveries, and numerous gas indications.

Earthquake risk is greatest in the Camden Bay area, 120 km east of Prudhoe Bay, he says, and some authorities are suggesting that structures should be designed to resist magnitude 6.0 earthquakes.

As for slumping and sliding, Dr. Grantz produced slides showing evidence of huge masses of seabed - up to 40 km long in some cases, and anything from 20 to 200 m deep, moving seaward along bedding planes.

Already one lease sale has been held for the nearshore region - last December (although it is currently held up by legal injunctions) - and another is tentatively scheduled for February 1983, quite possibly out to 200 m water depth.

The cost of studies into subsurface phenomena affecting these regions is small in comparison to potential disasters says Dr. Grantz.

No one took him up in discussion, possibly because we were all too awed by the forces of nature he was describing, but I believe the oil companies are in serious discussion with him and his colleagues.

On the other side of the continent, off the east coast of Baffin Island a further sea bottom phenomenon was described by Dr. Levy in his paper on natural petroleum seepage in this area.

Slicks and gas bubbles have been observed erupting off Scott Inlet and it is suggested that these are associated with submarine troughs.

Perhaps when the oil companies hear this they will start presenting a case for moving in there as soon as possible to drill for oil to relieve the pressure.

The behaviour of an oilspill in cold waters at the other end of the planet was described by Dr. Leonardo Guzman from Chile, in his paper on studies after the Metula supertanker oil spill.

It occurred in 1974 and is the largest such incident in South America (bigger than the Torrey Canyon) and oiled large areas of the coast of Tierra del Fuego.

A year later, 250 km of coast had been contaminated, with 18,000 t of oil assumed to have been washed ashore.

The spilled oil was not contained or dispersed, and slicks remained in the straits for over two months. Much mousse was stranded on the coast in thicknesses up to 200 mm.

Dr. Guzman spoke of the intertidal communities being hardest hit. Full impact on saltmarsh plantlife was not felt until two years after oiling.   Of the bird life, 3-4000 birds were killed in the first 6 months.

There are now some unconfirmed claims that bird populations are re-establishing, but one definite result of the spill, so far not reversed, was complete eradication of the tern colony.

Now turning to the main Canadian papers:

The history of Canada's move into the Arctic in the search for hydrocarbons was well summarised by Dr. Noel Boston.

Since the first well was drilled there in 1961, close to 650 m has been spent on exploration.

8 gas fields have been found, and the threshold level of reserves for the Polar Gas plan is near.

To the west, drilling from artificial islands in the Mackenzie Delta began in 1973, while to the east, it did not start until last year in the southern Davis Strait.

But he reminded us that though the Arctic is harsh, it bruises easily. "A few days disturbance at a birds breeding colony can delay nesting to the point where not enough of the season remains to complete the breeding that year".

And "the diminution of even one species can have a profound effect on all the others".

He outlined the intricate network of regulatory procedures designed to protect this environment. A typical time span between an application to drill, and its approval could be 18 months he said.

You may wish to comment on one point from his paper which he did not have time to make verbally.

This is the statement that: Marine transportation will become the next major environmental issue in the Canadian Arctic.

Both the Arctic Pilot project and the planned Beaufort Sea production schemes envisage ice-breaking tankers plying south continuously.

Dr. Kingham's paper also makes the point that accidents to tankers and pipelines are considered far more probable than blow-outs.

Studies for these schemes have now brought Canada to the forefront in ice knowledge and navigation, but there is still much to be deliberated.

In general the Canadian papers provided information on two main fronts: the engineering development work that is going ahead to evolve oil and gas production and transportation systems, and analyses of behaviour of spilled oil and the development of countermeasures.

The main paper providing examples of production schemes actually being studied came from Jean-Gérard Napoleoni.

For use in "Iceberg Alley" in the Labrador Sea he described both a floating platform arrangement and a completely underwater alternative.

To enable floating platforms to remain on station throughout normal winter sea ice, they would be equipped with rotating cutters to chew into the ice at the water line.

Such a severe environment does of course have its price - the cost of a Labrador production scheme is estimated at 70 per cent more than in the North Sea.

Also a seabed flowline may now and again have to be sacrificed to iceberg scour.

In the face of large icebergs such a platform would of course have to shut down temporarily and move out of the way.

Detailed knowledge of the habits of icebergs is therefore necessary and much study has already proceeded here.

Mr. Napoleoni conceded that, even under the scrutiny of intensely analytical minds, it will never be possible to arrive at a completely accurate means of predicting iceberg movement.

Although it can be established that, barring other factors, icebergs move exactly the same way as the water in which they float, they are subject to other unmeasurable influences.

Mr. Napoleoni mentioned the inconvenient fact that the sea is not flat, but slopes by a few mm in 1 km.

If you also add in a little wind, you have a series of iceberg pirouettes guaranteed to keep any platform crew on tenterhooks when an iceberg is near, even when diversionary towing is being attempted.

Noel Boston had already reminded us in an earlier paper, that an efficient crew can disconnect a platform in a matter of minutes. There are no inefficient crews he told us.

Techniques like sonar measurement from submersibles can establish berg keel depth to within 1 m.

All in all, Mr. Napoleoni concludes that coping with icebergs off Labrador is feasible by a combination of drift prediction and towing.

In all cases, he says, there is no doubt that the environment can always be protected.

It is unfortunate that we were unable to obtain an equivalent picture of the production schemes being considered at the other corner of Canada, in the Beaufort Sea.

Some details are however available about the main options being considered by Dome Petroleum - the company closest to going into production in the area.

Although oil might form a coat 5 to 10 mm thick under smooth ice, it does not form a continuous slick under moving ice.

It is likely to spread in very small concentrations over large areas, perhaps 10 x what might normally be expected.

Dr. Wadhams estimated the oiled area would be about 500 km long, although this was questioned.

There is some debate over whether oil would surface at all from beneath multi-year ice which lacks brine drainage channels through which the oil can rise.

Frozen into the ice, the oil would be kept toxic and unweathered.

"It is very important that a large scale field experiment be done in multi-year ice", says Dr. Wadhams.

His overall conclusions are: the extent of oil spread from a blowout, its persistence in a toxic state, and above all, the difficulty of cleanup, are all considerably greater than for a blowout in the open sea. The ice acts as blotting paper.

Some of the oil will not reappear for 2 or more years, during which time it will have travelled thousands of km across the Arctic Ocean.

Burning appears the only feasible option, but it must be carried out within a short time of the oil's appearance, and requires air droppable igniters.

It is questionable how much oil can be disposed of in this way.

The government's view on the state-of-the-art in oil spill research came from Dr. Kingham of Canada's Department of the Environment.

His paper details the studies of the ø 7 m five-year Arctic Marine Oilspill Programme started in 1977.

"It is universally recognised that the general state-of-the-art for dealing with major oil spills is inadequate and unsatisfactory" he says.

AMOP has now reached the stage where it is essential to try new countermeasures on actual spills in the north.

In the static landfast ice of the southern Beaufort Sea, in situ burning is held to be the answer, but, he says, this is not the case in the moving ice where exploration from drillships is now taking place.

In the Labrador Sea, the problem might be even more difficult to develop reasonably effective spill countermeasures. "It has as many spill control problems as any area in the Arctic".

All the evidence suggests that dispersants are not effective in cold waters.

From these three main papers on oil spills, various points of disagreement

Options include earthfill islands, huge concrete caissons, and monocone structures.

The final paper dealing with engineering projects came from Mr. Ollis Kaustinen, who gave the latest details of the Polar Gas pipeline project.

Only a week ago the group withdrew its original route application, and now intends a route which includes only two marine crossings.

However, these crossings are far more dramatic than their predecessors, particularly the 120 km long M'Clure Strait which reaches 500 m deep.

The construction method chosen here is to pull the pipe from shore along the seabed, at speeds of 6 m per minute using winches placed on the ice above.

Sections would be welded together on the seabed, preferably using a one-atmosphere chamber which has now been tested at 265 m depth.

The Polar Gas group is now 8 years old, and asked about the future timescale of the project, Mr. Kaustinen said that actual building should be completed on phase 1 (Mackenzie Delta) in 1989, and on Phase 2 (Arctic Island) in 1990.

Finally we come to the most crucial environmental issue affecting marine activity in the Arctic - oil spills and their clean-up.

The work being done by the oil companies was outlined by Mr. John Hnatiuk, speaking for the Arctic Petroleum Operators Association.

The APOA now has 23 member companies, is 10 years old and has spent nearly $ 30 m on over 150 research projects.

Experiments on burning oil in situ on the ice 5 years ago indicated that 90 per cent could be disposed of in this way if burnt not more than 2 to 3 weeks after exposure.

A further major study has been going ahead this winter, in the Beaufort Sea. Quantities of oil, with compressed air to simulate gas, have been released under various thicknesses of ice, with ice movement also simulated by moving the point of oil discharge on the seafloor.

In the final discussion you may care to ask if any preliminary results are becoming available from this programme.

A usefully precise and detailed description of the effects of a blowout under ice in the Beaufort Sea came from Dr. Peter Wadhams of the Scott Polar Research Institute.

The worst case would be a seabed blowout at the end of the open water season in October. A relief well could not then be drilled until the following July, by which time up to 1 million barrels of oil could have been released under heavy confused ice.

Seabed containment domes would be quite impractical where ice might scour the seabed.

774

and uncertainty emerge.

The most important is whether, when the oil released under ice finds its way to the surface, it will gather in large enough pools to burn.

The APOA appears to be fairly confident that it will, but Dr. Wadhams, for example, feels that the chances are low in multi-year ice particularly.

It must be the subject of future field experiments he says.

Taken together, the Polar activities papers form an impressively comprehensive summary of the current state of knowledge and activity offshore in this most sensitive of areas.

The degree of interest in the Arctic now, and the current scale of experimental work, ought to ensure that significant progress can be reported to any future conference on the subject.

It would also be most interesting to spread the net to include reports from the Soviet Union and Norway.

RAPPORTEUR'S SUMMARY OF SESSION 4 - TEMPERATE ZONE

Mr. G. Brockis
Manager Oil Spill Contingency Planning and Research
B.P. Co. Ltd., London, U. K.

Mr. Chairman, Ladies and Gentlemen,

In summarising my impressions of some of the salient features arising from the 13 papers, presentations and discussions during session 4 on The Temperate Zone, I shall no doubt do rough justice to some speakers and by sin of omission rather than, I hope, of commission, less than justice to others.

I have attempted to structure my remarks by taking the presentations in groups, beginning with the two opening papers on contingency planning and bringing in, from other presentations, comment which seems to me to be particularly relevant.

It was clear from these two papers - and, indeed, from plans discussed in other sessions, for example the Norwegian approach discussed in session 3 - that national contingency plans are designed to suit the political and social structures and the geographical limits within which they must operate. And, like oil spills themselves, there is a necessary uniqueness to each one.

Although there is no "best way" to design a contingency plan - any more than there is a "best way" to deal with spilled oil - there are important requirements which are common to all.

First, and with no implication of priority order, is the need for advance planning. Planning to utilise the available knowledge and expertise from all sources, involving close cooperation and exchange of information between government, industry, universities and organisations of all kinds, both internally and across national frontiers, is an essential prerequisite to any real success such plans might achieve. Several examples of advance planning were discussed and I mention but four as illustrative: The Manche Plan to which Admiral Stacey referred; the UNEP regional seas programmes about which Dr. Keckes spoke; the successful United States response to the protection of specific areas along the coast of Texas following the IXTOC 1 blowout, on which Mr. Biglane and Dr. Lindstedt-Siva commented; and the use of oil slick movement forecasting techniques, about which more in a moment.

Advance planning must also embrace the provision of adequate numbers of properly trained and equipped personnel, operating in a regularly exercised system.

I believe, though, that the papers and discussions have shown that while much has been and is being done we have not yet achieved the level of cooperation and integration between governments, industry and others that we should like to see - and I direct that comment both to the planning function and to the action function. More efforts are needed on both.

Regarding the action side of the business, if I may so refer to it, the importance of clear command structures and the relevance of good communications to the effective exercise of command were stressed. When an incident occurs who is in charge? Does the person in charge know he's in charge? And does everyone else concerned know that he is? Committee discussion and decisions are most important at preplanning stages. Authoritative individual decisions are necessary in the action phase.

Two more examples among many other useful points brought out with reference to contingency planning and action deserve mention. First, the urgent need for indentification of safe havens for vessels in distress. Ships and their crews must not be denied such sanctuary. Second, the "do nothing" option, the reasons for and aims of which should be more widely and overtly disseminated so that, when it is invoked, the reasons are not misconstrued by the public.

I noted from the impressive sequence of slides shown that the United States' plan covers response both to oil and to hazardous materials spills. Mr. Biglane predicted that the 1980's will be the decade of the hazardous materials spill. This prompts me to question whether plans to deal with oil spills and hazardous materials spills should be integrated in one response structure, or whether they might be best dealt with separately? I suggest that this is a matter to which further thought should be given.

Now to pass on quickly to Dr. Callaghan's presentation dealing with the probabilistic Sliktrak programme; and its integration with Oilsim, a deterministic programme, to produce the Slikforcast simulation both for oil spill emergency tracking and for longer term contingency planning as described to us by Dr. Audunson.

These programmes hold promise of useful applicability to most areas for which appropriate meteorological and other input data are available. Accurate predictions before and during spill events would directly assist in assessing clean-up response requirements, and their location, and materially improve the deployment of those resources to best strategic and economic advantage.

Assumptions have to be made in developing such programmes and, in addition, they can of course only be as good as the data base on which they rest. Some of the work of the API on the fate of oil spills at sea, as summarised in the paper given and references quoted by Dr. Lasday, provide good data for the oil types on which they worked and there is a wealth of other data on which to draw. But we have heard from Mr. Woollen and Mr. Tucker in their papers on the data buoy located in the Western approaches, and on the importance of oceanography to the offshore industry, of the difficulties and expense in obtaining high quality oceanographic and meteorological data in sufficient quantity. Wave directional spectra and accurate surface current measurements are, we have been told, not made on a routine basis; and it is sea surface roughness and surface currents which respectively break up and transport oil slicks. Further, near-surface currents in open waters appear to be well below 1 per cent of wind speed, although we did receive some assurance that the 3 per cent figure normally quoted and used in respect of surface movement is indeed a reasonable one. We know also that in the particular case of the IXTOC 1 blowout much oil, or water-in-oil emulsion, was floating in the water column as well as on the surface.

I don't wish to labour the point but do feel that the predictive programmes about which we have heard do need to be validated by testing them against measured events whenever the opportunity arises. Refinements and improvements based on such experience, together with increasing accuracy and breadth in the data base as more information becomes available, should lead to a tool which can be used with increasing confidence. I make the point that communication is important not only in response actions but in the planning function too. There seems to be a good case here for the oceanographers, meteorologists and others who can contribute to this valuable development to work even more closely together.

Meanwhile, we should be wary that in the early stages of using these programmes for advance contingency planning purposes, we are not misled into an unquestioning, and possibly misleading, acceptance of the predictions they provide.

The papers delivered by Mr. Hardy and by Dr. McAuliffe and the review of API-sponsored research by Dr. Lasday already referred to, presented a great deal of interesting and detailed data on monitoring of what happens to oil and to its many and varied components in the sea; on how it is dealt with by living organisms; and on what effects it, and its components, may or may not have on those organisms.

The recent API/EPA funded work on dispersants described by Dr. McAuliffe and the subsequent discussion on this and the other papers I have mentioned in this group have once again highlighted the immense interest in this subject and in the difficulties associated with it; in particular the proper interpretation and use of the results arising from the excellent laboratory and field work carried out. I could not summarise or interpret the wealth of data presented, even were I qualified to do so, better than the authors have done in their papers and will not make the attempt. There are, however, a few comments I should make.

It appeared to me from the discussion that some delegates had formed the impression that industry is attempting to formulate lower toxicity dispersants. That in my view is not a true reflection of the position. It is doubtful that much advance could be made in that direction. Industry is, however, attempting to formulate low toxicity dispersants which have even higher efficiencies. This includes work to try to extend to a capability to treat higher viscosity oils than can sensibly be treated at present - or, put another way, to be able to treat effectively, under lower ambient temperature conditions those oils already amenable to treatment under temperate conditions. To disperse the same quantity of oil using smaller quantities of dispersant would seem to me to make good environmental and economic sense.

Dispersants should of course be seen in context. They are but one weapon in the oil spill armoury. It is clear that there are circumstances in which dispersants should not be used. Equally, at the other end of the scale, circumstances arise when dispersants offer the best approach to a particular oil pollution problem. But between these two extremes and for many valid reasons there is an area, as the discussion showed, in which questions continue to arise and in which vociferous debate is likely to continue for some time yet.

As with the predictive programmes already summarised, the message seems to me to be that the way to eventually resolve these matters is to continue to acquire data and information in the field on the effects of oil spills, some of which will have been treated with dispersants and some not. More background data are desirable so that "before and after" data can be meaningfully compared.

On the subject of monitoring, it is worth recalling the UNEP view expressed by Dr. Keckes that monitoring of background pollution levels on a global scale is not technically justifiable and that their available financial resources must be directed to Regional problems.

The mention of field studies leads me to the paper on the Amoco Cadiz experience presented by Monsieur d'Ozouville. The visual aids used added force to the fairly recent concept of the Vulnerability Index and to the suggestion to extend the use of this Index as a planning tool.

It is, I believe, of interest to observe in connection with Mr. Berry's paper on produced water discharges that in the United Kingdom, as well as in the United States, as he described, the statistical approach to effluent control - that is, short period averaging and longer term averaging - is recognised. Also, that the latter is considered more relevant than a specific maximum; and that no adverse effects have been demonstrated at current effluent levels. Since no-one present at the Session offered comment on a query relating to the disposal of produced water under Arctic inshore winter conditions, this might suggest an area for potential study.

Session 4 was concluded with Mr. Angremond's paper on the development in Holland of a new vessel designed for both dredging and anti-pollution duties. This is an interesting approach which shows that the high, often prohibitive cost of dedicated anti oil pollution craft can be offset by designing vessels for dual purpose duty. We will wait with interest to see how she performs when she enters service later this year.

I am aware, Mr. Chairman, that many useful and important points raised in the papers and in the discussion on them, have been left unsaid. Authors and contributors who may feel aggrieved that their most important points have been omitted, or perhaps misrepresented, will I hope set the record straight in the discussion period which follows. May I thank the authors for their clear and concise presentations which were an enormous help in what I have found a fascinating, if daunting task of summarising their extensive work and the discussion it provoked.

RAPPORTEUR'S SUMMARY OF SESSION 5 - TROPICAL ZONE

Mr. R.I. Walker
General Manager - Engineering
British National Oil Corporation

Mr. Chairman, Authors, Ladies and Gentlemen,

When I was persuaded by the Conference Organiser, to act as rapporteur
I did not fully appreciate the task that I was undertaking; the programme
allowed quarter of an hour to the task and the eight papers took four
hours to read, I am therefore faced with the daunting task of
summarising each paper in less than two minutes. Rather than do this,
I hope the authors will allow me to delve into their papers, and
gather their thoughts at random in the manner that Field Marshall Lord
Wavell did in his anthology of poems entitled "The Garden of Other Men's
Flowers". My intention is to provoke thought and discussion for the
final session of this conference rather than summarise.

The role I assume for myself is akin to the retainer following at the
end of the Lord Mayor of London's Inaugural Procession. The man is
clothed in a cloth cap, wears overalls, pushes a hand cart and is equipped
with a shovel and broom. The pollution gathered by the procession
follower has other uses, so my gardening friends tell me.

Of the eight papers forming the 5th Session on the Tropical Zone, three
dealt with the effects of pollution, two deal with models and simulations
and one deals with the positive benefits of weather forecasting and its
positive spin-offs to the non oil related community. I would draw your
attention, Mr. Chairman, to the fact that there are no papers on the
prevention of pollution, its avoidance or its cure.

It is an unfortunate fact but as soon as the words Petroleum and Marine
environment are brought together in the same sentence, it invariably includes
    an allusion to pollution; this conference is no exception. At a future
session of this conference I hope the organisers will encourage papers
on the positive aspects as well as the negative one of pollution. Of course
the oil industry is responsible for a degree of pollution, I think it also
should be judged on the positive contribution that it has made.

In the first papers of the Session, Captain Hebden's paper "Gulf's Area Oil
Companies Mutual Aid Organisation" points to the frustration of getting an
agreed position and policy between an ever increasing number of individual
organisations with a common link of pooling resources to combat pollution.
That this organisation exists at all and continues to grow both in numbers
and areas of influence is a tribute to all concerned and is a recognition
of the very real needs it fulfils. Captain Hebden, as did many other speakers,
in all sessions of the Conference emphasised the need for effective
communications at all levels as the fundamental element for combating pollution.

Dr. W.G. Lehr's paper "Oil Slick Movement in the Arabian Gulf" provides a powerful tool to minimise the impact of pollution and, if used at the outset of any projects, should go a long way to reduce exposure.

Dr. Lehr's team have identified the sections of the Gulf coastline that have a high risk of exposure to pollution and the time of the year at which they are at risk. The Gulf Area Oil Companies Mutual Aid Organisation strategy, distribution, and the deployment of their pollution control equipment will benefit directly from this work. Dr. Lehr alludes to other work being undertaken by Dhahran University and its associated Research Institute of immediate interest to the Oil Industry.

Dr. Hibbert, when presenting his wide ranging paper on "The Gulf Weather Forecast Scheme 21 years on" stressed the positive contribution that the oil industry has made to the Gulf Area interests. Two hand-books covering historical wind and climatic data are available for purchase by all-comers; similar publications covering different aspects of climatic and oceano-graphic conditions are planned for the future. Dr. Hibbert highlighted the benefits that accrued to all users when the meteorological data was consciously gathered from a wide spread of location by amateur and professional observers and collated centrally for retransmission to users. An extreme of this approach being data from instrumented remote sensing satellites - synoptic wave charts could in the future be the outcome of such work. Dr. Hibbert's presentation raised many questions and comments from the assembled delegates.

The paper on "Tropical Oil Spill Contingency Plan" by Mr. Stoner makes a powerful and unanswerable case for oil spill contingency planning. As such it provides a check list and yardstick against which current and future oil spill contingency plans should be measured. Dr. Stoner concluded his paper with the observation that under the relatively favourable wind and sea conditions of the Tropics - as opposed to the onerous conditions of the temperate zones - that the current range of containment and clean up equipment can play a significant role on the cure of pollution; but that this does not absolve the Oil Industry from striving for further improvements.

There were two papers on the impact of oil pollution of the tropical environment. Dr. Baker's paper dealt with the impact of oil spills on coastal mangrove shoreline, the one by Dr. Ray covered its impact on coral reefs. Dr. Baker concluded that the only positive approach to deal with pollution on mangrove shorelines was to prevent the oil pollutant from reaching the shore-line. Dr. Baker also illustrated with words and slides the difficulty in obtaining meaningful scientific data on the effects of pollution in remote areas.

Dr. Ray's paper on coral reefs highlighted the paucity of reliable field data on the level of contamination leading to damage of coral communities. A substantial research has been carried out in controlled laboratory environments, but in all cases, the level of contamination was unrealistically high. Dr. Ray concluded that certain coral communities would adapt to occasional exposure by hydrocarbon and stressed the possible greater danger of damage to coral communities from hydrocarbon dispersants. Dr. Ray provided figures showing transportation and insidious run off from land masses are principal sources of marine pollution.

Dr. Jacques' paper entitled "The Development of a Model for Biological
Control of Shallow Tropical Waters Subjected to Petroleum Threats" does
not lend itself to a summary. It has to be taken in its entirety.
Dr. Jacques in his paper applies his model to the West African upwelling
between Senegal to Morocco.

The final paper of Session 5 contained an alarming set of statistics on
the Gulf of Mexico - Ixtoc I oil spill - Mr. Golov culled the figures
from weekly reports "Oil Spill Intelligence Report" covering the 295 days
of the blowout. Dr. Golov suggested that only 4.5% of the oil from Ixtoc I
had been recovered; he further suggested that figures released by relevant
agencies did not match actuals; the statistics qualify for the "Guinness Book of
Records". He concluded by expressing the fervent hope that was echoed
throughout the hall by all that the figures would never again be exceeded.

Dr. Golov's paper provoked many questions and comments from the delegates
especially on the subject of contingency plans and the absence of treaties
between Mexico and the U.S.A. (their shorelines were affected by this blowout)
and also on the lack of a collaborative US Government agency sponsored oil
spill research programme.

In conclusion, with your agreement, Mr. Chairman, I should like to make
some personal observations. It has worried me, while listening to this
the 5th Session of the Conference for which I am the Rapporteur, and in
all other sessions that the negative contribution of the Oil Industry -
pollution has been a major theme -   there are
many areas in which the industry has made a positive contribution, these
too should receive equal attention. In the 5th Session - that on weather
forecasting - there is an example of a positive contribution. There is a
further facet which received scant attention in the conference - the
prevention of pollution rather than the cure. I respectfully suggest that
the organising committee of the next Petromar Conference should encourage
papers in these areas.

I wish to conclude by expressing my thanks to the organisers of the
Conference, and to the authors of the papers especially, for their work
and their general contribution to the understanding of the marine environment.

GENERAL DISCUSSION - SESSION 6

This is a brief summary of the general discussion held after
the rapporteurs' reports. The summary has been divided into four
parts: pollution, cooperation, equipment and prevention.

Pollution

It is necessary to put oil spill environmental inpacts in the
perspective of the total marine oil pollution. Reference was made
to a study made by OECD in 1973 in which it was stated that 55 per cent
of oil in the marine environment starts from land e.g. industrial
waste and river run-off. As the estimates are rather old it was thought
that there are good reasons to believe that the pollution from land has
decreased due to construction of municipal sewage plants and due to the
fact that raw sewage is being treated to a greater extent.

Petroleum hydrocarbons are only a small part of complex products
entering the marine environment. Of major concern are those man-made
products entering the sea which are innately unstable. There is a
small percentage of petroleum in the marine environment which is sometimes not a critical issue e.g. there may be petroleum components in
the water columns which are barely noticeable but which may become
a major component in an organism.

However, the big difference between the land-based contribution
of hydrocarbons to the marine environment and oil spills at sea is
that the land-based sources are multi-point sources and discharged
over dispersed areas whereas when a tanker breaks up or an oil-well
blows up tens of thousands of gallons of oil are spilled in the same
place in a short period of time.

Cooperation

The importance of broad international cooperation was stressed not
only directly between countries but also between different international
organizations.

One example of the need for cooperation between governments and
industry is the need for contingency planning for oil spill clean-up
in areas where there is a concentration of oil transportation routes.
In many of these high risk areas the infrastructure facilities are
very scarce e.g. the Malacca Straits, South East Asia. Many of the
less developed countries were reported to be extremely concerned about

the hazards caused by petroleum transportation off their coastlines.

Regional cooperation should therefore be promoted. Through UNEP
an increased understanding has been reached between a number of
countries and industry.

Also stressed during the discussion was the need to establish
good relations and cooperation with the public as industry and
governments must realize that the public will play a more and more
important role.

Several speakers indicated the necessity for a future Petromar
conference in 2 to 3 years time with preparation work performed from
now on. They considered Petromar 80 as a very good forum for
exchanging knowledge and experience. Also mentioned was the
possibility of including workshops in the next conference.

Equipment

Part of the discussion dealt with the equipment for combatting oil
spills. It was stated that the efficiency of oil combatting does not
depend only on the various kinds of equipment available but on the
whole oil spill clean-up operation. The operation must be treated as
a system and it is the optimization of the system that is of importance
not the item of equipment.

Due to environmental conditions and the size of spills it was
stated that it is misleading to the public to state that we are today
in a position to effectively deal with oil spills. All too often the
authorities responsible make excuses as to why operations went wrong
when the truth was that there was a lack of capability from the
beginning. Therefore it is very important that we learn lessons from
each spill, and even if each spill is unique they have a lot of
aspects in common.

Concerning the special problems for less developed countries it
was said that the problem is not the lack of equipment but the fact that
no country developed or developing can deal with a big oil spill
effectively. Therefore it is necessary to organize and plan joint
international efforts so that equipment which is available nearby can be
utilized as effectively as possible.

The issue of chemical dispersants was raised and it was reported
that UNEP had organized a workshop on the Application and Environmental
Effects of Oil Spill Chemicals in November 1979. One of the major
conclusions was that there is a role for the use of chemical
dispersants when either physicial recovery is not feasible or the "do
nothing" option is not viable.

UNEP is in the process of drafting guidelines on the application
and environmental effects of oil spill chemicals. The guidelines will
be available in 1981.

Prevention

Several remarks were made about the lack of presentations dealing with the enormous amount of work that <u>is</u> being done on prevention. One referred to the fact that oil companies devote more time and effort to prevention than to clean-up and other efforts. More will be done in the field of prevention but it is necessary to strike a balance somehow.

The organization IMCO (Intergovernmental Maritime Consultative Organization) is working on prevention problems e.g. agreements concerning segregated ballast tanks. In 1978 the Tanker Safety and Pollution Prevention Protocol under IMCO required certain equipment to be installed on tankers for navigation and back-up systems for the power.

Concerning operational cleaning up of tankers at sea the hope was expressed that this problem would soon be eliminated because of the economic incentive to get every drop of oil out in port with oil prices at $ 30 a barrel.

Examples of prevention actions were given in connection with drilling in the Arctic where clean-up operations present very difficult problems. In Canada there are stringent regulations for operators e.g. dredging a hole for the blow-out preventor into the seabed, closing operations down at a certain time in the season to prevent blow-outs.

It was also stressed that most oil spills could have been prevented by better training or contingency planning or government regulations.

Prevention and safety actions taken or suggested in Norway and Sweden were also reported during the discussion.

OVERALL SUMMING-UP BY THE CONFERENCE CHAIRMAN

Dr. Hanns Kippenberger
Président d'EUROCEAN
Administrateur - Directeur Général
Banque Européenne de Crédit
Bruxelles, Belgique

Ladies and Gentlemen,

I think the following reasonable assumptions can be made:

- oil will remain a major raw material and most important
  individual source of energy during the coming decades

- oil consumption is estimated to increase by around 1.1 % p.a.
  during the period 1980-90 and by around 0.3 % during 1990-2000
  which is much less than the estimated growth of the world GNP for
  the same period (an increase of around 3-3.5 %)

- for political and commercial reasons we have to increase efforts to
  expand exploration, exploitation and transportation of offshore oil.

I would like to summarize my conclusions under the following headings:

1. Assessment of situation

In order to make preventive measures possible, more information and
knowledge about the importance and the existence of the problems seem
to be necessary.

- although there is a lot of information available, a more detailed
  collection of data appears necessary in order to better monitor the
  importance of chemical influences, waves, currents, etc., and also
  information on existing offshore drilling and production
  activities and on accidental or permanent oil discharges. A
  collection of case studies of accidents which have occurred would
  be very useful.

The public is not well informed on existing clean-up capabilities;
whilst regional pollution clean-up systems appear satisfactory, there is
a lack of such systems on a global basis. The warning of biologists of
the direct danger of clean-up operations especially in Arctic waters,
must be taken into account.

- basic scientific research on the effects of oil discharges on
  water, life, sediments, has to be extended.

- closer cooperation and coordination between governments, research
  institutions and centres appears advisable.

2.   Relationship between Government, Industry, the Public and
      Scientists

General standards, strategies and rules and regulations for the
protection of the marine environment have been prepared by:

-    Intergovernmental Maritime Consultative Organization (IMCO)

-    International Convention for the Prevention of Pollution of the Sea
     by Oil

-    a number of conventions: Oslo, Paris, London, Helsinki, Barcelona

-    specifically formed committees

Remaining problems:

-    improve national guidelines, laws and standards to respond better
     to the new phenomena. Such guidelines should be clear,
     meaningful and lasting

-    establish uniform international guidelines

-    find a way to enforce such guidelines:

     better training of people involved
     establish rules of conduct (tankers, industry, illegal dumping, etc.)
     better coordination of prevention measures
     clear definition of responsibilities in case of accident
     introduce legislation on damage compensation (industry insurance,
     funds, government).

3.   Development of new and improvement of existing techniques and
      procedures for offshore oil activities to prevent discharges of oil
      into the seas is in the interest of industry itself (traffic control,
      tankers with segregated ballast tanks, information on wind, waves,
      swell, current, etc.)

-    is technology for equipment used today adequate? - design, material,
     operational procedures

-    authorities and industry should coordinate their efforts even more
     and improve cooperation

-    who should bear the financial burden for improved equipment? -
     industry through price increases or the public through higher taxes?

-    at what point does the cost of development of a system outweigh the
     usefulness of the result?

4. Are we sufficiently prepared for emergency situations?

-   prevention is always better than taking risks

-   improvement of clean-up equipment (has to be produced on a larger scale)

-   availability of such equipment (dispersion, sedimentation, recovery)

-   improvement of detecting systems and contingency planning

-   improvement of management procedure (organization of intervention procedures and systems, human capacity, strike teams, industry, national and international cooperation).

5. Industry should give better information on existing prevention measures, equipment available, contingency plans and pollution expenditure.

FINAL CONCLUSIONS

-   there is no doubt that it is more intelligent and finally less expensive to install satisfactory or to improve existing prevention systems in order to reduce the number and importance of deliberate or accidental oil discharges. Better equipment, better planning and management would help considerably. Information and scientific research have to be improved

-   because of human insufficiency and material failures, accidents however will continue to occur. We are technically and financially able to avoid the disastrous consequences of major oil spills but we all have to become more conscious of the importance of environmental problems for the quality of our lives now and in the future and, we should try to improve the coordination of the necessary efforts between industry, national and international bodies

-   at the beginning of the conference, I was comparing pollution with inflation. I said that both were man-made and are consequences of technical and human insufficiencies, negligence, cost reasonings, indolence and lack of discipline

-   as a boy I was given a 1 billion Mark coin to remind me of the effects of inflation;

    If there were a similar symbol for pollution, I would give it to all those involved to remind us of the possible dangers and consequences of their actions.

But I would not like to end my summing-up remarks without thanking again:

-   the organizer of the conference - EUROCEAN - but particularly the Conference Committee led by Mr. Don Lennard for the very efficient work they have done so successfully

- the authors of the papers and the speakers for the high quality of their presentations

- the session chairmen for their efficiency and good coordination

- the interpreters for helping us to understand each other

- the supporters of this conference for all their help and fine cooperation

- the participants for their active and efficient contribution

- the government of Monaco for their hospitality and especially for putting the beautiful conference hall at our disposal

- but above all to His Serene Highness for having given us the honour and pleasure of supporting and addressing this conference.

I now officially declare the conference "Petromar 80" closed.